Second Edition

Discrete Mathematics
for
Computer Scientists
and
Mathematicians

Second Edition

Discrete Mathematics for Computer Scientists and Mathematicians

Joe L. Mott

Abraham Kandel

Theodore P. Baker

The Florida State University
Department of Mathematics and Computer Science

A Reston Book
Prentice-Hall
Englewood Cliffs, New Jersey 07632

Library of Congress Cataloging-in-Publication Data

Mott, Joe L.
  Discrete mathematics for computer scientists and
mathematicians.

  Rev. ed. of: Discrete mathematics for computer
scientists.   c1983.
  Bibliography: p.
  Includes index.
  1. Combinatorial analysis. 2. Graph theory.
3. Mathematics—Data processing.   I. Kandel, Abraham.
II. Baker, Theodore P.   III. Mott, Joe L. Discrete mathematics for computer
scientists.   IV. Title.
QA164.M68   1986   555′.6   85-14365
ISBN 0-8359-1391-0

A Reston Book
Published by Prentice-Hall
A Division of Simon & Schuster, Inc.
Englewood Cliffs, NJ 07632

10   9   8   7   6   5   4   3

Printed in the United States of America

Contents

Preface

This text is intended for use in a first course in discrete mathematics in an undergraduate computer science and mathematics curriculum. The level is appropriate for a sophomore or junior course, and the number of topics and the depth of analysis can be adjusted to fit a one-term or a two-term course. A computer science student can take this course concurrently with the first course in programming preliminary to the study of data structures and the design and analysis of algorithms. A mathematics student may take this course concurrently with the first calculus course.

No specific background is prerequisite outside of the material ordinarily covered in most college algebra courses. In particular, a calculus background is not required for Chapters 1 to 7. While it is not necessary, knowledge of limits would help in understanding the proof of one theorem in Chapter 7 and knowledge of integration would enhance understanding some of the discussions in Chapter 8. We have assumed that students will have had little or no programming experience, although it would be desirable.

Our assumption about background has dictated how we have written the text in certain places. For instance, in Chapter 3, we have avoided reference to the convergence of power series by representing the geometric series

$$\sum_{i=0}^{\infty} a^i X^i$$

as the multiplicative inverse of $1 - aX$; in other words, we have considered power series from a strictly algebraic rather than the analytical viewpoint. Likewise, in Chapter 4, we avoid reference to limits when we discuss the asymptotic behavior of functions and the "big O notation," but if students understand limits, then exercises 11 and 12 in Section 4.2.1 will greatly streamline the discussion.

The Association for Computing Machinery, CUPM, and others have recommended that a computer science curriculum include a discrete mathematics course that introduces the student to logical and algebraic structures and to combinatorial mathematics including enumeration methods and graph theory. This text is an attempt to satisfy that recommendation.

Furthermore, we expect that some of the teachers of this course will be mathematicians who are not computer scientists by profession or by training. Therefore, we have purposely suppressed writing many algorithms in computer programming language, although on occasions it would have been easier to do so.

We believe that a discrete mathematics course based on our book will meet several important needs of both computer science and mathematics majors. While the basic content of the book is mathematics, many applications are oriented toward computer science. Moreover, we have attempted to include examples from computer science that can be discussed without making presumptions about the reader's background in computer science.

Many apparently mathematical topics are quite useful for computer science students as well. In particular, computer science students need to understand graph theory, since many topics of graph theory will be applied in a data structures course. Moreover, they need mathematical induction as a proof technique and to understand recursion, Boolean algebra to prepare for digital circuit design, logic and other proof techniques to be able to prove correctness of algorithms, and recurrence relations to analyze algorithms. Besides that, computer science students need to see how some real life problems can be modeled with graphs (like minimal spanning trees in Section 5.4, scheduling problems and graph coloring in Section 5.11, and network flow problems in Chapter 7).

Mathematics majors, on the other hand, will use graphs as a modeling tool, and they will benefit from a study of recurrence relations to understand computer solutions of differential equations. But more than that, discrete mathematics provides a good training ground for the mathematics student to learn to solve problems and to make correct proofs. For this reason, mathematics majors should take discrete mathematics quite early in their program of studies, preferably before those courses that require many proofs.

Discrete mathematics embodies the spirit of mathematical and scientific research perhaps more than almost any other undergraduate mathematics course. In graph theory, for example, powerful concepts can be defined and grasped because they can be visualized and simple examples can be constructed easily. This feature and others make the subject both challenging and rewarding to student and teacher alike.

The text has evolved over a period of years and, in that time, our curriculum at Florida State University has changed significantly, especially for computer science students. Thus, not only has the list of topics changed, but also the order in which we discuss them. Consequently, we have written the text so that the chapters are more or less independent of each other.

The following diagram shows the basic logical relationship among the chapters.

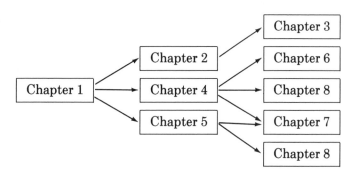

Chapter 1, of course, is introductory and as much or as little of it can be discussed as needed depending on the background of the students. Most students likely have been exposed to the material of Sections 1.1, 1.2, and 1.3 except possibly the definitions in Section 1.3 of equivalence relations, composition of relations, and one-to-one and onto functions.

We recommend covering, at the minimum, Section 1.7 (Methods of Proof of an Implication) and Section 1.10 (Induction). Sections 1.5 and 1.6 contain introductory material on logic and is the foundation upon which Section 1.7 is built. A thorough understanding of proof by induction is, in our opinion, absolutely essential.

Section 1.4 is a general discussion that can be assigned for reading. Section 1.9 (Rules of Inference for Quantified Propositions) may be omitted without injury.

Chapter 3 can be taught at any time after Chapter 2 is covered. In particular, in a curriculum that calls for an early introduction to trees and graph theory we recommend that Chapter 3 be postponed until after Chapter 5. Only elementary recurrences are used in Section 5.5, and in Section 5.6 there is only one use of a recurrence relation. But even this does not require any result from Chapter 3, as a solution can be obtained instead from Example 1.10.11 in Chapter 1.

Chapter 4 on directed graphs and Chapter 5 on nondirected graphs are related but may be treated as mutually independent chapters since definitions given in Chapter 4 for digraphs are repeated and illustrated for Chapter 5. In fact, Sections 4.1 and 4.2 can be taught concurrently with Sections 5.1 and 5.2.

We have made several significant changes from the first edition. First we have added two chapters, Chapter 7 on network flows and Chapter 8 on representation and manipulation of imprecision. Next, we have added several exercises in almost every section of the book. Moreover, we have consolidated two separate sections on partial orders into one in this second edition (Section 4.4), and we have removed the material on fuzzy sets from Chapter 1 of the first edition and incorporated that with other material on expert systems into Chapter 8. We have rewritten other sections including the section on methods of proof in Chapter 1, Section 3.6 on solutions of inhomogeneous recurrence relations in Chapter 3, and Sections 5.1 through 5.6 of Chapter 5 on graphs. The most notable change in Chapter 5 is that we have consolidated spanning trees and minimal spanning trees into one section and we have introduced breadth-first search and depth-first search spanning trees as well.

Finally, we have added chapter reviews at the end of each chapter. Chapter 5 has a review for Sections 5.1 to 5.6 and then one for Sections 5.7 to 5.12. These reviews contain questions and problems from actual classroom tests that we have given in our own classes.

There are several possible course syllabi. For mathematics students only, we suggest Chapters 1, 2, 3, 5, and 7. One for computer science majors alone could be Chapters 1, 2, 4, 5 (at least Sections 5.1 to 5.6), 7, and 8. Chapter 6 on Boolean algebras could replace Chapter 7 or 8 if preparation for a digital design course is needed.

At Florida State University our discrete classes contain both mathematics majors and computer science majors so we follow this syllabus:

Discrete I:
   Sections 1.5 to 1.10 of Chapter 1 (Section 1.9 is optional), Chapter 2, and Chapter 5 (at least Sections 5.1 to 5.6)
Discrete II:
   Chapters 3, 4, 7, and selected topics from sections 5.7 to 5.12 as time permits.

Exercises follow each section, and as a general rule the level of difficulty ranges from the routine to the moderately difficult, although some proofs may present a challenge. In the early chapters we include many worked-out examples and solutions to the exercises hoping to enable the student to check his work and gain confidence. Later in the book we make greater demands on the student; in particular, we expect the student to be able to make some proofs by the end of the text.

Acknowledgments

We express our appreciation to the Sloan Foundation for the grant awarded to the departments of Mathematics and Computer Science at Florida State University in 1983. The Sloan Foundation has played a major role in educating the academic community of the need for discrete mathematics in the curriculum, and we appreciate the support that the Foundation has given us.

To our colleagues and friends who have taught from an earlier version of the book and made suggestions for improvement we say a heartfelt thank you.

The editorial staff at Reston Publishing have been a great help and we thank them.

Portions of the material in Chapter 8 are based on recent work by Lofti A. Zadeh [50], Maria Zemankova-Leech and Abraham Kandel [52], L. Applebaum and E. H. Ruspini [46], and many researchers in the fields of fuzzy set theory and artificial intelligence. Special thanks are due Dalya and Peli Pelled, who provided the desk upon which A. Kandel wrote Chapter 8.

We wish to express our gratitude to several people who helped with the preparation of the manuscript. Sheila O'Connell and Pam Flowers read early versions and made several helpful suggestions while Sandy Robbins, Denise Khosrow, Lynne Pennock, Ruth Wright, Karen Serra, and Marlene Walker typed portions of the manuscript for the first edition. Robert Stephens typed most of the manuscript for the second edition.

Finally, we want to express our love and appreciation to our families for their patience and encouragement throughout the time we were writing this book.

A Note to the Reader

In each chapter of this book, sections are numbered by chapter and then section. Thus, section number 4.2 means that it is the second section of Chapter 4. Likewise theorems, corollaries, definitions, and examples are numbered by chapter, section, and sequence so that example 4.2.7 means that the example is the seventh example in section 4.2.

The end of every theorem proof is indicated by the symbol $\square$.

We acknowledge our intellectual debt to several authors. We have included at the end of the book a bibliography which references many, but not all, of the books that have been a great help to us. A bracket, for instance [25], means that we are referring to the article or book number 25 in the bibliography.

An asterisk (*) indicates that the problem beside which the asterisk appears is generally more difficult than the other problems of the section.

Joe L. Mott
Abraham Kandel
Theodore P. Baker

1

Foundations

1.1 BASICS

One of the important tools in modern mathematics is the theory of sets. The notation, terminology, and concepts of set theory are helpful in studying any branch of mathematics. Every branch of mathematics can be considered as a study of sets of objects of one kind or another. For example, algebra is concerned with sets of numbers and operations on those sets whereas analysis deals mainly with sets of functions. The study of sets and their use in the foundations of mathematics was begun in the latter part of the nineteenth century by Georg Cantor (1845–1918). Since then, set theory has unified many seemingly disconnected ideas. It has helped to reduce many mathematical concepts to their logical foundations in an elegant and systematic way and helped to clarify the relationship between mathematics and philosophy.

What do the following have in common?

- a crowd of people,
- a herd of animals,
- a bunch of flowers, and
- a group of children.

In each case we are dealing with a collection of objects of a certain type. Rather than use a different word for each type of collection, it is convenient to denote them all by the one word "set." Thus a **set** is a collection of well-defined objects, called the **elements** of the set. The elements (or **members**) of the set are said to belong to (or be contained in) the set.

1

It is important to realize that a set may itself be an element of some other set. For example, a line is a set of points; the set of all lines in the plane is a set of sets of points. In fact a set can be a set of sets of sets and so on. The theory dealing with the (abstract) sets defined in the above manner is called (**abstract** or **conventional**) **set theory,** in contrast to fuzzy set theory which will be introduced later in Chapter 8.

This chapter begins with a review of set theory which includes the introduction of several important classes of sets and their properties.

In this chapter we also introduce the basic concepts of relations and functions necessary for understanding the remainder of the material. The chapter also describes different methods of proof—including mathematical induction—and shows how to use these techniques in proving results related to the content of the text.

The material in Chapters 2–8 represents the applications of the concepts introduced in this chapter. Understanding these concepts and their potential applications is good preparation for most computer science and mathematics majors.

1.2   SETS AND OPERATIONS OF SETS

Sets will be denoted by *capital* letters $A, B, C, \ldots, X, Y, Z$. Elements will be denoted by *lower case* letters $a, b, c, \ldots, x, y, z$. The phrase "is an element of" will be denoted by the symbol $\in$. Thus we write $x \in A$ for "x is an element of A." In analogous situations, we write $x \notin A$ for "x is not an element of A."

There are five ways used to describe a set.

1. Describe a set by describing the properties of the members of the set.

2. Describe a set by listing its elements.

3. Describe a set A by its characteristic function, defined as

$$\mu_A(x) = 1 \text{ if } x \in A,$$

$$\mu_A(x) = 0 \text{ if } x \notin A,$$

for all x in U, where U is the universal set, sometimes called the "universe of discourse," or just the "universe," which is a fixed specified set describing the context for the duration of the discussion.

If the discussion refers to dogs only, for example, then the universe of discourse is the class of dogs. In elementary algebra or number theory, the universe of discourse could be numbers (rational, real, complex, etc.). The universe of discourse must be explicitly stated, because the truth value of a statement depends upon it, as we shall see later.

4. Describe a set by a recursive formula. This is to give one or more elements of the set and a rule by which the rest of the elements of the set may be generated. We return to this idea in Section 1.10 and in Chapter 3.

5. Describe a set by an operation (such as union, intersection, complement, etc.) on some other sets.

Example 1.2.1. Describe the set containing all the nonnegative integers less than or equal to 5.

Let A denote the set. Then the set A can be described in the following ways:

1. $A = \{x \mid x$ is a nonnegative integer less than or equal to 5$\}$.
2. $A = \{0,1,2,3,4,5\}$.
3. $\mu_A(x) = \begin{cases} 1 \text{ for } x = 0,1,\ldots,5, \\ 0 \text{ otherwise.} \end{cases}$
4. $A = \{x_{i+1} = x_i + 1, i = 0,1,\ldots,4, \text{ where } x_0 = 0\}$.
5. This part is left to the reader as an exercise to be completed once the operations on sets are discussed.

The use of braces and $\mid$ ("such that") is a conventional notation which reads: $\{x \mid \text{property of } x\}$ means "the set of all elements x *such that* x has the given property." Note that, for a given set, not all the five ways of describing it are always possible. For example, the set of real numbers between 0 and 1 cannot be described by either listing all its elements or by a recursive formula.

In this section, we shall introduce the fundamental operations on sets and the relations among these operations. We begin with the following definitions.

Definition 1.2.1. Let A and B be two sets. Then A is said to be a **subset** of B if every element of A is an element of B; A is said to be a proper subset of B if A is a subset of B and there is at least one element of B which is not in A.

If A is a subset of B, we say A is contained in B. Symbolically, we write $A \subseteq B$. If A is a proper subset of B, then we say A is strictly contained in

B, denoted by $A \subset B$. The containment of sets has the following properties. Let A, B, and C be sets.

1. $A \subseteq A$.
2. If $A \subseteq B$ and $B \subseteq C$, then $A \subseteq C$.
3. If $A \subseteq B$ and $B \subset C$, then $A \subset C$.
4. If $A \subseteq B$ and $A \nsubseteq C$, then $B \nsubseteq C$, where $\nsubseteq$ means "is not contained in."

The statement $A \subseteq B$ does not rule out the possibility that $B \subseteq A$. In fact, we have both $A \subseteq B$ and $B \subseteq A$ if and only if (abbreviated iff) A and B have the same elements. Thus we define the following:

Definition 1.2.2.   Two sets A and B are equal iff $A \subseteq B$ and $B \subseteq A$. We write $A = B$.

Therefore, we have the following principle.

Principle.   To show that two sets A and B are equal, we must show that each element of A is also an element of B, and conversely.

A set containing no elements is called the **empty set** or **null set,** denoted by $\varnothing$. For example, given the universal set U of all positive numbers, the set of all positive numbers x in U satisfying the equation $x + 1 = 0$ is an empty set since there are no positive numbers which can satisfy this equation. The empty set is a subset of every set. In other words, $\varnothing \subseteq A$ for every A. This is because there are no elements in $\varnothing$; therefore, every element in $\varnothing$ belongs to A. It is important to note that the sets $\varnothing$ and $\{\varnothing\}$ are very different sets. The former has no elements, whereas the latter has the unique element $\varnothing$. A set containing a single element is called a **singleton.**

We shall now describe three operations on sets; namely, complement, union, and intersection. These operations allow us to construct new sets from given sets. We shall also study the relationships among these operations.

Definition 1.2.3.   Let U be the universal set and let A be any subset of U. The **absolute complement** of A, $\overline{A}$, is defined as $\{x \mid x \notin A\}$ or, $\{x \mid x \in U \text{ and } x \notin A\}$. If A and B are sets, the **relative complement** of A with respect to B is as shown below.

$$B - A = \{x \mid x \in B \text{ and } x \notin A\}.$$

It is clear that $\overline{\varnothing} = U$, $\overline{U} = \varnothing$, and that the complement of the complement of A is equal to A.

Definition 1.2.4.   Let A and B be two sets. The **union** of A and B is $A \cup B = \{x \mid x \in A \text{ or } x \in B \text{ or both}\}$. More generally, if $A_1, A_2, \ldots, A_n$ are sets, then their union is the set of all objects which belong to at least one of them, and is denoted by

$$A_1 \cup A_2 \cup \cdots \cup A_n, \text{ or by } \bigcup_{j=1}^{n} A_j.$$

Definition 1.2.5. The **intersection** of two sets A and B is $A \cap B = \{x \mid x \in A \text{ and } x \in B\}$. The intersection of n sets $A_1, A_2, \ldots, A_n$ is the set of all objects which belong to every one of them, and is denoted by

$$A_1 \cap A_2 \cap \cdots \cap A_n, \text{ or } \bigcap_{j=1}^{n} A_j.$$

Some basic properties of union and intersection of two sets are as follows:

	Union	Intersection
Idempotent:	$A \cup A = A$	$A \cap A = A$
Commutative:	$A \cup B = B \cup A$	$A \cap B = B \cap A$
Associative:	$A \cup (B \cup C) = (A \cup B) \cup C$	$A \cap (B \cap C) = (A \cap B) \cap C$

It should be noted that, in general,

$$(A \cup B) \cap C \neq A \cup (B \cap C).$$

Definition 1.2.6. The **symmetrical difference** of two sets A and B is $A \triangle B = \{x \mid x \in A, \text{ or } x \in B, \text{ but not both}\}$. The symmetrical difference of two sets is also called the **Boolean sum** of the two sets.

Definition 1.2.7. Two sets A and B are said to be **disjoint** if they do not have a member in common, that is to say, if $A \cap B = \varnothing$.

We can easily show the following theorems from the definitions of union, intersection, and complement.

Theorem 1.2.1. (Distributive Laws). Let A, B, and C be three sets. Then,

$$C \cap (A \cup B) = (C \cap A) \cup (C \cap B),$$
$$C \cup (A \cap B) = (C \cup A) \cap (C \cup B).$$

Theorem 1.2.2.   (DeMorgan's Laws). Let A and B be two sets. Then,

$$\overline{(A \cup B)} = \overline{A} \cap \overline{B},$$

$$\overline{(A \cap B)} = \overline{A} \cup \overline{B}.$$

It is often helpful to use a diagram, called a Venn diagram [after John Venn (1834–1883)], to visualize the various properties of the set operations. The universal set is represented by a large rectangular area. Subsets within this universe are represented by circular areas. A summary of set operations and their Venn diagrams is given in Figure 1-1.

DeMorgan's laws can be established from the Venn diagram. If the area outside A represents $\overline{A}$ and the area outside B represents $\overline{B}$, the proof is immediate.

Let U be our universe; applying DeMorgan's laws, $A \cup B$ can be expressed as a union of disjoint sets:

$$A \cup B = \overline{(\overline{A} \cap \overline{B})} = U - (\overline{A} \cap \overline{B}) = (A \cap B) \cup (A \cap \overline{B}) \cup (\overline{A} \cap B).$$

Set Operation	Symbol	Venn Diagram
Set B is contained in set A	$B \subset A$	
The absolute complement of set A	$\overline{A}$	
The relative complement of set B with respect to set A	$A - B$	
The union of sets A and B	$A \cup B$	
The intersection of sets A and B	$A \cap B$	
The symmetrical difference of sets A and B	$A \,\Delta\, B$	

Figure 1-1. Venn diagram of set operations.

Example 1.2.2.

$$
\begin{aligned}
A - (A - B) &= A - (A \cap \overline{B}) &&\text{(by definition of } A - B), \\
&= A \cap \overline{(A \cap \overline{B})} &&\text{(by definition of } A - B), \\
&= A \cap (\overline{A} \cup B) &&\text{(by DeMorgan)}, \\
&= (A \cap \overline{A}) \cup (A \cap B) &&\text{(by distributive law)}, \\
&= \varnothing \cup (A \cap B) &&\text{(by } A \cap \overline{A} = \varnothing), \\
&= A \cap B &&\text{(by } \varnothing \cup X = X).
\end{aligned}
$$

Clearly, the elements of a set may themselves be sets. A special class of such sets is the **power set.**

Definition 1.2.8. Let A be a given set. The **power set** of A, denoted by $\mathcal{P}(A)$, is the family of sets such that $X \subseteq A$ iff $X \in \mathcal{P}(A)$. Symbolically, $\mathcal{P}(A) = \{X \mid X \subseteq A\}$.

Example 1.2.3. Let $A = \{a,b,c\}$. The power set of A is as follows:

$$\mathcal{P}(A) = \{\{\varnothing\},\{a\},\{b\},\{c\},\{a,b\},\{b,c\},\{c,a\},\{a,b,c\}\}.$$

Exercises for Section 1.2

1. List the elements in the following sets.
   (a) The set of prime numbers less than or equal to 31.
   (b) $\{x \mid x \in \mathbb{R}$ and $x^2 + x - 12 = 0\}$, where $\mathbb{R}$ represents the set of real numbers.
   (c) The set of letters in the word $SUBSETS$.
2. Russell's paradox: Show that set K, such that $K = \{S \mid S$ is a set such that $S \notin S\}$, does not exist.
3. Prove that the empty set is unique.
4. Cantor's paradox: Show that set A, such that $A = \{S \mid S$ is a set\}, does not exist.
5. Let $U = \{1,2,3,4,5\}$, $A = \{1,5\}$, $B = \{1,2,3,4\}$, and $C = \{2,5\}$. Determine the following sets.
   (a) $A \cap \overline{B}$.
   (b) $A \cup (B \cap C)$.
   (c) $(A \cup B) \cap (A \cup C)$.
   (d) $\overline{(A \cap B)} \cup \overline{(B \cup C)}$.
   (e) $\overline{A} \cup \overline{B}$.

6. Let A, B, and C be subsets of U. Prove or disprove:

$$(A \cup B) \cap (B \cup \overline{C}) \subset A \cap \overline{B}.$$

7. Use DeMorgan's laws to prove that the complement of

$$(\overline{A} \cap B) \cap (A \cup \overline{B}) \cap (A \cup C)$$

is

$$(A \cup \overline{B}) \cup (\overline{A} \cap (B \cup \overline{C})).$$

8. A_k are sets of real numbers defined as

$$A_o = \{a \mid a \leq 1\}$$

$$A_k = \{a \mid a < 1 + 1/k\}, k = 1, 2, \ldots.$$

Prove that

$$\bigcap_{k=1}^{\infty} A_k = A_o.$$

9. List the elements of the set $\{a/b: a$ and b are prime integers with $1 < a \leq 12$ and $3 < b < 9\}$.

10. Let A be a set. Define $\mathcal{P}(A)$ as the set of all subsets of A. List $\mathcal{P}(A)$, where $A = \{1,2,3\}$. If $\mathcal{P}(A)$ has 256 elements, how many elements are there in A?

11. If set A has k elements, formulate a conjecture about the number of elements in $\mathcal{P}(A)$.

12. The **Cartesian product** of the sets S and T, $(S \times T)$, is the set of all ordered pairs (s,t) where $s \in S$ and $t \in T$, with $(s,t) = (u,v)$ for $u \in S$, $v \in T$, iff $s = u$ and $t = v$. Prove that $S \times T$ is not equal to $T \times S$ unless $S = T$ or either S or T is $\varnothing$.

13. Prove that $B - A$ is a subset of $\overline{A}$.

14. Prove that $B - \overline{A} = B \cap A$.

15. Prove that $A \subset B$ implies $A \cup (B - A) = B$.

16. If $A = \{0,1\}$ and $B = \{1,a\}$, determine the sets
    (a) $A \times \{1\} \times B$.
    (b) $(B \times A) \times (B \times A)$.

Selected Answers for Section 1.2

2. It is observed that unrestricted freedom in using the concept of "set" must lead to contradiction. One of the paradoxes, exhibited by Bertrand Russell, may be formulated as follows. Most sets do not contain themselves as elements. For example, the set A of all integers contains as elements only integers; A, being itself not an integer but a *set of integers,* does not contain itself as element. Such a set we may call "ordinary." There may possibly be sets that do contain themselves as elements; for example, the set S defined as follows: "S contains as elements all sets definable by phrase of less than ten words" could be considered to contain itself as an element. Such sets we might call "extraordinary" sets. In any case, however, most sets will be ordinary, and we may exclude the erratic behavior of "extraordinary" sets by confining our attention to the *set of all ordinary sets.* Call this set C. Each element of the set C is itself a set; in fact an ordinary set. The question now arises, is C itself an ordinary set or an extraordinary set? It must be one or the other. If C is ordinary, it contains itself as an element, since C is defined as containing *all* ordinary sets. This being so, C must be extraordinary, since the extraordinary sets are those containing themselves as members. This is a contradiction. Hence C must be extraordinary. But then C contains as a member an extraordinary set (namely C itself), which contradicts the definition whereby C was to contain ordinary sets only. Thus in either case we see that the assumption of the mere existence of the set C has led us to a contradiction.

3. Suppose that there are two empty sets, $\varnothing_1$ and $\varnothing_2$. Since $\varnothing_1$ and $\varnothing_2$ are included in every set, $\varnothing_1 \subseteq \varnothing_2$ and $\varnothing_2 \subseteq \varnothing_1$, which implies that $\varnothing_1 = \varnothing_2$.

11. The power set of A, $\mathcal{P}(A)$ has 2^k elements if A has k elements. (A proof can be constructed using the binomial theorem discussed in Chapter 2.)

1.3 RELATIONS AND FUNCTIONS

In this section our main concern is sets whose elements are ordered pairs. By an **ordered pair** we mean that each set is specified by two objects in a prescribed order. The ordered pair of a and b, with first coordinate a and second coordinate b, is the set (a,b). We also define that $(a,b) = (c,d)$ iff $a = c$ and $b = d$. We are now in a position to define the Cartesian product of sets A and B.

Definition 1.3.1. Let A and B be two sets. The **Cartesian product** of A and B is defined as $A \times B = \{(a,b) \mid a \in A \text{ and } b \in B\}$. More generally, the Cartesian product of n sets $A_1, A_2, \ldots, A_n$ is defined as

$$A_1 \times A_2 \times \cdots \times A_n = \{(a_1, a_2, \ldots, a_n) \mid a_i \in A_i, i = 1, 2, \ldots, n\}.$$

The expression $(a_1, a_2, \ldots, a_n)$ is called an ordered n-tuple.

Example 1.3.1. Let $A = \{0,1,2\}$ and $B = \{a,b\}$. Then,

$$A \times B = \{(0,a), (0,b), (1,a), (1,b), (2,a), (2,b)\},$$

$$A \times A = \{(0,0)\ (0,1), (0,2), (1,0), (1,1), (1,2), (2,0), (2,1), (2,2)\}.$$

Example 1.3.2. Let R^1 be the set of real numbers. Then the Cartesian product $R^1 \times R^1 = \{(x,y) \mid x \text{ and } y \text{ are real numbers}\}$.

From the definition of the Cartesian product we have seen that any element (a,b) in a Cartesian product $A \times B$ is just an ordered pair. No relationship is required between the objects a and b for them to form an ordered pair. Thus, frequently we are not interested in the entire Cartesian product set but only in a certain portion of it which is in some way well defined.

Definition 1.3.2. A **(binary) relation** R from A to B is a subset of $A \times B$. If $A = B$, we say R is a (binary) relation on A. More generally, an **n-ary relation** is a subset of a Cartesian product of n sets $A_1, A_2, \ldots, A_n$. In case $n = 1$, a subset R of A is called a unary relation on A.

Example 1.3.3. Suppose it is desired to find all the points inside a unit circle whose center is at the origin. Then the set

$$R = \{(x,y) \mid x \text{ and } y \text{ are real numbers and } x^2 + y^2 < 1\}$$

is a relation on the set of real numbers.

Definition 1.3.3. Let R be a relation from A to B. The **domain** of R denoted by **dom R**, is defined:

$$\text{dom } R = \{x \mid x \in A \text{ and } (x,y) \in R \text{ for some } y \in B\}.$$

The **range** of R, denoted by **ran R**, is defined:

$$\text{ran } R = \{y \mid y \in B \text{ and } (x,y) \in R \text{ for some } x \in A\}.$$

Clearly, dom $R \subseteq A$ and ran $R \subseteq B$. Moreover, the domain of R is the set of first coordinates in R and the range of R is the set of second coordinates in R.

We sometimes write $(x,y) \in R$ as $x\ R\ y$ which reads **"x relates to y."**

Definition 1.3.4. Let R be a relation on A. R is an equivalence relation on A if the following conditions are satisfied:

1. xRx for all $x \in A$ (R is **reflexive**).
2. If xRy, then yRx, for all $x,y \in A$ (R is **symmetric**).
3. If xRy and yRz, then xRz for all $x,y,z \in A$ (R is **transitive**).

Example 1.3.4. Let N be the set of natural numbers, that is, $N = \{1,2,3,\ldots\}$. Define a relation R in N as follows:

$$R = \{(x,y) \mid x, y \in N \text{ and } x + y \text{ is even}\}.$$

R is an **equivalence relation** in N because the first two conditions are clearly satisfied. As to the third condition, if $x + y$ and $y + z$ are divisible by 2, then $x + (y + y) + z$ is divisible by 2. Hence $x + z$ is divisible by 2. In this equivalence relation all the odd numbers are equivalent and so are all the even numbers.

Example 1.3.5. Let $A = \{1,2,3,4\}$, and let $R = \{(1,3), (4,2), (2,4), (2,3), (3,1)\}$. Then, the relation R is not reflexive according to part (1) of definition 1.3.4, because $(1,1) \notin R$; similarly, R is not symmetric since $(2,3) \in R$, but $(3,2) \notin R$; and finally, R is not transitive since $(2,3) \in R$ and $(3,1) \in R$, but $(2,1) \notin R$.

On the other hand, the relation $S = \{(1,1), (2,2), (3,3), (4,4), (2,3), (3,1), (2,1)\}$ is reflexive and transitive but not symmetric as a relation on A.

A relation R on a set A is called an **antisymmetric** relation if $(a,b) \in R$ and $(b,a) \in R$ implies $a = b$.

The relation R in example 1.3.5 is not antisymmetric because $(1,3) \in R$ and $(3,1) \in R$, but certainly $1 \neq 3$.

Example 1.3.6. Let N be the set of natural numbers (that is, N is the set of positive integers). Let R be the relation on N defined by xRy iff x divides y. Thus, $(x,y) \in R$ iff $y = xz$, where z is some integer. Then R is reflexive since any integer x in N divides itself. Moreover, R is transitive since if r divides s and s divides t, then $s = ru$ and $t = sv$, where u and v are integers. But then $t = r\ (uv)$ so that r divides t. Finally, R is antisymmetric since if a divides b and b divides a, where a and b are integers, then in general $a = \pm b$, but, in this case, since a and b are positive integers, $a = b$.

Example 1.3.7. From time to time, we shall have occasion to refer to the equivalence relation known as **congruence modulo n,** where n is a positive integer. Let A be the set of integers and let n be a fixed positive integer. Define the relation R_n on A by aR_nb iff $a - b$ is an integral multiple of n. In other words, aR_nb iff $a - b = kn$, where k is some integer. The most common notation is to write $a \equiv b \bmod n$ instead of aR_nb, and we say that a *is congruent to b modulo n.* Thus, for example, $17 \equiv 2 \bmod 5$ since $17 - 2 = 3 \cdot 5$, but 17 is not congruent to 3 mod 5 since $17 - 3 = 14$ is not an integral multiple of 5.

We now check that R_n is an equivalence relation. Reflexivity: $a - a = 0 \cdot n$, so that $a \equiv a \bmod n$ for each integer a. Symmetry: If $a \equiv b \bmod n$, then $a - b = kn$, where k is some integer. But then $b - a = (-k)n$, so that $b \equiv a \bmod n$. Transitivity: If $a \equiv b \bmod n$, and $b \equiv c \bmod n$, then $a - b = kn$ and $b - c = mn$ where k and m are integers. Therefore, by adding these two equations we have $a - c = (k + m)n$, and we conclude $a \equiv c \bmod n$.

Now if a is a given integer, let $[a]$ denote the set of all integers b such that $a \equiv b \bmod n$. The set $[a]$ is called the **congruence class** (or equivalence class) containing the integer a and a is called a **representative** of the congruence class. Thus, $[a] = \{a + kn \mid k \text{ is any integer}\}$. In particular, if $n = 5$, then there are five different congruence classes for the relation congruence modulo 5, namely,

$$[0] = \{5k \qquad \mid k \in A\}$$
$$[1] = \{5k + 1 \mid k \in A\}$$
$$[2] = \{5k + 2 \mid k \in A\}$$
$$[3] = \{5k + 3 \mid k \in A\}$$
$$[4] = \{5k + 4 \mid k \in A\}$$

Each integer belongs to one and only one of these equivalence classes because the remainder is uniquely determined when an integer is divided by 5.

Functions

Definition 1.3.5. For any relation R from A to B and for any $a \in A$ and $b \in B$ define the sets

$$R^{-1} = \{(y,x) \in B \times A \mid (x,y) \in R\}$$
$$R(a) = \{y \in B \mid (a,y) \in R\}$$
$$R^{-1}(b) = \{x \in A \mid (x,b) \in R\}$$

The set R^{-1} is itself a relation from B to A where, by definition, y relates to x under R^{-1} iff x relates to y under R. Thus, if R is the relation defined on the set of real numbers by xRy iff $x < y$, then $cR^{-1}d$ means $c > d$. Likewise, if R is the relation on the set Z of integers defined by xRy iff x divides y, then R^{-1} is the relation where $cR^{-1}d$ means c is an integral multiple of d. The set $R(a)$ contains all elements of B to which the element a is related under R and can be viewed as the set of **images** of a. In general, $R(a)$ may be either empty or contain many elements. But if R is a **function** from A to B, then for each $a \in A$, not only is $R(a)$ nonempty, but, more than that, $R(a)$ contains precisely one element.

Let us reiterate this observation in another way in the following definition of function. The words **mapping, transformation, correspondence,** and **operator** are among these words that are sometimes used as synonyms for **function.**

Definition 1.3.6.   Let A and B be two nonempty sets. A *function,* denoted by f, from A to B is a relation from A to B such that:

1. dom $f = A$, that is, for each $a \in A$, $(a,b) \in f$ for some $b \in B$. In other words, f is defined at each $a \in A$.
2. If $(x,y) \in f$ and $(x,z) \in f$ then $y = z$. In this case, we say that f is **well-defined** or **single-valued.** Thus, no element of A is related to two elements of B. If $(x,y) \in f$, then we say that y is the **image of** x under f and we write $y = f(x)$.

Moreover, if f is a function from A to B, we write $f:A \to B$.

An alternate approach to the notion of function can be given in terms of a rule of assignment. In this version, a function from A to B is a rule that assigns to each element x in A a unique element y in B where $f(x) = y$ is written to denote this correspondence. This is essentially the same as the above definition since the term "rule" can be interpreted in such a way that each rule f determines the relation $\{(x,y) \in A \times B \mid f(x) = y\}$, and conversely. Thus, if the rule $y = 3x$ is defined on the set of real numbers, then the function is the set $\{(x,y) \mid x$ and y are real numbers and $y = 3x\}$.

To be sure, if f is a function from A to B, and if $b \in B$, then the set $f^{-1}(b)$ may be either empty or may contain several elements. This observation motivates the following two definitions.

Definition 1.3.7.   A function $f:A \to B$ is said to be **one-to-one** if $f(x_1) = y$ and $f(x_2) = y$ implies $x_1 = x_2$. Thus, f is one-to-one iff for each $b \in$ ran f, $f^{-1}(b)$ contains exactly one element.

If we describe $f^{-1}(b)$ as the set of **preimages** of b, then f is one-to-one if for each $b \in$ ran f, b has precisely one preimage.

Definition 1.3.8. A function $f{:}A \to B$ is **onto** iff ran $f = B$. In other words, f is onto iff $f^{-1}(b)$ is nonempty for each $b \in B$. To put it another way, f is onto if each $b \in B$ has some preimage in A.

If the function $f{:}A \to B$ is both one-to-one and onto, then the inverse relation f^{-1} is single-valued, and thus is a function from B to A. In this case, f^{-1} is called the **inverse function** of f.

A one-to-one, onto function $f{:}A \to B$ is usually called a **one-to-one correspondence** between A and B.

If a function $f{:}A \to B$ is not one-to-one, we call it a **many-to-one** function. If f is not necessarily onto B, then it is said to be **into** B.

Since a function is a set, two functions f and g from A into B are equal if they are equal as sets. In other words, $f = g$ iff $f(a) = g(a)$ for each $a \in A$.

Example 1.3.8. Let $A = \{r,s,t\}$, $B = \{1,2,3\}$, and $C = (r,s,t,u)$. $R = \{(r,1), (r,2), (t,2)\}$ is a relation from A to B but R is not a function since $R(r) = \{1,2\}$.

The set $f = \{(r,1), (s,2), (t,2)\}$ is a function from A to B but f is not one-to-one since $f^{-1}(2) = \{s,t\}$. Likewise, f is not onto B since $f^{-1}(3) = \varnothing$, the empty set.

The function $g = \{(r,1), (s,2), (t,3)\}$ is both a one-to-one and an onto function from A to B. Moreover, $g^{-1} = \{(1,r), (2,s), (3,t)\}$.

On the other hand, the function $h{:}C \to B$ defined as $h = \{(r,1), (s,1), (t,2), (u,3)\}$ is onto, but not one-to-one since $h^{-1}(1) = \{r,s\}$.

Definition 1.3.9. Let R be a relation from A to B and S a relation from B to C. The **composition** of R and S, denoted by $R \cdot S$ or simply RS, is the relation from A to C given by $aRSc$ iff there is an element $b \in B$ such that aRb and bSc. In other words, a relates to c under RS iff a relates to some $b \in B$ under R where b, in turn, relates to c under S.

The set B in the composition of $R \subseteq A \times B$ and $S \subseteq B \times C$ serves as an intermediary for establishing a correspondence between the sets A and C. For instance, suppose that A is a set of hospital patients, B is a list of symptoms, and C is a list of treatments. Suppose, moreover, that R relates a patient to his symptoms, and S relates symptoms to appropriate treatments. Then RS relates a patient to a treatment appropriate for his symptoms. Of course, a given patient may have several symptoms and each may have several treatments so that there may be several treatments associated with a patient under RS.

Example 1.3.9. If $A = \{1,2,3\}$, $B = \{4,5\}$, and $C = \{a,b,c\}$, let $R = \{(1,4), (1,5), (2,5)\}$ be a relation from A to B, and let $S = \{(4,a), (5,c)\}$ be a relation from B to C. Then $RS = \{(1,a), (1,c), (2,c)\}$ is a relation from A to C.

If $T = \{(a,1), (a,2), (b,1), (c,2), (c,3)\}$ is a relation from C to A, then $ST = \{(4,1), (4,2), (5,2), (5,3)\}$ is a relation from B to A.

Finally, for the relation $V = \{(1,1), (1,2), (2,3), (3,1)\}$ from A to A, then $V^2 = V \cdot V = \{(1,1), (1,2), (1,3), (2,1), (3,1), (3,2)\}$ is a relation from A to A.

If, in the definition of composition, the relations R and S are, in fact, functions, then to express the image of an element x under $R \cdot S$, we get into the peculiar notational anomaly $(R \cdot S)(x) = S(R(x))$ which means the value of x under R is determined first and then the value of $R(x)$ under S is determined. This anomaly has induced many writers to write $(x)(R \cdot S)$ instead of $(R \cdot S)(x)$, and, in general, for any function f to write xf for $f(x)$. Nevertheless, we shall stick to the classical notation and face the anomaly as it comes.

Example 1.3.10. If $R(x) = x + 1$ and $S(y) = y^2$ are functions defined on the set of real numbers, then $(R \cdot S)(x) = S(R(x)) = (x + 1)^2$ and $(S \cdot R)(x) = R(S(x)) = x^2 + 1$.

Exercises for Section 1.3

1. Let $A = \{1,2,3,4,5,6\}$. Construct pictorial descriptions of the relation R on A for the following cases.
   (a) $R = \{(j,k) \mid j \text{ divides } k\}$.
   (b) $R = \{(j,k) \mid j \text{ is a multiple of } k\}$.
   (c) $R = \{(j,k) \mid (j - k)^2 \in A\}$.
   (d) $R = \{(j,k) \mid j/k \text{ is a prime}\}$.

2. Let R be the relation from $A = \{1,2,3,4,5\}$ to $B = \{1,3,5\}$ which is defined by "x is less than y." Write R as a set of ordered pairs.

3. Let R be the relation in the natural numbers $N = \{1,2,3,\ldots\}$ defined by "$x + 2y = 10$," that is, let $R = \{(x,y) \mid x \in N, y \in N, x + 2y = 10\}$. Find
   (a) the domain and range of R
   (b) R^{-1}

4. Prove that if R is an antisymmetric relation so is R^{-1}.

5. Prove that if R is a symmetric relation, then $R \cap R^{-1} = R$.

6. Show that if R is an antisymmetric relation and R^* is an antisymmetric relation, so is $R \cap R^*$. What about $R \cup R^*$?

7. Let L be the set of lines in the Euclidean plane and let R be the relation in L defined by "x is parallel to y." Is R a symmetric relation? Why? Is R a transitive relation?

8. Replace the sentence "x is parallel to y" by the sentence "x is perpendicular to y" in Exercise 7. Is R a symmetric relation? Why? Is R a transitive relation?

9. Let D denote the diagonal line of $A \times A$, i.e., the set of all ordered pairs $(a,a) \in A \times A$. Prove that the relation R on A is antisymmetric if $R \cap R^{-1} \subseteq D$.

10. Can a relation R in a set A be both symmetric and antisymmetric?

11. Let $A = \{1,2,3\}$. Give an example of a relation R in A such that R is neither symmetric nor antisymmetric.

12. Show that when a relation R is symmetric, so is R^k for any $k > 0$, where R^k is the kth power of the relation R.

13. If $f:A \to B$ is a one-to-one correspondence, prove that $f^{-1}:B \to A$ is a one-to-one correspondence.

14. If $f:A \to B$ and $g:B \to C$ are both one-to-one correspondences, prove that $f \cdot g$ is a one-to-one correspondence.

15. Show that the function f from the reals into the reals defined by $f(x) = x^3 + 1$ is a one-to-one, onto function and find f^{-1}.

16. A function $b:A \times A \to A$ is called a binary operation on the set A. Moreover, a binary operation b on the set A is called **commutative** if $b(x,y) = b(y,x)$ for all $(x,y) \in A \times A$. Which of the following binary operations on the set of real numbers is commutative?
    (a) $b(x,y) = x - y$
    (b) $b(x,y) = x^2 + y^2$
    (c) $b(x,y) = \max\{x,y\}$, where $\max\{x,y\}$ denotes the larger of the two numbers x and y. (For example, $\max\{5,7\} = 7$, $\max\{-6,2\} = 2$.)
    (d) $b(x,y) = (x + y)/2$

17. A binary operation b on a set A is said to be **associative** if $b(x,b(y,z)) = b(b(x,y),z)$ for all x,y,z A. Which of the operations in exercise 16 are associative?

18. Which of the following relations are functions? Define R by xRy iff $x^2 + y^2 = 1$ where x and y are real numbers such that
    (a) $0 \le x \le 1, 0 \le y \le 1$
    (b) $-1 \le x \le 1, -1 \le y \le 1$
    (c) $-1 \le x \le 1, 0 \le y \le 1$
    (d) x arbitrary, $0 \le y \le 1$

19. Which of the following functions are one-to-one? Which are onto?
    (a) $f(x) = x + 1$ where $A = \{$positive integers$\}$, where $A = \{$integers$\}$.
    (b) $f(x) = x^2$ where $A = \{$positive integers$\}$, where $A = \{$integers$\}$, where $A = \{$real numbers$\}$.

20. If $A = \{1,2,3,4\}$ and $R = \{(1,2), (2,3), (3,4), (4,2)\}$ and $S = \{(1,3), (2,4), (4,2), (4,3)\}$, then compute $R \cdot S$, $S \cdot R$, and R^2.

21. If $A = \{1,2,3,4\}$ and $B = \{a,b,c,d\}$, determine if the following functions are one-to-one or onto.
    (a) $f = \{(1,a), (2,a), (3,b), (4,d)\}$
    (b) $g = \{(1,d), (2,b), (3,a), (4,a)\}$
    (c) $h = \{(1,d), (2,b), (3,a), (4,c)\}$

22. Suppose that a,b,c are integers, where $a \neq 0$. Suppose a divides b and a divides c. Prove that a divides $bx + cy$ where x and y are any integers.

23. Show that the following pairs of integers are congruent modulo 7.
    (a) 3 and 24
    (b) -31 and 11
    (c) -15 and -64

    Show that 25 and 12 are not congruent modulo 7.

24. Suppose that $a \equiv b \bmod n$ and d is a positive integer such that d divides n. Prove that $a \equiv b \bmod d$.

25. Show that the relation Q on the set $A = \{(a,b) \mid a,b$ are integers and $b \neq 0\}$ is an equivalence relation where Q is defined as $(a,b)Q(c,d)$ iff $ad = bc$.

Selected Answers for Section 1.3

3. (a) The solution set of $x + 2y = 10$ is $R = \{(2,4), (4,3), (6,2), (8,1)\}$ even though there is an infinite number of elements in N. Thus, the domain of R is $\{2,4,6,8\}$ and the range of R is $\{4,3,2,1\}$.
   (b) R^{-1} is found by interchanging x and y in the definition of R;

$$R^{-1} = \{(x,y) \mid x \in N, y \in N, 2x + y = 10\}$$

$$= \{(1,8), (2,6) (3,4), (4,2)\}.$$

10. Any subset of the "diagonal line" of $A \times A$, that is, any relation R in A in which $(a,b) \in R$ implies $a = b$ is both symmetric and anti-symmetric.

1.4   SOME METHODS OF PROOF AND PROBLEM-SOLVING STRATEGIES

We do not aspire to accomplish the ambitious task of presenting a discourse on all (or even a significant number) of the known methods of proof and problem-solving strategies. What we offer here are a few basic hints with some examples of the application of these hints. We expect you to return to these suggestions from time to time to refresh your memory as you are developing your ability to solve problems. Should you find this topic interesting, the books by G. Polya, *How to Solve it* [32], *Induction and Analogy in Mathematics* [33], *Patterns in Plausible Inference* [33], and *Mathematical Discovery* [34] are delightful reading, instructive, and most informative.

To some the word "prove" is a frightening word, but it should not be so; it only signifies "solve this problem," and solving a problem just means finding a way out of a difficulty, a way around an obstacle, a way of attaining a goal that otherwise was not immediately accessible. In other words, solving problems is a vital part of life, and, as in life, we may need to *improvise* and apply *ad hoc* techniques. But with each successful completion of a task, we become more capable of completing the next assignment.

Problem solving is part science and part art but is mostly hard work. It is a *science* in that there are several oft-repeated principles applied to varied types of problems. It is also an *art* because principles or rules cannot be applied mechanically but involve the skill of the student. Moreover, just as in playing an athletic game or a musical instrument, so also in problem solving this skill can *only* be learned by *imitation* and *practice*. We contend that *interest, effort, and experience are the primary factors to solving problems*. Thus, you the reader will be your own best teacher. Any solution that you have obtained by your own effort and insight, or one that you have followed with keen, intelligent concentration, may become a pattern for you, a model that you can imitate with success in solving similar problems.

We ask you to approach each problem with an attitude of research, to not only solve the problem at hand, but also to seek out the key ideas and techniques that made the solution possible. The general problem-solving methods taught in this section and throughout the book will *never never* compensate for lack of relevant knowledge or intelligent effort. Our discussion can take you only so far; you have to go the rest of the way on your own.

The solution of any problem worthy of the name will likely require some critical insight. Consequently, you simply *must* become familiar with the problem, gain information, and observe patterns to put yourself in a position to have the critical insight. *Without insight, method is largely useless*.

This insight is gained through hard work and hard work only. We agree with Thomas A. Edison who is supposed to have said that genius is two percent inspiration and ninety-eight percent perspiration.

We will get you started by introducing you to some general methods of approaching problems; specific tools like the inclusion-exclusion principle and the methods of recurrence relations and generating functions are all to be discussed in later chapters.

Several things must be clearly established *before* a solution of a problem should be attempted. For one thing, you must devise a plan of attack. We will give some more hints on this, but generally you will learn best by experience, so for a while your best plan may be to *imitate what you have seen others do.* It therefore becomes necessary for you to understand and record, for future reference, solutions suitable for imitation.

But before you can imitate you will have to be able to *analyze* what has been done. The solution of any problem consists of two major parts: the *discovery* of the solution and the *presentation* of the solution. There are two halves; one isn't complete without the other. Discovery is the cause, presentation is the effect; discovery is generally private while presentation is public. In fact, you usually confront solutions in their finished, polished form rather than while they are in the making, and therefore all the guesses, questions, false starts, and blind alleys have been eliminated. The beauty of this newborn baby need not reflect the birth pains that brought it into existence. Remember that you must not confuse the two halves of a solution and neither can you exclude either part.

The Basic Elements of an Argument

Generally speaking there are at least four basic elements of an argument that you will need to identify and analyze; these are: (1) goals, (2) grounds, (3) warrants, and (4) the frame of reference. Let us briefly explain what these kinds of elements are, and how they are connected together.

1. **Goals.** When we are asked to solve some problem or to present some argument, there is always some destination, some claim, some conclusion to which we are invited to arrive, and the first step in analyzing and criticizing is to make sure what the precise character of the destination is.

2. **Grounds.** Having clarified the goal, we must consider what kind of information is required to arrive at that goal. The term grounds refers to the *specific* facts relied on to support a given claim. If goals represent the destination, then grounds are the starting point. Even if we are analyzing someone else's solution or argument we must ask ourselves where we would begin and determine whether we can see how to take the same steps and so end by agreeing that the goal has been achieved.

Depending on what kind of goal is under discussion, these grounds may comprise matters of common knowledge, experimental observations, previously established claims, or other factual data that may or may not have been given in the problem.

3. **Warrants.** Knowing on what grounds an argument is founded is, nevertheless, only one step toward the destination. Next we must check whether these grounds really do provide genuine support for each individual assertion and are not just a lot of irrelevant information

having nothing to do with the assertion. The grounds may be true but have no *bearing* on the conclusion; the grounds may be based on correct information yet have no *relevancy* to the issue at hand.

Given the starting point and the goal, the question is: How do you justify the move from *these* grounds to *that* conclusion? The reasons or principles offered as justification are the warrants. In focusing on the warrants, the attention is not so much on the starting point or goal but on the correctness of each step along the way.

Thus there are two concerns about the warrants cited in an argument: Are they reliable and are they applicable? Frequently in a mathematical argument the warrant is some known formula, some commonly accepted fact, an established theorem, or a rule of logical inference. Thus, in this event, the reliability of the warrant may not be in question, but the question as to whether or not the warrant applies is all the more crucial. Moreover, even if the warrant is applicable, its correct application is still required. The laws of algebra may apply but there are restraints on their application; for instance, one cannot divide by zero.

4. **Frame of reference.** Aside from the particular fact, rule, theorem, formula, or principle that serves as grounds for an argument, we need to determine the frame of reference, that whole interlocking web of ideas, facts, definitions, theorems, principles, tacit assumptions, and methods that is presupposed by the warrant appealed to in the argument; otherwise, the warrant is likely to be meaningless.

The conclusions arrived at in an argument are *well founded* only if sufficient grounds of an appropriate and relevant kind can be offered in their support. The grounds must be connected to the conclusion by reliable, applicable warrants, which are capable of being justified by appeal to a relevant frame of reference.

Working Forward

The student probably has seen presentations of arguments in high school plane geometry that were patterned after the model of exposition presented in Euclid's *Elements,* written about 300 B.C. In the Euclidean exposition, all arguments proceed generally in the same direction: from the grounds toward the goals by way of reliable warrants. Any new assertion has to be correctly proved from the given hypothesis or from propositions correctly proved in foregoing steps. It is not enough that correct statements are listed, but they must be listed in logical order, each leading into the next. All statements should be connected together and organized into a well-adapted whole for "precept must be upon precept, precept upon precept, line upon line, line upon line." We call this method of presentation—*working forward*—and on occasion one can discover solutions following this pattern. We illustrate the general scheme by the following diagram:

Working Forward

Figure 1-2.

We emphasize that in *relating* a proof or a solution most writers follow the Euclidean model and work forward from the hypothesis to the conclusion. As a consequence, there may be a natural bias to work forward in *discovering* a solution, but this bias is often inappropriate in problem solving because the order in which we discover details is very often exactly *opposite* from the order in which we relate those details. Frequently, the critical insight is gained after focusing attention on the goal, focusing on the conclusion rather than the hypothesis.

The Aspects of Discovery

The discovery of the solution of a problem is likely to encompass at least four stages of development: (1) education, (2) experimentation, (3) incubation, and (4) revelation. In general terms, let us describe what we think each of these entail.

1. **Education.** Certainly you have little hope of solving a problem if you do not have a sound understanding of the statement of the problem. A first order of business then is to *obtain a precise statement of the problem in unambiguous language.* This probably will entail reviewing technical definitions, determining the general context and frame of reference, and identifying the different parts of the problem (like the goal and grounds).

2. **Experimentation.** In this stage you may want to examine special cases of the problem, replace conditions by equivalent ones, consider logical alternatives (like arguing by contrapositive or contradiction), or decompose the problem into parts and work on it case by case. In general terms, attempt to *reformulate* the problem and *reduce the complexity* of the problem.

3. **Incubation.** After becoming thoroughly familiar with the problem, it may take some time for these ideas to germinate. You may have to go

over steps 1 and 2 again and rethink the problem. As a general rule you probably should focus your attention on the goal and how to get there rather than focusing on the starting point.

4. **Revelation.** The critical insight may come at any moment, so be prepared to write it down and test it out as soon as possible.

Thus, our advice includes the following suggestions:

- Clarify the problem.
- Reformulate the problem.
- Reduce the complexity of the problem.
- Focus on the goal.

Perhaps a few more comments will be helpful to understand our suggestions.

Many times a problem can be translated from one in words to a mathematical problem by finding an equation to solve. The "word problems" in a college algebra course serve as an example.

The complexity of a problem may be reduced in a variety of ways; consider special cases, fewer variables, etc. For instance, suppose we are asked to count the number of elements in a set A. Then if A were the disjoint union of two sets B and C we need only count the number of elements in B and C separately and take the sum of these two numbers to discover the number of elements in A. This simple idea is one of the fundamental rules of counting called the *sum rule*. The difficulty of such counting problems depends on how difficult it is to spot a way of dividing the set A into such subsets which themselves can be counted with ease.

Inductive Reasoning

Another particular process is worthy of special attention, namely, *inductive* or *scientific reasoning*. To understand this process, we need to clarify the meaning of *specialization* and *generalization*.

Generalization is passing from the consideration of a restricted set (usually a small number of observations) to a more comprehensive set containing the original more restricted set. For instance, if a proposition holds for all triangles and rectangles, there may be a generalization holding for all polygons. Likewise, if something holds for the integers 2, 3, 5, 7, 29, and 59, it may hold for all prime integers.

Specialization reverses the process by changing the focus from a larger set to a smaller set of objects, say, for example, from the set of polygons down to the subset of triangles, from the set of prime integers down to a subset of one or more specific primes, or from a general integer n down to a specific value of n. In specialization, we examine special, more manageable cases. A good heuristic approach is to set any integer parameters equal to 1,2,3,4, ... in sequence and look for a pattern. Thus the magnitude of the problem may be reduced to simpler cases and certain patterns and relationships more easily observed.

It may be beneficial to make a diagram or tabulate several observations.

In particular, ordering data by one or several of their attributes into a table may help solve problems by facilitating pattern recognition. In addition to a list of successive observations, a list of differences between successive observations is sometimes useful.

As we observe some pattern emerging, we may suspect that this pattern is no mere coincidence, and therefore conjecture a generalization that will account for all observations and hopefully extend beyond the limits of our actual observations and hold true for all cases.

The process of reasoning by which one takes specific observations and formulates a general hypothesis or conjecture which accounts for these observations is called *inductive reasoning* or just plain *induction* (not to be confused with the principle of mathematical induction to be discussed later).

If the method of inductive reasoning is to be fruitful, then the evidence will have to be sufficient to make a conclusion, and the investigator will need to analyze and *interpret the evidence correctly.* Therefore, a word of caution is needed here. Remember that a conjecture is nothing more than a clearly formulated guess; *it is not proved yet.* You may have much confidence in your conjecture, but be careful—your guess could be wrong. A conjecture is merely tentative; it is an attempt to get at the truth, but without verification it should still be regarded with some wariness—a *proof is still required.* Be careful not to jump to conclusions.

We need to pay attention to later cases which could agree or not with the conjecture. A case in agreement makes the conjecture *more credible* but does not prove it. A conflicting case disproves the conjecture. If someone claims a conjecture is true for all cases and there is at least *one exception,* then his conjecture is false. This one exception is enough to refute any would-be rule or general statement. This one object that does not comply with the conjecture is called a *counterexample.*

Example 1.4.1. A few computations might lead us to conjecture that every odd integer greater than 3 can be written as the sum of a prime and a power of 2. Each new computation in agreement with the conjecture adds plausibility to the conjecture and we become more confident about the possibility of its validity. But confidence is not a proof. Indeed the fact that the conjecture holds for the first 63 odd integers does not guarantee validity, for, in fact, the conjecture fails for the integer 127.

You might ask how do we see that 127 is a counter example? We have at least two options: first, we could consider $f(p) = 127 - p$, for all primes p less than 127, and investigate if $f(p)$ is ever a power of 2 for some value of p. Or secondly, we could consider $g(n) = 127 - 2^n$, for all values of n so that 2^n is less than 127, and then investigate if $g(n)$ is prime for some value of n. Obviously, since there are more primes less than 127 than there are powers of 2, the second approach requires less effort. The following table shows that $g(n)$ is never prime for $n = 0,1,2,3,4,5,6$ and thus that 127 is indeed a counter example to the above conjecture.

n	0	1	2	3	4	5	6
$g(n) = 127 - 2^n$	2 (63)	5 (25)	3 (41)	7 (14)	3 (37)	5 (19)	7 (9)

Inductive reasoning does lead frequently to valid conjectures. For instance, consider the following example.

Example 1.4.2. Suppose we have the following problem: Conjecture a general formula for the sum

$$T_n = \frac{1}{2!} + \frac{2}{3!} + \ldots + \frac{n}{(n+1)!}$$

by specializing successive values of the positive integer n. (Recall that $n!$ is the product of $n(n-1)\ldots 2 \cdot 1$.)

If we allow n to be 1, 2, 3, or 4, we get the values of the sum $T_1 = 1/2$, $T_2 = 5/6$, $T_3 = 23/24$, $T_4 = 119/120$. We note that in each case the numerator is one less than the denominator. Moreover, the denominators are respectively 2!, 3!, 4!, and 5!. Is this coincidental? If not, then T_5 should be $(6! - 1)/6!$. In fact,

$$T_5 = T_4 + \frac{5}{6!} = \frac{119}{5!} + \frac{5}{6!} = \frac{719}{6!} = \frac{6! - 1}{6!};$$

We note one other feature: so far T_n has had the denominator $(n + 1)!$ for each observation. Now we are emboldened to conjecture that T_n is always equal to $[(n + 1)! - 1]/(n + 1)!$. Later we will discuss a method to prove this conjecture.

Analysis-Synthesis and Working Backward

We have suggested that you focus on the goal largely to guide the direction of your thoughts. You might do this in a variety of ways but two methods frequently prove beneficial.

We call the first method "*analysis-synthesis.*" Basically, we suppose the problem has a solution and then try to determine its characteristics.

Suppose that the problem is to prove a statement like: If A, then B. The method encompasses two stages. The first stage—the analysis—is the laboratory work so to speak; here the plan is devised. The second stage—the synthesis—is where the plan is actually carried out, where what was discovered in the analysis stage is applied. This stage actually constitutes the proof.

In the analysis stage, we start from what is to be concluded, take it for granted, and draw inferences from the conclusion until we reach something already known, admittedly true, or patently obvious when considered in relation to the given information A. In other words, we assume B, derive or conclude C, from C we derive D, and so on, until we arrive at statement Z that is obvious in conjunction with A. Now we are prepared for the second stage.

In the second stage, we simply reverse the process. Starting with the obvious statement Z, we work forward, following the Euclidean model of expositions and attempt to reverse each step of the derivations to conclude D and then C and then B. Of course, the success of the synthesis stage depends on whether or not each derivation is, in fact, reversible.

Example 1.4.3.   Suppose that we want to prove: If n is a positive integer such that $N = 6^{n+2} + 7^{2n+1}$ is divisible by 43, then $M = 6^{n+3} + 7^{2n+3}$ is divisible by 43. Here we may not have any idea how to proceed from the grounds that N is divisible by 43 to show that M is divisible by 43. However, we could attempt to employ analysis-synthesis and assume that M is divisible by 43. Then $M - N$, $M - 2N$, $M - 3N$, etc., are all divisible by 43. What we would hope is that in the analysis stage we could

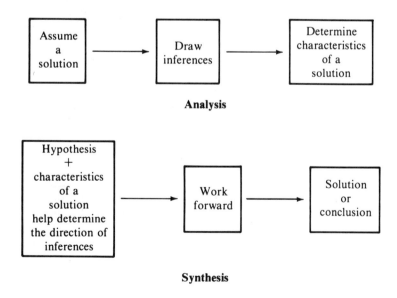

Figure 1-3. Diagram of analysis-synthesis method of problem solving.

discover some one of the numbers $M - kN$ that is *obviously* a multiple of 43, then we could reverse the process in the synthesis stage to prove the proposition. Let us consider

$$M - N = 6^{n+3} + 7^{2n+3} - (6^{n+2} + 7^{2n+1})$$
$$= 6^{n+2}(6 - 1) + 7^{2n+1}(7^2 - 1)$$
$$= 6^{n+2}(5) + 7^{2n+1}(48).$$

At first glance this is no simplification, but we notice first that the powers of 6 and 7 are the same as those in N.

Then if we write $48 = 5 + 43$ we see that

$$M - N = 6^{n+2}(5) + 7^{2n+1}(5) + 7^{2n+1}(43) = 5N + 7^{2n+1}(43).$$

Therefore, $M - 6N = 7^{2n+1}(43)$ is something obviously divisible by 43. Now we are prepared to present the synthesis stage.

Since $M - 6N = 7^{2n+1}(43)$ and N are divisible by 43, $M = 6N + 7^{2n+1}(43)$ is divisible by 43.

The method of *working backward* is similar to analysis-synthesis in that attention is focused on the goal. However, working backward differs in the way the goal is considered in relation to the given information. In the analysis stage of analysis-synthesis, the goal is considered to be part of the given information, and we attempt to derive consequences from the goal in conjunction with the givens. Thus, the direction of inference is from the goal statement to some new statements. In working backward, the goal is not *considered to be a piece of given information.* We start with the goal, but instead of drawing inferences from it, we try to *guess* a preceding statement or statements that, taken together, would imply the goal statement. Frequently, there are theorems or facts in the frame of reference that will give such statements that imply the goal statement. Thus, the person formulating a proof of an implication "If A, then B" is supposed to think like this: "I can prove B if I can prove C; I can prove C if I can prove D; I can prove D if I can prove E. But I can prove E from A."

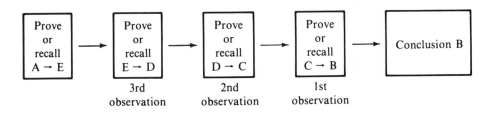

Figure 1-4. Working backward to prove "If A, then B".

Example 1.4.4. Prove the following statement by working backward.

If the right triangle T, with sides of length a and b and hypotenuse of length c, has area equal to $c^2/4$, then T is an isosceles triangle.

We must show $a = b$. But we can prove $a = b$ by showing the equivalent fact $a - b = 0$. To be sure, $a - b = 0$ iff $(a - b)^2 = a^2 - 2ab + b^2 = 0$. So we can prove $a - b = 0$ by showing $a^2 - 2ab + b^2 = 0$ or equivalently, that $a^2 + b^2 = 2ab$. The Pythagorean theorem states that $a^2 + b^2 = c^2$, so to prove $a^2 + b^2 = 2ab$ we need only show $c^2 = 2ab$ or, equivalently, that $c^2/4 = ab/2$. But recall that the area of T is $ab/2$. Thus, by hypothesis, $ab/2 = c^2/4$. Therefore, the proof is completed and we know that $a = b$. $\square$

Example 1.4.5. Recall that for integers a, b, and n, where n is positive, $a \equiv b \bmod n$ means $a - b = k \cdot n$, where k is some integer. Then prove that if $a \equiv b \bmod n$, and $c \equiv d \bmod n$, then $a + c \equiv b + d \bmod n$ by working backward.

We can prove $a + c \equiv b + d \bmod n$ if we can show $(a + c) - (b + d) = n \cdot x$ for some integer x. But $(a + c) - (b + d) = (a - b) + (c - d)$, and we know that $a - b = n \cdot k$ and $c - d = n \cdot l$ where k and l are integers, since $a \equiv b \bmod n$ and $c \equiv d \bmod n$. Thus, $(a + c) - (b + d) = n \cdot k + n \cdot l = n(k + l)$ and the integer $k + l$ is the integer x that we need. $\square$

Fallacies

What constitutes a proof is not always clear, nor is it obvious when an argument is convincing. In fact, mathematicians sometimes disagree among themselves as to whether an argument is sound. Their disagreement may be over whether to allow a particular warrant, such as the law of the excluded middle in logic, or their disagreement may occur when a purported proof is thought to contain some error. Thus, when a mathematical argument is presented, whether or not it is well founded it is usually determined by consensus; the mathematical community decides whether the argument is convincing. An argument is accepted as a valid proof if no one can perceive any flaws in it. Agreement in such matters is very good, but the process is by no means fail-safe. Examples exist of arguments that were widely accepted as proof for many years but were then shown to be fallacious by someone who discovered a possibility that had been overlooked in the original argument. It is frequently the case that the overlooked possibility provides grounds for refutation of the original assertion, but on occasion, the discovery results in a new argument being devised, which is then accepted as a proof.

While all are not agreed as to what makes an argument convincing, all agree that *all cases of an issue must be considered and justification must be given for every conclusion.* Moreover, all would agree that, both in the discovery and the presentation of a solution, *fallacies should be avoided at all costs.*

Fallacies are arguments that are persuasive but unsound, their persuasiveness comes from their superficial resemblance to sound arguments and this similarity serves to camouflage the deception. There are two main types of fallacies:

1. *Fallacies of Ambiguity* or arguments that are flawed because of ambiguities in their constituent terms.

2. *Fallacies of Unwarranted Assumptions* or arguments that involve an unacceptable or illicit step from grounds to conclusion because no appeal at all is made to a warrant, or the appeal is to warrants that are not valid or cannot be applied to the present argument.

Fallacies of ambiguity arise when some crucial term is being used in different senses. The ambiguity may be intentional but is often a result of the imprecise feature of language. Ambiguity should not be confused with vagueness, however. The question "Is there a pitcher in the room?" is ambiguous, because the pitcher in question could be a container or a baseball player. But the statement "I will come to see you sometime this afternoon" is vague, because it fails to tell us precisely what time the visit will take place.

There are many fallacies of ambiguity but we will mention just two: the *fallacy* of *equivocation* and the *fallacy* of *amphiboly.*

The fallacy of equivocation occurs when a word or phrase is used in more than one sense in a single argument with the result that its various senses are confused. You cannot switch from one sense to another in midstream, so to speak. The sentence "Our team needs a new pitcher so bring us one off the shelf in the kitchen" commits the fallacy of equivocation.

Another special kind of ambiguity gives rise to the fallacy of amphiboly. This fallacy occurs as a result of faulty grammar: omission of a comma or other punctuation, careless positioning of qualifying words or phrases, and the like. The *Reader's Digest* often prints humorous examples of such ambiguities.

Amphiboly occurs in mathematical problems because a problem is inadequately formulated, owing to some syntactical ambiguity. Thus the equation "$X = 3 \times 5 + 10$" is ambiguous as it stands, for lack of parentheses. The calculation may yield $X = 25$ or $X = 45$, depending on whether we insert parentheses as $X = (3 \times 5) + 10$ or as $X = 3 \times (5 + 10)$.

Practice identifying grammatical ambiguities can help us to avoid the pitfalls to which ambiguities can lead if their presence in an argument goes unrecognized. Once identified, such fallacies can be eliminated by rewriting and reformulating.

Likewise there are several fallacies of unwarranted assumptions, but we will discuss just four: (1) begging the question, (2) hasty generalization, (3) false cause, and (4) faulty inference.

Circular Reasoning

The fallacy of begging the question is also known as *circular reasoning.* We commit this fallacy when we make an assertion and then argue on its behalf by advancing what is purported to be grounds but whose meaning is simply equivalent to that of the original assertion. We make an assertion A and offer as grounds a statement B in support of the statement A, but actually A and B turn out to mean exactly the same things—though this fact may be concealed because they are phrased in different terms. The error here is that in attempting to prove a certain proposition, the desired proposition itself—perhaps in another form—is unwittingly assumed in the argument. Thus, the argument degenerates into an unconvincing assertion of the type "it's true because it's true," which has no more force than just the assumption of the proposition in question. From another standpoint, the argument might be called the "vicious circle" fallacy.

For example, in proving that the base angles of an isosceles triangle are equal, a student may assume that these angles are equal in order to prove two triangles are congruent and then go on to argue that the angles in question must be equal as they are corresponding parts of congruent triangles!

Example 1.4.6. The assertion that the integer 1987 is prime because it is not composite commits the fallacy of begging the question, as does the sentence "The square root of 2 is irrational because it is not rational." On the other hand, the statement "1987 is prime because 1987 is not divisible by any of the integers 2 through 1986" does not beg the question and is accepted as a fully documented proof, provided, of course, that indeed it has been verified that none of the integers 2 through 1986 divides 1987.

Question begging also occurs in definitions. The so-called circular definition actually begs the question by defining the word only in terms of synonyms. For instance, a dictionary may define the word "modal" as pertaining to "mode," "mode" as "a modality," and finally define "modality" as "the quality or state of being modal." Consider the following definitions:

A dog is a canine animal.

Horsemanship is an equestrian skill.

Distillation is the process of distilling.

Each of the formulations presupposes an understanding of the term to be defined. Anyone who does not already know what a dog is can hardly expect to have any idea of what it means to be canine. Nor, for that matter, is it likely that someone who has no idea of what distillation is will know what is meant by distilling.

We commit the fallacy of "jumping to conclusions" or hasty generalization when we (1) make a general conjecture based on too few specific instances or (2) draw a conclusion from examples that are not representative of the whole class.

For example, we might conclude that for every prime integer N, the integer $N + 2$ is also prime because we have considered the pairs of primes, 3 and 5, 5 and 7, 11 and 13, 17 and 19, and 29 and 31. Of course, if we consider the case where $N = 7$, then $N + 2 = 9$ is not prime and refutes the proposed conjecture.

The fallacy of false cause occurs when we take one event to be the cause of another simply because one event happened *before* the other. The Latin phrase that describes this fallacy is *post hoc ergo propter hoc* which literally means "after this, therefore on account of this." A political party, for example, may take credit for an economic upswing that took place after their party took office without indicating what policy brought about the upswing.

The fallacy of false cause also occurs when we are simply mistaken about a given phenomenon. The history of science abounds with such false attributions of causality; for example, the notion that the earth was flat was maintained for centuries, and even now it is held by some though this notion was dispelled when Magellan sailed around the world.

Exercises for Section 1.4

1. Show that $n = 40$ is a counterexample to the conjecture that $n^2 + n + 41$ is a prime for each integer n.

2. Find a counterexample to the statement "$n^2 - 79n + 1601$ yields a prime for each positive integer n."

3. Show that 509 is a counterexample to the conjecture that every odd integer greater than 3 is the sum of a prime and a power of 2.

4. Conjecture a general formula for the following 3 sums:

    (a) $\dfrac{1}{1 \cdot 2} + \dfrac{1}{2 \cdot 3} + \cdots + \dfrac{1}{n(n + 1)} =$

    (b) $\dfrac{1}{1 \cdot 3} + \dfrac{1}{3 \cdot 5} + \cdots + \dfrac{1}{(2n - 1)(2n + 1)} =$

    (c) $\dfrac{1}{1 \cdot 4} + \dfrac{1}{4 \cdot 7} + \cdots + \dfrac{1}{(3n - 2)(3n + 1)} =$

Conjecture a general formula that includes the results in (a), (b), and (c).

5. Conjecture a general formula for the sum $1 + 3 + 5 + \cdots + (2n - 1)$.

6. Guess the rule according to which the successive terms of the following sequence of numbers are chosen:

   $11, 31, 41, 61, 71, 101, 131, \ldots$.

7. Consider the expressions

$$1 = 0 + 1,$$
$$2 + 3 + 4 = 1 + 8,$$
$$5 + 6 + 7 + 8 + 9 = 8 + 27,$$
$$10 + 11 + 12 + 13 + 14 + 15 + 16 = 27 + 64.$$

   Guess the general law suggested by these examples and express it in suitable mathematical form.

8. The first three terms of the sequence $3, 13, 23, \ldots$ (numbers ending in 3) are all prime. Are the following terms of the sequence also prime integers?

9. Verify that if $11^{n+2} + 12^{2n+1}$ is divisible by 133 then $11^{n+3} + 12^{2n+3}$ is divisible by 133.

10. The table below lists for a few values of a positive integer n the number of positive divisors of n. We denote this number by $d(n)$. For example, $d(4) = 3$ because 1, 2, and 4 are the only positive divisors of 4.

n	1	2	3	4	5	6	7	8	9	10	11	12	13	14	15	16
$d(n)$	1	2	2	3	2	4	2	4	3	4	2	6	2	4	4	5

    Looking over the values of $d(n)$, we are struck by the frequency with which $d(n)$ is an even integer. Continue the table up to $n = 25$. Conjecture a general result.

11. Identify the type of fallacy in the following statements:
    (a) Pearls are soft.
       She is a pearl.
       Hence, she is soft.
    (b) David is telling the truth because he wouldn't lie to me.
    (c) After Melinda insulted her, Anne was mad.
       Mad people should be put in a hospital.
       Hence, Anne should be put in a hospital.

(d) The Democrats cannot blame the Republicans for the high prices because in 1980, when price controls were removed, the Republicans were not in office.

(e) People should do what is right.
People have the right to disregard good advice.
Therefore, people should disregard good advice.

(f) No designing persons are deserving of trust.
Architects are designers by profession.
Hence, architects are to be distrusted.

(g) Killing people is illegal.
Capital punishment is legalized killing.
Hence, capital punishment is illegal.

(h) People in the hospital are ill.
Hence, they should never have gone there.

(i) All members of the Marching Chiefs exercise daily.
If you exercise daily, you can be in the band.

(j) I will never see a doctor again. All my associates who were ill over the past winter went to doctors.

(k) Joe declared that he is insane. But he must be sane because the insane never admit they are insane.

(l) I walked under a ladder.
I was hit by a truck.
Walking under a ladder brings bad luck.

(m) The United Nations is necessary because it is vital that nations work together.

(n) Life makes sense because God exists.
God exists because if He didn't life would be nonsense.

(o) Governmental break-ins are legitimate if sufficient reason is given.
National security is a sufficient reason for a governmental break-in.
In matters of national security, the government should never reveal its reasons for a break-in.
Therefore, the governmental break-in of Ellsberg's office was legitimate, but the government cannot reveal the reason.

12. Make at least four conjectures concerning the values of $f(n) = n^2 - n + 11$ where n is a nonnegative integer. Prove or disprove these conjectures.

13. Find a counterexample to the conjecture that $M_p = 2^p - 1$ is a prime for each prime integer p.

14. Some integers are the sum of 2, 3, or 4 perfect squares. Use inductive reasoning to formulate a conjecture concerning which integers are the sum of the squares of (a) 2 integers, (b) 3 integers, (c) 4 integers. Hint: consider prime integers first.

Selected Answers for Section 1.4

> 11. (a) Ambiguity.
>   (b) Begs the question.
>   (c) Ambiguity.
>   (d) False cause.
>   (e) Ambiguity.
>   (f) Ambiguity.
>   (g) Begs the question.
>   (h) False cause.
>   (i) False cause.
>   (j) False cause.
>   (k) Begs the question.
>   (l) False cause.
>   (m) Begs the question.
>   (n) Begs the question.

1.5   FUNDAMENTALS OF LOGIC

Now before we can understand the fallacy of faulty inference we must discuss, in some detail at least, what we are assuming about logic and what we mean by *valid* inferences. It is not our intention here to discuss a complete course in logic but rather to introduce just enough concepts to formulate the statements of several commonly accepted rules of inference.

First of all, we shall restrict our attention only to those sentences that satisfy two fundamental assumptions that correspond to two famous "laws" of classical logic, the law of the excluded middle and the law of contradiction. We hasten to say that some logicians may not be in agreement with both of these assumptions, but they will serve our purpose.

Sentences are usually classified as declarative, exclamatory, interrogative, or imperative. We confine our attention to those declarative sentences to which it is meaningful to assign one and only one of the truth values "true" or "false." We call such sentences **propositions.**

Of course, not all sentences are propositions because, for one thing, not all sentences are declarative. But we also rule out certain semantical paradoxes like the sentence: "This sentence is false." For if we consider this sentence true, then we must determine from the content of the sentence that it is false, and likewise, if we consider it false, then it turns out to be true.

For definiteness let us list our assumptions about propositions.

Assumption 1.   The Law of the Excluded Middle. For every proposition p, either p is true or p is false.

Assumption 2.   The Law of Contradiction. For every proposition p, it is not the case that p is both true and false.

Five Basic Connectives

Propositions are combined by means of such connectives as *and, or, if . . . then* and *if and only if;* and they are modified by the word *not.* These five main types of connectives can be defined in terms of the three: *and, or,* and *not.*

We proceed to give the definitions of these connectives. If p is a proposition, then "p is not true" is a proposition, which we represent as $\sim p$, and refer to it as "not p", "the **negation** of p", or "the **denial** of p". Not p is a proposition that is true when p is false and false when p is true. The denial of p is accomplished by preceding p by the words "it is not the case that" or by the insertion of the word "not" in an appropriate place. For example, the negation of the proposition: Einstein is a genius, is the statement:

It is not the case that Einstein is a genius,

or

It is false that Einstein is a genius,

or

Einstein is not a genius.

While negation changes one proposition into another, other connectives combine two propositions to form a third. If p and q are propositions, then "p and q" is a proposition, which we represent in symbols as $p \wedge q$ and refer to it as the **conjunction** of p and q. The conjunction of p and q is true only when both p and q are true. On the other hand, the proposition "p or q," called the **disjunction** of p and q, and denoted by $p \vee q$, is true whenever at least one of the two propositions is true. Here we have defined *or* in the *inclusive* sense—either p is true or q is true or both are true so this "or" could be known as *inclusive or.* But, of course, "or" can be used in the exclusive sense—either p is true or q is true, but not both. Generally speaking in everyday usage the context determines which sense is meant, but, in our usage, if no indication is given, we shall mean the inclusive or.

The proposition "p implies q" or "if p then q" is represented as $p \rightarrow q$

and is called an **implication** or a **conditional.** In this setting, p is called the **premise, hypothesis,** or **antecedent** of the implication, and q is called the **conclusion** or **consequent** of the implication. We define $p \rightarrow q$ as a proposition that is false only when the antecedent p is true and the consequent q is false. It might be beneficial to emphasize the cases when the implication is true or false by the following **truth table** that contains all possible truth values for p and q separately and the corresponding truth values for $p \rightarrow q$. Let T denote "true" and F denote "false."

p	q	$p \rightarrow q$
T	T	T
T	F	F
F	T	T
F	F	T

As an example consider the following statement about geometrical objects: If two angles of a triangle are equal, then the triangle is a right triangle. This proposition, like many others that occur in geometry and in other branches of mathematics, is a conditional statement. The antecedent and the consequent are, respectively:

p: two angles of a triangle are equal,

q: the triangle is a right triangle.

According to our understanding of the English language this proposition seems to predict that whenever two angles of a triangle are equal then the triangle must be a right triangle. Of course, we know that there are cases when the proposition is true for there exist isosceles right triangles. But are there cases where this statement is false? Since a conditional is false only when the antecedent is true and the consequent is false, we should look for a triangle with two equal angles that is not a right triangle. This type of example is easy to find, in fact, an equilateral triangle has all three angles equal and is not a right triangle. Only one such example is needed to show that the proposition is not always true.

Note in particular that an implication $p \rightarrow q$ does not assert that its antecedent p *is* true; nor does it say that its consequent q *is* true. It only says that *if* the antecedent is true, then its consequent is true also. Moreover, this relationship between antecedent and consequent can be expressed by saying that the antecedent will be false if the consequent is false. Therefore, according to our definition of implication, there are two valid principles of implication that, nevertheless, are sometimes considered paradoxical.

(a) A false antecedent p implies any proposition q;
(b) A true consequent q is implied by any proposition p.

To illustrate (a), we are correct in claiming that the following is true:

If 1981 is a leap year, then Isaac Newton discovered America.

Here, if p denotes the proposition "1981 is a leap year" and q denotes "Isaac Newton discovered America," then since p is false, we know immediately that the implication $p \to q$ is true in spite of the falsity of q. To illustrate (b) it is correct to assert that the following implication is true:

If (p) Isaac Newton discovered America, then (q) there are seven days in a week.

Here, since q is true, we know that $p \to q$ is true, regardless of whether p is true or false.

Normally the English language uses implication to indicate a *causal* or *inherent relationship* between a premise and a conclusion. But the above illustrations emphasize that in the language of propositions *the premise need not be related to the conclusion in any substantive way.*

If such combinations are not allowed, then some rule must be given that will determine what propositions can be combined to form implications. Mathematicians have found that it is more difficult to set up such a rule than it is to allow the two paradoxical situations as described in (a) and (b) above. As a result mathematicians do not hesitate to combine any propositions by using any of the connectives.

When two propositions are joined by "if . . . then" *one must be careful not to confuse the truth or falsity of either of the propositions with the truth or falsity of the conditional* or to expect any cause and effect relationship between the antecedent and the consequent.

The proposition $p \to q$ may be expressed as:

p implies q

if p then q

p only if q

p is a sufficient condition for q

q is a necessary condition for p

q if p

q follows from p

q provided p

q is a consequence of p

q whenever p

The **converse** of $p \to q$ is the conditional $q \to p$ and the **biconditional** $p \leftrightarrow q$ is the conjunction of the conditionals $p \to q$ and $q \to p$. The biconditional can be formed with the words "if and only if." Thus, the symbol $p \leftrightarrow q$ is read "p if and only if q." By consulting the truth table of $p \to q$ and $q \to p$, we see that $p \leftrightarrow q$ is true when p and q have the same truth values, and is false otherwise.

Notice that we are talking about mappings here: if we think of p and q not as sentences but as variables taking the value T or F, then the different connectives are functions from either $\{T,F\}$ or $\{T,F\}^2 = \{T,F\} \times \{T,F\}$ into $\{T,F\}$. Negation $\sim$ is a function from $\{T,F\}$ into $\{T,F\}$ while, for example, disjunction $\vee$ is a function from $\{T,F\}^2$ into $\{T,F\}$. There are four elements in $\{T,F\}^2$, namely, (T,T), (T,F), (F,T) and (F,F), and any function from $\{T,F\}^2$ into $\{T,F\}$ must assign a value—either T or F—to each of these four elements. Since each of these four elements may take either of the two values independently, there are $2 \times 2 \times 2 \times 2 = 16$ possible functions from $\{T,F\}^2$ into $\{T,F\}$.

A **propositional function** is a function whose variables are propositions. Thus, there are only 16 propositional functions of two variables. But there can be propositional functions of several variables involving many connectives. For example, the proposition $[(p \wedge q) \vee \sim r] \leftrightarrow p$ may be viewed as a function of the three variables p, q, and r and involves the four connectives $\sim$, $\vee$, $\wedge$, and $\leftrightarrow$. Then as a function of three variables $(p \wedge q) \vee (\sim r) \leftrightarrow p$ maps the 8 points of $\{T,F\}^3$ into $\{T,F\}$. We list all the values of this function in the following truth table:

p	q	r	$p \wedge q$	$\sim r$	$(p \wedge q) \vee (\sim r)$	$[p \wedge q) \vee \sim r) \leftrightarrow p$
F	F	F	F	T	T	F
F	F	T	F	F	F	T
F	T	F	F	T	T	F
F	T	T	F	F	F	T
T	F	F	F	T	T	T
T	F	T	F	F	F	F
T	T	F	T	T	T	T
T	T	T	T	F	T	T

In constructing truth tables, it is useful to follow these two conventions.

1. Place all propositional variables in the left-most columns.

2 Assign truth values to the variables according to the following pattern: Let 0 represent F and 1 represent T, then assign the values of the variable by counting in binary numbers from 0 to $2^k - 1$ where k is the number of propositional variables. (For example, the assignment of the values $T F T$ to p, q, and r in the above example corresponds to the binary number 1 0 1.)

Two functions are the same if they have the same domain and range and take on the same values at the same points. Thus, in the context of propositions, since a truth table lists all the values of a propositional function we then see that two propositional functions are the same if they have the same identical truth tables. Now it is frequently the case that mathematicians say, rather, that then two propositional functions are **logically equivalent** or just **equivalent,** for short. Thus, if P and Q are propositional functions, then P and Q are equivalent if they have the same truth tables, and we write $P \equiv Q$ to mean P is equivalent to Q. Moreover, sad to say, mathematicians frequently become somewhat imprecise here and drop the word "functions" and say that two propositions P and Q are equivalent if P and Q have the same truth table.

A **tautology** is a propositional function whose truth value is *true* for all possible values of the propositional variables; for example, $p \vee \sim p$ is a tautology. A **contradiction** or **absurdity** is a propositional function whose truth value is always *false,* such as $p \wedge \sim p$. A propositional function that is neither a tautology nor a contradiction is called a **contingency.**

Abbrievated Truth Table

Properties of a propositional function can sometimes be determined by constructing an abbreviated truth table. For instance, if we wish to show that a propositional function is a contingency, it is enough to exhibit two lines of the truth table, one which makes the proposition true and another that makes it false. To determine if a propositional function is a tautology, it is only necessary to check those lines of the truth table for which the proposition could be false; or to show that two propositional functions are equivalent we need only check those lines where each function can be false.

Example 1.5.1. To show that $p \rightarrow q$ and $(\sim p) \vee q$ are equivalent we will use an abbreviated truth table. Since $p \rightarrow q$ is false only when p is true and q is false, we need only consider the one line of the truth table of $p \rightarrow q$. But likewise there is only one line of the truth table of $(\sim p) \vee q$ where $(\sim p) \vee q$ is false and that is when $\sim p$ is false and q is false or, in other words, when p is true and q is false. Thus, $(p \rightarrow q) \equiv [(\sim p) \vee q]$.

Example 1.5.2.   Prove that $[(p \wedge \sim q) \to r] \to [p \to (q \vee r)]$ is a tautology.

Since this proposition involves 3 variables, p, q, r the complete truth table would require 8 lines, but we will appeal to an abbreviated truth table. The one case where an implication is false is the case where the antecedent is true but the consequent is false. In the above, the consequent $[p \to (q \vee r)]$ is false when p is true and q and r are both false. Then, the abbreviated truth table addresses this one case:

pqr	$[p \to (q \vee r)]$	$p \wedge \sim q$	$[(p \wedge \sim q) \to r]$	$[(p \wedge \sim q) \to r] \to [p \to (q \vee r)]$
TFF	F	T	F	T

Since the last column has the value T in this case, the proposition must be a tautology.

Propositional Functions of Two Variables

Let us now list the 16 propositional functions of 2 variables.

p	q	1	2	3	4	5	6	7	8	9	10	11	12	13	14	15	16
F	F	T	F	T	F	T	F	T	F	T	F	T	F	T	F	T	F
F	T	T	T	F	F	T	T	F	F	T	T	F	F	T	T	F	F
T	F	T	T	T	T	F	F	F	F	T	T	T	T	F	F	F	F
T	T	T	T	T	T	T	T	T	F	F	F	F	F	F	F	F	F

Examining this table, we readily find the representation of the connectives we have already defined: disjunction ($\vee$) is in column 2, conjunction ($\wedge$) is in column 8, p is represented in column 4, $\sim p$ is in column 13, q is represented in column 6, and $\sim q$ in column 11; column 1 is "universally true" and so is a tautology, and column 16 is "universally false" and thus represents a contradiction. Moreover, $p \to q$ is represented in column 5 while the converse of $p \to q$, or $q \to p$, is represented in column 3.

What of the remaining 6 columns? It would seem plausible that these *also* represent ways of combining propositions. That is indeed the case, and in fact some are just the negation of columns already mentioned. Thus, as column 2 represents the disjunction $p \vee q$, column 15 represents nondisjunction or $\sim(p \vee q)$.

Column 7 represents $p \leftrightarrow q$, and column 10 represents the negation of $p \leftrightarrow q$ and is at the same time the truth table for the exclusive or. We denote $\sim(p \leftrightarrow q)$ by $p \leftarrow\!/\!\to q$. Columns 12 and 14 are the negations of columns 5 and 3, respectively. We shall write $p —\!/\!\to q$ for $\sim(p \to q)$ and $q —\!/\!\to p$ for $\sim(q \to p)$.

Thus, you should see that an understanding for $\sim$, $\vee$, $\wedge$, and $\rightarrow$ gives an understanding of all 16 propositional functions of 2 variables, and since $p \rightarrow q$ is equivalent to $(\sim p) \vee q$, in fact, $\sim$, $\vee$, and $\wedge$ generate all 16 functions. Let us list these functions now with the assignments we have made.

The Connectives of 2 Propositions

"True"	$p \vee q$	$q \rightarrow p$	p	$p \rightarrow q$	q	$p \leftrightarrow q$	$p \wedge q$
T	T	T	T	T	T	T	T
T	T	T	T	F	F	F	F
T	T	F	F	'T	T	F	F
T	F	T	F	T	F	T	F

$\sim(p \wedge q)$	$p \leftarrow/\rightarrow q$	$\sim q$	$p-/\rightarrow q$	$\sim p$	$q-/\rightarrow p$	$\sim(p \vee q)$	"False"
F	F	F	F	F	F	F	F
T	T	T	T	F	F	F	F
T	T	F	F	T	T	F	F
T	F	T	F	T	F	T	F

Now, using all the connectives between 2 propositions, it is possible to generate many more than 16 such propositions. It follows, then, that there are several equivalences. Let us list some of the equivalences.

1. $\sim(p \vee q) \equiv (\sim p) \wedge (\sim q)$ ⎫ (DeMorgan's laws)
2. $\sim(p \wedge q) \equiv (\sim p) \vee (\sim q)$ ⎭
3. $p \equiv \sim(\sim p)$ (Law of double negation)
4. $(p \rightarrow q) \equiv (\sim p) \vee q$ (Law of implication)
5. $(p \rightarrow q) \equiv (\sim q \rightarrow \sim p)$ (Law of contrapositive)

DeMorgan's laws are useful in forming negations of disjunctions and conjunctions. For example, suppose p and q are the following propositions:

p: God makes little green apples.

q: It rains in Indianapolis in the summer time.

The negation of $p \vee q$ is: It is false that God makes little green apples or it rains in Indianapolis in the summer time. By DeMorgan's law, since $\sim(p \vee q) \equiv (\sim p) \wedge (\sim q)$, we see that the negation of $p \vee q$ is also: God doesn't make little green apples and it doesn't rain in Indianapolis in the summer time.

Example 1.5.3. Verify the first DeMorgan law. $\sim(p \vee q)$ is true precisely when $p \vee q$ is false or when both p and q are false. In this case $(\sim p) \wedge (\sim q)$ is true.

Conversely, if $(\sim p) \wedge (\sim q)$ is true then p and q are false and $p \vee q$ is false so that $\sim(p \vee q)$ is true.

Thus, these two propositional functions have the same truth table and are therefore equivalent.

If $p \to q$ is a proposition, then $\sim q \to \sim p$ is called the **contrapositive** of $p \to q$. The law of contrapositive states that $p \to q$ and its contrapositive are equivalent propositions.

The contrapositive of $(\sim q) \to (\sim p)$ is $\sim(\sim p) \to \sim(\sim q)$ and since $\sim(\sim p) \equiv p$ and $\sim(\sim q) \equiv q$, we see that the contrapositive of $(\sim q) \to (\sim p)$ is equivalent to $p \to q$.

The **converse** of $p \to q$ is $q \to p$. The **opposite** of $p \to q$ (sometimes called the **inverse** of $p \to q$) is the proposition $(\sim p) \to (\sim q)$, which is the contrapositive of $q \to p$, and thus equivalent to $q \to p$. The following truth table shows that $p \to q$ and $(\sim q) \to (\sim p)$ are equivalent and $q \to p$ and $(\sim p) \to (\sim q)$ are equivalent.

p	q	$\sim p$	$\sim q$	$p \to q$	$(\sim q) \to (\sim p)$	$q \to p$	$(\sim p) \to (\sim q)$
F	F	T	T	T	T	T	T
F	T	T	F	T	T	F	F
T	F	F	T	F	F	T	T
T	T	F	F	T	T	T	T

The meanings of contrapositive, converse, and opposite of an implication will become apparent upon examining the following diagrams.

Implication	Converse
$p \to q$. If p, then q. p is sufficient for q. q is necessary for p.	$q \to p$. If q, then p. q is sufficient for p. p is necessary for q.
Opposite	Contrapositive
$(\sim p) \to (\sim q)$. If not p, then not q (equivalent to the converse).	$(\sim q) \to (\sim p)$. If not q, then not p (equivalent to the implication).

Theorem	Converse
$p \rightarrow q$	$q \rightarrow p$
If triangle I and triangle II are similar, then the corresponding sides of triangles I and II are proportional. True	If the corresponding sides of triangles I and II are proportional, then triangle I and triangle II are similar. True
Opposite	Contrapositive
$\sim p \rightarrow \sim q$	$\sim q \rightarrow \sim p$
If triangle I and triangle II are *not* similar, then the corresponding sides of triangles I and II are *not* proportional. True	If the corresponding sides of triangles I and II are *not* proportional, then triangle I and II are not similar. True

Theorem	Converse
$p \rightarrow q$	$q \rightarrow p$
If the quadrilateral ABCD is a square, then the sides of quadrilateral ABCD are equal. True	If the sides of quadrilateral ABCD are equal, then the quadrilateral ABCD is a square. False
Opposite	Contrapositive
$\sim p \rightarrow \sim q$	$\sim q \rightarrow \sim p$
If the quadrilateral ABCD is *not* a square, then the sides of quadrilateral ABCD are *not* equal. False	If the sides of quadrilateral ABCD are *not* equal, then the quadrilateral ABCD is *not* a square. True

The latter diagram shows that the converse of a theorem is not necessarily true just because the theorem is true!

Summary

From the definition of $\sim$, $\vee$, $\wedge$, $\rightarrow$, and $\leftrightarrow$, let us remember the following:

1. Either p is true or $\sim p$ is true (not both).
2. It is not possible for both p and $\sim p$ to be true simultaneously. (Thus, if an assumption leads to a situation where p and $\sim p$ are both found to be true, then we say that the assumption has led to a contradiction and the assumption must be false.)
3. If $p \wedge q$ is true, then p must be true and q must be true.

4. If $p \wedge q$ is false, then at least one of p, q must be false. Hence, in particular, if $p \wedge q$ is false and p is true, then q must be false; and, similarly, if $p \wedge q$ is false and q is true, then p must be false.

5. If $p \vee q$ is false, then p must be false and so must q.

6. If $p \vee q$ is true, then at least one of p, q is true. Hence, in particular, if $p \vee q$ is true and p is false, then q must be true; and, similarly, if $p \vee q$ is true and q is false, then p must be true. (Note, however, that if $p \vee q$ is true and if p is true, we cannot conclude that q is true or that it is false.)

7. If $p \rightarrow q$ is true and p is true, then q must be true. (Note, however, if $p \rightarrow q$ is true and q is true, p could be true or false.)

8. If $p \rightarrow q$ is true and q is false, then p must be false.

9. If $p \rightarrow q$ is true, then either p is false or q is true, or both. (Note that we cannot in this case conclude explicitly that p is false and q is true.)

10. If $p \rightarrow q$ is false, then p must be true and q must be false.

11. If $p \leftrightarrow q$ is true, p and q must have the same truth value.

12. If $p \leftrightarrow q$ is false, then p and q have opposite truth values.

Exercises for Section 1.5

1. Construct truth tables for the following:
   (a) $[(p \vee q) \wedge (\sim r)] \leftrightarrow q$.
   (b) $(p \vee q) \wedge ((\sim p) \vee (\sim r))$.
   (c) $\{(p \wedge q) \vee (\sim p \wedge r)\} \vee (q \wedge r)$.
   (d) $[(p \vee q) \wedge (\sim r)] \leftrightarrow (q \rightarrow r)$.

2. Prove the following are tautologies:
   (a) $\sim(p \vee q) \vee [(\sim p) \wedge q] \vee p$.
   (b) $[(p \rightarrow q) \wedge (r \rightarrow s) \wedge (p \vee r)] \rightarrow (q \vee s)$.
   (c) $[(p \rightarrow r) \wedge (q \rightarrow r)] \rightarrow [(p \vee q) \rightarrow r]$.
   (d) $\{[p \rightarrow (q \vee r)] \wedge (\sim q)\} \rightarrow (p \rightarrow r)$.
   (e) $\{[p \vee q) \rightarrow r] \wedge (\sim p)\} \rightarrow (q \rightarrow r)$.

3. Consider the propositions:
   p: David is playing pool.
   q: David is inside.
   r: David is doing his homework.
   s: David is listening to music.
   Translate the following sentences into symbolic notation using $p, q, r, s, \sim, \vee, \wedge$, and parentheses only.
   (a) Either David is playing pool or he is inside.
   (b) Neither is David playing pool, nor is he doing his homework.

(c) David is playing pool and not doing his homework.

(d) David is inside doing his homework, not playing pool.

(e) David is inside doing his homework while listening to music, and he is not playing pool.

(f) David is not listening to music, nor is he playing pool, neither is he doing his homework.

4. Using the specifications of p, q, r, and s of Exercise 3 translate the following propositions into acceptable English.

(a) $(\sim p) \wedge (\sim q)$.

(b) $p \vee (q \wedge r)$.

(c) $\sim((\sim p) \wedge r)$.

(d) $[(\sim p) \vee q] \wedge [\sim r \vee s]$.

(e) $[(\sim p) \wedge q] \vee [(\sim r) \wedge s]$.

5. Restate the following implications, $p \rightarrow q$, in the equivalent form, $(\sim p) \vee q$.

(a) If he fails to follow orders, he will lose his commission.

(b) If the work is not finished on time, then I am in trouble.

(c) If triangle ABC is isosceles, then the base angles A and B are equal.

(d) If K-Mart does not refund the money, I will not shop there anymore.

(e) If lines AB and CD are parallel, then the alternate interior angles are equal.

6. Restate the following as implications "If . . ., then . . .":

(a) A necessary condition that a given quadrilateral $ABCD$ be a rectangle is that it be a parallelogram.

(b) A sufficient condition that $ABCD$ be a rectangle is that it be a square.

(c) A necessary condition that a given integer n be divisible by 9 is that it is divisible by 3.

(d) A sufficient condition that a given integer n be divisible by 9 is that it be divisible by 18.

(e) A sufficient condition for n to be divisible by 9 is that the sum of its digits be divisible by 9.

7. State the converse, opposite, and contrapositive to the following:

(a) If triangle ABC is a right triangle, then $|AB|^2 + |BC|^2 = |AC|^2$.

(b) If the triangle is equiangular, then it is equilateral.

Selected Answers for Section 1.5

1. (a)

p	q	r	$p \vee q$	$\sim r$	$(p \vee q) \wedge \sim r$	$[(p \vee q) \wedge \sim r] \leftrightarrow q$
F	F	F	F	T	F	T
F	F	T	F	F	F	T
F	T	F	T	T	T	T
F	T	T	T	F	F	F
T	F	F	T	T	T	F
T	F	T	T	F	F	T
T	T	F	T	T	T	T
T	T	T	T	F	F	F

3. (a) $p \vee q$
   (b) $(\sim p) \wedge (\sim r)$
   (c) $p \wedge (\sim r)$
   (d) $(\sim p) \wedge q \wedge r$

5. (d) K-Mart refunds the money or I will not shop there anymore.
   (e) Lines AB and CD are not parallel or the alternate interior angles are equal.

7. (a) (Converse) If $|AB|^2 + |BC|^2 = |AC|^2$, then triangle ABC is a right triangle.
   (Opposite) If triangle ABC is not a right triangle, then $|AB|^2 + |BC|^2 \neq |AC|^2$.
   (Contrapositive) If $|AB|^2 + |BC|^2 \neq |AC|^2$, then triangle ABC is not a right triangle.

1.6   LOGICAL INFERENCES

We have said that a well-founded proof of a theorem is a sequence of statements which represent an argument that the theorem is true. Some statements appear as grounds, some as warrants, and some are known as part of the frame of reference. Other statements may be given as part of the hypothesis of the theorem, assumed to be true in the argument. But some assertions must be *inferred* from those that have occurred earlier in the proof.

Thus, to construct proofs, we need a means of drawing conclusions or deriving new assertions from old ones; this is done using **rules of inference.** Rules of inference specify which conclusions may be inferred legitimately from assertions known, assumed, or previously established.

Now in this context let us define **logical implication.** A proposition *p logically implies* a proposition q, and q is a **logical consequence** of p, if the implication $p \to q$ is true for all possible assignments of the truth values of p and q, that is, if $p \to q$ is a tautology. Much care must be taken *not* to confuse *implication* (or *conditional*) with *logical implication*. The conditional is only a way of connecting the two propositions p and q, whereas if p logically implies q then p and q *are related* to the extent that whenever p has the truth value T then so does q. We do note that every logical implication is an implication (conditional), but not all implications are logical implications.

If we examine the truth table of the conditional again, we recall that whenever the antecedent is false, the conditional is true. Moreover, the only time the implication is false is when the antecedent is true and the consequent is false. This allows us, then, to shorten the work involved in checking whether a conditional is or is not a logical implication. All we need do is check all possible cases in which the antecedent is true to see if the consequent is also true. If this is the case, then the implication $p \to q$ is, in fact, a logical implication. (Of course, we would draw the same conclusion if we checked all possible cases where the consequent is false and determined in those cases that the antecedent is also false.)

The word *inference* will be used to designate a set of premises accompanied by a suggested conclusion regardless of whether or not the conclusion is a logical consequence of the premises. Thus, there are **faulty inferences** and **valid inferences.** Each inference can be written as an implication as follows:

(conjunction of premises) $\to$ (conclusion).

We say that this inference is **valid** if the implication is a tautology, that is, if the implication is a logical implication or, in other words, if the conclusion is a logical consequence of the conjunction of all the premises. Otherwise, we say that the inference is faulty or invalid.

The important fact to realize is that *in a valid inference, whenever all the premises are true so is the conclusion* (the case where one or more of the premises are false automatically gives the implication a truth value T, regardless of the truth value of the conclusion). Therefore we reiterate: to check that an inference is valid or not, it is sufficient to check only those rows of the truth table for which all the premises are true, and see if the conclusion also is true there.

It is important not to confuse the validity of an inference with the truth of its conclusion. The fact that the conclusion of a valid inference is not necessarily true can be seen from the following argument:

If Joe reads *The Daily Worker,* then he is a communist.
Joe reads *The Daily Worker.* Therefore, he is a communist.

If p represents the statement "Joe reads *The Daily Worker*" and q is substituted for "he is communist," then the above inference is valid because $[p \land (p \to q)]$ logically implies q (we need only check the truth table when p and $p \to q$ is true). But the conclusion of the above argument need not be *true,* for Joe may be a propaganda interpreter for some government agency or a political science instructor.

This is an example of a valid inference but an unsound argument since we define an argument to be *sound* if the inference is *valid* and the premises and conclusion are *true.*

The most familiar type of proof uses two fundamental rules of inference.

Fundamental Rule 1.   If the statement in p is assumed as true, and also the statement $p \to q$ is accepted as true, then, in these circumstances, we must accept q as true.

Symbolically we have the following pattern, where we use the familiar symbol $\therefore$ to stand for "hence" or "therefore":

$$\frac{\begin{array}{c} p \\ p \to q \end{array}}{\therefore q} .$$

In this tabular presentation of an argument, the assertions above the horizontal line are the **hypotheses** or **premises;** the assertion below the line is the **conclusion.** (Observe that the premises are not accompanied by a truth value, we *assume* they are true.) The rule depicted is known as *modus ponens* or the *rule of detachment.* The rule of detachment is a valid inference because $[p \land (p \to q)] \to q$ is a tautology.

For example, suppose that we know the following two statements:

It is 11:00 o'clock in Tallahassee.

If it is 11:00 o'clock in Tallahassee, then it is 10:00 o'clock in New Orleans.

Then, by the rule of detachment we must conclude:

It is 10:00 o'clock in New Orleans.

The distinction between implication and inference is worth emphasizing. It is essentially this: the truth of an implication $p \to q$ does not guarantee the truth of either p or q. But the truth of *both* p and $p \to q$ does guarantee the truth of q.

It is understood that in place of the propositions p, q in the statement of the rule of detachment we can substitute propositional functions of any degree of complexity. For example, the rule of detachment would permit us to make the following inference, where we assume the proposition above the horizontal line as premises:

$$\frac{\begin{array}{l} p \wedge (q \vee r) \\ [p \wedge (q \vee r)] \to s \wedge [(\sim t) \vee u] \end{array}}{\therefore\ s \wedge [(\sim t) \vee u]}$$

The following line of argument is typical in geometry:

Since (p) two sides and the included angle of triangle ABC are equal, respectively, to two sides and the included angle of triangle $A'B'C'$, (q) triangle ABC is congruent to triangle $A'B'C'$.

The argument as thus stated is abbreviated; there is more to it than the words that appear. Using p and q as indicated in parentheses, the completed argument would appear as follows:

Since p is true and $p \to q$ is true (by a special case of a previous theorem established in the framework of geometry), hence q is true (by the rule of detachment).

Another rule of inference is commonly called the **law of hypothetical syllogism** or the **transitive rule.**

Fundamental Rule 2.   Whenever the two implications $p \to q$ and $q \to r$ are accepted as true, we must accept the implication $p \to r$ as true.

In pattern form, we write:

$$\frac{\begin{array}{l} p \to q \\ q \to r \end{array}}{\therefore p \to r}\ .$$

This rule is a valid rule of inference because the implication

$$(p \to q) \wedge (q \to r) \to (p \to r)$$

is a tautology.

The transitive rule can be extended to a larger number of implications as follows:

$$\frac{\begin{array}{l} p \to q \\ q \to r \\ r \to s \end{array}}{\therefore p \to s}\ .$$

Most arguments in mathematics are based on the two fundamental rules of inference, with occasional uses of the law of contrapositive and DeMorgan's laws. Therefore, we suggest that the student become thoroughly versed in understanding at least these rules.

There are other valid inferences; we list some of the more important ones in the following table. We do not discuss them because of space limitations. You will notice that some of these other rules of inference are nothing more than a reinterpretation of the two fundamental rules in the light of the law of contraposition.

Rules of Inference Related to the Language of Propositions

Rule of Inference	Tautological Form	Name
1. p $\therefore p \vee q$	$p \rightarrow (p \vee q)$	addition
2. $p \wedge q$ $\therefore p$	$(p \wedge q) \rightarrow p$	simplification
3. p $p \rightarrow q$ $\therefore q$	$[p \wedge (p \rightarrow q)] \rightarrow q$	*modus ponens*
4. $\sim q$ $p \rightarrow q$ $\therefore \sim p$	$[\sim q \wedge (p \rightarrow q)] \rightarrow \sim p$	*modus tollens*
5. $p \vee q$ $\sim p$ $\therefore q$	$[(p \vee q) \wedge \sim p] \rightarrow q$	disjunctive syllogism
6. $p \rightarrow q$ $q \rightarrow r$ $\therefore p \rightarrow r$	$[(p \rightarrow q) \wedge (q \rightarrow r)] \rightarrow [p \rightarrow r]$	hypothetical syllogism
7. p q $\therefore p \wedge q$		conjunction
8. $(p \rightarrow q) \wedge (r \rightarrow s)$ $p \vee r$ $\therefore q \vee s$	$[(p \rightarrow q) \wedge (r \rightarrow s) \wedge (p \vee r)] \rightarrow [q \vee s]$	constructive dilemma
9. $(p \rightarrow q) \wedge (r \rightarrow s)$ $\sim q \vee \sim s$ $\therefore \sim p \vee \sim r$	$[(p \rightarrow q) \wedge (r \rightarrow s) \wedge (\sim q \vee \sim s)] \rightarrow [\sim p \vee \sim r]$	destructive dilemma

Since most of the rules in the above table follow from the two fundamental rules, DeMorgan's laws, and the law of contraposition, we list them as fundamental rules.

Fundamental Rule 3:   DeMorgan's laws.

Fundamental Rule 4:   Law of contrapositive.

We can summarize the above table of rules of inferences by saying these rules constitute valid methods of reasoning—valid arguments—valid inferences. Nevertheless, the truth of the conclusion need not follow from the validity of the argument unless the premises are also true. A faulty inference, on the other hand, may sometimes include a true conclusion.

Fallacies

There are three forms of faulty inferences that we will now discuss:

1. The fallacy of affirming the consequent (or affirming the converse).
2. The fallacy of denying the antecedent (or assuming the opposite).
3. The *non sequitur* fallacy.

The fallacy of affirming the consequent is presented in the following form:

$$\frac{\begin{array}{c} p \to q \\ q \end{array}}{\therefore\ p}\ . \qquad \text{Fallacy}$$

Consider the following argument:

If the price of gold is rising, then inflation is surely coming. Inflation is surely coming. Therefore, the price of gold is rising.

This argument is faulty because the conclusion can be false even though $p \to q$ and q are true, that is, the implication $[(p \to q) \land q] \to p$ is *not* a tautology: the cause of inflation may not have been related to the price of gold at all but perhaps to the price of oil, the overall increase in wages, or some other cause.

The fallacy of denying the antecedent takes the form:

$$\frac{\begin{array}{c} p \to q \\ \sim p \end{array}}{\therefore\ \sim q}\ . \qquad \text{Fallacy}$$

Since the opposite of $p \to q$ is $\sim p \to \sim q$, this fallacy is the same as affirming the opposite. The converse $q \to p$ of $p \to q$ need not hold if $p \to q$ is true, and since the contrapositive of $q \to p$ is equivalent to $q \to p$ we see that affirming the opposite is equivalent to affirming the converse.

In a sense, the fallacies of assuming the converse or opposite, and perhaps all logical errors, are special cases of the *non sequitur* fallacy. *Non sequitur* means "it does not follow." A typical pattern of a *non sequitur* error is the following:

$$\frac{p}{\therefore q}.$$

This is like the pattern of the law of detachment with the premise $p \to q$ omitted. Of course, if this premise is known to be correct, the argument is valid, though abbreviated. But conceivably the premise $p \to \sim q$ could hold, in which case the correct conclusion would be $\sim q$ instead of q.

For example, consider the argument:

If Socrates is a man, then Socrates is mortal.

Socrates is a man.

Therefore, Socrates is mortal.

This argument is valid because it follows the pattern of *modus ponens*. However, consider the argument:

Socrates is a man.

Therefore, Socrates is mortal.

Here the conclusion may be thought to follow from the premise, but it does so only because of the meanings of "man" and "mortal," not by mere inference.

Let us put the arguments in symbolic form. The first argument has the form:

$$\begin{array}{c} p \to q \\ \underline{p} \\ \therefore q \end{array}.$$

The second has the form:

$$\frac{p}{\therefore q}.$$

It is the *form* of the first which makes it valid; any other argument with the same form would also be valid. However, the second argument does not share this quality. There are many arguments of the second form which we would not regard as valid. For example:

A triangle has three sides.

Therefore, a triangle is a square.

Here this argument has the form:

$$\frac{p}{\therefore q},$$

but we would not consider it a valid argument.

Notice that what remains when arguments are symbolized in this way is the bare logical bones, the mere *form* of the argument which many arguments may have in common regardless of the content of the sentences. It is precisely this form that enables us to analyze the inference, for deduction has more to do with the forms of the propositions in an argument than with their meanings.

To illustrate these ideas, let us determine whether or not the following arguments are valid.

If a baby is hungry, then the baby cries. If the baby is not mad, then he does not cry. If a baby is mad, then he has a red face. Therefore, if a baby is hungry, then he has a red face.

The basic statements may be represented with the following symbols:

h: a baby is hungry,
c: a baby cries,
m: a baby is mad, and
r: a baby has a red face.

Then the argument takes the following form:

$$\frac{\begin{array}{l} h \rightarrow c \\ \sim m \rightarrow \sim c \\ m \rightarrow r \end{array}}{\therefore h \rightarrow r}.$$

We see that this is a valid inference because $\sim m \rightarrow \sim c$ is the contrapositive of $c \rightarrow m$. Thus, with this replacement, we have:

$$\frac{\begin{array}{l} h \rightarrow c \\ c \rightarrow m \\ m \rightarrow r \end{array}}{\therefore h \rightarrow r}.$$

In this final form, the form of the argument is nothing more than Fundamental Rule 2 and therefore is valid.

Consider the following argument:

If Nixon is not reelected, then Tulsa will lose its air base.
Nixon will be reelected if and only if Tulsa votes for him.
If Tulsa keeps its air base, Nixon will be reelected.
Therefore, Nixon will be reelected.

Let us make the following representations:

R: Nixon will be reelected,
T: Tulsa votes for Nixon, and
A: Tulsa keeps its air base.

The form of the argument is:

$$\sim R \rightarrow \sim A$$
$$R \leftrightarrow T$$
$$A \rightarrow R$$
$$\therefore R$$

Now $\sim R \rightarrow \sim A$ and $A \rightarrow R$ are equivalent so that actually the argument can be simplified to:

$$R \leftrightarrow T$$
$$A \rightarrow R$$
$$\therefore R$$

We suspect that the *non sequitur* fallacy has been committed since neither A nor T is a premise. Of course, we could consider the truth table of $[(A \rightarrow R) \wedge (R \leftrightarrow T)] \rightarrow R$ to see that we do not have a tautology and thus that the inference is invalid. Nevertheless, a valid inference would have been $A \rightarrow T$, or in words:

If Tulsa keeps its air base, then Tulsa votes for Nixon.

Consider the argument:

If a pair of angles A and B are right angles, then they are equal.
The angles A and B are equal.
Hence, the angles A and B must be right angles.
Represent the statements as follows:

R: a pair of angles A and B are right angles,
E: the angles A and B are equal.

We therefore have an argument of the form:

$$R \to E$$
$$\underline{E}$$
$$\therefore R$$

Obviously, this argument is faulty; in fact, it demonstrates the fallacy of affirming the consequent.

Exercises for Section 1.6

I. Complete the blanks in the following sets of propositions, so that each set is in conformity with the rule of detachment. (In each case the first two propositions are to be assumed as premises, so that no question as to the actual truth or falsity of any of the prepositions is to be considered.)

1. (a) If the year N is a leap year, then N is a multiple of four.
   (b) The number 1984 is a multiple of four.
   (c) Hence, . . .
2. (a) If high interest rates are to be continued, then the housing industry will be hurt.
   (b) High interest rates are to be continued.
   (c) Hence, . . .
3. (a) If today is Thursday, ten days from now will be Sunday.
   (b) Today is Thursday.
   (c) Hence, . . .
4. (a) If today is Thursday, ten days from now will be Monday.
   (b) Today is Thursday.
   (c) Hence, . . .
5. (a) . . . .
   (b) 1984 is a leap year.
   (c) Hence 1984 is a presidential election year.
6. (a) If today is Sunday, then I will go to church.
   (b) . . . .
   (c) Therefore, I will go to church.
7. (a) If two triangles are congruent, then the triangles are mutually equiangular.
   (b) The two triangles are congruent.
   (c) Hence, . . .
8. (a) If the triangle is isosceles, then the triangle has two equal angles.
   (b) . . . .
   (c) Hence, the triangle has two equal angles.

9. (a) If triangles *ABC and A'B'C'* are congruent, then angle $A =$ angle A'.
   (b) Triangles *ABC* and *A'B'C'* are congruent.
   (c) Hence, . . .
10. (a) . . . .
   (b) Price controls are to be adopted.
   (c) Hence, the country will be saved from inflation.

II. Complete the blanks in the following sets of propositions so that each set is in conformity with the transitive rule.

1. (a) Triangle *ABC* is equilateral implies triangle *ABC* is equiangular.
   (b) Triangle *ABC* is equiangular implies angle A = 60°.
   (c) Hence, . . .
2. (a) If x is greater than y, than u is less than v.
   (b) If u is less than v, then z is greater than w.
   (c) Hence, . . .
3. (a) If 1960 was a leap year, then 1964 was a leap year.
   (b) . . . .
   (c) If 1960 was a leap year, then 1968 was a leap year.
4. (a) If New York time is five hours slower than London time, then Denver time is two hours slower than New York time.
   (b) If Denver time is two hours slower than New York time, then San Francisco time is three hours slower than New York time.
   (c) Hence, . . .
5. (a) Since "*X* is guilty" implies "*Y* is innocent," and (b) "*Y* is innocent" implies "*Z* is under suspicion," hence, (c) . . .

III. Determine whether each of the following inferences is valid or faulty. If the inference is valid, produce some evidence which will confirm its validity. If the inference is faulty, produce a combination of truth values that will confirm a fallacy, or indicate a fallacy.

1. If today is David's birthday, then today is January 24.
   Today is January 24.
   Hence, today is David's birthday.
2. If the client is guilty, then he was at the scene of the crime.
   The client was not at the scene of the crime.
   Hence, the client is not guilty.
3. The days are becoming longer.
   The nights are becoming shorter if the days are becoming longer.
   Hence, the nights are becoming shorter.

4. If angle α = angle β, then the lines AB and BC are equal.
   We know $AB = BC$.
   Hence, angle α = angle β.
5. The earth is spherical implies that the moon is spherical.
   The earth is not spherical.
   Hence, the moon is not spherical.
6. If David passes the final exam, then he will pass the course.
   David will pass the course.
   Hence, he will pass the final exam.
7. If the patient has a virus, he must have a temperature above 99°.
   The patient's temperature is not above 99°.
   Hence, the patient does not have a virus.
8. If diamonds are not expensive, then gold is selling cheaply.
   Gold is not selling cheaply.
   Hence, diamonds are expensive.
9. AB is parallel to EF or CD is parallel to EF.
   AB is parallel to EF.
   Hence, CD is not parallel to EF.
10. AB is parallel to EF or CD is parallel to EF.
    CD is not parallel to EF.
    Hence, AB is parallel to EF.
11. A is not guilty and B is not telling the truth.
    Hence, it is false that "A is guilty or B is telling the truth."
12. Either Mack is not guilty or Mike is telling the truth.
    Mike is not telling the truth.
    Hence, Mack is not guilty.
13. If Lowell is studying for the ministry, then he is required to take theology and Greek.
    Lowell is not required to take Greek.
    Hence, Lowell is not studying for the ministry.
14. The governor will call a special session only if the Senate cannot reach a compromise.
    If a majority of the Cabinet are in agreement, then the governor will call a special session.
    The Senate cannot reach a compromise.
    Hence, a majority of the Cabinet are in agreement.
15. The governor will call a special session only if the Senate cannot reach a compromise.
    If a majority of the Cabinet are in agreement, then the governor will call a special session.
    The Senate can reach a compromise.
    Hence, a majority of the Cabinet are in agreement.
16. The governor will call a special session only if the Senate cannot reach a compromise.

If a majority of the Cabinet are in agreement, then the governor will call a special session.

The Senate can reach a compromise.

Hence, a majority of the Cabinet are not in agreement.

17. The governor will call a special session only if the Senate cannot reach a compromise.

If a majority of the Cabinet are in agreement, then the governor will call a special session.

The Senate can reach a compromise.

Hence, the governor will not call a special session.

18. The new people in the neighborhood have a beautiful boat.

They also have a nice car.

Hence, they must be nice people.

IV. Fill in the blanks in the following arguments by using the law of contraposition and/or *modus tollens*.

1. If it is not raining, the sun will come out.

   The sun will not come out.

   Hence, . . .

2. If lines AB and CD are parallel, then alternate interior angles α and β are equal.

   But angles α and β are not equal.

   Hence, . . .

3. If the graphs are isomorphic, then their degree spectrum will be the same.

   Their degree spectra are different.

   Hence, . . .

4. If the graph G is bipartite, then G is two-colorable.

   The graph G is not two-colorable.

   Hence, . . .

5. If C is on the perpendicular bisector of the line segment AB, then C is equidistant from A and B.

   Hence, if C is not . . .

6. If Joe does not pass the language requirement, then he does not graduate.

   Hence, if Joe . . .

7. If the graphs are isomorphic, then they have the same number of edges.

   The graphs have different numbers of edges.

   Hence, . . .

8. If Melinda is late, then she will be placed on restrictions.

   Hence, if . . .

V. Determine whether each of the following inference patterns is valid or invalid. If the inference pattern is invalid, indicate a combination of truth values which will produce a counterexample. If the inference pattern is valid, produce some evidence which will confirm its validity.

1. $r \to s$
    $\underline{\sim s}$.
    $\therefore \sim r$

2. $r \to s$
    $p \to q$
    $\underline{r \lor p}$.
    $\therefore s \lor q$

3. $\underline{\quad q \quad}$.
    $\therefore (p \land q)$

4. $p \to (r \to s)$
    $\sim r \to \sim p$
    $\underline{p}$.
    $\therefore s$

5. $(p \land q) \to \sim t$
    $w \lor r$
    $w \to p$
    $\underline{r \to q}$.
    $\therefore (w \lor r) \to \sim t$

6. $\sim t \to \sim r$
    $\sim s$
    $t \to w$
    $\underline{r \lor s}$.
    $\therefore w$

7. $\sim r \to (s \to \sim t)$
    $\sim r \lor w$
    $\sim p \to s$
    $\underline{\sim w}$.
    $\therefore t \to p$

8. p
    $p \to q$
    $\underline{q \to r}$.
    $\therefore r$

9. $\sim r$
    $p \to q$
    $\underline{q \to r}$.
    $\therefore \sim p$

10. $\sim p$
    $p \to q$
    $\underline{q \to r}$.
    $\therefore \sim r$

11. r
    $p \to q$
    $\underline{q \to r}$.
    $\therefore p$

12. If Tallahassee is not in Florida, then golf balls are not sold in Chicago.
    Golf balls are not sold in Chicago.
    Hence, Tallahassee is in Florida.
13. If the cup is styrofoam, then it is lighter than water.
    If the cup is lighter than water, then Joe can carry it.
    Hence, if the cup is styrofoam, then Joe can carry it.
14. If wages are raised, buying increases.
    If there is a depression, wages cannot be raised.
    Thus, if there is a depression, buying cannot increase.
15. The given triangles are similar.
    If the given triangles are mutually equiangular, then they are similar.
    Therefore, the triangles are mutually equiangular.

16. If Joe or Abe needs a vacation, then Ted deserves an assistant.
Hence, if Joe needs a vacation, then Ted deserves an assistant.

VI. Verify that the following argument is valid by translating into symbols and using truth tables to check for tautologies:

If Joe is a mathematician, then he is ambitious.

If Joe is an early riser, then he does not like oatmeal.

If Joe is ambitious, then he is an early riser.

Hence, if Joe is a mathematician, then he does not like oatmeal.

VII. Verify that the following argument is valid by using the rules of inference:

If Clifton does not live in France, then he does not speak French.

Clifton does not drive a Datsun.

If Clifton lives in France, then he rides a bicycle.

Either Clifton speaks French, or he drives a Datsun.

Hence, Clifton rides a bicycle.

Selected Answers for Section 1.6

I. 1. (c) No conclusion.

   2. (c) Hence, the housing industry will be hurt.

   5. (a) If 1984 is a leap year, then it is a presidential election year.

   8. (b) The triangle is isosceles.

II. 2. (c) Hence if x is greater than y, then z is greater than w.

III. 1. Fallacy of affirming the consequent.

   5. Fallacy of denying the antecedent.

   12. Valid; disjunctive syllogism.

   18. Nonsequitur.

IV. 3. Hence the graphs are not isomorphic.

   8. Hence, if Melinda will not be placed on restrictions, then she is not late.

V. 1. Valid

    (1) $r \rightarrow s$ premise

    (2) $\sim s \rightarrow \sim r$ Rule 4

    (3) $\sim s$ premise

    (4) $\sim r$ by (2), (3), Rule 1.

5. Invalid. Let the statements have one of the following sets of truth-values. Then the premises are true but the conclusion is false.

p	q	w	r	t	~t
T	F	T	F	T	F
F	T	F	T	T	T

7. Valid
    (1) $\sim r \vee w$ premise
    (2) $\sim w$ premise
    (3) $\sim r$ by (1) and (2), disjunctive syllogism
    (4) $(\sim r) \rightarrow (s \rightarrow \sim t)$ premise
    (5) $s \rightarrow \sim t$ by (3), (4), and Rule 1.
    (6) $\sim p \rightarrow s$ premise
    (7) $\sim p \rightarrow \sim t$ by (5), (6), and Rule 2.
    (8) $t \rightarrow p$ by (7) and Rule 4.

10. invalid.

1.7 METHODS OF PROOF OF AN IMPLICATION

In the previous two sections we introduced truth tables to be able to define the several logical connectives: $\vee$, $\wedge$, $\sim$, $\rightarrow$, and $\leftrightarrow$. Moreover, truth tables are used to determine which propositional functions are, in fact, tautologies. Tautologies then, in turn, are the foundations upon which valid inferences are based.

Now in this section, we apply valid inference patterns in two basic ways—first, to validate nine common methods for proving implications, and, second, to supply warrants and justification for the conclusions drawn in each step of a proof.

The nine methods of proof that we list are so common that they are frequently referred to by name, but these are by no means all of the known methods of proof—they just provide a good foundation upon which to build.

We must emphasize that what we describe in general terms is just the skeleton outline form that proofs usually take. In developing a proof the student should follow the skeleton outline of some indicated method of proof; but, more than that, the student will have to add flesh to the bones of the skeleton by supplying the details and documentation for each deduction. Every conclusion must be made according to valid inference patterns.

Brief descriptions of the nine methods of proof are listed below:

1. **Trivial proof** of $p \rightarrow q$. If it is possible to establish that q is true, then, regardless of the truth value of p, the implication $p \rightarrow q$ is true. Thus, the construction of a trivial proof of $p \rightarrow q$ requires showing that the truth value of q is true.

2. **Vacuous proof** of $p \rightarrow q$. If p is shown to be false, then the implication $p \rightarrow q$ is true for any proposition q.

3. **Direct proof** of $p \rightarrow q$. The construction of a direct proof of $p \rightarrow q$ begins by assuming p is true and then, from available information from the frame of reference, the conclusion q is shown to be true by valid inference.

4. **Indirect proof** of $p \rightarrow q$ (direct proof of contrapositive). The implication $p \rightarrow q$ is equivalent to the implication $\sim q \rightarrow \sim p$. Consequently, we can establish the truth of $p \rightarrow q$ by establishing $(\sim q) \rightarrow (\sim p)$. Of course, this last implication is likely to be shown to be true by a direct proof proceeding from the assumption that $\sim q$ is true. Thus, an indirect proof of $p \rightarrow q$ proceeds as follows:

   (a) Assume q is false.
   (b) Prove on the basis of that assumption and other available information from the frame of reference that p is false.

5. **Proof** of $p \rightarrow q$ **by contradiction.** This method of proof exploits the fact [derived from DeMorgan's laws and the equivalence of $p \rightarrow q$ to $(\sim p) \vee q$] that $p \rightarrow q$ is true iff $p \wedge (\sim q)$ is false. Thus, a proof by contradiction is constructed as follows:

   (a) Assume $p \wedge (\sim q)$ is true.
   (b) Discover on the basis of that assumption some conclusion that is patently false or violates some other fact already established in the frame of reference.
   (c) Then the contradiction discovered in step (b) leads us to conclude that the assumption in step (a) was false and therefore that $p \wedge (\sim q)$ is false, so that $p \rightarrow q$ is true.

Frequently the contradiction one obtains in a proof by contradiction is the proposition $p \wedge (\sim p)$. Hence, in this case, one could have given a proof by contrapositive just as easily.

6. **Proof** of $p \rightarrow q$ **by cases.** If p is in the form $p_1 \vee p_2 \vee \cdots \vee p_n$, then $(p_1 \vee p_2 \vee \cdots \vee p_n) \rightarrow q$ can be established by proving separately the different implications:

$$p_1 \rightarrow q, p_2 \rightarrow q, \ldots, \text{and } p_n \rightarrow q.$$

The method of proof by cases is valid when $n = 2$, for example, because the statement $(p_1 \lor p_2) \to q$ and $(p_1 \to q) \land (p_2 \to q)$ are equivalent.

7. **Proof by elimination of cases.** Frequently in solving a problem or in constructing a proof we are confronted with two alternatives: either p has to be true or q has to be true. If for some other reasons we go on to verify that p is, in fact, false, then obviously we must conclude that q is true.

To be sure, this method of proof is nothing more than the law of disjunctive syllogism: $[(p \lor q) \land (\sim p)] \to q$, and, of course, it can be extended to any finite number of cases:

$$[\{(p_1 \lor p_2 \lor \cdots \lor p_n) \lor q\} \land (\sim p_1) \land (\sim p_2) \land \cdots (\sim p_n)] \to q$$

Likewise, a proof by elimination of cases can take another form based on the following tautology:

$$\{[p \to (q \lor r)] \land (\sim q)\} \to (p \to r).$$

In other words, if we are given that p implies two possible conclusions q or r, and if one of the conclusions q is definitely false, then, in fact, we may conclude that p implies the other conclusion r.

Finally, there is a third form of a proof by elimination of cases. In this situation, a statement of the form $p \to (q \lor r)$ is proved by proving instead the equivalent statement $(p \land \sim q) \to r$.

The second proposition $(p \land \sim q) \to r$ is at least potentially simpler to prove, since one would then have two premises p and $\sim q$ with which to work instead of just the one premise p.

Suppose, for example, that we wish to prove the following statement for a real number X: If $X^2 - 5X + 6 = 0$, then $X = 3$ or $X = 2$. Then, we could accomplish this by assuming $X^2 - 5X + 6 = 0$ and $X \neq 3$ and then demonstrating that $X = 2$.

We might observe that this method of proof of $p \to (q \lor r)$ by elimination of cases is quite similar in form to proof by contradiction; however, a proof by contradiction of $p \to (q \lor r)$ would go one step further and assume $p \land (\sim q) \land (\sim r)$ (since $\sim(q \lor r) = (\sim q) \lor (\sim r)$ by DeMorgan's law), and then attempt to find a contradiction.

The next method of proof is actually just another form of elimination of cases, but we separate it out for better recall.

8. **Conditional proof.** The two propositions $p \to (q \to r)$ and $(p \land q) \to r$ are equivalent. Therefore, a proof of the conditional

$p \to (q \to r)$ can be constructed as follows:

(a) Combine the two antecedents p and q.
(b) Then prove r on the basis of these assumptions and other available information.

9. **Proof of equivalence.** To prove that a proposition p is true if and only if a proposition q is true frequently we break the proof into two halves: we prove $p \to q$, and then we prove $q \to p$. We may, of course, choose any method of proof to prove either half.

Having given several basic patterns of proof in outline form, let us now give some examples of proof following these basic patterns. Since we are trying to illustrate the methods of proof, we will not be so concerned with the importance of the facts that we are proving; we view them primarily as instruments for practice.

Examples of direct proof. First we list some elementary examples of direct proofs.

Example 1.7.1. Suppose that we wish to prove the statement: If X is a number such that $X^2 - 5X + 6 = 0$, then $X = 3$ or $X = 2$. Now since we are discussing numbers, the frame of reference should include all the rules of algebra. A direct proof of the statement proceeds as follows: Assume $X^2 - 5X + 6 = 0$. Using the rules of algebra, we have $X^2 - 5X + 6 = (X - 3)(X - 2) = 0$. It is known (a fact from the frame of reference) that if the product of two numbers is zero then one or the other of the two factors must be zero. Hence, $X - 3 = 0$ or $X - 2 = 0$. But $X - 3 = 0$ implies $X = 3$, and $X - 2 = 0$ implies $X = 2$. Thus, $X = 3$ or $X = 2$. $\square$

Example 1.7.2. If an integer a is such that $a - 2$ is divisible by 3, then $a^2 - 1$ is divisible by 3.

Again we give a direct proof. Here we let p represent the statement "a is an integer such that $a - 2$ is divisible by 3" and we let q represent "$a^2 - 1$ is divisible by 3." Next, we translate what "divisible by 3" means. Since $a - 2$ is divisible by 3, we know that $a - 2 = 3k$, where k is some integer. Therefore, $a + 1$ is divisible by 3 since $a + 1 = (a - 2) + 3 = 3(k + 1)$, and then $a^2 - 1 = (a + 1)(a - 1) = 3(k + 1)(a - 1)$ is divisible by 3. $\square$

Example 1.7.3. If a and b are odd integers, then $a + b$ is an even integer.

Let us recall that an even integer has the form $2k$, where k is some integer. Likewise, an odd integer has the form $2m + 1$, where m is an integer. Therefore, since in a direct proof we can assume the antecedent, we know that $a = 2k + 1$ and $b = 2m + 1$, for integers k and m, respectively. But then, $a + b = (2k + 1) + (2m + 1) = 2(k + m + 1)$ is an even integer since $k + m + 1$ is an integer.   □

Before giving some more examples of proofs, let us recall a well-known result called

> **The Division Algorithm.** If a and b are integers where b is positive, then there are (unique) integers q and r such that $a = bq + r$, where $0 \le r < b$. The integers q and r are called, respectively, the *quotient* and the *remainder*.

Of course, a is divisible by b if and only if the remainder r given by the division algorithm is zero.

Example of a proof of equivalence.

Example 1.7.4.   Two integers a and b have the same remainder when divided by the positive integer n iff the integer $a - b$ is divisible by n.

We are to prove the statement $p \leftrightarrow q$, where p is the statement "a and b have the same remainder when divided by the positive integer n" and q is the statement "the integer $a - b$ is divisible by n." Thus, we make two proofs, namely, $p \rightarrow q$ and $q \rightarrow p$.

First, let us give an interpretation of what p means. If we apply the division algorithm, then $a = nq_1 + r_1$, where q_1 and r_1 are integers and $0 \le r_1 < n$. Likewise, $b = nq_2 + r_2$, where q_2 and r_2 are integers with $0 \le r_2 < n$. The statement p just asserts that the remainders r_1 and r_2 are equal.

Now, let us give a direct proof of $p \rightarrow q$. Thus, we assume $r_1 = r_2$ and attempt to prove that $a - b$ has n as a factor. But

$$a - b = (nq_1 + r_1) - (nq_2 + r_2) = n(q_1 - q_2) + (r_1 - r_2) = n(q_1 - q_2)$$

since $r_1 - r_2 = 0$. Therefore, $a - b$ is divisible by n since $q_1 - q_2$ is an integer.

Next, let us give a direct proof of the converse implication $q \rightarrow p$. Thus, we assume that $a - b$ is divisible by n or, in other words, $a - b = nk$, where k is an integer. Then, if $a = nq_1 + r_1$ and $b = nq_2 + r_2$, we are to prove that $r_1 = r_2$. In the equation $a - b = nk$, we substitute the expression $nq_2 + r_2$ for b and then transpose to get the equation $a = nq_2 + r_2 + nk = n(q_2 + k) + r_2$. But since $0 \le r_2 < n$ and quotients and remainders are unique when we apply the division algorithm, we conclude that the remainder r_1 must equal r_2.   □

Example 1.7.5. If a and b are integers such that a has remainder 2 and b has remainder 8, respectively, when divided by 10, then $a + b$ is divisible by 10.

We know that $a = (10)q_1 + 2$ and $b = (10)q_2 + 8$, where q_1 and q_2 are integers. Therefore, $a + b = (10)(q_1 + q_2) + (2 + 8) = (10)(q_1 + q_2 + 1)$, and we conclude $a + b$ is divisible by 10 since $q_1 + q_2 + 1$ is an integer. $\square$

Examples of proofs by contrapositive. We wish now to illustrate some simple proofs by contrapositive. Our first example uses the familiar concepts of even and odd integers.

Example 1.7.6. If the product of two integers a and b is even, then either a is even or b is even.

Let p be the proposition "ab is even" and let q be "a is even or b is even." Then we are to prove $p \rightarrow q$, but by contrapositive we prove instead $\sim q \rightarrow \sim p$. By DeMorgan's law, $\sim q$ is the statement "a is odd and b is odd." Therefore, $a = 2m + 1$ and $b = 2n + 1$, where m and n are integers. But then, $ab = 4mn + 2m + 2n + 1 = 2(2mn + m + n) + 1$ is odd. Thus, the assumption $\sim q$ has led to the conclusion $\sim p$, and the proof is complete. $\square$

Example 1.7.7. If n is the product of two positive integers a and b, then either $a \le n^{1/2}$ or $b \le n^{1/2}$.

Let q_1 be the statement "$a \le n^{1/2}$" and q_2 the statement "$b \le n^{1/2}$." Then by DeMorgan's law $\sim(q_1 \vee q_2) \equiv (\sim q_1) \wedge (\sim q_2)$. Thus, let us assume that $a > n^{1/2}$ and $b > n^{1/2}$. But then from properties of inequalities, we know that $ab > n^{1/2}n^{1/2} = n$. Hence, n is not the product of a and b. $\square$

Example 1.7.8. The following fact is also readily verified by a contrapositive proof. If n is an integer, then its smallest factor $p > 1$ must be a prime.

This is true since, if we assume otherwise, p would have a factor that would be a smaller factor of n. $\square$

The above two facts go together to provide the foundation for an algorithm to find the smallest prime divisor of a given positive integer n and hence determine whether or not n is itself a prime. We assume that a list of all primes less than or equal to the square root of n is available.

Algorithm 1.7.1. The smallest prime divisor of n.

Input: a positive integer n.

Output: The smallest prime divisor of n.

Method:
1. Consider the integer n. If n is even, 2 is the smallest prime divisor of n. If n is odd, go to Step 2.
2. Find the largest integer s less than or equal to the square root of n. Let P be the set of all primes less than or equal to s. Go to Step 3.
3. For each $x \in P$ considered in order, determine if x is a divisor of n. If the currently considered x is a divisor of n, then output x. If no $x \in P$ is a divisor of n, output n.

For instance, 167 is a prime because 12 is the largest integer less than or equal to the square root of 167. The primes less than or equal to 12 are 2, 3, 5, 7, 11, and none of these are divisors of 167.

Examples of proofs by contradiction.

Example 1.7.9.   In a room of 13 people, 2 or more people have their birthdays in the same month.

We prove this proposition by contradiction by assuming that the room has 13 people and *no pair* of people have their birthdays in the same month. But then since each person is born in some month, and since we are assuming that no two people were born in the *same* month, there must be 13 months represented as the birth months of the people in the room. This conclusion is in violation of the well-known fact that there are only 12 months. Thus, the proposition is true. $\square$

Example 1.7.10.   Suppose that the 10 integers 1,2,...,10 are randomly positioned around a circular wheel. Show that the sum of some set of 3 consecutively positioned numbers is at least 17.

Let us give a proof by contradiction.
Let X_i represent the integer positioned at position i on the wheel. Then we are to prove:

either
$$X_1 + X_2 + X_3 \geq 17,$$
$$X_2 + X_3 + X_4 \geq 17,$$
$$\vdots$$

or
$$X_{10} + X_1 + X_2 \geq 17$$

(since the tenth and the first and second positions are consecutive on a circular wheel).

If, on the contrary, we assume that the conclusion is false, then by DeMorgan's laws we assume the conjunction of the following statements:

$$X_1 + X_2 + X_3 < 17,$$
$$X_2 + X_3 + X_4 < 17,$$
$$\vdots$$

and
$$X_{10} + X_1 + X_2 < 17.$$

Moreover, since each X_i is an integer, in fact, we can say more; namely, each of the above inequalities is ≤ 16 rather than < 17. Now taking the sum of all of these inequalities, we discover that

$$3(X_1 + X_2 + \cdots + X_{10}) \leq (10)(16).$$

But the sum $X_1 + X_2 + \cdots + X_{10}$ is just the sum of the first 10 positive integers, which we can observe is equal to 55. Therefore, the last inequality becomes

$$165 = 3(55) \leq (10)(16) = 160.$$

Clearly, this is a contradiction.   □

Example 1.7.11.   If 41 balls are chosen from a collection of red, white, blue, garnet, and gold colored balls, then there are at least 12 red balls, 15 white balls, 4 blue, 10 garnet, or 4 gold balls chosen.

Let X_1, X_2, X_3, X_4, X_5 represent, respectively, the number of red, white, blue, garnet, and gold balls chosen. We are to prove that either $X_1 \geq 12$, $X_2 \geq 15$, $X_3 \geq 4$, $X_4 \geq 10$, or $X_5 \geq 4$. Suppose, on the contrary, that $X_1 \leq 11$, $X_2 \leq 14$, $X_3 \leq 3$, $X_4 \leq 9$, and $X_5 \leq 3$. Then $X_1 + X_2 + \cdots + X_5 \leq 11 + 14 + 3 + 9 + 3 = 40$. But, on the other hand, the sum $X_1 + X_2 + \cdots + X_5 = 41$, since this is the total number of balls chosen. Thus, we have arrived at the contradiction $41 \leq 40$ and, as a result, the conclusion is verified.   □

Example 1.7.12.   The same idea of proof can be used to prove the **pigeonhole principle.**

Let $m_1, m_2, \ldots, m_n$ be positive integers. If $m_1 + m_2 + \cdots + m_n - n + 1$ objects are put into n boxes, then either the first box contains at least m_1 objects, or the second box contains at least m_2 objects, $\ldots$, or the nth box contains at least m_n objects.

Let q_1 represent the statement "the ith box contains at least m_i objects," and let p represent the statement "$m_1 + m_2 + \cdots + m_n - n + 1$ objects are put into n boxes." Then we are asked to prove: $p \rightarrow (q_1 \vee q_2 \vee \cdots \vee q_n)$. We do this by contradiction, that is, we assume p and $\sim(q_1 \vee q_2 \vee \cdots \vee q_n)$.

By DeMorgan's laws, $\sim(q_1 \vee q_2 \vee \cdots \vee q_n) \equiv (\sim q_1) \wedge (\sim q_2) \wedge \cdots \wedge (\sim q_n)$. Thus we are assuming $p \wedge (\sim q_1) \wedge (\sim q_2) \wedge \cdots \wedge (\sim q_n)$. Now $\sim q_i$ means that the ith box contains less than m_i objects; in other words, the ith box contains *at most* $m_i - 1$ objects. But since we are assuming the conjunction of all the statements $\sim q_i$, we are assuming that the first box has at most $m_1 - 1$ objects, *and* the second box contains at most $m_2 - 1$ objects, ..., *and* the nth box contains at most $m_n - 1$ objects. But then all n boxes contain at most $(m_1 - 1) + (m_2 - 1) + \cdots + (m_n - 1)$ objects. Since this last number is equal to $m_1 + m_2 + \cdots + m_n - n$, we see that all n boxes contain at most $m_1 + m_2 + \cdots + m_n - n$ objects, and hence the statement p is contradicted in that not all of the $m_1 + m_2 + \cdots + m_n - n + 1$ objects have been distributed. This contradiction proves the pigeonhole principle is valid. $\square$

This principle is known by its name because we often think of the objects as pigeons and the boxes as pigeonholes. What the principle says is that if we distribute a large number of pigeons into a specified number of holes, then we can be assured that some hole contains a certain number of pigeons or more.

Example 1.7.13   Another form of the pigeonhole principle. If A is the average number of pigeons per hole, then some pigeonhole contains at least A pigeons and some pigeonhole contains at most A pigeons.

The proof is almost the same as above. If n is the number of pigeonholes and m_i is the number of pigeons in the ith hole, then we prove that either $m_1 \geq A$ or $m_2 \geq A$,. . .,or $m_n \geq A$. (The proof that some $m_j \leq A$ is similar.)

Let us assume the contrary, namely, that $m_1 < A$ and $m_2 < A$,. . .,and $m_n < A$. But then the sum $m_1 + m_2 + \cdots + m_n < n A$ = the total number of pigeons. This clearly is a contradiction since $m_1 + m_2 + \cdots + m_n$ also equals the total number of pigeons.

Now the number of pigeons in a pigeonhole is necessarily an integer, but the average A need not be an integer. Let us use the notation (called the ceiling of A) $\lceil A \rceil$ to mean the smallest integer $x \geq A$ and $\lfloor A \rfloor$ (called the floor of A) to mean the largest integer $x \leq A$. Thus, $\lceil 9.7 \rceil = 10$ and $\lfloor 9.7 \rfloor = 9$. In this terminology, the second version of the pigeonhole principle can be stated:

> *If A is the average number of pigeons per hole, then some pigeonhole contains at least $\lceil A \rceil$ pigeons and some pigeonhole contains at most $\lfloor A \rfloor$ pigeons.*

Applications of the pigeonhole principle.

1. If $n + 1$ pigeons are distributed among n pigeonholes, then some hole contains at least 2 pigeons. If $2n + 1$ pigeons are distributed among n pigeonholes, then some hole contains at least 3 pigeons. In general, if k is an integer and $kn + 1$ pigeons are distributed among n pigeonholes, then some hole contains at least $k + 1$ pigeons. This follows since the average number of pigeons per hole is $k + 1/n$ and $\lceil k + 1/n \rceil = k + 1$.

2. In any group of 367 people there must be at least one pair with the same birthday.

3. If 4 different pairs of socks are scrambled in a drawer, one need only select 5 individual socks in order to guarantee finding a matching pair. Here the pairs determine 4 pigeonholes and 5 individual socks in 4 holes implies a matching pair.

4. In a group of 61 people at least 6 people were born in the same month.

5. If 401 letters were delivered to 50 apartments, then some apartment received at most 8 letters.

6. Suppose 50 chairs are arranged in a rectangular array of 5 rows and 10 columns. Suppose that 41 students are seated randomly in the chairs (one student per chair, of course). Then some row contains at least 9 students, some column contains at least 5 students, some row contains at most 8 students, and some column contains at most 4 students. The result follows from the pigeonhole principle because the average number of students per row is 8.2 and the average number per column is 4.1.

7. Suppose that a patient is given a prescription of 45 pills with the instructions to take at least one pill per day for 30 days. Then prove that there must be a period of consecutive days during which the patient takes a total of exactly 14 pills.

To see why, let a_i be the number of pills the patient has taken through the end of the ith day. Since the patient takes at least one pill per day and at most 45 pills in 30 days, we have $1 \le a_1 < a_2 < \cdots a_{30} \le 45$. Also, adding 14 to each of these inequalities gives $a_1 + 14 < a_2 + 14 < \cdots < a_{30} + 14 \le 45 + 14 = 59$. We now have 60 integers: $a_1, a_2, \ldots, a_{30}, a_1 + 14$, $a_2 + 14, \ldots, a_{30} + 14$. Moreover, these numbers all lie in the range between 1 and 59. Thus, we have 60 pigeons in 59 pigeonholes, so there must be 2 of these numbers that are equal. Since $a_1, a_2, \ldots, a_{30}$ are all different and

$a_1 + 14, a_2 + 14, \ldots, a_{30} + 14$ are all different, it must be that one of a_1, $a_2, \ldots, a_{30}$ is equal to one of the integers $a_1 + 14, a_2 + 14, \ldots, a_{30} + 14$. In other words, there are i and j such that $a_i = a_j + 14$. Thus, between days i and j, the patient takes exactly 14 pills.

8. If $x_1, x_2, \ldots, x_8$ are 8 distinct integers, then there is some pair of these integers with the same remainder when divided by 7.

If each integer is divided by 7 and their remainders are recorded, the only possibilities for these remainders are 0, 1, 2, 3, 4, 5, and 6. Thus, we have 8 remainders to be distributed among the 7 possibilities from 0 to 6. In other words, it is like distributing 8 pigeons among 7 pigeonholes. We conclude that there are at least 2 pigeons in some hole or in this framework, there are at least 2 of the remainders which are equal.

Let us rephrase what we have proved in yet another way. The division algorithm implies that the set of integers can be partitioned into 7 disjoint subsets (called congruence classes) which we denote by [0],[1],[2],[3],[4],[5],[6]. Moreover, an integer x is in $[r]$ iff r is the remainder when x is divided by 7.

Therefore, in the above example, we have 8 integers which are distributed among 7 congruence classes so that, by the pigeonhole principle, two of these integers must be in the same congruence class. By the definition of congruence class, these 2 integers must have the same remainder when divided by 7.

This latter approach, while in fact the same as the former, nevertheless requires a slightly deeper technical understanding because of the appeal to congruences. The concept of congruence class is useful in the next example.

9. Given any set of 7 distinct integers, there must exist 2 integers in this set whose sum or difference is divisible by 10.

If we apply what we learned in (8) above, we see that we need 6 pigeonholes for which if 2 integers a and b are in the same pigeonhole, then we must conclude that either $a + b$ or $a - b$ is divisible by 10.

Our first thought is to let pigeonholes be the sets $[r] = \{$integers x such that x has remainder r when divided by 10$\}$. This choice for pigeonholes would have the feature that if 2 integers a and b are in the same hole, then their difference $a - b$ is divisible by 10. But we have too many pigeonholes using this choice for we have 10 pigeonholes (congruence classes) determined by $r = 0, 1, 2, \ldots, 9$. We need to reduce the number of holes and also incorporate the extra feature about sums being divisible by 10. We use the following ploy: group the 10 congruence classes into a group of 6 pigeonholes according to the following scheme: $\{[0]\}, \{[1],[9]\}, \{[2],[8]\}, \{[3],[7]\}, \{[4],[6]\}, \{[5]\}$. Now we have our 6 pigeonholes. Moreover, since we have 7 integers, there is some one of these holes which contains at least 2 of these 7 integers. Say a and b are in the same pigeonhole. If a and b are in [0] or [5], then both $a + b$ and $a - b$ are divisible by 10. However, if a and b are in some one of the other 4 pigeonholes, $\{[2],[8]\}$ for example, then it may be that a and b are in [2], in which case $a - b$ is divisible by 10, but their sum is not divisible by 10. Or it may be that a is [2] and b is in [8]. In this case, $a + b$ is divisible by 10 but their difference is not. Nevertheless, we have accomplished what we set out to prove.

Examples of proofs by cases.   Before we give the next example, we need to recall the definition of the absolute value of a real number. If x is a real number, then the absolute value of x, denoted by $|x|$, is the number x itself if x is nonnegative; but if x is negative, then the absolute value of x is the number $-x$.

Example 1.7.14.   If a real number x is such that $|x| > 4$, then $x^2 > 16$.

At first glance, this proof may not appear to require cases, but the cases are hidden in the definition of absolute value. Note $|x| > 4$ means $x > 4$ or $-x > 4$.

Therefore, if $x > 4$, then $x^2 > 4^2 = 16$. Likewise, if $-x > 4$, then $x^2 = (-x)(-x) > 4^2 = 16$.   □

Example 1.7.15.   If the integer a is such that $a - 1$ or $a - 2$ is divisible by 3, then $a^2 - 1$ is divisible by 3.

First, we prove: If the integer a is such that $a - 1$ is divisible by 3, then $a^2 - 1$ is divisible by 3. In this case, since $a - 1 = 3m$, where m is an integer, it follows that $a^2 - 1 = (a + 1)(a - 1) = (a + 1)(3m)$ is divisible by 3 since $m(a + 1)$ is an integer.

Next, we prove: If the integer a is such that $a - 2$ is divisible by 3, then $a^2 - 1$ is divisible by 3. We have already proved this in example 1.7.2.   □

Example 1.7.16.   An example of a proof by elimination of cases. Suppose that we are to prove: If p is an odd prime, then p has the form $6n + 1$ or $6n + 5$ or $p = 3$.

By the division algorithm, we see that if we divide 6 into p, then we have only 6 possibilities for remainders, namely, 0, 1, 2, 3, 4, or 5. Therefore, we conclude that either $p = 6n$, $p = 6n + 1$, $p = 6n + 2$, $p = 6n + 3$, $p = 6n + 4$, or $p = 6n + 5$. Nevertheless, some of these cases may be eliminated since p is an odd prime. The cases $p = 6n, 6n + 2, 6n + 4$ can be eliminated since p is odd, and the case $p = 6n + 3$ can be eliminated except for the special case $n = 0$ or $p = 3$, for in all other cases p would be divisible by 3 and not prime.   □

As a final example, let us illustrate the use of the method of conditional proof.

Example 1.7.17. An example of conditional proof. Suppose that we are to prove:

(i) If a is a prime integer, then if a divides the product of two integers bc, then a divides b or a divides c. $[p \rightarrow (q \rightarrow r \vee s)]$
By conditional proof (i) can be changed to:
(ii) If a is a prime integer and a divides the product of two integers bc, then a divides b or a divides c. $[(p \wedge q) \rightarrow (r \vee s)]$
Not only is this so, but also by elimination of cases (ii) can be reduced to proving:
(iii) If a is a prime integer such that a divides the product of two integers bc, and a does not divide b, then a divides c. $[(p \wedge q \wedge \sim r) \rightarrow s]$.

Now we have used valid inference patterns to change the form of the proposition (i) to another equivalent form (iii). But we have not proved anything yet; we have just reformulated the problem. We anticipate (iii) will be easier to prove since we can use the three premises, but nevertheless, additional facts and details still have to be supplied in order to actually prove (iii).

Let us complete the proof of (iii) assuming the reader has in his arsenal the fact that if 1 is the greatest common divisor of two integers d and e, then 1 can be written as a linear combination $dx + ey$, where x and y are integers. (This fact will be proved in Chapter 4.)

With this extra weapon, we prove (iii). Since a is prime and a does not divide b, the greatest common divisor of a and b is 1. Hence, there are integers x and y such that $1 = ax + by$. But then if we multiply this equation by c, we have $c = acx + bcy$ and clearly a divides acx since a is a factor. Moreover, a divides bc since this is one of the premises of the proposition. Thus, a divides acx and bcy, so a divides their sum—or in other words, a divides c—and this is what we were to prove. □

Exercises for Section 1.7

1. Use the division algorithm and proof by cases or elimination of cases to prove the following:
   (a) Every odd integer is either of the form $4n + 1$ or $4n + 3$.
   (b) The square of any odd integer is of the form $8n + 1$.
   (c) The square of any integer is either of the form $3n$ or $3n + 1$.
   (d) The cube of any integer is either of the form $9n$, $9n + 1$, or $9n + 8$.
   (e) The equation $x^3 - 117y^3 = 5$ has no integer solutions.
   (f) Any integer $n > 0$ is either of the form $6k, 6k + 1, \ldots, 6k + 5$. Conclude that $n(n + 1)(2n + 1)/6$ is an integer.

(g) If $p = p_1^2 + p_2^2 + p_3^2$, where p, p_1, p_2, and p_3 are primes, then one of them is the prime 3. Hint: use 1(c).

(h) If x is an integer, then $x^3 - x$ is divisible by 3.

(i) If p is a prime > 3, then p^2 has the form $12k + 1$, where k is an integer.

(j) If an integer is simultaneously a square and a cube (as is the case with $64 = 8^2 = 4^3$), verify that the integer must be of the form $7n$ or $7n + 1$.

2. Give direct proofs for the following: For integers a, b, c, and d, prove that

(a) a divides 0; the integer 1 divides a; and a divides a.

(b) a divides 1 iff $a = 1$ or $a = -1$.

(c) If a divides b and c divides d, then ac divides bd.

(d) a divides b and b divides a iff $a = b$ or $a = -b$.

(e) If a divides b and b is not zero, then $|a| \leq |b|$.

(f) If a divides b and a divides c, then a divides $bx + cy$ for arbitrary integers x and y.

(g) If a divides b, then a divides bc.

(h) If a divides b and a divides c, then a^2 divides bc.

(i) a divides b iff ac divides bc, where $c \neq 0$.

3. Prove or disprove: if a divides $(b + c)$, then a divides b or a divides c.

4. Prove for integers a and b:

(a) If a and b are even, then $a + b$ is even and ab is even.

(b) If a and b are odd, then $a + b$ is even and ab is odd.

(c) If a is even and b is odd, then $a + b$ is odd and ab is even.

(d) If a^2 is even, then a is even.

(e) If $a + b$ is odd and b is even, then a is odd.

(f) If a^2 is odd, then a is odd.

5. Prove for an integer a,

(a) 2 divides $(a)(a + 1)$, and

(b) 3 divides $(a)(a + 1)(a + 2)$.

(c) If a is an odd integer, then 24 divides $a(a^2 - 1)$. Hint: the square of an odd integer is of the form $8n + 1$.

(d) If a and b are odd integers, then 8 divides $(a^2 - b^2)$.

6. (a) If the sum of 2 real numbers is less than 100, prove that one of the numbers is less than 50.

(b) If the sum of 4 real numbers is less than 100, prove that one of the numbers is less than 25.

(c) If the sum of 11 real numbers is greater than 100, prove that one of the numbers is greater than 9.

(d) If the sum of $n > 1$ real numbers is less than r, where r is a real number, then at least one of the numbers is less than r/n.

(e) If the sum of $n > 1$ real numbers is greater than r, then at least one of the numbers is greater than r/n.

7. Prove by contradiction: If the product of a certain 2-digit decimal integer n by 5 is a 2-digit number, then the tens digit of n is 1.

8. (a) If m and n are each integers > 2, prove by contradiction that $mn > m + n$.
   (b) If m and n are positive integers, give a direct proof that $mn \geq n$.

9. (a) Prove by contrapositive: If a is an odd integer, then there are no integral roots for the polynomial $f(X) = X^2 - X - a$.
   (b) Prove by contradiction: There do not exist 3 consecutive integers such that the cube of the largest is equal to the sum of the cubes of the 2 other integers.
   (c) Prove by contradiction: If n has the form $4k + 3$, where k is an integer, then the equation $x^2 + y^2 = n$ has no integral solutions for x and y.

10. Use the pigeonhole principle to prove the following:
    (a) Given 10 distinct integers, then some pair of them have the same remainder when divided by 9 and so their difference is divisible by 9.
    (b) Given $n + 1$ distinct integers, then there is some pair of them such that their difference is divisible by the positive integer n.
    (c) Given 37 positive integers, then there must be at least 4 of them that have the same remainder when divided by 12.
    (d) Among 13 different integral powers of the integer 5, there must be at least 2 of them that have the same remainder when divided by 12.
    (e) Among 61 different integral powers of the integer 5, there are at least 6 of them that have the same remainder when divided by 12.
    (f) Among $n + 1$ different integral powers of an integer a, there are at least 2 of them that have the same remainder when divided by the positive integer n.

11. Suppose that a man hiked 6 miles the first hour and 4 miles the twelfth hour and hiked a total of 71 miles in 12 hours. Prove that he must have hiked at least 12 miles within a certain period of two consecutive hours. (Hint: prove by contradiction.)

12. Suppose that the circumference of a circular wheel is divided into 50 sectors and that the numbers 1 and 50 are randomly assigned to these sectors.
    (a) Show that there are 3 consecutive sectors whose sum of assigned numbers is at least 77.
    (b) Show that there are 5 consecutive sectors whose sum of assigned numbers is at least 122.

13. Show that among $n + 1$ positive integers less than or equal to $2n$ there are 2 consecutive integers.

14. Show that for an arbitrary integer N, there is a multiple of N that contains only the digits 0 and 5. (Hint: Consider $M_1 = 5$, $M_2 = 55$, $M_3 = 555, \ldots, M_N = 555 \ldots 5 = (5)10^n + (5)10^{n-1} + \cdots + 5 \cdot 10 + 5$—the decimal expansion of M_N has N 5s. Then apply the pigeonhole principle.)

15. A typewriter is used for 102 hours over a period of 12 days. Show that on some pair of consecutive days, the typewriter was used for at least 17 hours.

16. Give a direct proof that if x and y are numbers such that $5x + 15y = 116$, then either x or y is not an integer.

17. If n is a positive integer, an integer d is a *proper* divisor if $0 < d < n$ and d divides n. A positive integer n is *perfect* if n is the sum of its proper divisors. Give a contrapositive argument of the following: A perfect integer is not a prime.

18. Give a contradiction proof that the square root of 2 is not a rational number. (Hint: Use x^2 is even implies x is even, if x is an integer.)

19. Use a contradiction argument to verify the following valid inferences:

(a) $q \rightarrow t$
$s \rightarrow r$
$q \lor s$
$\therefore t \lor r$.

(b) $\sim p \rightarrow (q \rightarrow \sim w)$
$\sim s \rightarrow q$
$\sim t$
$\sim p \lor t$
$\therefore w \rightarrow s$.

20. Use a contrapositive argument to verify the following valid inference:

$$w \rightarrow (r \rightarrow s)$$
$$\therefore (w \land r) \rightarrow s.$$

21. Use the pigeon-hole principle to show that one of any n consecutive integers is divisible by n.

22. Use the pigeonhole principle to show that the decimal expansion of a rational number must, after some point, become periodic.

23. The circumference of two concentric disks is divided into 200 sections each. For the outer disk, 100 of the sections are painted red and 100 of the sections are painted white. For the inner disk the sections are painted red or white in an arbitrary manner. Show that it is possible to align the two disks so that 100 or more of the sections on the inner disk have their colors matched with the corresponding sections on the outer disk.

24. Given 20 French, 30 Spanish, 25 German, 20 Italian, 50 Russian, and 17 English books, how many books must be chosen to guaran-

tee that at least

(a) 10 books of one language were chosen?

(b) 6 French, 11 Spanish, 7 German, 4 Italian, 20 Russian, or 8 English were chosen?

25. If there are 104 different pairs of people who know each other at a party of 30 people, then show that some person has 6 or fewer acquaintances.

26. Show that given any 52 integers, there exist two of them whose sum, or else whose difference, is divisible by 100.

27. From the integers 1,2,3,. . .,200, 101 integers are chosen. Show that among the integers chosen there are two such that one of them is divisible by the other.

28. A student has 37 days to prepare for an examination. From past experience she knows that she will require no more than 60 hours of study. She also wishes to study at least 1 hour per day. Show that no matter how she schedules her study time (a whole number of hours per day), there is a succession of days during which she will have studied exactly 13 hours.

29. Prove that in a group of n people there are two who have the same number of acquaintances in the group.

30. Given the information that no human being has more than 300,000 hairs on his head, and that the state of Florida has a population of 10,000,000, observe that there are at least two persons in Florida with the same number of hairs on their heads. What is the largest integer that can be used for n in the following assertion? There are n persons in Florida with the same number of hairs on their heads.

31. (a) Prove that 421 and 2477 are prime integers.

(b) Write 91001 as a product of prime integers.

32. (a) *Twin primes* are primes a and b such that $b = a + 2$. Thus 5 and 7 are twin primes. Find 4 other pairs of twin primes. Whether or not there are infinitely many twin primes is not known.

(b) Three primes a,b,c such that $a + 4 = b + 2 = c$ might be called *triple primes*. Find a set of triple primes and prove it is unique.

(c) *Adjacent primes* are primes a and b such that $b = a + 1$. Find a pair of adjacent primes and prove that no other pair exists.

(d) *Quadruple primes* are primes a,b,c,d where $a + 8 = b + 6 = c + 2 = d$. Thus, 11,13,17,19 are a set of quadruple primes. Find another set of quadruple primes.

(e) Show that in a set of quadruple primes a,b,c,d if the number $a + 4$ is as large as 15, then in fact $a + 4$ is a multiple of 15.

33. (a) Give a direct argument that if n is a perfect integer then the sum of the reciprocals of all divisors of n is 2.
    (b) Show that 28 and 496 and 8128 are perfect integers.

34. (a) If the sum of 5 integers x_1, x_2, x_3, x_4, x_5 is 14, and if each $0 \le x_i \le 3$, give a direct proof that the 5 integers are 3,3,3,3, and 2.
    (b) If counters are placed on 14 squares of a bingo board (which consists of 5 parallel rows and 5 columns of squares), then prove that some row or some column contains 2 adjacent squares with counters.

35. An ordered triple of positive integers (a,b,c) is called a *Pythagorean triple* iff $a^2 + b^2 = c^2$. (Therefore, the integers a and b can represent the length of two sides of a right triangle and c the length of the hypotenuse.) A *primitive triple* is an ordered triple of integers $(2mn, m^2 - n^2, m^2 + n^2)$ where m and n are integers and $0 < n < m$. Give a direct proof that any primitive triple is a Pythagorean triple.

36. Given n pigeons to be distributed among k pigeonholes:
    (a) What is a necessary and sufficient condition on n and k that, in every distribution, at least two pigeonholes must contain the same number of pigeons?
    (b) Suppose that the pigeons are distributed so that each hole is nonempty. How many pigeons must we choose in order to be sure that we have chosen the entire contents of at least one box? Hint: Use the pigeonhole principle in reverse form; that is, if fewer than n pigeons are distributed among n pigeonholes, then some pigeonhole is unoccupied.

37. Let S be a square of side length 1 unit. Choose any five points $P_1, \ldots, P_5$ in the interior of S. Let d_{ij} be the distance from P_i to P_j. Demonstrate that for at least one pair of the points P_i and P_j where $i \ne j$, that $d_{ij} < 1/\sqrt{2}$. Is the statement true if $1/\sqrt{2}$ is replaced by a smaller number?

38. Let $a_1, a_2, \ldots, a_n$ be n (not necessarily distinct) integers. Show that for some pair of integers k and m where $1 \le k \le m \le n$, $a_k + a_{k+1} + \cdots + a_m$ is divisible by n.

39. Let p be a prime different from 2 or 5. Show that there is some power of p whose last two decimal digits are 01. Hint: Consider p^i for $i = 1, 2, \ldots, 100$.

40. (a) Let A be the set of all integers that can be written as a sum of two squares of integers. Prove that if a and b are in A, then so is ab.
    (b) Prove that no integer of the form $8n + 7$ is a sum of 3 squares. Hint: Look at possible squares modulo 8.

41. If a is an integer, let $D(a)$ be the set of all positive integers k where k is a divisor of a, that is, where $a = kl$, where l is an integer. Then if a and b are integers $D(a) \cap D(b)$ is the set of all positive divisors common to both a and b.
    (a) Give a direct proof that if a or b is not zero, then $D(a) \cap D(b)$ is bounded above by some integer M. That is, for each $x \in D(a) \cap D(b)$, $x \le M$.
    (b) Suppose that $a = bq + r$, where a,b,q, and r are integers. Then give a direct proof that $D(a) \cap D(b) = D(b) \cap D(r)$.
    (c) Let us define the greatest common divisor of a and b, where a and b are not both zero as the largest integer in $D(a) \cap D(b)$.
        Prove then that the greatest common divisor of a and b (where a and b are not both zero) is that unique positive integer d such that
        (i) d divides both a and b, and
        (ii) if d' divides both a and b, then d' divides d.
    (d) Prove that if $a = bq + r$ where a,b,q, and r are integers and $b \ne 0$, then the greatest common divisor of a and b is the same as the greatest common divisor of b and r. (This is the basis for the Euclidean algorithm for finding the greatest common divisor.)

42. In any calendar year, determine how many Friday the thirteenths there can be. What is the smallest number possible?

43. If the sum of 10 nonnegative integers is 74, prove that the sum of some set of 4 of these integers is at least 30, and the sum of some set of 5 of them is at least 37.

Selected Answers for Section 1.7

11. Let x_i = distance hiked in the ith hour.

$$x_1 + x_{12} = 10$$
$$x_1 + x_2 + \cdots + x_{12} = 71$$

implies $x_2 + \cdots + x_{11} = 61$. If

$$x_1 + x_2 < 12$$
$$x_2 + x_3 < 12$$
$$\vdots$$
$$x_{11} + x_{12} < 12.$$

Then their sum

$$x_1 + x_{12} + 2(x_2 + \cdots + x_{11}) < 12 \cdot 11 = 132$$
$$2(x_2 + \cdots + x_{11}) < 122$$

which implies $122 < 122$. Contradiction.

15. We actually prove a stronger result. Let x_i = the number of hours the typewriter is used on day i. Suppose

$$x_1 + x_2 < 17, \text{ and}$$
$$x_3 + x_4 < 17$$
$$x_5 + x_6 < 17$$
$$\cdots \cdots$$
$$x_{11} + x_{12} < 17$$

Then $102 = x_1 + x_2 + \cdots + x_{12} < 6 \cdot 17 = 102$. This contradiction shows that either

$$x_1 + x_2 \geq 17 \text{ or}$$
$$x_3 + x_4 \geq 17 \text{ or}$$
$$\vdots$$
$$x_{11} + x_{12} \geq 17.$$

23. Let's hold the outer disk fixed and rotate the inner disk through the 200 possible alignments. For each alignment, let us count the number of matches. The sum of the counts for the 200 possible alignments must be 20,000, because each of the 200 sections on the inner disk will match its corresponding section on the outer disk in exactly 100 of the alignments. Thus, we have 20,000 pigeons to place in the 200 holes. We conclude that some hole must have at least 100 pigeons.

25. If each person has 7 or more acquaintances, there are at least $1/2(7)(30) = 105$ pairs.

1.8 FIRST ORDER LOGIC AND OTHER METHODS OF PROOF

To this point we have analyzed sentences and arguments, breaking them down into constituent simple propositions and regarding these simple propositions as building blocks. By this means we were able to discover something of what makes a valid argument. Nevertheless, there are arguments that are not susceptible to such a treatment. For example, let us consider the following argument:

> All mathematicians are rational.
>
> Joe is a mathematician.
>
> Therefore, Joe is rational.

We would intuitively regard this argument as a valid argument, but if we try to symbolize the form of the argument as we have been doing, we get an argument of the form:

$$\frac{\begin{array}{l} p \\ q \end{array}}{\therefore\ r}\ .$$

According to what we have learned thus far, this is not a valid argument form.

But, in fact, the argument is valid and the validity depends, in this case, not upon the form of the argument, but upon *relationships* between parts of the sentences and upon the form of the sentences themselves; in short, upon the *content* of the sentences.

First-order logic is that part of logic which emphasizes the *content* of the sentences involved in arguments as well as the *form* of arguments.

From a purely grammatical point of view, simple declarative sentences must involve a subject and a predicate, each of which may consist of a single word, a short phrase, or a whole clause. Putting it very roughly, the subject is the thing about which the sentence is making an assertion, and the predicate refers to a "property" that the subject has.

From a mathematical point of view, it is convenient to represent predicates by capital letters and subjects by small letters and thereby to symbolize sentences in such a way as to reflect a subject-predicate relationship. For example, the sentence "Florida is a state" could be symbolized as $S(f)$ where f represents Florida and S represents the predicate "is a state." Likewise, the symbols $M(j)$ could be used to represent the sentence "Joe is a mathematician." Moreover, the sentences "Joe is a gossip," "Joe gossips," and "Joe is gossipy" all have the same meaning and can be symbolized as $G(j)$, where j represents "Joe" and G represents "is gossipy."

In case our predicate is a negation we have a choice. For instance, in the sentence "The number whose square is -1 is not real," we could let S represent the predicate "is not real" and let i represent "the number whose square is -1," and then the sentence could be symbolized as $S(i)$. On the other hand, we could let R represent "is real" and then the sentence can be symbolized as $\sim R(i)$.

It should be clear that compound sentences can also be translated into symbols just by symbolizing all the constituent simple sentences.

But we still need something more than subject-predicate analysis to symbolize sentences like "All mathematicians are rational." We might, for example, attempt to write sentences of this type in the form $p \rightarrow q$. We could say:

1. In all cases, if a person is a mathematician, then that person is rational.

2. Always "a person is a mathematician" implies "that person is rational."

At first glance, we may think these statements are not unlike the form $p \rightarrow q$. However, there are two differences. First, the word "always" or the phrase "in all cases" indicates that more is being asserted than just an implication. Second, the p and q here could not themselves be propositions according to our definition of proposition, for p would be "a person is a mathematician" and q "that person is rational." How can we determine whether it is true or false that an unspecified person is a mathematician, baker, or candlestick maker? Likewise how can we determine whether or not that unspecified person is rational?

If we write the sentence "a person is a mathematician" as "x is a mathematician," then we realize that x is an unspecified variable. Moreover, this sentence is such that once the variable is specified the sentence becomes a proposition. For example, "Carl F. Gauss is a mathematician" is a true proposition. We refer to the sentence "x is a mathematician" as an example of an *open proposition* in one variable. Of course, we can have open propositions of more than one variable as the following definition shows.

Definition 1.8.1. An **open proposition** (or **predicate**) in n variables from a set U is a function $f:U^n \rightarrow \{T,F\}$, where U^n denotes the Cartesian product of n copies of the set U and T and F respectively stand for true and false. The set U is called the **universe of discourse** (universe, for short) of the open proposition f.

Thus, generally speaking, an open proposition is a declarative sentence which

1. contains one or more variables,
2. is not a proposition (except in the trivial case when the set U is a singleton set), and
3. produces a proposition when each of its variables is replaced by a specific element from the set U.

Some examples of open propositions are:

1. x is a rational number.
2. $y > 5$.
3. $x + y = 5$.
4. x climbed Mount Everest.
5. He is a lawyer and she is a computer scientist.

We have not specified the universe of discourse in any of the above but we would presumably choose sets of numbers for the universe in (1), (2), and (3) to avoid meaningless assertions such as "Joe > 5."

Just as in our subject-predicate analysis of sentences, let us introduce functional notation to emphasize that open propositions are functions of variables, and that when we assign specific values to the variables we obtain "values" of this function, the latter "values" of function being propositions that are either true or false. We might adopt the following notation for the indicated propositions:

$R(x)$: x is a rational number.
$G(y)$: $y > 5$.
$S(x,y)$: $x + y = 5$.
$E(x)$: x climbed Mount Everest.
$L(x)$: x is a lawyer.
$C(y)$: y is a computer scientist.

Then we see that $R(\sqrt{2})$ is false, $R(3/4)$ is true, $G(4)$ is false, and $G(7)$ is true. Likewise, $S(2,3)$ is true, while $S(4,3)$ is false.

Open propositions can be combined with logical connectives just as propositions are. For example, the sentence "x is a rational number or y is greater than 5" can be symbolized as $R(x) \lor G(y)$. Likewise the sentence "If x is a rational number, then x is greater than 5" can be symbolized as $R(x) \rightarrow G(x)$.

But still we have not been able to completely analyze the content of sentences like "all mathematicians are rational." Let us now discuss the role of the word "all" in these sentences.

Certain declarative sentences involve words that indicate quantity such as *all, some, none* or *one*. These words help determine the answer to the question "How many?". Since such words indicate quantity they are called *quantifiers*.

Consider the following statements:

1. All isosceles triangles are equiangular.
2. Some parallelograms are squares.
3. There are some real numbers that are not rational numbers.

4. Not all prime integers are odd.
5. Some birds cannot fly.
6. Not all vegetarians are healthy persons.
7. All smokers are flirting with danger.
8. There is one and only one even prime integer.
9. Each rectangle is a parallelogram.
10. Not every angle can be trisected by ruler and compass.

After some thought, we realize that there are two main quantifiers: all and some, where some is interpreted to mean at least one. For example, (1) uses "all"; (2) can be restated as "there is at least one parallelogram that is a square"; (4) means that there is at least one prime integer that is not odd, and (10) can be restated to say "there is at least one angle that cannot be trisected by ruler and compass."

The quantifier "all" is called the **universal quantifier,** and we shall denote it by $\forall x$, which is an inverted A followed by the variable x. It represents each of the following phrases, since they all have essentially the same meaning.

For all x,   All x are such that
For every x,   Every x is such that
For each x,   Each x is such that

The quantifier "some" is the **existential quantifier,** and we shall denote it by $\exists x$, which is a reversed E followed by x. It represents each of the following phrases:

There exists an x such that. . . .

There is an x such that. . . .

For some x. . . .

There is at least one x such that. . . .

Some x is such that. . . .

For a given open proposition $F(x)$ (like, for example, x is a mathematician) we can write "$\forall x$, $F(x)$" meaning "for each x in the universe of discourse, $F(x)$ is true," or we can write "$\exists x$, $F(x)$" meaning "there is at least one x in the universe of discourse such that $F(x)$ is true." The symbol $\exists! x$ is read "there is a unique x such that" or "there is one and only one x such that." For example, the sentence "there is one and only one even prime" can be written "$\exists! x$, [x is an even prime]." Moreover, if we had already designated $P(x)$ as the open proposition "x is an even prime integer," then the above sentence could be written even more cryptically: $\exists! x, P(x)$.

Now then if $M(x)$ denotes the sentence "x is a mathematician" and $R(x)$ denotes the sentence "x is rational" we can write the sentence "All mathematicians are rational" as "For all x, if x is a mathematician, then x is rational" and then symbolically as For all x, $M(x) \rightarrow R(x)$, or as $\forall x$, $[M(x) \rightarrow R(x)]$.

A rephrasing of the sentence "some parallelograms are squares" would be "there is at least one parallelogram that is a square" or "there is at least one object x in the universe such that x is a parallelogram and x is a square." The sentence may now be represented as $\exists x$, $[P(x) \wedge S(x)]$, where $P(x)$ and $S(x)$ mean "x is a parallelogram" and "x is a square," respectively.

In translating sentences with quantifiers into symbols we find a common, but not universal, pattern. The universal quantifier is very often followed by an implication because a universal statement is most often of the form "given any x, if it has property A, then it also has property B." The existential quantifier, on the other hand, is very often followed by a conjunction, because an existential statement is most often of the form "there exists an x with property A that also satisfies property B."

Still speaking generally, when we are considering the content of sentences, we should pay attention to at least five elements:

1. the subject,
2. the predicate,
3. the quantifiers,
4. the quality, and
5. the universe of discourse.

By the quality we mean that we determine whether or not the subject satisfies the property described in the predicate. For example, in a sentence like "all S [subject] is P [predicate]" the quality is affirmative, whereas in the sentence "all S is not P" the quality is negative.

The universe plays a significant role when analyzing sentences. In fact, for a given open proposition P the *truth set* of P is defined as the subset of the universe consisting of all x such that $P(x)$ is true. Then to say that the quantified statement "$\forall x$, $P(x)$" is true is the same as asserting that the truth set of P is equal to the *entire universe*. The statement "$\exists x$, $P(x)$" is true if the truth set is nonempty. The sentence, for example, "$\forall x$, $x > 2$" is true if, in fact, the universe consists of numbers all greater than 2, but this sentence is false if there is some object in the universe that is not greater than 2. Thus, if the universe includes the number 0, then the sentence, "$\forall x$, $x > 2$" is false. Likewise The sentence "$\exists x$, $(x > 2)$" is true if the universe includes, say, the number 5, but the sentence is false if the universe consists of, say, only negative integers.

We see then from these examples that open propositions become propositions once the variables are quantified but the truth value of the quantified proposition depends heavily upon the universe.

Of course, there is the possibility that a given open proposition $F(x)$ is never true for any value of x in the universe and then we use the symbol "$\sim[\exists x, F(x)]$" to mean "there do not exist any values of x in the universe such that $F(x)$ is true." Later, we will discuss the negation of quantifiers in greater detail.

Of course, using the modifier $\sim$ and the quantifiers $\forall$ and $\exists$, we can form eight different expressions involving the open proposition $F(x)$.

For example, $\forall x, [\sim F(x)]$ means "for each x in the universe, $F(x)$ is false" or in abbreviated form "all false." Likewise $\sim[\forall x, F(x)]$ means "it is false that for each x, $F(x)$ is true" that is, this says that $F(x)$ is not always true or in abbreviated form "not all true." Similarly, $\sim\{\exists x, [\sim F(x)]\}$ means "none false" while $\sim\{\exists x, F(x)\}$ means "none true". Let us list these eight quantified statements and their abbreviated meaning in the following list:

Sentence	Abbreviated Meaning
$\forall x, F(x)$	all true
$\exists x, F(x)$	at least one true
$\sim[\exists x, F(x)]$	none true
$\forall x, [\sim F(x)]$	all false
$\exists x, [\sim F(x)]$	at least one false
$\sim\{\exists x, [\sim F(x)]\}$	none false
$\sim\{\forall x, [F(x)]\}$	not all true
$\sim\{\forall x, [\sim F(x)]\}$	not all false

Now after some thought we conclude that

"all true" means the same as "none false,"

"all false" means the same as "none true,"

"not all true" means the same as "at least one false," and

"not all false" means the same as "at least one true."

Thus, the eight expressions can be grouped into four groups of two each, where the two have the same meaning. We list these four types as equivalences:

"all true" $\{\forall x, F(x)\} \equiv \{\sim[\exists x, \sim F(x)]\}$ "none false"

"all false" $\{\forall x, [\sim F(x)]\} \equiv \{\sim[\exists x, F(x)]\}$ "none true"

"not all true" $\{\sim[\forall x, F(x)]\} \equiv \{\exists x, [\sim F(x)]\}$ "at least one false"

"not all false" $\{\sim[\forall x, \{\sim F(x)\}]\} \equiv \{\exists x, F(x)\}$ "at least one true"

The equivalences also provide information about the negation of this type of quantified statement. In the first statement we have $\forall x, F(x)$, its negation $\sim[\forall x, F(x)]$ occurs in the third statement and is equivalent to $\exists x, [\sim F(x)]$. Thus, the negation of "all true" is "at least one false." The second statement is "all false" and its negation is the fourth statement which is equivalent to "at least one true."

We list these facts as follows:

Statement			Negation
"all true"	$\forall x, F(x)$	$\exists x, [\sim F(x)]$	"at least one false"
"at least one false"	$\exists x, [\sim F(x)]$	$\forall x, F(x)$	"all true"
"all false"	$\forall x, [\sim F(x)]$	$\exists x, F(x)$	"at least one true"
"at least one true"	$\exists x, F(x)$	$\forall x, [\sim F(x)]$	"all false"

We see that *to form the negation of a statement involving one quantifier* we need only *change the quantifier from universal to existential,* or *from existential to universal,* and *negate the statement which it quantifies.*

DeMorgan's Laws Revisited

Let us record the following observations: An open proposition $F(x)$ quantified by the universal quantifier $\forall$ is really just the conjunction of many propositions. Likewise, the sentence "$\exists x, F(x)$" is just the disjunction of many propositions. In other words, "$\forall x, F(x)$" and "$\exists x, F(x)$" are just the conjunction and, respectively, the disjunction of all the propositions $F(x)$ where x runs through all the elements of the universe.

For instance, if the universe consists of only $a, b, c,$ and d, then $\forall x, F(x)$ means the same as $F(a) \wedge F(b) \wedge F(c) \wedge F(d)$.

In the light of these comments, we conclude that the rule for negating a quantified open proposition is nothing more than a glorified version of DeMorgan's laws. Again, if the universe $U = \{a,b,c,d\}$, then $\forall x, F(x) = F(a) \wedge F(b) \wedge F(c) \wedge F(d)$, and, therefore, $\sim[\forall x, F(x)] \equiv \sim[F(a) \wedge F(b) \wedge F(c) \wedge F(d)]$. But, by DeMorgan's law, this last expression is the same as $[\sim F(a)] \vee [\sim F(b)] \vee [\sim F(c)] \vee [\sim F(d)]$, which, in turn, is the same as $\exists x, \sim F(x)$. Likewise, if $U = \{a,b,c,d\}$, then $\exists x, F(x) \equiv F(a) \vee F(b) \vee F(c) \vee F(d)$, so that

$$\sim[\exists x, F(x)] \equiv \sim[F(a) \vee F(b) \vee F(c) \vee F(d)]$$
$$\equiv [\sim F(a)] \wedge [\sim F(b)] \wedge [\sim F(c)] \wedge [\sim F(d)]$$
$$\equiv \forall x, [\sim F(x)] \text{ (where the third expression is}$$
$$\text{obtained from DeMorgan's law).}$$

We have seen that there are four main types of statements involving a single quantifier; namely $\forall x, F(x)$, $\exists x, F(x)$, $\forall x,[\sim F(x)]$, and $\exists x,[\sim F(x)]$. The following chart shows when each main type of quantified proposition is true and when it is false. For example, the third entry of the three columns gives the following information about the sentence $\forall x,[\sim F(x)]$; the sentence $\forall x,[\sim F(x)]$ is true if for all c, $F(c)$ is false, but the sentence is false if for at least one c, $F(c)$ is true.

Let c represent an object in the universe of the quantified proposition.

The Statement	Is True	Is False
$\forall x, F(x)$	if for all c, $F(c)$ is true.	if for at least one c, $F(c)$ is false.
$\exists x, F(x)$	if for at least one c, $F(c)$ is true.	if for all c, $F(c)$ is false.
$\forall x,[\sim F(x)]$	if for all c, $F(c)$ is false.	if for at least one c, $F(c)$ is true.
$\exists x,[\sim F(x)]$	if for at least one c, $F(c)$ is false.	if for all c, $F(c)$ is true.

From this chart we observe more proof techniques:

1. **Proof by example.** To show $\exists x, F(x)$ is true, it is sufficient to show $F(c)$ is true for some c in the universe. This type is the only situation where an example proves anything.

2. **Proof by exhaustion.** A statement of the form $\forall x, [\sim F(x)]$, that $F(x)$ is false for all x (all false) or, equivalently, that there are no values x for which $F(x)$ is true (none true) will have been proven after all the objects in the universe have been examined and none found with property $F(x)$.

3. **Proof by counterexample.** To show that $\forall x, F(x)$ is false, it is sufficient to exhibit a specific example c in the universe such that $F(c)$ is false. This one value c is called a **counterexample** to the assertion $\forall x, F(x)$. In actual fact, a counterexample *disproves* the statement $\forall x, F(x)$.

Proof of assertions of the form $\exists x, F(x)$ are referred to as **existence proofs** and existence proofs are classified as either **constructive** or **nonconstructive.** Constructive proofs actually exhibit a value c for which $F(c)$ is true or sometimes, rather than exhibiting c, the proof specifies a process (algorithm) for obtaining such a value.

A nonconstructive existence proof establishes the assertion $\exists x, F(x)$ without indicating how to find a value c such that $F(c)$ is true. Such a proof most commonly involves a proof by contradiction; it shows that the assumption that $\sim\{\exists x, F(x)\}$ is true leads to an absurdity or the negation of some previous result.

For example, there is a theorem that asserts that any polynomial with real coefficients and odd degree must have a real root. But the proof of this theorem does not say how to find this real root. Thus, the proof of this theorem is nonconstructive.

Lagrange's interpolation formula shows how to construct a polynomial with specific values at specified points. Suppose $x_1, x_2, \ldots, x_n$ and $y_1, y_2, \ldots, y_n$ are specified real numbers, where the x_i's are distinct. Then the polynomial

$$P(X) = \sum_{1 \le i \le n} \left[\prod_{i \ne j} \frac{(X - x_j)}{x_i - x_j} \right] y_i = \frac{(X - x_2)(X - x_3) \cdots (X - x_n)}{(x_1 - x_2)(x_1 - x_3) \cdots (x_1 - x_n)} y_1$$

$$+ \frac{(X - x_1)(X - x_3) \cdots (X - x_n)}{(x_2 - x_1)(x_2 - x_3) \cdots (x_2 - x_n)} y_2$$

$$+ \cdots + \frac{(X - x_1)(X - x_2) \cdots (X - x_{n-1})}{(x_n - x_1)(x_n - x_2) \cdots (x_n - x_{n-1})} y_n$$

is the unique polynomial such that $P(x_i) = y_i$ for each i.

For example, suppose $x_1 = 1$, $x_2 = 3$, $x_3 = 4$ and $y_1 = 5$, $y_2 = -6$, $y_3 = 2$. Then,

$$P(X) = \frac{(X - 3)(X - 4)}{(1 - 3)(1 - 4)}\,(5) + \frac{(X - 1)(X - 4)}{(3 - 1)(3 - 4)}\,(-6)$$

$$+ \frac{(X - 1)(X - 3)}{(4 - 1)(4 - 3)}\,(2)$$

$$= (X^2 - 7X + 12)\left(\frac{5}{6}\right) + (X^2 - 5X + 4)(3)$$

$$+ (X^2 - 4X + 3)\left(\frac{2}{3}\right)$$

$$= \frac{9}{2} X^2 - \frac{47}{2} X + 24$$

is a polynomial such that $P(1) = 5$, $P(3) = -6$, and $P(4) = 2$. Thus, once we know Lagrange's interpolation formula we can show the existence of a polynomial that attains specific values at specified points by construction.

To show how to make a **proof by exhaustion,** we consider the following proposition:

There are no rational roots to the polynomial
$$P(X) = 2X^8 - X^7 + 8X^4 + X^2 - 5.$$

To prove this by exhaustion we cannot consider *all* rational numbers, we need some theorem that will enable us to consider only a finite set of rational numbers. This is provided by the **rational roots theorem.**

If $P(X) = a_0 + a_1X + \cdots + a_nX^n$ is a polynomial with integer coefficients, then any rational root of $P(X)$ has the form a/b where a and b are integers such that a divides a_0 and b divides a_n.

Thus, this theorem enables us to consider as universe for the above proposition the set $U = \{\pm 1, \pm 5, \pm 1/2, \pm 5/2\}$. Since we can determine that $P(c) \neq 0$ for each c in U, we have proved the proposition by exhaustion.

An example of a proof by counterexample. Let n be a positive integer and define $p(n)$ to be the number of partitions of n; that is, the number of different ways to write n as a sum of positive integers, disregarding order. Since 5 can be written as

$$1 + 1 + 1 + 1 + 1, 2 + 1 + 1 + 1, 2 + 2 + 1, 3 + 1 + 1,$$
$$3 + 2, 4 + 1, \text{ and } 5,$$

we have $p(5) = 7$. In fact, it is easy to establish that

$$p(1) = 1, p(2) = 2, p(3) = 3, p(4) = 5, p(5) = 7,$$

and attempt to prove the conjecture: For each positive integer n, $p(n)$ is a prime integer.

To test this conjecture, we calculate $p(6)$ and the observation that $p(6) = 11$ adds credence to the conjecture, (but does not prove it!). However, the calculation that $p(7) = 15$ provides a counterexample. Thus, the conjecture has been proved to be false by counterexample.

Sentences with multiple quantifiers. So far we have considered only those sentences in which the universal and existential quantifiers appear singly. We shall now consider cases in which the quantifiers occur in combinations. These combinations become particularly important in the case of sentences involving more than one variable. For example, the fact that the product of two real numbers is a real number can be written:

$$(\forall x)\,(\forall y)\,[x \in \mathbb{R} \wedge y \in \mathbb{R} \to xy \in \mathbb{R}].$$

In general, if $P(x,y)$ is any predicate involving the two variables x and y, then the following possibilities exist:

$$(\forall x)\,(\forall y)\,P\,(x,y) \qquad (\forall x)\,(\exists y)\,P\,(x,y)$$
$$(\exists x)\,(\forall y)\,P\,(x,y) \qquad (\exists x)\,(\exists y)\,P\,(x,y)$$
$$(\forall y)\,(\forall x)\,P\,(x,y) \qquad (\exists y)\,(\forall x)\,P\,(x,y)$$
$$(\forall y)\,(\exists x)\,P\,(x,y) \qquad (\exists y)\,(\exists x)\,P\,(x,y)$$

If a sentence involves both the universal and the existential quantifiers, one must be careful about the order in which they are written. (One always works from left to right.) For instance, of the two sentences concerning real numbers:

$$(\forall x)\,(\exists y)\,[x + y = 5], \qquad (\exists y)\,(\forall x)\,[x + y = 5],$$

the first is true, while the second is false. The first sentence says that if x is any real number, then there exists a real number y such that the sum of x and y is 5. Of course, y is just $5 - x$. The second sentence says that every real number x is equal to the same number, $5 - y$, where y is some fixed real number. Thus, the second sentence says, in effect, that all real numbers are equal.

There are logical relationships between sentences with two quantifiers if the same predicate is involved in each sentence. We depict these relationships in the following diagram:

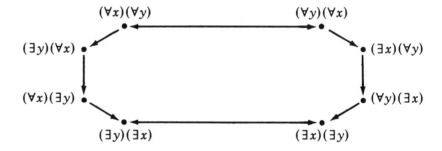

Figure 1-5. Graphical representation of relationships among sentences involving two quantifiers.

What this diagram tells us, for example, is that the sentences $(\forall x)$ $(\forall y)\,P\,(x,y)$ and $(\forall y)\,(\forall x)\,P\,(x,y)$ are logically equivalent while $(\forall x)$ $(\forall y)\,P\,(x,y)$ logically implies $(\exists y)\,(\forall x)\,P\,(x,y)$. Let us restate all the above relationships in the following list:

$$(\forall x)\,(\forall y)\,P\,(x,y) \leftrightarrow (\forall y)\,(\forall x)\,P\,(x,y)$$
$$(\forall x)\,(\forall y)\,P\,(x,y) \rightarrow (\exists y)\,(\forall x)\,P\,(x,y)$$
$$(\forall y)\,(\forall x)\,P\,(x,y) \rightarrow (\exists x)\,(\forall y)\,P\,(x,y)$$
$$(\exists y)\,(\forall x)\,P\,(x,y) \rightarrow (\forall x)\,(\exists y)\,P\,(x,y)$$
$$(\exists x)\,(\forall y)\,P\,(x,y) \rightarrow (\forall y)\,(\exists x)\,P\,(x,y)$$
$$(\forall x)\,(\exists y)\,P\,(x,y) \rightarrow (\exists y)\,(\exists x)\,P\,(x,y)$$
$$(\forall y)\,(\exists x)\,P\,(x,y) \rightarrow (\exists x)\,(\exists y)\,P\,(x,y)$$
$$(\exists y)\,(\exists x)\,P\,(x,y) \leftrightarrow (\exists x)\,(\exists y)\,P\,(x,y)$$

The negation of any sentence involving more than one quantifier can be accomplished by systematically applying the rule for negating a sentence with only one quantifier. Let us illustrate. Suppose your were asked to prove by contradiction a sentence that has the following form:

$$(\forall x)\,(\exists y)\,[F(x,y) \rightarrow G(x,y) \vee H(x,y)].$$

Thus, you must show that the negation of this sentence implies some false sentence such as $r \wedge \sim r$. Letting $F = F(x,y)$, etc., we find the negation as follows:

$$\sim[(\forall x)\,(\exists y)\,(F \rightarrow G \vee H)] \equiv (\exists x)\,[\sim(\exists y)\,(F \rightarrow G \vee H)]$$
$$\equiv (\exists x)\,(\forall y)\,[\sim(F \rightarrow G \vee H)]$$
$$\equiv (\exists x)\,(\forall y)\,[F \wedge \sim(G \vee H)]$$
$$\equiv (\exists x)\,(\forall y)\,[F \wedge (\sim G) \wedge (\sim H)]$$

Therefore, with the knowledge of a few tautologies and rules of logic, the work of negating a complicated sentence becomes almost mechanical.

Exercises for Section 1.8

1. Translate each of the following statements into symbols, using quantifiers, variables, and predicate symbols.
   (a) All birds can fly.
   (b) Not all birds can fly.
   (c) All babies are illogical.
   (d) Some babies are illogical.
   (e) If x is a man, then x is a giant.
   (f) Some men are giants.
   (g) Some men are not giants.
   (h) All men are giants.
   (i) No men are giants.
   (j) There is a student who likes mathematics but not history.

    (k) x is an odd integer and x is prime.

    (l) For all integers x, x is odd and x is prime.

    (m) For each integer x, x is odd and x is prime.

    (n) There is an integer x such that x is odd and x is prime.

    (o) Not every actor is talented who is famous.

    (p) Some numbers are rational.

    (q) Some numbers are not rational.

    (r) Not all numbers are rational.

    (s) Not every graph is planar.

    (t) If some students are lazy, then all students are lazy.

    (u) x is rational implies that x is real.

2. Let the universe consist of all integers and let

    $P(x)$: x is a prime,

    $Q(x)$: x is positive,

    $E(x)$: x is even,

    $N(x)$: x is divisible by 9,

    $S(x)$: x is a perfect square, and

    $G(x)$: x is greater than 2.

    Then express each of the following in symbolic form.

    (a) x is even or x is a perfect square.

    (b) x is a prime and x is divisible by 9.

    (c) x is a prime and x is greater than 2.

    (d) If x is a prime, then x is greater than 2.

    (e) If x is a prime, then x is positive and not even.

3. Translate each of the following sentences into symbols, first using no existential quantifier, and second using no universal quantifier.

    (a) Not all cars have carburetors.

    (b) Some people are either religious or pious.

    (c) No dogs are intelligent.

    (d) All babies are illogical.

    (e) Every number either is negative or has a square root.

    (f) Some numbers are not real.

    (g) Every connected and circuit-free graph is a tree.

    (h) Not every graph is connected.

4. Determine the truth or falsity of the following sentences where the universe U is the set of integers.

    (a) $\forall x, [x^2 - 2 \geq 0]$.

    (b) $\forall x, [x^2 - 10x + 21 = 0]$.

    (c) $\exists x, [x^2 - 10x + 21 = 0]$.

    (d) $\forall x, [x^2 - x - 1 \neq 0]$.

    (e) $\exists x, [2x^2 - 3x + 1 = 0]$.

    (f) $\exists x, [15x^2 - 11x + 2 = 0]$.

    (g) $\exists x, [x^2 - 3 = 0]$.

(h) $\exists x, [x^2 - 9 = 0]$.
(i) $\exists x, [\{x^2 > 10\} \wedge \{x \text{ is even}\}]$.
(j) $\forall x, \{\exists y, [x^2 = y]\}$.
(k) $\exists x, \{\forall y, [x^2 = y]\}$.
(l) $\forall y, \{\exists x, [x^2 = y]\}$.
(m) $\exists y, \{\forall x, [x^2 = y]\}$.

5. Write the negations (as universal or existential propositions) of sentences (a) through (i) in Exercise 4.

6. Write the negations of the following sentences by changing quantifiers.
   (a) For each integer x, if x is even, then $x^2 + x$ is even.
   (b) There is an integer x such that x is even and x is prime.
   (c) Every complete bipartite graph is not planar.
   (d) There is no integer x such that x is prime and $x + 6$ is prime.
   (e) For each integer x, $x^2 + 3 > 5$ or $x < 2$.
   (f) For each integer x, either $x, x - 1, x - 2$, or $x - 3$ is divisible by 4.
   (g) For each integer x, if x^2 is even, then x is even.
   (h) $\forall x, x^2 = 25$ or x is negative.
   (i) $\exists x, x^2 = 25$ and $x > 0$.
   (j) There is an integer x such that $x^2 = 9$.

7. Consider the open propositions over the universe $U = \{-5,0,1,2,3,4,5,6,7,8,9,10\}$.
   $P(x)$: $x^2 < 5$.
   $Q(x)$: $x \geq 3$.
   $R(x)$: x is a multiple of 2.
   $S(x)$: $x^2 = 25$.

   Find the truth sets of:
   (a) $P(x) \vee Q(x)$.
   (b) $P(x) \wedge R(x)$.
   (c) $[\sim P(x)] \vee Q(x)$.
   (d) $P(x) \wedge [\sim Q(x)]$.
   (e) $\sim\{[\sim P(x)] \wedge [\sim Q(x)]\}$.
   (f) $[\sim P(x)] \vee \{Q(x) \wedge [\sim R(x)]\}$.
   (g) $S(x)$.
   (h) $S(x) \wedge Q(x)$.
   (i) $S(x) \wedge [\sim Q(x)]$.
   (j) $[P(x) \wedge Q(x)] \wedge S(x)$.

8. Which of the following are propositions, open propositions, or neither?
   (a) $x < 2$.
   (b) $1 < 2$.
   (c) He is a baseball player.

(d) Reggie Jackson is a baseball player.

(e) $2 + 3 = 3 + 2$.

(f) This sentence is false.

(g) WOW!

(h) There is an integer x such that $x^2 - 25 = 0$.

(i) For each integer x, there is a integer y such that $x + y = 5$.

9. Using the Lagrange interpolation formula, construct a polynomial $P(x)$ such that $P(1) = 360$, $P(2) = 420$, $P(3) = 360$, $P(4) = 195$, and $P(5) = 0$.

10. Use Lagrange interpolation to construct a polynomial whose graph passes through the points $(-1,1)$, $(0,1)$, $(1,1)$, and $(2,-5)$.

11. If $x_1, x_2, \ldots, x_n$ are distinct real numbers and if

$$L_i(X) = \frac{(X - x_1)(X - x_2)\ldots(X - x_{i-1})(X - x_{i+1})\ldots(X - x_n)}{(X_i - x_1)(X_i - x_2)\ldots(X_i - x_{i-1})(X_i - x_{i+1})\ldots(X_i - x_n)}$$

then the Lagrange interpolation formula states that

$$P(X) = \sum_{i=1}^{n} L_i(X)y_i \text{ is the unique polynomial such that}$$

$P(x_i) = y_i$ for each i.

(a) Prove that, in general, $\Sigma_{i=1}^{n} L_i(X) = 1$. Hint: use uniqueness.

This fact in (a) enables one to compute $P(X)$ more quickly in many examples. For instance, using the data of exercise 10, $(x_1,y_1) = (-1,1)$, $(x_2,y_2) = (0,1)$, $(x_3,y_3) = (1,1)$, and $(x_4,y_4) = (2,-5)$, the polynomial $P(X)$ can be computed two ways: first,

$$P(X) = L_1(X) + L_2(X) + L_3(X) - 5L_4(X)$$

by the interpolation formula, and second, $P(X) = 1 - 6L_4(X)$ by (a) since $L_1(X) + L_2(X) + L_3(X) + L_4(X) = 1$. But then $P(X)$ can be computed easily by computing only $L_4(X)$. Since

$$L_4(X) = \frac{(X + 1)(X)(X - 1)}{(2 + 1)(2)(2 - 1)} = \frac{X^3 - X}{6}, \text{ we see that}$$

$$P(X) = 1 - \frac{6(X^3 - X)}{6} = -X^3 + X + 1$$

(b) Using the fact in (a) compute the Lagrange interpolation formula polynomial $P(X)$ for the following table of values:

x_i	1	2	3	4
y_i	1	1	1	4

(c) Compute $P(X)$ as in (b) for the following table of values:

x_i	1	2	3	4
y_i	1	1	1	-8

(d) Derive $P(X)$ as in (b) for the following table of values:

x_i	1	3	5	7
y_i	1	1	1	4

(e) If $x_i = i$ for $i = 1,2,\ldots,n$, prove that

$$\sum_{i-1}^{n} iL_i(X) = X \text{ and}$$
$$\sum_{i-1}^{n} i^2 L_i(X) = X^2.$$

Can you generalize?

12. Derive $P(X)$ for the following table:

x_i	1	2	3	4	5
y_i	1	1	1	1	v

Thus observe that a string of ones need not continue; any value v can be justified as the next value of the interpolation polynomial $P(X)$. Note that the special choice $v = 1$ substantially reduces the degree of $P(X)$.

13. Derive $P(X)$ for the following table:

x_i	1	2	3	4	5
y_i	1	2	3	4	v

Thus conclude that a sequence of values beginning 1, 2, 3, 4 need not continue with 5; any value v can be justified as the next value. However, $v = 5$ substantially reduces the degree of $P(X)$.

14. Find all rational roots of the following polynomials:
    (a) $X^3 - 2X + 1$
    (b) $2X^3 + 6X^2 + 4X + 1$
    (c) $X^3 - 3X - 1$
    (d) $2X^3 - 11X^2 + 17X - 6$
    (e) $3X^3 - X^2 - 6X + 2$

15. (Magic Squares Problem) Suppose that we have a 3×3 square and nine 1×1 tiles containing the digits 1 through 9. The problem is to place the tiles in the squares so that the sum of the numbers across each row, down each column, and along each diagonal is the same. Such an arrangement is called a *magic square*.

(a) First discover what each row sum (and therefore what each column and diagonal sum) must be.

(b) Observe that 9 cannot be placed in the second row and second column. Neither can 9 be placed in any corner position. Why? Thus there are only 4 legitimate positions for 9. Moreover, in any legitimate location for 9, there is a limited number of positions for 6, 7, and 8. Next observe that 5 must occupy the second row and second column position. Finally, observe that 1, 5, and 9 must occur in the same row or column.

(c) Use proof by exhaustion to find all 8 magic squares using the digits 1 through 9. Note that any one of these can be transformed into any other by either a rotation or a reflection.

(d) Change the magic squares problem to use the digits 0 through 8. Indicate how to find all such magic squares.

(e) Indicate how to find all 3×3 magic squares using the integers 3 through 11.

(f) Find all 3×3 magic squares using any 9 consecutive integers.

(g) Indicate how to generate 5×5 magic squares using the first 25 integers as labels on the individual tiles. Hint: consider the middle 3×3 square using the entries 9 through 17. Can you generalize to an $n \times n$ square where n is odd?

(h) In general, an $n \times n$ magic square using the integers $1, 2, ..., n^2$ is an $n \times n$ array of these integers in such a way that the sum s of the integers in each row, in each column, and in each of the two diagonals is the same. The number s is called the magic sum of the magic square.

Show that the magic sum of an $n \times n$ magic square using the integers $1, 2, ..., n^2$ is $n(n^2 + 1)/2$. Observe that there are no 2×2 magic squares. Find a 4×4 magic square using the integers $1, 2, ..., 16$.

16. Define a Latin square as an $n \times n$ table that uses the integers 1 through n as entries, and does so in such a way that no integer appears more than once in the same row or column.

(a) Observe that an $n \times n$ table of entries 1 through n is a Latin square if and only if each integer appears exactly once in each row and column.

(b) Find both 2×2 Latin squares.

(c) Use proof by exhaustion to find all 12 3×3 Latin squares.

(d) Find one of the 576 4×4 Latin squares.

(e) Find a 5×5 Latin square.

Selected Answers for Section 1.8

1. (a) $\forall x, [B(x) \rightarrow F(x)]$.

   (b) $\sim[\forall x, (B(x) \rightarrow F(x))]$ or $\exists x, (B(x) \wedge [\sim F(x)])$.

   (d) $\exists x, [B(x) \wedge I(x)]$.

   (e) $M(x) \rightarrow G(x)$.

   (f) $\exists x, [M(x) \wedge G(x)]$.

   (j) $\exists x, [S(x) \wedge M(x) \wedge \sim H(x)]$.

   (k) $O(x) \wedge P(x)$.

   (l) Let the universe be the set of integers $\forall x, (O(x) \wedge P(x))$.

   (n) $\exists x, [O(x) \wedge P(x)]$.

   (p) $\exists x, [N(x) \wedge R(x)]$.

   (r) $\sim[\forall x, [N(x) \rightarrow R(x)]]$ or $\exists x, [N(x) \wedge \sim R(x)]$.

   (s) $\sim[\forall x, (G(x) \rightarrow P(x))]$.

3. (c) $\forall x, [D(x) \rightarrow \sim I(x)]$; $\sim[\exists x, \{D(x) \wedge I(x)\}]$.

   (d) $\forall x, [B(x) \rightarrow I(x)]$; $\sim[\exists x, \{B(x) \wedge \sim I(x)\}]$.

   (f) $\sim[\forall x, [N(x) \rightarrow R(x)]]$; $\exists x, [N(x) \wedge \sim R(x)]$.

6. (c) Write the sentence as follows: Let U be the universe of graphs.

$$\forall x, [C(x) \wedge B(x) \rightarrow \sim P(x)]$$

where

$C(x)$: x is complete,

$B(x)$: x is bipartite, and

$P(x)$: x is planar.

The negation is:

$$\exists x, [C(x) \wedge B(x) \wedge (P(x)]$$

1.9   RULES OF INFERENCE FOR QUANTIFIED PROPOSITIONS

Additional rules of inference are necessary to prove assertions involving open propositions and quantifiers. A careful treatment of these rules is beyond our scope, but we will illustrate some of the techniques. The following four rules describe when the universal and existential quantifiers can be added to or deleted from an assertion. We continue our list to include the four rules of inference we have already discussed:

Fundamental Rule 5.   Universal Specification. If a statement of the form $\forall x, P(x)$ is assumed to be true, then the universal quantifier can be dropped to obtain $P(c)$ is true for an arbitrary object c in the universe. This rule may be represented as

$$\frac{\forall x, P(x)}{\therefore P(c) \text{ for all } c} .$$

Thus, suppose the universe is the set of humans, and suppose that $M(x)$ denotes the statement "x is mortal," then if we can establish the truth of the sentence "$\forall x, M(x)$," that is, "all men are mortal," then the rule of universal specification allows us to conclude "Socrates is mortal."

Informally stated the next rule says that what is true for arbitrary objects in the universe is true for all objects. This rule permits the universal quantification of assertions.

Fundamental Rule 6. Universal Generalization. If a statement $P(c)$ is true for each element c of the universe, then the universal quantifier may be prefixed to obtain $\forall x, P(x)$. In symbols, this rule is

$$\frac{P(c) \text{ for all } c}{\therefore \forall x, P(x)} .$$

This rule holds provided we know P(c) is true for each element c in the universe.

The next rule, informally stated, says that if a statement is true of some object then we may refer to this object by assigning it a name.

Fundamental Rule 7. Existential Specification. If $\exists x, P(x)$ is assumed to be true, then there is an element c in the universe such that $P(c)$ is true. This rule takes the form

$$\frac{\exists x, P(x)}{\therefore P(c) \text{ for some } c} .$$

Note that the element c is not arbitrary (as it was in Rule 5), but must be one for which P(x) is true. It follows from the truth of $\exists x, P(x)$ that at least one such element must exist, but nothing more is guaranteed. This places constraints on the proper use of this rule. For example, if we know that $\exists x, P(x)$ and $\exists x, Q(x)$ are both true, then we can conclude $P(c) \wedge Q(d)$ is true for some elements c and d of the universe, but as a general rule we *cannot* conclude that $P(c) \wedge Q(c)$ is true. For example, suppose that the universe is the set of integers and $P(x)$ is the sentence "x is even" while $Q(x)$ is the sentence "x is odd." Then $\exists x, P(x)$ and $\exists x, Q(x)$ are both true, but $P(c) \wedge Q(c)$ is false for every c in the universe of integers.

Fundamental Rule 8. Existential Generalization. If $P(c)$ is true for some element c in the universe, then $\exists x, P(x)$ is true. In symbols, we have

$$\frac{P(c) \text{ for some } c}{\therefore \exists x, P(x)}.$$

When quantifiers are involved, construction of proofs is more complicated because of the care required in the application of the rules of inference. An exploration into the subtleties of proofs involving quantifiers is beyond our intention, but we shall give a few simple examples to illustrate the application of the above rules.

Generally speaking, *in order to draw conclusions from quantified premises, we need to remove quantifiers properly, argue with the resulting propositions, and then properly prefix the correct quantifiers.*

Example 1.9.1. Consider the argument

All men are fallible.

All kings are men.

Therefore, all kings are fallible.

Let $M(x)$ denote the assertion "x is a man," $K(x)$ denote the assertion "x is a king," and $F(x)$ the sentence "x is fallible." Then the above argument is symbolized:

$$\frac{\begin{array}{l} \forall x, [M(x) \rightarrow F(x)] \\ \forall x, [K(x) \rightarrow M(x)] \end{array}}{\therefore \forall x, [K(x) \rightarrow F(x)]}.$$

A formal proof is as follows:

Assertion	Reasons
1. $\forall x, [M(x) \rightarrow F(x)]$	Premise 1
2. $M(c) \rightarrow F(c)$	Step 1 and Rule 5
3. $\forall x, [K(x) \rightarrow M(x)]$	Premise 2
4. $K(c) \rightarrow M(c)$	Step 3 and Rule 5
5. $K(c) \rightarrow F(c)$	Steps 2 and 4 and Rule 2
6. $\forall x, [K(x) \rightarrow F(x)]$	Step 5 and Rule 6

Example 1.9.2.   Symbolize the following argument and check for its validity:

Lions are dangerous animals.

There are lions.

Therefore, there are dangerous animals.

Represent $L(x)$ and $D(x)$ as "x is a lion" and "x is dangerous," respectively. Then the argument takes the form

$$\forall x, [L(x) \to D(x)]$$
$$\underline{\exists x, L(x)}$$
$$\therefore \exists x\, D(x).$$

A formal proof is as follows:

Assertion	Reasons
1. $\exists x, L(x)$	Premise 2
2. $L(a)$	Step 1 and Rule 7
3. $\forall x, [L(x) \to D(x)]$	Premise 1
4. $L(a) \to D(a)$	Step 3 and Rule 5
5. $D(a)$	Steps 2 and 4, Rule 1
6. $\exists x, D(x)$	Step 5 and Rule 8

Exercises for Section 1.9

1. Obtain a conclusion in the following:
   (a) If there are any rational roots to the equation $X^2 - 2 = 0$, then either ± 1 or ± 2 are roots.
   It is not the case that ± 1 or ± 2 are roots of the equation.
   Hence, ...
   (b) Some negative numbers are rational numbers.
   No rational numbers are imaginary.
   Hence, ...
   (c) Some politicians are corrupt.
   All corrupt persons should be sentenced to prison.
   Hence, ...
   (d) All Democrats are not conservative.
   Hence, all conservatives are ...
   (e) All squares are rectangles.
   All rectangles are parallelograms.
   All parallelograms are quadrilaterals.
   Hence, ...

2. Prove or disprove the validity of the following arguments:
   (a) Every living thing is a plant or an animal.
       David's dog is alive and it is not a plant.
       All animals have hearts.
       Hence, David's dog has a heart.
   (b) No mathematicians are ignorant.
       All ignorant people are haughty.
       Hence, some haughty people are not mathematicians.
   (c) Babies are illogical.
       Nobody is despised who can manage a crocodile.
       Illogical people are despised.
       Hence, babies cannot manage crocodiles. [Lewis Carroll]
   (d) Students of average intelligence can do arithmetic.
       A student without average intelligence is not a capable student.
       Your students cannot do arithmetic.
       Therefore, your students are not capable.
   (e) All integers are rational numbers.
       Some integers are powers of 2.
       Therefore, some rational numbers are powers of 2.
   (f) Some rational numbers are powers of 3.
       All integers are rational numbers.
       Therefore, some integers are powers of 3.
   (g) All clear explanations are satisfactory.
       Some excuses are unsatisfactory.
       Hence, some excuses are not clear explanations. [Lewis Carroll]
   (h) Some dogs are animals.
       Some cats are animals.
       Therefore, some dogs are cats.
   (i) All dogs are carnivorous.
       Some animals are dogs.
       Therefore, some animals are carnivorous.

3. The following propositions involve predicates that define sets. Use the properties to conclude relationships between these sets. Use Venn diagrams to check the validity of the arguments.
   (a) All cigarettes are hazardous to health.
       All Smokums are cigarettes.
       Hence, all Smokums are hazardous to health.
   (b) Some scientists are not engineers.
       Some astronauts are not engineers.
       Hence, some scientists are not astronauts.
   (c) All astronauts are scientists.
       Some astronauts are engineers.
       Hence, some engineers are scientists.

     (d) Some humans are vertebrates.
         All humans are mammals.
         Therefore, some mammals are vertebrates.

     (e) No mothers are males.
         Some males are politicians.
         Hence, some politicians are not mothers.

     (f) Some females are not mothers.
         Some politicians are not females.
         Hence, some politicians are not mothers.

     (g) All doctors are college graduates.
         Some doctors are not golfers.
         Hence, some golfers are not college graduates.

     (h) All fathers are males.
         Some students are fathers.
         Hence, all students are males.

     (i) All fathers are males.
         Some students are fathers.
         Hence, some students are males.

Selected Answers for Section 1.9

2. (a) Let the universe consist of all living things, and let

$$P(x):\ x \text{ is a plant.}$$
$$A(x):\ x \text{ is an animal}$$
$$H(x):\ x \text{ has a heart}$$
$$a:\ \text{David's dog}$$

Then the inference pattern is:

$$\forall x,\ [P(x) \vee A(x)].$$
$$\sim P(a).$$
$$\forall x,\ [A(x) \rightarrow H(x)].$$

Hence, $H(a)$.
The proof of validity is the following:

(1) $\forall x,\ [P(x) \vee A(x)]$ Premise

(2) $\sim P(a)$ Premise

(3) $P(a) \vee A(a)$ From (1) and Universal Specification

(4) $A(a)$ From (2) and (3) and disjunctive syllogism

(5) $\forall x,\ [A(x) \rightarrow H(x)]$ Premise

(6) $A(a) \rightarrow H(a)$ (5) and Specification
(7) $H(a)$ (4), (6), and Inference Rule (1)

   3. (b), (f), (g), and (h) are invalid.

1.10   MATHEMATICAL INDUCTION

In mathematics, as in science there are two main aspects of inquiry whereby we can discover new results: deductive and inductive. As we have said the deductive aspect involves accepting certain statements as premises and axioms and then deducing other statements on the basis of valid inferences. The inductive aspect, on the other hand, is concerned with the search for facts by observation and experimentation—we arrive at a conjecture for a general rule by inductive reasoning. Frequently we may arrive at a conjecture that we believe to be true *for all positive integers n*. But then before we can put any confidence in our conjecture we need to verify the truth of the conjecture. There is a proof technique that is useful in verifying such conjectures. Let us describe that technique now.

The Principle of Mathematical Induction.   Let $P(n)$ be a statement which, for each integer n, may be either true or false. To prove $P(n)$ is true for all integers $n \geq 1$, it suffices to prove:

1. $P(1)$ is true.
2. For all $k \geq 1$, $P(k)$ implies $P(k + 1)$.

If one replaces (1) and (2) by (1') $P(n_0)$ is true, and (2') For all $k \geq n_0$, $P(k)$ implies $P(k + 1)$, then we can prove $P(n)$ is true for all $n \geq n_0$, and the starting point n_0, or *basis of induction,* may be any integer—positive, negative, or zero. Normally we expect to prove $P(k) \rightarrow P(k + 1)$ directly so there are 3 steps to a proof using the principle of mathematical induction:

  (i) (Basis of induction) Show $P(n_0)$ is true.
  (ii) (Inductive hypothesis) Assume $P(k)$ is true for $k \geq n_0$.
  (iii) (Inductive step) Show that $P(k + 1)$ is true on the basis of the inductive hypothesis.

We emphasize that the inductive hypothesis is not tantamount to assuming what is to be proved; it is just part of proving the implication

$$P(k) \rightarrow P(k + 1).$$

Now the principle of mathematical induction is a reasonable method of proof for part (1) tells us that $P(1)$ is true. Then using (2) and the fact that part (1) tells us that $P(1)$ is true, we conclude $P(2)$ is true. But then (2) implies that $P(2 + 1) = P(3)$ is true, and so on. Continuing in this way we would ultimately reach the conclusion that $P(n)$ is true for any fixed positive integer n. The principle of mathematical induction is much like the game we played as children where we would stand up dominos so that if one fell over it would collide with the next domino in line. This is like part (2) of the principle. Then we would tip over the first domino (this is like part (1) of the principle). Then what would happen? All the dominos would fall down—like the conclusion that $P(n)$ is true for all positive integers n.

Example 1.10.1. Let us use this approach on the problem of determining a formula for the sum of the first n positive integers. Let $S(n) = 1 + 2 + 3 + \cdots + n$. Let us examine a few values for $S(n)$ and list them in the following table:

n	1	2	3	4	5	6	7
$S(n)$	1	3	6	10	15	21	28

The task of guessing a formula for $S(n)$ may not be an easy one and there is no sure-fire approach for obtaining a formula. Nevertheless, one might observe the following pattern:

$$2\,S(1) = \ 2 = 1 \cdot 2$$
$$2\,S(2) = \ 6 = 2 \cdot 3$$
$$2\,S(3) = 12 = 3 \cdot 4$$
$$2\,S(4) = 20 = 4 \cdot 5$$
$$2\,S(5) = 30 = 5 \cdot 6$$
$$2\,S(6) = 42 = 6 \cdot 7$$

This leads us to conjecture that

$$2\,S(n) = n(n + 1) \text{ or that } S(n) = \frac{n(n + 1)}{2}.$$

Now let us use mathematical induction to prove this formula. Let $P(n)$ be the statement: the sum $S(n)$ of the first n positive integers is equal to $n(n + 1)/2$.

1. **Basis of Induction.** Since $S(1) = 1 = 1(1 + 1)/2$, the formula is true for $n = 1$.

2. **Inductive Hypothesis.** Assume the statement $P(n)$ is true for $n = k$, that is, that $S(k) = 1 + 2 + \cdots + k = k(k + 1)/2$.

3. **Inductive Step.** Now show that the formula is true for $n = k + 1$, that is, show that $S(k + 1) = (k + 1)(k + 2)/2$ follows from the inductive hypothesis. To do this, we observe that $S(k + 1) = 1 + 2 + \cdots + (k + 1) = S(k) + (k + 1)$.

Since $S(k) = k(k + 1)/2$ by the inductive hypothesis, we have

$$S(k + 1) = S(k) + (k + 1) = \frac{k}{2}(k + 1) + (k + 1) = (k + 1)\left(\frac{k}{2} + 1\right)$$

$$= \frac{(k + 1)(k + 2)}{2},$$

and the formula holds for $k + 1$. So, by assuming the formula was true for k, we have been able to prove the formula holds for $k + 1$, and the proof is complete by the principle of mathematical induction.

The principle of mathematical induction is based on a result that may be considered one of the axioms for the set of positive integers. This axiom is called the *well-ordered property* of the positive integers; its statement is the following: *Any nonempty set of positive integers contains a least positive integer.*

Example 1.10.2. Find and prove a formula for the sum of the first n cubes, that is, $1^3 + 2^3 + \cdots + n^3$.

We consider the first few cases:

$$1^3 = 1 = 1^2$$
$$1^3 + 2^3 = 9 = 3^2$$
$$1^3 + 2^3 + 3^3 = 36 = 6^2$$
$$1^3 + 2^3 + 3^3 + 4^3 = 100 = 10^2$$

From this meager information we expect that $1^3 + 2^3 + 3^3 + 4^3 + 5^3$ to be a perfect square. But the square of what integer? After computing we find that it is 15^2. Still we may not see the pattern at first, but by comparing the table for $S(n)$ in Example 1.11.1 we see that we have obtained thus far,

$[S(1)]^2 = 1^2$, $[S(2)]^2 = 3^2$, $[S(3)]^2 = 6^2$, $[S(4)]^2 = 10^2$, and $[S(5)]^2 = 15^2$.

We conjecture then that $1^3 + 2^3 + \cdots + n^3 = [n(n + 1)/2]^2$. Let us verify this formula by mathematical induction:

   1. **Basis of Induction.** Since $1^3 = [1 (1 + 1)/2]^2$ the formula holds for $n = 1$.

   2. **Inductive Hypothesis.** Suppose the formula holds for $n = k$. Thus, suppose $1^3 + 2^3 + \cdots + k^3 = [k(k + 1)/2]^2$.

   3. **Inductive Step.** Show the formula holds for $n = k + 1$; that is, show $1^3 + 2^3 + \cdots + k^3 + (k + 1)^3 = [(k + 1) (k + 2)/2]^2$.

   Now $1^3 + 2^3 + \cdots + k^3 + (k + 1)^3$ is nothing more that the sum of $1^3 + 2^3 + \cdots + k^3$ and $(k + 1)^3$, so we use the inductive hypothesis to replace $1^3 + 2^3 + \cdots + k^3$ by $[k (k + 1)/2]^2$. Thus,

$$1^3 + 2^3 + \cdots + k^3 + (k + 1)^3 = \left[\frac{k(k + 1)}{2}\right]^2 + (k + 1)^3$$

$$= (k + 1)^2 \left[\left(\frac{k}{2}\right)^2 + k + 1\right] = (k + 1)^2 \left[\frac{k^2}{4} + k + 1\right]$$

$$= (k + 1)^2 \left[\frac{k^2 + 4k + 4}{4}\right]$$

$$= (k + 1)^2 \left[\frac{k + 2}{2}\right]^2$$

$$= \left[\frac{(k + 1) (k + 2)}{2}\right]^2.$$

Hence, the formula holds for $k + 1$ and thus by the principle of mathematical induction for all positive integers n.

   Example 1.10.3. Prove by mathematical induction that $6^{n+2} + 7^{2n+1}$ is divisible by 43 for each positive integer n.

   1. First we show that $6^{1+2} + 7^{2+1} = 6^3 + 7^3$ is divisible by 43. But this follows because $6^3 + 7^3 = 559 = 43(13)$.

   2. Next we suppose that $6^{k+2} + 7^{2k+1} = 43x$ for some integer x.

   3. Then we show that, on the basis of the inductive hypothesis, $6^{k+3} + 7^{2(k+1)+1} = 6^{k+3} + 7^{2k+3}$ is divisible by 43.

   We showed in Section 1.4 that $6^{k+3} + 7^{2k+3} = 6(6^{k+2} + 7^{2k+1}) + 43(7^{2k+1})$. Thus, $6^{k+3} + 7^{2k+3} = 6(43x) + 43(7^{2k+1}) = 43[6x + 7^{2k+1}]$ or $6^{k+3} + 7^{2k+3} = 43(y)$ where y is an integer.

   Hence, $6^{k+3} + 7^{2k+3}$ is divisible by 43, and by the principle of mathematical induction $6^{n+2} + 7^{2n+1}$ is divisible by 43 for each positive integer n.

Example 1.10.4.   For each positive integer n, there are more than n prime integers.

Let $P(n)$ be the proposition: there are more than n prime integers.

1. $P(1)$ is true since 2 and 3 are primes.
2. Assume $P(k)$ is true.
3. Let $a_1, a_2, \ldots, a_k, a_{k+1}$ be $k + 1$ distinct prime integers whose existence is guaranteed since $P(k)$ is true. Form the integer

$$N = a_1 a_2 \ldots a_k \, a_{k+1} + 1 = \prod_{i=1}^{k+1} a_i + 1.$$

Now N is not divisible by any of the primes a_i. But N is either a prime or is divisible by a new prime a_{k+2}. In either case there are more than $k + 1$ primes.

Example 1.10.5.   Suppose the Postal Department prints only 5- and 9-cent stamps. Prove that it is possible to make up any postage of n-cents using only 5- and 9-cent stamps for $n \geq 35$.

1. First, we see that postage of exactly 35 cents can be made up with seven 5-cent stamps.
2. Assume that n-cents postage can be made up with 5- and 9-cent stamps where $n \geq 35$.
3. Now consider postage of $n + 1$ cents. There are two possibilities to consider:
   (a) The n cents postage is made up with only 5-cent stamps, or
   (b) there is at least one 9-cent stamp involved in the makeup of n-cents postage.

In case (a), the number of 5-cent stamps is at least seven since $n \geq 35$. Thus, we can replace those seven 5-cent stamps by four 9-cent stamps and make up $n + 1$ cents postage.

In case (b), the n cents postage includes at least one 9-cent stamp. Therefore, if we replace that one 9-cent stamp by two 5-cent stamps we can make up $n + 1$ cents postage.

Therefore, in either case we have shown how to make up $n + 1$ cents postage in terms of only 5- and 9-cent stamps.

Example 1.10.6. Prove that for all integers $n \geq 4$, $3^n > n^3$.

Let $P(n)$ be the statement: $3^n > n^3$.

1. First, $P(4)$ is true because $3^4 = 81 > 4^3 = 64$.
2. Assume $P(n)$ is true for $n \geq 4$, that is, assume $3^n > n^3$ for $n \geq 4$.
3. We prove $P(n + 1)$ is true on the basis of our assumption. Thus, we must prove $3^{n+1} > (n + 1)^3$. Let us rewrite $(n + 1)^3 = n^3 + 3n^2 + 3n + 1 = n^3(1 + 3/n + 3/n^2 + 1/n^3)$. Since the inductive hypothesis gives us that $3^n > n^3$, we would be done if we could also prove that $3 > 1 + 3/n + 3/n^2 + 1/n^3$ for $n \geq 4$. We now prove this. Observe that the function $f(n) = 1 + 3/n + 3/n^2 + 1/n^3$ decreases as n increases, so then $f(n)$ is largest when n is smallest. In other words $f(4)$ is the largest value of $f(n)$ when n ranges over the integers greater than or equal to 4. Since

$$f(4) = 1 + 3/4 + 3/4^2 + 1/4^3 = 125/64$$

is obviously less than 3, we have for any integer $n \geq 4$, $3 > 1 + 3/n + 3/n^2 + 1/n^3$. Thus, combining the two facts: $3 > 1 + 3/n + 3/n^2 + 1/n^2$ and $3^n > n^3$ for $n \geq 4$, we can multiply and obtain

$$3^{n+1} > 3n^3 > n^3(1 + 3/n + 3/n^2 + 1/n^3)$$
$$= (n + 1)^3, \text{ and the proof is complete.}$$

Recursion

In computer programming the evaluation of a function or the execution of a procedure is usually achieved at machine-language level by the use of a subroutine. The idea of a subroutine which itself calls another subroutine is common; however, it is frequently beneficial to have a subroutine that contains a call to itself. Such a routine is called a *recursive subroutine*. Informally speaking, we give the name *recursion* to the technique of defining a function, a set, or an algorithm *in terms of itself* where it is generally understood that the definition will be in terms of "previous" values. Thus, a recursive subroutine applied to a list of objects would be defined in terms of applying the subroutine to proper sublists.

Moreover, a function f from the set N of nonnegative integers is *defined recursively* if the value of f at 0 is given and for each positive integer n the value of f at n is defined in terms of the values of f at k where $0 \leq k < n$.

Conceivably, the mechanism we are describing may not actually define a function. Thus, if the object defined by the recursive definition is, in fact, a function we say that the function is *well-defined* by the definition. Hence, when a function is defined recursively it is necessary to *prove* that the function is, in fact, well-defined.

The sequence $1, 3, 9, 27, \ldots, 3^n, \ldots$, for example, can be defined explicitly by the formula $T(n) = 3^n$ for all integers $n \geq 0$, but the same function can be defined recursively as follows:

(i) $T(0) = 1$
(ii) $T(n + 1) = 3T(n)$ for all integers $n \geq 0$.

Here part (ii) embodies the salient feature of recursion, namely, the feature of "self-reference".

It is clear that $T(n) = 3^n$ satisfies the conditions (i) and (ii), but it may not be clear that the two conditions alone are enough to define T.

The property of self-reference is the area of concern—we have "defined" $T(n + 1) = 3T(n)$ provided that $T(n)$ is defined, but we normally expect $T(n)$ to be defined only in the case that the function T is itself already defined. This state of affairs makes the recursive definition vulnerable to the charge of circularity.

The recursive definition for T is, in fact, not circular, but we must clarify what we mean when we say that T is well-defined by (i) and (ii). We mean two things: first, that there *exists* a function from the set N of nonnegative integers into the set of integers satisfying (i) and (ii), and second, that there is *only one* such function. Therefore, any proof that a function is well-defined by a recursive definition must involve two proofs: one of *existence* and one of *uniqueness*.

The existence causes us no problem in the present case for we have a candidate, namely, $T(n) = 3^n$. Moreover, the fact that there is only one function satisfying (i) and (ii) can be verified by mathematical induction. (We leave the proof as an exercise.) Thus, the recursive definition for T is, in fact, well-defined.

A thorough discussion of recursive definitions (and all the machinery necessary to prove that certain functions are well-defined by recursive definitions) is beyond our intentions for this book. We shall be content to mention only the following theorem; this theorem can be used to verify that many functions are well-defined by recursive definitions even in cases where no explicit formula is known.

The Recursion Theorem. Let F be a given function from a set S into S. Let s_0 be a fixed element of S, and let N denote the set of nonnegative integers. Then there is a unique function $f: N \rightarrow S$ satisfying

1. $f(0) = s_0$, and
2. $f(n + 1) = F(f(n))$ for all integers $n \in N$.

The interested reader can find a proof of this theorem in an excellent article written by Leon Henkin [16] or on page 74 of [15]. Another interesting discussion of recursive definitions and induction may be found in the article by R. C. Buck [7].

The condition (1) of the Recursion Theorem is called the *initial condition* and condition (2) is called the *recurrence relation* or *generating rule*. Both parts are necessary for the conclusion of the theorem.

Example 1.10.7.   Let us show how to apply the Recursion Theorem to obtain the existence of a function h satisfying

(i) $h(0) = 9$
(ii) $h(n + 1) = 5h(n) + 24$

for all $n \geq 0$.

Let $s_0 = 9$ and let $F(k) = 5k + 24$ for all k.

In many cases it is possible to obtain from a recurrence relation an explicit formula for the general term of a sequence; in fact Chapter 3 is devoted to developing techniques to obtain explicit formulas. But even if one cannot obtain a formula, a recurrence relation provides a powerful computational tool. Indeed, from a strictly computational point of view, a formula may not be as valuable as a recurrence relation.

Let us list several recursively defined functions some of which are quite familiar. For example, the recursively defined function

1. $f(0) = 1$
2. $f(n + 1) = (n + 1)f(n)$ for all $n \geq 0$

is just the factorial function $f(n) = n!$.

Likewise, if a and d are given numbers then the recursively defined function

1. $A(0) = a$
2. $A(n + 1) = A(n) + d$ for all $n \geq 0$

is just the function $A(n) = a + nd$.

Moreover, the sequence of numbers $\{A(n)\}_{n=0}^{\infty}$ is usually called the **arithmetic progression with initial term a** and **common difference d.** On the other hand, if multiplication is used in the above definition instead of addition we get the **geometric progression** $G(n) = ad^n$, and in this case d is called the **common ratio.**

The famous Fibonacci sequence is defined recursively:

1. $F_0 = 1 = F_1$
2. $F_{n+1} = F_n + F_{n-1}$ for all integers $n \geq 1$.

To find a new Fibonacci number simply add the last two:

$$F_2 = F_1 + F_0 = 2$$
$$F_3 = F_2 + F_1 = 3$$
$$F_4 = F_3 + F_2 = 5, \text{ etc.}$$

In this example we have *two* initial conditions which are required because the recurrence relation for F_{n+1} is defined in terms of both n and $n - 1$. A stronger form of the Recursion Theorem is needed to prove the existence and uniqueness of a function satisfying these conditions. We discuss this sequence in greater detail in Chapter 3.

To prove properties of sequences like the Fibonacci sequence we need another form of the principle of mathematical induction. This principle is actually equivalent to the principle of mathematical induction but, nevertheless, we shall call it strong mathematical induction.

Strong Mathematical Induction: Let $P(n)$ be a statement which, for each integer n, may be either true or false. Then $P(n)$ is true for all positive integers if there is an integer $q \geq 1$ such that

1. $P(1), P(2), \ldots, P(q)$ are all true.
2. When $k \geq q$, the assumption that $P(i)$ is true for all integers $1 \leq i \leq k$ implies that $P(k + 1)$ is true.

As in the case of the principle of mathematical induction, this form can be modified to apply to statements in which the starting value is an integer different from 1.

Thus, just as before, there are 3 steps to proofs by strong mathematical induction.

1. **Basis of Induction.** Show $P(1), P(2), \ldots, P(q)$ are all true.
2. **Strong Inductive Hypothesis.** Assume $P(i)$ is true for all integers i such that $1 \leq i \leq k$, where $k \geq q$.
3. **Inductive Step.** Show that $P(k + 1)$ is true on the basis of the strong inductive hypothesis.

In a proof using the principle of mathematical induction we are allowed to assume $P(k)$ in order to establish $P(k + 1)$. But in using strong mathematical induction we assume not only $P(k)$ but also $P(k - 1)$, $P(k - 2),\ldots,P(1)$ as well, to establish $P(k + 1)$.

Strong mathematical induction is a natural choice for proofs in which the properties of elements in the $(n + 1)$th step depend on the properties of elements generated in several previous steps.

To illustrate proofs by strong mathematical induction, let us consider the following examples.

Example 1.10.8. Prove that for each positive integer n, the nth Fibonacci number F_n is less than $(7/4)^n$.

Let $P(n)$ be the sentence: $F_n < (7/4)^n$. Then clearly $P(1)$ and $P(2)$ are true since $F_1 = 1 < 7/4$ and $F_2 = 2 < (7/4)^2$. Assume that $P(i)$ is true for all $1 \le i \le k$, where $k \ge 2$, that is, suppose $F_i < (7/4)^i$ for each $1 \le i \le k$. Then, show $F_{k+1} < (7/4)^{k+1}$ on the basis of the strong inductive hypothesis. Since $k \ge 2$, $k - 1$ is a positive integer and thus $F_k < (7/4)^k$ and $F_{k-1} < (7/4)^{k-1}$. Hence, $F_{k+1} = F_k + F_{k-1} < (7/4)^k + (7/4)^{k-1} = (7/4)^{k-1}(7/4 + 1) = (7/4)^{k-1}(11/4) < (7/4)^{k-1}(7/4)^2 = (7/4)^{k+1}$ since $(11/4) = 44/16 < (7/4)^2$.

Example 1.10.9. Prove that the function $b(n) = 2(3^n) - 5$ is the unique function defined by

(1) $b(0) = -3, b(1) = 1$, and
(2) $b(n) = 4b(n - 1) - 3b(n - 2)$ for $n \ge 2$.

First, it is easy to check that $b(n) = 2(3^n) - 5$ does, in fact, satisfy the relations (1) and (2).

Next, we claim that if $a(n)$ is any other function satisfying relations (1) and (2), then $a(n) = b(n)$ for all n. Let $P(n)$ be the statement: $a(n) = b(n)$. We prove $P(n)$ is a true statement for all nonnegative integers n by strong induction.

1. **Basis of induction.** Show $P(0)$ and $P(1)$ is true. This is immediate since we know $b(0) = -3$ and $b(1) = 1$ and we are assuming $a(0) = -3$ and $a(1) = 1$. Thus, $a(0) = b(0)$ and $a(1) = b(1)$.

2. **Strong inductive hypothesis.** Assume $P(i)$ is true for all $0 \le i \le k$ where $k \ge 1$. In other words, assume $a(i) = b(i)$ for all integers $0 \le i \le k$ where $k \ge 1$.

3. **Induction step.** Show that $P(k + 1)$ is true or that $a(k + 1) = b(k + 1)$. By (2), $a(k + 1) = 4a(k) - 3a(k - 1)$. But by the strong inductive hypothesis, $a(k) = b(k)$ and $a(k - 1) = b(k - 1)$ since $k \ge 1$ and $k - 1 > 0$.

$$\text{Thus, } a(k+1) = 4(2(3^k) - 5) - 3(2(3^{k-1}) - 5)$$
$$= (8)(3^k) - (6)(3^{k-1}) - 5$$
$$= (8)(3^k) - 2(3^k) - 5$$
$$= (6)(3^k) - 5 = 2(3^{k+1}) - 5$$
$$= b(k+1)$$

and the result is proved by induction.

Example 1.10.10. Prove that if F_n is the nth Fibonacci number, then

$$F_n = \frac{1}{\sqrt{5}} \left[\left(\frac{1 + \sqrt{5}}{2} \right)^{n+1} - \left(\frac{1 - \sqrt{5}}{2} \right)^{n+1} \right]$$

for all integers $n \geq 0$.

Proof: **Basis of induction.** When $n = 0$,

$$F_0 = 1 = \frac{1}{\sqrt{5}} \left[\left(\frac{1 + \sqrt{5}}{2} \right) - \left(\frac{1 - \sqrt{5}}{2} \right) \right]$$

Likewise when $n = 1$,

$$F_1 = 1 = \frac{1}{\sqrt{5}} \left[\left(\frac{1 + \sqrt{5}}{2} \right)^2 - \left(\frac{1 - \sqrt{5}}{2} \right)^2 \right]$$

So the formula is valid for $n = 0$ and $n = 1$.

Strong inductive hypothesis. Now for $n \geq 1$, assume

$$F_k = \frac{1}{\sqrt{5}} \left[\left(\frac{1 + \sqrt{5}}{2} \right)^{k+1} - \left(\frac{1 - \sqrt{5}}{2} \right)^{k+1} \right]$$

for each integer k where $0 \leq k \leq n$.

Inductive step. Prove that

$$F_{n+1} = \frac{1}{\sqrt{5}} \left[\left(\frac{1 + \sqrt{5}}{2} \right)^{n+2} - \left(\frac{1 - \sqrt{5}}{2} \right)^{n+2} \right]$$

To expedite the proof, let

$$a = \frac{1 + \sqrt{5}}{2} \quad \text{and} \quad b = \frac{1 - \sqrt{5}}{2}.$$

Then since we know that $F_{n+1} = F_n + F_{n-1}$ and

$$F_n = \frac{1}{\sqrt{5}} [a^{n+1} - b^{n+1}] \quad \text{and}$$

$$F_{n-1} = \frac{1}{\sqrt{5}} [a^n - b^n]$$

by the inductive hypothesis, it follows that

$$F_{n+1} = \frac{1}{\sqrt{5}} [a^{n+1} - b^{n+1} + a^n - b^n]$$

$$= \frac{1}{\sqrt{5}} [a^n(a + 1) - b^n(b + 1)].$$

Now

$$a + 1 = \frac{1 + \sqrt{5}}{2} + 1 = \frac{3 + \sqrt{5}}{2} \quad \text{and} \quad a^2 = \frac{3 + \sqrt{5}}{2}$$

so that $a + 1 = a^2$. Likewise, $b + 1 = b^2$. Therefore,

$$F_{n+1} = \frac{1}{\sqrt{5}} [a^n(a + 1) - b^n(b + 1)]$$

$$= \frac{1}{\sqrt{5}} [a^{n+2} - b^{n+2}] \text{ and the proof is complete.}$$

Example 1.10.11. In section 5.6 we shall have occasion to consider the function $V(n)$ where $V(0) = 1$ and $V(1) = 2$ and $V(n + 1) = V(n) + V(n - 1) + 1$ for all $n \geq 1$. We wish to show that $V(n) = F_{n+2} - 1$ for all integers $n \geq 0$.

Basis of induction. When $n = 0$, $V(0) = 1$ and $F_2 - 1 = 2 - 1 = 1$. Likewise, $V(1) = 2$ and $F_3 = 3 - 1 = 2$.

Strong inductive hypothesis. For $n \geq 1$, assume $V(k) = F_{k+2} - 1$ for all $0 \leq k \leq n$.

Inductive step. Prove $V(n + 1) = F_{n+3} - 1$. Now $V(n + 1) = V(n) + V(n - 1) + 1 = (F_{n+2} - 1) + (F_{n+1} - 1) + 1 = F_{n+2} + F_{n+1} - 1 = F_{n+3} - 1$ by definition of F_{n+3}.

Now since we have shown $V(n) = F_{n+2} - 1$ and we know a closed expression for F_{n+2} we can get one for $V(n)$, namely,

$$V(n) = F_{n+2} - 1 = \frac{1}{\sqrt{5}}\left[\left(\frac{1 + \sqrt{5}}{2}\right)^{n+3} - \left(\frac{1 - \sqrt{5}}{2}\right)^{n+3}\right] - 1.$$

Exercises for Section 1.10

I. Use mathematical induction to prove that each of the following statements is true for all positive integers n.

1. $11^{n+2} + 12^{2n+1}$ is divisible by 133.

2. If $a_n = 5a_{n-1} - 6a_{n-2}$ for $n \geq 2$ and $a_0 = 12$ and $a_1 = 29$, then $a_n = 5(3^n) + 7(2^n)$.

3. $1^2 + 2^2 + 3^2 + \cdots + n^2 = n(n + 1)(2n + 1)/6$ for $n \geq 1$.

4. $1^2 + 3^2 + 5^2 + \cdots + (2n - 1)^2 = n(2n - 1)(2n + 1)/3$.

5. $1/(1)(2) + 1/(2)(3) + \cdots + 1/n(n + 1) = n/(n + 1)$.

6. $1 \cdot 3 + 3 \cdot 5 + 5 \cdot 7 + \cdots + (2n - 1)(2n + 1) = n(4n^2 + 6n - 1)/3$.

7. $1^3 + 3^3 + 5^3 + \cdots + (2n - 1)^3 = n^2(2n^2 - 1)$.

8. $1 \cdot 2 + 2 \cdot 3 + 3 \cdot 4 + \cdots + n(n + 1) = n(n + 1)(n + 2)/3$.

9. $1 \cdot 2 \cdot 3 + 2 \cdot 3 \cdot 4 + \cdots + n(n + 1)(n + 2) = n(n + 1)(n + 2)(n + 3)/4$.

10. $1^2 - 2^2 + 3^2 - 4^2 \cdots (-1)^{n-1}n^2 = (-1)^{n-1}n(n + 1)/2$.

11. $x - y$ is a factor of the polynomial $x^n - y^n$.

12. $x + y$ is a factor of the polynomial $x^{2n+1} + y^{2n+1}$.

13. $n(n^2 + 5)$ is an integer multiple of 6.

14. $n(n^2 - 1)(3n + 2)$ is an integer multiple of 24.

15. $3n^5 + 5n^3 + 7n$ is divisible by 15 for each positive integer n.

16. $a_n = 5(2^n) + 1$ is the unique function defined by
    (1) $a_0 = 6$,
        $a_1 = 11$,
    (2) $a_n = 3a_{n-1} - 2a_{n-2}$ for $n \geq 2$.

17. For any real number $x > -1$, $(1 + x)^n \geq 1 + nx$.

18. Show that the sum of the first n terms of an arithmetic progression with initial term a and common difference d is $n/2[2a + (n - 1)d]$.

19. Show that the sum of the first n terms of a geometric progression with initial term a and common ratio $r \neq 1$ is a $[(r^n - 1)/(r - 1)] = a[(1 - r^n)/(1 - r)]$.

20. Let D_n be the number of diagonals of an n-sided convex polygon. Make a table of values of D_n for $n \geq 3$, and then conjecture a formula for D_n in terms of n. Prove that this formula is valid by mathematical induction. (Hint: Search for a pattern describing $2 D_n$.)

21. For each integer $n \geq 4$, $n! > 2^n$.

22. For each integer $n \geq 5$, $2^n > n^2$.

23. Suppose that we have a system of currency that has \$3 and \$5 bills. Show that any debt of \$$n$ can be paid with only \$3 and \$5 bills for each integer $n \geq 8$. Do the same problem for \$2 and \$7 bills and $n \geq 9$.

24. Show that any integer composed of 3^n identical digits is divisible by 3^n. (For example, 222 and 555 are divisible by 3, while 222, 222, 222 and 555, 555, 555 are divisible by 9.)

25. For every positive integer $n \geq 2$, the number of lines obtained by joining n distinct points in the plane, no three of which are collinear, is $n(n - 1)/2$.

26. $g(n) = (3)5^n + (7)2^n$ is the unique function defined by
    1. $g(0) = 10$
       $g(1) = 29$
    2. $g(n + 1) = 7g(n) - 10g(n - 1)$ for $n \geq 1$.

27. For each integer $n \geq 2$,
$$\frac{1}{\sqrt{1}} + \frac{1}{\sqrt{2}} + \cdots + \frac{1}{\sqrt{n}} > \sqrt{n}$$

28. For each integer $n \geq 10$, $2^n > n^3$.

29. For each integer $n \geq 4$, $3^n > 2^n + 64$.

30. For each integer $n \geq 2$, $n^3 + 1 > n^2 + n$.

31. For each integer $n \geq 5$, $4^n > n^4$.

32. For each integer $n \geq 17$, $2^n > n^4$.

33. For each integer $n \geq 9$, $n! > 4^n$.

34. For each integer $n \geq 1$, the nth Fibonacci number F_n is less than

$$\left(\frac{13}{8}\right)^n.$$

35. $a_n = 2^n - 1$ is the unique function defined by
    (1) $a_0 = 0$
        $a_1 = 1$
    (2) $a_n = 3a_{n-1} - 2a_{n-2}$ for $n \geq 2$

36. Suppose that b_n is the function defined by $b_1 = 1$, $b_2 = 2$, $b_3 = 3$, and $b_n = b_{n-1} + b_{n-2} + b_{n-3}$ for all integers $n \geq 4$. Prove that $b_n < 2^n$ for all positive integers n.

37. Let $a_1 = 1$ and $a_n = \sqrt{3a_{n-1} + 1}$ for $n > 1$. Prove that $a_n < 7/2$ for all positive integers n.

38. Let $a_0 = a_1 = a_2 = 1$ and $a_n = a_{n-2} + a_{n-3}$ for $n \geq 3$. Prove that $a_n \leq (4/3)^n$ for each integer $n \geq 0$.

II. Apply the Recursion Theorem to verify that the following recursive definitions do in fact define functions.

1. $g(0) = 1$
   $g(n + 1) = 3g(n)^2 + 7$ for $n \geq 0$.

2. $h(0) = 3$
   $h(n + 1) = 7\,h(n)^3 - 3$ for $n \geq 0$.

3. $k(0) = 1$
   $k(n + 1) = \sqrt{3k(n)^2 + 7\,k(n) - 3}$ for $n \geq 0$.

III. Perhaps the oldest recorded nontrivial algorithm is known as the *Euclidean Algorithm*. This algorithm computes the greatest common divisor of 2 nonnegative integers. If a and b are nonnegative integers, then $gcd(a,b)$ is defined as the largest positive integer d such that d divides both a and b. If $a > b \geq 0$, then the Euclidean algorithm is based upon the following facts:

(a) $gcd(a,b) = a$ if $b = 0$
(b) $gcd(a,b) = gcd(b,r)$ if $b \neq 0$

and $a = b\,q + r$ where $0 \leq r < b$.

Thus the greatest common division of 22 and 8 can be found by applying the above facts recursively as follows:

$$gcd(22,8) = gcd(8,6) = gcd(6,2) =$$
$$gcd(2,0) = 2.$$

Find the greatest common divisors of the following pairs of integers.
   (a) 81 and 36
   (b) 144 and 118
   (c) 1317 and 56
   (d) 10,815 and 6489
   (e) 510 and 374.

IV. Number Theory

1. (a) In example 1.10.4 we used the fact that every integer $a > 1$ is either a prime or is divisible by a prime. Give a contradiction proof of this fact by using the well-ordered property of the positive integers.

   (b) Extend the proof in (a) to conclude that every integer $a > 1$ is a product of prime integers.

   (c) Write 1235 and 5124 as products of primes.

   (d) (*The Fundamental Theorem of Arithmetic.*) Prove that the factorization of an integer $a > 1$ into a product of primes is unique except for rearrangement of factors. Hint: use contradiction and the well-ordered property.

   (e) Show that if $p_1, p_2, \ldots, p_k$ are distinct primes and $e_1, e_2, \ldots, e_k$ are positive integers, then the number of positive divisors of $n = p_1^{e_1} p_2^{e_2} \ldots p_k^{e_k}$ is $(e_1 + 1)(e_2 + 1) \ldots (e_k + 1)$. Hint: Note that each prime p_i will appear 0 or 1 or 2 or ... or e_i times as a factor of any positive division of n.

2. If p is a prime, let $E(p)$ be 1 added to the product of all primes up to and including p, that is, $E(p) = (2\,3\,5\,7\ldots p) + 1$. For lack of a better word, call p a *Euclidean prime* if $E(p)$ is a prime. (The proof of Example 1.10.4 was first published by Euclid.)

   (a) Show that 13 is not a Euclidian prime.

   (b) Find all Euclidian primes less than or equal to 21.

3. For each integer $n > 2$, prove that there is a prime p such that $n < p < n!$. Hint: consider $q = n! - 1$, and apply exercise 1(a)).

4. (a) Show that each prime is of the form $4n + 1$ or $4n + 3$ for some integer n.

   (b) Show that the product of 2 integers of the form $4n + 1$, where n is an integer, is again of that form.

   (c) Modify the proof of Example 1.10.4 to show that there are infinitely many primes of the form $4n + 3$, that is, show that there are more than k primes of the form $4n + 3$ for any positive integer k. Hint: Suppose $p_1, p_2, \ldots, p_k$ are k primes of the form $4n + 3$. Consider $4p_1 p_2 \ldots p_k - 1$ and use exercises 1(a), 4(a), and 4(b).

   (d) Show that each prime is of the form $6n + 1$ or $6n + 5$.

   (e) Show that the product of two integers of the form $6n + 1$, where n is an integer, is again of that form.

   (f) Modify the proof of Example 1.10.4 to show that there are infinitely many primes of the form $6n + 5$.

   (g) Show that there are infinitely many primes p such that $p - 2$ is not a prime. Hint: use exercise 4(f).

5. The *Goldbach conjecture* is a famous unsolved problem which conjectures that every even integer larger than 4 is the sum of two odd primes. For instance, $6 = 3 + 3$, $8 = 5 + 3$, $10 = 5 + 5$, and so on. Verify the Goldbach conjecture for all even integers between 12 and 50.

6. Let $m > 1$ be an integer; prove that exactly one integer in any sequence of m consecutive integers is divisible by m.

7. (a) Observe that the polynomial function $f_2(n) = n^2 - n + 2$ takes on only prime values for $n = 0,1$.
   (b) Observe that $f_3(n) = n^2 - n + 3$ takes on only prime values for $n = 0,1,2$.
   (c) Define an integer m to be "lucky" if $f_m(n) = n^2 - n + m$ takes on only prime values for $n = 0,1,2,\ldots,m - 1$. We have observed in (a) and (b) that 2 and 3 are lucky numbers. Show that 5, 11, 17, and 41 are also lucky numbers.
   (d) Prove that a lucky integer m must also be prime.
   (e) Show that 13 is not a lucky integer. In fact, show that if p is a lucky integer > 10, then the units digit of p must be 1 or 7. Hint: consider the units digit of several values of $n^2 - n$.
   (f) If m is a lucky number, prove that $g(n) = n^2 + n + m$ takes on only prime values for $n = 0,1,2,\ldots,m - 2$.
   (g) If m is a lucky number, show that $f_m(m) = m^2 - m + m$ is not a prime. Likewise, show that $f_m(m + 1)$ is not prime.
   (h) Call a lucky prime p super-lucky if $n^2 - n + p$ is a prime for $n = p + 2$. Check that of the six lucky primes 2, 3, 5, 11, 17, 41 all are super-lucky except 2.
       Make a similar investigation for $n = p + 3$.
       Show that $n^2 - n + p$ is always composite $n = p + 4$.
   (i) One might be tempted to search for general formulas that generate only prime integers. First, we observe that the polynomial $h(n) = 2n + 1$ takes on prime values for infinitely many values of n. But, of course, $h(n)$ is not prime for infinitely many values of n also.
       Prove that the quadratic polynomial $g(n) = n^2 + an + b$, where a and b are integers, cannot take on prime values for every value of n. Hint: observe that there is an integer n_0 such that $m = g(n_0) > 1$ and $g(n) > m$ for all values of $n > n_0$. Then observe that $g(n_0 + m) - g(n_0)$ is divisible by m.
       The same ideas of proof prove that for any nonconstant polynomial $g(x)$ with integer coefficients there is a value k such that $g(k)$ is not prime. The only other technical fact required to complete the proof of this fact is the Binomial Theorem of Section 2.7. For instance, use the Binomial Theo-

rem to observe that for any integers m and n_0, $(n_0 + m)^i - n_0^i$ is always divisible by m. Then conclude that $g(n_0 + m) - g(n_0)$ is divisible by m. Choose n_0 according to the hint given for quadratic polynomials.

8. A positive integer that is not prime is said to be **composite.** There can be arbitrarily long sequences of consecutive composite integers. For instance, 24, 25, 26, 27, 28 is a sequence of 5 consecutive composite integers.

   (a) Find a sequence of 7 consecutive composite integers.

   (b) Given a positive integer n, find an integer a (depending on n) such that none of a, $a + 1$, $a + 2$,...,$a + n - 1$ is a prime integer. In other words, find a sequence of n consecutive composite integers. Hint: consider $a = (n + 1)! + 2$, where $(n + 1)!$ means the product $(n + 1)(n)(n - 1) \ldots (2)(1)$.

9. (a) Find four primes of the form $2^n - 1$ for some integer n.

   (b) Prove that if $2^n - 1$ is prime then n is prime. Hint: use contrapositive and exercise 11 of part I. The primes of the form $2^n - 1$, for n an integer, are called *Mersenne primes.*

   (c) Prove that the converse of part (b) is false by showing that 23 divides $2^{11} - 1$.

   (d) Prove that if a is a positive integer such that $a^n - 1$ is prime, then $a = 2$.

   (e) Prove that if $2^n - 1$ is prime, then $2^{n-1}(2^n - 1) = N$ is perfect. (Thus, N is the sum of all positive divisions of N different from N.)

10. (a) Find four primes of the form $2^n + 1$ for an integer n.

    (b) Prove that if $2^k + 1$ is prime, then $n = 0$ or $n = 2^k$ for some integer k. Hint: use exercise 12 of part I.

    (c) The primes of the form $F_k = 2^{2^k} + 1$ are called *Fermat primes.* Show that F_5 is not prime. Hint: 641 is a divisor of F_5. Show that F_1, F_2, F_3, and F_4 are primes.

11. (a) Find a prime of the form $n^2 + 1$ for some integer n.

    (b) Find a prime of the form $n^2 - 1$ for some integer n.

    (c) Prove that the prime of (b) is unique.

    (d) Prove that the prime you found in (a) is not unique.

Selected Answers for Section 1.10

3. Inductive step:

$$1^2 + 2^2 + \cdots + n^2 + (n + 1)^2 = \frac{n(n + 1)(2n + 1)}{6} + (n + 1)^2$$

$$= \frac{n(n + 1)(2n + 1) + 6(n + 1)^2}{6}$$

$$= \frac{(n + 1)[(n)(2n + 1) + 6(n + 1)]}{6}$$

$$= \frac{(n + 1)[2n^2 + 7n + 6]}{6}$$

$$= \frac{(n + 1)[(n + 2)(2n + 3)]}{6}$$

5. Inductive step:

$$\frac{1}{1 \cdot 2} + \frac{1}{2 \cdot 3} + \cdots + \frac{1}{n(n + 1)} + \frac{1}{(n + 1)(n + 2)}$$

$$= \frac{n}{n + 1} + \frac{1}{(n + 1)(n + 2)}$$

$$= \frac{n(n + 2) + 1}{(n + 1)(n + 2)} = \frac{n^2 + 2n + 1}{(n + 1)(n + 2)} = \frac{(n + 1)(n + 1)}{(n + 1)(n + 2)}$$

$$= \frac{n + 1}{n + 2}$$

13. True for $n = 1$

Inductive hypothesis: Suppose $n(n^2 + 5) = 6t$, where t is some integer.

Inductive step:

$$(n + 1)((n + 1)^2 + 5) = (n + 1)(n^2 + 2n + 6)$$

$$= (n + 1)(n^2 + 5 + 2n + 1)$$

$$= n(n^2 + 5) + 3n^2 + 3(n + 2)$$

$$= 6t + 3[(n)(n + 1) + 2] = 6(t + s),$$

where $3n^2 + 3(n + 2) = 3[(n)(n + 1) + 2] = 6s$. Note that for any integer n, $n(n + 1) + 2$ is even so that $3[(n)(n + 1) + 2]$ is a multiple of 6.

26. Suppose that $f(n)$ is any sequence satisfying (1) and (2). We show that $f(n) = g(n)$ for all integers $n \geq 0$ by strong mathematical induction.

   (a) *Basis of induction.* Clearly $g(0) = 10 = f(0)$ and $g(1) = 29 = f(1)$.

   (b) *Inductive hypothesis.* Suppose that $g(i) = f(i)$ for all integers $0 \leq i \leq k$ where $k \geq 1$.

   (c) *Induction Step.* Show that $g(k + 1) = f(k + 1)$.

   Since $k \geq 1$, $k + 1+ \geq 2$ so that $f(k + 1) = 7f(k) - 10f(k - 1)$. But by the inductive hypothesis

$$f(k) = g(k) = (3)5^k + (7)2^k$$
$$\text{and} \quad f(k - 1) = g(k - 1)$$
$$= (3)5^{k-1} + (7)2^{k-1}.$$

   But then
$$f(k + 1) = 7(3 \cdot 5^k + 7 \cdot 2^k) - 10(3 \cdot 5^{k-1} + 7 \cdot 2^{k-1})$$
$$= 5^{k-1}(7 \cdot 3 \cdot 5 - 10 \cdot 3) + 2^{k-1}(7^2 \cdot 2 - 10 \cdot 7)$$
$$= 5^{k-1}(3 \cdot 5^2) + 2^{k-1}(2^2 \cdot 7)$$
$$= 3 \cdot 5^{k+1} + 7 \cdot 2^{k+1} = g(k + 1),$$

   and the proof is complete.

III. (c) Since

$$1317 = (23)(56) + 29$$
$$56 = (1)(29) + 27$$
$$29 = (1)(27) + 2$$
$$27 = (13)(2) + 1$$
$$2 = (2)(1) + 0$$

$$gcd(1317,56) = gcd(56,29) = gcd(29,27)$$
$$= gcd(27,2) = gcd(2,1) = gcd(1,0)$$
$$= 1.$$

REVIEW FOR CHAPTER ONE

1. Write the converse and the contrapositive of the statements.
   (a) If $x \leq y$ and $y \leq x$, then $x = y$.
   (b) If p is an odd prime, then p is congruent to 1 modulo 6 or congruent to 5 modulo 6.

2. For each statement, write an equivalent statement in "If...
   then..." form.
   (a) x is an integer or y is an integer.
   (b) In order for a relation to be asymmetric, it is necessary that it
       be antisymmetric.
   (c) In order for an integer n to be divisible by 6, it is sufficient that
       it be divisible by 2 and 3.

3. Write a nontrivial negation of each of the following statements:
   (a) Some relations are not transitive.
   (b) No prime numbers are even.
   (c) $x > 2$ or y is negative.
   (d) If $\sqrt{2} = m/n$, then m and n are both even integers.

4. Disprove the following conjecture by finding a counterexample:
   (a) $3^n > n!$ for all positive integers n.
   (b) $f(n) = n^2 - n + 17$ is prime for all positive integers n.

5. Prove by mathematical induction that
   (a) $4^n > 3^n$ for all integers $n \geq 1$.
   (b) $2^{n-1}(3^n + 4^n) > 7^n$ for all integers $n \geq 2$.
   (c) $5^n > n^5$ for integers $n \geq 6$.

6. Suppose a function f is defined recursively as follows: $f(0) = 7$,
   $f(1) = 26$, and $f(n) = 7f(n-1) - 12f(n-2)$ for all integers
   $n \geq 2$. Prove by mathematical induction that $f(n) = 2(3^n) + 5(4^n)$
   for all integers $n \geq 0$.

7. Use the Euclidian algorithm to find the GCD (greatest common
   divisor) of 924 and 4410.

8. Suppose there are 59 different pairs of people who know each other
   at a party of 15 people. Prove that at least one person at the party
   knows 7 or fewer other persons at the party.

9. A bag of M&M's contains brown, tan, yellow, green, and orange
   M&M's. How many M&M's must you grab from the bag to
   guarantee that you will have 8 brown or 5 tan or 6 yellow or 10
   green or 12 orange M&M's?

10. Tell whether each argument below is valid or invalid.
    (a) If there is a legal holiday in March, then the banks will close in
        March. There is no legal holiday in March. Therefore, the
        banks will not close in March.
    (b) If R is an equivalence relation, then R is transitive and
        reflexive. R is reflexive but not transitive. Therefore, R is not
        an equivalence relation.
    (c) If $xy = 0$, then $x = 0$ or $y = 0$. $x = 0$. Therefore, $y = 0$.

12. Supply a valid conclusion for the following argument:

    If the positive integer n is even and perfect, then the units digit of n is 6 or 8.

    If the units digit of the positive integer n is 6 or 8, then the units digit of n^2 is 4 or 6.

    The units digit of n^2 is not 4 and not 6. Therefore, . . .

13. Suppose that the circumference of a circular wheel is divided into 40 sectors and that the integers 1 through 40 are randomly assigned to these sectors. Prove that there is at least one group of 4 consecutive sectors whose sum of assigned numbers is 82 or more.

14. Suppose that no pine tree has more than 9000 pine cones and that there are 6,381,400 pine trees in Leon County. What is the largest integer that can be used for n in the following assertion? There are at least n pine trees in Leon County with the same number of pine cones.

15. Construct a truth table for the propositional function
    $(p \leftrightarrow \sim r) \wedge [(r \to q) \vee \sim q]$

16. Prove that $\{[(p \vee q) \to r] \wedge (\sim p)\} \to (q \to r)$ is a tautology by appeal to an abbreviated truth table.

17. Find the error in the following "proof" by induction that $1 + 2 + \cdots + n = (1/8)(2n + 1)^2$ for all positive integers n. "If $1 + 2 + \cdots + n = (1/8)(2n + 1)^2$, then $1 + 2 + \cdots + n + n + 1 = (1/8)(2n + 1)^2 + (n + 1) = (1/8)(4n^2 + 4n + 1 + 8n + 8) = (1/8)(4n^2 + 12n + 9) = (1/8)(2n + 3)^2 = (1/8)(2(n + 1) + 1)^2$. Thus, the truth of the statement for n implies its truth for $n + 1$."

2

Elementary Combinatorics

INTRODUCTION

For most applications of computers to problems, one normally needs to know, at least approximately, how much storage will be required and about how many operations are necessary. A major component of estimating the storage needed may be determining the number of items of a particular type that have to be stored. Similarly, a knowledge of how many operations the computation involves will help in assessing the length of program execution time, and thereby aid in determining the potential cost of the computation. Being able to answer such questions of the form "How many?", is important if one attempts to compare different methods of computation or even to decide whether or not a given computation is feasible.

For example, we will be able to determine by the methods and concepts in this chapter that there are $(n - 1)! = (n - 1)(n - 2) \cdots (3)(2)(1)$ different ways of visiting each of n cities exactly once by starting and finishing each trip at a given city. Furthermore, the most straightforward way of finding the shortest round trip would be to list all $(n - 1)!$ routes and calculate the total distance associated with each route. Such a process of "complete enumeration" or "exhaustive searching" has the virtue of being easily programmed, but the problems of using such an algorithm become apparent if the number of cities is not small. For instance, finding the total distance for a single route requires n additions and since there are $(n - 1)!$ different routes, the total number of additions is $n!$. Thus, if there are 50 cities, 50! is approximately equal to 3×10^{64}, and even if the computer performs 10^9 additions per second, it will take more than 10^{47} *years* just to perform the additions required by this algorithm.

Therefore, we need to find algorithms better than mere exhaustive searching. If on the other hand, another algorithm could be found that required only n^2 operations, then for $n = 50$ cities, a shortest route could be found after only 2500 operations and these could be performed in less than 1 second by the same computer that required 10^{47} years to perform 50! operations.

The basic ideas, techniques, and concepts necessary for one to make an assessment of the amount of storage and work that algorithms entail is the topic of this chapter. We will use very elementary settings such as counting the number of license plates of a certain form, the number of n-digit numbers of a certain type, the number of words of prescribed form, etc., but the reader should keep in mind that a list of such license plates, numbers, or words would require that amount of storage in a computer.

2.1   BASICS OF COUNTING

If X is a set, let us use $|X|$ to denote the number of elements in X.

Two Basic Counting Principles

Two elementary principles act as "building blocks" for all counting problems. The first principle essentially says that the whole is the sum of its parts; it is at once immediate and elementary, we need only be clear on the details.

Sum Rule: The principle of disjunctive counting. If a set X is the union of disjoint nonempty subsets $S_1, \ldots, S_n$, then $|X| = |S_1| + |S_2| + \cdots + |S_n|$.

We emphasize that the subsets $S_1, S_2, \ldots, S_n$ must have no elements in common. Moreover, since $X = S_1 \cup S_2 \cup \cdots \cup S_n$, each element of X is in *exactly* one of the subsets S_i. In other words, $S_1, S_2, \ldots, S_n$ is a *partition* of X.

If the subsets $S_1, S_2, \ldots, S_n$ were allowed to overlap, then a more profound principle will be needed—the principle of inclusion and exclusion. We will discuss this principle later in Section 2.8.

Frequently, instead of asking for the number of elements in a set *per se*, some problems ask for how many ways a certain event can happen.

The difference is largely in semantics, for if A is an event, we can let X be the set of ways that A can happen and count the number of elements in X. Nevertheless, let us state the sum rule for counting events.

If $E_1, \ldots, E_n$ are mutually exclusive events, and E_1 can happen e_1 ways, E_2 can happen e_2 ways, $\ldots, E_n$ can happen e_n ways, then E_1 or E_2 or $\ldots$ or E_n can happen $e_1 + e_2 + \cdots + e_n$ ways.

Again we emphasize that mutually exclusive events E_1 and E_2 mean that E_1 or E_2 can happen but both cannot happen simultaneously.

The sum rule can also be formulated in terms of choices: If an object can be selected from a reservoir in e_1 ways and an object can be selected from a separate reservoir in e_2 ways, then the selection of one object from *either* one reservoir *or* the other can be made in $e_1 + e_2$ ways.

Example 2.1.1. In how many ways can we draw a heart or a spade from an ordinary deck of playing cards? A heart or an ace? An ace or a king? A card numbered 2 through 10? A numbered card or a king?

Since there are 13 hearts and 13 spades we may draw a heart or a spade in $13 + 13 = 26$ ways; we may draw a heart or an ace in $13 + 3 = 16$ ways since there are only 3 aces that are not hearts. We may draw an ace or a king in $4 + 4 = 8$ ways. There are 9 cards numbered 2 through 10 in each of 4 suits, clubs, diamonds, hearts, or spades, so we may choose a numbered card in 36 ways. (Note: we are counting aces as distinct from numbered cards.) Thus, we may choose a numbered card or a king in $36 + 4 = 40$ ways.

Example 2.1.2. How many ways can we get a sum of 4 or of 8 when two distinguishable dice (say one die is red and the other is white) are rolled? How many ways can we get an even sum?

Let us label the outcome of a 1 on the red die and a 3 on the white die as the ordered pair $(1,3)$. Then we see that the outcomes $(1,3)$, $(2,2)$, and $(3,1)$ are the only ones whose sum is 4. Thus, there are 3 ways to obtain the sum 4. Likewise, we obtain the sum 8 from the outcomes $(2,6)$, $(3,5)$, $(4,4)$, $(5,3)$, and $(6,2)$. Thus, there are $3 + 5 = 8$ outcomes whose sum is 4 or 8. The number of ways to obtain an even sum is the same as the number of ways to obtain either the sum 2, 4, 6, 8, 10, or 12. There is 1 way to obtain the sum 2, 3 ways to obtain the sum 4, 5 ways to obtain 6, 5 ways to obtain an 8, 3 ways to obtain a 10, and 1 way to obtain a 12. Therefore, there are $1 + 3 + 5 + 5 + 3 + 1 = 18$ ways to obtain an even sum.

Perhaps at this stage it is worthwhile to discuss the semantical differences between the words *distinct* and *distinguishable*. In the above example, the two dice were distinct because we had two dice and not just one die; moreover, the two dice were distinguishable by color so that their outcome $(1,5)$ could be differentiated from the outcome $(5,1)$. Yet if the two distinct dice had no distinguishing characteristics (such as size, color, weight, feel, or smell) then we could not make such a differentiation.

Example 2.1.3. How many ways can we get a sum of 8 when two *indistinguishable* dice are rolled? An even sum?

Had the dice been distinguishable, we would obtain a sum of 8 by the outcomes (2,6), (3,5), (4,4), (5,3), and (6,2), but since the dice are similar, the outcomes (2,6) and (6,2) and, as well, (3,5) and (5,3) cannot be differentiated and thus we obtain the sum of 8 with the roll of two similar dice in only 3 ways. Likewise, we can get an even sum in $1 + 2 + 3 + 3 + 2 + 1 = 12$ ways. (Recall from Example 2.1.2 that we could get an even sum from 2 distinguishable dice in 18 ways.)

Now let us state the other basic counting rule.

Product Rule: the principle of sequential counting. If $S_1, \ldots, S_n$ are nonempty sets, then the number of elements in the Cartesian product $S_1 \times S_2 \times \cdots \times S_n$ is the product $\Pi_{i=1}^n |S_i|$. That is,

$$|S_1 \times S_2 \times \cdots \times S_n| = \Pi_{i=1}^n |S_i|.$$

Let us illustrate $S_1 \times S_2$ by a tree diagram (see Figure 2-1) where

$$S_1 = \{a_1, a_2, a_3, a_4, a_5\} \qquad \text{and} \qquad S_2 = \{b_1, b_2, b_3\}.$$

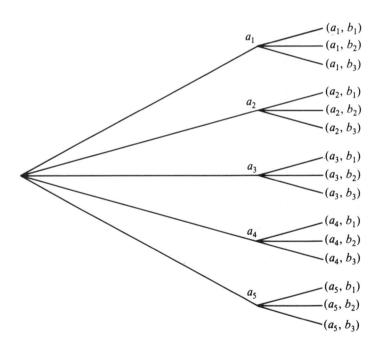

Figure 2-1

Observe that there are 5 branches in the first stage corresponding to the 5 elements of S_1 and to each of these branches there are 3 branches in the second stage corresponding to the 3 elements of S_2 giving a total of 15 branches altogether. Moreover, the Cartesian product $S_1 \times S_2$ can be partitioned as $(a_1 \times S_2) \cup (a_2 \times S_2) \cup (a_3 \times S_2) \cup (a_4 \times S_2) \cup (a_5 \times S_2)$, where $(a_i \times S_2) = \{(a_i,b_1),(a_i,b_2),(a_i,b_3)\}$. Thus, for example, $(a_3 \times S_2)$ corresponds to the third branch in the first stage followed by each of the 3 branches in the second stage.

More generally, if $a_1,\ldots,a_n$ are the n distinct elements of S_1 and $b_1,\ldots,b_m$ are the m distinct elements of S_2, then $S_1 \times S_2 = \cup_{i=1}^{n} (a_i \times S_2)$. For if x is an arbitrary element of $S_1 \times S_2$, then $x = (a,b)$ where $a \in S_1$ and $b \in S_2$. Thus, $a = a_i$ for some i and $b = b_j$ for some j. Thus, $x = (a_i,b_j) \in (a_i \times S_2)$ and therefore $x \in \cup_{i=1}^{n} (a_i \times S_2)$. Conversely, if $x \in \cup_{i=1}^{n} (a_i \times S_2)$, then $x \in (a_i \times S_2)$ for some i, and thus $x = (a_i,b_j)$ where b_j is some element of S_2. Therefore, $x \in S_1 \times S_2$.

Next observe that $(a_i \times S_2)$ and $(a_j \times S_2)$ are disjoint if $i \neq j$ since if $x \in (a_i \times S_2) \cap (a_j \times S_2)$ then $x = (a_i,b_k)$ for some k and $x = (a_j,b_l)$ for some l. But then $(a_i,b_k) = (a_j,b_l)$ implies that $a_i = a_j$ and $b_k = b_l$. But since $i \neq j$, $a_i \neq a_j$.

Thus, we conclude that $S_1 \times S_2$ is the disjoint union of the sets $(a_i \times S_2)$. Furthermore $|a_i \times S_2| = |S_2|$ since there is obviously a one-to-one correspondence between the sets $a_i \times S_2$ and S_2, namely, $(a_i,b_j) \rightarrow b_j$. Then by the sum rule $|S_1 \times S_2| = \Sigma_{i=1}^{n} |a_i \times S_2| = (n$ summands$)$ $|S_2| + |S_2| + \cdots + |S_2| = n|S_2| = nm$.

Therefore, we have proven the product rule for two sets. The general rule follows by mathematical induction.   □

We can reformulate the product rule in terms of events. If events $E_1,E_2,\ldots,E_n$ can happen $e_1,e_2,\ldots,$ and e_n ways, respectively, then the sequence of events E_1 first, followed by $E_2,\ldots,$ followed by E_n can happen $e_1 \cdot e_2 \cdot \ldots \cdot e_n$ ways.

In terms of choices, the product rule is stated thus: If a first object can be chosen e_1 ways, a second e_2 ways, $\ldots,$ and an nth object can be chosen e_n ways, then a choice of a first, second, $\ldots,$ and an nth object can be made in $e_1 e_2 \cdots e_n$ ways.

Example 2.1.4. If 2 distinguishable dice are rolled, in how many ways can they fall? If 5 distinguishable dice are rolled, how many possible outcomes are there? How many if 100 distinguishable dice are tossed?

The first die can fall (event E_1) in 6 ways and the second can fall (event E_2) in 6 ways. Thus, there are $6 \cdot 6 = 6^2 = 36$ outcomes when 2 dice are rolled. Also the third, fourth, and fifth die each have 6 possible outcomes so there are $6 \cdot 6 \cdot 6 \cdot 6 \cdot 6 = 6^5$ possible outcomes when all 5 dice are tossed. Likewise there are 6^{100} possible outcomes when 100 dice are tossed.

Example 2.1.5. Suppose that the license plates of a certain state require 3 English letters followed by 4 digits. (a) How many different plates can be manufactured if repetition of letters and digits are allowed? (b) How many plates are possible if only the letters can be repeated? (c) How many are possible if only the digits can be repeated? (d) How many are possible if no repetitions are allowed at all?

Answers. (a) $26^3 \cdot 10^4$ since there are 26 possibilities for each of the 3 letters and 10 possibilities for each of 4 digits. (b) $26^3 \cdot 10 \cdot 9 \cdot 8 \cdot 7$. (c) $26 \cdot 25 \cdot 24 \cdot 10^4$. (d) $26 \cdot 25 \cdot 24 \cdot 10 \cdot 9 \cdot 8 \cdot 7$.

Example 2.1.6. (a) How many 3-digit numbers can be formed using the digits $1, 3, 4, 5, 6, 8$, and 9? (b) How many can be formed if no digit can be repeated?

There are 7^3 such 3-digit numbers in (a) since each of the 3 digits can be filled with 7 possibilities. Likewise, the answer to question (b) is $7 \cdot 6 \cdot 5$ since there are 7 possibilities for the hundreds digit but once one digit is used it is not available for the tens digit (since no digit can be repeated in this problem). Thus there are only 6 possibilities for the tens digit, and then for the same reason there are only 5 possibilities for the units digit.

Frequently the solution of combinatorial problems call for the application of *both* the sum rule and the product rule, perhaps even a repeated and intermixed application of both principles. Generally speaking, the best way to approach any counting problem is by using these two principles to produce a thorough case by case decomposition into small, manageable subproblems. Then frequently the counting itself is easy once one has decided exactly what to count.

Let us illustrate with a few examples.

Example 2.1.7. (a) How many different license plates are there that involve 1, 2, or 3 letters followed by 4 digits?

We can form plates with 1 letter followed by 4 digits in $26 \cdot 10^4$ ways, plates with 2 letters followed by 4 digits in $26^2 \cdot 10^4$ ways, and plates with 3 letters followed by 4 digits in $26^3 \cdot 10^4$ ways. These separate events are mutually exclusive so we can apply the sum rule to conclude that there are $26 \cdot 10^4 + 26^2 \cdot 10^4 + 26^3 \cdot 10^4 = (26 + 26^2 + 26^3)10^4$ plates with 1, 2, or 3 letters followed by 4 digits. Now we can use what we have learned to solve the next question.

(b) How many different plates are there that involve 1, 2, or 3 letters followed by 1, 2, 3, or 4 digits?

Following the pattern of the solution to (a) we see that there are $(26 + 26^2 + 26^3)10$ ways to form plates of 1, 2, or 3 letters followed by 1 digit, $(26 + 26^2 + 26^3)10^2$ plates of 1, 2, or 3 letters follwed by 2 digits; $(26 + 26^2 + 26^3)10^3$ plates of 1, 2, or 3 letters followed by 3 digits, and $(26 + 26^2 + 26^3)10^4$ plates of 1, 2, or 3 letters followed by 4 digits. Thus, we can apply the sum rule to conclude that there are $(26 + 26^2 + 26^3)10 + (26 + 26^2 + 26^3)10^2 + (26 + 26^2 + 26^3)10^3 + (26 + 26^2 + 26^3)10^4 = (26 + 26^2 + 26^3)(10 + 10^2 + 10^3 + 10^4)$ ways to form plates of 1, 2, or 3 letters followed by 1, 2, 3, or 4 digits.

(c) Having seen the above explanations, can you now conjecture how many plates there are that involve from 1 to 10 letters followed by from 1 to 10 digits?

Example 2.1.8. (a) How many 2-digit or 3-digit numbers can be formed using the digits 1, 3, 4, 5, 6, 8, and 9 if no repetition is allowed?

We have already seen in Example 2.1.6 (b) that there are $7 \cdot 6 \cdot 5$ three-digit numbers possible. Likewise, we can apply the product rule to see that there are $7 \cdot 6$ possible 2-digit numbers. Hence, there are $7 \cdot 6 + 7 \cdot 6 \cdot 5$ possible two-digit or three-digit numbers.

(b) How many numbers can be formed using the digits 1, 3, 4, 5, 6, 8, and 9 if no repetitions are allowed?

The number of digits are not specified in this problem so we can form one-digit numbers, two-digit numbers, or three-digit numbers, etc. But since no repetitions are allowed and we have only the 7 integers to work with, the maximum number of digits would have to be 7. Applying the product rule, we see that we may form 7 one-digit numbers, $7 \cdot 6 = 42$ two-digit numbers, $7 \cdot 6 \cdot 5$ three-digit numbers, $7 \cdot 6 \cdot 5 \cdot 4$ four-digit numbers, $7 \cdot 6 \cdot 5 \cdot 4 \cdot 3$ five-digit numbers, $7 \cdot 6 \cdot 5 \cdot 4 \cdot 3 \cdot 2$ six-digit numbers, and $7 \cdot 6 \cdot 5 \cdot 4 \cdot 3 \cdot 2 \cdot 1$ seven-digit numbers.

The events of forming one-digit numbers, two-digit numbers, three-digit numbers, etc., are mutually exclusive events so we apply the sum rule to see that there are $7 + 7 \cdot 6 + 7 \cdot 6 \cdot 5 + 7 \cdot 6 \cdot 5 \cdot 4 + 7 \cdot 6 \cdot 5 \cdot 4 \cdot 3 + 7 \cdot 6 \cdot 5 \cdot 4 \cdot 3 \cdot 2 + 7 \cdot 6 \cdot 5 \cdot 4 \cdot 3 \cdot 2 \cdot 1$ different numbers we can form under the restrictions of this problem.

Example 2.1.9. How many three-digit numbers are there which are even and have no repeated digits? (Here we are using all digits 0 through 9.)

For a number to be even it must end in 0,2,4,6, or 8. There are two cases to consider. First, suppose that the number ends in 0; then there are 9 possibilities for the first digit and 8 possibilities for the second since no digit can be repeated. Hence there are $9 \cdot 8$ three-digit numbers that end in 0. Now suppose the number does not end in 0. Then there are 4 choices for the last digit (2,4,6, or 8); when this digit is specified, then there are only 8 possibilities for the first digit, since the number cannot begin with 0. Finally, there are 8 choices for the second digit and therefore there are $8 \cdot 8 \cdot 4$ numbers that do not end in 0. Accordingly since these two cases are mutually exclusive, the sum rule gives $9 \cdot 8 + 8 \cdot 8 \cdot 4$ even three-digit numbers with no repeated digits.

Indirect Counting

It is sometimes beneficial to solve some combinatorial problems by *counting indirectly*, that is, by counting the complement of a set. We will discuss this more when we consider the principle of inclusion and exclusion but for now let us list a few examples.

Example 2.1.10. Let us determine, by counting indirectly, the number of nonnegative integers less than 10^9 that contain the digit 1. First, we determine the integers less than 10^9 that *do not* contain the digit 1. We are considering 1-digit, or 2-digit, . . . , up to 9-digit numbers. Of course, a representation like 000002578 is actually a 4-digit number so we can consider 9 positions to be filled with any of the digits 0, 2, 3, 4, 5, 6, 7, 8, or 9. There are 9^9 such integers that do not contain the digit 1. Thus, there are $10^9 - 9^9 = 612{,}579{,}511$ integers less than 10^9 that do contain the digit 1.

Example 2.1.11 Suppose that we draw a card from a deck of 52 cards and replace it before the next draw. In how many ways can 10 cards be drawn so that the tenth card is a repetition of a previous draw?

We answer this by counting indirectly. First we count the number of ways we can draw 10 cards so that the 10th card is not a repetition. We analyze this as follows. First, choose what the 10th card will be. This can be done in 52 ways. If the first 9 draws are different from this, then each of the 9 draws can be chosen from 51 cards. Thus, there are 51^9 ways to draw the first 9 cards different from the 10th card. Hence there are $(51^9)(52)$ ways to choose 10 cards with the 10th card different from any of the previous 9 draws. Hence, there are $52^{10} - (51^9)(52)$ ways to draw 10 cards where the 10th is a repetition since there are 52^{10} ways to draw 10 cards with replacements.

Example 2.1.12. In how many ways can 10 people be seated in a row so that a certain pair of them are not next to each other?

There are 10! ways of seating all 10 people. Thus, by indirect counting, we need only count the number of ways of seating the 10 people where the certain pair of people (say, A and B) are seated next to each other. If we treat the pair $A B$ as one entity, then there are 9 total entities to arrange in 9! ways. But A and B can be seated next to each other in 2 different orders, namely $A B$ and $B A$. Thus, there are $(2)(9!)$ ways of seating all 10 people where A and B are next to each other. The answer to our problem then is $10! - (2)(9!)$.

One-to-One Correspondence

There is, finally, another technique that is often used in counting. In this technique the problem at hand is replaced by another problem where it is observed that there are exactly the same number of objects of the first type as there are of the second type. In other words, a one-to-one correspondence is set up between the objects of the first type with those of the second. (A one-to-one correspondence between 2 sets A and B is just a one-to-one function from A onto B.) Of course, we would hope that in the new context we can see how to count the objects of the second type more readily than those of the first type.

Example 2.1.13. Suppose that there are 101 players entered in a single elimination tennis tournament. In such a tournament, any player who loses a match must drop out, and every match ends in a victory for some player—there are no ties. In each round of the tournament, the players remaining are matched into as many pairs as possible, but if there is an odd number of players left someone receives a bye (which means an automatic victory for this player in this round). Enough rounds are played until a single player remains who wins the tournament. The problem is: how many matches must be played in total?

There are two approaches to this problem. The straightforward approach is to analyze each round of the tournament as follows. The 50 winners and bye will go into the second round and pair into 25 matches and a bye. After this the 25 winners and the bye will go into the third round where there will be exactly 13 matches. The fourth round will have six matches and a bye; the fifth round, 3 matches and a bye; the sixth, 2 matches; the seventh will have 1 match and the winner of the seventh round wins the entire tournament. In total, there must be $50 + 25 + 13 + 6 + 2 + 1 = 100$ matches.

Nevertheless, there is a better way to solve this problem. Observe that there is a one-to-one correspondence between the number of matches and the number of losers. Each match has one and only one loser and each loser was eliminated in one and only one match. Therefore, the total number of matches is the same as the total number of losers. At the start there are 101 players and at the end there is only one undefeated player. Consequently, there must have been 100 losers and hence 100 matches are required to determine a winner.

One of the nice features of the second approach is that the problem and its solution can be generalized. Any single elimination tournament similar to the one above that starts with n contestants will require $n - 1$ matches in order to determine a winner.

Example 2.1.14. Determine the number of subsets of a set with n elements.

Let S_n be the number of subsets of a set with n elements. Then we calculate the value of S_n for a few values of n to determine a conjecture.

n	0	1	2	3	4
S_n	1	2	4	8	16

For example, if $n = 3$ where the set contains the 3 elements a,b,c, then there is a total of 8 subsets including the empty set, the 3 singleton sets: $\{a\},\{b\},\{c\}$, 3 subsets with 2 elements each: $\{a,b\},\{a,c\},\{b,c\}$, and finally, the entire set $\{a,b,c\}$. From this meager information, we conjecture that, in general, $S_n = 2^n$ (a fact perhaps already familiar to the reader).

Now let us give a proof that $S_n = 2^n$. We could prove the result by mathematical induction, but we shall apply the product rule and the idea of one-to-one correspondence. Let $V = \{x_1, x_2, \ldots, x_n\}$ denote the entire set. Then if T is any subset of V, assign the n-digit binary sequence $(y_1, y_2, \ldots, y_n)$ where $y_i = 1$ if $x_i \in T$ and $y_i = 0$ if $x_i \notin T$. In this manner we associate a unique n-digit binary sequence to each subset of V. For instance, if $T = \{x_1, x_3, x_5\}$, then the associated n-digit binary sequence is $(1,0,1,0,1,0,0,\ldots,0)$ indicating $x_1 \in T$, $x_3 \in T$, $x_5 \in T$ but that the other $n - 3$ elements are not in T. Moreover, to each n-digit binary sequence there is a unique subset of V. For example, the binary sequence $(0,1,1,0,0,1,0,\ldots,0)$ corresponds to the subset $\{x_2, x_3, x_6\}$. We understand then that we have established a one-to-one correspondence between the collection of subsets of V and the collection of all n-digit binary sequences. There are clearly 2^n n-digit binary sequences so that there are, likewise, 2^n subsets of V.

Applications to Computer Science

A 2-valued Boolean function of n variables is defined by the assignment of a value of either 0 or 1 to each of the 2^n n-digit binary numbers. How many Boolean functions of n variables are there?

Since there are 2 ways to assign a value to each of the 2^n binary n-tuples, by the rule of product there are

$$\underbrace{2 \cdot 2 \cdot \ \cdots \ \cdot 2}_{2^n \text{ factors}} = 2^{2^n}$$

ways to assign all the values, and therefore 2^{2^n} different Boolean functions of n variables.

A 2-valued Boolean function can be represented in tabular form where the n-digit binary numbers and their values are given in the table below. Such a tabular form is also known as the truth table of a 2-valued Boolean function. For example, the following table is a truth table of a 2-valued Boolean function of four variables:

Four-Digit Binary Number	Value
0000	0
0001	1
0010	1
0011	0
0100	1
0101	0
0110	0
0111	0
1000	1
1001	0
1010	0
1011	0
1100	0
1101	0
1110	0
1111	0

A *self-dual 2-valued Boolean function* is one which will remain unchanged after all the 0's and 1's in the truth table are interchanged.

How many self-dual 2-valued Boolean functions of n variables are there?

Partition the set of 2^n binary n-tuples into 2^{n-1} blocks, each block containing an n-tuple and its 1's complement. In constructing a self-dual function, assigning a value to either member of a block fixes the value that must be given to the other member. So, independent value assignments may be made for only 2^{n-1} of the 2^n n-tuples. Thus, there are $2^{2^{n-1}}$ different self-dual Boolean functions of n variables.

The applications of the 2-valued Boolean functions, and self-dual Boolean functions is quite important to computer scientists who study the nature and applications of switching functions and logic design. It is

therefore important to understand their properties as well as to enumerate them. Further discussion of this subject is given in Chapter 6.

Factorials

Frequently it is useful to have a simple notation for products such as

$$4 \cdot 3 \cdot 2 \cdot 1, \quad 6 \cdot 5 \cdot 4 \cdot 3 \cdot 2 \cdot 1, \quad \text{or} \quad 7 \cdot 6 \cdot 5 \cdot 4.$$

Definition 2.1.1. For each positive integer we define $n! = n \cdot (n-1)(n-2) \cdots 3 \cdot 2 \cdot 1 = $ the product of all integers from 1 to n.

Also define $0! = 1$. Note that $1! = 1$

Thus,
$$4! = 4 \cdot 3 \cdot 2 \cdot 1,$$
$$6! = 6 \cdot 5 \cdot 4 \cdot 3 \cdot 2 \cdot 1,$$

and
$$7 \cdot 6 \cdot 5 \cdot 4 = \frac{7!}{3!} = \frac{7 \cdot 6 \cdot 5 \cdot 4 \cdot 3 \cdot 2 \cdot 1}{3 \cdot 2 \cdot 1}.$$

We read $n!$ as "n factorial."

It is true that $4! = 24$ and $6! = 720$ but frequently we leave our answers in factorial form rather than evaluating the factorials. Nevertheless, the relation $n! = n[(n-1)!]$ enables us to compute the values of $n!$ for small n fairly quickly. For example:

$0! = 1,$	$1! = 1,$	$2! = 2,$
$3! = 6,$	$4! = 24,$	$5! = 120,$
$6! = 720,$	$7! = 5,040,$	$8! = 40,320,$
$9! = 362,880,$	$10! = 3,628,800,$	$11! = 39,916,800.$

Exercises for Section 2.1

1. How many possible telephone numbers are there when there are seven digits, the first two of which are between 2 and 9 inclusive, the third digit between 1 and 9 inclusive, and each of the remaining may be between 0 and 9 inclusive?

2. Suppose that a state's license plates consist of three letters followed by three digits: How many different plates can be manufactured (repetitions are allowed)?

3. A company produces combination locks, the combinations consist of three numbers from 0 to 39 inclusive. Because of the construc-

tion no number can occur twice in a combination. How many different combinations for locks can be attained?

4. (a) How many 4-digit numbers can be formed using the digits 2, 3, 5, 6, 8, and 9 if repetitions are allowed?
   (b) How many if no repetitions are allowed?
   (c) How many if those in (b) are even numbers?
   (d) How many of those numbers in (b) are greater than 4000?
   (e) How many of those in (b) are divisible by 5?

5. How many 3-letter words can be formed using the letters a, b, c, d, e, and f and using a letter only once if:
   (a) the letter a is to be used?
   (b) either a or b or both a and b are used?
   (c) the letter a is not used?

6. How many ways are there to pick a man and a woman who are not married from 30 married couples?

7. (a) How many ways are there to select 2 cards (without replacement) from a deck of 52?
   How many ways are there to select the 2 cards such that:
   (b) the first card is an ace and the second card is a king?
   (c) the first card is an ace and the second is not a king?
   (d) the first card is a heart and the second is a club?
   (e) the first card is a heart and the second is a king?
   (f) the first card is a heart and the second is not a king?
   (g) neither card is an ace?
   (h) at least one of the cards drawn is an ace?

8. How many ways are there to roll two distinguishable dice to yield a sum that is divisible by 3?

9. How many integers between 1 and 10^4 contain exactly one 8 and one 9?

10. How many different license plates are there (allowing repetitions):
    (a) involving 3 letters and 4 digits if the 3 letters must appear together either at the beginning or at the end of the plate?
    (b) involving 1, 2, or 3 letters and 1, 2, 3, or 4 digits if the letters must occur together?

11. How many 5-letter words are there where the first and last letters:
    (a) are consonants?
    (b) are vowels?
    (c) are vowels and the middle letters are consonants?
    (d) How many 5-letter words are there if vowels can only appear (if at all) as the first or last letter?

(e) Do (a)–(d) assuming no repetitions are allowed.

12. (a) How many 7-digit numbers are there (leading zeroes are not allowed like 0123456)?
    (b) How many 7-digit numbers are there with exactly one 5?

13. A palindrome is a word that reads the same forward or backward. How many 9-letter palindromes are possible using the English alphabet?

14. There are five different roads from City A to City B, three different roads from City B to City C, and three different roads that go directly from A to C.
    (a) How many different ways are there to go from A to C via B?
    (b) How many different ways are there from A to C altogether?
    (c) How many different ways are there from A to C and then back to A?
    (d) How many different trips are there from A to C and back again to A that visit B both going and coming?
    (e) How many different trips are there that go from A to C via B and return directly from C to A?
    (f) How many different trips are there that go directly from A to C and return to A via B?
    (g) How many different trips are there from A to C and back to A that visit B at least once?
    (h) Suppose that once a road is used it is closed and cannot be used again. Then how many different trips are there from A to C via B and back to A again via B?
    (i) Using the assumption in (h) how many different trips are there from A to C and back to A again?

15. Find the total number of positive integers that can be formed from the digits 1, 2, 3, 4, and 5 if no digit is repeated in any one integer.

16. A newborn child can be given 1, 2, or 3 names. In how many ways can a child be named if we can choose from 300 names (and no name can be repeated)?

17. There are 9 positions on a baseball team. If the baseball coach takes 25 players on a road trip,
    (a) how many different teams can he field?
    (b) how many different batting orders can he make if he has 10 pitchers and he always places the pitcher in the 9th position of his batting order?
    (c) how many teams can the coach field if he has 4 catchers, 10 pitchers, 7 infielders, and 4 outfielders?

18. In a certain programming language, an identifier is a sequence of a certain number of characters where the first character must be a

letter of the English alphabet and the remaining characters may be either a letter or a digit.

(a) How many identifiers are there of length 5?

(b) In particular, in some implementations of Pascal an identifier is a sequence of from 1 up to 8 characters with the above restrictions. How many Pascal identifiers are there?

19. (a) There are 10 telegrams and 2 messenger boys. In how many different ways can the telegrams be distributed to the messenger boys if the telegrams are distinguishable?

(b) In how many different ways can the telegrams be distributed to the messenger boys and then delivered to 10 different people if the telegrams are distinguishable?

(c) Rework (a) under the assumption that the telegrams are indistinguishable?

20. A shoe store has 30 styles of shoes. If each style is available in 12 different lengths, 4 different widths, and 6 different colors, how many kinds of shoes must be kept in stock?

21. A chain letter is sent to 10 people in the first week of the year. The next week each person who received a letter sends letters to 10 new people, and so on.

(a) How many people have received letters after 10 weeks?

(b) at the end of the year?

22. A company has 750 employees. Explain why there must be at least 2 people with the same pair of initials.

23. A tire store carries 10 different sizes of tires, each in both tube and tubeless variety, each with either nylon, rayon cord, or steel-belted, and each with white sidewalls or plain black. How many different kinds of tires does the store have?

24. How many integers between 10^5 and 10^6

(a) have no digits other than 2, 5, or 8?

(b) have no digits other than 0, 2, 5, or 8?

25. In how many different orders can 3 men and 3 women be seated in a row of 6 seats if

(a) anyone may sit in any of the seats?

(b) the first and last seats must be filled by men?

(c) men occupy the first 3 seats and women occupy the last three seats?

(d) all members of the same sex are seated in adjacent seats?

(e) men and women are seated alternately?

26. Find the sum of all 4-digit numbers that can be obtained by using (without repetition) the digits 2, 3, 5, and 7.

27. A new state flag is to be designed with 6 vertical stripes in yellow, white, blue, and red. In how many ways can this be done so that no 2 adjacent stripes have the same color?

28. How many ways can one right and one left shoe be selected from 10 pairs of shoes without obtaining a pair?

29. (a) If 6 men intend to speak at a convention, in how many orders can they do so with B speaking immediately before A?
    (b) How many orders are there with B speaking after A?

30. Given 8 different English books, 12 different German books, and 5 different Russian books, determine how many ways to arrange these books on a shelf where
    (a) all books of the same language are grouped together.
    (b) the 8 English books are on the left.
    (c) the 8 English books are on the left and the 12 German books are on the right.

    Determine the number of ways to select two books from different subjects.

31. Of the integers 50 to 500 inclusive, how many integers
    (a) are there in total?
    (b) are even?
    (c) are odd?
    (d) contain the digit 7?
    (e) are greater than 100?
    (f) are greater than 100 and do not contain the digit 7?
    (g) are divisible by 5?
    (h) have their digits in strictly increasing order?

32. Prove by mathematical induction that one-half of the 6^n outcomes of rolling n distinguishable dice have an even sum.

33. (a) In how many ways can we choose a black square and a white square on an 8×8 chess board?
    (b) In how many ways can we choose a black square and a white square on a chess board if the two squares must not belong to the same row or column?

34. (a) In how many ways can 4 cards of different suits be selected from an ordinary deck of 52 cards?
    (b) In (a) how many ways can the 4 cards be selected where the diamond selection has the same value as the hearts selection and the spade has the same value as the club?

35. Each of n different light posts can be made to beam a red, a yellow, or a green light. How many different signals can be beamed by the array of light posts?

36. Twenty athletes compete in a contest. Each of 3 judges assign 20 different ratings to the 20 athletes. For an athlete to be named winner he must be given the highest rating by at least 2 judges. Compute the fraction of cases for which a winner is named.

37. If the population of a certain city is 25,000 and each resident has 3 initials, determine whether or not there must be 2 citizens with the same 3 initials.

38. A railroad compartment has 10 seats, 5 facing the engine and 5 facing away from the engine. Of 10 passengers, 4 prefer to face the engine, 3 prefer to face away from the engine, and 3 have no preference. In how many ways can the passengers be seated according to their preference?

39. A mother distributes 5 different apples among 8 children.
    (a) How many ways can this be done if each child receives at most one apple?
    (b) How many ways can this be done if there is no restriction on the number of apples a child can receive?

40. How many ways can a committee of k people be chosen from 10 people if k can be 1,2,3,..., or 10?

Selected Answers for Section 2.1

1. $8 \cdot 8 \cdot 9 \cdot 10^4$.

2. $26^3 \cdot 10^3$.

7. (a) $52 \cdot 51$.
   (b) $4 \cdot 4$.
   (c) $4 \cdot 47$.
   (f) $1 \cdot 48 + 12 \cdot 47$.
   (g) $48 \cdot 47$.
   (h) $52 \cdot 51 - 48 \cdot 47$.

9. $4 \cdot 3 \cdot 8^2$.

10. (a) $2 \cdot 26^3 \cdot 10^4$.
    (b) $(2 \cdot 10 + 3 \cdot 10^2 + 4 \cdot 10^3 + 5 \cdot 10^4)(26 + 26^2 + 26^3)$.

11. (a) $21^2 \cdot 26^3$.
    (b) $5^2 \cdot 26^3$.
    (c) $5^2 \cdot 21^3$.
    (d) $26^2 \cdot 21^3 = 21^5 + 2 \cdot 5 \cdot 21^4 + 5^2 \cdot 21^3$.
    (e) $32 \cdot 20 \cdot 24 \cdot 23 \cdot 22$; $5 \cdot 4 \cdot 24 \cdot 23 \cdot 22$; $5 \cdot 4 \cdot 21 \cdot 20 \cdot 19$; $5 \cdot 4 \cdot 21 \cdot 20 \cdot 19 + 5 \cdot 21 \cdot 20 \cdot 19 \cdot 18 + 5 \cdot 21 \cdot 20 \cdot 19 \cdot 18 + 21 \cdot 20 \cdot 19 \cdot 18 \cdot 17$.

14. (a) $5 \cdot 3 = 15$.
    (b) $15 + 3 = 18$.
    (c) 18^2.
    (d) 15^2.
    (e) $15 \cdot 3 = 45$.
    (f) $3 \cdot 15 = 45$.
    (g) $15 \cdot 3 + 3 \cdot 15 + 15^2 = 18^2 - 3^2 = 15 \cdot 18 + 3 \cdot 15$.
    (h) $15 \cdot 8$.
    (i) $15 \cdot 8 + 15 \cdot 3 + 3 \cdot 15 + 3 \cdot 2$.

15. $5 + 5 \cdot 4 + 5 \cdot 4 \cdot 3 + 5 \cdot 4 \cdot 3 \cdot 2 + 5 \cdot 4 \cdot 3 \cdot 2 \cdot 1$.

16. $300 + 300 \cdot 299 + 300 \cdot 299 \cdot 298$.

17. (a) $25!/16! \cdot$
    (b) $(15!/7!) \cdot 10$.
    (c) $4 \cdot 10 \cdot 7 \cdot 6 \cdot 5 \cdot 4 \cdot 4 \cdot 3 \cdot 2$.

18. In (a) the identifiers are of length 5 and the first character may be one of 26 letters, while the remaining 4 characters can be any of the 26 letters a through z or 10 digits 0 through 9. Thus, each of the remaining 4 characters can be filled in 36 ways. Hence there are $26 \cdot 36^4$ identifiers of length 5.
    (b) Here we are asked to count the number of identifiers of length 1, or 2 etc., up to 8. There are 26 of length 1, $26 \cdot 36$ of length 2, ..., and $26 \cdot 36^7$ of length 8. Hence there are $26 + 26 \cdot 36 + 26 \cdot 36^2 + \cdots + 26 \cdot 36^7$ Pascal identifiers.

19. (a) 2^{10}.
    (b) $2^{10} (10!)$.
    (c) 11; give the first boy 0, 1, 2, ..., 10 telegrams and the rest to the second boy.

20. $30 \cdot 12 \cdot 4 \cdot 6$.

21. (a) $10 + 10^2 + 10^3 + \cdots + 10^{10}$.
    (b) $10 + 10^2 + 10^3 + \cdots + 10^{52}$.

23. $10 \cdot 3 \cdot 2 \cdot 2$.

24. (a) 3^6.
    (b) $3 \cdot 4^5$.

25. (a) $6! = 720$.
    (b) $3 \cdot 2 \cdot 4! = 144$.
    (c) $(3!)^2 = 36$.
    (d) $2(3!)^2 = 72$.
    (e) $2(3!)^2 = 2 \cdot 36 = 72$.

26. There are $24 = 4!$ such numbers. Each digit occurs 6 times in every one of the 4 positions. The sum of the digits is 17. Hence the sum of these 24 numbers is $(6)(17) \cdot (1111)$.

27. Choose the color for the first stripe in 4 ways. The second stripe can be one of the 3 remaining colors. Then the third stripe can be any of 3 colors excluding the color of the second stripe, and so on. The number is $4 \cdot 3^5$.

28. $10 \cdot 9$

29. (a) $5!$

    (b) $6!/2$

2.2   COMBINATIONS AND PERMUTATIONS

Definition 2.2.1.   A combination of n objects taken r at a time (called an r-combination of n objects) is an **unordered selection** of r of the objects.

A permutation of n objects taken r at a time (also called an r-permutation of n objects) is an **ordered selection** or **arrangement** of r of the objects.

Some remarks will help clarify these definitions. Note that we are simply defining the terms r-combinations and r-permutations here and have not mentioned anything about the properties of the n objects. For example, these definitions say nothing about whether or not a given element may appear more than once in the list of n objects. In other words, it may be that the n objects do not constitute a set in the normal usage of the word.

Example 2.2.1.   Suppose that the 5 objects from which selections are to be made are: a,a,a,b,c. Then the 3-combinations of these 5 objects are: aaa,aab,aac,abc. The 3-permutations are:

$$aaa, aab, aba, baa, aac, aca, caa,$$
$$abc, acb, bac, bca, cab, cba.$$

Neither do these definitions say anything about any rules governing the selection of the r-objects: on one extreme, objects could be chosen where all repetition is forbidden, or on the other extreme, each object may be chosen up to r times, or then again there may be some rule of selection between these extremes; for instance, the rule that would allow a given object to be repeated up to a certain specified number of times.

We will use expressions like $\{3 \cdot a, 2 \cdot b, 5 \cdot c\}$ to indicate either (1) that we have $3 + 2 + 5 = 10$ objects including 3 a's, 2 b's, and 5 c's, or (2) that we have 3 objects a,b,c where selections are constrained by the conditions that a can be selected at most three times, b can be selected at most twice, and c can be chosen up to five times.

The numbers 3, 2, and 5 in this example will be called *repetition numbers*.

Example 2.2.2.   The 3-combinations of $\{3 \cdot a, 2 \cdot b, 5 \cdot c\}$ are:

$$aaa, aab, aac, abb, abc,$$
$$ccc, ccb, cca, cbb.$$

Example 2.2.3.   The 3-combinations of $\{3 \cdot a, 2 \cdot b, 2 \cdot c, 1 \cdot d\}$. are:

$$aaa, aab, aac, aad, bba, bbc, bbd,$$
$$cca, ccb, ccd, abc, abd, acd, bcd.$$

In order to include the case where there is no limit on the number of times an object can be repeated in a selection (except that imposed by the size of the selection) we use the symbol ∞ as a repetition number to mean that an object can occur an infinite number of times.

Example 2.2.4.   The 3-combinations of $\{\infty \cdot a, 2 \cdot b, \infty \cdot c\}$ are the same as in Example 2.2.2 even though a and c can be repeated an infinite number of times. This is because, in 3-combinations, 3 is the limit on the number of objects to be chosen.

If we are considering selections where each object has ∞ as its repetition number then we designate such selections as selections with *unlimited repetitions*. In particular, a selection of r objects in this case will be called *r-combinations with unlimited repetitions* and any ordered arrangement of these r objects will be an *r-permutation with unlimited repetitions*.

Example 2.2.5.   The 3-combinations of a,b,c,d with unlimited repetitions are the 3-combinations of $\{\infty \cdot a, \infty \cdot b, \infty \cdot c, \infty \cdot d\}$. There are 20 such 3-combinations, namely:

$$aaa, aab, aac, aad,$$
$$bbb, bba, bbc, bbd,$$
$$ccc, cca, ccb, ccd,$$
$$ddd, dda, ddb, ddc,$$
$$abc, abd, acd, bcd.$$

Moreover, there are $4^3 = 64$ of 3-permutations with unlimited repetitions since the first position can be filled 4 ways (with a, b, c, or d), the second position can be filled 4 ways, and likewise for the third position.

We leave it to the student to make a list of all 64 3-permutations of a,b,c,d with unlimited repetitions.

The 2-permutations of $\{\infty \cdot a, \infty \cdot b, \infty \cdot c, \infty \cdot d\}$ do not present such a formidable list and so we tabulate them in the following table.

2-Combinations With Unlimited Repetitions	2-Permutations With Unlimited Repetitions
aa	aa
ab	ab, ba
ac	ac, ca
ad	ad, da
bb	bb
bc	bc, cb
bd	bd, db
cc	cc
cd	cd, dc
dd	dd
Total Number 10	16

Of course, these are not the only constraints that can be placed on selections; the possibilities are endless. We list some more examples just for concreteness. We might, for example, consider selections of $\{\infty \cdot a, \infty \cdot b, \infty \cdot c\}$ where b can be chosen only an even number of times. Thus, 5-combinations with these repetition numbers and this constraint would be those 5-combinations with unlimited repetitions and where b is chosen 0, 2, or 4 times.

Example 2.2.6. The 3-combinations of $\{\infty \cdot a, \infty \cdot b, 1 \cdot c, 1 \cdot d\}$ where b can be chosen only an even number of times are the 3-combinations of a,b,c,d where a can be chosen up to 3 times, b can be chosen 0 or 2 times, and c and d can be chosen at most once. The 3-combinations subject to these constraints are:

$$aaa, aac, aad, bba, bbc, bbd, acd.$$

As another example, we might be interested in, say, selections of $\{\infty \cdot a,$ $3 \cdot b, 1 \cdot c\}$ where a can be chosen a prime number of times. Thus, the 8-combinations subject to these constraints would be all those 8-combinations where a can be chosen 2, 3, 5, or 7 times, b can be chosen up to 3 times, and c can be chosen at most once.

There are, as we have said, an infinite variety of constraints one could place on selections. You can just let your imagination go free in conjuring up different constraints. Nevertheless, any selection of r objects, regardless of the constraints on the selection, would constitute an r-combination according to our definition. Moreover, any arrangement of these r objects would constitute an r-permutation.

While there may be an infinite variety of constraints, we are primarily interested in two major types: one we have already described—combinations and permutations with unlimited repetitions, the other we now describe.

If the repetition numbers are all 1, then selections of r objects are called *r-combinations without repetitions* and arrangements of the r objects are *r-permutations without repetitions*. We remind you that r-combinations without repetitions are just subsets of the n elements containing exactly r elements. Moreover, we shall often drop the repetition number 1 when considering r-combinations without repetitions. For example, when considering r-combinations of $\{a,b,c,d\}$ we will mean that each repetition number is 1 unless otherwise designated, and, of course, we mean that in a given selection an element need not be chosen at all, but, if it is chosen, then in this selection this element cannot be chosen again.

Example 2.2.7. Suppose selections are to be made from the four objects a,b,c,d.

2-Combinations Without Repetitions	2-Permutations Without Repetitions
ab	ab, ba
ac	ac, ca
ad	ad, da
bc	bc, cb
bd	bd, db
cd	cd, dc

Total Number: 6   12

There are six 2-combinations without repetitions and to each there are two 2-permutations giving a total of twelve 2-permutations without repetitions.

Note the total number of 2-combinations with unlimited repetitions in Example 2.2.5 included the six 2-combinations without repetitions of Example 2.2.7 and as well 4 other 2-combinations where repetitions actually occur. Likewise, the sixteen 2-permutations with unlimited repetitions included the twelve 2-permutations without repetitions.

3-Combinations Without Repetitions	3-Permutations Without Repetitions
abc	abc, acb, bac, bca, cab, cba
abd	abd, adb, bad, bda, dab, dba
acd	acd, adc, cad, cda, dac, dca
bcd	bcd, bdc, cbd, cdb, dbc, dcb
Total Number 4	24

Note that to each of the 3-combinations without repetitions there are 6 possible 3-permutations without repetitions. Momentarily, we will show that this observation can be generalized.

Exercises for Section 2.2

1. List all 5-combinations of $\{\infty \cdot a, \infty \cdot b, \infty \cdot c\}$, where b is chosen an even number of times.

2. List all 64 3-permutations of $\{\infty \cdot a, \infty \cdot b, \infty \cdot c, \infty \cdot d\}$.

3. List all 3-combinations and 4-combinations of $\{2 \cdot a, b, 3 \cdot c\}$.

4. Determine the number of 5-combinations of $\{1 \cdot a, \infty \cdot b, \infty \cdot c, 1 \cdot d\}$. More generally, develop a formula for the number of r-combinations of a collection of letters $a_1, a_2, \ldots, a_k$ whose repetition numbers are each either 1 or ∞.

2.3   ENUMERATION OF COMBINATIONS AND PERMUTATIONS

General formulas for enumerating combinations and permutations will now be presented. At this time, we will only list formulas for combinations and permutations without repetitions or with unlimited repetitions. We will wait until later to use generating functions to give general techniques for enumerating combinations where other rules govern the selections.

Let $P(n,r)$ denote the number of r-permutations of n elements without repetitions.

Theorem 2.3.1. (Enumerating r-permutations without repetitions).

$$P(n,r) = n(n-1) \cdots (n-r+1) = \frac{n!}{(n-r)!}.$$

Proof. Since there are n distinct objects, the first position of an r-permutation may be filled in n ways. This done, the second position can be filled in $n-1$ ways since no repetitions are allowed and there are $n-1$ objects left to choose from. The third can be filled in $n-2$ ways and so on until the rth position is filled in $n-r+1$ ways (see Figure 2-2). By applying the product rule, we conclude that

$$P(n,r) = n(n-1)(n-2) \cdots (n-r+1).$$

From the definition of factorials, it follows that

$$P(n,r) = \frac{n!}{(n-r)!}. \quad \square$$

When $r = n$, this formula becomes

$$P(n,n) = \frac{n!}{0!} = n! \,.$$

When explicit reference to r is not made, we assume that all the objects are to be arranged; thus when we talk about the *permutations of n objects* we mean the case $r = n$.

Corollary 2.3.1. There are $n!$ permutations of n distinct objects.

Example 2.3.1. There are $3! = 6$ permutations of $\{a,b,c\}$. There are $4! = 24$ permutations of $\{a,b,c,d\}$. The number of 2-permutations of

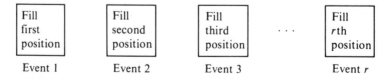

Figure 2-2

$\{a,b,c,d,e\}$ is $P(5,2) = 5!/(5 - 2)! = 5 \cdot 4 = 20$. The number of 5-letter words using the letters a,b,c,d, and e at most once is $P(5,5) = 120$.

Example 2.3.2. There are $P(10,4) = 5{,}040$ 4-digit numbers that contain no repeated digits since each such number is just an arrangement of four of the digits $0,1,2,3,\ldots,9$ (leading zeroes are allowed). There are $P(26,3)$ 3-letter words formed from the English alphabet with no repeated letters. Thus, there are $P(26,3)$ $P(10,4)$ license plates formed by 3 distinct letters followed by 4 distinct digits.

Example 2.3.3. In how many ways can 7 women and 3 men be arranged in a row if the 3 men must always stand next to each other?

There are 3! ways of arranging the 3 men. Since the 3 men always stand next to each other, we treat them as a single entity, which we denote by X. Then if $W_1, W_2, \ldots, W_7$ represents the women, we next are interested in the number of ways of arranging $\{X, W_1, W_2, W_3, \ldots, W_7\}$. There are 8! permutations of these 8 objects. Hence there are $(3!)$ $(8!)$ permutations altogether (of course, if there has to be a prescribed order of an arrangement on the 3 men then there are only 8! total permutations).

Example 2.3.4. In how many ways can the letters of the English alphabet be arranged so that there are exactly 5 letters between the letters a and b?

There are $P(24,5)$ ways to arrange the 5 letters between a and b, 2 ways to place a and b, and then 20! ways to arrange any 7-letter word treated as one unit along with the remaining 19 letters. The total is $P(24,5)$ $(20!)$ (2).

Example 2.3.5. How many 6-digit numbers without repetition of digits are there such that the digits are all nonzero and 1 and 2 do not appear consecutively in either order?

We are asked to count certain 6-permutations of the 9 integers $1,2,\ldots,9$. In the following table we separate these 6-permutations into 4 disjoint classes and count the number of permutations in each class.

Class	Number of Permutations in the Class
(i) Neither 1 nor 2 appears as a digit	7!
(ii) 1, but not 2, appears as a digit	6 $P(7,5)$
(iii) 2, but not 1, appears	6 $P(7,5)$
(iv) Both 1 and 2 appear	$(2)(7)(4)\,P(6,3) + (4)(7)(6)(3)\,P(5,2)$
Total	7! $+$ $(2)(6)\,P(7,5)$ $+$ (56) P (6,3) $+$ (504) P (5,2)

Let us explain how to count the elements in class (iv).

1. The hundred thousands digit is 1 (and thus the ten thousands digit is not 2). The second digit can be chosen in 7 ways. Choose the position for 2 in 4 ways; then fill the other 3 positions $P(6,3)$ ways. Hence, there are $(7)4\,P(6,3)$ numbers in this category.
2. The units digit is 1 (and hence the tens digit is not 2). Likewise, there are $(7)4\,P(6,3)$ numbers in this category.
3. The integer 1 appears in a position different from the hundred thousands digit and the units digit. Hence, 2 cannot appear immediately to the left or to the right of 1. Since 1 can be any one of the digits from the tens digit up to the ten thousands digit, 1 can be placed in 4 ways. The digit immediately to the left of 1 can be filled in 7 ways, while the digit immediately to the right of 1 can be filled in 6 ways. The integer 2 can be placed in any of the remaining positions in 3 ways and then the other 2 digits are a 2-permutation of the remaining 5 integers. Hence, there are $(4)(7)(6)(3)\,P(5,2)$ numbers in this category.

Thus, there are

$$(2)(7)(4)\,P(6,3) + (4)(7)(6)(3)\,P(5,2)$$

numbers in class (iv). Then, by the sum rule, there are

$$P(7,6) + (2)(6)\,P(7,5) + (56)\,P(6,3) + (504)\,P(5,2)$$

elements in all four classes. (Look for the shorter indirect counting solution.)

The permutations we have been considering are more properly called **linear permutations** for the objects are being arranged in a line. If instead of arranging objects in a line, we arrange them in a circle, then the number of permutations decreases.

Example 2.3.6. In how many ways can 5 children arrange themselves in a ring?

Here, the 5 children are not assigned to particular places but are only arranged relative to one another. Thus, the arrangements (see Figure 2-3) are considered the same if the children are in the same order clockwise. Hence, the position of child C_1 is immaterial and it is only the position of the 4 other children relative to C_1 that counts. Therefore, keeping C_1 fixed in position, there are 4! arrangements of the remaining children. This can be generalized to conclude:

Theorem 2.3.2. There are $(n-1)!$ permutations of n distinct objects in a circle.

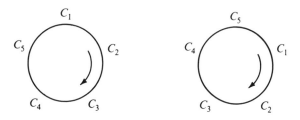

Figure 2-3

As we did before discussing circular permutations, we will use the word "permutation" for "linear permutation" unless ambiguity would result.

Notation. Let $C(n,r)$ denote the number of r-combinations of n distinct objects where 1 is the repetition number for each element. Thus, if S is a set with n elements, $C(n,r)$ is the number of subsets of S with exactly r elements. Or in the terminology of the previous section, $C(n,r)$ is the number of r-combinations of n elements without repetitions.

We read $C(n,r)$ as "n choose r" to emphasize that selections are being made. Frequently, the terms $C(n,r)$ are called binomial coefficients because of their role in the expansion of $(x + y)^n$. Moreover, it is common practice to use the notation $\binom{n}{r}$ instead of $C(n,r)$; nevertheless, we shall use $C(n,r)$ most of the time so that lines of the text will not have to be widely spaced. We will study the binomial theorem and identities involving binomial coefficients in Section 2.6.

Now let us discuss a formula for enumerating r-combinations of n objects without repetition. Any r-permutation of n objects without repetition can be obtained by first choosing the r elements [this can be done in $C(n,r)$ ways] and then arranging the r elements in all possible orders (this can be done in $r!$ ways). Thus, it follows that $P(n,r) = r!C(n,r)$. We know a formula for $P(n,r)$ from Theorem 2.3.1 and thus obtain a formula for $C(n,r)$ by dividing by $r!$

Theorem 2.3.3. (Enumerating r-combinations without repetitions).

$$C(n,r) = \frac{P(n,r)}{r!} = \frac{n!}{r!(n-r)!}.$$

Example 2.3.7. In how many ways can a hand of 5 cards be selected from a deck of 52 cards?

Each hand is essentially a 5-combination of 52 cards. Thus there are

$$C(52,5) = \frac{52!}{5!\,47!} = \frac{52 \cdot 51 \cdot 50 \cdot 49 \cdot 48}{5 \cdot 4 \cdot 3 \cdot 2 \cdot 1}$$

$$= 52 \cdot 51 \cdot 10 \cdot 49 \cdot 2 = 2{,}598{,}960 \text{ such hands.}$$

Example 2.3.8. (a) How many 5-card hands consist only of hearts?

Since there are 13 hearts to choose from, each such hand is a 5-combination of 13 objects. Thus, there is a total of

$$C(13,5) = \frac{13!}{5!\,8!} = \frac{13 \cdot 12 \cdot 11 \cdot 10 \cdot 9}{5 \cdot 4 \cdot 3 \cdot 2 \cdot 1}$$

$$= 13 \cdot 11 \cdot 9 = 1{,}287.$$

(b) How many 5-card hands consist of cards from a single suit?

For each of the 4 suits, spades, hearts, diamonds, or clubs, there are $C(13,5)$ 5-card hands. Hence, there are a total of $4\,C(13,5)$ such hands.

(c) How many 5-card hands have 2 clubs and 3 hearts?

Answer. $C(13,2)\,C(13,3)$.

(d) How many 5-card hands have 2 cards of one suit and 3 cards of a different suit?

For a fixed choice of 2 suits there are $2C(13,2)\,C(13,3)$ ways to choose 2 from one of the suits and 3 from the other. We can choose the 2 suits in $C(4,2)$ ways. Thus, there are $2C(13,2)\,C(13,3)\,C(4,2)$ such 5-card hands.

Recall that two of a kind means 2 aces, 2 kings, 2 queens, etc. Similarly, 3 tens are called three of a kind. Thus there are 13 "kinds" in a deck of 52 cards.

(e) How many 5-card hands contain 2 aces and 3 kings?

Answer. $C(4,2)\,C(4,3)$.

(f) How many 5-card hands contain exactly 2 of one kind and 3 of another kind?

Choose the first kind 13 ways, choose 2 of the first kind $C(4,2)$ ways, choose the second kind 12 ways and choose 3 of the second kind in $C(4,3)$ ways. Hence there are $(13)\,C(4,2)\,(12)\,C(4,3)$ 5-card hands with 2 of one kind and 3 of another kind.

Example 2.3.9. (a) In how many ways can a committee of 5 be chosen from 9 people?

Answer. $C(9,5)$ ways.

(b) How many committees of 5 or more can be chosen from 9 people?

Answer. $C(9,5) + C(9,6) + C(9,7) + C(9,8) + C(9,9)$.

(c) In how many ways can a committee of 5 teachers and 4 students be chosen from 9 teachers and 15 students?
The teachers can be selected in $C(9,5)$ ways while the students can be chosen in $C(15,4)$ ways so that the committee can be formed in $C(9,5)$ $C(15,4)$ ways.

(d) In how many ways can the committee in (c) be formed if teacher A refuses to serve if student B is on the committee?
We answer this question by counting indirectly. First we count the number of committees where both A and B are on the committee. Thus, there are only 8 teachers remaining from which 4 teachers are to be chosen. Likewise, there are only 14 students remaining from which 3 more students are to be chosen. There are $C(8,4)$ $C(14,3)$ committees containing both A and B, and hence there are

$$C(9,5)\ C(15,4) - C(8,4)\ C(14,3)$$

committees that do not have both A and B on the committee.

Example 2.3.10. There are 21 consonants and 5 vowels in the English alphabet. Consider only 8-letter words with 3 different vowels and 5 different consonants. (a) How many such words can be formed?

Answer. $C(5,3)$ $C(21,5)$ 8! (Choose the vowels, choose the consonants, and then arrange the 8 letters.)

(b) How many such words contain the letter a? $C(4,2)$ $C(21,5)$8!
(c) How many contain the letters a and b? $C(4,2)$ $C(20,4)$8!
(d) How many contain the letters b and c? $C(5,3)$ $C(19,3)$8!
(e) How many contain the letters $a,b,$ and c? $C(4,2)$ $C(19,3)$8!
(f) How many begin with a and end with b? $C(4,2)$ $C(20,4)$6!
(g) How many begin with b and end with c? $C(5,3)$ $C(19,3)$6!

Example 2.3.11. There are 30 females and 35 males in the junior class while there are 25 females and 20 males in the senior class. In how many ways can a committee of 10 be chosen so that there are exactly 5 females and 3 juniors on the committee?

Let us draw a chart illustrating the possible male-female and junior-senior constitution of the committee.

Juniors		Seniors		
Female	Male	Female	Male	Number of Ways of Selecting
0	3	5	2	$C(30,0)\,C(35,3)\,C(25,5)\,C(20,2)$
1	2	4	3	$C(30,1)\,C(35,2)\,C(25,4)\,C(20,3)$
2	1	3	4	$C(30,2)\,C(35,1)\,C(25,3)\,C(20,4)$
3	0	2	5	$C(30,3)\,C(35,0)\,C(25,2)\,C(20,5)$

Thus, the total number of ways is the sum of the terms in the last column:

$$C(30,0)\,C(35,3)\,C(25,5)\,C(20,2) + C(30,1)\,C(35,2)\,C(25,4)\,C(20,3) +$$
$$C(30,2)\,C(35,1)\,C(25,3)\,C(20,4) + C(30,3)\,C(35,0)\,C(25,2)\,C(20,5).$$

Exercises for Section 2.3

1. Compute $P(8,5)$ and $C(6,3)$.

2. How many ways are there to distribute 10 different books among 15 people if no person is to receive more than 1 book?

3. (a) In how many ways can 6 boys and 5 girls sit in a row?
   (b) In how many ways can they sit in a row if the boys and the girls are each to sit together?
   (c) In how many ways can they sit in a row if the girls are to sit together and the boys do not sit together?

4. Solve Exercise 3 in the case of m boys and g girls (leave your answers in factorial form).

5. (a) Find the number of ways in which 5 boys and 5 girls can be seated in a row if the boys and girls are to have alternate seats.
   (b) Find the number of ways to seat them alternately if boy A and girl B are to sit in adjacent seats.
   (c) Find the number of ways to seat them alternately if boy A and girl B must not sit in adjacent seats.

6. Find the number of ways in which 5 children can ride a toboggan if 1 of the 3 oldest children must drive?

7. (a) How many ways are there to seat 10 boys and 10 girls around a circular table?
   (b) If boys and girls alternate, how many ways are there?

8. In how many ways can the digits 0, 1, 2, 3, 4, 5, 6, 7, 8, and 9 be arranged so that
   (a) 0 and 1 are adjacent?
   (b) 0 and 1 are adjacent and in the order 01?
   (c) 0, 1, 2, and 3 are adjacent.

9. A group of 8 scientists is composed of 5 psychologists and 3 sociologists.
   (a) In how many ways can a committee of 5 be formed?
   (b) In how many ways can a committee of 5 be formed that has 3 psychologists and 2 sociologists?

10. A bridge hand consists of 13 cards dealt from an ordinary deck of 52 cards.
    (a) How many possible bridge hands are there?
    (b) In how many ways can a person get exactly 6 spades and 5 hearts?

11. How many 4-digit telephone numbers have one or more repeated digits?

12. In how many ways can 10 people arrange themselves
    (a) In a row of 10 chairs?
    (b) In a row of 7 chairs?
    (c) In a circle of 10 chairs?

13. A collection of 100 light bulbs contains 8 defective ones.
    (a) In how many ways can a sample of 10 bulbs be selected?
    (b) In how many ways can a sample of 10 bulbs be selected which contain 6 good bulbs and 4 defective ones?
    (c) In how many ways can the sample of 10 bulbs be selected so that either the sample contains 6 good ones and 4 defective ones or 5 good ones and 5 defective ones?

14. (a) How many binary sequences are there of length 15?
    (b) How many binary sequences are there of length 15 with exactly six 1's.

15. A farmer buys 3 cows, 8 pigs, and 12 chickens from a man who has 9 cows, 25 pigs, and 100 chickens. How many choices does the farmer have?

16. Suppose there are 15 red balls and 5 white balls. Assume that the balls are distinguishable and that a sample of 5 balls is to be selected.
    (a) How many samples of 5 balls are there?

(b) How many samples contain all red balls?

(c) How many samples contain 3 red balls and 2 white balls?

(d) How many samples contain at least 4 red balls?

17. A multiple choice test has 15 questions and 4 choices for each answer. How many ways can the 15 questions be answered so that
    (a) exactly 3 answers are correct?
    (b) at least 3 answers are correct?

18. Suppose there are 50 distinguishable books including 18 English books, 17 French books, and 15 Spanish books.
    (a) How many ways can 2 books be selected?
    (b) How many ways can 3 books be selected so that there is 1 book from each of the 3 languages?
    (c) How many ways are there to select 3 books where exactly 1 language is missing?

19. In how many ways can a team of 5 be chosen from 10 players so as to
    (a) include both the strongest and the weakest player?
    (b) include the strongest but exclude the weakest player?
    (c) exclude both the strongest and the weakest player?

20. Consider the set $S = \{a,e,i,b,c,d,f,g,h,m,n,p\}$. How many 5-letter words containing 2 different vowels and 3 different consonants
    (a) can be formed from the letters in S?
    (b) contain the letter b?
    (c) begin with a?
    (d) begin with b?
    (e) begin with a and contain b?

21. Find the number of ways in which 5 different English books, 6 French books, 3 German books, and 7 Russian books can be arranged on a shelf so that all books of the same language are together.

22. How many 9-letter words can be formed that contain 3, 4, or 5 vowels,
    (a) Allowing repetition of letters?
    (b) Not allowing repetition?

23. A committee is to be chosen from a set of 9 women and 5 men. How many ways are there to form the committee if the committee has,
    (a) 6 people, 3 women, and 3 men?
    (b) any number of people but equal numbers of women and men?
    (c) 6 people and at least 3 are women?
    (d) 6 people including Mr. A?

(e) 6 people but Mr. and Mrs. *A* cannot both be on the committee?

(f) 6 people, 3 of each sex, and Mr. and Mrs. *A* cannot both be on the committee?

24. A man has 15 close friends of whom 6 are women.

(a) In how many ways can he invite 3 or more of his friends to a party?

(b) In how many ways can he invite 3 or more of his friends if he wants the same number of men (including himself) as women?

25. The dean of a certain college has a pool of 10 chemists, 7 psychologists, and 3 statisticians from which to form a committee.

(a) How many 5-member committees can the dean appoint from this pool?

(b) How many 5-member committees can be formed that have at least 1 statistician member?

(c) The chemist, Professor *C*, and the psychologist, Professor *P*, do not get along. How many 5-member committes can be formed that have at most 1 of Professor *C* and Professor *P* on the committee?

(d) How many 5-member committees can be formed so that the number of chemists is greater than or equal to the number of psychologists and the number of psychologists is greater than or equal to the number of statisticians?

26. How many 5-card poker hands have

(a) 4 aces?

(b) 4 of a kind?

(c) exactly 2 pairs?

(d) a full house (that is, 3 of a kind and 2 of another kind)?

(e) a straight (a set of 5 consecutive values)? (Here the ace can be the highest or lowest card.)

(f) a pair of aces (and no other pairs)?

(g) exactly 1 pair?

(h) no cards of the same kind?

(i) a flush (a set of 5 cards in one suit)?

(j) a straight flush (a set of 5 consecutive cards in one suit)?

(k) a royal flush (a straight flush with ace-high card)?

(l) a spade flush?

(m) an ace-high spade flush?

(n) a full house with 3 aces and another pair?

(o) 3 of a kind (without another pair)?

(p) 3 aces (and no other ace or pair)?

27. In how many ways can a 6-card hand have
    (a) exactly one pair (no 3 of a kind or 2 pairs)?
    (b) one pair, or 2 pairs, or 3 of a kind, or 4 of a kind?
    (c) at least 1 card of each suit?
    (d) at least 1 of each of the 4 honor cards: ace, king, queen, and jack?
    (e) the same number of diamonds as clubs?
    (f) at least one diamond and at least one club and the values of the diamonds are all greater than the values of the clubs?

28. How many ways can 5 days be chosen from each of the 12 months of an ordinary year of 365 days?

29. Compute the number of 6-letter combinations of the letters of the English alphabet if no letter is to appear in a combination more than 2 times.

30. In how many ways can 30 distinguishable books be distributed among 3 people $A,B,$ and C so that
    (a) A and B together receive exactly twice as many books as C?
    (b) C receives at least 2 books, B receives at least twice as many books as C, and A receives at least 3 times as many books as B?

31. How many 13-card bridge hands from a 52-card deck are there that have
    (a) all hearts?
    (b) all cards in the same suit?
    (c) exactly 2 suits represented?
    (d) 4 spades, 3 hearts, 3 clubs, and 3 diamonds?
    (e) 4 cards each of spades, hearts, and diamonds and 1 club?
    (f) 4 cards of 3 suits and 1 card of the fourth suit?
    (g) 3 cards of 3 suits and 4 cards of the fourth suit?
    (h) 4 cards of two suits, 3 cards in a third suit, and 2 cards in the fourth suit?
    (i) 4 cards in two suits, 5 in a third suit, and a void in the fourth suit?
    (j) 5 cards in one suit, 4 in another, 3 in a third, and 1 in a fourth suit?
    (k) no face cards (a face card is a 10, jack, queen, king, or ace)?
    (l) all four kings?
    (m) 10 cards in one suit and 3 cards in another suit?
    (n) 10 cards in one suit?
    (o) exactly 3 suits?

32. Determine whether or not the following solutions are correct.
    (a) There are $C(4,2)\,C(50,3)$ 5-card hands with at least 2 aces.
    (b) There are $C(52,5) - C(36,5)$ 5-card hands with at least one of each of the honor cards (ace, king, queen, and jack).

(c) There are $C(39,5)$ 5-card hands that contain only spades, hearts, and diamonds. For any choice of 3 suits, there are $C(39,5)$ 5-card hands that contain cards from at most these 3 suits. Thus, there are $C(4,3)$ $C(39,5)$ 5-card hands that contain cards from at most three suits.

(d) There are $C(4,2)$ (48) (44) (40) 5-card hands that contain exactly one pair of kings and no other matching cards because we can choose the 2 kings in $C(4,2)$ ways, remove all kings from the deck, choose one card in 48 ways, remove all cards of that kind, choose another card from the remaining 44 cards in 44 ways, remove all cards of that kind, and, finally, choose a last card from the remaining cards in 40 ways.

(e) A team can either win, lose, or tie each game it plays. Thus, there are $C(20,7)$ 2^{13} different team records with exactly 7 wins in a 20-game season since we can choose the games for wins in $C(20,7)$ ways and then there are 2 possibilities (lose or tie) in the other 13 games. Therefore, there are $C(20,7)$ 3^{13} team records with at least 7 wins.

33. A man has 5 female and 7 male friends and his wife has 7 female and 5 male friends. In how many ways can they invite 6 males and 6 females if husband and wife are to invite 6 friends each?

34. A jar contains 10 counters numbered 1,2,...,10. Someone removes 3 of the counters from the jar. How many ways will the sum of the numbers on the 3 counters be
(a) exactly 9?
(b) at least 9?

35. A bag contains 20 distinguishable balls of which 6 are red, 6 are white, and 8 are blue. We draw out 5 balls with at least one red ball, replace them, and then draw 5 balls with at most one white one. How many ways can this be done?

36. How many ways can 3 integers be selected from the integers 1,2,...,30 so that their sum is even?

37. How many ways can a person invite 3 of his 6 friends to lunch every day for 20 days?

38. A student is to answer 12 of 15 questions on an examination. How many choices does the student have
(a) in all?
(b) if he must answer the first two questions?
(c) if he must answer the first or the second question but not both?
(d) if he must answer exactly 3 of the first 5 questions?
(e) if he must answer at least 3 of the first 5 questions?

39. Four people can sit on each side of a boat. How many ways can one select a crew for the boat if out of 30 candidates 10 prefer the port side, 12 prefer the starboard side, and 8 have no preference?

40. A coin is tossed 12 times. How many outcomes are possible
    (a) in all?
    (b) with exactly 3 heads?
    (c) with at most 3 heads?
    (d) with a head on the fifth and seventh toss?
    (e) with the same number of heads as tails?

Selected Answers for Section 2.3

2. $P(15,10)$.

3. (a) $11!$.
   (b) $2!6!5!$.
   (c) $5 \cdot 6!5!$.

6. $3 \cdot 4!$.

7. (a) $19!$.
   (b) $10!9!$.

9. (a) $C(8,5)$.
   (b) $C(5,3)\, C(3,2)$.

10. (a) $C(52,13)$
    (b) $C(13,6)\, C(13,5)\, C(26,2)$.

11. $10^4 - P(10,4)$.

13. (a) $C(100,10)$
    (b) $C(92,6)\, C(8,4)$
    (c) $C(92,6)\, C(8,4) + C(92,5)\, C(8,5)$.

14. (a) 2^{15}.
    (b) $C(15,6)$.

15. $C(9,3)\, C(25,8)\, C(100,12)$.

16. (a) $C(20,5)$.
    (b) $C(15,5)$.
    (c) $C(15,3)\, C(5,2)$.

17. (a) $C(15,3)3^{12}$.
    (b) $4^{15} - (3^{15} + 3^{14}C(15,1) + 3^{13}C(15,2))$.

18. (a) $C(50,2)$.
    (b) $18 \cdot 17 \cdot 15$.
    (c) $\binom{17}{2}\binom{15}{1} + \binom{17}{1}\binom{15}{2} + \binom{18}{2}\binom{15}{1} + \binom{18}{1}\binom{15}{2} + \binom{18}{2}\binom{17}{1} + \binom{18}{1}\binom{17}{2}$. Recall $C(n,r) = \binom{n}{r}$.

19. (a) $C(8,3)$.
   (b) $C(8,4)$.
   (c) $C(8,5)$.

21. $4!5!6!3!7!$.

24. (a) $2^{15} - 1 - \binom{15}{1} - \binom{15}{2} = \binom{15}{3} + \binom{15}{4} + \cdots + \binom{15}{15}$.

   (b) $\binom{9}{1}\binom{6}{2} + \binom{9}{2}\binom{6}{3} + \binom{9}{3}\binom{6}{4} + \binom{9}{4}\binom{6}{5} + \binom{9}{5}\binom{6}{6}$.

25. (a) $C(20,5)$.

   (b) $\binom{17}{4}\binom{3}{1} + \binom{17}{3}\binom{3}{2} + \binom{17}{2}\binom{3}{3} = \binom{20}{5} - \binom{17}{5}$.

   (c) $\binom{20}{5} - \binom{18}{3} = \binom{18}{4} + \binom{18}{4} + \binom{18}{5}$.

   (d) $\binom{10}{5} + \binom{10}{4}\binom{7}{1} + \binom{10}{3}\binom{7}{2} + \binom{10}{3}\binom{7}{1}\binom{3}{1} + \binom{10}{2}\binom{7}{2}\binom{3}{1}$.

26. (a) $\binom{4}{4} \cdot 48$.

   (b) $\binom{13}{1}\binom{4}{4} \cdot 48$.

   (c) $\binom{13}{2}\binom{4}{2}\binom{4}{2} \cdot 44$.

   (d) $13\binom{4}{3} \cdot 12\binom{4}{2}$.

   (e) $10 \cdot 4^5$ (choose the top card in 10 ways).

   (f) $\binom{4}{2}\binom{12}{3} \cdot 4^3$.

   (g) $\binom{13}{1}\binom{4}{2}\binom{12}{3}(4)^3$ (choose the kind, choose the pair, choose the other 3 kinds, choose the 3 cards).

   (h) $\binom{13}{5}(4)^5 = 52 \cdot 48 \cdot 44 \cdot 40 \cdot 36/5!$

   (i) $\binom{4}{1}\binom{13}{5}$ (choose the suit, choose 5 cards in the suit).

   (j) $\binom{4}{1}(10)$.

   (k) $\binom{4}{1} = 4$.

   (l) $\binom{13}{5}$

   (m) $\binom{12}{4}$

   (n) $\binom{4}{3}\binom{12}{1}\binom{4}{2}$ (choose the 3 aces $\binom{4}{3}$ ways, choose the kind for the pair 12 ways choose the 2 of that kind $\binom{4}{2}$ ways).

(o) $\binom{13}{1}\binom{4}{3}\binom{12}{2}4 \cdot 4$ (choose a kind, 3 of that kind, choose 2 other kinds, pick the 4th and 5th cards).

(p) $\binom{4}{3}\binom{12}{2}$ (4) (4) (choose 3 aces, choose 2 other kinds, pick 4th and 5th cards).

28. $\binom{30}{5}^4 \binom{31}{5}^7 \binom{28}{5}$.

29. There are $C(26,6)$ combinations with all 6 letters distinct; $C(26,1)$ $C(25,4)$ combinations with 1 pair of letters and 4 other letters, etc. Total: $C(26,6) + C(26,1) C(25,4) + C(26,2) C(24,2) + C(26,3)$.

30. (a) $C(30,10)2^{20}$.

(b)

Number of Books C Has	Number of Books B Has	Number of Books A Has
2	4	24
2	5	23
2	6	22
2	7	21
3	6	21

$$C(30,2) C(28,4) + C(30,2) C(28,5) + C(30,2) C(28,6) +$$
$$C(30,2) C(28,7) + C(30,3) C(27,6).$$

2.4 ENUMERATING COMBINATIONS AND PERMUTATIONS WITH REPETITIONS

Now let us turn our attention to counting permutations and combinations with unlimited repetitions. Let $U(n,r)$ denote the number of r-permutations of n objects with unlimited repetitions and let $V(n,r)$ denote the number of r-combinations of n objects with unlimited repetitions. That is, if $a_1,a_2, \ldots ,a_n$ are the n objects, we are counting r-combinations and r-permutations of $\{\infty \cdot a_1, \infty \cdot a_2, \ldots , \infty \cdot a_n\}$.

We have already used $U(n,r)$ (without so designating) in examples 2.1.5, 2.1.6, 2.1.10, 2.1.11, and 2.2.5.

Theorem 2.4.1   (Enumerating r-permutations with unlimited repetitions).

$$U(n,r) = n^r$$

Proof. Each of the r positions can be filled in n ways and so by the product rule, $U(n,r) = n^r$.

Example 2.4.1. There are 25 true or false questions on an examination. How many different ways can a student do the examination if he or she can also choose to leave the answer blank?

Answer: 3^{25}.

Example 2.4.2. The results of 50 football games (win, lose or, tie) are to be predicted. How many different forecasts can contain exactly 28 correct results?

Answer. Choose 28 correct results $C(50,28)$ ways. Each of the remaining 22 games has 2 wrong forecasts. Thus, there are $C(50,28) \cdot 2^{22}$ forecasts with exactly 28 correct predictions.

Example 2.4.3. A telegraph can transmit two different signals: a dot and a dash. What length of these symbols is needed to encode the 26 letters of the English alphabet and the ten digits $0,1,\ldots,9$?

Answer. Since there are two choices for each character, the number of different sequences of length k is 2^k. The number of nontrivial sequences of length n or less is $2 + 2^2 + 2^3 + \cdots + 2^n = 2^{n+1} - 2$. If $n = 4$ this total is 30, which is enough to encode the letters of the English alphabet, but not enough to also encode the digits. To encode the digits we need to allow sequences of length up to 5 for then there are possibly $2^{5+1} - 2 = 62$ total sequences. (This is why in Morse code all letters are transmitted by sequences of four or fewer characters while all digits are transmitted by sequences of length 5.)

Of course, sequences of dots and dashes are in one-to-one correspondence with sequences of 0's and 1's, the so-called *binary* sequences. Thus, we conclude there are 2^k binary sequences of length k, or, in other words, there are 2^k k-digit *binary numbers*. Moreover, there are $2^{n+1} - 2$ binary sequences of positive length n or less. (There are $2^{n+1} - 1$ if we include the sequence of length 0.)

Example 2.4.4. How many 10-digit binary numbers are there with exactly six 1's?

Answer. The key to this problem is that we can specify a binary number by choosing the subset of 6 positions where the 1's go (or the subset of 4 positions for the 0's). Thus, there are $C(10,6) = C(10,4) = 210$ such binary numbers.

We might recall that the formula for $C(n,r)$ was obtained by dividing $P(n,r)$ by $r!$ since each r-combination without repetitions gave rise to $r!$ permutations. The formula for $V(n,r)$ (when unlimited repetitions is allowed) is more difficult to obtain—we cannot simply divide the permutation result for unlimited repetition, $U(n,r) = n^r$, by an appropriate factor since different combinations with repetition will not in general give rise to the same number of permutations. For example, the 3-combination aab gives rise to 3 different permutations while the 3-combination abc gives rise to 6 permutations.

To give what we believe is an understandable explanation of the formula for enumerating combinations with unlimited repetitions, we will reformulate the problem in several different ways.

Let the distinct objects be $a_1, a_2, \ldots, a_n$ so that selections are made from $\{\infty \cdot a_1, \infty \cdot a_2, \ldots, \infty \cdot a_n\}$. Any r-combination will be of the form $\{x_1 \cdot a_1, x_2 \cdot a_2, \ldots, x_n \cdot a_n\}$ where $x_1, \ldots, x_n$ are the repetition numbers, each x_i is nonnegative, and $x_1 + x_2 + \cdots + x_n = r$. Conversely, any sequence of nonnegative integers $x_1, x_2, \ldots, x_n$ where $x_1 + x_2 + \cdots + x_n = r$ corresponds to an r-combination $\{x_1 \cdot a_1, x_2 \cdot a_2, \ldots, x_n \cdot a_n\}$.

First observation: *The number of r-combinations of $\{\infty \cdot a_1, \infty \cdot a_2, \ldots, \infty \cdot a_n\}$ equals the number of solutions of $x_1 + x_2 + \cdots + x_n = r$ in nonnegative integers.*

We believe the next formulation makes it easier to conceptualize the problem.

Second observation.   *The number of nonnegative integral solutions of $x_1 + x_2 + \cdots + x_n = r$ is equal to the number of ways of placing r indistinguishable balls in n numbered boxes.* We see this by just interpreting that the kth box contains x_k balls.

Third observation: *The number of ways of placing r indistinguishable balls in n numbered boxes is equal to the number of binary numbers with $(n - 1)$ 1's and r 0's.* We see this as follows: If there are x_1 balls in box number 1, x_2 balls in box number 2, $\ldots$, x_n balls in box number n, then in a corresponding binary number let there be x_1 0's to the left of the first 1, x_2 0's between the first and second 1, x_3 0's between the second and third, $\ldots$, and finally x_n 0's to the right of the last 1. (Two consecutive 1's mean that there were no balls in that box.) Conversely, to any such binary number with $(n - 1)$ 1's and r 0's we associate a distribution of r balls into n boxes by reversing the above process.

Perhaps an example will be instructive.

Suppose $r = 7$ and $n = 10$ in the above, that is, we are interested in 7-combinations of $\{\infty \cdot a_1, \infty \cdot a_2, \ldots, \infty \cdot a_{10}\}$. To the 7-combination $a_1a_1a_1a_4a_4a_8a_8$ we associate the solution $(3,0,0,2,0,0,0,2,0,0)$ of $x_1 + x_2 + \ldots + x_{10} = 7$. Then to the solution $(3,0,0,2,0,0,0,2,0,0)$ we associate the distribution of 3 balls in box 1, 0 balls in boxes 2, 3, 5, 6, 7, 9, and 10, 2 balls in box 4, and 2 balls in box 8. Then to this distribution of balls associate the binary number 0001110011110011.

We reverse the process in one more example for further clarification. The binary number 1100011010111001 signifies that there are no balls in boxes 1, 2, 4, 7, 8, or 10, 3 balls in box 3, 2 balls in box 9, and 1 ball each in boxes 5 and 6. To this distribution of balls is associated the solution $(0,0,3,0,1,1,0,0,2,0)$ of $x_1 + x_2 + \ldots + x_{10} = 7$. Then to this solution is associated the 7-combination $\{3 \cdot a_3, 1 \cdot a_5, 1 \cdot a_6, 2 \cdot a_9\}$.

Fourth observation. *The number of binary numbers with $n - 1$ 1's and r 0's is $C(n - 1 + r, r)$.* For just as in Example 2.4.4, we have $n - 1 + r$ positions and we need only choose which r positions will be occupied by a 0, and then the remaining $n - 1$ positions are filled by 1's.

We summarize:

Theorem 2.4.2.   (Enumerating r-combinations with unlimited repetitions).

$V(n,r)$ = the number of r-combinations of n distinct objects with unlimited repetitions
   = the number of nonnegative integral solutions to $x_1 + x_2 + \ldots + x_n = r$
   = the number of ways of distributing r similar balls into n numbered boxes
   = the number of binary numbers with $n - 1$ one's and r zeros.
   = $C(n - 1 + r, r) = C(n - 1 + r, n - 1)$
   = $(n + r - 1)!/[r!(n - 1)!]$.

Remark.   Of course, the number of r-combinations of $\{\infty \cdot a_1, \infty \cdot a_2, \ldots, \infty \cdot a_n\}$ is the same as the number of r-combinations of $\{r \cdot a_1, r \cdot a_2, \ldots, r \cdot a_n\}$.

The following examples will clarify the conclusions of Theorem 2.4.2.

Example 2.4.5.   (a) The number of 4-combinations of $\{\infty \cdot a_1, \infty \cdot a_2, \infty \cdot a_3, \infty \cdot a_4, \infty \cdot a_5\}$ is $C(5 - 1 + 4,4) = C(8,4) = 70$.

(b) The number of 3-combinations of 5 objects with unlimited repetitions is $C(5 - 1 + 3,3) = C(7,3) = 35$.

(c) The number of nonnegative integral solutions to $x_1 + x_2 + x_3 + x_4 + x_5 = 50$ is $C(5 - 1 + 50,50) = C(54,50) = 54!/4!50! = 27 \cdot 53 \cdot 17 \cdot 13 = 316{,}251$.

(d) The number of ways of placing 10 similar balls in 6 numbered boxes is $C(6 - 1 + 10,10) = C(15,10) = 3{,}003$.

(e) The number of binary numbers with ten 1's and five 0's is $C(10 + 5,5) = C(15,5) = 3{,}003$.

Other problems though couched in different settings can be solved by Theorem 2.4.2.

Example 2.4.6.   (a) How many different outcomes are possible by tossing 10 similar coins?

Answer.   This is the same as placing 10 similar balls into two boxes labeled "heads" and "tails." $C(2 - 1 + 10,10) = C(11,10) = 11$.

(b) How many different outcomes are possible from tossing 10 similar dice?

Answer.   This is the same as placing 10 similar balls into 6 numbered boxes. Therefore there are $C(15,10) = 3{,}003$ possibilities.

(c) How many ways can 20 similar books be placed on 5 different shelves?

Answer. $C(5 - 1 + 20,20) = C(24,20)$.

(d) Out of a large supply of pennies, nickels, dimes, and quarters, in how many ways can 10 coins be selected?

Answer. $C(4 - 1 + 10,10) = C(13,10)$ since this is equivalent to placing 10 similar balls in 4 numbered boxes labeled "pennies," "nickels," "dimes," and "quarters."

(e) How many ways are there to fill a box with a dozen doughnuts chosen from 8 different varieties of doughnuts?

Answer.   First, we observe that relative positions in the box are immaterial so that order does not count. Therefore, this is a combination problem. Secondly, a box might consist of a dozen of one variety of doughnut, so that we see that this problem allows unlimited repetitions. The answer then is $C(8 - 1 + 12,12) = C(19,12)$.

Now let us consider a slight variation of the above examples.

Example 2.4.7. (a) Enumerate the number of ways of placing 20 indistinguishable balls into 5 boxes where each box is nonempty.

We analyze this problem as follows: First, place one ball in each of the 5 boxes. Then we must count the number of ways of distributing the 15 remaining balls into 5 boxes with unlimited repetitions. By Theorem 2.4.2, we can do this in $C(5 - 1 + 15,15) = C(19,15)$ ways.

Of course, we can also model this problem as a solution-of-an-equation problem. If x_i represents the number of balls in the ith box, then we are asked to enumerate the number of integral solutions to $x_1 + x_2 + x_3 + x_4 + x_5 = 20$ where each $x_i > 0$. After distributing 1 ball into each of the 5 boxes, we then are to enumerate the number of integral solutions of $y_1 + y_2 + y_3 + y_4 + y_5 = 15$ where each $y_i \geq 0$.

Likewise we can solve the following:

(b) How many integral solutions are there to $x_1 + x_2 + x_3 + x_4 + x_5 = 20$ where each $x_i \geq 2$?

Here we can model this problem as a distribution-of-similar-balls problem whereby first we place 2 balls in each of 5 boxes and then enumerate the number of ways of placing the remaining 10 balls in 5 boxes with unlimited repetitions. In other words the number of integral solutions to $x_1 + x_2 + x_3 + x_4 + x_5 = 20$ where each $x_i \geq 2$ is the same as the number of integral solutions of $y_1 + y_2 + y_3 + y_4 + y_5 = 10$ where each $y_i \geq 0$. We know that there are $C(5 - 1 + 10,10) = C(14,10)$ such solutions.

(c) How many integral solutions are there to $x_1 + x_2 + x_3 + x_4 + x_5 = 20$ where $x_1 \geq 3$, $x_2 \geq 2$, $x_3 \geq 4$, $x_4 \geq 6$, and $x_5 \geq 0$?

First, distribute 3 balls in box 1, 2 balls in box 2, 4 balls in box 3, 6 balls in box 4, and 0 balls in box 5. That leaves 5 balls to be distributed into 5 boxes with unlimited repetition. That is, we now wish to count the number of integral solutions of $y_1 + y_2 + y_3 + y_4 + y_5 = 5$ where each $y_i \geq 0$. There are $C(5 - 1 + 5,5) = C(9,5)$ such solutions.

(d) How many integral solutions are there to $x_1 + x_2 + x_3 + x_4 + x_5 = 20$ where $x_1 \geq -3$, $x_2 \geq 0$, $x_3 \geq 4$, $x_4 \geq 2$, $x_5 \geq 2$?

Here we interpret placing -3 balls in box 1 as actually increasing the total number of balls from 20 to 23. Then placing 4 in box 3, and 2 in each of boxes 4 and 5, leaves only 15 balls. Thus, we have $C(5 - 1 + 15,15) = C(19,15)$ solutions of $y_1 + y_2 + y_3 + y_4 + y_5 = 15$ where each $y_i \geq 0$.

These ideas can be incorporated to prove the following theorem.

Theorem 2.4.3.   The number of integral solutions of $x_1 + x_2 + \ldots + x_n = r$ where each $x_i > 0$

> $=$ the number of ways of distributing r similar balls into n numbered boxes with at least one ball in each box
> $= C(n - 1 + (r - n), r - n) = C(r - 1, r - n)$
> $= C(r - 1, n - 1)$.

Likewise, suppose that $r_1, r_2, \ldots, r_n$ are integers. Then the number of integral solutions of $x_1 + x_2 + \ldots + x_n = r$ where $x_1 \geq r_1, x_2 \geq r_2, \ldots$, and $x_n \geq r_n$

> $=$ the number of ways of distributing r similar balls into n numbered boxes where there are at least r_1 balls in the first box, at least r_2 balls in the second box, $\ldots$, and at least r_n balls in the nth box
> $= C(n - 1 + r - r_1 - r_2 - \ldots - r_n, r - r_1 - r_2 - \ldots - r_n)$
> $= C(n - 1 + r - r_1 - r_2 - \ldots - r_n, n - 1)$.

Finally consider the following example.

Example 2.4.8.   Enumerate the number of nonnegative integral solutions to the inequality $x_1 + x_2 + x_3 + x_4 + x_5 \leq 19$.

Of course, this asks for the number of nonnegative integral solutions to 20 equations $x_1 + x_2 + x_3 + x_4 + x_5 = k$ where k can be any integer from 0 to 19. By repeated application of Theorem 2.4.2 we see that there are $C(5 - 1 + 0,4) + C(5 - 1 + 1,4) + \ldots + C(5 - 1 + 19,4)$ such solutions.

From this point of view we have 19 similar balls and we are counting the number of ways of distributing either 0 or 1, or 2, $\ldots$, or 19 of these balls into the 5 boxes.

But there is also an alternate way to approach the problem. If k is some integer between 0 and 19, then for every distribution of k balls into 5 boxes, one could distribute the remaining $19 - k$ balls into a *sixth* box. Hence the number of nonnegative integral solutions of $x_1 + x_2 + x_3 + x_4 + x_5 \leq 19$ is the same as the number of nonnegative integral solutions of $y_1 + y_2 + y_3 + y_4 + y_5 + y_6 = 19$ (note we have one more variable, y_6). By

Theorem 2.4.2, there are $C(6 - 1 + 19,5) = C(24,5) = C(24,19)$ such solutions. Hence from what we have already seen, we conclude that

$$C(24,5) = \sum_{k=0}^{19} C(4 + k,k) = \sum_{k=0}^{19} C(k + 4,4).$$

Exercises for Section 2.4

1. A quarterback of a football team has a repertoire of 20 plays and runs 60 plays in the course of a game. The coach is interested in the frequency distribution of the play-calling showing how many times each of the various plays were called. How many such frequency distributions are there?

2. In how many ways can 5 similar books be placed on 3 different shelves?

3. How many outcomes are obtained from rolling n indistinguishable dice?

4. How many dominos are there in a set which are numbered from
   (a) double blank to double six?
   (b) double blank to double nine?

5. In how many ways can 5 glasses be filled with 10 different kinds of Kool-aid if no mixing is allowed and the glasses are
   (a) indistinguishable?
   (b) distinguishable?

6. How many solutions are there to the equation $x_1 + x_2 + x_3 + x_4 + x_5 = 50$ in nonnegative integers?

7. Find the number of distinct triples (x_1, x_2, x_3) of nonnegative integers satisfying $x_1 + x_2 + x_3 < 15$.

8. How many integers between 1 and 1,000 inclusive have a sum of digits
   (a) equal to 7?
   (b) less than 7?

9. Find all $C(5,3)$ integral solutions of $y_1 + y_2 + y_3 + y_4 = 2$ where each $y_i \geq 0$. Then list all integral solutions to $x_1 + x_2 + x_3 + x_4 = 22$ where each $x_i \geq 5$.

10. Find all integral solutions to $y_1 + y_2 + y_3 = 3$ where each $y_i \geq 0$. Then list all integral solutions to $x_1 + x_2 + x_3 = 8$ where $x_1 \geq 3, x_2 \geq -2$, and $x_3 \geq 4$.

11. Find the number of nonnegative integral solutions to $x_1 + x_2 + x_3 + x_4 + x_5 = 10$.

12. Find the number of integral solutions to $x_1 + x_2 + x_3 + x_4 = 50$, where $x_1 \geq -4$, $x_2 \geq 7$, $x_3 \geq -14$, $x_4 \geq 10$.

13. Find the number of distinct triples (x_1, x_2, x_3) of nonnegative integers satisfying the inequality $x_1 + x_2 + x_3 < 6$.

14. How many integers between 1 and 1,000 inclusive have a sum of digits
    (a) equal to 10?
    (b) less than 10?

15. For what values of r is it true that $x_1 + x_2 + x_3 = r$ has no integral solutions where $2 \leq x_1$, $5 \leq x_2$, and $4 \leq x_3$?

16. For what values of r is it true that $x_1 + x_2 + x_3 + x_4 = r$ has no integral solutions with $7 \leq x_1$, $8 \leq x_2$, $9 \leq x_3$, $10 \leq x_4$?

17. A bag of coins contains 10 nickels, 8 dimes, and 7 quarters. Assuming that the coins of any one denomination are indistinguishable, in how many ways can 6 coins be selected from the bag?

18. How many ways are there to make a selection of coins from $1.00 worth of identical pennies, $1.00 worth of identical nickels, and $1.00 worth of identical dimes if a total of
    (a) 10 coins are selected?
    (b) 20 coins are selected?
    (c) 25 coins are selected?

19. How many ways are there to arrange a deck of 52 cards with no adjacent hearts?

20. How many ways are there to place 20 identical balls into 6 different boxes in which exactly 2 boxes are empty?

21. (a) How many ways are there to distribute 20 chocolate doughnuts, 12 cherry-filled doughnuts, and 24 cream-filled doughnuts to 4 different students?

    (b) How many ways can the different kinds of doughnuts be distributed to the students if each student receives at least 2 of each kind of doughnut?

22. How many integral solutions are there of $x_1 + x_2 + x_3 + x_4 + x_5 = 30$ where for each i
    (a) $x_i \geq 0$;
    (b) $x_i \geq 1$;
    (c) $x_1 \geq 2$, $x_2 \geq 3$, $x_3 \geq 4$, $x_4 \geq 2$, $x_5 \geq 0$;
    (d) $x_i > i$.

23. Six distinct symbols are transmitted through a communication channel. A total of 12 blanks are to be inserted between the symbols with at least 2 blanks between every pair of symbols. In how many ways can the symbols and blanks be arranged?

24. A teacher wishes to give an examination with 10 questions. In how many ways can the test be given a total of 30 points if each question is to be worth 2 or more points?

25. In how many ways can a lady wear 5 rings on 4 fingers of her hand?

26. Let m and n be positive integers where $m \le n$. In how many ways can n be written as a sum of m positive integers where order is taken into account? (Note that $4 = 1 + 1 + 2 = 1 + 2 + 1 = 2 + 1 + 1$ so that 4 can be written as a sum of 3 positive integers in 3 ways.)

27. In how many ways can we partition 12 similar coins into 5 numbered nonempty batches?

28. (a) How many ways can 6 (identical) apples, 1 pear, 1 plum, 1 pomegranate, 1 persimmon, 1 peach, and 1 quince be distributed among 3 people?
    (b) What is the answer to (a) if each person receives exactly 4 fruits?

29. In how many ways can we place 4 red balls, 4 white balls, and 4 blue balls in 6 numbered boxes?

30. How many ways can we distribute 12 white balls and 2 black balls
    (a) into 9 numbered boxes?
    (b) into 9 numbered boxes where each box contains at least one white ball?

31. How many integers between 1 and 10,000,000 have the sum of digits equal to 18?

32. How many ways can 5 (identical) apples and 5 (identical) oranges be distributed among 5 people such that each person receives exactly 2 fruits?

Selected Answers for Section 2.4

1. $C(60 + 20 - 1,60)$ (60 balls into 20 numbered boxes).

3. $C(n + 6 - 1,n)$.

4. (a) $C(8,2)$.
   (b) $C(11,2)$.

5. (a) $C(10 - 1 + 5,5)$.
   (b) 10^5.

9. For a solution (y_1, y_2, y_3, y_4) where $y_1 + y_2 + y_3 + y_4 = 2$, let $x_i - 5 = y_i$ or $x_i = y_i + 5$. Then $x_1 + x_2 + x_3 + x_4 = 22$.

10. For a solution $y_1 + y_2 + y_3 = 3$, let $y_1 = x_1 - 3$, $y_2 = x_2 + 2$, $y_3 = x_3 - 4$, then $x_1 + x_2 + x_3 = 8$.

11. $C(10 + 5 - 1, 5 - 1) = C(14,4)$.

12. $C(50 + 4 - 7 + 14 - 10 + 4 - 1, 4 - 1) = C(54,3)$.

13. Count the number of solutions for $x_1 + x_2 + x_3 = n$ where $n = 0, 1, 2, 3, 4, 5$ and sum.

14. Let x_1 = units digit, x_2 = tens digit, and x_3 = hundreds digit.
    (a) Count the number of nonnegative integral solutions to $x_1 + x_2 + x_3 = 10$, and exclude the 3 cases where $x_i = 10$.
    (b) Count the number of nonnegative solutions for $x_1 + x_2 + x_3 = n$ where $n = 0, 1, 2, 3, 4, 5, 6, 7, 8, 9$ and sum.

19. Let nonhearts be dividers to determine 40 cells. Choose 13 of these cells in which to place a heart. Then arrange the hearts and the nonhearts $C(40,13)$ 13!39! ways, or use hearts as dividers to form 14 cells of which the first and last may be empty but others are nonempty.

20. Choose the 2 boxes to be empty; place 1 ball in each of 4 remaining boxes. Then distribute the remainder. $\binom{6}{2} \binom{20 - 4 + 4 - 1}{16}$.

21. (a) $\binom{20 + 4 - 1}{20} \binom{12 + 4 - 1}{12} \binom{24 + 4 - 1}{24}$.

    (b) $\binom{12 + 4 - 1}{12} \binom{4 + 4 - 1}{4} \binom{16 + 4 - 1}{16}$.

22. (a) $\binom{34}{30}$.

    (b) $\binom{29}{25}$.

    (c) $\binom{23}{19}$.

    (d) $\binom{14}{10}$.

23. Fill 5 boxes with 2 or more blanks in $C(5 - 1 + 2, 2) = \binom{6}{2}$ ways. Then arrange the 6 symbols 6! ways. Total 6! $\binom{6}{2}$.

24. $\binom{19}{10}$

2.5 ENUMERATING PERMUTATIONS WITH CONSTRAINED REPETITIONS

There are, of course, intermediate cases between selections with no repetitions and selections with unlimited repetition of the objects.

Suppose that we are given a particular selection of r objects where there are some repetitions. What we desire is a formula for the number of permutations on this given selection of r objects.

Perhaps an example is in order. Recall that in Example 2.2.3 we listed all 3-combinations of $\{3 \cdot a, 2 \cdot b, 2 \cdot c, 1 \cdot d\}$. In the following table we will list all permutations of each of these 3-combinations:

3-combinations of $\{3 \cdot a, 2 \cdot b, 2 \cdot c, 1 \cdot d\}$	The number of 3-permutations
aaa	1
aab	3 (*aab, aba, baa*)
aac	3
aad	3
bba	3
bbd	3
bbc	3
cca	3
ccb	3
ccd	3
abc	6
abd	6
acd	6
bcd	6

Total number	14	52

(See Example 2.2.7 for a list of the last 24 permutations)

We note that in the above table there corresponded 3 permutations to each 3-combination where one object was repeated twice and another was repeated once. While, on the other hand, there were 6 permutations corresponding to each 3-combination where 3 distinct objects were selected. We ask: is there a rule here that holds in general? The answer is yes, and we begin to explain why by considering the following example.

Example 2.5.1   How many different arrangements are there of the letters $a, a, a, b,$ and c?

This is asking for the number of 5-permutations of the particular 5-combination $\{3 \cdot a, 1 \cdot b, 1 \cdot c\}$. Let x be the number of such permutations.

Now consider a particular permutation, for example, *aabca*. If the letters a were distinct, that is, if they were written as a_1, a_2, a_3, then this permutation would give rise to 3! different permutations; namely;

$$a_1 a_2 bc a_3 \qquad a_2 a_1 bc a_3$$
$$a_2 a_3 bc a_1 \qquad a_3 a_2 bc a_1$$
$$a_1 a_3 bc a_2 \qquad a_3 a_1 bc a_2$$

These correspond to the different ways of arranging the 3 letters a_1, a_2, a_3. Likewise each permutation of $\{3 \cdot a, 1 \cdot b, 1 \cdot c\}$ will give rise to 3! permutations where the letters a are replaced by distinct letters. Thus there are $3!x$ permutations of the letters $\{a_1, a_2, a_3, b, c\}$. But there are 5! permutations of the letters $\{a_1, a_2, a_3, b, c\}$. Thus $3!x = 5!$ and $x = 5!/3! = 20$. Hence there are 20 permutations of $\{3 \cdot a, b, c\}$.

One more example will be enough to see the general pattern.

Example 2.5.2   How many 10-permutations are there of $\{3 \cdot a, 4 \cdot b, 2 \cdot c, 1 \cdot d\}$?

Let x be the number of such permutations. Reasoning as in Example 2.5.1, we see that there are $3!x$ permutations if we replace the a's by a_1, a_2, and a_3. Likewise there are $(4!)\,(3!x)$ permutations if we also replace the b's by b_1, b_2, b_3, and b_4 corresponding to the number of ways of arranging b_1, b_2, b_3, b_4. Continuing we see that there are $(2!)\,(4!)\,(3!)x$ permutations if we also replace the c's by c_1 and c_2. But then we know that there are 10! permutations of $\{a_1, a_2, a_3, b_1, b_2, b_3, b_4, c_1, c_2, d\}$. Hence,

$$(2!)\,(4!)\,(3!)x = 10! \text{ or } x = \frac{10!}{2!3!4!} = 12{,}600.$$

Let us give an alternate solution. Note there are 10 letters with 3 alike (the a's), 4 alike (the b's), 2 alike (the c's), and 1 alike (the letter d). Thus, we have 10 positions to be filled with these letters to give the various permutations. From the 10 positions first choose the 3 positions for the a's; then from the remaining 7 positions, choose the 4 positions for the b's; from the remaining 3 positions, choose the 2 positions for the c's; and finally, choose the last position for the letter d. This can be done in

$$C(10,3)\,C(7,4)\,C(3,2)\,C(1,1) = \frac{10!}{3!7!}\frac{7!}{4!3!}\frac{3!}{2!1!}\frac{1!}{1!0!} = \frac{10!}{3!4!2!1!}$$

(Note the cancellation of certain factorials.)

Now let us introduce the following notation. Suppose that $q_1, q_2, \ldots, q_t$ are nonnegative integers such that $n = q_1 + q_2 + \ldots + q_t$. Suppose, moreover, that $a_1, \ldots, a_t$ are t distinct objects. Let $P(n; q_1, q_2, \ldots, q_t)$

denote the number of n-permutations of the n-combination $\{q_1 \cdot a_1, q_2 \cdot a_2, \ldots, q_t \cdot a_t\}$.

Armed with the above two examples and this notation we prove the following theorem.

Theorem 2.5.1   (Enumerating n-permutations with constrained repetitions).

$$P(n; q_1, \ldots, q_t) = \frac{n!}{q_1! q_2! \ldots q_t!}$$
$$= C(n, q_1)\, C(n - q_1, q_2)\, C(n - q_1 - q_2, q_3)$$
$$\ldots C(n - q_1 - q_2 \ldots - q_{t-1}, q_t)$$

Proof.   Let $x = P(n; q_1, q_2, \ldots, q_t)$.

If the q_1 a_1's were all different there would be $(q_1!)x$ permutations since each old permutation would give rise to $q_1!$ new permutations corresponding to the number of ways of arranging the q_1 distinct objects in a row. If the q_2 a_2's were all replaced by distinct objects, then by similar reasoning there would be $(q_2!)\,(q_1!)x$ permutations. If we repeat this procedure until all the objects are distinct we will have $(q_t!) \ldots (q_2!)\,(q_1!)x$ permutations.

However, we know that there are $n!$ permutations of n distinct objects. Equating these two quantities and solving for x gives the first equality of the theorem.

The second equality is obtained as follows. First choose the q_1 positions for the a_1's; then from the remaining $n - q_1$ positions, choose q_2 positions for the a_2's and so on. Note that at the last we will have left $n - q_1 - q_2 \ldots - q_{t-1} = q_t$ positions to fill with the q_t a_t's, so $C(n - q_1 - q_2 \ldots -q_{t-1}, q_t) = C(q_t, q_t)$.

The last equality of the theorem follows because both numbers represent the same number of permutations, or we can obtain it by canceling factorials as in Example 2.5.2. $\square$

Example 2.5.3.   The number of arrangements of letters in the word T A L L A H A S S E E is

$$P(11; 3, 2, 2, 2, 1, 1) = \frac{11!}{3! 2! 2! 2! 1! 1!}$$

since this equals the number of permutations of $\{3 \cdot A, 2 \cdot E, 2 \cdot L, 2 \cdot S, 1 \cdot H, 1 \cdot T\}$. The number of arrangements of these letters that begin with T and end with E is $9!/3!1!2!2!1!$.

Example 2.5.4. In how many ways can 23 different books be given to 5 students so that 2 of the students will have 4 books each and the other 3 will have 5 books each?

Answer. Choose the 2 students to receive 4 books each in $C(5,2)$ ways. Then to each such choice the 23 books can be distributed in $P(23;4,4,5,5,5) = 23!/4!4!5!5!5!$ ways. Thus there are $C(5,2)$ $(23!/4!^2 5!^3)$ total distributions.

Example 2.5.5. Find the number of 5-combinations and the number of 5-permutations of $\{5 \cdot a, 3 \cdot b, 2 \cdot c, 3 \cdot d, 2 \cdot e, 1 \cdot f, 4 \cdot g\}$. The different ways of selecting 5 letters may be classified as in the following table. There may be several combinations of each type, but each one will give rise to the same number of permutations so we can compute that number by applying Theorem 2.5.1.

Types of Selection	Number of 5-Combinations	Number of Arrangements From Each Selection	Number of 5-Permutations
(1) All 5 alike	1	$\dfrac{5!}{5!} = 1$	1
(2) 4 alike and 1 different	12	$\dfrac{5!}{4!1!} = 5$	$12 \cdot 5 = 60$
(3) 3 alike, 2 others alike	20	$\dfrac{5!}{3!2!} = 10$	$20 \cdot 10 = 200$
(4) 3 alike, 2 others different	$4C(6,2) = 60$	$\dfrac{5!}{3!1!1!} = 20$	$60 \cdot 20 = 1{,}200$
(5) 2 alike, 2 others alike, and 1 different	$C(6,2)5 = 75$	$\dfrac{5!}{2!2!1!} = 30$	$75 \cdot 30 = 2{,}250$
(6) 2 alike and 3 different	$6C(6,3) = 120$	$\dfrac{5!}{2!1!1!1!} = 60$	$120 \cdot 60 = 7{,}200$
(7) All 5 different	$C(7,5) = 21$	$\dfrac{5!}{1!1!1!1!1!} = 120$	$21 \cdot 120 = 2{,}520$
Total	309		13,431

The table should be self-explanatory except possibly how we arrived at the numbers in column two. Let us explain.

Selection (1) can be made in only 1 way [namely by the selection $\{5 \cdot a\}$].

Selection (2) can be made in 12 ways; choose the 4 alike in 2 ways [either $\{4 \cdot a\}$ or $\{4 \cdot g\}$], and then choose the 1 different letter in 6 ways.

Selection (3) can be made in 20 ways; choose the 3 alike from the a's, b's, d's, or g's, and once one of these is chosen there are 5 choices for the 2 alike (since the c's and e's can now be chosen).

Selection (4) can be made in $4C(6,2)$ ways since there are 4 ways to choose 3 alike letters [either $\{3 \cdot a\}$, $\{3 \cdot b\}$, $\{3 \cdot d\}$, or $\{3 \cdot g\}$], and once these have been chosen there are 6 different letters from which 2 must be chosen.

Make selection (5) as follows. First, note that there are only 6 letters with repetition numbers ≥ 2 (only f has repetition number 1). Choose the 2 different letters that are each to have repetition number 2 in $C(6,2)$ ways. Now after choosing these 2 letters there are 5 letters that remain (f can now be included) from which to choose the 1 different letter. This can be done in 5 ways. Thus, selection (5) can be made in $C(6,2)5 = 75$ ways.

Selection (6) is explained in much the same way as was selection 5. Choose the letter that is to be repeated twice in 6 ways. There are 6 distinct letters remaining from which 3 are to be chosen. Thus, selection (6) can be made in $6C(6,3) = 120$ ways.

Selection (7) can be made in $C(7,5) = 21$ ways because there are 7 distinct letters from which 5 are to be chosen.

The total number of 5-combinations is $1 + 12 + 20 + 60 + 75 + 120 + 21 = 309$, and the total number of 5-permutations is $1 + 60 + 200 + 1{,}200 + 2{,}250 + 7{,}200 + 2{,}520 = 13{,}431$.

Ordered and Unordered Partitions

The very essence of combinatorial mathematics is reformulation of problems. Let us, therefore, interpret Theorem 2.5.1 from a different perspective. The following discussion will be suggestive.

A carpet manufacturer has 1,000 rugs in his warehouse for sale. Upon investigation, he finds that they differ in quality. He decides to classify them in three quality grades, 1, 2, and 3. Each rug is inspected, and a tag is attached bearing one of the numbers 1, 2, or 3.

Now let us give a mathematical description of what has been accomplished.

Let S denote the set of 1,000 rugs. By attaching a tag bearing one of the numbers 1, 2, or 3 to each of the rugs, the manufacturer has defined a function f whose domain is S and whose range is $\{1,2,3\}$. This function defines the following three subsets of S:

$$A_1 = \{a \in S \mid f(a) = 1\},$$
$$A_2 = \{a \in S \mid f(a) = 2\}, \text{ and}$$
$$A_3 = \{a \in S \mid f(a) = 3\}.$$

The sets A_1, A_2, A_3 have the following properties:

(a) $A_1 \cup A_2 \cup A_3 = S$; their union equals S.
(b) $A_i \cap A_j = \varnothing$ for $i \neq j$; they are disjoint.

That is to say, the sets A_1, A_2, A_3 form an (unordered) *partition of S*. But, more than this, there can be a definite ordering on the sets themselves, for the manufacturer may want to charge more for the higher quality rugs. Thus, the ordered triple of sets (A_1, A_2, A_3) form a (3-part) *ordered partition of S*. Of course, the ordered triple (A_2, A_1, A_3) gives rise to a different ordered partition of S (for example the manufacturer may want to charge more for rugs of quality grade 2), even though this would constitute the same unordered partition of S.

Definition 2.5.1. Let S be a set with n distinct elements, and let t be a positive integer. A *t-part partition* of the set S is a set $\{A_1, \ldots, A_t\}$ of t subsets of S, $A_1, \ldots, A_t$ such that

$$S = A_1 \cup A_2 \cup \cdots \cup A_t$$
$$A_i \cap A_j = \varnothing \quad \text{for} \quad i \neq j.$$

The subsets A_i are called *parts or cells* of S. Frequently, we will suppress the words "*t-part*" and occasionally we will use the term unordered partition to emphasize the distinction between this definition and the following one.

An **ordered partition** of S is first of all a partition of S but, secondly, there is a specified order on the subsets. Thus, an ordered t-tuple of sets $(A_1, A_2, \ldots, A_t)$ is a t-part ordered partition of S if the sets $A_1, \ldots, A_t$ form a partition of S.

Example 2.5.6. $A_1 = \{a,b\}$, $A_2 = \{c\}$, $A_3 = \{d\}$ form a 3-part partition of $S = \{a,b,c,d\}$ whereas $(A_1,A_2,A_3),(A_2,A_1,A_3),(A_2,A_3,A_1),(A_3,A_2,A_1),$ $(A_3,A_1,A_2),(A_1,A_3,A_2)$ form 6 different ordered partitions of S using these same 3 subsets.

Of course, there are other 3-part partitions of S, for example, $B_1 = \{a,c\}$, $B_2 = \{b\}$, and $B_3 = \{d\}$. There is nothing in our definition to exclude the empty set as one of the subsets, so $C_1 = \{a,b,c\}$, $C_2 = \{d\}$, and $C_3 = \varnothing$ is another 3-part partition of S.

We are interested in ordered partitions of certain types. For this reason, we usually specify the numbers of elements of the subsets in the ordered partition. Thus, by an ordered partition of S of type $(q_1, q_2, \ldots, q_t)$, we mean an ordered partition $(A_1, \ldots, A_t)$ of S where $|A_i| = q_i$ for

A_1	A_2		A_{t-1}	A_t
q_1	q_2		q_{t-1}	q_t
elements	elements		elements	elements

Figure 2-4

each i. Since S has n elements, clearly we must have $n = q_1 + q_2 + \cdots + q_t$.

We might depict a partition of S of type $(q_1, \ldots, q_t)$ as illustrated in Figure 2-4.

Example 2.5.7. List all ordered partitions of $S = \{a,b,c,d\}$ of type $(1,1,2)$.

$$(\{a\},\{b\},\{c,d\}) \qquad (\{b\},\{a\},\{c,d\})$$
$$(\{a\},\{c\},\{b,d\}) \qquad (\{c\},\{a\},\{b,d\})$$
$$(\{a\},\{d\},\{b,c\}) \qquad (\{d\},\{a\},\{b,c\})$$
$$(\{b\},\{c\},\{a,d\}) \qquad (\{c\},\{b\},\{a,d\})$$
$$(\{b\},\{d\},\{a,c\}) \qquad (\{d\},\{b\},\{a,c\})$$
$$(\{c\},\{d\},\{a,b\}) \qquad (\{d\},\{c\},\{a,b\}).$$

Of course, we have learned that most often we are interested in "how many" rather than "a list of all."

Theorem 2.5.2. (Enumerating ordered partitions of a set). The number of ordered partitions of a set S of type $(q_1, q_2, \ldots, q_t)$ where $|S| = n$ is

$$P(n;q_1, \ldots ,q_t) = \frac{n!}{q_1!q_2!\ldots q_t!}.$$

Proof. We see this by choosing the q_1 elements to occupy the first subset in $C(n,q_1)$ ways; the q_2 elements for the second subset in $C(n - q_1,q_2)$ ways, etc. Then the number of ordered partitions of type $(q_1,q_2, \ldots ,q_t)$ is $C(n,q_1)C(n - q_1,q_2) \cdots C(n - q_1 - q_2 - \cdots - q_{t-1},q_t)$ and we know that this equals $P(n;q_1, \ldots ,q_t)$. $\square$

Thus in Example 2.5.7 we could use Theorem 2.5.2 to compute the number of ordered partitions of $S = \{a,b,c,d\}$ of type $(1,1,2)$ for here $n =$

4, $q_1 = 1$, $q_2 = 1$, and $q_2 = 2$. Thus, there are $4!/1!1!2! = 12$ such ordered partitions.

Example 2.5.8.   In the game of bridge, four players (usually called North, East, South, and West) seated in a specified order are each dealt a "hand" of 13 cards.

(a) How many ways can the 52 cards be dealt to the four players?

Answer. $52!/13!^4$ (Here order counts.)

(b) In how many ways will one player be dealt all four kings?

Answer.   Choose the player to receive the kings in 4 ways; then partition the remaining cards. There are $4(48!/9!13!^3) = 4C(48,9)$ $C(39,13)\, C(26,13)\, C(13,13)$ ways.

(c) In how many deals will North be dealt 7 hearts and South the other 6 hearts?

Answer.   Choose the 7 hearts for North (and automatically give the other 6 hearts to South) in $C(13,7)$ ways; then partition the remaining cards. $C(13,7)\, 39!/6!7!13!^2$ hands.

(d) In how many ways will North and South have together all four kings?

Number of kings for North	South	Number of hands
0	4	$\dfrac{48!}{9!13!^3}$
1	3	$C(4,1)\,\dfrac{48!}{12!10!13!^2}$
2	2	$C(4,2)\,\dfrac{48!}{11!^2 13!^2}$
3	1	$C(4,1)\,\dfrac{48!}{12!10!13!^2}$
4	0	$\dfrac{48!}{9!13!^3}$

Answer. The total number of deals is

$$2 \cdot \frac{48!}{9!13!^3} + 2C(4,1)\frac{48!}{12!10!13!^2} + C(4,2)\frac{48!}{11!^213!^2}.$$

Determining the number of unordered partitions is a much more complex matter. We will give a formula only in the case that all subsets have the same number of elements.

Theorem 2.5.3. (Enumerating unordered partitions of equal cell size). Let S be a set with n elements where $n = q \cdot t$. Then the number of unordered partitions of S of type $(q,q,\ldots,q)$ is $1/t!\,(n!/(q!)^t)$. Here recall that t equals the number of subsets.

This follows immediately from the fact that each unordered partition of the t subsets gives rise to $t!$ ordered partitions. There are $1/4!\,(52!/13!^4)$ bridge deals disregarding dealing order.

Example 2.5.9. (a) In how many ways can 14 men be partitioned into 6 teams where the first team has 3 members, the second team has 2 members, the third team has 3 members, and the fourth, fifth, and sixth teams each have 2 members?

Answer. This calls for the number of ordered partitions of type $(3,2,3,2,2,2)$; there are

$$P(14;3,2,3,2,2,2) = \frac{14!}{3!2!3!2!2!2!} = \frac{14!}{3!^22!^4}$$

such ways.

(b) In how many ways can 12 of the 14 people be distributed into 3 teams where the first team has 3 members, the second has 5, and the third has 4 members?

Answer. First count the number of ways to choose the 12 people to be placed into teams, then count the number of ordered partitions of type $(3,5,4)$. There are $C(14,12)\,(12!/3!5!4!)$ such ways.

(c) In how many ways can 12 of the 14 people be distributed into 3 teams of 4 each?

Answer. First, count the number of ways to choose the 12 people; then count the number of unordered partitions of type $(4,4,4)$. There are $C(14,12)\,(12!/4!^33!)$ such ways.

(d) In how many ways can 14 people be partitioned into 6 teams when the first and second teams have 3 members each and the third, fourth, fifth, and sixth teams have 2 members each?

Answer. $14!/(3!^2 2!^4)$ (Count the number of ordered partitions.)

(e) In how many ways can 14 people be partitioned into 6 teams where two teams have 3 each and 4 teams have 2 each?

Answer. $14!/(2!4!3!^2 2!^4)$

(We divide by 2!4! because each unordered partition gives rise to 2! arrangements of the two teams with 3 each and 4! arrangements of the 4 teams with 2 each.)

(f) In how many ways can 14 people be distributed into 6 teams where *in some order* 2 teams have 3 each and 4 teams have 2 members each? Let us be clear how this problem differs from (d) and from (e). Problem (d) calls for counting the number of ordered partitions of type (3,3,2,2,2,2), that is, there is a *specified order*. There are of course other types of ordered partitions with these same occupancy numbers, (2,3,2,3,2,2), for example. Indeed, there are $C(6,2) = 6!/2!4!$ different types of orderings of these 6 numbers including four 2's and two 3's. For each selection of a type of partition there are $14!/3!^2 2!^4$ ordered partitions of that type. Hence there are $(6!/2!4!)$ $(14!/3!^2 2!^4)$ such partitions in some order. Note this answer is 6! times the answer in (e). Perhaps one more example of this kind will make the concept clear.

(g) In how many ways can 14 people be partitioned into 7 teams where in some order 2 teams have 3 members each, 3 teams have 2 each, and 2 teams have 1 member each?

Answer. $(7!/2!3!2!)$ $(14!/3!^2 2!^3 1!1!)$

Some Hints

Basically we have focused our attention on these types of problems: permutation, combination, and partition problems.

We have discussed two major subtopics to each of these types of problems: permutations with or without repetitions, combinations with or without repetitions, and ordered or unordered partitions. We have seen in our examples that these problems can be formulated in all sorts of settings. What we need are some clues that will help determine whether the problem is calling for counting r-permutations, r-combinations, or partitions.

First, we suggest looking for key words—the key word in the definition of combination is *selection* while the key word for permutation is *arrangement*. Nevertheless, not all permutation or combination problems use these words; some for example, may use phrases that suggest that *order counts* (a permutation problem) while others make no reference to order (probably a combination problem).

But these clues are not fail-safe, for sometimes arrangements and order are only implied by the context rather than being mentioned explicitly. Thus, we need additional clues. Frequently in combination and permutation problems *two sets of objects* are involved either explicitly or implicitly. It is in this context where many problems are phrased as *distribution problems*. The general idea is that we have r objects and we want to enumerate the number of ways in which these objects can be assigned or distributed to n cells.

For example, we may wish to count the number of ways of assigning r balls to n boxes, r cards to n hands, or r players to n teams.

In some applications the objects are indistinguishable; in others they are distinguishable; the cells may be distinguishable, (maybe they are numbered or equivalently ordered in some way), or the cells may be identical.

Order Counts	Set of r Objects	Type of Repetition Allowed	Name	Number	Theorem Reference
Yes	Distinguishable	None	r-permutation	$P(n,r)$	2.3.1
No	Indistinguishable	None	r-combination	$C(n,r)$	2.3.3
Yes	Distinguishable	Unlimited	r-permutation with unlimited repetition	n^r	2.4.1
No	Indistinguishable	Unlimited	r-combination with unlimited repetition	$C(n+r-1,r)$	2.4.2
Yes		Constrained	Ordered partition	$P(n;q_1,\ldots,q_t)$	2.5.1 & 2.5.2
No		Constrained	Unordered partition		No general formula

Figure 2-5

Generally speaking, if the objects and the cells are distinguishable, that is, if elements in *both* sets are distinguishable, we suspect a permutation problem. But, on the other hand, if the objects are indistinguishable and the cells are distinguishable (or in other words, the elements of only one of the sets are distinguishable), then we suspect a combination problem.

Ordered partitions are really permutation problems; nevertheless, frequently the clue here is the mention of two or more subsets of a set to which elements are to be assigned.

After determining whether the problem calls for counting permutations or combinations, determine next whether or not repetitions are allowed. Or, in the case of partitions determine whether to count ordered or unordered partitions.

The number of ways of choosing/arranging r objects from n objects is illustrated in Figure 2-5.

Exercises for Section 2.5

1. Use Theorem 2.5.2. to calculate the following expressions:
   (a) $P(10;4,3,2,1)$
   (b) $P(10;3,3,2,2)$
   (c) $P(16;4,7,0,3,2)$.

2. A store has 25 flags to hang along the front of the store to celebrate a special occasion. If there are 10 red flags, 5 white flags, 4 yellow flags, and 6 blue flags, how many distinguishable ways can the flags be displayed?

3. In how many ways can 8 students be divided into
   (a) 4 teams of 2 each?
   (b) 2 teams of 4 each?
   (c) 3 teams one with 1 student, one with 2 students, and one with 5 students?

4. From 200 automobiles 40 are selected to test whether they meet the antipollution requirements. Also 50 automobiles are selected from the same 200 autos to test whether or not they meet the safety requirements.
   (a) In how many ways can the selections be made?
   (b) In how many ways can the selections be made so that there are exactly 10 automobiles that undergo both tests?

5. Suppose that Florida State University has a residence hall that has 5 single rooms, 5 double rooms, and 3 rooms for 3 students each. In how many ways can 24 students be assigned to the 13 rooms?

6. Suppose that a set S has n distinct elements. How many n-part ordered partitions $(A_1, A_2, \ldots, A_n)$ are there in which each set A_i has exactly 1 element?

7. In how many ways can 3 boys share 15 different sized apples,
   (a) if each takes 5?
   (b) if the youngest boy gets 7 apples and the other two boys get 4 each?

8. A child has blocks of 6 different colors.
   (a) If the child selects one block of each color, in how many ways can these be arranged in a line?
   (b) In how many ways can the 6 blocks be arranged in a circle?
   (c) If the child selects 4 blocks of each color, in how many ways can these 24 blocks be arranged in a line?
   *(d) In how many ways can the 24 blocks be arranged in a circle?

9. How many different 8-digit numbers can be formed by arranging the digits 1,1,1,1,2,3,3,3?

10. Find the number of arrangements of the letters of
    (a) Mississippi.
    (b) Tennessee.

11. How many anagrams (arrangements of the letters) are there of $7 \cdot a, 5 \cdot c, 1 \cdot d, 5 \cdot e, 1 \cdot g, 1 \cdot h, 7 \cdot i, 3 \cdot m, 9 \cdot n, 4 \cdot o, 5 \cdot t$?

12. How many ways are there to distribute 10 balls into 6 boxes with at most 4 balls in the first 2 boxes (that is, if x_i = the number of balls in box i, then $x_1 + x_2 \leq 4$) if:
    (a) the balls are indistinguishable?
    (b) the balls are distinguishable?

13. How many arrangements are there of $\{8 \cdot a, 6 \cdot b, 7 \cdot c\}$ in which each a is on at least one side of another a?

14. How many n-digit binary numbers are there without any pair of consecutive digits being the same?

15. Compute the number of rows of 6 Americans, 7 Mexicans, and 10 Canadians in which an American invariably stands between a Mexican and a Canadian and in which a Mexican and a Canadian never stand side by side.

16. (a) Compute the number of 10-digit numbers which contain only the digits 1, 2, and 3 with the digit 2 appearing in each number exactly twice.
    (b) How many of these numbers are divisible by 9? (Recall that an integer is divisible by 9 iff the sum of its digits is divisible by 9.)

17. (a) In how many ways can we choose 3 of the numbers from 1 to 100 so that their sum is divisible by 3?

(b) In how many ways can we choose 3 out of $3n$ successive positive integers so that their sum is divisible by 3?

18. In how many ways can we distribute 10 red balls, 10 white balls, and 10 blue balls into 6 different boxes (any box may be left empty)?

19. A chess player places black and white chess pieces (2 knights, 2 bishops, 2 rooks, 1 queen, and 1 king of each color) in the first two rows of a chessboard. In how many ways can this be done?

20. (a) In how many ways can we package 20 books into 5 packages of 4 books each?
    (b) Solve problem (a) if 2 packages contain 5 books each, 2 other packages contain 3 each, and 1 package has 4 books.

21. In how many ways can 5 different messages be delivered by 3 messenger boys if no messenger boy is left unemployed? The order in which a messenger delivers his message is immaterial.

22. How many ways can 12 white pawns and 12 black pawns be placed on the black squares of an 8×8 chess board?

23. Given the integers $1,2,3,\ldots,15$, two groups are selected; the first group contains 5 integers and the second group contains 2 integers. In how many ways can the selection be made if (a) unlimited repetition is allowed or (b) repetition is allowed but a group contains either all odd integers or all even integers and, moreover, if one group contains even integers, then the other group contains only odd integers, and vice versa or (c) no repetition is allowed, and the smallest number of the first group is larger than the largest number of the second group?

24. A shop sells 20 different flavors of ice cream. In how many ways can a customer choose 4 ice cream cones (one dip of ice cream per cone) if they
    (a) are all of different flavors?
    (b) are not necessarily of different flavors?
    (c) contain only 2 or 3 flavors?
    (d) contain 3 different flavors?

25. How many bridge deals are there in which North and South get all the spades?

26. In how many ways can 10 people be divided into disjoint committees where each committee must contain at least 2 people? (A division into committees of 3, 3, and 4 is considered the same as the division into committees of 3, 4, and 3.)

27. Consider the word TALLAHASSEE. How many arrangements are there
    (a) altogether?

     (b) where no two letters A appear together?

     (c) where the letters S are together and the letters E are together?

     (d) of 4 of the letters taken from TALLAHASSEE?

28. Consider the word TRIANNUAL.

     (a) How many arrangements are there of these 9 letters?

     (b) How many 9-letter words are there with the letters T, I, and U separated by exactly 2 of the other letters?

     (c) How many 6-letter words can be formed from the letters of TRIANNUAL with no N's?

29. Suppose there are n houses in a housing project, and to make them look different r of them are painted white, s of them are painted yellow, and t of them are painted blue. In how many ways can colors be assigned to the houses?

30. How many ways can we distribute 52 cards among 4 players where each player gets 3 cards of each of 3 suits and 4 cards of the fourth suit?

31. Suppose that we place black and white chess pieces (2 knights, 2 bishops, 2 rooks, 1 queen, and 1 king of each of the two colors) on an 8×8 chess board.

     (a) How many ways can 16 pieces be placed on the first two rows of the chess board?

     (b) How many ways can the pieces be placed on the entire board?

32. How many 6-letter words can be formed if the letters are taken from a set of 20 different letters and no letter can appear more than twice in a word?

33. (a) Compute the number of 10-digit integers that contain only the digits 1, 2, 3 with the digit 3 appearing exactly twice in each integer.

     (b) How many of the integers in (a) are divisible by 9?

34. Given 3 yellow balls, 1 red ball, 1 white ball, and 1 blue ball, compute the number of different rows of 4 of these balls.

35. How many ways can we divide a deck of 52 cards into 2 halves if each half is to contain 2 kings?

Selected Answers for Section 2.5

2. $\dfrac{25!}{10!6!5!4!}$.

5. $P(24;1,1,1,1,1,2,2,2,2,2,3,3,3)$.

6. $P(n;1,1,\ldots,1) = n!$.

7. (a) $\dfrac{15!}{5!^3}$

(b) $\dfrac{15!}{7!4!^2}$

12. (a) For $k = 0,1,2,3,4$ there are $\binom{k+1}{k} = \binom{k+1}{1}$ ways to place k balls in the first 2 boxes. $\binom{10-k+4-1}{3}$ ways to distribute the remaining balls into the remaining boxes. The total is

$$\sum_{k=0}^{4} (k + 1) \binom{13-k}{3}.$$

(b) $x_1 + x_2 = k$, $k = 0,1,2,3,4$. Choose k balls to go into the first 2 boxes $\binom{10}{k}$ ways. For each of these there are 2 arrangements (box 1 and box 2). The remaining balls have 4 choices each.

$$\text{Total} = \sum_{k=0}^{4} \binom{10}{k} 2^k 4^{10-k}.$$

13. $\dfrac{13!}{6!7!} \left[\binom{14}{4} + 3\binom{14}{3} + 3\binom{14}{3} + 5\binom{14}{2} + \binom{14}{1} \right].$

15. $6!7!10! \left[\binom{6}{3}\binom{9}{2} + \binom{6}{2}\binom{9}{3} \right].$

16. (a) $2^8 C(10,2)$.
    (b) $P(10;2,3,5)$.

17. (a) $2\binom{33}{3} + \binom{34}{3} + \binom{34}{1}\binom{33}{1}^2$.
    (b) $3\binom{n}{3} + \binom{n}{1}^3$.

18. $\binom{10+6-1}{10}^3$.

19. $P(16;2,2,2,2,2,2,1,1,1,1) = \dfrac{16!}{2!^6}.$

20. (a) $\dfrac{20!}{5!(4!)^5}$

(b) $\dfrac{20!}{2!^2 3!^2 4! 5!^2}.$

21. The messages are partitioned into 3 cells of sizes 3,1,1 or 2,2,1. Hence $3(5!/3!1!1!) + 3(5!/2!2!1!)$

22. $\dfrac{32!}{(12!)^2 8!}$

(Note: there are 32 black squares so 8 will not have a pawn on them.)

23. (a) $C(15 - 1 + 5,5) \, C(15 - 1 + 2,2) = \binom{19}{5}\binom{16}{2}$.

(b) There are 8 odd and 7 even integers to choose from, and either there are to be 5 odd and 2 even or 5 even and 2 odd to be chosen in $\binom{12}{5}\binom{8}{2} + \binom{11}{5}\binom{9}{2}$ ways.

(c) $\binom{15}{7}$, since the 2 groups are determined once the 7 integers are selected.

24. (a) $\binom{20}{4}$

(b) $\binom{20 - 1 + 4}{4} = \binom{23}{4}$

(c) $\binom{23}{4} - \binom{20}{4} - 20$

(d) There are $\binom{20}{3}$ ways to choose 3 flavors times 3 ways to fill 4 cones with exactly 3 flavors $= 3\binom{20}{3}$.

25. $2\binom{13}{0}\left(\dfrac{39!}{13!^3}\right) + 2\binom{13}{1}\left(\dfrac{39!}{1!12!13!^2}\right) + \cdots + 2\binom{13}{6}\left(\dfrac{39!}{6!7!13!^2}\right) = \binom{39}{13}\binom{26}{13}\binom{26}{13}$

2.6 BINOMIAL COEFFICIENTS

In this section we will present some basic identities involving binomial coefficients. In formulas arising from the analysis of algorithms in computer science, the binomial coefficients occur over and over again, so that a facility for manipulating them is a necessity. Moreover, different approaches to problems often give rise to formulas that are different in appearance yet identities of binomial coefficients reveal that they are, in fact, the same expressions.

The study of identities is itself a major field of study in combinatorial mathematics and we will of necessity merely scratch the surface. For a more thorough study we recommend Riordan's book [36]. In this section we only hope to get a good idea of the type of identities involved and an idea as to the methods by which these identities are obtained.

Combinatorial Reasoning

The symbol $C(n,r)$ has two meanings: the combinatorial and the factorial. In other words, $C(n,r)$ represents the number of ways of choosing r objects from n distinct objects (the combinatorial meaning) and, as well, $C(n,r)$ equals n!/r!(n-r)! (the factorial or algebraic meaning). Therefore as a general rule, all theorems and identities about factorials and binomial coefficients can be viewed as two kinds of statements for which two kinds of proofs can be given—a combinatorial proof and an algebraic proof. Roughly speaking, a combinatorial proof will be based on decomposing a set into subsets in a certain prescribed manner and then counting the number of ways of selecting these subsets, while an algebraic proof will be patterned mainly on the manipulation of factorials. We feel, as a general rule, that combinatorial proofs are preferable in that they are intuitive, instructive, and easy to remember. While algebraic proofs are more formal than combinatorial proofs, they have an advantage in that verifications can be made even when understanding of the combinatorial meaning is missing. Nevertheless, an awareness of both proofs is probably necessary for a thorough understanding of the meaning of an identity.

Let us give an example of how combinatorial reasoning can be used.

Example 2.6.1.   Prove that $(n^2)!/(n!)^{n+1}$ is an integer. The first hint we have is that this number is reminiscent of the conclusion of Theorem 2.5.3. In fact, this is the clue to the solution, for the number $(n^2)!/(n!)^n$ enumerates the ordered n-part partitions of a set S containing n^2 elements where each cell contains n elements. On the other hand, $(n^2)!/(n!)^{n+1}$ enumerates such unordered n-part partitions of S.

Moreover, it is often the case that the very form of one's solution reflects the combinatorial reasoning used to solve the problem.

Example 2.6.2.   Suppose that there are a different roads from A to B, b different roads from B to C and c different roads directly from A to C. How many different trips are there from A to C and back to A that visit B at least once? The answer $(ab)c + c(ab) + (ab)^2$ could be arrived at by counting first, the abc trips from A to C via B that return directly to A; then counting the trips directly from A to C and that return via B and, finally, counting those $(ab)^2$ trips that go and return via B.

Of course, there are a total of $ab + c$ trips from A to C and thus $(ab + c)^2$ from A to C and back to A, and there are c^2 trips that go directly to C and return directly from C back to A. Thus the difference $(ab + c)^2 - c^2$ represents the number of trips that visit B at least once.

Another approach could observe that there are $(ab)(ab + c)$ trips from A to C via B that return to A anyway. Moreover, there are $c(ab)$ trips to go directly to C and return via B. Thus, there are $(ab)(ab + c) + c(ab)$ trips in all that visit B at least once.

Now, of course, simple algebra will show that all 3 expressions: $(ab)c + c(ab) + (ab)^2$, $(ab + c)^2 - c^2$, and $(ab)(ab + c) + c(ab)$ are the same. What we wish to point out here is that the *form* of the expression suggests the combinatorial reasoning used to obtain the solution.

This point of view will be very beneficial when you are called upon to verify some identities involving binomial coefficients.

Some Examples of Combinatorial Identities

(1) Representation by factorials.

$$C(n,r) = \frac{n!}{r!(n-r)!}$$

for every pair of integers n and r where $n \geq r \geq 0$. This identity was proved in Theorem 2.3.4.

(2) Symmetry property: $C(n,r) = C(n,n-r)$.

A combinatorial proof of this identity is easy to see because when we choose r objects from n objects there are $n - r$ objects left. These $n - r$ objects can be considered as an $(n - r)$-combination. Hence to every r-combination automatically there is an associated $(n - r)$-combination and conversely. In other words there are precisely the same number of r-combinations as $(n - r)$-combinations which is just what identity (2) states.

Alternatively, a proof using factorials follows from the factorial representation because $C(n,r) = n!/(n - r)!r!$ while

$$C(n,n - r) = \frac{n!}{(n - (n - r))!(n - r)!}.$$

But since $n - (n - r) = r$, we have $(n - (n - r))! = r!$ and $C(n,r) = C(n,n - r)$.

(3) Newton's Identity: $C(n,r)\, C(r,k) = C(n,k)\, C(n - k,r - k)$ for integers $n \geq r \geq k \geq 0$.

The left-hand side counts the number of ways of selecting two sets: first a set A of r objects and then from A, a set B of k objects. For example, we may be counting the number of ways to select a committee of r people and then to select a subset of k leaders from this committee. On the other hand, the right-hand side counts the number of ways we could select the group of k leaders from the n people first, and then select the remaining $r - k$ people for the committee from the remaining $n - k$ people.

A special case of this identity is:

(3a) $C(n,r)r = nC(n - 1,r - 1)$.

Here just let $k = 1$; in other words, choose only one leader of the committee of r.

Then, of course, if $r \neq 0$, we can rearrange to give a rule for the removal of constants from binomial coefficients:

(3b) $(n/r) \, C(n - 1, r - 1) = C(n, r)$.

Another special case of identity (3) is:

(3c) $C(n, r + 1) \, (r + 1) = (n - r) \, C(n, r)$.

Here replace r and k in identity (3) by $r + 1$ and 1, respectively. Then, of course, we have:

(3d) $C(n, r + 1) = [(n - r)/(r + 1)] \, C(n, r)$ for integers $n \geq r + 1 \geq 1$.

Sir Isaac Newton (1646–1727) discovered the importance of this identity: it shows how to compute $C(n, r+1)$ from $C(n, r)$.

It might be instructive at this point to list some formulas for r-permutations along with a combinatorial interpretation.

(4) $P(n, r) = nP(n - 1, r - 1)$

This identity holds because in arranging n objects we can fill the first position n ways and then arrange the remaining $n - 1$ objects in $r - 1$ positions.

On the other hand, we observe that any r-permutation of n objects can also be attained by first arranging $r - 1$ of the objects in some order and then filling the rth position. The first $r - 1$ positions can be filled in $P(n, r - 1)$ ways while the rth position can be filled in $n - (r - 1) = n - r + 1$ ways. Thus, we see that

(4a) $P(n, r) = P(n, r - 1) \, (n - r + 1)$.

The following result is very useful; it is commonly associated with Blaise Pascal (1623–1662), although an equivalent version was known by M. Stifel (1486–1567).

(5) Pascal's Identity.

$$C(n, r) = C(n - 1, r) + C(n - 1, r - 1).$$

Let us give a combinatorial proof; we leave the easy factorial proof as an exercise. Let S be a set of n objects. Distinguish one of the objects, say $x \in S$. (For example, S might consist of $n - 1$ women and 1 male.) The r-combinations of S can be divided into two classes:

(A) those selections that include x and
(B) those selections that do not include x.

In (A), we need merely to choose $r - 1$ objects from the remaining $n - 1$ objects in $C(n - 1, r - 1)$ ways. In (B), we choose r objects from the

Row Number

$n = 0$ $C(0, 0)$

$n = 1$ $C(1, 0)$ $C(1, 1)$

$n = 2$ $C(2, 0)$ $C(2, 1)$ $C(2, 2)$

$n = 3$ $C(3, 0)$ $C(3, 1)$ $C(3, 2)$ $C(3, 3)$

$n = 4$ $C(4, 0)$ $C(4, 1)$ $\boxed{C(4, 2)}$ $\boxed{C(4, 3)}$ $C(4, 4)$

$n = 5$ $C(5, 0)$ $C(5, 1)$ $C(5, 2)$ $\boxed{C(5, 3)}$ $C(5, 4)$ $C(5, 5)$

Figure 2-6. Pascal's triangle.

remaining $n - 1$ objects (excluding x) in $C(n - 1, r)$ ways. Since these two classes are disjoint, we can apply the sum rule to obtain the identity.

Thus, the number of committees of r people is the sum of the number of committees that contain a given person and the number of committees that do not contain that person.

This identity gives us an alternate method for determining the numerical values of $C(n,r)$. For example, if we know $C(4,0)$, $C(4,1)$, $C(4,2)$, $C(4,3)$, and $C(4,4)$, we can determine $C(5,1)$, $C(5,2)$, $C(5,3)$, and $C(5,4)$ simply by addition.

Using identity (5) and the fact that $C(n,0) = C(n,n) = 1$ for all nonnegative integers n, we can build successive rows in the table of binomial coefficients, called Pascal's triangle (Figure 2-6).

What Pascal's identity says is that the numbers on the rth row are found by adding the two nearest binomial coefficients in the row above it. For instance, $C(5,3)$ is the sum of the two circled binomial coefficients above it. Since $C(4,2) = 6$ and $C(4,3) = 4$ we see that $C(5,3) = C(4,2) + C(4,3) = 6 + 4 = 10$.

Let's list Pascal's triangle again (Figure 2-7) with the numerical values of the binomial coefficients entered. The number on the nth row along the rth diagonal is $C(n,r)$ or, in other words, the rth number on the nth row is $C(n,r)$.

We can construct the row corresponding to $n = 8$ by using row 7 and Pascal's identity as follows:

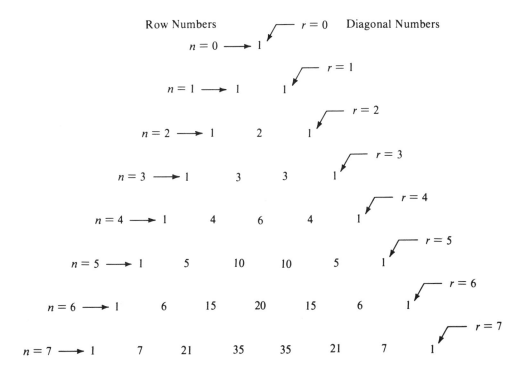

Figure 2-7. Pascal's triangle with the numerical values of the binomial coefficients entered.

We might note that the diagonal corresponding to $r = 0$ on the left has all 1's and likewise the opposite diagonal on the extreme right has only 1's. This is because of,

(6) Boundary Conditions.

$$C(n,0) = 1 = C(n,n).$$

Likewise the diagonal corresponding to $r = 1$ always has the row number 1, 2, 3, 4, 5, 6, etc., and the same is true for the opposite diagonal second from the extreme right. This is because,

(7) Secondary Conditions.

$$C(n,1) = n = C(n,n - 1).$$

Indeed there is a symmetry property of the triangle: on any row as we proceed from the left to right the numbers are the same as we proceed

Figure 2-8 Table of Binomial Coefficients

n	$C(n,0)$	$C(n,1)$	$C(n,2)$	$C(n,3)$	$C(n,4)$	$C(n,5)$	$C(n,6)$	$C(n,7)$	$C(n,8)$	$C(n,9)$	$C(n,10)$
0	1										
1	1	1									
2	1	2	1								
3	1	3	3	1							
4	1	4	6	4	1						
5	1	5	10	10	5	1					
6	1	6	15	20	15	6	1				
7	1	7	21	35	35	21	7	1			
8	1	8	28	56	70	56	28	8	1		
9	1	9	36	84	126	126	84	36	9	1	
10	1	10	45	120	210	252	210	120	45	10	1
11	1	11	55	165	330	462	462	330	165	55	11
12	1	12	66	220	495	792	924	792	495	220	66
13	1	13	78	285	715	1287	1716	1716	1287	715	286
14	1	14	91	364	1001	2002	3003	3432	3003	2002	1001
15	1	15	105	455	1365	3003	5005	6435	6435	5005	3003
16	1	16	120	560	1820	4368	8008	11440	12870	11440	8008
17	1	17	136	680	2380	6188	12376	19448	24310	24310	19448
18	1	18	153	816	3060	8568	18564	31824	43758	48620	43758
19	1	19	171	969	3876	11628	27132	50388	75582	92378	92378
20	1	20	190	1140	4845	15504	38760	77520	125970	167960	184756

For coefficients missing from the Table use the relation $C(n,r) = C(n,n-r)$
$C(20,11) = C(20,9) = 167,960$

from the extreme right to left. This follows because $C(n,r) = C(n,n-r)$. Thus, we need only know approximately one half the values of the binomial coefficients in Pascal's triangle as the other values are known from symmetry. Thus, if we have computed $C(20,0)$, $C(20,1)$, ..., and $C(20,10)$ then by symmetry we know $C(20,11) = C(20,9)$, $C(20,12) = C(20,8)$, $C(20,13) = C(20,7)$, etc., and the rest of the binomial coefficients on row 20. Thus, we can list more in a tabular form by omitting those values of $C(n,r)$ that we can obtain by symmetry (see Figure 2-8).

Note one other thing about Pascal's triangle or the table of Figure 2-8: the sum of all numbers on a diagonal (proceeding downward from left to right) is the number immediately below the last number on a diagonal. For example, take the entries of the diagonal of Figure 2-8 starting with the first 1 in row 3. Then we have, say, $1 + 4 + 10 + 20 + 35$ equals the 70 in row 8 directly below the 35 of row 7. Check this out with other diagonals. For instance, by adding more terms of that same diagonal we have $1 + 4 + 10 + 20 + 35 + 56 + 84 = 210$ (the number directly below the 84 in the next column). Is there a mathematical explanation for this? The answer is yes as shown in Figure 2-8.

(8) Diagonal Summation.

$$C(n,0) + C(n + 1,1) + C(n + 2,2) + \cdots + C(n + r,r)$$
$$= C(n + r + 1,r)$$

Let us give a combinatorial proof by counting the number of ways to distribute r indistinguishable balls into $n + 2$ numbered boxes. This can be done in $C(n + r + 1,r) = C(n + 2 + r - 1,r)$ ways. But the balls may also be distributed as follows: For each $0 \le k \le r$, distribute k of the balls in the first $n + 1$ boxes, and then the remainder in the last box. This can be done in $\Sigma_{k=0}^{r} C(n + k,k)$ ways.

An alternate proof can be made by repeated applications of Pascal's identity (5). In this proof we start with $C(n + r + 1,r) = C(n + r,r) + C(n + r,r - 1)$ and then decompose $C(n + r,r - 1)$ into $C(n + r - 1,r - 1) + C(n + r - 1,r - 2)$. This gives $C(n + r + 1,r) = C(n + r,r) + C(n + r - 1,r - 1) + C(n + r - 1,r - 2)$. Again decompose the last term by Pascal's identity and combine to get $C(n + r + 1,r) = C(n + r,r) + C(n + r - 1,r - 1) + C(n + r - 2,r - 2) + C(n + r - 2,r - 3)$. This process can be continued until the last term is $C(n,0)$. The sum is the one desired.

Thus in several ways, we can compute more entries of the table—for example, for row 21, we know $C(21,0) = 1$, $C(21,1) = 21$, but then we compute $C(21,2)$ by using identity (8), by computing the sum of the diagonal $1 + 19 + 190$ or using identity (5), $C(21,2) = 20 + 190 = C(20,1) + C(20,2)$. Thus, $C(21,2) = 210$. Likewise, $C(21,3) = 1 + 18 + 171 + 1140 = 190 + 1140 = C(20,2) + C(20,3)$; $C(21,4) = 1140 + 4845 = C(20,3) + C(20,4) = 5985$, etc.

The sum of the numbers of the nth row of Pascal's triangle gives the following identity.

(9) Row Summation.

$$C(n,0) + C(n,1) + \cdots + C(n,r) + \cdots + C(n,n) = 2^n.$$

This just means that there is a total of 2^n subsets of a set S with n elements and this number is also the sum of the number of all the subsets of S respectively with 0 elements, 1 element, 2 elements, ..., r elements, ..., and n elements.

Another way of interpreting this identity is that we have n people to be put into two buses. We can place none in the first bus and all n in the second bus in $C(n,0)$ ways; we can put 1 in the first bus and all $n - 1$ in the second bus in $C(n,1)$ ways, and so on, until we place all n people in the first bus and none in the second in $C(n,n)$ ways. There is a total of 2^n ways to do this since each of the people can be placed two ways.

(10) Row Square Summation.

$$C(n,0)^2 + C(n,1)^2 + \cdots + C(n,r)^2 + \cdots + C(n,n)^2 = C(2n,n)$$

for each positive integer n.

This just says that the sum of the squares of the nth row of Pascal's triangle is the middle number in the $2n$th row.

To verify this, let S be a set with $2n$ elements. Then the right-hand side of identity (10) counts the n-combinations of S. Now partition S into two subsets A and B of S where A and B have n elements each. (We might have a set of n men and n women, for example.) Then an n-combination of S is a union of an r-combination of A and an $(n - r)$-combination of B for $r = 0, 1, \ldots, n$. We might be choosing r men and $n - r$ women. Then for a given r, there are $C(n,r)$ r-combinations of A and $C(n,n - r)$ $(n - r)$-combinations of B. Thus by the product rule there are $C(n,r)$ $C(n,n - r)$ n-combinations which are unions of an r-combination of A and an $(n - r)$-combination of B. By symmetry, $C(n,r) \times C(n,n - r) = C(n,r)^2$. Hence by the sum rule the number of n-combinations of S equals $\Sigma_{r-0}^n C(n,r)^2$. But we have already observed that this is also equal to $C(2n,n)$, hence the identity is proved.

After observing the proof of this identity we see that we could use the same ideas to obtain a more general identity.

(10a) $C(m,0)C(n,0) + C(m,1)C(n,1) + \cdots + C(m,n)C(n,n) = C(m + n,n)$ for integers $m \geq n \geq 0$.

Here we suppose that we have a set S including a subset A of m men and a subset B of n women. We can choose n people from this set. Any n-combination of S is a union of an r-combination of A and an $(n - r)$-combination of B for $r = 0, 1, \ldots, n$. Thus, for a given r, there are $C(m,r)$ r-combinations of A (since A has m members), and $C(n,n - r) = C(n,r)$ $(n - r)$-combinations of B (since B has n members). Thus, for each r, there are $C(m,r)$ $C(n,r)$ n-combinations of S which are the union of an r-combination of A and an $(n - r)$-combination of B. Apply the sum rule as $r = 0, 1, \ldots, n$ to get the identity.

The sum of all numbers including a given binomial coefficient and all numbers above it in a column in Pascal's triangle (or in Figure 2-8) is the number in the next column and in the next row.

(11) Column Summation.

$$C(r,r) + C(r + 1,r) + \cdots + C(n,r) = C(n + 1,r + 1)$$

for any positive integer $n \geq r$.

Following the pattern of Example 2.4.8, we see that the above identity counts in 2 different ways the number of nonnegative solutions of the inequality

$$x_1 + x_2 + \cdots + x_{r+1} \le n - r$$

We can use Pascal's identity to obtain a corresponding identity for permutations.

(12) $P(n,r) = r\,P(n - 1, r - 1) + P(n - 1, r).$

This can be obtained easily from identity (5) just by multiplying by $r!$. For $P(n,r) = r!\,C(n,r) = r![C(n - 1, r - 1) + C(n - 1, r)] = r!\,C(n - 1, r - 1) + r!\,C(n - 1, r) = r[(r - 1)!\,C(n - 1, r - 1)] + r!\,C(n - 1, r) = rP(n - 1, r - 1) + P(n - 1, r).$

Of course, there is a combinatorial meaning for identity (12). Recall that identity (5) was obtained by dividing all r-combinations into two classes: (a) those selections that included a fixed element x or (b) those selections that did not include x. Likewise, all r-permutations can be obtained by arranging each selection of class A by placing x in one of r positions and then arranging the other $r - 1$ elements chosen from $n - 1$ elements [this is done in $rP(n - 1, r - 1)$ ways], or by arranging the r elements chosen in each selection of class B [this is done $P(n - 1, r)$ ways].

There are many more interesting properties of Pascal's triangle, but for now let us move on to other topics.

Let us show how to obtain other identities from combinatorial identities.

Example 2.6.3. Evaluate the sum $1 + 2 + 3 + \ldots + n$.

Note that $k = C(k,1)$ so that $\Sigma_{k=1}^{n} k = \Sigma_{k=1}^{n} C(k,1) = C(n + 1, 2) = n(n + 1)/2$ by identity (11).

Example 2.6.4. Evaluate the sum $1^2 + 2^2 + 3^2 + \ldots + n^2$.

Here we observe that $k^2 = k(k - 1) + k = 2C(k,2) + C(k,1)$. Therefore, $\Sigma_{k=1}^{n} k^2 = 2\Sigma_{k=1}^{n} C(k,2) + \Sigma_{k=1}^{n} C(k,1) = 2C(n + 1, 3) + C(n + 1, 2)$ by (11). Here we have used the convention that $C(k,r) = 0$ if $k < r$.

Exercises for Section 2.6

1. (a) Show that $k^3 = k(k - 1)(k - 2) + 3k^2 - 2k = C(k,1) + 6\,C(k,2) + 6\,C(k,3).$

(b) Evaluate $1^3 + 2^3 + \cdots + n^3$.

(c) Using the fact that $k^4 = k(k-1)(k-2)(k-3) + 6k^3 - 11k^2 + 6k$, derive a formula for k^4 like (a) in terms of binomial coefficients.

2. Use the column summation identity (11) and $r = 1, 2$, and 3 to derive the formulas

(a) $1 + 2 + \cdots + n = n(n+1)/2$,

(b) $1 \cdot 2 + 2 \cdot 3 + 3 \cdot 4 + \cdots + n(n+1) = n(n+1)(n+2)/3$, and

(c) $1 \cdot 2 \cdot 3 + (2 \cdot 3 \cdot 4) + \cdots + (n)(n+1)(n+2) = n(n+1)(n+2)(n+3)/4$.

3. Derive the column summation identity (11) from the diagonal summation identity (8).

4. Use identity (11) to verify that

(a) $C(n+1,r) + \cdots + C(n+m,r) = C(n+m+1,r+1) - C(n+1,r+1)$.

(b) In particular,

$$\sum_{k=1}^{n-1} C(k+2,2) = C(n+2,3) - 1$$

or

$$\sum_{k=0}^{n-1} C(k+2,2) = C(n+2,3).$$

(c) Use (b) to obtain a proof that

$$1 \cdot 2 + 2 \cdot 3 + 3 \cdot 4 + \cdots + n(n+1) = \frac{1}{3}(n)(n+1)(n+2).$$

5. We wish to make triples (x,y,z) from the integers $\{1,2,\ldots,(n+1)\}$ such that z is larger than either x or y.

(a) Prove that if z is $k+1$, then the number of such triples is k^2.

(b) These triples can be classified into 3 types:

(1) $x = y$,

(2) $x < y$, and

(3) $x > y$.

Show that there are $C(n+1,2)$ of the first type and $C(n+1,3)$ of each of the other 2 types.

6. Solve for the unknowns in the following:

(a) $C(10,4) + C(10,3) = C(n,r)$;

(b) $C(50,20) - C(49,19) = C(n,r)$;

(c) $P(n,2) = 90$;

(d) $C(10,5)\, C(5,3) = C(10,3)\, C(n,r)$;

(e) $C(5,0)^2 + C(5,1)^2 + C(5,2)^2 + C(5,3)^2 + C(5,4)^2 + C(5,5)^2 = C(n,r)$;

(f) $C(5,5) + C(6,5) + C(7,5) + C(8,5) + C(9,5) = C(n,r)$;

(g) $C(n,0) + C(n,1) + \cdots + C(n,n) = 128$.

7. Use the binomial identities to evaluate the sum

$$1 \cdot 2 \cdot 3 + 2 \cdot 3 \cdot 4 + \cdots + (n - 2)\,(n - 1)\,n.$$

8. (a) Show that

$$C(r + m + n, r)\, C(m + n, m) = \frac{(r + m + n)!}{r!\,m!\,n!}$$

(b) Show that

$$\sum_{r=0}^{n} \frac{(2n)!}{(r!)^2\,(n - r)!^2} = C(2n,n)^2.$$

9. Give a combinatorial argument to explain why

(a) $P(n,n) = P(n,n - 1)$,

(b) $P(n,n) = 2P(n,n - 2)$,

(c) $(3n)!/3!(n!)^3$ is an integer,

(d) $(3n)!/3!^n$ and $(3n)!/3^n$ are integers,

(e) $[(n!)^2]!/[(n!)!]^{n!+1}$ is an integer, and

(f) $(n!)!/\{[(n - 1)!]!\}^n$ is an integer.

10. Show by a combinatorial argument that:

(a) $C(2n,2) = 2C(n,2) + n^2$. (Hint: Consider a set of n men and n women.)

(b) $(n - r)\, C(n,r) = nC(n - 1,r)$.

(c) $C(3n,3) = 3C(n,3) + 3 \cdot 2C(n,2)\, C(n,1) + C(n,1)^3$.

(d) $C(3n,3) = C(2n,3) + C(n,3) + C(2n,1)\, C(n,2) + C(2n,2)\, C(n,1)$.

(e) $C(4n,4) = C(n,1)^4 + C(4,1)\, C(n,2)\, C(3,2)\, C(n,1)^2 + C(4,2)\, C(n,2)^2 + C(4,1)\, C(3,1)\, C(n,1)\, C(n,3) + C(4,1)\, C(n,4)$.

11. Show by a factorial argument that

$$(n - r)\, C(n + r - 1, r)\, C(n,r) = nC(n + r - 1, 2r)\, C(2r,r)$$

Selected Answers for Section 2.6

1. (b) $\sum_{k=1}^{n} k^3 = C(n + 1,2) + 6C(n + 1,3) + 6C(n + 1,4)$.

7. Since $(k - 2)\,(k - 1)\,k = 3!\,C(k,3)$

$$\sum_{k=1}^{n} (k - 2)\,(k - 1)k = 6 \sum_{k=1}^{n} C(k,3) = 6C(n + 1,4) \text{ by (11).}$$

9. (a) Designate one of the n distinct objects as a special object. There are $P(n,n - 1)$ ways to arrange the $n - 1$ objects into n

positions. For each of these, there is only one way to place the special object.

(b) Place $n - 2$ of the objects. Then there are 2 ways to arrange the remaining 2 objects in the 2 vacant places.

(c) Count the number of unordered 3-part partitions of $3n$ objects where each part contains n elements.

(e) Count the number of unordered $n!$-part partitions of $(n!)^2$ elements where each part has $n!$ elements.

2.7 THE BINOMIAL AND MULTINOMIAL THEOREMS

Any sum of two unlike symbols, such as $x + y$, is called a **binomial**. The binomial theorem is a formula for the powers of a binomial. The first few cases of this theorem should be familiar to the reader. We list these first cases of the binomial theorem in triangular form to suggest the correspondence with Pascal's triangle:

$$
\begin{array}{ll}
1 & = (x + y)^0 \\
x + y & = (x + y)^1 \\
x^2 + 2xy + y^2 & = (x + y)^2 \\
x^3 + 3x^2y + 3xy^2 + y^3 & = (x + y)^3 \\
x^4 + 4x^3y + 6x^2y^2 + 4xy^3 + y^4 & = (x + y)^4
\end{array}
$$

If we focus on the coefficients alone we find Pascal's triangle again. That is just what the binomial theorem says.

Theorem 2.7.1. (The Binomial Theorem). Let n be a positive integer. Then for all x and y,

$$(x + y)^n = C(n,0)\, x^n + C(n,1)\, x^{n-1}y + C(n,2)\, x^{n-2}y^2 + \cdots$$

$$+ C(n,r)x^{n-r}y^r + \cdots + C(n,n)\, y^n = \binom{n}{0} x^n + \binom{n}{1} x^{n-1}y$$

$$+ \binom{n}{2} x^{n-2}y^2 + \cdots$$

$$+ \binom{n}{r} x^{n-r}y^r + \cdots + \binom{n}{n} y^n = \sum_{r=0}^{n} C(n,r)x^{n-r}y^r.$$

The binomial coefficients $C(n,r) = \binom{n}{r}$ receive their name from their appearance in the expansion of powers of a binomial.

First Proof: Write $(x + y)^n$ as a product of n factors $(x + y)$ $(x + y) \cdots (x + y)$. Then expand this product until no parentheses remain. We can do this in many ways. One way is to select an x or a y from each factor, multiply and arrange these into a term of the form $x^{n-r}y^r$ for $r = 0, 1, \ldots, n$. The collection of all terms with the same exponents on x and y will determine the coefficients in the expansion of $(x + y)^n$. Thus for any given r we need only determine the number of terms of the form $x^{n-r}y^r$ obtained as described. But such a term is obtained by selecting y from r of the factors and then x from the remaining $n - r$ factors. The number of such terms is therefore the number of ways of choosing r of the n factors from which to choose the y. Since this can be done in $C(n,r)$ ways the coefficient of $x^{n-r}y^r$ is $C(n,r)$ as stated in the theorem.   □

Example 2.7.1.

$$(x + y)^8 = C(8,0)x^8 + C(8,1)x^7y + C(8,2)x^6y^2$$
$$+ C(8,3)x^5y^3 + C(8,4)x^4y^4 + C(8,5)x^3y^5$$
$$+ C(8,6)x^2y^6 + C(8,7)x\,y^7 + C(8,8)\,y^8 \qquad (2.7.1)$$
$$= x^8 + 8x^7y + 28x^6y^2 + 56x^5y^3 + 70x^4y^4$$
$$+ 56x^3y^5 + 28x^2y^6 + 8xy^7 + y^8;$$

$$(2a + 5b)^6 = C(6,0)(2a)^6 + C(6,1)(2a)^5(5b) + C(6,2)(2a)^4(5b)^2$$
$$+ C(6,3)(2a)^3(5b)^3 + C(6,4)(2a)^2(5b)^4 \qquad (2.7.2)$$
$$+ C(6,5)(2a)(5b)^5 + C(6,6)(5b)^6.$$

Let $x = 2a$ and $y = 5b$ in Theorem 2.7.1. Then

$$(2a + 5b)^6 = 2^6a^6 + 6 \cdot 2^5 \cdot 5a^5b + 15 \cdot 2^4 \cdot 5^2a^4b^2 + 20 \cdot 2^3 \cdot 5^3a^3b^3$$
$$+ 15 \cdot 2^2 \cdot 5^4a^2b^4 + 6 \cdot 2 \cdot 5^5ab^5 + 5^6b^6.$$

Second Proof.   This proof is by mathematical induction on n. If $n = 1$, the formula becomes $(x + y)^1 = C(1,0)x + C(1,1)y = x + y$, and this is clearly true. We now assume the formula is true for a positive integer n and prove that it is true when n is replaced by $n + 1$. We write $(x + y)^{n+1} = (x + y)(x + y)^n$ and by the inductive hypothesis this becomes

$$(x+y)\left(\sum_{r=0}^{n} C(n,r)x^{n-r}y^r\right) = x\left(\sum_{r=0}^{n} C(n,r)x^{n-r}y^r\right) + y\left(\sum_{r=0}^{n} C(n,r)x^{n-r}y^r\right)$$

$$= C(n,0)x^{n+1} + \sum_{r=1}^{n} C(n,r)x^{n-r+1}y^r$$

$$+ \sum_{r=0}^{n-1} C(n,r)x^{n-r}y^{r+1} + C(n,n)y^{n+1}.$$

If we set $r = k - 1$ in the third term above, then as r runs from 0 to $n - 1$, k runs from 1 to n, $\sum_{r=0}^{n-1} C(n,r) x^{n-r} y^{r+1}$ becomes $\sum_{k=1}^{n} C(n,k-1) x^{n+1-k} y^k$. Now the letter of the dummy variable is immaterial so now replace k by r and the third term becomes $\sum_{r=1}^{n} C(n,r-1) x^{n+1-r} y^r$ and

$$(x + y)^{n+1} = x^{n+1} + \sum_{r=1}^{n} [C(n,r) + C(n,r-1)] x^{n+1-r} y^r + y^{n+1}.$$

But using Pascal's identity, we then have

$$(x + y)^{n+1} = x^{n+1} + \sum_{r=1}^{n} C(n+1,r) x^{n+1-r} y^r + y^{n+1}$$

or

$$(x + y)^{n+1} = \sum_{r=0}^{n+1} C(n+1,r) x^{n+1-r} y^r.$$

Thus the formula is true for $n + 1$ and the theorem is proved by mathematical induction. $\square$

The binomial theorem can be written in several other equivalent forms:

$$(x + y)^n = \sum_{r=0}^{n} C(n, n-r) x^{n-r} y^r = \sum_{r=0}^{n} C(n,r) x^r y^{n-r}$$
$$= \sum_{r=0}^{n} C(n, n-r) x^r y^{n-r}$$

The first of these follows from Theorem 2.7.1 and the symmetry property $C(n,r) = C(n, n-r)$ for $r = 0, 1, \ldots, n$. The other two follow by interchanging x with y.

The case $y = 1$ occurs frequently enough to warrant recording it as a special case.

Corollary 2.7.1.   Let n be a positive integer. Then for all x,

$$(x + 1)^n = \sum_{r=0}^{n} C(n,r) x^{n-r} = \sum_{r=0}^{n} C(n, n-r) x^{n-r} = \sum_{r=0}^{n} C(n,r) x^r.$$

Replacing x by $-x$ we have

$$(1 - x)^n = \sum_{r=0}^{n} C(n,r)(-x)^r = \sum_{r=0}^{n} C(n,r)(-1)^r x^r.$$

Some More Identities

The identity $C(n,0) + C(n,1) + \cdots + C(n,n) = 2^n$ has already been proved by a combinatorial argument in identity (9) but it also follows from the binomial theorem by setting $x = y = 1$.

If we set $x = 1$ and $y = -1$ in the binomial theorem, then we see that

(13) $C(n,0) - C(n,1) + C(n,2) + \cdots + (-1)^n C(n,n) = 0.$

This says that the alternating sum of the members of any row of Pascal's triangle is zero.

We can also write this as

$$C(n,0) + C(n,2) + C(n,4) \cdots = C(n,1) + C(n,3) \cdots.$$

Let S be the common total of these two sums. Add the right-hand side to the left. By identity (9), this is 2^n. Thus $2S = 2^n$ or $S = 2^{n-1}$. Therefore, we have

(14) $C(n,0) + C(n,2) + C(n,4) \cdots = C(n,1) + C(n,3) \cdots = 2^{n-1}.$

This identity has the following combinatorial interpretation. If S is a set with n elements, then the number of subsets of S with an even number of elements is 2^{n-1} and this equals the number of subsets of S with an odd number of elements.

(15) $1C(n,1) + 2C(n,2) + 3C(n,3) + \cdots nC(n,n) = n2^{n-1}$ for each positive integer n.

To see this we use Newton's identity (3) and identity (9). By identity (3a), $rC(n,r) = nC(n-1,r-1)$, so $1C(n,1) + 2C(n, 2) + \cdots + nC(n,n) = nC(n-1,0) + nC(n-1,1) + \cdots + nC(n-1,n-1) = n[C(n-1,0) + \cdots + C(n-1,n-1)] = n2^{n-1}$ by identity (9). Likewise we can apply identities (3b) and (9) to obtain the identity:

(15a) $C(n-1,0)/1 + C(n-1,1)/2 + C(n-1,2)/3 + \cdots + C(n-1,n-1)/n = (2^n - 1)/n.$

(16) Vandermonde's Identity. $C(n+m,r) = C(n,0) C(m,r) + C(n,1) C(m,r-1) + \cdots + C(n,r) C(m,0)$ for integers $n \geq r \geq 0$ and $m \geq r \geq 0$.

We give a proof using the binomial theorem; we leave a combinatorial proof as an exercise.

First, consider the coefficient of x^r in $(1 + x)^{n+m}$. By the binomial theorem that coefficient is $C(n + m,r)$. But $(1 + x)^{n+m}$ can also be written as $(1 + x)^n (1 + x)^m$, and each of these factors can be expanded by the binomial theorem: $(1 + x)^n = C(n,0) + C(n,1)x + \cdots + C(n,n)x^n$ and $(1 + x)^m = C(m,0) + C(m,1)x + \cdots + C(m,m)x^m$. Now in the product the coefficient of x^r is obtained by summing over $k = 0, 1, \ldots, r$, the products of a term of degree k, $C(n,k)x^k$, from the first factor and a term of degree $r - k$, $C(m,r - k)x^{r-k}$, from the second factor, so the coefficient of x^r in the product is

$$\sum_{k=0}^{r} C(n,k)\, C(m,r - k) = C(n,0)\, C(m,r) +$$

$$C(n,1)\, C(m,r - 1) + \cdots + C(n,r)\, C(m,0).$$

But, as we have already observed, this coefficient is also $C(n + m,r)$, so the identity follows.

In summary we have obtained combinatorial identities in a variety of ways including the use of

1. combinatorial reasoning;
2. representation of binomial coefficients by factorials;
3. Pascal's identity for binomial coefficients;
4. mathematical induction; and
5. the binomial theorem.

The Multinomial Theorem

The sum of two unlike things $x_1 + x_2$ is a *binomial,* the sum of three unlike things is a **trinomial,** and, more generally, the sum of t unlike things, $x_1 + x_2 + \cdots + x_t$, is a **multinomial.**

The binomial theorem provides a formula for $(x_1 + x_2)^n$ when n is a positive integer. This formula can be extended to give a formula for powers of trinomials $(x_1 + x_2 + x_3)^n$ or more generally for powers of multinomials $(x_1 + x_2 + \cdots + x_t)^n$. In this theorem the role of the binomial coefficients is replaced by the numbers

$$P(n;q_1, q_2, \ldots, q_t) = \frac{n!}{q_1!q_2! \cdots q_t!}$$

where $q_1, q_2, \ldots, q_t$ are nonnegative integers with $q_1 + q_2 + \cdots + q_t = n$. It is legitimate to name these numbers **multinomial coefficients** and to denote them by $(q_1, \overset{n}{\ldots}, q_t)$. Recall that they enumerate the ordered partitions of a set of n elements of type $(q_1, q_2, \ldots, q_t)$.

Before stating the general theorem let us first consider some special cases. If we multiply out $(x_1 + x_2 + x_3)^3$, we get $x_1^3 + x_2^3 + x_3^3 + 3x_1^2 x_2 + 3x_1^2 x_3 + 3x_1 x_2^2 + 3x_1 x_3^2 + 3x_2 x_3^2 + 6x_1 x_2 x_3$.

The coefficient of $x_2 x_3^2$, for example, can be discovered as we did in the proof of Theorem 2.7.1 by choosing x_2 from one factor and x_3 from the remaining 2 factors. In other words we could choose this in $C(3,1)$ $C(2,2) = 3$ ways.

Perhaps another example will be more instructive. Suppose that we wish to find the coefficient of $x_1^4 x_2^5 x_3^6 x_4^3$ in $(x_1 + x_2 + x_3 + x_4)^{18}$. (Note the exponents add up to 18.) This product will occur in the multinomial expansion as often as x_1 can be chosen from 4 of the 18 factors, x_2 from 5 of the remaining 14 factors, x_3 from 6 of the remaining 9 factors, and x_4 then taken from the last 3 factors. We see that the coefficient of $x_1^4 x_2^5 x_3^6 x_4^3$ must be

$$C(18,4)\ C(14,5)\ C(9,6)\ C(3,3) = \frac{18!}{4!5!6!3!}\,.$$

This is not surprising for we can formulate the problem in another way. We are calculating the number of ways of arranging the following 18 letters: $\{4 \cdot x_1, 5 \cdot x_2, 6 \cdot x_3, 3 \cdot x_4\}$. Moreover, we know that the number of such arrangements is

$$P(18;4,5,6,3) = \frac{18!}{4!5!6!3!}\,.$$

More generally, we can say that $(x_1 + x_2 + x_3 + x_4)^{18}$ is the sum of all terms of the form $P(18;q_1,q_2,q_3,q_4)$ where q_1, q_2, q_3, and q_4 range over all possible sets of nonnegative integers such that $q_1 + q_2 + q_3 + q_4 = 18$. Further generalization is apparent and we state it as follows:

Theorem 2.7.2 (The Multinomial Theorem). Let n be a positive integer. Then for all $x_1, x_2, \ldots, x_t$ we have $(x_1 + x_2 + \cdots x_t)^n = \Sigma P(n;q_1, \ldots, q_t) x_1^{q_1} x_2^{q_2} \cdots x_t^{q_t}$ where the summation extends over all sets of nonnegative integers $q_1, q_2, \ldots, q_t$ where $q_1 + q_2 + \cdots + q_t = n$. There are $C(n + t - 1, n)$ terms in the expansion of $(x_1 + x_2 + \cdots + x_t)^n$.

Proof. The coefficient of $x_1^{q_1} x_2^{q_2} \cdots x_t^{q_t}$ is the number of ways of arranging the n letters $\{q_1 \cdot x_1, q_2 \cdot x_2, \ldots, q_t \cdot x_t\}$, therefore, it is $P(n;q_1,q_2,\ldots,q_t)$.

The number of terms is determined as follows: each term of the form $x_1^{q_1} x_2^{q_2} \cdots x_t^{q_t}$ is a selection of n objects with repetitions from t distinct types. Hence there are $C(n + t - 1, n)$ ways to do this. $\square$

Example 2.7.2. (a) In $(x_1 + x_2 + x_3 + x_4 + x_5)^{10}$ the coefficient of $x_1^2 x_3 x_4^3 x_5^4$ is

$$P(10;2,0,1,3,4) = \frac{10!}{2!0!1!3!4!} = 12,600.$$

There are $C(10 + 5 - 1,10) = C(14,10) = 1,001$ terms in the expansion $(x_1 + x_2 + x_3 + x_4 + x_5)^{10}$.

(b) In $(2x - 3y + 5z)^8$, we let $x_1 = 2x$, $x_2 = -3y$, $x_3 = 5z$, and then the coefficient of $x_1^3 x_2^3 x_3^2$ is $P(8;3,3,2) = 560$. Thus, the coefficient of $x^3 y^3 z^2$ is $2^3(-3)^3(5)^2 P(8;3,3,2) = (2^3)(-3)^3(5^2)(560)$.

(c) $(x - 2y + z)^3 = P(3;3,0,0)x^3 + P(3;0,3,0)(-2)^3 y^3 + P(3;0,0,3)z^3 + P(3;2,0,1)x^2 z + P(3;2,1,0)x^2(-2y) + P(3;1,2,0) x (-2y)^2 + P(3;1,0,2) xz^2 + P(3;0,2,1) (-2y)^2 z + P(3;0,1,2) (-2y)z^2 + P(3;1,1,1) x (-2y)z = x^3 - 8y^3 + z^3 + 3x^2 z - 6x^2 y + 12xy^2 + 3xz^2 + 12y^2 z - 6yz^2 - 12xyz.$

Corollary 2.7.2. For any positive integer t, we have $t^n = \Sigma\, P(n;q_1, q_2, \ldots, q_t)$ where the summation extends over all sets of nonnegative integers $q_1, q_2, \ldots, q_t$ where $q_1 + q_2 + \cdots + q_t = n$.

Proof. Just let $1 = x_1 = x_2 = \cdots = x_t$ in Theorem 2.7.2. $\square$

Corollary 2.7.2 states that there are t^n t-part ordered partitions of a set S with n elements.

Example 2.7.3. Find the number of 3-part unordered partitions of a set S with n distinct elements. We know the number of 3-part ordered partitions of S is 3^n.

Let $P_n(t)$ denote the number of t-part unordered partitions of a set with n elements. We are asked to find $P_n(3)$ in this example.

One 3-part unordered partition of S is $\{S,\phi,\phi\}$ of which there are 3 orderings: (S,ϕ,ϕ), (ϕ,S,ϕ), and (ϕ,ϕ,S). From each of the other $P_n(3) - 1$ unordered partitions there are 3! orderings. Thus, $3!(P_n(3) - 1) + 3 = 3^n$, and $P_n(3) = (3^{n-1} + 1)/2$.

Exercises for Section 2.7

1. What is the coefficient of $x^3 y^7$ in $(x + y)^{10}$? in $(2x - 9y)^{10}$?

2. Using Figure 2-8 complete the 21st row of Pascal's triangle.

3. (a) Use the binomial theorem to prove that $3^n = \Sigma_{r=0}^n C(n,r)\, 2^r$.

(b) Generalize to find the sum $\Sigma_{r=0}^{n} C(n,r)\, t^r$ for any real number t.

(c) Likewise prove that $2^n = \Sigma_{r=0}^{n} (-1)^r\, C(n,r)\, 3^{n-r}$.

4. Use the multinomial theorem to expand $(x_1 + x_2 + x_3 + x_4)^4$.

5. (a) Determine the coefficient of $x_1^3 x_2^2 x_3^2 x_5^3$ in $(x_1 + x_2 + x_3 + x_4 + x_5)^{10}$.

   (b) Determine the coefficient of $x^5 y^{10} z^5 w^5$ in $(x - 7y + 3z - w)^{25}$.

   (c) Determine the number of terms in the expansion of $(x - 7y + 3z - w)^{25}$.

   (d) Determine the coefficient of x^5 in $(a + bx + cx^2)^{10}$.

6. What is the sum of all numbers of the form $12!/q_1!q_2!q_3!$ where q_1, q_2, q_3 range over all sets of nonnegative integers such that $q_1 + q_2 + q_3 = 12$?

7. (a) Use Pascal's identity to prove that $C(2n + 2, n + 1) = C(2n, n + 1) + 2C(2n,n) + C(2n, n - 1)$.

   (b) Therefore $C(2n + 2, n + 1) = 2[C(2n,n) + C(2n, n - 1)]$.

   (c) Consider a set of $2n$ men and 2 women and give a combinatorial argument for the equation in (a).

8. Obtain relations by equating the coefficients of x^k in the following:

   (a) $(1 + x)^{n+1} = (1 + x)\,(C(n,0) + C(n,1)x + \cdots + C(n,n)x^n)$;

   (b) $(1 + x)^{n+2} = (1 + 2x + x^2)\,(C(n,0) + C(n,1)x + \cdots + C(n,n)x^n)$;

   (c) $(1 + x)^{n+3} = (1 + 3x + 3x^2 + x^3)\,(\Sigma_{r=0}^{n} C(n,r)x^r)$.

9. Prove that:
$$[C(n,0) + C(n,1) + \cdots + C(n,n)]^2 = C(2n,0) + C(2n,1) \\ + \cdots + C(2n,2n).$$

10. (a) Show that for $n \geq 2$, $C(n,1) - 2C(n,2) + 3C(n,3) \cdots + (-1)^{n-1} nC(n,n) = 0$.

    (b) Conclude that $C(n,1) + 3C(n,3) + 5C(n,5) \cdots = 2C(n,2) + 4C(n,4) \cdots = n2^{n-2}$; [here the last term is $nC(n,n)$].

    (c) What is the value of $C(n,0) - 2C(n,1) + 3C(n,2) \cdots + (-1)^n (n + 1) C(n,n)$?

    (d) Verify that $C(n,0) + 3C(n,1) + 5C(n,2) + \cdots + (2n + 1) C(n,n) = (n + 1) 2^n$.

    (e) Verify that $C(n,2) + 2C(n,3) + 3C(n,4) + \cdots + (n - 1) C(n,n) = 1 + (n - 2) 2^{n-1}$.

11. (a) Consider $(1 + 2x)^n$ to prove $C(n,0) + 2C(n,1) + 2^2 C(n,2) + \cdots + 2^n C(n,n) = 3^n$.

    (b) Verify $C(n,0) - 2C(n,1) + 2^2 C(n,2) + \cdots + (-1)^n 2^n C(n,n) = (-1)^n$.

(c) Verify a formula for $C(n,0) + 3C(n,1) + 3^2C(n,2) + \cdots + 3^nC(n,n)$ and for $C(n,0) - 3C(n,1) + 3^2C(n,2) + \cdots + (-1)^n3^nC(n,n)$.

12. Give a factorial proof of (a) Pascal's identity and, (b) Newton's identity.

13. (a) Evaluate the sum $1 + 2\,C(n,1) + \cdots + (r + 1)\,C(n,r) + \cdots + (n + 1)\,C(n,r)$ by breaking this sum into 2 sums, each of which is an identity in this section.

    (b) Evaluate the sum $C(n,0) + 2C(n,1) + C(n,2) + 2C(n,3).\ldots$

14. (a) Observe that
$$\frac{(1 + x)^n + (1 - x)^n}{2} = C(n,0) + C(n,2)x^2 + \cdots + C(n,q)x^q,$$

    where

$$q = \begin{cases} n \text{ if } n \text{ is even} \\ n - 1 \text{ if } n \text{ is odd.} \end{cases}$$

    (b) Then verify that

$$C(n,0) + C(n,2) + \cdots + C(n,q) = \begin{cases} 2^{n-1} \text{ for } n > 0 \\ 1 \text{ for } n = 0 \end{cases}$$

    for $q = \begin{cases} n \text{ if } n \text{ is even} \\ n - 1 \text{ if } n \text{ is odd.} \end{cases}$

15. Give a combinatorial proof of Vandermonde's identity. (Hint: Let S be the union set of m men and n women.)

16. Show that the product of k consecutive integers is divisible by $k!$. (Hint: consider the number of ways of selecting k objects from $n + k$ objects.)

17. Show that $P(n;q_1,q_2) = C(n,q_1) = C(n,q_2)$.

18. Give a combinatorial argument that $t^n = \Sigma\,P(n;q_1,q_2, \ldots, q_t)$ where the summation is taken over all sets of nonnegative integers $q_1,q_2, \ldots, q_t$ where $n = q_1 + q_2 + \cdots + q_t$.

19. (a) Among $2n$ objects, n of them are indistinguishable. Find the number of ways to select n of these $2n$ objects. (Hint: first select r objects from the n distinguishable objects; then select $n - r$ objects from the indistinguishable objects.)

    (b) Among $3n + 1$ objects, n of them are indistinguishable. Find the number of ways to select n of these $3n + 1$ objects.

20. The Stirling number of the second kind, $S(n,t)$, denotes the number of t-part unordered partitions of a set with n distinct elements where each cell is nonempty (that is, each $q_i > 0$). Show

by an argument similar to the verification of Pascal's identity (5) that $S(n,t) = S(n-1, t-1) + tS(n-1,t)$. Observe the boundary conditions $S(n,1) = S(n,n) = 1$. Then following the pattern of Pascal's triangle list the values of $S(n,t)$ for $n = 1,2,3,4,5$.

21. Let $T(n,t)$ denote the number of ordered t-part partitions of a set of n distinct elements where each cell is nonempty. Show that $T(n,t) = t[T(n-1), t-1) + T(n-1,t)]$. List the values of $T(n,t)$ for $n = 1,2,3,4,5$. Note that $t! \, S(n,t) = T(n,t)$.

22. Let $P(n)$ denote the number of unordered partitions of a set with n distinct elements where the number of cells is not specified, that is $P(n)$ is the number of t-part unordered partitions for all possible values of t where $1 \le t \le n$. Thus, $P(n) = \Sigma_{t-1}^n S(n,t)$.

Let $Q(n)$ denote the number of ordered t-part partitions for all values of t where $1 \le t \le n$. Then, $Q(n) = \Sigma_{t-1}^n T(n,t)$. Use the results of Problem 20 and 21 to compute $P(n)$ and $Q(n)$ for $n = 1,2,3,4,5$.

23. In how many ways can 7 distinguishable balls be placed in 4 distinguishable boxes if the first box contains 2 balls (and the order of the balls in a box is immaterial)?

24. Verify that
    (a) $C(n+2,r) - 2C(n+1,r) + C(n,r) = C(n,r-2)$;
    (b) $C(n+3,r) - 3C(n+2,r) + 3C(n+1,r) - C(n,r) = C(n,r-3)$; and
    (c) $\Sigma_{j-0}^q (-1)^j C(q,j) C(n+q-j,r) = C(n,r-q)$ where q is a positive integer.

25. (a) Give a combinatorial proof and an algebraic proof for the identity $P(n;q_1,q_2,q_3) = P(n-1;q_1-1,q_2,q_3) + P(n-1; q_1,q_2-1,q_3) + P(n;q_1,q_2,q_3-1)$.
    (b) State a similar formula for $P(n;q_1,q_2,\ldots,q_t)$.

26. Derive identity (10) for row square summation by computing the coefficient of x^n in $(1+x)^{2n}$.

27. Prove by mathematical induction the following two formulas for Stirling numbers of the second kind:
    (a) $S(n,2) = 2^{n-1} - 1$ for $n \ge 1$.
    (b) $S(n,n-1) = C(n,2)$ for $n \ge 1$.

28. (a) Prove that if p is a prime integer and k is an integer such that $0 < k < p$, then p divides $C(p,k)$.
    (b) Use (a) to conclude that the prime p divides $2^p - 2$.

Selected Answers for Section 2.7

3. (a) Consider the binomial expansion of $(1+2)^n$.

7. (c) Choose a committee of $n+1$ from the $2n+2$ people by

choosing $n + 1$ men, or n men and 1 woman, or $n - 1$ men and 2 women.

8. (a) $C(n + 1,k) = C(n,k - 1) + C(n,k)$.
   (b) $C(n + 2,k) = C(n,k) + 2C(n,k - 1) + C(n,k - 2)$.
   (c) $C(n + 3,k) = C(n,k) + 3C(n,k - 1) + 3C(n,k - 2) + C(n,k - 3)$.

9. Recall that $(1 + x)^{2n} = C(2n, 0) + C(2n, 1) x + \ldots + C(2n, 2n) x^{2n}$ $= [(1 + x)^n]^2 = [C(n, 0) + C(n,1)x + \cdots + C(n, n)x^n]^2$. Then in this expression let $x = 1$.

10. (a) $C(n, 1) - 2C(n, 2) + 3C(n, 3) + \cdots + (-1)^{n-1}nC(n, n) = n[C(n - 1, 0) - C(n - 1, 1) + C(n - 1, 2) + \cdots + (-1)^{n-1}C(n - 1, n - 1)] = n \cdot 0 = 0$ by identity (13).
    (c) $C(n, 0) - 2C(n, 1) + 3C(n, 2) \cdots (-1)^n(n + 1)C(n,n) = C(n,0) - C(n,1) - C(n,2) \cdots (-1)^{n-1} C(n, n) - [C(n,1) - 2C(n,2) \cdots (-1)^n n C(n,n)] = 0$ by identity (13) and 10 (a).

14. $(1 + x)^n = C(n,0) + C(n,1)x + \cdots + C(n,n)x^n$
    $(1 - x)^n = C(n,0) - C(n,1)x + \cdots + (-1)^n C(n,n)x^n$

$$\frac{(1 + x)^n + (1 - x)^n}{2} = C(n,0) + C(n,2)x^2 + \cdots$$

$$+ C(n,q)x^q \text{ where}$$

$$q = \begin{cases} n \text{ if } n \text{ is even} \\ n - 1 \text{ if } n \text{ is odd} \end{cases}$$

Let $x = 1$ in the above expression and we get

$$C(n,0) + C(n,2) + \cdots + C(n,q)$$

$$\begin{cases} 2^{n-1} \text{ for } n > 0 \\ 1 \text{ for } n = 0. \end{cases}$$

24. See exercise 8.

2.8 THE PRINCIPLE OF INCLUSION-EXCLUSION

In Section 2.1 we discussed the sum rule by which we can count the number of elements in the union of **disjoint** sets. However, if the sets are not disjoint we must refine the statement of the sum rule to a rule commonly called the **principle of inclusion-exclusion** (it is sometimes called the **sieve method**).

First Statement: If A and B are subsets of some universe set U, then

$$|A \cup B| = |A| + |B| - |A \cap B|. \qquad (2.8.1)$$

This is fairly clear from a Venn diagram illustrated in Figure 2–9 since in counting the elements of A and the elements of B we have counted the elements of $A \cap B$ twice.

But it is also clear that $A \cup B$ is the union of the 3 disjoint sets

$$A \cap \overline{B},\ A \cap B \text{ and } \overline{A} \cap B, \text{ so that by the sum rule,}$$
$$|A \cup B| = |A \cap \overline{B}| + |A \cap B| + |\overline{A} \cap B| \qquad (2.8.2)$$

since $(A \cap \overline{B}) \cup (A \cap B) = A$ and $B = (\overline{A} \cap B) \cup (A \cap B)$, we see that $|A| = |A \cap \overline{B}| + |A \cap B|$ and $|B| = |\overline{A} \cap B| + A \cap B|$. The sum of these two equations is

$$|A| + |B| = |A \cap \overline{B}| + |A \cap B| + |\overline{A} \cap B| + A \cap B|. \qquad (2.8.3)$$

The combination of Equations (2.8.2) and (2.8.3) gives the desired result:

$$|A \cup B| = |A| + |B| - |A \cap B|.$$

Of course, if $A \cap B = \phi$, then this is just the sum rule.

Since sets are frequently defined in terms of properties, we translate this equation into the following statement:

The number of elements with either of the properties A or B equals the number of elements with property A plus the

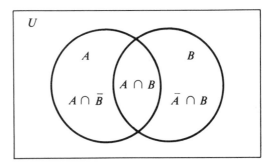

Figure 2-9

number of elements with property B minus the number of elements that satisfy *both* properties A and B.

A simple example will suffice to illustrate this statement.

Example 2.8.1. Suppose that 200 faculty members can speak French and 50 can speak Russian, while only 20 can speak both French and Russian. How many faculty members can speak either French or Russian?

If F is the set of faculty who speak French and R is the set of faculty who speak Russian, then we know that $|F| = 200, |R| = 50$ and $|F \cap R| = 20$. We are asked to compute $|F \cup R|$ which by the formula is $200 + 50 - 20 = 230$.

The principle of inclusion-exclusion offers an alternative method of solving some combinatorial problems described below.

Example 2.8.2. From a group of 10 professors how many ways can a committee of 5 members be formed so that at least one of Professor A and Professor B will be included?

We will give 3 solutions; one using the sum rule, one by counting indirectly, and one using the principle of inclusion-exclusion.

The number of committees including *both* Professor A and Professor B is $C(8,3) = 56$. The number of committees including Professor A but excluding Professor B is $C(8,4) = 70$, (since Professor A fills one position there are only 4 more positions to fill and since Professor B is excluded there are only 8 people from which to choose). Likewise $C(8,4)$ is the number of committees including Professor B and excluding Professor A. Consequently by the *sum rule* the total number of ways of selecting a committee of 5 including Professor A *or* Professor B is $C(8,3) + 2C(8,4) = 56 + 2 \cdot 70 = 196$. By counting indirectly we obtain a second solution; we just need the observation that the total number of committees excluding *both* Professor A and B is $C(8,5) = 56$ and the total number of committees is $C(10,5)$. Thus, we see that $C(10,5) - C(8,5) = 252 - 56 = 196$ is the number of committees including at least one of the professors.

A third solution uses the principle of inclusion-exclusion. Among the 252 committees of 5 members, let A_1 and A_2 be the set of committees that *include* Professor A and Professor B, respectively. Since $|A_1| = C(9,4) = 126 = |A_2|$ and $|A_1 \cap A_2| = C(8,3) = 56$, it follows that $|A_1 \cup A_2| = 126 + 126 - 56 = 196$.

Likewise we can obtain a statement of the principle of inclusion-exclusion for 3 sets.

If A, B, C are any 3 subsets of the universal set U, we can find $|A \cup B \cup C|$ by examining the Venn diagram illustrated in Figure 2–10. We see that

$$A = (A \cap \overline{B} \cap \overline{C}) \cup (A \cap B \cap \overline{C}) \cup (A \cap \overline{B} \cap C) \cup (A \cap B \cap C),$$

$$B = (\overline{A} \cap B \cap \overline{C}) \cup (A \cap B \cap \overline{C}) \cup (\overline{A} \cap B \cap C) \cup (A \cap B \cap C),$$

$$C = (\overline{A} \cap \overline{B} \cap C) \cup (A \cap \overline{B} \cap C) \cup (\overline{A} \cap B \cap C) \cup (A \cap B \cap C).$$

Therefore since these are disjoint unions we can use the sum rule to compute:

$$|A| = |A \cap \overline{B} \cap \overline{C}| + A \cap B \cap \overline{C}| + |A \cap \overline{B} \cap C| + |A \cap B \cap C|,$$
(2.8.4)

$$|B| = |\overline{A} \cap B \cap \overline{C}| + |A \cap B \cap \overline{C}| + |\overline{A} \cap B \cap C| + |A \cap B \cap C|,$$
(2.8.5)

$$|C| = |\overline{A} \cap \overline{B} \cap C| + |A \cap \overline{B} \cap C| + |\overline{A} \cap B \cap C| + |A \cap B \cap C|,$$
(2.8.6)

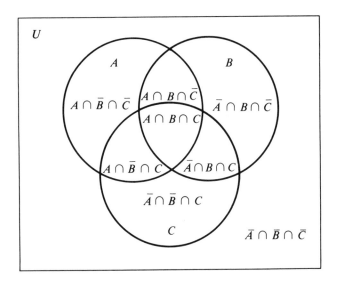

Figure 2-10

Adding Equations (2.8.4), (2.8.5), and (2.8.6), we have:

$$|A| + |B| + |C| = |A \cap \overline{B} \cap \overline{C}| + |A \cap B \cap \overline{C}| + |A \cap \overline{B} \cap C|$$
$$+ |A \cap B \cap C| + |\overline{A} \cap B \cap \overline{C}| + |\overline{A} \cap B \cap C|$$
$$+ |\overline{A} \cap \overline{B} \cap C| + |A \cap B \cap \overline{C}| + |A \cap B \cap C|$$
$$+ |A \cap \overline{B} \cap C| + |A \cap B \cap C| + |\overline{A} \cap B \cap C|.$$

The first 7 of these sets make up $A \cup B \cup C$, the next 2 make up $A \cap B$, and the next 2 give $A \cap C$. Thus, we have $|A| + |B| + |C| = |A \cup B \cup C| + |A \cap B| + |A \cap C| + |\overline{A} \cap B \cap C|$.

By rearranging terms, we have $|A \cup B \cup C| = |A| + |B| + |C| - |A \cap B| - |A \cap C| - |\overline{A} \cap B \cap C|$.

But we would like an expression free of complements. We note that $|\overline{A} \cap B \cap C| + |A \cap B \cap C| = |B \cap C|$ so that we have the following theorem.

Theorem 2.8.1.   If A, B, and C are finite sets, then

$$|A \cup B \cup C| = |A| + |B| + |C| - |A \cap B| - |A \cap C|$$
$$- |B \cap C| + |A \cap B \cap C| \qquad (2.8.7)$$

Example 2.8.3.   If there are 200 faculty members that speak French, 50 that speak Russian, 100 that speak Spanish, 20 that speak French and Russian, 60 that speak French and Spanish, 35 that speak Russian and Spanish, while only 10 speak French, Russian, and Spanish, how many speak either French or Russian or Spanish?

Let F be the set of faculty who speak French, R be the set of faculty who speak Russian, and S be the set of faculty that speak Spanish. We know from example 2.8.1. that $|F| = 200, |R| = 50, |S| = 100, |F \cap R| = 20$, $|F \cap S| = 60, |R \cap S| = 35$ and $|F \cap R \cap S| = 10$. Thus, $|F \cup R \cup S| = 200 + 50 + 100 - 20 - 60 - 35 + 10 = 245$.

Frequently it has been beneficial to count indirectly and there is a form of the principle of inclusion-exclusion that encompasses counting complements and DeMorgan's laws. Let us explain this version for 2 sets and we will give the general version later.

If A and B are subsets of a universal set U, then $\overline{A} \cap \overline{B} = \overline{A \cup B}$ by

DeMorgan's laws, and then by Equation (2.8.1), we have

$$
\begin{aligned}
|\overline{A} \cap \overline{B}| = |\overline{A \cup B}| &= |U| - |A \cup B| \\
&= |U| - \{|A| + |B| - |A \cap B|\} \qquad (2.8.8) \\
&= |U| - |A| - |B| + |A \cap B|.
\end{aligned}
$$

Example 2.8.4. If in Example 2.8.1 there are 1,000 faculty altogether, then there are $1{,}000 - |F| - |R| + |F \cap R| = 1{,}000 - 200 - 50 + 20 = 1{,}000 - |F \cup R| = 1{,}000 - 230 = 770$ people who speak neither French nor Russian.

Likewise, if A, B, and C are subsets of U, we can apply DeMorgan's laws and Equation (2.8.7) to get

$$
\begin{aligned}
|\overline{A} \cap \overline{B} \cap \overline{C}| = |\overline{A \cup B \cup C}| &= |U| - |A \cup B \cup C| \\
&= |U| - \{|A| + |B| + |C| - |A \cap B| \\
&\quad - |A \cap C| - |B \cap C| + |A \cap B \cap C|\} \qquad (2.8.9) \\
&= |U| - |A| - |B| - |C| + |A \cap B| \\
&\quad + |A \cap C| + |B \cap C| - |A \cap B \cap C|.
\end{aligned}
$$

Thus in Example 2.8.2 there are $1{,}000 - 245 = 755 = 1{,}000 - 200 - 50 - 100 + 20 + 60 + 35 - 10$ faculty who do not speak either of the 3 languages.

Example 2.8.5. In a survey of students at Florida State University the following information was obtained: 260 were taking a statistics course, 208 were taking a mathematics course, 160 were taking a computer programming course, 76 were taking statistics and mathematics, 48 were taking statistics and computer programming, 62 were taking mathematics courses and computer programming, 30 were taking all 3 kinds of courses, and 150 were taking none of the 3 courses.
   Let

$$
\begin{aligned}
S &= \{\text{students taking statistics}\}, \\
M &= \{\text{students taking mathematics}\}, \\
C &= \{\text{students taking computer programming}\}.
\end{aligned}
$$

(a) How many students were surveyed?
(b) How many students were taking a statistics and a mathematics course but not a computer programming course?
(c) How many were taking a statistics and a computer course but not a mathematics course?

(d) How many were taking a computer programming and a mathematics course but not a statistics course?

(e) How many were taking a statistics course but not taking a course in mathematics or in computer programming?

(f) How many were taking a mathematics course but not taking a statistics course or a computer programming course?

(g) How many were taking a computer programming course but not taking a course in mathematics or in statistics?

The Venn diagram illustrated in Figure 2-11 will also be helpful in our analysis.

We know that $|S| = 260, |M| = 208, |C| = 160, |S \cap M| = 76, |S \cap C| = 48, |M \cap C| = 62, |S \cap M \cap C| = 30$, and $|\overline{S \cup M \cup C}| = 150$; so we can immediately insert the number of students in 2 of the 8 regions of the Venn diagram, the regions $S \cap M \cap C$ and $\overline{S \cup M \cup C}$.

(a) The total number of students surveyed:

$$
\begin{aligned}
|U| &= |S \cup M \cup C| + |\overline{S \cup M \cup C}| \\
&= |S| + |M| + |C| - |S \cap M| - |S \cap C| - |M \cap C| \\
&\quad + |S \cap M \cap C| + |\overline{S \cup M \cup C}| \\
&= 260 + 208 + 160 - 76 - 48 - 62 + 30 + 150 = 622.
\end{aligned}
$$

(b) We are asked to find:

$$
\begin{aligned}
|S \cap M \cap \overline{C}| &= |S \cap M - S \cap M \cap C| = |S \cap M| - |S \cap M \cap C| \\
&= 76 - 30 = 46.
\end{aligned}
$$

(c) $|S \cap C \cap \overline{M}| = |S \cap C| - |S \cap C \cap M| = 48 - 30 = 18.$

(d) $|M \cap C \cap \overline{S}| = |M \cap C| - |M \cap C \cap S| = 62 - 30 = 32.$

Figure 2-11

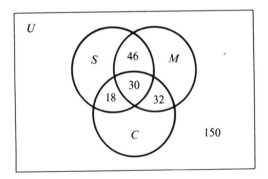

Figure 2-12

Thus, we can insert some more numbers in the appropriate regions of the Venn diagram (see Figure 2-12).

Only the computation of the number of students taking courses in exactly one of the 3 subjects remains, and that is precisely the content of parts (e), (f), and (g).

(e) $|S \cap \overline{M} \cap \overline{C}| = |S| - |S \cap M \cap \overline{C}| - |S \cap C \cap \overline{M}| - |S \cap M \cap C| = 260 - 46 - 18 - 30 = 166.$

(f) $|M \cap \overline{S} \cap \overline{C}| = |M| - |M \cap S \cap \overline{C}| - |M \cap C \cap \overline{S}| - |M \cap S \cap C| = 100$

(g) $|C \cap \overline{S} \cap \overline{M}| = |C| - |C \cap S \cap \overline{M}| - |C \cap \overline{S} \cap M| - |C \cap S \cap M| = 80$

Thus, we can fill in the numbers for all 8 regions in the Venn diagram illustrated in Fig 2-13.

Let us now proceed to give a general formulation of the principle of inclusion-exclusion. If $P_1, P_2, \cdots, P_n$ are properties that elements of a universal set U may or may not satisfy, then for each i, let A_i be the set of those elements of U that satisfy the property P_i. Then $A_1 \cup A_2 \cup \cdots \cup A_n$ is the set of all elements of U that satisfy at least one of the properties $P_1, P_2, \cdots, P_n$, while $A_i \cap A_j$ is the set of elements that satisfy both the properties P_i and P_j, $A_i \cap A_j \cap A_k$ is the set of elements that satisfy the 3 properties P_i, P_j, and P_k. The set $A_1 \cap A_2 \cap \cdots \cap A_n$ is the set of elements that satisfy all of the properties $P_1, P_2, \cdots, P_n$. The set of elements that do not satisfy P_i is $\overline{A}_i$ and $\overline{A}_1 \cap \overline{A}_2 \cap \cdots \cap \overline{A}_n$ is the set of elements that satisfy none of the properties.

Let us list a few facts that will be useful. Suppose A, B, A_i are subsets of U.

1. $|\overline{A}| = |U| - |A|$ (counting indirectly).
2. $|B - A| = |B \cap \overline{A}| = |B - (A \cap B)| = |B| - |A \cap B|$ (counting relative complements).

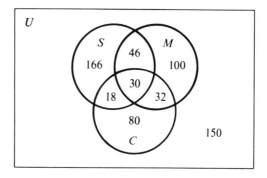

Figure 2-13

3. $|\overline{A_1} \cap \overline{A_2} \cap \cdots \cap \overline{A_n}| = |\overline{A_1 \cup A_2 \cup \cdots \cup A_n}| = |U| - |A_1 \cup A_2 \cup \cdots \cup A_n|$ (counting by DeMorgan's law).

Theorem 2.8.1. (General statement of the principle of inclusion-exclusion). If A_i are finite subsets of a universal set U, then

$$|A_1 \cup A_2 \cup \cdots \cup A_n| = \sum_{i=1}^{n} |A_i| - \sum_{i,j} |A_i \cap A_j|$$

$$+ \sum_{i,j,k} |A_i \cap A_j \cap A_k| \qquad (2.8.10)$$

$$+ \cdots +$$

$$(-1)^{n-1} |A_1 \cap A_2 \cdots \cap A_n|,$$

where the second summation is taken over all 2-combinations $\{i,j\}$ of the integers $\{1,2,\ldots,n\}$, the third summation is taken over all 3-combinations $\{i,j,k\}$ of $\{1,2,\ldots,n\}$, and so on.

For $n = 4$ there are $4 + C(4,2) + C(4,3) + 1 = 2^4 - 1 = 15$ terms and the theorem states that

$$|A_1 \cup A_2 \cup A_3 \cup A_4| = |A_1| + |A_2| + |A_3| + |A_4|$$
$$- |A_1 \cap A_2| - |A_1 \cap A_3| - |A_1 \cap A_4|$$
$$- |A_2 \cap A_3| - |A_2 \cap A_4| - |A_3 \cap A_4|$$
$$+ |A_1 \cap A_2 \cap A_3| + |A_1 \cap A_2 \cap A_4|$$
$$+ |A_1 \cap A_3 + \cap A_4| + |A_2 \cap A_3 \cap A_4|$$
$$- |A_1 \cap A_2 \cap A_3 \cap A_4|.$$

In general there are $\binom{n}{1} + \binom{n}{2} + \binom{n}{3} + \cdots + \binom{n}{n} = 2^n - 1$ terms on the right-hand side of Equation (2.8.10).

Proof. The theorem will be proved by induction on the number n of subsets A_i. The theorem is obviously true for $n = 1$, and we have indicated why the theorem holds for $n = 2$ and $n = 3$. Assume the theorem holds for any n subsets of U. Suppose, then, that we have $n + 1$ sets $A_1, A_2, \ldots, A_n, A_{n+1}$. We show the formula holds with n replaced by $n + 1$. We will use the result for 2 sets repeatedly in the proof.

Consider $A_1 \cup A_2 \cup \ldots \cup A_n \cup A_{n+1}$ as the union of the sets $A_1 \cup A_2 \cup \cdots \cup A_n$ and A_{n+1}. Then

$$
\begin{aligned}
|A_1 \cup A_2 \cup \cdots \cup A_n \cup A_{n+1}| &= |(A_1 \cup \cdots \cup A_n) \cup A_{n+1}| \\
&= |A_1 \cup A_2 \cup \cdots \cup A_n| + |A_{n+1}| \quad (2.8.11) \\
&\quad - |(A_1 \cup A_2 \cup \cdots \cup A_n) \cap A_{n+1}|
\end{aligned}
$$

Use the fact that intersection distributes over unions to get that

$$
\begin{aligned}
|(A_1 \cup A_2 \cup \cdots \cup A_n) \cap A_{n+1}| \\
= |(A_1 \cap A_{n+1}) \cup (A_2 \cap A_{n+1}) \cup \cdots \cup (A_n \cap A_{n+1})|.
\end{aligned}
$$

Thus, 2.8.11 becomes

$$
\begin{aligned}
|A_1 \cup A_2 \cup \cdots \cup A_n| + |A_{n+1}| \\
- |(A_1 \cap A_{n+1}) \cup (A_2 \cap A_{n+1}) \cup \cdots \cup (A_n \cap A_{n+1})|. \quad (2.8.12)
\end{aligned}
$$

We can apply the inductive hypothesis to 2 of the 3 sets in equation (2.8.12), namely,

$$
\begin{aligned}
|A_1 \cup A_2 \cup \cdots \cup A_n| = \sum_{i=1}^{n} |A_i| - \sum_{i,j \leq n} |A_i \cap A_j| \\
+ \sum_{i,j,k \leq n} |A_i \cap A_j \cap A_k| \quad (2.8.13) \\
+ \cdots + (-1)^{n-1} \\
|A_1 \cap A_2 \cap \cdots \cap A_n|
\end{aligned}
$$

and

$$
\begin{aligned}
|(A_1 \cap A_{n+1}) \cup (A_2 \cap A_{n+1}) \cup \cdots \cup (A_n \cap A_{n+1})| = \\
\sum_{i=1}^{n} |A_i \cap A_{n+1}| - \sum_{i,j} |A_i \cap A_{n+1} \cap (A_j \cap A_{n+1})| \\
+ \sum_{i,j,k \leq n} |(A_i \cap A_{n+1}) \cap (A_j \cap A_{n+1}) \cap (A_k \cap A_{n+1})| \\
\quad (2.8.14) \\
+ \cdots + (-1)^{n-1} |(A_1 \cap A_{n+1}) \cap (A_2 \cap A_{n+1}) \cap \cdots \\
\cap (A_n \cap A_{n+1})|.
\end{aligned}
$$

Substituting Equations (2.8.13) and (2.8.14) into (2.8.12) and making simplifications like $(A_i \cap A_{n+1}) \cap (A_j \cap A_{n+1}) \cap (A_k \cap A_{n+1}) = A_i \cap A_j \cap A_k \cap A_{n+1}$, we have

$$|A_1 \cup A_2 \cup \cdots \cup A_n \cup A_{n+1}| =$$

$$\sum_{i=1}^{n}|A_i| - \sum_{i,j \leq n}|A_i \cap A_j| + \sum_{i,j,k \leq n}|A_i \cap A_j \cap A_k| + \cdots$$

$$+ (-1)^{n-1}|A_1 \cap A_2 \cap \cdots \cap A_n| + |A_{n+1}| - \sum_{i=1}^{n}|A_i \cap A_{n+1}| \qquad (2.8.15)$$

$$+ \sum_{i,j \leq n}|A_i \cap A_j \cap A_{n+1}| + \cdots +$$

$$(-1)^n|A_1 \cap A_2 \cap \cdots \cap A_n \cap A_{n+1}|.$$

Now observe that $(\Sigma_{i,j \leq n}|A_i \cap A_j| + \Sigma_{i=1}^{n}|A_i \cap A_{n+1}|)$, where the first sum is taken over I, the 2-combinations $\{i,j\}$ of $\{1,2,\ldots,n\}$, and the second sum is taken over J, the 2-combinations of the form $\{i, n+1\}$ where $i \in \{1,2,\ldots,n\}$, can be simplified to $\Sigma A_i \cap A_j$ where this sum is taken over all 2-combinations of $\{1,2,\ldots,n,n+1\}$ since $I \cup J$ is the set of all 2-combinations of $\{1,2,\ldots,n, n+1\}$. Likewise the two sums

$$\sum_{i,j,k \leq n}|A_i \cap A_j \cap A_k| + \sum_{i,j \leq n}|A_i \cap A_j \cap A_{n+1}|,$$

where the first sum is taken over all 3-combinations of $\{1,2,\ldots,n\}$ and the second is taken over all 3-combinations of the form $\{i, j, n+1\}$ of $\{1,2,\ldots,n,n+1\}$, can be simplified to $\Sigma_{i,j,k}|A_i \cap A_j \cap A_k|$ where this sum is taken over all 3-combinations of $\{1,2,\ldots,n,n+1\}$. Other similar simplifications can also be made. Thus, Equation (2.8.15) becomes

$$|A_1 \cup A_2 \cup \cdots \cup A_n \cup A_{n+1}| =$$

$$\sum_{i=1}^{n+1}|A_i| - \sum_{i,j}|A_i \cap A_j| + \sum_{i,j,k}|A_i \cap A_j \cap A_k| \qquad (2.8.16)$$

$$+ \cdots + (-1)^n|A_1 \cap A_2 \cdots \cap A_{n+1}|,$$

where the second sum is taken over all 2-combinations of $\{1,2,\ldots,n+1\}$, the third sum is taken over all 3-combinations of $\{1,2,\cdots,n+1\}$, and so on. In other words, Equation (2.8.16) is just Equation (2.8.10) with n replaced by $n+1$.

The theorem, then, is proved by mathematical induction. $\square$

Let us give an alternate proof of the theorem for additional clarity. We must show that every element of $A_1 \cup A_2 \cup \cdots \cup A_n$ is counted exactly once in the right-hand side of Equation (2.8.10).

Suppose that an element $x \in A_1 \cup A_2 \cup \cdots \cup A_n$ is in exactly m of the sets, for definiteness, say $x \in A_1$, $x \in A_2, \ldots$, $x \in A_m$, and $x \notin A_{m+1}, \ldots, x \notin A_n$. Then x will be counted in each of the terms $|A_i|$ for $i = 1$, $2, \ldots, m$, in other words, x will be counted $\binom{m}{1}$ times in the $\Sigma_{i=1}^{n} |A_i|$.

Furthermore, x will be counted $C(m,2)$ times in $\Sigma |A_i \cap A_j|$ since there are $C(m,2)$ pairs of sets A_i, A_j where x is in both A_i and A_j.

Likewise, x is counted $C(m,3)$ times in $\Sigma |A_i \cap A_j \cap A_k|$ since there are $C(m,3)$ 3-combinations A_i, A_j, A_k where $x \in A_i$, $x \in A_j$, and $x \in A_k$ (namely 3-combinations of the sets $A_1, A_2, \ldots, A_m$). Continuing in like manner, we see that on the right side, x is counted

$$C(m,1) - C(m,2) + C(m,3) + \cdots + (-1)^{m-1} C(m,m) \quad \text{times.} \quad (2.8.17)$$

Now we must show that this last expression is 1. Expanding $(1 - 1)^m$ by the binomial theorem yields $0 = (1 - 1)^m = C(m,0) - C(m,1) + C(m,2) + \cdots + (-1)^m C(m,m)$. Use the fact that $C(m,0) = 1$, and transpose all the other terms to the left-hand side of the last equation, and change signs to see that Equation (2.8.17) is equal to 1. $\square$

Corollary 2.8.1

$$|\overline{A}_1 \cap \overline{A}_2 \cap \cdots \cap \overline{A}_n| = |U| - |A_1 \cup A_2 \cup \cdots \cup A_n|$$

$$= |U| - \sum_{i=1}^{n} |A_i| + \sum_{i,j} |A_i \cap A_j| \qquad (2.8.18)$$

$$- \sum_{i,j,k} |A_i \cap A_j \cap A_k| + \cdots$$

$$+ (-1)^n |A_1 \cap A_2 \cap \cdots \cap A_n|.$$

In general, the principle of inclusion-exclusion can be used together with Theorem 2.4.2 to count the number of integral solutions of an equation $x_1 + x_2 + \cdots + x_n = r$ where for each i, the solution x_i are bounded above and below by integers b_i and c_i. Let us give an example.

Example 2.8.6. Count the number of integral solutions to (2.8.19)

$$x_1 + x_2 + x_3 = 20 \quad \text{where} \quad 2 \le x_1 \le 5,$$
$$4 \le x_2 \le 7 \quad \text{and} \quad -2 \le x_3 \le 9. \qquad (2.8.19)$$

Let U be the set of solutions (x_1, x_2, x_3) where $2 \le x_1$, $4 \le x_2$, $-2 \le x_3$. We know that $|U| = C(20 - 2 - 4 + 2 + 3 - 1, 3 - 1) = C(18,2)$.
Let

$$A_1 = \{(x_1, x_2, x_3) \in U \mid x_1 \ge 6\}$$

$$A_2 = \{(x_1, x_2, x_3) \in U \,|\, x_2 \geq 8\}$$
$$A_3 = \{(x_1, x_2, x_3) \in U \,|\, x_3 \geq 10\}.$$

We wish to count the number of elements in $\overline{A}_1 \cap \overline{A}_2 \cap \overline{A}_3 = \overline{(A_1 \cup A_2 \cup A_3)}$.

By the principle of inclusion-exclusion $|\overline{(A_1 \cup A_2 \cup A_3)}| = |U| - |A_1 \cup A_2 \cup A_3| = |U| - \{|A_1| + |A_2| + |A_3| - |A_1 \cap A_2| - |A_1 \cap A_3| - |A_2 \cap A_3| + |A_1 \cap A_2 \cap A_3|\}$.

Now A_1 is the set of solutions of Equation (2.8.19) where $6 \leq x_1, 4 \leq x_2, -2 \leq x_3$.

Thus, $|A_1| = C(20 - 6 - 4 + 2 + 3 - 1, 3 - 1) = C(14,2)$.

Similarly, A_2 is the set of solutions of Equation (2.8.19) where $2 \leq x_1, 8 \leq x_2, -2 \leq x_3$.

Therefore, $|A_2| = C(20 - 2 - 8 + 2 + 3 - 1, 3 - 1) = C(14,2)$.

Likewise, $|A_3| = C(20 - 2 - 4 - 10 + 3 - 1, 3 - 1) = C(6,2)$.

Now $A_1 \cap A_2$ is the set of solutions where $6 \leq x_1, 8 \leq x_2, -2 \leq x_3$.

Thus $|A_1 \cap A_2| = C(20 - 6 - 8 + 2 + 3 - 1, 3 - 1) = C(10,2)$.

Likewise $|A_1 \cap A_3| = C(20 - 6 - 4 - 10 + 3 - 1, 3 - 1) = C(2,2)$, and $|A_2 \cap A_3| = C(20 - 2 - 8 - 10 + 3 - 1, 3 - 1) = C(2,2)$.

Moreover, $|A_1 \cap A_2 \cap A_3| = 0$ since 20 does not exceed $6 + 8 + 10$.

Therefore, $|\overline{(A_1 \cup A_2 \cup A_3)}| = C(18,2) - 2C(14,2) - C(6,2) + C(10,2) + 2C(2,2)$.

Example 2.8.7. In how many ways can the letters $\{5 \cdot a, 4 \cdot b, 3 \cdot c\}$ be arranged so that all the letters of the same kind are not in a single block?

Let U be the set of $12!/[5!4!3!]$ permutations of these letters. Let A_1 be the arrangements of the letters where the 5 a's are in a single block, A_2 the arrangements where the 4 b's are in a single block, and A_3 the arrangements where the 3 c's are in one block. Then,

$$|A_1| = \frac{8!}{4!3!}, |A_2| = \frac{9!}{5!3!}, |A_3| = \frac{10!}{5!4!},$$

$$|A_1 \cap A_2| = \frac{5!}{3!}, |A_1 \cap A_3| = \frac{6!}{4!}, |A_2 \cap A_3| = \frac{7!}{5!},$$

$$|A_1 \cap A_2 \cap A_3| = 3! \,.$$

Thus,

$$|\overline{A}_1 \cap \overline{A}_2 \cap \overline{A}_3| = \frac{12!}{5!4!3!} - \left(\frac{8!}{4!3!} + \frac{9!}{5!3!} + \frac{10!}{5!4!}\right) + \frac{5!}{3!} + \frac{6!}{4!} + \frac{7!}{5!} - 3!.$$

Example 2.8.8. (The sieve of Eratosthenes). One of the great mysteries of mathematics is the distribution of prime integers among the positive integers. Sometimes primes are separated by only one integer like 17 and 19, 29 and 31, but at other times they are separated by arbitrarily large gaps. A method developed by the Greek mathematician Eratosthenes who lived in Alexandria in the third century B.C. gives a way of listing all primes between 1 and n. His procedure is the following: Remove all multiples of 2 other than 2. Keep the first remaining integer exceeding 2, namely, the prime 3. Remove all multiples of 3 except 3 itself. Keep the first remaining integer exceeding 3, namely, the prime 5. Remove all the multiples of 5 except 5, and so on. The retained numbers are the primes. This method is called the "the sieve of Eratosthenes."

We now compute how many integers between 1 and 1,000 are not divisible by 2, 3, 5, or 7, that is, how many integers remain after the first 4 steps of Eratosthenes' sieve method. The problem is solved using the principle of inclusion-exclusion.

Let U be the set of integers x such that $1 \le x \le 1,000$. Let A_1, A_2, A_3, A_4 be the set of elements of U divisible by 2, 3, 5, and 7, respectively. Thus, $A_1 \cap A_2$ denotes those positive integers $\le 1,000$ that are divisible by 6, $A_1 \cap A_3$ those divisible by 10, $A_2 \cap A_3 = \{$all integers $x \mid 1 \le x \le 1,000$ and x is divisible by 15$\}$ and so on. We wish to compute $|\overline{A_1} \cap \overline{A_2} \cap \overline{A_3} \cap \overline{A_4}|$. We know that

$$|A_1| = \frac{1,000}{2} = 500, \qquad |A_2| = \left\lfloor \frac{1,000}{3} \right\rfloor$$

(where $\lfloor x \rfloor$ means the greatest integer $\le x$),

$$|A_3| = \frac{1,000}{5} = 200,$$

$$|A_4| = \left\lfloor \frac{1,000}{7} \right\rfloor = 142,$$

$$|A_1 \cap A_2| = \left\lfloor \frac{1,000}{6} \right\rfloor = 166$$

$$|A_1 \cap A_3| = \frac{1,000}{10} = 100,$$

$$|A_1 \cap A_4| = \left\lfloor \frac{1,000}{14} \right\rfloor = 71,$$

$$|A_2 \cap A_3| = \left\lfloor \frac{1,000}{15} \right\rfloor = 66,$$

$$|A_2 \cap A_4| = \left\lfloor \frac{1,000}{21} \right\rfloor = 47,$$

$$|A_3 \cap A_4| = \left\lfloor \frac{1,000}{35} \right\rfloor = 28,$$

$$|A_1 \cap A_2 \cap A_3| = \left\lfloor \frac{1,000}{30} \right\rfloor = 33,$$

$$|A_1 \cap A_2 \cap A_4| = \left\lfloor \frac{1,000}{42} \right\rfloor = 23,$$

$$|A_1 \cap A_3 \cap A_4| = \left\lfloor \frac{1,000}{70} \right\rfloor = 14,$$

$$|A_2 \cap A_3 \cap A_4| = \left\lfloor \frac{1,000}{105} \right\rfloor = 9,$$

$$|A_1 \cap A_2 \cap A_3 \cap A_4| = \left\lfloor \frac{1,000}{210} \right\rfloor = 4.$$

Then,

$$|A_1 \cup A_2 \cup A_3 \cup A_4| = 500 + 333 + 200 + 142 - 166$$
$$- 100 - 71 - 66 - 47 - 28$$
$$+ 33 + 23 + 14 + 9 - 4 = 772.$$

Thus,

$$|\overline{A}_1 \cap \overline{A}_2 \cap \overline{A}_3 \cap \overline{A}_4| = 1,000 - 772 = 228.$$

Example 2.8.9. (Euler's ϕ-function). Two positive integers are said to be relatively prime if 1 is the only common positive divisor. If n is a positive integer, $\phi(n)$ is the number of integers x such that $1 \leq x \leq n$ and such that n and x are relatively prime. For example, $\phi(30) = 8$ because 1, 7, 11, 13, 17, 19, 23, and 29 are the only positive integers less than 30 and relatively prime to 30. Let $U = \{1, 2, \ldots, n\}$, and suppose $P_1, \ldots, P_k$ are the distinct prime divisors of n. Let A_i denote the subset of U consisting of those integers divisible by P_i. The integers in U relatively prime to n are those in none of the subsets $A_1, A_2, \ldots, A_k$, so $\phi(n) = |\overline{A}_1 \cap \overline{A}_2 \cap \cdots \cap \overline{A}_n| = |U| - |A_1 \cup A_2 \cup \cdots \cup A_k|$. If d divides n, then there are n/d multiples of d in U. Hence

$$|A_i| = \frac{n}{P_i}, |A_i \cap A_j| = \frac{n}{P_i P_j}, \ldots, |A_1 \cap A_2 \cap \ldots \cap A_k| = \frac{n}{P_1 P_2 \ldots P_k}.$$

Thus, by the principle of inclusion-exclusion,

$$\phi(n) = n - \sum_{i=1}^{k} \frac{n}{P_i} + \sum_{1 \le i < j \le k} \frac{n}{P_i P_j} + \cdots + (-1)^k \frac{n}{P_1 P_2 \ldots P_k}.$$

This last expression can be seen to equal the product

$$n \left[1 - \frac{1}{P_1}\right]\left[1 - \frac{1}{P_2}\right] \cdots \left[1 - \frac{1}{P_k}\right]$$

Thus, in this formula since $30 = 2 \cdot 3 \cdot 5$, $\phi(30) = 30[1 - (1/2)] [1 - (1/3)] [1 - (1/5)] = 30(1/2) (2/3) (4/5) = 8$.

In many applications of the principle of inclusion-exclusion there is a symmetry about the properties so that all the sets A_i have the same number of elements, the intersection of any pair of sets have the same number of elements, and so on. That is,

$$|A_1| = |A_2| = \cdots = |A_i| = \cdots = |A_n|,$$
$$|A_1 \cap A_2| = |A_1 \cap A_3| = \cdots = |A_i \cap A_j| = \cdots = |A_{n-1} \cap A_n|,$$
$$|A_1 \cap A_2 \cap A_3| = \cdots = |A_i \cap A_j \cap A_k| \ldots$$

Then since there are $C(n,1) = n$ 1-combinations of the sets, $C(n,2)$ 2-combinations, etc., we see that

$$\begin{aligned}
|A_1 \cup A_2 \cup \cdots \cup A_n| &= n|A_1| - C(n,2)|A_1 \cap A_2| \\
&+ C(n,3)|A_1 \cap A_2 \cap A_3| + \cdots \\
&+ (-1)^{n-1}|A_1 \cap A_2 \cap \cdots \cap A_n|,
\end{aligned} \tag{2.8.20}$$

and

$$\begin{aligned}
|\overline{A}_1 \cap \overline{A}_2 \cap \cdots \cap \overline{A}_n| &= |U| - n|A_1| \\
&+ C(n,2)|A_1 \cap A_2| - C(n,3)|A_1 \cap A_2 \cap A_3| + \cdots \\
&+ (-1)^n|A_1 \cap A_2 \cap \cdots \cap A_n|.
\end{aligned} \tag{2.8.21}$$

Example 2.8.10. (Derangements). Among the permutations of $\{1,2,\ldots,n\}$ there are some, called derangements, in which none of the n integers appears in its natural place. Thus, $(i_1,i_2,\ldots,i_n)$ is a derangement if $i_1 \neq 1, i_2 \neq 2,\ldots$, and $i_n \neq n$. Let D_n be the number of derangements of $\{1,2,\ldots,n\}$.

As illustrations we note that $D_1 = 0$, $D_2 = 1$ because there is one derangement, namely (2,1); and $D_3 = 2$ because (2,3,1) and (3,1,2) are the only derangements.

We want to derive a formula for D_n that is valid for each positive integer n. This can be achieved by use of the principle of inclusion-exclusion.

Let U be the set of $n!$ permutations of $\{1,2,\ldots,n\}$. For each i, let A_i be the permutations $(b_1,b_2,\ldots,b_n)$ of $\{1,2,\ldots,n\}$ such that $b_i = i$. Then the set of derangements is precisely the set $\overline{A}_1 \cap \overline{A}_2 \cap \cdots \cap \overline{A}_n$. Therefore $D_n = |\overline{A}_1 \cap \overline{A}_2 \cap \cdots \cap \overline{A}_n|$. The permutations in A_1 are all of the form $(1,b_2,\ldots,b_n)$ where $(b_2,\ldots,b_n)$ is a permutation of $\{2,3,\ldots,n\}$. Thus $|A_1| = (n - 1)!$; similarly $|A_i| = (n - 1)!$. Likewise, $A_1 \cap A_2$ is the set of permutations of the form $(1,2,b_3,\ldots,b_n)$ so that $|A_1 \cap A_2| = (n - 2)!$. In a similar way we see that $|A_i \cap A_j| = (n - 2)!$. For any integer k where $1 \le k \le n$, the permutations in $A_1 \cap A_2 \cap \cdots \cap A_k$ are of the form $(1,2,\ldots,k,b_{k+1},\ldots,b_n)$ where $(b_{k+1},\ldots,b_n)$ is a permutation of $\{k + 1,\ldots,n\}$. Thus, $|A_1 \cap A_2 \cap \cdots \cap A_k| = (n - k)!$, and more generally, $|A_{i_1} \cap A_{i_2} \cap \cdots \cap A_{i_k}| = (n - k)!$ for $\{i_1,i_2,\ldots,i_k\}$ a k-combination of $\{1,2,\ldots,n\}$. Thus, we have the conditions prevailing in (2.8.20), so that

$$|U| - |A_1 \cup A_2 \cup \cdots \cup A_n| = n! - C(n,1)(n - 1)!$$
$$+ C(n,2)(n - 2)! + \cdots + (-1)^n C(n,n)$$

$$= n! - \frac{n!}{1!} + \frac{n!}{2!} - \frac{n!}{3!} + \cdots + (-1)^n \frac{n!}{n!}.$$

Thus, $$D_n = n!\left[1 - \frac{1}{1!} + \frac{1}{2!} - \frac{1}{3!} + \cdots + \frac{(-1)^n}{n!}\right]. \qquad (2.8.22)$$

In particular,

$$D_5 = 5!\left[1 - \frac{1}{1!} + \frac{1}{2!} - \frac{1}{3!} + \frac{1}{4!} - \frac{1}{5!}\right] = 44.$$

Example 2.8.11. Let n books be distributed to n students. Suppose that the books are returned and distributed to the students again later on. In how many ways can the books be distributed so that no student will get the same book twice?

Answer: The first time the books are distributed $n!$ ways, the second time D_n ways. Hence, the total number of ways is given by

$$n!D_n = (n!)^2\left[1 - \frac{1}{1!} + \frac{1}{2!} + \cdots + (-1)^n\frac{1}{n!}\right].$$

Example 2.8.12. Find the number of derangements of the integers from 1 to 10 inclusive, satisfying the condition that the set of elements in the first 5 places is:

(a) 1,2,3,4,5, in some order,
(b) 6,7,8,9,10, in some order.

Answer. (a) The integers 1, 2, 3, 4, and 5 can be placed into the first 5 places in D_5 ways; the last 5 integers 6, 7, 8, 9, and 10 can be placed in the last 5 places in D_5 ways. Hence, the answer is $D_5 \cdot D_5 = 1936$. (b) Any arrangement of 6, 7, 8, 9, and 10 in the first 5 places is a derangement so there are 5! possibilities; the same is true for the integers 1, 2, 3, 4, and 5 in the last 5 places. Hence, there are $(5!)^2 = 14{,}400$ such derangements.

Exercises for Section 2.8

1. A certain computer center employs 100 computer programmers. Of these 47 can program in FORTRAN, 35 in Pascal and 23 can program in both languages. How many can program in neither of these 2 languages?

2. Suppose that, in addition to the information given in Exercise 1, there are 20 employees that can program in COBOL, 12 in COBOL and FORTRAN, 11 in Pascal and COBOL and 5 in FORTRAN, Pascal, and COBOL. How many can program in none of these 3 languages?

3. An insurance company claimed to have 900 new policy holders of which
   796 bought auto insurance,
   402 bought life insurance,
   667 bought fire insurance,
   347 bought auto and life insurance,
   580 bought auto and fire insurance,
   291 bought life and fire insurance, and
   263 bought auto, life, and fire insurance.
   Explain why the state insurance commission ordered an audit of the company's records.

4. An advertising agency has 1,000 clients. Suppose that T is the set of clients that use television advertising, R is the set of clients that use radio advertising, and N is the set of clients who use newspaper advertising. Suppose that $|T| = 415$, $|R| = 350$, $|N| = 280$, 100

clients use all 3 types of advertising, 175 use television and radio, 180 use radio and newspapers, and $|T \cap N| = 165$.

(a) Find $|T \cap R \cap \overline{N}|$.

(b) How many clients use radio and newspaper advertising but not television?

(c) How many use television but do not use newspaper advertising and do not use radio advertising?

(d) Find $|\overline{T} \cap \overline{R} \cap \overline{N}|$.

5. In a survey of 800 voters, the following information was found: 300 were college educated, 260 were from high-income families, 325 were registered Democrats, 184 were college educated and from high-income families, 155 were college educated and registered Democrats, 165 were from high-income families and were registered Democrats, 94 were college educated, from high-income families, and were registered Democrats. Let

$$E = \{\text{voters who were college educated}\}$$
$$I = \{\text{voters who were from high-income families}\}$$
$$D = \{\text{voters who were registered Democrats}\}.$$

Draw a Venn diagram and list the number of elements in the 8 different regions of the diagram.

6. How many integral solutions are there of $x_1 + x_2 + x_3 + x_4 = 20$ if $1 \le x_1 \le 6$, $1 \le x_2 \le 7$, $1 \le x_3 \le 8$, and $1 \le x_4 \le 9$?

7. How many integral solutions are there of $x_1 + x_2 + x_3 + x_4 = 20$ if $2 \le x_1 \le 6$, $3 \le x_2 \le 7$, $5 \le x_3 \le 8$, and $2 \le x_4 \le 9$?

8. How many integers from 1 to 10^6 inclusive are neither perfect squares, perfect cubes, nor perfect fourth powers?

9. Find the number of integers between 1 and 1,000 inclusive that are divisible by none of 5, 6, and 8. Note that the intersection of the set of integers divisible by 6 with the set of integers divisible by 8 is the set of integers divisible by 24.

10. Find the number of permutations of the integers 1 to 10 inclusive

(a) such that exactly 4 of the integers are in their natural positions (that is, exactly 6 of the integers are deranged).

(b) such that 6 or more of the integers are deranged.

(c) that do not have 1 in the first place, nor 4 in the fourth place, nor 7 in the seventh place.

(d) such that no odd integer will be in the natural position.

(e) that do not begin with a 1 and do not end with 10.

11. Suppose that S is the set of integers 1 to n inclusive and that A is a subset of r of these integers. Show that the number of permutations of S in which the elements of A are the only elements that are deranged is $n! - C(r,1)(n-1)! + C(r,2)(n-2)! - \cdots + (-1)^r C(r,r)(n-r)!$.

12. A simple code is made by permuting the letters of the alphabet with every letter being replaced by a distinct letter. How many different codes can be made in this way?

13. Prove that $D_n - nD_{n-1} = (-1)^n$ for $n \geq 2$.

14. Eight people enter an elevator at the first floor. The elevator discharges passengers on each successive floor until it empties on the fifth floor. How many different ways can this happen?

15. In how many ways can the letters $\{4 \cdot a, 3 \cdot b, 2 \cdot c\}$ be arranged so that all the letters of the same kind are not in a single block?

16. A bookbinder is to bind 10 different books in red, blue, and brown cloth. In how many ways can he do this if each color of cloth is to be used for at least one book?

17. At a theater 10 men check their hats. In how many ways can their hats be returned so that
    (a) no man receives his own hat?
    (b) at least 1 of the men receives his own hat?
    (c) at least 2 of the men receive their own hats?

18. How many ways are there to select a 5-card hand from a deck of 52 cards such that the hand contains at least one card in each suit?

19. How many 13-card bridge hands have at least
    (a) one card in each suit?
    (b) one void suit?
    (c) one of each honor card (honor cards are aces, kings, queens, and jacks)?

20. How many ways are there to assign 20 different people to 3 different rooms with at least 1 person in each room?

21. (a) How many integers between 1 and 10^6 inclusive include all of the digits 1, 2, 3, and 4?
    (b) How many of the numbers between 1 and 10^6 inclusive consist of the digits 1, 2, 3, 4 alone?

22. Three Americans, 3 Mexicans, and 3 Canadians are to be seated in a row. How many ways can they be seated so that,

(a) no 3 countrymen sit together?

(b) no 2 countrymen may sit together?

23. Thirty students take a quiz. Then for the purpose of grading, the teacher asks the students to exchange papers so that no one is grading his own paper. How many ways can this be done?

24. In how many ways can each of 10 people select a left glove and a right glove out of a total of 10 pairs of gloves so that no person selects a matching pair of gloves?

25. The squares of a chessboard are painted 8 different colors. The squares of each row are painted all 8 colors and no 2 consecutive squares in one column can be painted the same color. In how many ways can this be done?

26. How many arrangements are there of the letters a, b, c, d, e, and f with either a before b, or b before c, or c before d? (By "before," we mean anywhere before, not just immediately before.)

27. How many arrangements are there of the letters of the word MATHEMATICS with both T's before both A's, or both A's before both M's, or both M's before the E?

28. How many integers between 1 and 50 are relatively prime to 50? (The integer a is relatively prime to the integer b iff 1 is the only common positive divisor.)

*29. How many arrangements are there of MISSISSIPPI with no pair of consecutive letters the same?

30. Suppose that a person with 10 friends invites a different subset of 3 friends to dinner every night for 10 days. How many ways can this be done so that all friends are included at least once?

31. How many arrangements are there of 3 a's, 3 b's, and 3 c's
    (a) without 3 consecutive letters the same?
    (b) having no adjacent letters the same?

32. Same problem as 31 for 3 a's, 3 b's, 3 c's, and 3 d's.

33. A secretary types n letters and their corresponding envelopes; in a fit of temper, she then puts the letters into the envelopes at random. How many ways could she have placed the letters so that no letter is in its correct envelope?

34. Given $2n$ letters of the alphabet, 2 of each of n types, how many arrangements are there with no pair of consecutive letters the same?

35. Use the principle of inclusion exclusion to count the number of primes between 41 and 100 inclusive.

36. How many derangements of the integers 1 to 20 inclusive are there in which the even integers must be deranged (odd integers may or may not occupy their natural position)?

37. How many arrangements of the 26 letters of the alphabet are there which contain none of the patterns LEFT, TURN, SIGN, or CAR?

38. Each of thirty students is taking an examination in two different subjects. One teacher examines the students in one subject and another in the other subject, and each teacher takes 5 minutes to examine a student in a subject. In how many ways can the examinations be scheduled without a student being required to appear before both examiners at the same time?

39. How many 6-digit decimal numbers contain exactly three different digits?

40. How many n-digit decimal numbers contain exactly k different digits?

41. How many 4-digit numbers can be composed of the digits in the number 123,143?

42. How many 5-digit numbers can be composed of the digits in the number 12,334,233?

43. How many 6-digit numbers can be composed of the digits in the number 1,223,145,345 if the same digit must not appear twice in a row?

44. How many 5-digit numbers can be composed of the digits of the number 12,123,334 if the digit 3 must not appear three times in a row?

45. (a) In how many ways can we arrange the digits in the number 11,223,344 if the same digit must not appear twice in a row?
    (b) Solve part (a) for the number 12,234,455.

46. Use the sieve of Eratosthenes to compute all prime integers $\leq$ 200.

Selected Answers for Section 2.8

1. $100 - 47 - 35 + 23 = 41$.

2. $100 - 47 - 35 - 20 + 23 + 12 + 11 - 5 = 39$.

4. (a) 75.
   (b) $|R \cap N \cap \overline{T}| = 80$
   (c) $|T \cap \overline{N} \cap \overline{R}| = 175$.
   (d) $|\overline{T} \cap \overline{R} \cap \overline{N}| = |\overline{T \cup R \cup N}| = 1,000 - 625 = 375$.

5.

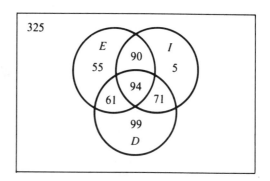

Figure 2-14

6. $\binom{19}{3} - \binom{13}{3} - \binom{12}{3} - \binom{11}{3} - \binom{10}{3} + \binom{6}{3} + \binom{5}{3} + \binom{4}{3} + \binom{4}{3} + \binom{3}{3}$
$= 217.$

9. $1{,}000 - (200 + 166 + 125) + (33 + 25 + 41) - 8 = 600.$

10. (a) $C(10,6)D_6.$

    (b) $\binom{10}{6}D_6 + \binom{10}{7}D_7 + \binom{10}{8}D_8 + \binom{10}{9}D_9 + \binom{10}{10}D_{10}$

    (c) $10! - (3)9! + (3)8! - 7!.$

    (d) $10! - \binom{5}{1}9! + \binom{5}{2}8! - \binom{5}{3}7! + \binom{5}{4}6! - \binom{5}{5}5!.$

    (e) $10! - (2)9! + 8!.$

12. $D_{26}.$

14. $4^8 - \binom{4}{1}3^8 + \binom{4}{2}2^8 - \binom{4}{3}1^8 = 40{,}824.$

15. Let A_1 be the arrangements where the 4 a's are one block, A_2 the arrangements with the 3 b's in a single block, and A_3 the arrangements with the 2 c's in a block. Then,

$$|\overline{A}_1 \cap \overline{A}_2 \cap \overline{A}_3| = \frac{9!}{4!3!2!} - \left(\frac{6!}{3!2!} + \frac{7!}{4!2!} + \frac{8!}{4!3!}\right)$$

$$+ \frac{4!}{2!} + \frac{5!}{3!} + \frac{6!}{4!} - 3! = 871.$$

16. $3^{10} - (3)2^{10} + 3$ ways.

17. (a) $D_{10}.$

    (b) $10! - D_{10}.$

    (c) $10! - D_{10} - 10D_9.$

18. $\binom{52}{5} - 4\binom{39}{5} + 6\binom{26}{5} - 4\binom{13}{5}.$

19. (a) $\binom{52}{13} - \binom{4}{1}\binom{39}{13} + \binom{4}{2}\binom{26}{13} - \binom{4}{3}$.

20. $3^{20} - 3 \cdot 2^{20} + 3$.

21. (a) $10^6 - 4 \cdot 9^6 + 6 \cdot 8^6 - 4 \cdot 7^6 + 6^6$.

(b) $4 + 4^2 + 4^3 + 4^4 + 4^5 + 4^6 = (4^7 - 4)/3$.

22. There are 9! permutations of the 9 people. If A_1, A_2, and A_3 is the set of permutations with 3 Americans together, 3 Mexicans together, and 3 Canadians together, respectively, then $|A_i| = 3!7!$, $|A_i \cap A_j| = 3!3!5!$, $|A_1 \cap A_2 \cap A_3| = (3!)^4$. Thus, $|\overline{A}_1 \cap \overline{A}_2 \cap \overline{A}_3| = 9! - 3 \cdot 3!7! + 3(3!)^2 5! - (3!)^4$.

(b) $9! - 9 \cdot 2!8! + 27(2!)^2 7! + 3 \cdot 3!7! - 27(2!)^3 6! - 18 \cdot 3!2!6! + 3(3!)^2 5! + 27 \cdot 3!(2!)^2 5! - 9(3!)^2 2!4! + (3!)^4$.

23. D_{30}.

24. $(10!)D_{10}$.

25. The first row can be painted 8! ways. Each row after the first can be painted D_8 ways. Hence the number of ways is $8!(D_8)^7$.

REVIEW FOR CHAPTER 2

1. How many license plates are there (with repetitions allowed) if
   (a) there is a letter followed by 3 digits followed by 3 letters followed by a letter or a digit?
   (b) there are 1, 2, or 3 digits followed by 1, 2, or 3 letters followed by a letter or a digit?
   (c) there are 1, 2, or 3 digits and 1, 2, or 3 letters and the letters must occur together?

2. How many ways are there to arrange the letters of the word MATHEMATICS?

3. There are 21 consonants and 5 vowels in the English alphabet. Consider only 10-letter words with 4 different vowels and 6 different consonants.
   (a) How many such words can be formed?
   (b) How many contain the letter a?
   (c) How many begin with b and end with c?

4. Give a combinatorial argument to explain why $(5n)!/5!^n n!$ is an integer.

5. In a class of 10 girls and 6 boys, how many
   (a) ways can a committee of 5 students be selected?
   (b) ways can a committee of 5 girls and 3 boys be chosen?
   (c) ways can a committee be selected that contain 3, 5 or 7 students?

(d) committees will contain 3 or more girls?

(e) committees of 2 or more can be chosen that have twice as many girls as boys?

(f) ways can the students be divided into teams where 2 teams have 5 members each and 2 teams have 3 each?

(g) ways can the girls be divided into teams with 2 members each?

(h) ways can the students be divided where the first team and the second team each have 4 students, the third and fourth teams have 3 each and the fifth team has 2 members?

(i) committees of 6 students can be formed with 3 of each sex but girl G and boy B cannot both be on the committee?

6. A test with 20 questions is a multiple choice test with 5 answers for each question but only 1 correct answer to each question. How many ways are there to have
   (a) exactly 6 correct answers?
   (b) at least 6 correct answers?

7. How many integral solutions are there of $x_1 + x_2 + x_3 + x_4 = 20$ where $2 \leq x_1, 3 \leq x_2, 0 \leq x_3, 5 \leq x_4$?

8. How many ways can 20 indistinguishable books be arranged on 5 different shelves?

9. How many 5-card hands from a deck of 52 have
   (a) 5 cards in 1 suit?
   (b) exactly 3 aces and no other pair?
   (c) exactly 1 pair?

10. (a) State Pascal's identity.
    (b) Use the binomial theorem to prove

$$\binom{n}{0} - \binom{n}{1} + \binom{n}{2} - \binom{n}{3} \cdots + (-1)^n \binom{n}{n} = 0.$$

   (c) Use the binomial theorem to prove

$$\left[\binom{n}{0} + \binom{n}{1} + \cdots + \binom{n}{n} \right]^2 = \sum_{k=0}^{2n} \binom{2n}{k}.$$

   (d) Give a combinatorial proof that

$$(n + 1) \binom{n}{r} = \binom{n + 1}{r + 1} (r + 1).$$

(e) Prove that

$$C(n + 4,r) = C(n,r) + 4C(n,r - 1) + 6C(n,r - 2)$$
$$+ 4C(n,r - 3) + C(n,r - 4).$$

11. Use the principle of inclusion-exclusion to determine the number of prime integers less than 400.

12. Find the number of 7-card hands that have at least one card in each suit.

13. Find the number of permutations of the letters of TOMTOM so that the same letters do not appear together.

14. Use the notation $\left\langle \begin{matrix} n \\ r \end{matrix} \right\rangle$ to stand for $\left(\begin{matrix} n - 1 + r \\ r \end{matrix} \right)$. Then prove

(a) $\left\langle \begin{matrix} n \\ r \end{matrix} \right\rangle = \left\langle \begin{matrix} n - 1 \\ r - 1 \end{matrix} \right\rangle + \left\langle \begin{matrix} n - 1 \\ r \end{matrix} \right\rangle$

(b) $\left\langle \begin{matrix} n \\ r \end{matrix} \right\rangle = \dfrac{n}{r} \left\langle \begin{matrix} n + 1 \\ r - 1 \end{matrix} \right\rangle = \dfrac{n + r - 1}{r} \left\langle \begin{matrix} n - 1 \\ r - 1 \end{matrix} \right\rangle.$

15. How many different arrangements are there of the letters of the word CONNECTICUT
    (a) With no two C's adjacent?
    (b) If there are only four letters in each arrangement?
    (c) If there is exactly one consonant between each pair of vowels?

16. Suppose that we have a 15-letter alphabet. How many 6-letter words have exactly 3 consecutive letters the same?

3

Recurrence
Relations

3.1   GENERATING FUNCTIONS OF SEQUENCES

The objects of interest in this chapter are sequences of real numbers $(a_0, a_1, a_2, \ldots, a_r, \ldots)$, that is, functions whose domain is the set of nonnegative integers and whose range is the set of real numbers. We shall use expressions like $A = \{a_r\}_{r=0}^{\infty}$ to denote such sequences.

Example 3.1.1. The sequence $A = \{2^r\}_{r=0}^{\infty}$ is the sequence $(1,2,4,8,16,\ldots,2^r,\ldots)$; the sequence $B = \{b_r\}_{r=0}^{\infty}$ where

$$b_r = \begin{cases} 0 & \text{if } 0 \le r \le 4 \\ 2 & \text{if } 5 \le r \le 9 \\ 3 & \text{if } r = 10 \\ 4 & \text{if } 11 \le r \end{cases}$$

is the sequence where $b_0 = b_1 = b_2 = b_3 = b_4 = 0$, $b_5 = b_6 = b_7 = b_8 = b_9 = 2$, $b_{10} = 3$, and $b_r = 4$ for all subscripts $r \ge 11$ thus, $B = (0,0,0,0,0,2,2,2,2,2,3,4,4,\ldots)$. The sequence $C = \{C_r\}_{r=0}^{\infty}$, where $C_r = r + 1$ for each value of r, is the sequence $(1,2,3,4,5,\ldots)$, and the sequence $D = \{d_r\}_{r=0}^{\infty}$ where, for each r, $d_r = r^2$ is the sequence $(0,1,4,9,16,25,\ldots)$.

The letter we use for the subscript has no particular significance, another will do just as easily; in other words, there is no difference in the sequence denoted by $\{a_i\}_{i=0}^{\infty}$ and that denoted by $\{a_r\}_{r=0}^{\infty}$. Normally we will be interested in sequences $A = \{a_r\}_{r=0}^{\infty}$ where a_r is the number of ways to select r objects in some procedure. For example, let a_r be the number of nonnegative integral solutions to the equation

$$x_1 + x_2 + \cdots + x_n = r \qquad\qquad (3.1.1)$$

where n is a fixed positive integer and each x_i is subject to certain constraints. Of course, you will recall from Section 2.4 that a_r is also described as the number of ways of distributing r similar balls into n numbered boxes, where the occupancy numbers for the different boxes are subject to certain constraints.

If, for instance, the constraints are only that each $x_i \geq 0$, then we know that $a_r = C(n - 1 + r, r)$. In other words, $a_0 = C(n - 1, 0) = 1$, $a_1 = C(n,1) = n$, $a_2 = C(n + 1, 2)$, and so on. In this case, there will be infinitely many nonzero terms of the sequence A.

However, if each x_i is restricted so that $0 \leq x_i \leq 1$, then in particular $a_r = 0$ if $r \geq n + 1$ because we cannot distribute more than n balls into n boxes if each box can contain at most 1 ball. Hence, in this case,

$$a_r = \begin{cases} C(n,r) & \text{if } 0 \leq r \leq n \\ 0 & \text{if } n + 1 \leq r. \end{cases}$$

Likewise if $0 \leq x_i \leq 2$, then $a_r = 0$ if $r \geq 2n + 1$, but a general expression for a_r if $r \leq 2n$ is not immediate using the techniques that we have developed thus far. By comparing this problem with Example 2.8.6 in the previous chapter you might realize that the principle of inclusion-exclusion will be required. In fact, the reader may want to verify that, in case $n = 3$,

$$a_0 = 1, a_1 = 3, a_2 = C(4,2), a_3 = C(5,3) - 3, a_4 = C(6,4) - 9, a_5$$
$$= C(7,5) - 3C(4,2), a_6 = C(8,6) - 3C(5,3) + 3.$$

Of course, we could place all sorts of complicated restrictions on the values for x_i and you might well imagine that the difficulty for giving an expression for each a_r could become insurmountable. We don't deny that this would be the case using the methods developed up to this point. Nevertheless, it is our intention to introduce a method in this section that will handle problems like the last case fairly easily.

Oddly enough, the clue to the new method is found in something quite old and familiar: multiplication of polynomials. Since polynomials only involve finitely many nonzero terms and sequences can involve infinitely many nonzero terms, we introduce the concept of **generating function,** a generalization of the concept of polynomial, to allow for that eventuality. The use of generating functions will be the most abstract technique used in this text to solve combinatorial problems, but once this method is mastered it will be the easiest method to apply to a broad spectrum of problems.

To the sequence $A = \{a_r\}_{r=0}^{\infty}$, we assign the symbol $A(X) = a_0 + a_1 X + \ldots + a_n X^n \ldots = \sum_{r=0}^{\infty} a_r X^r$. The expression $A(X)$ is called a **formal power series**, a_i is the **coefficient** of X^i, the term $a_i X^i$ is the **term of degree** i, and the term $a_0 X^0 = a_0$ is called the **constant term**. The coefficients are really what are of interest; the symbol X^i is simply a device for locating the coefficient a_i, and for this reason, the formal power series $A(X) = \sum_{n=0}^{\infty} a_r X^r$ is called an (ordinary) generating function for the sequence $A = \{a_r\}_{r=0}^{\infty}$.

The words "generating function" are used because, in some sense, $A(X)$ generates its coefficients. The word "ordinary" is used to denote the fact that powers of X are used; other kinds of generating functions could use, by contrast, other functions like $X^r/r!$, $\sin(rX)$, or $\cos(rX)$ in place of X^r. For the most part we will suppress the word "ordinary" in our usage. We use the word "formal" to distinguish between the abstract symbol $A(X) = \sum_{r=0}^{\infty} a_r X^r$ and the concept of power series some students may have seen in calculus courses. We emphasize that in our concept X will *never* be assigned a numerical value if there are infinitely many nonzero coefficients in the sequence generated by $A(X)$. Therefore, we avoid having to discuss such topics as convergence and divergence of power series, topics often discussed in calculus courses.

If all of the coefficients are zero from some point on, $A(X)$ is just a *polynomial*. If $a_k \neq 0$ and $a_i = 0$ for $i \geq k + 1$, then $A(X)$ is a polynomial of degree k.

Example 3.1.2. The generating functions

$$A(X) = \sum_{r=0}^{\infty} 2^r X^r,$$

$$B(X) = 2X^5 + 2X^6 + 2X^7 + 2X^8 + 2X^9 + 3X^{10}$$
$$+ 4X^{11} + 4X^{12} + \cdots$$

$$C(X) = \sum_{r=0}^{\infty} (r+1) X^r,$$

$$D(X) = \sum_{r=0}^{\infty} (r^2) X^r$$

generate the sequences A, B, C, and D of Example 3.1.1.

Definition 3.1.1. Let $A(X) = \sum_{r=0}^{\infty} a_r X^r$, $B(X) = \sum_{s=0}^{\infty} b_s X^s$ be 2 formal power series. We then define the following concepts.

Equality: $A(X) = B(X)$ iff $a_n = b_n$ for each $n \geq 0$.

Multiplication by a scalar number C: $CA(X) = \Sigma_{r=0}^{\infty} (Ca_r) X^r$.

Sum: $A(X) + B(X) = \Sigma_{n=0}^{\infty} (a_n + b_n) X^n$.

Product: $A(X) B(X) = \Sigma_{n=0}^{\infty} P_n X^n$, where $P_n = \Sigma_{j+k=n} a_j b_k$.

Let us take some time to discuss the definition of product of 2 formal power series. The sum $\Sigma_{j+k=n} a_j b_k$ means take the sum of all possible products $a_j b_k$ where the sum of the subscripts j and k is n. Since these subscripts correspond to the exponents on X, we see that the term $P_n X^n$ in the product $A(X) B(X)$ is obtained by taking the sum of all possible products of one term $a_j X^j$ from $A(X)$ and one term $b_k X^k$ from $B(X)$ such that the sum of exponents $j + k = n$.

Of course, this can be accomplished very systematically by starting with the exponent 0 and thus with the constant term a_0 of $A(X)$ and multiply it by the coefficient b_n of X^n in $B(X)$; then proceed to the coefficient a_1 of X in $A(X)$ and multiply it by the coefficient b_{n-1} of X^{n-1} in $B(X)$, and so on, using the coefficients of increasing powers of X in $A(X)$ and the coefficients of corresponding decreasing powers of X in $B(X)$ as follows:

$$P_n = a_0 b_n + a_1 b_{n-1} + a_2 b_{n-2} + \cdots + a_{n-1} b_1 + a_n b_0 = \sum_{i=0}^{n} a_i b_{n-i}.$$

Thus,

$$A(X) B(X) = a_0 b_0 + (a_0 b_1 + a_1 b_0)X + (a_0 b_2 + a_1 b_1 + a_2 b_0)X^2$$
$$+ \cdots + (a_0 b_n + a_1 b_{n-1} + a_2 b_{n-2} + \cdots + a_n b_0)X^n + \cdots$$

Example 3.1.3. If $A(X) = a_0 + a_3 X^3 + a_4 X^4 + a_8 X^8$ and $B(X) = b_0 + b_4 X^4 + b_5 X^5 + b_8 X^8$ (where for the moment we are not assigning any values to the coefficients—we just mean that the coefficients are zero for the missing powers of X), then the coefficient of X^r in $A(X) B(X)$ is found by considering the powers $\{X^0, X^3, X^4, X^8\}$ from the first factor and the powers $\{X^0, X^4, X^5, X^8\}$ in the second factor such that their sum is r. For instance, the coefficient of X^8 can be obtained by using X^0 in the first factor and X^8 in the second; X^3 in the first and X^5 in the second; X^4 in the first and X^4 in the second; or X^8 in the first and X^0 in the second factor. Thus, the coefficient of X^8 in the product $A(X) B(X)$ is such that $P_8 = a_0 b_8 + a_3 b_5 + a_4 b_4 + a_8 b_0$, because $(0,8),(3,5),(4,4)$ and $(8,0)$ are the only pairs of exponents of $A(X)$ and $B(X)$ whose sum is 8.

Likewise the coefficient of X^5 in the product is $a_0 b_5$ because there is only one pair of exponents of $A(X)$ and $B(X)$, namely $(0,5)$, whose sum is 5.

Thus, if $a_0 = 2$, $a_3 = -5$, $a_4 = 7$, and $a_8 = 3$, $b_0 = 3$, $b_4 = -6$, $b_5 = 8$, and $b_8 = 3$, then $P_8 = (2)(3) + (-5)(8) + (7)(-6) + (3)(3) = -67$, and $P_5 = (2)(8) = 16$. Of course, we can complete all the computations of coefficients to see that with these values of a_i's and b_j's, then

$$A(X)B(X) = 6 - 15X^3 + 9X^4 + 16X^5 + 30X^7 - 67X^8$$
$$+ 56X^9 - 15X^{11} + 3X^{12} + 24X^{13} + 9X^{16}.$$

However, the case where all the nonzero coefficients of $A(X)$ and $B(X)$ are 1 is of special interest. Note that in this case, $P_8 = a_0 b_8 + a_3 b_5 + a_4 b_4 + a_8 b_0 = 4$, and $P_5 = 1$. In particular, in this case, the coefficient of X^8 in the product is just the *number* of pairs of exponents whose sum is 8, that is, the coefficient of X^8 in the product $(1 + X^3 + X^4 + X^8)(1 + X^4 + X^5 + X^8)$ is just the number of integral solutions to the equation $e_1 + e_2 = 8$, where e_1 and e_2 represent the exponents of $A(X)$ and $B(X)$, respectively. Hence e_1 can only be 0,3,4, or 8 and $e_2 = 0,4,5$, or 8. Likewise there is only one solution to $e_1 + e_2 = 5$ with these restrictions on e_1 and e_2 so the coefficient of X^5 in $(1 + X^3 + X^4 + X^8)(1 + X^4 + X^5 + X^8)$ is 1.

We have stumbled upon a clue here: the coefficient of X^r in the product $(1 + X^3 + X^4 + X^8)(1 + X^4 + X^5 + X^8)$ is the number of integral solutions to the equation $e_1 + e_2 = r$ subject to the constraints $e_1 = 0,3,4,8$ and $e_2 = 0,4,5,8$. Note that the *exponents* of the factors in the product reflect the *constraints* in the equation. Note also that we can view this clue in two ways. We can compute the coefficient of X^r by algebra and then discover the number of integral solutions to the equation $e_1 + e_2 = r$ subject to the constraints; or we can compute all solutions of the equation subject to the constraints and then discover the coefficient of X^r.

Generating Function Models

Thus, if we wanted to count the number of nonnegative integral solutions to an equation $e_1 + e_2 = r$ with certain constraints on e_1 and e_2, we would expect that we need only find the coefficient of X^r in the product of generating functions $A(X)B(X)$, where the exponents of $A(X)$ reflect the constraints on e_1 and the exponents of $B(X)$ reflect the constraints on e_2. Thus, if e_1 can only be 0, 1, or 9, then let $A(X) = 1 + X + X^9$. If e_2 can only be even and $0 \le e_2 \le 8$, then $B(X) = 1 + X^2 + X^4 + X^6 + X^8$. On the other hand, if e_1 can be any nonnegative integer value, then we let $A(X) = 1 + X + X^2 \ldots$ [in this case, $A(X)$ has an infinite number of terms]. Likewise, if e_2 can only take on the integral values, say, that are multiples of 5, then we let $B(X) = 1 + X^5 + X^{10} + \ldots$. No doubt you see that the possibilities are endless.

Moreover, we can extend what we have said to equations with more than two variables e_i because the definition of product of formal power series extends to more than two factors.

Let us illustrate for 3 factors. Let

$$A(X) = \sum_{i=0}^{\infty} a_i X^i,$$

$$B(X) = \sum_{j=0}^{\infty} b_j X^j,$$

$$C(X) = \sum_{k=0}^{\infty} c_k X^k.$$

Then

$$A(X)\, B(X)\, C(X) = \sum_{r=0}^{\infty} P_r X^r$$

where

$$P_r = \sum_{i+j+k=r} a_i b_j c_k,$$

that is, the term $P_r X^r$ in the product is obtained by taking any one term $a_i X^i$ from $A(X)$, any one term $b_j X^j$ from $B(X)$, and any one term $c_k X^k$ from $C(X)$ such that the sum of exponents $i + j + k = r$. Of course, the reader can extend this idea to products of several factors. Thus, here is our rule:

> Assume each nonzero coefficient of each formal power series $A_i(X)$ is 1. Then the coefficient of X^r in the product $A_1(X)$ $A_2(X) \cdots A_n(X)$ can be interpreted as the number of nonnegative integral solutions to an equation $e_1 + e_2 + \cdots + e_n = r$ where constraints on each e_i are determined by the exponents of the ith factor $A_i(X)$.

Of course, this line of reasoning can be reversed. Given a problem to count the number of nonnegative integral solutions to an equation $e_1 + e_2 \cdots + e_n = r$ with constraints on each e_i, then we can build a generating function $A_1(X)\, A_2(X) \cdots A_n(X)$ whose coefficient of X^r is the answer to the problem.

Example 3.1.4.   The coefficient of X^9 in $(1 + X^3 + X^8)^{10}$ is $C(10,3)$ because the only solutions of $e_1 + e_2 + \cdots + e_{10} = 9$ where each $e_i = 0,3,8$ are those solutions where 3 values are equal to three, and the remaining values are 0. Likewise the coefficient of X^{25} is $10!/3!2!5!$ because each solution to the equation will involve three 3's, two 8's, and five 0's.

Example 3.1.5. Find a generating function for a_r = the number of nonnegative integral solutions of $e_1 + e_2 + e_3 + e_4 + e_5 = r$ where $0 \leq e_1 \leq 3, 0 \leq e_2 \leq 3, 2 \leq e_3 \leq 6, 2 \leq e_4 \leq 6, e_5$ is odd, and $1 \leq e_5 \leq 9$. Let $A_1(X) = A_2(X) = 1 + X + X^2 + X^3, A_3(X) = A_4(X) = X^2 + X^3 + X^4 + X^5 + X^6$, and $A_5(X) = X + X^3 + X^5 + X^7 + X^9$. Thus, the generating function we want is

$$A_1(X) A_2(X) A_3(X) A_4(X) A_5(X) = (1 + X + X^2 + X^3)^2$$
$$(X^2 + X^3 + X^4 + X^5 + X^6)^2$$
$$(X + X^3 + X^5 + X^7 + X^9).$$

Example 3.1.6. Find a generating function for a_r = the number of nonnegative integral solutions to $e_1 + e_2 + \cdots + e_n = r$ where $0 \leq e_i \leq 1$. Let $A_i(X) = 1 + X$ for each $i = 1, 2, \ldots, n$. Thus, the generating function we want is $A_1(X) A_2(X) \cdots A_n(X) = (1 + X)^n$. The binomial theorem gives all the coefficients and thus we know the number of solutions to the above equation is $C(n,r)$.

Example 3.1.7. Find a generating function for a_r = the number of nonnegative integral solutions to $e_1 + e_2 + \cdots + e_n = r$ where $0 \leq e_i$ for each i.

Here since there is no upper bound constraint on the e_i's, we let $A_1(X) A_2(X) \cdots A_n(X) = (1 + X + X^2 + \cdots + X^k \ldots)^n$. Using Theorem 2.4.2, we know that $\sum_{r=0}^{\infty} C(n - 1 + r, r) X^r$ must be another expression for this same generating function, that is,

$$\left(\sum_{k=0}^{\infty} X^k \right)^n = \sum_{r=0}^{\infty} C(n - 1 + r, r) X^r.$$

In particular,

$$\left(\sum_{k=0}^{\infty} X^k \right)^2 = \sum_{r=0}^{\infty} (r + 1) X^r,$$

and

$$\left(\sum_{k=0}^{\infty} X^k \right)^3 = \sum_{r=0}^{\infty} \frac{(r + 2)(r + 1)}{2} X^r$$

since for $n = 2$, $C(n - 1 + r,r) = r + 1$ and for $n = 3$, $C(n - 1 + r,r) = (r + 2)(r + 1)/2$.

Example 3.1.8. Find a generating function for $a_r = $ the number of ways of distributing r similar balls into n numbered boxes where each box is nonempty.

First we model this problem as an integral-solution-of-an-equation problem; namely, we are to count the number of integral solutions to $e_1 + e_2 + \cdots + e_n = r$, where each $e_i \geq 1$.

Then, in turn, we build the generating function $(X + X^2 + \cdots)^n = (\Sigma_{r-1}^{\infty} X^r)^n$, which by Theorem 2.4.3 must equal $\Sigma_{r-n}^{\infty} C(r - 1,n - 1)X^r$. (The reader should give a combinatorial explanation as to why the coefficients of X^r are all zero if $0 \leq r \leq n - 1$).

Example 3.1.9. Find a generating function for $a_r = $ the number of ways the sum r can be obtained when:

(a) 2 distinguishable dice are tossed.
(b) 2 distinguishable dice are tossed and the first shows an even number and the second shows an odd number.
(c) 10 distinguishable dice are tossed and 6 specified dice show an even number and the remaining 4 show an odd number.

In (a) we are to count the number of integral solutions to $e_1 + e_2 = r$ where $1 \leq e_i \leq 6$. Then a_r is the coefficient of X^r in the generating function $(X + X^2 + X^3 + X^4 + X^5 + X^6)^2$.

In (b) we are looking for the coefficient of X^r in $(X^2 + X^4 + X^6)(X + X^3 + X^5)$ since $1 \leq e_1 \leq 6$ and e_1 is even while $1 \leq e_2 \leq 6$ and e_2 is odd.

Likewise, the generating function called for in (c) is $(X^2 + X^4 + X^6)^6(X + X^3 + X^5)^4$.

Example 3.1.10. Find a generating function to count the number of integral solutions to $e_1 + e_2 + e_3 = 10$ if for each i, $0 \leq e_i$.

Here we can take two approaches. Of course we are looking for the coefficient of X^{10} in $(1 + X + X^2 + X^3 + \cdots)^3$. But since the equation is a model for the distribution of 10 similar balls into 3 boxes we see that each $e_i \leq 10$ for we cannot place more than 10 balls in each box. Thus we could also interpret the problem as one where we are to find the coefficient of X^{10} in $(1 + X + X^2 + \cdots + X^{10})^3$.

Exercises for Section 3.1

1. Build a generating function for a_r = the number of integral solutions to the equation $e_1 + e_2 + e_3 = r$ if:
   (a) $0 \le e_i \le 3$ for each i.
   (b) $2 \le e_i \le 5$ for each i.
   (c) $0 < e_i$ for each i.
   (d) $0 \le e_1 \le 6$ and e_1 is even; $2 < e_2 \le 7$ and e_2 is odd; $5 \le e_3 \le 7$.

2. Write a generating function for a_r when a_r is
   (a) the number of ways of selecting r balls from 3 red balls, 5 blue balls, 7 white balls.
   (b) the number of ways of selecting r coins from an unlimited supply of pennies, nickels, dimes and quarters.
   (c) the number of r-combinations formed from n letters where the first letter can appear an even number of times up to 12, the second letter can appear an odd number of times up to 7, the remaining letters can occur an unlimited number of times.
   (d) the number of ways of obtaining a total of r upon tossing 50 distinguishable dice. [Which coefficient do we want in (d) if we want to know the number of ways of obtaining a total of 100 upon tossing the 50 dice?]
   (e) the number of integers between 0 and 999 whose sum of digits is r.

3. In $(1 + X^5 + X^9)^{10}$ find
   (a) the coefficient of X^{23}.
   (b) the coefficient of X^{32}.

4. Find the coefficient of X^{16} in $(1 + X^4 + X^8)^{10}$.

5. (a) Find a generating function for the number of ways to distribute 30 balls into 5 numbered boxes where each box contains at least 3 balls and at most 7 balls.
   (b) Factor out X^{15} from the above functions, and interpret this revised generating function combinatorially.

6. Find a generating function for a_r = the number of ways of distributing r similar balls into 7 numbered boxes where the second, third, fourth, and fifth boxes are nonempty.

7. (a) Find a generating function for the number of ways to select 6 nonconsecutive integers from $1, 2, \ldots, n$.
   (b) Which coefficient do we want to find in case $n = 20$?
   (c) Which coefficient do we want for general n?

8. Build a generating function for a_r = the number of ways to distribute r similar balls into 5 numbered boxes with
   (a) at most 3 balls in each box.
   (b) 3, 6, or 8 balls in each box.

(c) at least 1 ball in each of the first 3 boxes and at least 3 balls in each of the last 2 boxes.

(d) at most 5 balls in box 1, at most 7 balls in the last 4 boxes.

(e) a multiple of 5 balls in box 1, a multiple of 10 balls in box 2, a multiple of 25 balls in box 3, a multiple of 50 balls in box 4, and a multiple of 100 balls in box 5.

9. Find a generating function for the number of r-combinations of $\{3 \cdot a, 5 \cdot b, 2 \cdot c\}$

10. Build a generating function for determining the number of ways of making change for a dollar bill in pennies, nickels, dimes, quarters, and half-dollar pieces. Which coefficient do we want?

11. Find a generating function for the sequence $A = \{a_r\}_{r=0}^{\infty}$ where

$$a_r = \begin{cases} 1 & \text{if } 0 \leq r \leq 2 \\ 3 & \text{if } 3 \leq r \leq 5 \\ 0 & \text{if } r \geq 6 \end{cases}$$

Selected Answers for Section 3.1

1. (a) $(1 + X + X^2 + X^3)^3$.
   (b) $(X^2 + X^3 + X^4 + X^5)^3$.
   (c) $(X + X^2 + \cdots)^3$.
   (d) $(1 + X^2 + X^4 + X^6)(X^3 + X^5 + X^7)(X^5 + X^6 + X^7)$.

2. (a) $(1 + X + X^2 + X^3)(1 + X + \cdots + X^5)(1 + X + \cdots + X^7)$.
   (b) $(1 + \cdots + X^n + \cdots)^4$.
   (c) $(1 + X^2 + \cdots + X^{12})(X + X^3 + X^5 + X^7)(1 + \cdots + X^n \cdots)^{n-2}$.
   (d) $(X + \cdots + X^6)^{50}; a_{100}$.
   (e) $(1 + X + \cdots + X^9)^3$.

3. (a) Solve $e_1 + e_2 + \cdots + e_{10} = 23$ where $e_i = 0,5,9$. This can be done only with one 5, two 9's and seven 0's. Hence the coefficient is $10!/1!2!7!$.
   (b) 32 can be otained only with three 9's, one 5, and 6 0's. Thus the coefficient of X^{32} is $10!/3!1!6!$.

4. The only solutions to $e_1 + e_2 + \cdots + e_{10} = 16$ where $e_i = 0,4,8$ are those with four 4's, no 8's, and six 0's; two 8's, no 4's, and eight 0's; or two 4's, one 8, and seven 0's. Thus the coefficient is

$$\binom{10}{4} + \binom{10}{2} + 8\binom{10}{2} = \binom{10}{4} + 9\binom{10}{2}.$$

6. $(1 + X \cdots)^3 (X + X^2 \cdots)^4$.

7. (a) $(1 + X \cdots)^2 (X + X^2 \cdots)^5$. Think of the 6 integers chosen as dividers for 7 boxes where the first and last box can be empty and the other 5 boxes are nonempty.
   (b) Coefficient of X^{14}.
   (c) Coefficient of X^{n-6}.

9. $(1 + X + X^2 + X^3)(1 + X + \cdots + X^5)(1 + X + X^2)$.

10. Find the coefficient of X^{100} in the product of

$$(1 + X + X^2 + \cdots + X^{100})$$

$$(1 + X^5 + X^{10} + \cdots + X^{100})$$

$$(1 + X^{10} + X^{20} + \cdots + X^{100})$$

$$(1 + X^{25} + X^{50} + X^{75} + X^{100})$$

$$(1 + X^{50} + X^{100}).$$

11. $1 + X + X^2 + 3X^3 + 3X^4 + 3X^5$.

3.2   CALCULATING COEFFICIENTS OF GENERATING FUNCTIONS

Up to this point we have been interested primarily in building generating functions to determine solutions to combinatorial problems. We now develop algebraic techniques for calculating the coefficients of generating functions.

The most important concept we introduce is that of **division of formal power series.** First let us discuss the meaning of $1/A(X)$.

Definition 3.2.1. If $A(X) = \Sigma_{n=0}^{\infty} a_n X^n$ is a formal power series, then $A(X)$ is said to have a multiplicative inverse if there is a formal power series $B(X) = \Sigma_{k=0}^{\infty} b_k X^k$ such that $A(X)B(X) = 1$.

In particular, if $A(X)$ has a multiplicative inverse, then we see that $a_0 b_0 = 1$, so that a_0 must be nonzero. The converse is also true. In fact, if $a_0 \neq 0$, then we can determine the coefficients of $B(X)$ by writing down the coefficients of successive powers of X in $A(X)B(X)$ from the definition of product of 2 power series, and then equating these to the coefficients of like powers of X in the power series 1. Therefore, we have:

$$a_0 b_0 = 1$$
$$a_0 b_1 + a_1 b_0 = 0$$
$$a_0 b_2 + a_1 b_1 + a_2 b_0 = 0$$
$$a_0 b_3 + a_1 b_2 + a_2 b_1 + a_3 b_0 = 0$$

$$\vdots$$

$$a_0 b_n + a_1 b_{n-1} + \cdots + a_n b_0 = 0,$$

and so on.

From the first equation, we can solve for $b_0 = 1/a_0$; from the second, we find

$$b_1 = \frac{-a_1 b_0}{a_0} = \frac{-a_1}{a_0^2};$$

in the third equation, we get

$$b_2 = \frac{-a_1 b_1 - a_2 b_0}{a_0} = \frac{a_1^2 - a_2 a_0}{a_0^3};$$

from the fourth, we find

$$b_3 = \frac{-a_1 b_2 - a_2 b_1 - a_3 b_0}{a_0}.$$

We can substitute into this expression for b_0, b_1, and b_2 to obtain an expression for b_3 involving only the coefficients of $A(X)$. Continuing in this manner, we can solve for each coefficient of $B(X)$.

Thus, we established that a *formal power series* $A(X) = \Sigma_{n=0}^{\infty} a_n X^n$ has a *multiplicative inverse* iff *the constant* term a_0 *is different from* zero.

The reader is doubtless familiar with division of polynomials and, in fact, the discovery of the coefficients of $B(X) = 1/A(X)$ as above is nothing more than an extension of that idea.

Definition 3.2.2. If $A(X)$ and $C(X)$ are power series, we say that $A(X)$ *divides* $C(X)$ if there is a formal power series $D(X)$ such that $C(X) = A(X) D(X)$, and we write $D(X) = C(X)/A(X)$.

Of course, for arbitrary formal power series, $A(X)$ and $C(X)$, it need not be the case that $A(X)$ divides $C(X)$. However, if $A(X) = \Sigma_{n=0}^{\infty} a_n X^n$ is such that $a_0 \neq 0$, then $A(X)$ has a multiplicative inverse $B(X) = 1/A(X)$

and then $A(X)$ divides any $C(X)$—just let $D(X) = C(X) B(X) = (C(X)$
$1/A(X))$.

If $A(X) = \sum_{n=0}^{\infty} a_n X^n$, and $a_0 = 0$, but some coefficient of $A(X)$ is not
zero, then let a_k be the first nonzero coefficient of $A(X)$, and $A(X) =$
$X^k A_1(X)$, where a_k, the constant term of $A_1(X)$, is nonzero. Then in
order for $A(X)$ to divide $C(X)$ it must be true that X^k is also a factor of
$C(X)$, that is, $C(X) = X^k C_1(X)$ where $C_1(X)$ is a formal power series. If
this is the case, then cancel the common powers of X from both $A(X)$ and
$C(X)$ and then we can find $C(X)/A(X) = C_1(X)/A_1(X)$ by using the
multiplicative inverse of $A_1(X)$.

Geometric Series

Let us use what we have learned to determine the multiplicative
inverse for $A(X) = 1 - X$. Let $B(X) = 1/A(X) = \sum_{n=0}^{\infty} b_n X^n$. Solving
successively for $b_0, b_1, \ldots$, as above, we see that

$$b_0 = \frac{1}{a_0} = 1,$$

$$b_1 = \frac{-a_1 b_0}{a_0} = \frac{(-1)(1)}{(1)} = 1,$$

$$b_2 = \frac{-a_1 b_1 - a_2 b_0}{a_0} = \frac{-(-1)(1) - (0)(1)}{1} = 1,$$

$$b_3 = \frac{-a_1 b_2 - a_2 b_1 - a_3 b_0}{a_0} = 1,$$

and so on. We see that each $b_i = 1$ so that we have an expression for the
geometric series:

$$\frac{1}{1 - X} = \sum_{r=0}^{\infty} X^r. \tag{3.2.1}$$

If we replace in the above expression X by aX where a is a real number,
then we see that:

$$\frac{1}{1 - aX} = \sum_{r=0}^{\infty} a^r X^r, \tag{3.2.2}$$

the so called geometric series (with common ratio a).

In particular, let $a = -1$; then we get

$$\frac{1}{1 + X} = \sum_{r=0}^{\infty} (-1)^r X^r = 1 - X + X^2 - X^3 \cdots, \qquad (3.2.3)$$

the so called **alternating geometric series.** Likewise,

$$\frac{1}{1 + aX} = \sum_{r=0}^{\infty} (-1)^r a^r X^r \qquad (3.2.4)$$

Suppose that n is a positive integer. If $B_1(X), B_2(X), \cdots$, and $B_n(X)$ are the multiplicative inverses of $A_1(X), A_2(X), \cdots$, and $A_n(X)$, respectively, then $B_1(X) B_2(X) \cdots B_n(X)$ is the multiplicative inverse of $A_1(X)A_2(X) \cdots A_n(X)$—just multiply $A_1(X)A_2(X) \cdots A_n(X)$ by $B_1(X)B_2(X) \cdots B_n(X)$ and use the facts that $A_i(X)B_i(X) = 1$ for each i. In particular, if $B(X)$ is the multiplicative inverse of $A(X)$, then $(B(X))^n$ is the multiplicative inverse of $(A(X))^n$. Let us apply this observation to $A(X) = 1 - X$.

For n a positive integer,

$$\frac{1}{(1 - X)^n} = \left(\sum_{k=0}^{\infty} X^k \right)^n = \sum_{r=0}^{\infty} C(n - 1 + r, r) X^r. \qquad (3.2.5)$$

The first equality follows from the above comments and the fact that $\sum_{k=0}^{\infty} X^k$ is the multiplicative inverse of $1 - X$; we have already observed the second equality in Example 3.1.7. The equality $1/(1 - X)^n = \sum_{r=0}^{\infty} C(n - 1 + r, r) X^r$ could also be proved by mathematical induction and use of the identity $C(n - 1, 0) + C(n, 1) + C(n + 1, 2) + \cdots + C(n + r - 1, r) = C(n + r, r)$.

By replacing X by $-X$ in the above we get the following identity: For n a positive integer,

$$\frac{1}{(1 + X)^n} = \sum_{r=0}^{\infty} C(n - 1 + r, r)(-1)^r X^r. \qquad (3.2.6)$$

Following this pattern, replace X by aX in (3.2.5) and (3.2.6) to obtain

$$\frac{1}{(1 - aX)^n} = \sum_{r=0}^{\infty} C(n - 1 + r, r)a^r X^r, \qquad (3.2.7)$$

$$\frac{1}{(1 + aX)^n} = \sum_{r=0}^{\infty} C(n - 1 + r, r)(-a)^r X^r. \qquad (3.2.8)$$

Likewise, replace X by X^k in (3.2.1) to get for k a positive integer,

$$\frac{1}{1 - X^k} = \sum_{r=0}^{\infty} X^{kr} = 1 + X^k + X^{2k} + \cdots, \qquad (3.2.9)$$

and

$$\frac{1}{1 + X^k} = \sum_{r=0}^{\infty} (-1)^r X^{kr}. \qquad (3.2.10)$$

If a is a nonzero real number,

$$\frac{1}{a - X} = \frac{1}{a}\left(\frac{1}{1 - X/a}\right) = \frac{1}{a}\sum_{r=0}^{\infty} \frac{X^r}{a^r} \qquad (3.2.11)$$

and

$$\frac{1}{X - a} = -\frac{1}{a - X} = -\frac{1}{a}\sum_{r=0}^{\infty} \frac{X^r}{a^r}. \qquad (3.2.12)$$

Other identities that we will use frequently are:
If n is a positive integer,

$$1 + X + X^2 + \cdots + X^n = \frac{1 - X^{n+1}}{1 - X}. \qquad (3.2.13)$$

If n is a positive integer,

$$(1 + X)^n = 1 + \binom{n}{1} X + \binom{n}{2} X^2 + \cdots + \binom{n}{n} X^n \qquad (3.2.14)$$

$$(1 + X^k)^n = 1 + \binom{n}{1} X^k + \binom{n}{2} X^{2k} + \cdots + \binom{n}{n} X^{nk} \qquad (3.2.15)$$

$$(1 - X)^n = 1 - \binom{n}{1} X + \binom{n}{2} X^2 + \cdots + (-1)^n \binom{n}{n} X^n \qquad (3.2.16)$$

$$(1 - X^k)^n = 1 - \binom{n}{1} X^k + \binom{n}{2} X^{2k} + \cdots + (-1)^n \binom{n}{n} X^{nk} \qquad (3.2.17)$$

The formulas (3.2.14)–(3.2.17) are all special cases of the binomial theorem.

Use of Partial Fraction Decomposition

If $A(X)$ and $C(X)$ are polynomials, we show how to compute $C(X)/A(X)$ by using the above identities and partial fractions. The reader will recall from algebra that if $A(X)$ is a product of linear factors, $A(X) = a_n(X - \alpha_1)^{r_1}(X - \alpha_2)^{r_2} \cdots (X - \alpha_k)^{r_k}$, and if $C(X)$ is any polynomial of degree less than the degree of $A(X)$, then $C(X)/A(X)$ can be written as the sum of elementary fractions as follows:

$$\frac{C(X)}{A(X)} = \frac{A_{11}}{(X - \alpha_1)^{r_1}} + \frac{A_{12}}{(X - \alpha_1)^{r_1 - 1}} + \cdots + \frac{A_{1r_1}}{(X - \alpha_1)}$$

$$+ \frac{A_{21}}{(X - \alpha_2)^{r_2}} + \frac{A_{22}}{(X - \alpha_2)^{r_2 - 1}} + \cdots + \frac{A_{2r_2}}{(X - \alpha_2)}$$

$$+ \cdots + \frac{A_{k1}}{(X - \alpha_k)^{r_k}} + \frac{A_{k2}}{(X - \alpha_k)^{r_k - 1}} + \cdots + \frac{A_{kr_k}}{(X - \alpha_k)}.$$

To find the numbers $A_{11}, \ldots A_{kr_k}$ we multiply both sides of the last equation by $(X - \alpha_1)^{r_1}(X - \alpha_2)^{r_2} \cdots (X - a_k)^{r_k}$ to clear of denominators and then we equate coefficients of the same powers of X. Then the required coefficients can be solved from the resulting system of equations.

A few examples will illustrate the method and refresh your memory.

Example 3.2.1.   Calculate $B(X) = \Sigma_{r=0}^{\infty} b_r X^r = 1/(X^2 - 5X + 6)$.

Since $X^2 - 5X + 6 = (X - 3)(X - 2)$, we see that $1/(X^2 - 5X + 6) = A/(X - 3) + B/(X - 2)$. Thus, $A(X - 2) + B(X - 3) = 1$. Let $X = 2$ and we find $B = -1$. Let $X = 3$ and we see that $A = 1$. Thus $1/(X^2 - 5X + 6) = 1/(X - 3) - 1/(X - 2)$. Then we use (3.2.11) and (3.2.12) to see that

$$\frac{1}{X^2 - 5X + 6} = -\frac{1}{3 - X} + \frac{1}{2 - X} = -\frac{1}{3(1 - X/3)} + \frac{1}{2(1 - X/2)}$$

$$= -\frac{1}{3}\sum_{r=0}^{\infty}\left(\frac{1}{3}\right)^r X^r + \frac{1}{2}\sum_{r=0}^{\infty}\left(\frac{1}{2}\right)^r X^r$$

$$= \sum_{r=0}^{\infty}\left(-\frac{1}{3^{r+1}} + \frac{1}{2^{r+1}}\right) X^r = B(X).$$

Therefore, for each r, $b_r = -1/3^{r+1} + 1/2^{r+1}$.

Thus,

$$\frac{X^5}{X^2 - 5X + 6} = X^5 \sum_{r=0}^{\infty}\left(-\frac{1}{3^{r+1}} + \frac{1}{2^{r+1}}\right) X^r = \sum_{r=0}^{\infty}\left(-\frac{1}{3^{r+1}} + \frac{1}{2^{r+1}}\right) X^{r+5}$$

and if we make the substitution $k = r + 5$ we see that

$$\frac{X^5}{X^2 - 5X + 6} = \sum_{k=5}^{\infty} \left(-\frac{1}{3^{k-4}} + \frac{1}{2^{k-4}} \right) X^k = \sum_{k=0}^{\infty} d_k X^k$$

and what this final equality says is that

$$d_0 = d_1 = d_2 = d_3 = d_4 = 0$$

$$d_5 = -\frac{1}{3} + \frac{1}{2}$$

$$d_6 = -\frac{1}{3^2} + \frac{1}{2^2}$$

$$d_k = -\frac{1}{3^{k-4}} + \frac{1}{2^{k-4}}$$

if $k \geq 5$ and so on.

Example 3.2.2. Compute the coefficients of

$$\sum_{r=0}^{\infty} d_r X^r = \frac{X^2 - 5\ X + 3}{X^4 - 5\ X^2 + 4}.$$

Since $X^4 - 5\ X^2 + 4 = (X^2 - 1)\ (X^2 - 4) = (X - 1)\ (X + 1)\ (X - 2)$ $(X + 2)$ we can write

$$\frac{X^2 - 5X + 3}{X^4 - 5X^2 + 4} = \frac{A}{X - 1} + \frac{B}{X + 1} + \frac{C}{X - 2} + \frac{D}{X + 2}.$$

Multiplication by $X^4 - 5X^2 + 4$ gives

$$X^2 - 5X + 3 = A(X + 1)(X - 2)(X + 2) + B(X - 1)(X - 2)(X + 2)$$
$$+ C(X - 1)(X + 1)(X + 2) + D(X - 1)(X + 1)(X - 2).$$

Let $X = 1$, then all terms of the right-hand side that involve the factor $X - 1$ vanish, and we have $-1 = -6A$ or $A = 1/6$.

Similarly putting $X = -1$, $X = 2$, and $X = -2$, we find $B = 3/2$, $C = -1/4$, and $D = -17/12$.

Thus,

$$\frac{X^2 - 5X + 3}{X^4 - 5X^2 + 4} = \frac{1}{6(X - 1)} + \frac{3}{2(X + 1)} - \frac{1}{4(X - 2)} - \frac{17}{12(X + 2)}$$

$$= \frac{1}{2}\left[-\frac{1}{3(1 - X)} + \frac{3}{1 + X} + \frac{1}{4(1 - X/2)} - \frac{17}{12(1 + X/2)}\right]$$

$$= \frac{1}{2}\left[-\frac{1}{3}\sum_{r=0}^{\infty} X^r + 3 \sum_{r=0}^{\infty} (-1)^r X^r + \frac{1}{4}\sum_{r=0}^{\infty} \left(\frac{1}{2^r}\right) X^r \right.$$

$$\left. - \frac{17}{12}\sum_{r=0}^{\infty} \left(-\frac{1}{2}\right)^r X^r \right]$$

$$= \frac{1}{2}\sum_{r=0}^{\infty} \left[\left(-\frac{1}{3}\right) + 3 (-1)^r + \frac{1}{4}\frac{1}{2^r} - \frac{17}{12}\left(-\frac{1}{2}\right)^r\right] X^r$$

Therefore,

$$d_r = \frac{1}{2}\left[-\frac{1}{3} + 3 (-1)^r + \frac{1}{2^{r+2}} - \frac{17}{3}(-1)^r \frac{1}{2^{r+2}}\right]$$

which can be simplified to

$$d_r = \begin{cases} \dfrac{1}{2}\left[-\dfrac{1}{3} + 3 + \dfrac{1}{2^{r+2}}\left(1 - \dfrac{17}{3}\right)\right] = \dfrac{1}{3}\left(4 - \dfrac{14}{2^{r+3}}\right) & \text{if } r \text{ is even} \\[3mm] \dfrac{1}{2}\left[-\dfrac{1}{3} - 3 + \dfrac{1}{2^{r+2}}\left(1 + \dfrac{17}{3}\right)\right] = \dfrac{1}{3}\left(-5 + \dfrac{5}{2^{r+1}}\right) & \text{if } r \text{ is odd.} \end{cases}$$

After doing these examples we see that it is desirable to write $C(X)/A(X)$ in the form,

$$\frac{B_{11}}{[1 - (X/\alpha_1)]^{r_1}} + \frac{B_{12}}{[1 - (X/\alpha_1)]^{r_1-1}}$$

$$+ \cdots + \frac{B_{1r_1}}{[1 - (X/\alpha_1)]} + \cdots + \frac{B_{k1}}{[1 - (X/\alpha_k)]^{r_k}}$$

$$+ \frac{B_{k2}}{[1 - (X/\alpha_k)]^{r_k-1}} + \cdots + \frac{B_{kr_k}}{[1 - (X/\alpha_k)]}$$

where $$A(X) = a_n(X - \alpha_1)^{r_1} (X - \alpha_2)^{r_2} \cdots (X - \alpha_k)^{r_k}$$

and then solve for the constants $B_{11}, \ldots, B_{1r_1}, \ldots, B_{k1}, \ldots, B_{kr_k}$ by algebraic techniques. This is desirable because in this form we can readily apply the formulas (3.2.1) through (3.2.8) without having to resort to the intermediate step of applying (3.2.11) and (3.2.12).

Example 3.2.3. Find the coefficient of X^{20} in $(X^3 + X^4 + X^5 \cdots)^5$.

Simplify the expression by extracting X^3 from each factor. Thus,

$$(X^3 + X^4 + X^5 \cdots)^5 = [X^3(1 + X + \cdots)]^5 = X^{15}\left(\sum_{r=0}^{\infty} X^r\right)^5$$

$$= \frac{X^{15}}{(1 - X)^5} = X^{15}\sum_{r=0}^{\infty} C(5 - 1 + r, r)X^r.$$

The coefficient of X^{20} in the original expression becomes the coefficient of X^5 in $\sum_{r=0}^{\infty} C(4 + r, r)X^r$ (cancel X^{15} from the above expression). Thus, the coefficient we seek is when $r = 5$ in the last power series, that is, the coefficient is $C(4 + 5, 5) = C(9, 5)$.

Example 3.2.4. Calculate the coefficient of X^{15} in $A(X) = (X^2 + X^3 + X^4 + X^5)(X + X^2 + X^3 + X^4 + X^5 + X^6 + X^7)(1 + X + \cdots + X^{15})$ Note that we can rewrite the expression for $A(X)$ as

$$X^2(1 + X + X^2 + X^3)(X)(1 + X + \cdots + X^6)$$

$$(1 + X + \cdots + X^{15}) = X^3 \frac{(1 - X^4)}{(1 - X)}\frac{(1 - X^7)}{1 - X}\frac{(1 - X^{16})}{1 - X}$$

$$= X^3 \frac{(1 - X^4)(1 - X^7)(1 - X^{16})}{(1 - X)^3}$$

The coefficient of X^{15} in $A(X)$ is the same as the coefficient of X^{12} in

$$\frac{(1 - X^4)(1 - X^7)(1 - X^{16})}{(1 - X)^3}$$

$$= (1 - X^4)(1 - X^7)(1 - X^{16})\left(\sum_{r=0}^{\infty} C(r + 2, r)X^r\right).$$

Since the coefficient of X^{12} in a product of several factors can be obtained by taking one term from each factor so that the sum of their exponents equals 12, we see that the term X^{16} in the third factor and all terms of degree greater than 12 in the last factor need not be considered. Hence we look for the coefficient of X^{12} in

$$(1 - X^4)(1 - X^7)\sum_{r=0}^{12} C(r + 2, r)X^r$$

$$= (1 - X^4 - X^7 + X^{11})\sum_{r=0}^{12} C(r + 2, r)X^r.$$

Using the successive terms of the first factors and the corresponding terms of the second factor so that the sum of their exponents is 12, we see that $C(14,12) - C(10,8) - C(7,5) + C(3,1)$ is the coefficient we seek.

Note that $A(X)$ is a generating function such that the coefficient of X^r counts the number of ways of distributing r similar balls into 3 numbered boxes where the first box can have any number of balls between 2 and 5 inclusive, the second box can contain any number between 1 and 7 inclusive, and the third box can contain any number up to 15 balls. Factoring X^2 out of the first factor and X out of the second factor amounted to the combinatorial strategy of placing 2 balls in the first box and 1 in the second to begin with. Then finding the coefficient of X^{12} in $A(X)/X^3$ amounted to counting the number of ways of distributing 12 balls into 3 boxes where the first box could contain up to 3 balls, the second box could contain up to 6 balls, and the last up to 15 balls. Had we done this problem in Chapter 2 we would have used the principle of inclusion-exclusion, and the form of the answer suggests that that is precisely what is going on behind the scenes in all the algebraic manipulation. This is the major reason for using generating functions: *the algebraic techniques automatically do the combinatorial reasoning for us.*

To illustrate this point let us solve one problem with techniques from Chapter 2 and compare those techniques with generating function techniques.

Example 3.2.5. Find the number of ways of placing 20 similar balls into 6 numbered boxes so that the first box contains any number of balls between 1 and 5 inclusive and the other 5 boxes must contain 2 or more balls each. The integer-solution-of-an-equation-model is: count the number of integral solutions to $e_1 + e_2 + e_3 + e_4 + e_5 + e_6 = 20$ where $1 \le e_1 \le 5$ and $2 \le e_2, e_3, e_4, e_5, e_6$.

First, we will count the solutions where $1 \le e_1$ and $2 \le e_i$ for $i = 2,3,4,5,6$. We do this by placing 1 ball in box number one, 2 balls each in the other 5 boxes, and then counting the number of ways to distribute the remaining 9 balls into 6 boxes with unlimited repetition. There are $C(14,9)$ ways to do this.

But then we wish to discard the number of solutions for which $6 \le e_1$ and $2 \le e_i$ for $i = 2,3,4,5,6$. There are $C(9,4)$ of these. Hence the total number of solutions subject to the stated constraints is $C(14,9) - C(9,4)$.

Now let us solve this problem with generating functions. The generating function we consider is:

$$(X + X^2 + X^3 + X^4 + X^5)(X^2 + X^3 \cdots)^5$$
$$= X(1 + X + X^2 + X^3 + X^4)[X^2(1 + X + X^2 \cdots)]^5$$
$$= X(1 + X + X^2 + X^3 + X^4)(X^{10})(1 + X + X^2 \cdots)^5$$
$$= X^{11}(1 + X + X^2 + X^3 + X^4)(1 + X + X^2 \cdots)^5.$$

We desire to compute the coefficient of X^{20} in this last product, but we need only compute the coefficient of X^9 in $(1 + X + X^2 + X^3 + X^4)$ $(1 + X + X^2 \cdots)^5$, which can be rewritten as:

$$\left(\frac{1 - X^5}{1 - X}\right)\left(\frac{1}{1 - X}\right)^5 = (1 - X^5)\left(\frac{1}{1 - X}\right)^6 = (1 - X^5)\left(\sum_{r=0}^{\infty} C(r + 5, r)X^r\right).$$

Thus, the coefficient of X^9 in this last product is $C(14,9) - C(9,4)$. Note again that the algebra did the combinatorial reasoning for us.

Linearity

Ordinary generating functions possess an important transformation property, namely *linearity*. Linearity implies that, if any two sequences of numbers $\{a_r\}_{r=0}^{\infty}$ and $\{b_r\}_{r=0}^{\infty}$ and scalars D and E are given, then sequence $\{c_r\}_{r=0}^{\infty}$ formed by the linear combination $c_r = Da_r + Eb_r$ has the generating function

$$C(X) = \sum_{r=0}^{\infty} (Da_r + Eb_r)X^r = DA(X) + EB(X),$$

where

$$A(X) = \sum_{r=0}^{\infty} a_r X^r \quad \text{and} \quad B(X) = \sum_{r=0}^{\infty} b_r X^r$$

are the generating functions for $\{a_r\}_{r=0}^{\infty}$ and $\{b_r\}_{r=0}^{\infty}$, respectively.

Example 3.26. Since $1/(1 - 2X)$ generates $\{2^r\}_{r=0}^{\infty}$

and $\dfrac{1}{(1 - 5X)^3}$ generates $\left\{\dfrac{(r + 2)(r + 1)}{2} 5^r\right\}_{r=0}^{\infty}$,

$$\frac{3}{1 - 2X} - \frac{7}{(1 - 5X)^3}$$ generates the sequence

$$\left\{3 \cdot 2^r - 7\frac{(r + 2)(r + 1)}{2} 5^r\right\}_{r=0}^{\infty}.$$

Exercises for Section 3.2

1. Using the equations of Definition 3.2.1, find the coefficients b_0,b_1,b_2,b_3,b_4,b_5 for the following generating functions $B(X)$:

   (a) $\dfrac{1}{1 - 11X + 28X^2}$,

   (b) $\dfrac{1}{1 + 3X + X^2}$,

   (c) $\dfrac{1 + 2X}{1 + 3X + X^2}$.

2. Write the formal power series expression for the following:

   (a) $\dfrac{1}{1 - 5X}$,   (d) $\dfrac{1}{3 - X}$,   (g) $\dfrac{1}{(1 - 5X)^3}$,

   (b) $\dfrac{1}{1 + 5X}$,   (e) $\dfrac{1}{(1 - 5X)^2}$,   (h) $\dfrac{1}{(3 + X)^2}$,

   (c) $\dfrac{1}{3 + X}$,   (f) $\dfrac{1}{(1 - X)^5}$,   (i) $\dfrac{1}{(3 + X)^4}$.

3. Use partial fractions to compute:

   (a) $\dfrac{1}{1 - 7X + 12X^2}$,

   (b) $\dfrac{1}{1 - 7X + 10X^2}$,

   (c) $\dfrac{X + 21}{(X - 5)(2X + 3)}$,

   (d) $\dfrac{7X^2 + 3X + 2}{(X - 2)(X + 1)^2}$,

   (e) $\dfrac{1 - 7X + 3X^2}{(1 - 3X)(1 - 2X)(1 + X)}$.

4. Write the generating function for the sequence $\{a_r\}_{r=0}^{\infty}$ defined by
   (a) $a_r = (-1)^r$,
   (b) $a_r = (-1)^r 3^r$,
   (c) $a_r = 5^r$,
   (d) $a_r = r + 1$,
   (e) $a_r = 6(r + 1)$,
   (f) $a_r = C(r + 3, r)$,
   (g) $a_r = (r + 3)(r + 2)(r + 1)$,
   (h) $a_r = \dfrac{(-1)^r (r + 2)(r + 1)}{2!}$,

   (i) $a_r = 5^r + (-1)^r 3^r + 8C(r + 3, r)$,
   (j) $a_r = (r + 1) 3^r$,
   (k) $a_r = (r + 2)(r + 1) 3^r$.

5. Write an expression for a_r where a_r is the coefficient of X^r in the following generating functions $A(X)$:

   (a) $\dfrac{1}{1-X} + \dfrac{5}{1+2X} + \dfrac{7}{(1-X)^5}$,

   (b) $\dfrac{3}{(1-X)^2} - \dfrac{7}{(1-2X)^3} + \dfrac{8}{3+2X}$,

   (c) $\dfrac{8}{(3+2X)^2} + \dfrac{1}{(5+X)^3}$,

   (d) $\dfrac{-1}{X-1} - \dfrac{4}{3(X+1)} + \dfrac{13}{12(X-2)} + \dfrac{9}{4(2+X)}$.

6. Find the coefficient of X^{10} in
   (a) $(1 + X + X^2 + \cdots)^2$,

   (b) $\dfrac{1}{(1-X)^3}$,

   (c) $\dfrac{1}{(1-X)^5}$,

   (d) $\dfrac{1}{(1+X)^5}$,

   (e) $(X^3 + X^4 + \cdots)^2$,
   (f) $X^4(1 + X + X^2 + X^3)(1 + X + X^2 + X^3 + X^4)(1 + X + X^2 + \cdots + X^{12})$.

7. Find the coefficient of X^{12} in

$$\frac{1 - X^4 - X^7 + X^{11}}{(1-X)^5}.$$

8. Find the coefficient of X^{14} in
   (a) $(1 + X + X^2 + X^3)^{10}$,
   (b) $(1 + X + X^2 + X^3 + X^4 + \cdots + X^8)^{10}$,
   (c) $(X^2 + X^3 + X^4 + X^5 + X^6 + X^7)^4$.

9. Find the coefficient of X^{20} in

$$(X + X^2 + X^3 + X^4 + X^5)(X^2 + X^3 + X^4 \cdots)^5.$$

10. (a) Find the coefficient of X^{50} in $(X^{10} + X^{11} + \cdots + X^{25})(X + X^2 + \cdots + X^{15})(X^{20} + X^{21} + \cdots + X^{45})$.

    (b) Find the coefficient of X^{25} in

$$(X^2 + X^3 + X^4 + X^5 + X^6)^7.$$

11. Find the coefficient of X^{12} in

    (a) $\dfrac{X^2}{(1 - X)^{10}}$,

    (b) $\dfrac{X^5}{(1 - X)^{10}}$,

    (c) $(1 - X)^{20}$,

    (d) $(1 + X)^{20}$,

    (e) $(1 + X)^{-20}$,

    (f) $(1 - 4X)^{-5}$,

    (g) $(1 - 4X)^{15}$,

    (h) $(1 + X^3)^{-4}$,

    (i) $\dfrac{X^2 - 3X}{(1 - X)^4}$

    (j) $(1 - 2X)^{19}$.

12. Let a_r be the number of ways the sum r can be obtained by tossing 50 distinguishable dice. Write a generating function for the sequence $\{a_r\}_{r=0}^{\infty}$. Then find the number of ways to obtain the sum of 100, that is, find a_{100}.

13. Let a_r be the number of nonnegative integral solutions to $X_1 + X_2 + X_3 = r$.
    (a) Find a_{10} if $0 \le X_i \le 4$ for each i.
    (b) Find a_{50} where $2 \le X_1 \le 50, 0 \le X_2 \le 50, 5 \le X_3 \le 25$.

14. Find a_{10} in the exercise numbers listed below from Section 3.1:
    (a) 5(a)
    (b) 6
    (c) 7(a)
    (d) 8(a)
    (e) 8(c)
    (f) 8(d)
    (g) 8(e)

15. Use generating functions to find the number of ways to select 10 balls from a large pile of red, white, and blue balls if
    (a) the selection has at least 2 balls of each color,
    (b) the selection has at most 2 red balls, and
    (c) the selection has an even number of blue balls.

16. How many ways are there to place an order for 12 chocolate sundaes if there are 5 types of sundaes, and at most 4 sundaes of one type are allowed?

17. How many ways are there to paint 20 identical rooms in a hotel with 5 colors if there is only enough blue, pink, and green paint to paint 3 rooms?

18. Write a generating function for a_n, the number of ways of obtaining the sum n when tossing 9 distinguishable dice. Then find a_{25}.

19. How many solutions are there for the equation $X_1 + X_2 + X_3 + X_4 = 12$ if each X_i must have one of values 1, 2, 3, or 4?

20. Write a generating function for a_r, the number of ways of selecting r letters from $\{5 \cdot a, 6 \cdot b, 8 \cdot c\}$ if each selection must include at least one a, at least one b, and at least two c's. Find a_8.

21. Find the generating function $A(X)$ for a_r, the number of ways to select r objects from n distinguishable objects, where unlimited repetition is allowed and each kind of object must appear an even number of times. Derive a single series expression for $A(X)$ and find an expression for a_{12}.

22. Solve exercise 21 in case each object must appear a multiple of 3 times.

23. Find a generating function for determining a_r, the number of r-combinations of the letters M, A, T, R, I, X in which M, A, and T can appear any number of times but R, I, and X appear at most once. Find a_{10}.

24. Let a_r denote the number of ways to partition r identical marbles into 4 distinct piles so that each pile has an odd number of marbles that is at least 3.
    (a) Determine a generating function for a_r.
    (b) Determine a closed form expression for a_r.
    (c) Find a_{20}.

25. A set of r balls is selected from an infinite supply of red, white, blue, and gold balls. A selection must satisfy the condition that either the number of red balls is even and the number of white balls is odd, or the number of blue balls is even and the number of gold balls is odd. Let a_r denote the number of such selections.
    (a) Determine the generating function for a_r.
    (b) Determine an expression for a_r.
    (c) Find a_{13}.

26. (a) In how many ways can $3r$ balls be selected from $2r$ garnet balls, $2r$ gold balls, and $2r$ blue balls?
    (b) How many ways can $4r$ balls be chosen?

27. Let a_r denote the number of ways to select r balls from 20 garnet balls, 20 gold balls, 30 green balls, and 30 blue balls with the constraints that the number of garnet balls is not equal to 2, the number of gold balls is not 3, the number of green balls is not 4, and the number of blue balls is not 5. Determine the generating function for a_r.

Selected Answers for Section 3.2

2. (c) $\dfrac{1}{3+X} = \dfrac{1}{3\left(1 + \dfrac{X}{3}\right)} = \dfrac{1}{3}\sum_{r=0}^{\infty}(-1)^r \left(\dfrac{1}{3}\right)^r X^r = \dfrac{1}{3}\sum_{r=0}^{\infty}\left(-\dfrac{1}{3}\right)^r X^r$

(h) $\dfrac{1}{(3 + X)^2} = \dfrac{1}{3^2} \left(1 + \dfrac{X}{3}\right)^{-2} = \dfrac{1}{9} \sum_{r=0}^{\infty} (-1)^r (r + 1) \left(\dfrac{1}{3}\right)^r X^r$

(i) $\dfrac{1}{(3 + X)^4} = \dfrac{1}{3^4} \left(1 + \dfrac{X}{3}\right)^{-4} = \dfrac{1}{81} \sum_{r=0}^{\infty} (-1)^r C(r + 3,r) \left(\dfrac{1}{3}\right)^r X^r$

3. (a) $\dfrac{1}{1 - 7X + 12X^2} = \dfrac{1}{(1 - 3X)(1 - 4X)} = \dfrac{-3}{1 - 3X} + \dfrac{4}{1 - 4X}$

$= -3 \sum_{r=0}^{\infty} 3^r X^r + 4 \sum_{r=0}^{\infty} 4^r X^r$

$= \sum_{r=0}^{\infty} (4^{r+1} - 3^{r+1}) X^r$

(c) $A(X) = \dfrac{X + 21}{(X - 5)(2X + 3)} = \dfrac{2}{X - 5} - \dfrac{3}{2X + 3}$

$= \dfrac{-2}{5\left(1 - \dfrac{X}{5}\right)} - \dfrac{1}{\left(1 + \dfrac{2}{3}X\right)}$

$= \dfrac{-2}{5} \sum_{r=0}^{\infty} \left(\dfrac{1}{5}\right)^r X^r - \sum_{r=0}^{\infty} \left(-\dfrac{2}{3}\right)^r X^r$

$= \sum_{r=0}^{\infty} \left[\left(-\dfrac{2}{5}\right)\left(\dfrac{1}{5}\right)^r - \left(-\dfrac{2}{3}\right)^r\right] X^r$

$\therefore a_r = \left(-\dfrac{2}{5}\right)\left(\dfrac{1}{5}\right)^r - \left(-\dfrac{2}{3}\right)^r$ for $r \geq 0$.

(d) $A(X) = \dfrac{7X^2 + 3X + 2}{(X - 2)(X + 1)^2} = \dfrac{3}{X + 1} - \dfrac{2}{(X + 1)^2} + \dfrac{4}{X - 2}$

$= \dfrac{3}{1 + X} - \dfrac{2}{(1 + X)^2} - \dfrac{2}{\left(1 - \dfrac{X}{2}\right)}$

$= 3 \sum_{r=0}^{\infty} (-1)^r X^r - 2 \sum_{r=0}^{\infty} (r + 1)(-1)^r X^r - 2 \sum_{r=0}^{\infty} \left(\dfrac{1}{2}\right)^r X^r$

$= \sum_{r=0}^{\infty} \left[3(-1)^r - 2(r + 1)(-1)^r - 2\left(\dfrac{1}{2}\right)^r\right] X^r;$

simplify.

(e) $a_n = (-9/4)(3^n) + (7/3)(2^n) + (11/12)(-1)^n$

5. (b) $a_r = 3C(r + 1,r) - 7 (2)^r C(r + 2,r) + 8/3 (-1)^r (2/3)^r$

(d) $1 - (4/3) (-1)^n - 13/24 (1/2)^n + 9/8 (-1/2)^n$

6. (a) $\dfrac{1}{(1 - X)^2} = \sum\limits_{r=0}^{\infty} C(r + 1, r) X^r = \sum\limits_{r=0}^{\infty} (r + 1) X^r$;

   coefficient of X^{10} is 11.

   (b) $\dfrac{1}{(1 - X)^3} = \sum\limits_{r=0}^{\infty} C(r + 2, r) X^r = \sum\limits_{r=0}^{\infty} \dfrac{(r + 2)(r + 1)}{2} X^r$;

   coefficient of X^{10} is $(12)(11)/2$.

   (c) $C(14, 10)$

   (d) $(-1)^{10} C(14, 10) = C(14, 10)$

   (e) $[X^3(1 + X + X^2 + \cdots)]^2 = X^6 [1/(1 - X)^2]$; coefficient of X^{10} is the coefficient of X^4 in $1/(1 - X)^2 = \sum_{r=0}^{\infty} C(r + 1, r) X^r$; coefficient $= 5$

   (f) $C(8, 6) - C(4, 2) - C(3, 1)$

8. (a) $(1 + X + X^2 + X^3)^{10} = \left(\dfrac{1 - X^4}{1 - X}\right)^{10} = \dfrac{(1 - X^4)^{10}}{(1 - X)^{10}}$

   $= (1 - X^4)^{10} \sum\limits_{r=0}^{\infty} C(r + 9, r) X^r$

   $= [1 - C(10,1) X^4 + C(10,2) X^8 - C(10,3) X^{12} + \cdots + X^{40}]$

   $\sum\limits_{r=0}^{\infty} C(r + 9, r) X^r$;

   coefficient of X^{14} is $C(23,9) - C(10,1) C(19,10)$
   $+ C(10,2) C(15,6) - C(10,3) C(11,2)$

   (b) $C(23,14) - 10 C(14,5)$

9. $C(14,9) - C(9,4)$

10. (a) $C(21,19) - C(6,4) - C(5,3)$

    (b) $C(17,11) - 7 C(12,6) + C(7,2) C(7,1)$

11. (a) $C(19,10)$

    (b) $C(16,7)$

    (c) $C(20,12)$

    (d) $C(20,12)$

    (e) $C(31,12)$

    (g) $4^{12} C(15,12)$

    (j) $(-2)^{12} C(19,12)$

12. $(X + X^2 + X^3 + X^4 + X^5 + X^6)^{50} = X^{50} (1 - X^6)^{50} 1/(1 - X)^{50}$;

    coefficient of X^{100} is $C(99,50) - C(44 + 49,44) C(50,1) + C(49 + 38,38) C(50,2) - C(49 + 32,32) C(50,3) \cdots$

13. (a) $C(12,10) - 3C(7,5) + 3$
    (b) $C(45,43) - C(24,22)$

15. (a) $C(6,4)$
    (b) $C(12,10) - C(9,7)$

16. $C(16,12) - C(5,1) C(11,7) + C(5,2) C(6,2)$

17. The coefficient of X^{20} in $(1 + X + X^2 + X^3)^3 (\Sigma_{r=0}^{\infty} X^r)^2$ is $C(24,20) - 3C(20,16) + 3C(16,12) - C(12,8)$

18. $A(X) = (X + X^2 + X^3 + X^4 + X^5 + X^6)^9 = X^9 (1 + X + \cdots + X^5)^9 = X^9 (1 - X^6)^9 (1 - X)^{-9}$.

    Find the coefficient of X^{16} in $(1 - X^6)^9 (1 - X)^{-9}$;
    $$a_{25} = C(24,16) - 9C(18,10) + C(9,2) C(12,4)$$

3.3   RECURRENCE RELATIONS

The expressions for permutations, combinations, and partitions developed in Chapter 2 are the most fundamental tools for counting the elements of finite sets. Nevertheless, these expressions often prove inadequate for many combinatorial problems that the computer scientist must face. An important alternate approach uses **recurrence relations** (sometimes called **difference equations**) to define the terms of a sequence. We desire to demonstrate how many combinatorial problems can be modeled with recurrence relations, and then we will discuss methods of solving several common types of recurrence relations.

A formal discussion of recurrence relation beyond that of Section 1.10 is somewhat difficult within the scope of this book but the concept of recurrence relations is straightforward. Many combinatorial problems can be solved by reducing them to analogous problems involving a smaller number of objects, and the salient feature of recurrence relations is the specification of one term of a collection of numbers as a function of preceding terms of the collection. Using a recurrence relation we can reduce a problem involving n objects to one involving $n - 1$ objects, then to one involving $n - 2$ objects, and so on. By successive reductions of the number of objects involved, we hope to eventually end up with a problem that can be solved easily. Perhaps an example will be instructive.

A Computer Science Application

Suppose that in a given programming language we wish to count the number of valid expressions using only the ten digits 0,1, . . . ,9, and the four arithmetic operation symbols $+, -, \div, \times$. Assume that the syntax of this language requires that each valid expression end in a digit, and that 2 valid expressions can be combined by using the 4 arithmetic operations. Therefore, a valid expression is a sequence of one or more digits or of the form $A \circ B$ where A and B are valid expressions and the operator $\circ$ is one of the 4 arithmetic operations. Thus, for instance, $1 + 2$ is a valid expression as is 3×45, and then $1 + 2 - 3 \times 45$ is also a valid expression, but $1 + + 2$ is not. The problem then is: how many such valid expressions of length n are there in this language?

First of all, we note that the answer is not $(14^{n-1})10$ because we do not allow expressions like $1 + + 2$, that is, 2 successive arithmetic symbols are not allowed so the first $n - 1$ entries cannot be filled arbitrarily. Thus, our analysis needs to be a bit more sophisticated. We attempt to use the idea of recurrence relations.

Let a_n be the number of valid expressions of length n, and let us consider a particular valid expression of length n. Focus attention on the entry in position $n - 1$. This symbol may be a digit, in which case the first $n - 1$ symbols form a valid expression of length $n - 1$. Or this symbol may be one of the 4 arithmetic symbols, in which case the preceeding $n - 2$ symbols form a valid expression.

The number of valid expressions in the first class is $10a_{n-1}$ since there are 10 digits that can be appended to a valid expression of length $n - 1$. Likewise, there are $40a_{n-2}$ expressions in the second class since each one of 4 arithmetic symbols may be followed by any one of 10 digits and then both appended to a valid expression of length $n - 2$. Thus, we can determine a_n from a_{n-1} and a_{n-2} according to the relation $a_n = 10a_{n-1} + 40a_{n-2}$. This recurrence relation is valid for $n \geq 2$, but clearly $a_0 = 0$ and $a_1 = 10$ since the number of valid expressions of length 1 is just the number of digits. Thus, we can determine that

$$a_2 = 10a_1 + 40a_0 = 10(10) + 40(0) = 100,$$
$$a_3 = 10a_2 + 40a_1 = 10(100) + 40(10)$$
$$= 1400,$$
$$a_4 = 10a_3 + 40a_2 = 10(1400) + 40(100)$$
$$= 18{,}000, \text{ and so on.}$$

Pascal's identity (see Chapter 2) is another example of a recurrence relation: $C(n,r) = C(n - 1,r) + C(n - 1, r - 1)$. Here a term in the n-th row of Pascal's triangle is determined by 2 terms in a preceding row. But this example is an example of a recurrence relation involving the 2 integer variables n and r. By and large we shall restrict our attention to recurrence relations that involve only one integer variable, so let us adopt the following working definition of recurrence relation.

Definition 3.3.1. A **recurrence relation** is a formula that relates for any integer $n \geq 1$, the n-th term of a sequence $A = \{a_r\}_{r=0}^{\infty}$ to one or more of the terms $a_0, a_1, \ldots, a_{n-1}$.

Examples of recurrence relations. If s_n denotes the sum of the first n positive integers, then (1) $s_n = n + s_{n-1}$. Similarly if d is a real number, then the nth term of an arithmetic progression with common difference d satisfies the relation (2) $a_n = a_{n-1} + d$. Likewise if p_n denotes the nth term of a geometric progression with common ratio r, then (3) $p_n = rp_{n-1}$. We list other examples as:

(4) $a_n - 3a_{n-1} + 2a_{n-2} = 0$.

(5) $a_n - 3a_{n-1} + 2a_{n-2} = n^2 + 1$.

(6) $a_n - (n-1)a_{n-1} - (n-1)a_{n-2} = 0$.

(7) $a_n - 9a_{n-1} + 26a_{n-2} - 24a_{n-3} = 5n$.

(8) $a_n - 3(a_{n-1})^2 + 2a_{n-2} = n$.

(9) $a_n = a_0 a_{n-1} + a_1 a_{n-2} + \cdots + a_{n-1} a_0$.

(10) $a_n^2 + (a_{n-1})^2 = -1$.

Definition 3.3.2. Suppose n and k are nonnegative integers. A recurrence relation of the form $c_0(n)\, a_n + c_1(n)a_{n-1} + \cdots + c_k(n)\, a_{n-k} = f(n)$ for $n \geq k$, where $c_0(n), c_1(n), \ldots, c_k(n)$, and $f(n)$ are functions of n is said to be a **linear** recurrence relation. If $c_0(n)$ and $c_k(n)$ are not identically zero, then it is said to be a linear recurrence relation of *degree k*. If $c_0(n), c_1(n), \ldots, c_k(n)$ are constants, then the recurrence relation is known as a linear recurrence relation with constant coefficients. If $f(n)$ is identically zero, then the recurrence relation is said to be **homogeneous;** otherwise, it is **inhomogeneous.**

Thus, all the examples above are linear recurrence relations except (8), (9), and (10); the relation (8), for instance, is not linear because of the squared term. The relations in (3), (4), (5), and (7) are linear with constant coefficients. Relations (1), (2), and (3) have degree 1; (4), (5), and (6) have degree 2; (7) has degree 3. Relations (3), (4), and (6) are homogeneous.

There are no general techniques that will enable one to solve all recurrence relations. There are, nevertheless, techniques that will enable us to solve linear recurrence relations with constant coefficients.

Solutions of Recurrence Relations

In elementary algebra solving an equation like $X^2 - 7X + 10 = 0$ was defined to mean that we find all those values of X which, when substituted into the quadratic equation, made the equation a true statement. By factoring or by use of the quadratic formula, we determine that $X = 2$ and $X = 5$ are the only 2 values that solve the equation.

Suppose now that we are given the recurrence relation $a_n - 5a_{n-1} = 0$ for $n \geq 1$ and are asked to solve it. We first ask: what is meant by a solution of this recurrence relation? We answer this by recalling that a sequence $A = \{a_n\}_{n=0}^{\infty}$ is a function from the nonnegative integers into the real numbers. What the recurrence relation does is describe a relation between the values of this function at n and at $n - 1$. We ask then: is there a function, defined with domain the set of nonnegative integers, which makes this equation true for every value of n? The answer is yes, as is shown by the function $A = \{a_n\}_{n=0}^{\infty}$ where $a_n = 5^n$ for $n \geq 0$. For this function we have $a_n - 5a_{n-1} = 5^n - 5(5^{n-1}) = 0$ for $n \geq 1$, so that this function satisfies the recurrence relation. However, it is one of many solutions; as a matter of fact, if c is any constant the function $\{a_n\}_{n=0}^{\infty}$ where $a_n = c5^n$ for $n \geq 0$ also satisfies the same recurrence relation because $a_n - 5a_{n-1} = c5^n - 5c5^{n-1} = 0$ for $n \geq 1$.

Just as in algebra, recurrence relations may have no solution. Equation (10) above is an example of this since there are no *real*-valued functions f such that $(f(n))^2 + (f(n-1))^2 = -1$ since the squares of real numbers are always nonnegative.

Definition 3.3.3. Suppose that S is a subset of the nonnegative integers. Then a sequence $A = \{a_n\}_{n=0}^{\infty}$ is a **solution** to a recurrence relation over S if the values a_n of A make the recurrence relation a true statement for every value of n in S. If the sequence $A = \{a_n\}_{n=0}^{\infty}$ is a solution of a recurrence relation, then it is said to *satisfy* the relation.

Example 3.3.1. (a) $A = \{a_n\}_{n=0}^{\infty}$ where $a_n = 2^n$ satisfies the recurrence relation $a_n = 2a_{n-1}$ over the set S of integers $n \geq 1$. In fact for any constant c, the sequence $\{c2^n\}_{n=0}^{\infty}$ satisfies the same recurrence relation. More generally, if a and c are any real numbers, then $a_n = ca^n$ satisfies the recurrence relation: $a_n = aa_{n-1}$ for $n \geq 1$. (b) If c_1 and c_2 are arbitrary constants, then $a_n = c_1 2^n + c_2 5^n$ satisfies the recurrence relation: $a_n - 7a_{n-1} + 10a_{n-2} = 0$ over the set S of integers $n \geq 2$. For by substituting this expression for a_n into the recurrence relation, we have

$$
\begin{aligned}
a_n - 7a_{n-1} + 10a_{n-2} &= (c_1 2^n + c_2 5^n) - 7(c_1 2^{n-1} + c_2 5^{n-1}) \\
&\quad + 10(c_1 2^{n-2} + c_2 5^{n-2}) \\
&= c_1 2^n - 7c_1 2^{n-1} + 10c_1 2^{n-2} + c_2 5^n - 7c_2 5^{n-1} + 10c_2 5^{n-2} \\
&= 2^{n-2} c_1 [2^2 - 7(2) + 10] + 5^{n-2} c_2 [5^2 - 7(5) + 10] \\
&= 2^{n-2} c_1 (0) + 5^{n-2} c_2 (0) = 0.
\end{aligned}
$$

(c) Similarly for arbitrary constants c_1 and c_2, $a_n = c_1 5^n + c_2 n5^n$ satisfies the recurrence relation $a_n - 10a_{n-1} + 25a_{n-2} = 0$. (d) Likewise for arbitrary constants c_1, c_2, and c_3, $a_n = c_1 2^n + c_2 5^n + c_3 n5^n$ satisfies the recurrence relation $a_n - 12a_{n-1} + 45a_{n-2} - 50a_{n-3} = 0$. We leave this verification as an exercise.

Note that in each of the above examples there are infinitely many different solutions, one for each specific value of the constants. Suppose we are asked to find a solution of the recurrence relation in (b) for which $a_0 = 10$ and $a_1 = 41$. These so-called *boundary conditions* are requirements that must be satisfied in addition to that of satisfying the recurrence relation.

Let us see whether there is such a solution among those already listed of the form $a_n = c_1 2^n + c_2 5^n$. If we set $n = 0$ and $n = 1$, then $10 = a_0 = c_1 2^0 + c_2 5^0 = c_1 + c_2$ and $41 = a_1 = c_1 2^1 + c_2 5^1 = 2 c_1 + 5 c_2$. Thus, the constants c_1 and c_2 satisfy the equations

$$10 = c_1 + c_2 \qquad \text{and} \qquad 41 = 2 c_1 + 5 c_2.$$

Solving these two equations for c_1 and c_2, respectively, we find that $c_1 = 3$ and $c_2 = 7$. Thus, $a_n = (3) 2^n + (7) 5^n$ is a solution of the recurrence relation that satisfies the boundary conditions.

If we are given a recurrence relation describing the nth term of a sequence $A = \{a_r\}_{r=0}^{\infty}$ as a function of the terms $a_0, a_1, \ldots, a_{n-1}$, what we desire is a closed-form expression for a_n in terms of n alone as in (b) above. But even if we do not have such a closed-form expression the recurrence relation is still very useful in computation. For we can compute a_n in terms of $a_{n-1}, \ldots, a_1, a_0$; then compute a_{n+1} in terms of a_n, $a_{n-1}, \ldots, a_0$; and so on, provided the value of the sequence at one or more points is given so that the computation can be initiated. That is why we need the boundary conditions.

The above example is a linear recurrence relation of degree 2 and the 2 boundary conditions for the value of a_0 and a_1 gave rise to a unique solution. In general, if a linear recurrence relation of degree k has constant coefficients, and if there are fewer than k boundary conditions, then there will not be a unique solution. But even if k values of the sequence are stipulated these may not, as a rule, guarantee a unique solution except in the case that the k values are consecutive. In other words, if there is some integer n_0 such that the values for $a_{n_0}, a_{n_0+1}, \ldots, a_{n_0+k-1}$ are given, then there will be a unique solution of the linear recurrence relation of degree k satisfying these boundary conditions. Usually the values for $a_0, a_1, \ldots, a_{k-1}$ are given and then it would be appropriate to call these *initial conditions*.

The Fibonacci Relation

In a book published in 1202 A.D. Leonardo of Pisa, also known as Fibonacci, posed a problem of determining how many pairs of rabbits are born of one pair in one year. The problem posed by Fibonacci is the following. Initially, suppose that there is only one pair of rabbits, male and female, just born, and suppose, further, that every month each pair of rabbits that are over one month old produce a new pair of offspring of opposite sexes. Find the number of rabbits after 12 months and after n months.

We start with one pair of newly born rabbits. After one month we still have only one pair of rabbits since they are not yet mature enough to reproduce. After 2 months we have 2 pairs of rabbits because the first pair has now reproduced. After 3 months we have 3 pairs of rabbits since those born just last month cannot reproduce yet, but the original pair has reproduced again. After 4 months we have 5 pairs of rabbits because the first pair is continuing to reproduce, the second pair has produced a new pair, and the third pair is still maturing. For each integer $n \geq 0$, let F_n denote the number of pairs of rabbits alive at the end of the nth month. Here we mean that $F_0 = 1$, the original number of pairs of rabbits. Then what we have said is that $F_0 = 1 = F_1$, $F_2 = 2$, $F_3 = 3$, and $F_4 = 5$.

Note that F_n is formed by starting with F_{n-1} pairs of rabbits alive last month and adding the babies that can only come from the F_{n-2} pairs alive 2 months ago. Thus, $F_n = F_{n-1} + F_{n-2}$ is the recurrence relation and $F_0 = F_1 = 1$ are the initial conditions.

Using this relation and the values for F_2, F_3, F_4 already computed we see that

$$
\begin{aligned}
F_5 &= F_4 + F_3 = 5 + 3 = 8, \\
F_6 &= F_5 + F_4 = 8 + 5 = 13, \\
F_7 &= F_6 + F_5 = 13 + 8 = 21, \\
F_8 &= F_7 + F_6 = 21 + 13 = 34, \\
F_9 &= F_8 + F_7 = 34 + 21 = 55, \\
F_{10} &= F_9 + F_8 = 55 + 34 = 89, \\
F_{11} &= F_{10} + F_9 = 89 + 55 = 144, \\
F_{12} &= F_{11} + F_{10} = 144 + 89 = 233.
\end{aligned}
$$

Thus, after 12 months there are 233 pairs of rabbits alive. We could continue this process to compute F_{36}, and so on. Indeed this numerical approach, even for more complicated recurrence relations is quite practical especially if an electronic computer is used.

Shortly we will show how to obtain an explicit solution of this recurrence relation. The relation $F_n = F_{n-1} + F_{n-2}$ is called the **Fibonacci relation** and the numbers F_n generated by the Fibonacci relation with the initial conditions $F_0 = 1 = F_1$ are called the **Fibonacci numbers** and the sequence of Fibonacci numbers $\{F_n\}_{n=0}^{\infty}$ is the **Fibonacci sequence.** Fibonacci numbers arise quite naturally in many combinatorial settings. There is even a scientific journal, *Fibonacci Quarterly,* devoted primarily to research involving the Fibonacci relation and Fibonacci numbers.

The Fibonacci relation comes up again in the following stair-climbing example.

Example 3.3.2. (a) In how many ways can a person climb up a flight of n steps if the person can skip at most one step at a time?

Let a_n = the number of ways the person can climb n steps for $n \geq 1$. Note $a_1 = 1$, and $a_2 = 2$ (since one can proceed one step at a time or take 2 steps in one stride). Let us solve for a_n in terms of a fewer number of steps. Suppose the person takes only 1 step on the first stride, there then are left $n - 1$ steps to climb for which there are a_{n-1} ways to climb them. If, on the other hand, the person took 2 steps in the first stride, there are $n - 2$ steps left for which there are a_{n-2} ways to climb. Since there are only these 2 possibilities and these events are mutually exclusive we apply the sum rule to get $a_n = a_{n-1} + a_{n-2}$.

(b) Suppose we change the conditions of the above example and assume that the person may take either 1, 2, or 3 steps in each stride. Find a recurrence relation for the number of ways the person can climb n steps.

Again let a_n = the number of ways to climb n steps. Then it should be clear that $a_n = a_{n-1} + a_{n-2} + a_{n-3}$ where each summand is determined by whether the first stride takes 1, 2, or 3 steps.

Some Properties of Fibonacci Numbers

Let us examine some immediate consequences of the Fibonacci relation. First we attempt to find a compact formula for the sum $S_n = F_0 + F_1 + \cdots + F_n$. The following table shows that $S_0 = 1 = F_2 - 1$; $S_1 = 2 = F_3 - 1$; $S_2 = 4 = F_4 - 1$:

n	0	1	2	3	4	5	6	7
F_n	1	1	2	3	5	8	13	21
S_n	1	2	4	7	12	20	33	54

This leads us to conjecture that:

(1) The sum of the first $n + 1$ Fibonacci numbers is one less than F_{n+2}, that is, $F_0 + F_1 + F_2 + \cdots + F_n = F_{n+2} - 1$.

The proof is straightforward; it could be certified by mathematical induction, but the following observation makes the proof immediate. Write the numbers in an array as follows:

$$F_0 = F_2 - F_1$$
$$F_1 = F_3 - F_2$$
$$F_2 = F_4 - F_3$$
$$F_3 = F_5 - F_4$$
$$\vdots$$
$$F_n = F_{n+2} - F_{n+1}$$

If we add all of these equations, the telescopic property of the right-hand side causes all but $-F_1$ and F_{n+2} to vanish. Thus, $F_0 + F_1 + F_2 + \cdots + F_n = F_{n+2} - F_1$ but $F_1 = 1$.

Likewise we have:

(2) $F_0 + F_2 + F_4 + \cdots + F_{2n} = F_{2n+1}$, and

(3) $F_0^2 + F_1^2 + F_2^2 + \cdots + F_n^2 = F_n F_{n+1}$.

Theorem 3.3.1. (General Solution of the Fibonacci Relation). If F_n satisfies the Fibonacci relation $F_n = F_{n-1} + F_{n-2}$ for $n \geq 2$, then there are constants C_1 and C_2 such that

$$F_n = C_1 \left(\frac{1 + \sqrt{5}}{2} \right)^n + C_2 \left(\frac{1 - \sqrt{5}}{2} \right)^n,$$

where the constants are completely determined by the initial conditions.

Proof. Let $F(X) = \sum_{n=0}^{\infty} F_n X^n$ be the generating function for sequence $\{F_n\}_{n=0}^{\infty}$. Then note that

$$F(X) = F_0 + F_1 X + F_2 X^2 + F_3 X^3 + \cdots + F_n X^n + \cdots$$
$$XF(X) = F_0 X + F_1 X^2 + F_2 X^3 + \cdots + F_{n-1} X^n + \cdots$$
$$X^2 F(X) = F_0 X^2 + F_1 X^3 + \cdots + F_{n-2} X^n + \cdots$$

Subtracting the last 2 equations from the first gives:

$$\begin{aligned}
F(X) - XF(X) - X^2 F(X) &= F_0 + (F_1 - F_0)X + (F_2 - F_1 - F_0)X^2 \\
&\quad + (F_3 - F_2 - F_1)X^3 + \cdots \\
&\quad + (F_n - F_{n-1} - F_{n-2})X^n + \cdots \\
&= F_0 + (F_1 - F_0)X + 0\,X^2 + 0\,X^3 \cdots \\
&= F_0 + (F_1 - F_0)X.
\end{aligned}$$

Thus,

$$F(X) = \frac{F_0 + (F_1 - F_0)X}{1 - X - X^2} = \frac{F_0 + (F_1 - F_0)\,X}{[1 - \dfrac{(1 + \sqrt{5})}{2}X]\,[1 - \dfrac{(1 - \sqrt{5})}{2}X]}$$

Thus, for whatever initial conditions on F_0 and F_1, the method of partial fractions applies to give

$$F(X) = \frac{C_1}{1 - (1 + \sqrt{5})X/2} + \frac{C_2}{1 - (1 - \sqrt{5})X/2}.$$

Using the identities for geometric series we see that if $a = (1 + \sqrt{5})/2$ and $b = (1 - \sqrt{5})/2$ then,

$$\begin{aligned}
F(X) = \frac{C_1}{1 - aX} + \frac{C_2}{1 - bX} &= C_1 \sum_{n=0}^{\infty} a^n X^n + C_2 \sum_{n=0}^{\infty} b^n X^n \\
&= \sum_{n=0}^{\infty} (C_1 a^n + C_2 b^n) X^n = \sum_{n=0}^{\infty} F_n X^n.
\end{aligned}$$

In other words,

$$F_n = C_1 a^n + C_2 b^n = C_1 \left(\frac{1 + \sqrt{5}}{2}\right)^n + C_2 \left(\frac{1 - \sqrt{5}}{2}\right)^n$$

for each $n \geq 0$. $\square$

Of course if we are given the initial conditions that $F_0 = 1 = F_1$, then we can find

$$C_1 = \frac{1}{\sqrt{5}}\left(\frac{1 + \sqrt{5}}{2}\right) \qquad \text{and} \qquad C_2 = \frac{-1}{\sqrt{5}}\left(\frac{1 - \sqrt{5}}{2}\right)$$

so that, in this case, the nth Fibonacci number is

$$F_n = \frac{1}{\sqrt{5}} \left[\left(\frac{1 + \sqrt{5}}{2} \right)^{n+1} - \left(\frac{1 - \sqrt{5}}{2} \right)^{n+1} \right].$$

One consequence of this last observation is the relative size of the Fibonacci numbers for large n. Since $b = (1 - \sqrt{5})/2$ is approximately -0.618, b^{n+1} gets very small for large n so that F_n is approximated by $(1/\sqrt{5}) \, a^{n+1}$ for large n.

The Fibonacci numbers occur frequently in combinatorial problems, and in fact, there is an interesting property of Pascal's triangle that states that the sum of the elements lying on the diagonal running upward from the left are Fibonacci numbers. We illustrate this as follows:

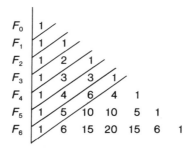

Figure 3-1

In particular, we have the identity $F_n = C(n,0) + C(n - 1,1) + C(n - 2,2) + \cdots + C(n - k,k)$ where $k = \lfloor n/2 \rfloor$ = the greatest integer in $n/2$.

To prove this we define $q_n = C(n,0) + C(n - 1,1) + \cdots + C(n - k,k)$ for $n \geq 0$ and $k = \lfloor n/2 \rfloor$.

If we adopt the convention that $C(m,r) = 0$ if $r > m$, then we can write $q_n = C(n,0) + C(n - 1,1) + \cdots + C(n - k,k) + C(n - k - 1,k+1) + \cdots + C(0,n)$. By Theorem 3.3.1, we need only show q_n satisfies the Fibonacci relation and that $q_0 = 1 = q_1$. But $q_0 = C(0,0) = 1$ and $q_1 = C(1,0) + C(0,1) = 1$. Then, using Pascal's identity, we see that for $n \geq 2$,

$$\begin{aligned}
q_{n-1} + q_{n-2} &= C(n - 1,0) + C(n - 2,1) + \cdots + C(0, n - 1) + C(n - 2,0) \\
&\quad + C(n - 3,1) + \cdots + C(0,n - 2) \\
&= C(n - 1,0) + [C(n - 2,1) + C(n - 2,0)] + [C(n - 3,1) \\
&\quad + C(n - 3,2)] + \cdots \\
&\quad + [C(0,n - 1) + C(0,n - 2)] \\
&= C(n - 1,0) + C(n - 1,1) \\
&\quad + C(n - 2,2) + \cdots + C(1,n - 1) \\
&= C(n,0) + C(n - 1,1) + C(n - 2,2) + \cdots + C(1,n - 1) \\
&\quad + C(0,n) = q_n.
\end{aligned}$$

Other Recurrence Relation Models

Compound interest problems can be described in terms of recurrence relations.

Example 3.3.3. Let P represent the principal borrowed from a bank, let r equal the interest rate per period, and let a_n represent the amount due after n periods. Then $a_n = a_{n-1} + r\, a_{n-1} = (1 + r)\, a_{n-1}$. In particular, $a_0 = P$, $a_1 = (1 + r)\, P$, $a_2 = (1 + r)\, a_1 = (1 + r)^2\, P$, and so on, so that $a_n = (1 + r)^n\, P$.

Example 3.3.4. The number of derangements satisfy a recurrence relation. Recall that a derangement of $\{1,2,\ldots,n\}$ is a permutation $(i_1, i_2, \ldots, i_n)$ where $i_1 \neq 1$, $i_2 \neq 2$, $\ldots$, and $i_n \neq n$. Let D_n = the number of derangements of $\{1,2,\ldots,n\}$. (We derived a formula for D_n in Section 2.8.)

These derangements can be partitioned into two classes. Consider an arbitrary derangement $(i_1, i_2, \ldots, i_n)$. Then either 1 is in the i_1th position (that is, 1 and i_1 have changed places) or 1 is not in the i_1th position. In the first case, the remaining $n - 2$ numbers form a derangement of those numbers and since i_1 can be chosen in $(n - 1)$ ways, there are $(n - 1)D_{n-2}$ derangements of this kind. Now let us count the derangements where 1 is not in the i_1th position. First choose i_1 in $n - 1$ ways. Now take any derangement of $\{2,3,\ldots,n\}$. We can form a derangement of $\{1,2,\ldots,n\}$ as follows: first replace i_1 in this derangement by 1 and then put i_1 in front of this derangement. Since originally i_1 did not occupy its natural position now 1 is not in the i_1th position. This process will produce all derangements of the second kind. Thus, there are $(n - 1)\,D_{n-1}$ of these derangements, and hence by adding both numbers we get $D_n = (n - 1)\,D_{n-1} + (n - 1)\,D_{n-2}$, for $n \geq 3$.

Example 3.3.5. Find a recurrence relation for a_n the number of different ways to distribute either a \$1 bill, a \$2 bill, a \$5 bill, or a \$10 dollar bill on successive days until a total of n dollars has been distributed.

We use the same kind of analysis as in the stair-climbing example. If, on the first day, we distribute a \$1 bill, then we are to distribute $n - 1$ dollars on the succeeding days and there are a_{n-1} ways to do that. If, on the other hand, the first days distribution was a \$2 bill, there remains the problem of distributing $n - 2$ dollars; this can be done a_{n-2} ways and so on. Thus, we see that $a_n = a_{n-1} + a_{n-2} + a_{n-5} + a_{n-10}$.

Let us follow this analysis in one more example.

Example 3.3.6. Suppose that a school principal decides to give a prize away each day. Suppose further that the principal has 3 different kinds of prizes worth $1 each and 5 different kinds of prizes worth $4 each. Find a recurrence relation for a_n = the number of different ways to distribute prizes worth n dollars.

If, on the first day, a $1 prize is given, the prize could have been chosen from one of the 3 different kinds of $1 prizes, and then there are $n - 1$ dollars worth of prizes to be given away later. Thus there are $3\, a_{n-1}$ ways to do this. If, on the other hand, the prize given on the first day was a $4 prize then there are 5 different ways to choose the prize and a_{n-4} ways to distribute the remaining prizes. Thus, $a_n = 3\, a_{n-1} + 5\, a_{n-4}$.

Many problems in the biological, management, and social sciences lead to sequences which satisfy recurrence relations. In some cases the problems lead, as in the examples above, to one recurrence relation for which the sequence is a solution. In other cases the principles which describe economic forces or the interaction between two or more populations can be formulated as a system of linear recurrence relations. For example, the growth of one population may affect the growth of another population either favorably or unfavorably because they may compete for food, one may prey on the other, or each population may prey on the other, and so on.

An example will illustrate the nature of this kind of problem.

Example 3.3.7. (The Lancaster Equations of Combat). Two armies engage in combat. Each army counts the number of men still in combat at the end of each day. Let a_0 and b_0 denote the number of men in the first and second armies, respectively, before combat begins, and let a_n and b_n denote the number of men in the two armies at the end of the nth day. Thus $a_{n-1} - a_n$ represents the number of soldiers lost by the first army during the battle on the nth day. Likewise $b_{n-1} - b_n$ represents the number of soldiers lost by the second army on the nth day.

Suppose it is known that the decrease in the number of soldiers in each army is proportional to the number of soldiers in the other army at the beginning of each day. Thus we have constants A and B such that $a_{n-1} - a_n = Ab_{n-1}$ and $b_{n-1} - b_n = Ba_{n-1}$. These constants measure the effectiveness of the weapons of the different armies. Of course, we can rewrite these so that we have

$$a = a_{n-1} - Ab_{n-1} \qquad \text{and} \qquad b_n = -Ba_{n-1} + b_{n-1},$$

a system very much reminiscent of two-linear equations in two unknowns.

Exercises for Section 3.3

1. In each of the following a recurrence relation and a function are given. In each case, show that the function is a solution of the given recurrence relation.
   (a) $a_n - a_{n-1} = 0$; $a_n = C$.
   (b) $a_n - a_{n-1} = 1$; $a_n = n + C$.
   (c) $a_n - a_{n-1} = 2$; $a_n = 2n + C$.
   (d) $a_n - 3a_{n-1} + 2a_{n-2} = 0$; $a_n = C_1 + C_2 2^n$.
   (e) $a_n - 3a_{n-1} + 2a_{n-2} = 1$; $a_n = C_1 + C_2 2^n - n$.
   (f) $a_n - 7a_{n-1} + 12a_{n-2} = 0$; $a_n = C_1 3^n + C_2 4^n$.
   (g) $a_n - 7a_{n-1} + 12a_{n-2} = 1$; $a_n = C_1 3^n + C_2 4^n + 1/6$.

2. For each of the recurrence relations in Exercise 1, we will give initial conditions. Of the solutions given, find the unique solution satifying the initial conditions.
   (a) $a_0 = 5$.
   (b) $a_0 = 6$.
   (c) $a_0 = 6$.
   (d) $a_0 = 5$, $a_1 = 6$.
   (e) $a_0 = 4$, $a_1 = 6$.
   (f) $a_0 = 4$, $a_1 = 6$.
   (g) $a_0 = 19/6$, $a_1 = 31/6$.

3. (a) Consider a $1 \times n$ chessboard. Suppose we can color each square of the chessboard either red or white. Let a_n = the number of ways of coloring the chessboard in which no 2 red squares are adjacent. Find a recurrence relation that a_n satisfies.
   (b) Suppose now that each square can be colored either red, white, or blue. Let b_n be the number of ways of coloring the n squares so that no two adjacent squares are colored red. Find a recurrence relation satisfied by b_n.

4. Find a recurrence relation for the number of ways to arrange flags on a flagpole n feet tall using 4 types of flags: red flags 2 feet high, or white, blue, and yellow flags each 1 foot high.

5. Find a recurrence relation for the number of ways to arrange vehicles in a row with n spaces if we are to park Volkswagens, Hondas, Toyotas, Fiestas, Buicks, Cadillacs, Continentals, Mack trucks, and Greyhound buses. Each of the Volkswagens, Hondas, Toyotas, and Fiestas requires one parking space; the Buicks, Cadillacs, and Continentals require 2 spaces each, and the Mack trucks and Greyhound buses require 4 spaces each.

6. Find a recurrence relation for the number of ways to make a pile of n chips using garnet, gold, red, white, and blue chips such that no two gold chips are together.

7. Let P_n be the number of permutations of m letters taken n at a time with repetitions but no 3 consecutive letters being the same. Derive a recurrence relation connecting P_n, P_{n-1}, and P_{n-2}.

8. (a) Find a recurrence relation for the number of n-digit binary sequences with no triple of consecutive 1's.

   (b) Repeat for n-digit ternary sequences. (Ternary sequences use only 0, 1, or 2 as digits).

9. (a) Find a recurrence relation for the number of n-digit ternary sequences that have an even number of 0's.

   (b) Repeat for n-digit quaternary sequences. Quaternary sequences use only 0, 1, 2, 3 for digits.

10. Suppose that a circular disk is divided into n sectors, like one would cut a pie, with the boundaries of all sectors meeting at a point in the center of the disk. Suppose, further, that we have 10 different colored paints with which we are required to paint the sectors on this disk in such a way that no adjacent sectors have the same color paint on them. Let a_n be the number of ways to paint the n sectors of the disk with the 10 different paints. Find a recurrence relation satisfied by a_n.

11. (Gambler's Ruin) Suppose we repeatedly bet \$1 on the toss of a coin; heads you win, tails I win. Each of us has a probability of 1/2 of winning on each flip of the coin. Suppose, further, that you have \$100 and I have \$200 to begin with. Let P_n be the probability that you win all \$300 when you have n dollars. Find a recurrence relation involving P_n.

12. (a) Suppose a coin is flipped until 2 heads appear and then the experiment stops. Find a recurrence relation for the number of experiments that end on the nth flip or sooner.

   (b) Repeat assuming the experiment stops only after 3 heads appear.

   (c) Repeat, assuming the experiment stops after 4 heads appear.

13. If \$1,000 is invested in a money market fund earning 16% a year, how much money is in the account after n years?

14. A bacteriologist wants the number of bacteria in a certain solution S_0. The bacteriologist proceeds as follows: one tenth of the solution S_0 is taken and diluted to form a new solution S_1; one tenth of S_1 is taken and diluted to form a new solution S_2; and so on. Let a_n denote the number of bacteria in S_n. Suppose it is determined that $a_4 = 10$. Find a_0.

15. A student starts a chain letter by writing to 4 of his friends and asking that each of them write to 4 others and so on. Let a_n denote the number of letters written at the nth stage of the chain letter. Find and solve a recurrence relation involving a_n. (Note: $a_0 = 4$.)

16. From Theorem 3.3.1 find the constants C_1 and C_2 so that F_n satisfies the following initial conditions.
    (a) $F_0 = 2, F_1 = 1$ (the Lucas sequence).
    (b) $F_0 = 0, F_1 = 1$.

17. Let $F_0 = 1 = F_1, F_2, \ldots$ be the Fibonacci sequence. By evaluating each of the following expressions for small values of n, conjecture a general formula and then prove it.
    (a) $F_1 + F_3 + \cdots + F_{2n-1}$.
    (b) $F_0 - F_1 + F_2 + \cdots + (-1)^n F_n$.
    (c) $F_n F_{n+2} + (-1)^n$.

18. The relation $F_n = F_{n-1} + F_{n-2}$ holds for all integers $n \geq 2$, explain how that relation can be translated to the relation $F_{n+2} = F_{n+1} + F_n$ for all integers n. Also observe that $F_{n+3} = F_{n+2} + F_{n+1}$.
    (a) Combine these to give an expression for F_{n+3} in terms of F_{n+1} and F_n.
    (b) Express F_{n+4} in terms of F_{n+1} and F_n.
    (c) Express F_{n+3} in terms of F_{n-1} and F_{n-2} for $n \geq 2$.

19. (a) Find r, given that $F_r = 2F_{101} + F_{100}$.
    (b) Find s, given that $F_s = 3F_{200} + 2F_{199}$.
    (c) Find t, given that $F_t = 5F_{317} + 3F_{316}$.

20. Let $L_0, L_1, \ldots$ be the Lucas sequence. That is, $L_n = L_{n-1} + L_{n-2}$ for $n \geq 2$ and $L_0 = 2, L_1 = 1$.
    (a) Prove that $L_0 + L_1 + L_2 + \cdots + L_n = L_{n+2} - 1$.
    (b) Derive a formula for $L_1 + L_3 + L_5 + \cdots + L_{2n+1}$.

21. Conjecture and prove formulas for the Lucas sequence:
    (a) $L_0 + L_3 + L_6 + L_9 + \cdots + L_{3n}$.
    (b) $L_1 + L_4 + L_7 + \cdots + L_{3n+1}$.
    (c) $L_2 + L_5 + L_8 + \cdots + L_{3n+2}$.

22. Find a recurrence relation that counts the number of ways of making a selection from the numbers $1, 2, 3, \ldots, n$ without taking a pair of consecutive numbers (counting the empty set as a selection).

23. Find a recurrence relation for the number of permutations of the integers $1, 2, 3, \ldots, n$ such that no integer is more than one place removed from its position in the natural order.

24. Find a recurrence relation for the number of n-digit quarternary sequences with at least one 1 and the first 1 occurring before the first 0 (there may be no 0's).

25. Find a recurrence relation for the number of n-digit binary sequences that have the pattern 010 occurring at the nth digit. For example, the pattern 010 occurs at the fourth and the ninth digits in the sequence 1010000100011, but not at the eighth and thirteenth digits. After the pattern occurs, scanning starts all over

again to search for the second occurrence so that in 110101010101 the pattern occurs at the fifth and ninth digits but not at the seventh and eleventh digits.

26. Find a recurrence relation for the number of n-digit ternary sequences that have the pattern 012 occurring:
    (a) for the first time at the end of the sequence.
    (b) for the second time at the end of the sequence.

27. Find a recurrence relation for the number of n-digit binary sequences with an even number of 0's and an even number of 1's.

28. Find a system of recurrence relations for the number of n-digit binary sequences which:
    (a) do not contain 2 consecutive 0's.
    (b) contain exactly one pair of consecutive 0's.

29. Find a system of recurrence relations for the number of n-digit binary sequences that contain the pattern 010 for the first time at the end.

30. Find a recurrence relation for the number of ways to pair off $2n$ people for tennis matches.

Selected Answers for Section 3.3

3. (a) $a_n = a_{n-1} + a_{n-2}$
   (b) $b_n = 2b_{n-1} + 2b_{n-2}$

4. Let a_n = the number of ways to arrange the flags on the flagpole n-feet tall. If the first flag is 1 foot high, then there are a_{n-1} ways to arrange the flags on the other $n - 1$ feet of the pole. Since there are 3 colors for the first flag, in this case, there are $3a_{n-1}$ ways to arrange the pole. In case the first flag is red, and hence, 2 feet high, there are a_{n-2} ways to arrange the flags on the other $n - 2$ feet of the pole. Thus, $a_n = 3a_{n-1} + a_{n-2}$.

5. $a_n = 4a_{n-1} + 3a_{n-2} + 2a_{n-4}$.

6. If the first chip is gold there are 4 colors for the second chip and then a_{n-2} ways to make a pile of the other $n - 2$ chips. There are, on the other hand, 4 colors for the first chip to be other than a gold one, and then a_{n-1} ways to make a pile of the other $n - 1$ chips. Thus $a_n = 4a_{n-1} + 4a_{n-2}$.

7. The first 2 letters may be the same or different; if the same, the remaining $n - 2$ letters must be a permutation of the specified type where the third letter cannot be like the first 2 letters. Hence we may append to a "good" $(n - 2)$-permutation two repeated letters different from the first letter of the $(n - 2)$-permutation. Similarly for the other case, the second letter can be chosen from $m - 1$ other letters and the other $n - 1$ letters can be arranged P_{n-1}

ways. Hence the recurrence relation is $P_n = (m - 1) [P_{n-1} + P_{n-2}]$.

9. (a) If the first digit is not 0, then there are $2a_{n-1}$ $(n - 1)$-digit such ternary sequences. If the first digit is 0, then we must count the number of $(n - 1)$-digit ternary sequences that have an odd number of 0's. Since there are 3^{n-1} total $(n - 1)$-digit ternary sequences and a_{n-1} with an even number of 0's, there are $3^{n-1} - a_{n-1}$ with an odd number of 0's. Hence there are $3^{n-1} - a_{n-1}$ n-digit sequences with an even numbers of 0's and that start with 0. Thus, $a_n = 2a_{n-1} + 3^{n-1} - a_{n-1} = a_{n-1} + 3^{n-1}$.

   (b) Here $a_n = 2a_{n-1} + 4^{n-1}$.

10. Number the n sectors. Then either sectors 1 and 3 have the same color or not. In the first case we could remove sector number 2 and imagine sectors 1 and 3 coalesced to leave a disk with $n - 2$ sectors painted according to the rules. Thus, in this case, the sector number 2 could be painted 9 different colors, so that the total number of ways to paint the n sectors in this way is $9a_{n-2}$. In the other case since sectors 1 and 3 are colored differently if we remove sector 2 we are left with $n - 1$ sectors of a kind we are considering. Since sector 2 can have only 8 different colors for it, the total number of ways to color disks of this type is $8a_{n-1}$. Hence the recurrence relation is $a_n = 8a_{n-1} + 9a_{n-2}$.

11. $P_n = \frac{1}{2} P_{n-1} + \frac{1}{2} P_{n+1}$.

12. (a) $a_n = a_{n-1} + (n - 1)$.
    (b) $a_n = a_{n-1} + C(n - 1,2)$.
    (c) $a_n = a_{n-1} + C(n - 1,3)$.

17. (a) $F_{2n} - 1$.
    (b) $(-1)^n F_{n-1} + 1$.
    (c) $F_{n-1}F_{n+3} + 3$.

19. (a) $r = 103$.
    (c) $t = 321$.

20. (b) $L_{2n+2} - 2$.

3.4 SOLVING RECURRENCE RELATIONS BY SUBSTITUTION AND GENERATING FUNCTIONS

We shall consider four methods of solving recurrence relations in this and the next two sections:

1. substitution (also called iteration),
2. generating functions,
3. characteristic roots, and
4. undetermined coefficients.

In the substitution method the recurrence relation for a_n is used repeatedly to solve for a general expression for a_n in terms of n. We desire that this expression involve no other terms of the sequence except those given by boundary conditions.

The mechanics of this method are best described in terms of examples. We used this method in Example 3.3.4. Let us also illustrate the method in the next three examples.

Example 3.4.1. Solve the recurrence relation $a_n = a_{n-1} + f(n)$ for $n \geq 1$ by substitution.

$$a_1 = a_0 + f(1)$$

$$a_2 = a_1 + f(2) = (a_0 + f(1)) + f(2)$$

$$a_3 = a_2 + f(3) = (a_0 + f(1) + f(2)) + f(3)$$

$$\vdots$$

$$a_n = a_0 + f(1) + f(2) + \cdots + f(n)$$

$$= a_0 + \sum_{k=1}^{n} f(k).$$

Thus, a_n is just the sum of the $f(k)$'s plus a_0.

More generally, if c is a constant then we can solve $a_n = ca_{n-1} + f(n)$ for $n \geq 1$ in the same way:

$$a_1 = ca_0 + f(1)$$

$$a_2 = ca_1 + f(2) = c(ca_0 + f(1)) + f(2)$$

$$= c^2 a_0 + cf(1) + f(2)$$

$$a_3 = ca_2 + f(3) = c(c^2 a_0 + cf(1) + f(2)) + f(3)$$

$$= c^3 a_0 + c^2 f(1) + cf(2) + f(3)$$

$$\vdots$$

$$a_n = ca_{n-1} + f(n) = c(c^{n-1}a_0 + c^{n-2}f(1) + \cdots + cf(n-2)$$

$$+ f(n-1)) + f(n)$$

$$= c^n a_0 + c^{n-1}f(1) + c^{n-2}f(2) + \cdots + cf(n-1) + f(n).$$

or

$$a_n = c^n a_0 + \sum_{k=1}^{n} c^{n-k} f(k).$$

Example 3.4.2. (The Towers of Hanoi Problem). There are 3 pegs and n circular disks of increasing diameter on one peg, with the largest disk on the bottom. These disks are to be transferred one at a time onto another peg with the provision that at no time is one allowed to put a larger disk on one with smaller diameter. The problem is to determine the number of moves for the transfer.

Let a_n be the number of moves required to transfer n disks. Clearly $a_0 = 0$, $a_1 = 1$, and $a_2 = 3$. Let us find a recurrence relation that a_n satifies. To transfer n disks to another peg we must first transfer the top $n - 1$ disks to a peg, transfer the largest disk to the vacant peg, and then transfer the $n - 1$ disks to the peg which now contains the largest peg. Thus $a_n = 2 a_{n-1} + 1$ for $n \geq 1$.

Now we use $c = 2$ and $f(k) = 1$ for each k in the formula of Example 3.4.1. Then $a_n = 2^n a_0 + 2^{n-1} + \cdots + 2^2 + 2 + 1$. But since $a_0 = 0$, we have $a_n = 2^{n-1} + \cdots + 2^2 + 2 + 1$, and using the formula for the sum of terms in a geometric progression, we find that

$$a_n = \frac{1-2^n}{1-2} = 2^n - 1 \text{ for } n \geq 0.$$

Example 3.4.3. (Analysis of the Mergesort Algorithm).

A sequence of numbers $(b_1, b_2, \ldots, b_n)$ is *sorted* if the numbers are arranged according to the nondecreasing order, that is, if $b_1 \leq b_2 \leq b_3 \leq \cdots \leq b_n$. Then b_1 is the *head* of the sequence and b_n is the *tail*. By *merging* two sorted lists we mean to combine them into one sorted list. One way to merge two sorted lists, LIST 1 and LIST 2, is the following procedure: Since the smaller one of the heads of the two lists must be the smallest of all the numbers in the two lists, we can remove this number from the list it is in and place it as the first number in the merged list. We shall label this new list as LIST. Now we can compare the two heads of the two lists of the remaining numbers and place the smaller one of the two as the second number in LIST. This process is repeated until one of the lists is empty, at which time the remainder of the nonempty list is appended (concatenated is the usual computer terminology) to the tail of LIST.

Since each comparison of an element of LIST 1 with an element of LIST 2 results in an element being removed from one of these lists and added to LIST, there can be no more than $m + k$ comparisons where LIST 1 and LIST 2 have m and k numbers respectively. Moreover, since no comparison can be made when either of the lists is empty, there can be, in fact, at most $m + k - 1$ comparisons. Thus, if n is the total number of numbers in the two lists, then there are at most $n - 1$ comparisons to merge the two sorted lists.

Mergesort is a sorting algorithm that splits an unsorted input list into two "halves", sorts each half recursively, and then merges the resulting sorted lists into a single sorted list. This final sorted list is the output of this algorithm. (A more formal description of Mergesort is given in Chapter 4.)

An illustration of the subdivisions and the merges necessary in sorting the input list {5,4,0,9,3,2,8,6,23,21} is shown in Fig. 3-2. The top half of the diagram exhibits the subdivision of the lists; the bottom half illustrates the merging of lists successively as described above.

An estimate of the cost of sorting an input list by Mergesort can be obtained by counting the maximum number of comparisons that are necessary. To do this let us make the simplifying assumption that the number n of items to be sorted is a power of 2. Then an upper bound a_n on the number of comparisons required to sort n items is given by the

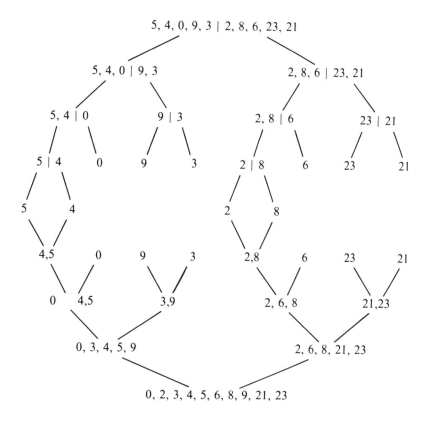

Figure 3-2

recurrence relation:

$$a_n = 2\,a_{n/2} + (n - 1) \text{ for } n \geq 2, \text{ where } a_1 = 0$$

The first of these two equations expresses the fact that the number of comparisons required to sort a large list must be, at most, the sum of the number of comparisons, $2a_{n/2}$, required to sort both halves and the number of comparisons, $(n - 1)$, required to merge the halves. The second equation expresses the fact that sorting a single element does not require any comparisons.

We can solve this recurrence relation for $n = 2^k$ where $k \geq 1$ by repeatedly using the basic relation to get:

$$2a_{n/2} = 2\left[2a_{n/4} + \frac{n}{2} - 1\right]$$

$$= 2^2 a_{n/4} + n - 2$$

$$2^2\,a_{n/4} = 2^2[2a_{n/8} + n/4 - 1]$$

$$= 2^3\,a_{n/8} + n - 2^2,$$

etc., so that we have the following sequence of equations:

$$a_n - 2a_{n/2} = n - 1$$

$$2a_{n/2} - 2^2 a_{n/4} = n - 2$$

$$2^2 a_{n/4} - 2^3 a_{n/8} = n - 2^2$$

$$\vdots$$

$$2^{k-1} a_{n/2^{k-1}} - 2^k a_{n/2^k} = n - 2^{k-1}.$$

Summing both sides of this sequence of equations, cancelling appropriate summands, and noting that $a_{n/2^k} = a_1 = 0$, we have

$$a_n = n - 1 + n - 2 + n - 2^2 + \cdots + n - 2^{k-1}$$

$$= kn - (1 + 2 \ldots + 2^{k-1})$$

$$= kn - (2^k - 1)$$

$$= kn - (n - 1)$$

$$= n\log_2(n) - (n - 1).$$

Divide-and-Conquer Relations

The above recurrence relation is a special case of a so-called "divide-and-conquer" relation. Frequently these relations arise in the analysis of recursive computer algorithms and usually take the form:

$a_n = ca_{n/d} + f(n)$ where c and d are constants and $f(n)$ is some function of n.

Usually these relations can be solved by substituting a power of d for n. We will discuss another method of solution for divide-and-conquer relations in Section 3.6.

Solutions By Generating Functions

We showed how to solve the Fibonacci relation with generating functions and that approach works for arbitrary linear recurrence relations with constant coefficients.

To describe this method in detail we need to understand the following basic property.

The shifting properties of generating functions. If $A(X) = \sum_{n=0}^{\infty} a_n X^n$ generates the sequence $(a_0, a_1, a_2, \ldots)$, then $XA(X)$ generates the sequence $(0, a_0, a_1, a_2, \ldots)$; $X^2 A(X)$ generates $(0, 0, a_0, a_1, a_2, \ldots)$, and, in general $X^k A(X)$ generates $(0, 0, \ldots, 0, a_0, a_1, a_2, \ldots)$ where there are k zeros before a_0.

Thus, if $A(X)$ is the generating function for the sequence $(a_0, a_1, \ldots)$, then multiplying $A(X)$ by X amounts to shifting the sequence one place to the right and inserting a zero in front. Multiplying $A(X)$ by X^k amounts to shifting the sequence k positions to the right and inserting k zeros in front.

This process is described by a change in the dummy variable in the formal power series expressions as follows:

$$X^k A(X) = X^k \sum_{n=0}^{\infty} a_n X^n = \sum_{n=0}^{\infty} a_n X^{n+k}.$$

In the last expression replace $n + k$ by r, and then $n = r - k$ and the expression $\sum_{n=0}^{\infty} a_n X^{n+k}$ becomes $\sum_{r-k}^{\infty} a_{r-k} X^r$. In this form the expression $\sum_{r-k}^{\infty} a_{r-k} X^r$ signifies that it generates the sequence $\{b_r\}_{r=0}^{\infty}$ where $0 = b_0 = b_1 = \cdots = b_{k-1}$, $b_k = a_0$, $b_{k+1} = a_1$, and in general $b_r = a_{r-k}$ if $r \geq k$. Thus, the nth term in the new sequence is obtained from the old sequence by replacing a_n by a_{n-k} if $n \geq k$ and by 0 if $n < k$.

For instance, we know that $1/(1 - X) = \sum_{n=0}^{\infty} X^n$ generates the sequence $(1, 1, 1, \ldots)$, that is, the sequence $\{a_n\}_{n=0}^{\infty}$ where $a_n = 1$ for each $n \geq 0$.

Thus,

$$\frac{X}{(1-X)} = \sum_{n=0}^{\infty} X^{n+1} = \sum_{r=1}^{\infty} X^r$$

generates $(0,1,1,1,\ldots)$, and

$$\frac{X^2}{(1-X)} = \sum_{n=0}^{\infty} X^{n+2} = \sum_{r=2}^{\infty} X^r$$

generates $(0,0,1,1,1,\ldots)$. Similarly,

$$\frac{1}{(1-X)^2} = \sum_{n=0}^{\infty} C(n+1,n) X^n = \sum_{n=0}^{\infty} (n+1) X^n$$

generates the sequence $(1,2,3,4,\ldots)$ so that

$$\frac{X}{(1-X)^2} = \sum_{n=0}^{\infty} (n+1) X^{n+1} = \sum_{r=1}^{\infty} rX^r$$

generates the sequence $\{r\}_{r=0}^{\infty} = (0,1,2,3,4\ldots)$. Note that the expression $\sum_{r=1}^{\infty} rX^r$ describes that the coefficient of X^0 is 0 because the sum is taken from $r = 1$ to ∞, but the form of the coefficients would give the same conclusion even if r is allowed to equal zero; hence we can write $X/(1-X)^2$ two ways: as $\sum_{r=1}^{\infty} rX^r$ and as $\sum_{r=0}^{\infty} rX^r$, and both expressions mean that the coefficient of X^0 is zero. Likewise,

$$\frac{X^2}{(1-X)^2} = \sum_{n=0}^{\infty} (n+1) X^{n+2} = \sum_{r=2}^{\infty} (r-1) X^r$$

generates the sequence $(0,0,1,2,3,4,\ldots)$, that is, the sequence $\{b_r\}_{r=0}^{\infty}$, where $b_r = r - 1$ if $r \geq 2$, but $0 = b_0 = b_1$. Since the expression $b_r = r - 1$ equals zero when $r = 1$, it happens that $X^2/(1-X)^2 = \sum_{r=2}^{\infty} (r-1) X^r$ also can be written as $\sum_{r=1}^{\infty} (r-1) X^r$.

Following this line of thought further, we see that

$$\frac{1}{(1-X)^3} = \sum_{n=0}^{\infty} C(n+2,n)X^n = \sum_{n=0}^{\infty} \frac{(n+2)(n+1)}{2} X^n$$

generates the sequence

$$\left\{\frac{(n+2)(n+1)}{2}\right\}_{n=0}^{\infty} = \left(\frac{1 \cdot 2}{2}, \frac{2 \cdot 3}{2}, \frac{3 \cdot 4}{2}, \ldots\right),$$

and, therefore, $2/(1 - X)^3 = \sum_{n=0}^{\infty} (n + 2)(n + 1)X^n$ generates $\{(n + 2)(n + 1)\}_{n=0}^{\infty} = (1 \cdot 2, 2 \cdot 3, 3 \cdot 4, \ldots)$. But then

$$\frac{2X}{(1 - X)^3} = \sum_{n=0}^{\infty} (n + 2)(n + 1)X^{n+1} = \sum_{r=1}^{\infty} (r + 1)(r) X^r$$

generates the sequence $(0, 1 \cdot 2, 2 \cdot 3, 3 \cdot 4, \ldots)$. Now since $b_r = (r + 1)(r)$ equals 0 when $r = 0$, we see that we can write

$$\frac{2X}{(1 - X)^3} = \sum_{r=1}^{\infty} (r + 1)(r)X^r = \sum_{r=0}^{\infty} (r + 1)(r)X^r,$$

so that $2X/(1 - X)^3$ generates $\{(r + 1)r\}_{r=0}^{\infty}$. Likewise

$$\frac{2X^2}{(1 - X)^3} = \sum_{n=0}^{\infty} (n + 2)(n + 1)X^{n+2} = \sum_{r=2}^{\infty} (r)(r - 1)X^r = \sum_{r=0}^{\infty} (r)(r - 1)X^r$$

generates the sequence $(0, 0, 1 \cdot 2, 2 \cdot 3, 3 \cdot 4, \ldots)$, and the last sum can be taken from 0 to ∞ because the coefficient $r(r - 1)$ is 0 when $r = 0, 1$.

We can combine these results to discover generating functions for other sequences. For instance,

$$\frac{2X}{(1 - X)^3} - \frac{X}{(1 - X)^2} = \frac{X(1 + X)}{(1 - X)^3}$$

generates the sequence $\{(r + 1)(r) - r\}_{r=0}^{\infty} = \{r^2\}_{r=0}^{\infty} = (0, 1, 4, 9, \ldots)$.

No doubt the reader can verify that

$$\frac{1}{(1 - X)^4} = \sum_{n=0}^{\infty} C(n + 3, n)X^n = \sum_{n=0}^{\infty} \frac{(n + 3)(n + 2)(n + 1)}{6} X^n$$

generates $\{(n + 3)(n + 2)(n + 1)/6\}_{n=0}^{\infty}$; $6/(1 - X)^4$ generates $\{(n + 3)(n + 2)(n + 1)\}_{n=0}^{\infty}$;

$$\frac{6X}{(1 - X)^4} = \sum_{n=0}^{\infty} (n + 3)(n + 2)(n + 1)X^{n+1}$$

$$= \sum_{r=1}^{\infty} (r + 2)(r + 1)(r)X^r$$

$$= \sum_{r=0}^{\infty} (r + 2)(r + 1)(r)X^r$$

generates $\{(r + 2)(r + 1)(r)\}_{r=0}^{\infty}$; and

$$\frac{6X^2}{(1-X)^4} = \sum_{n=0}^{\infty} (n+3)(n+2)(n+1)X^{n+2}$$

$$= \sum_{r=2}^{\infty} (r+1)(r)(r-1)X^r = \sum_{r=0}^{\infty} (r+1)(r)(r-1)X^r$$

generates $\{(r+1)(r)(r-1)\}_{r=0}^{\infty}$.

Since $(r+3)(r+2)(r+1) = r^3 + 6r^2 + 11r + 6$ then $r^3 = (r+3)(r+2)(r+1) - 6r^2 - 11r - 6$ so that $\{r^3\}_{r=0}^{\infty}$ is generated by

$$\frac{6}{(1-X)^4} - \frac{6(X)(1+X)}{(1-X)^3} - 11\frac{X}{(1-X)^2} - \frac{6}{(1-X)} = \frac{X(1+4X+X^2)}{(1-X)^4}.$$

In a similar manner we can find generating functions for the sequences $\{r^4\}_{r=0}^{\infty}$, $\{r^5\}_{r=0}^{\infty}$, and so on.

Now there are other shifting properties of generating functions. If $A(X) = \sum_{n=0}^{\infty} a_n X^n$ generates $(a_0, a_1, a_2, \ldots)$, then $A(X) - a_0 = \sum_{n=1}^{\infty} a_n X^n$ generates $(0, a_1, a_2, \ldots)$; $A(X) - a_0 - a_1 X = \sum_{n=2}^{\infty} a_n X^n$ generates $(0, 0, a_2, a_3, \ldots)$; and, in general, $A(X) - a_0 - a_1 X - \cdots - a_{k-1} X^{k-1} = \sum_{n=k}^{\infty} a_n X^n$ generates $(0, 0, \ldots, 0, a_k, a_{k+1} \ldots)$, where there are k zeros before a_k.

But then dividing by powers of X shifts the sequence to the left. For instance, $(A(X) - a_0)/X = \sum_{n=1}^{\infty} a_n X^{n-1}$ generates the sequence $(a_1, a_2, a_3, \ldots)$; $(A(X) - a_0 - a_1 X)/X^2$ generates $(a_2, a_3, a_4, \ldots)$; and, in general, for $k \geq 1$, $(A(X) - a_0 - a_1 X - \cdots - a_{k-1} X^{k-1})/X^k$ generates $(a_k, a_{k+1}, a_{k+2}, \ldots)$.

Again this shifting property can be described by a change in the dummy variable in the power series expressions. If we replace $n - 1$ by r, then $(A(X) - a_0)/X = \sum_{n=1}^{\infty} a_n X^{n-1}$ becomes $\sum_{r=0}^{\infty} a_{r+1} X^r$, which signifies that the coefficient a_n in the original sequence is replaced by a_{n+1}, that is, the sequence has been shifted one place to the left.

Likewise replace $n - k$ by r, and the expression $(A(X) - a_0 - a_1 X - \cdots - a_{k-1} X^{k-1})/X^k = \sum_{n=k}^{\infty} a_n X^{n-k}$ becomes $\sum_{r=0}^{\infty} a_{r+k} X^r$ which generates the sequence $(a_k, a_{k+1}, a_{k+2}, \ldots)$. In other words, the term a_n in the original sequence is replaced by a_{n+k} for each n, indicating that the sequence has been shifted k places to the left. Thus, for instance, $A(X) = 1/(1-X)^2$ generates $(1, 2, 3, \ldots) = \{n+1\}_{n=0}^{\infty}$ so that $[1/(1-X)^2 - 1]/X$ generates $(2, 3, 4, \ldots) = \{n+2\}_{n=0}^{\infty}$, and $[1/(1-X)^2 - 1 - 2X]/X^2$ generates $(3, 4, 5, \ldots) = \{n+3\}_{n=0}^{\infty}$ similarly $2/(1-X)^3$ generates $\{(n+2)(n+1)\}_{n=0}^{\infty}$ so that $[2/(1-X)^3 - 2]/X$ generates $\{(n+3)(n+2)\}_{n=0}^{\infty}$.

Let us combine these results on the shifting property of generating functions together with the different identities for geometric series and other series to obtain generating functions for a few familiar sequences (see Table 3-1).

Table 3-1. Table of Generating Functions

	Sequence a_n	Generating Function $A(X)$
(1)	$C(k,n)$	$(1 + X)^k$
(2)	1	$\dfrac{1}{1 - X}$
(3)	a^n	$\dfrac{1}{1 - aX}$
(4)	$(-1)^n$	$\dfrac{1}{1 + X}$
(5)	$(-1)^n a^n = (-a)^n$	$\dfrac{1}{1 + aX}$
(6)	$C(k - 1 + n, n)$ k is a fixed positive integer	$\dfrac{1}{(1 - X)^k}$
(7)	$C(k - 1 + n, n))a^n$	$\dfrac{1}{(1 - aX)^k}$
(8)	$C(k - 1 + n, n)(-a)^n$	$\dfrac{1}{(1 + aX)^k}$
(9)	$n + 1$	$\dfrac{1}{(1 - X)^2}$
(10)	n	$\dfrac{X}{(1 - X)^2}$
(11)	$(n + 2)(n + 1)$	$\dfrac{2}{(1 - X)^3}$
(12)	$(n + 1)(n)$	$\dfrac{2X}{(1 - X)^3}$
(13)	n^2	$\dfrac{X(1 + X)}{(1 - X)^3}$
(14)	$(n + 3)(n + 2)(n + 1)$	$\dfrac{6}{(1 - X)^4}$
(15)	$(n + 2)(n + 1)(n)$	$\dfrac{6X}{(1 - X)^4}$
(16)	n^3	$\dfrac{X(1 + 4X + X^2)}{(1 - X)^4}$
(17)	$(n + 1)a^n$	$\dfrac{1}{(1 - aX)^2}$
(18)	na^n	$\dfrac{aX}{(1 - aX)^2}$
(19)	$n^2 a^n$	$\dfrac{(aX)(1 + aX)}{(1 - aX)^3}$
(20)	$n^3 a^n$	$\dfrac{(aX)(1 + 4aX + a^2 X^2)}{(1 - aX)^4}$

In solving recurrence relations by generating functions we encounter these shifting properties so frequently that we list some equivalent expressions for ready reference.

Table of Equivalent Expressions for Generating Functions

If $A(X) = \displaystyle\sum_{n=0}^{\infty} a_n X^n$, then

$$\sum_{n=k}^{\infty} a_n X^n = A(X) - a_0 - a_1 X - \cdots - a_{k-1} X^{k-1},$$

$$\sum_{n=k}^{\infty} a_{n-1} X^n = X(A(X) - a_0 - a_1 X - \cdots - a_{k-2} X^{k-2}),$$

$$\sum_{n=k}^{\infty} a_{n-2} X^n = X^2(A(X) - a_0 - a_1 X - \cdots - a_{k-3} X^{k-3}),$$

$$\sum_{n=k}^{\infty} a_{n-3} X^n = X^3(A(X) - a_0 - a_1 X - \cdots - a_{k-4} X^{k-4}),$$

$$\vdots$$

$$\sum_{n=k}^{\infty} a_{n-k} X^n = X^k(A(X)).$$

Now we are prepared to describe how to solve linear recurrence relations with constant coefficients by the use of generating functions. The process is best illustrated by examples.

Example 3.4.4.   Solve the recurrence relation

$$a_n - 7 a_{n-1} + 10 a_{n-2} = 0 \text{ for } n \geq 2.$$

We number the steps of the procedure.

1. Let $A(X) = \sum_{n=0}^{\infty} a_n X^n$.
2. Next multiply each term in the recurrence relation by X^n and sum from 2 to ∞:

$$\sum_{n=2}^{\infty} a_n X^n - 7 \sum_{n=2}^{\infty} a_{n-1} X^n + 10 \sum_{n=2}^{\infty} a_{n-2} X^n = 0.$$

3. Replace each infinite sum by an expression from the table of equivalent expressions:

$$[A(X) - a_0 - a_1 X] - 7X[A(X) - a_0] + 10X^2[A(X)] = 0.$$

4. Then simplify:

$$A(X)(1 - 7X + 10X^2) = a_0 + a_1 X - 7a_0 X$$

or

$$A(X) = \frac{a_0 + (a_1 - 7a_0)X}{1 - 7X + 10X^2} = \frac{a_0 + (a_1 - 7a_0)X}{(1 - 2X)(1 - 5X)}.$$

5. Decompose $A(X)$ as a sum of partial fractions:

$$A(X) = \frac{C_1}{1 - 2X} + \frac{C_2}{1 - 5X},$$

where C_1 and C_2 are constants, as yet undetermined.
6. Express $A(X)$ as a sum of familiar series:

$$A(X) = \frac{C_1}{1 - 2X} + \frac{C_2}{1 - 5X} = C_1 \sum_{n=0}^{\infty} 2^n X^n + C_2 \sum_{n=0}^{\infty} 5^n X^n.$$

7. Express a_n as the coefficient of X^n in $A(X)$ and in the sum of the other series:

$$a_n = C_1 2^n + C_2 5^n.$$

(Thus, we see that the suggested solutions in Example 3.3.1(b) were the only possible solutions to this recurrence relation).
8. Now the constants C_1 and C_2 are uniquely determined once values for a_0 and a_1 are given. For example, if $a_0 = 10$ and $a_1 = 41$, we may use the form of the general solution $a_n = C_1 2^n + C_2 5^n$, and let $n = 0$ and $n = 1$ to obtain the two equations

$$C_1 + C_2 = 10 \quad \text{and} \quad 2C_1 + 5C_2 = 41,$$

which determine the values $C_1 = 3$ and $C_2 = 7$. Thus, in this case the unique solution of the recurrence relation is $a_n = (3)\, 2^n + (7)\, 5^n$.

Example 3.4.5.   Solve the recurrence relation $a_n - 9a_{n-1} + 26a_{n-2} - 24a_{n-3} = 0$ for $n \geq 3$.

As above let $A(X) = \sum_{n=0}^{\infty} a_n X^n$. Then multiply by X^n and sum from 3 to ∞ since $n \geq 3$. Thus,

$$\sum_{n=3}^{\infty} a_n X^n - 9 \sum_{n=3}^{\infty} a_{n-1} X^n + 26 \sum_{n=3}^{\infty} a_{n-2} X^n - 24 \sum_{n=3}^{\infty} a_{n-3} X^n = 0.$$

Replace the infinite sums by equivalent expressions:

$$(A(X) - a_0 - a_1 X - a_2 X^2) - 9X(A(X) - a_0 - a_1 X)$$
$$+ 26X^2(A(X) - a_0) - 24X^3 A(X) = 0.$$

Simplify:

$$A(X)(1 - 9X + 26X^2 - 24X^3) = a_0 + a_1 X + a_2 X^2$$
$$- 9a_0 X - 9a_1 X^2 + 26a_0 X^2$$

or

$$A(X) = \frac{a_0 + (a_1 - 9a_0)X + (a_2 - 9a_1 + 26a_0)X^2}{1 - 9X + 26X^2 - 24X^3}.$$

Now $1 - 9X + 26X^2 - 24X^3 = (1 - 2X)(1 - 3X)(1 - 4X)$ so that there are constants C_1, C_2, C_3 such that

$$A(X) = \frac{C_1}{1 - 2X} + \frac{C_2}{1 - 3X} + \frac{C_3}{1 - 4X}$$

$$= C_1 \sum_{n=0}^{\infty} 2^n X^n + C_2 \sum_{n=0}^{\infty} 3^n X^n + C_3 \sum_{n=0}^{\infty} 4^n X^n$$

$$= \sum_{n=0}^{\infty} (C_1 2^n + C_2 3^n + C_3 4^n) X^n.$$

Thus, $a_n = C_1 2^n + C_2 3^n + C_3 4^n$ and C_1, C_2, and C_3 can be determined once the initial conditions for a_0, a_1, and a_2 are specified.

For illustration let us assume that the initial conditions are $a_0 = 0$, $a_1 = 1$, and $a_2 = 10$. Then

$$A(X) = \frac{X + X^2}{(1 - 2X)(1 - 3X)(1 - 4X)}$$

$$= \frac{C_1}{(1 - 2X)} + \frac{C_2}{(1 - 3X)} + \frac{C_3}{(1 - 4X)}$$

and $C_1(1 - 3X)(1 - 4X) + C_2(1 - 2X)(1 - 4X) + C_3(1 - 2X)(1 - 3X) = X + X^2$. Let $X = 1/2, 1/3, 1/4$, and find that $C_1 = 3/2$, $C_2 = -4$, and $C_3 = 5/2$. Thus, in this case, $a_n = 3/2\,(2^n) - 4\,(3^n) + 5/2\,(4^n)$.

Example 3.4.6. Solve $a_n - 8a_{n-1} + 21a_{n-2} - 18a_{n-3} = 0$ for $n \geq 3$.

Here we see that if $A(X) = \sum_{n=0}^{\infty} a_n X^n$, then

$$\sum_{n=3}^{\infty} a_n X^n - 8 \sum_{n=3}^{\infty} a_{n-1} X^n + 21 \sum_{n=3}^{\infty} a_{n-2} X^n - 18 \sum_{n=3}^{\infty} a_{n-3} X^n = 0,$$

$$(A(X) - a_0 - a_1 X - a_2 X^2) - 8X(A(X) - a_0 - a_1 X)$$
$$+ 21X^2(A(X) - a_0) - 18X^3 A(X) = 0,$$

or

$$A(X) = \frac{a_0 + (a_1 - 8a_0)X + (a_2 - 8a_1 + 21a_0)X^2}{1 - 8X + 21X^2 - 18X^3}.$$

But since $1 - 8X + 21X^2 - 18X^3 = (1 - 2X)(1 - 3X)^2$ we see that there are constants C_1, C_2, C_3 such that $A(X) = C_1/(1 - 2X) + C_2/(1 - 3X) + C_3/(1 - 3X)^2$. Then $A(X) = \sum_{n=0}^{\infty} [C_1 2^n + C_2 3^n + C_3 3^n C(n + 1, n)]X^n$ or $a_n = C_1 2^n + C_2 3^n + C_3(n + 1)3^n$.

We are beginning to discover some things about this method of solving linear recurrence relations by generating functions. For one thing, in each of the Examples 3.4.4–3.4.6 $A(X)$ can be written $P(X)/Q(X)$ where the coefficients of the denominator $Q(X)$ has a definite relationship with the coefficients of the recurrence relation. Moreover note the relationship of the powers of X in $Q(X)$ with the subscripts in the recurrence relation.

For instance, in Examples 3.4.4–3.4.6, the relations $a_n - 7a_{n-1} + 10a_{n-2} = 0$, $a_n - 9a_{n-1} + 26a_{n-2} - 24a_{n-3} = 0$ and $a_n - 8a_{n-1} + 21a_{n-2} - 18a_{n-3} = 0$ gave rise to $A(X) = P(X)/Q(X)$, where the denominator $Q(X)$ was equal to $1 - 7X + 10X^2$, $1 - 9X + 26X^2 - 24X^3$, and $1 - 8X + 21X^2 - 18X^3$, respectively.

Let us note the form of $P(X)$ in each of these examples; they are $a_0 + (a_1 - 7a_0)X$, $a_0 + (a_1 - 9a_0)X + (a_2 - 9a_1 + 26a_0)X^2$, and $a_0 + (a_1 - $

$8a_0)X + (a_2 - 8a_1 + 21a_0)X^2$, respectively. Here, too, the coefficients of the recurrence relation and the values a_0, a_1, etc., determine $P(X)$.

Example 3.4.7. Solve $a_n - 6a_{n-1} + 12a_{n-2} - 8a_{n-3} = 0$ by generating functions.

Following the ideas expressed above, we expect that

$$A(X) = \sum_{n=0}^{\infty} a_n X^n = \frac{P(X)}{Q(X)}$$

$$= \frac{a_0 + (a_1 - 6a_0)X + (a_2 - 6a_1 + 12a_0)X^2}{1 - 6X + 12X^2 - 8X^3}.$$

But since $1 - 6X + 12X^2 - 8X^3 = (1 - 2X)^3$ we use partial fractions to conclude that there are constants C_1, C_2, C_3 such that

$$A(X) = \frac{C_1}{1 - 2X} + \frac{C_2}{(1 - 2X)^2} + \frac{C_3}{(1 - 2X)^3}$$

$$= C_1 \sum_{n=0}^{\infty} 2^n X^n + C_2 \sum_{n=0}^{\infty} \binom{n+1}{n} 2^n X^n + C_3 \sum_{n=0}^{\infty} \binom{n+2}{n} 2^n X^n$$

$$= \sum_{n=0}^{\infty} \left[C_1 2^n + C_2(n+1)2^n + C_3 \frac{(n+2)(n+1)}{2} 2^n \right] X^n,$$

so that

$$a_n = C_1 2^n + C_2(n+1)2^n + C_3 \frac{(n+2)(n+1)}{2} 2^n.$$

Outline of the Method of Generating Functions

What we have discovered in these examples holds in general. Let us outline the method of generating functions as follows.

1. A linear recurrence relation with constant coefficients of degree k is given, which without loss of generality, we assume has the following form: $a_n + c_1 a_{n-1} + \cdots + c_k a_{n-k} = 0$, where $c_1, c_2, \ldots, c_k$ are constants, $c_k \neq 0$, and $n \geq k$.
2. Let $A(X) = \sum_{n=0}^{\infty} a_n X^n$, multiply each term of the recurrence relation by X^n, sum from k to ∞, and replace all infinite sums by equivalent expressions. Thereby, the recurrence relation is transformed into an algebraic equation: $A(X) = P(X)/Q(X)$, where $P(X) = a_0 + (a_1 +$

$c_1 a_0)X + (a_2 + c_1 a_1 + c_2 a_0)X^2 + \cdots + (a_{k-1} + c_1 a_{k-2} + \cdots + c_{k-1} a_0)X^{k-1}$ and $Q(X) = 1 + c_1 X + c_2 X^2 + \cdots + c_k X^k$.

3. Then, knowing $P(X)$ and $Q(X)$, transform $A(X)$ back to get the coefficients a_n (call this performing the inverse transformation). This can be accomplished in one of several ways. We shall describe two.

If the factorization of $Q(X)$ is known, say $Q(X) = (1 - q_1 X)(1 - q_2 X) \cdots (1 - q_k X)$, then use partial fractions and the different identities for familiar generating functions to get $A(X)$ as a sum of familiar series, and hence that a_n is the sum of the coefficients of known series.

However, even when we cannot factor $Q(X)$ if we are given initial conditions we can solve for as many coefficients of $A(X)$ as we desire by long division of $Q(X)$ into $P(X)$ [or by finding the multiplicative inverse of $Q(X)$ since the constant term of $Q(X)$ is nonzero].

Actually the process we have described is reversible and we affirm this in the following theorem.

Theorem 3.4.1. If $\{a_n\}_{n=0}^{\infty}$ is a sequence of numbers which satisfy the linear recurrence relation with constant coefficients $a_n + c_1 a_{n-1} + \cdots + c_k a_{n-k} = 0$, where $c_k \neq 0$, and $n \geq k$, then the generating function $A(X) = \sum_{n=0}^{\infty} a_n X^n$ equals $P(X)/Q(X)$, where $P(X) = a_0 + (a_1 + c_1 a_0)X + \cdots + (a_{k-1} + c_1 a_{k-2} + \cdots + c_{k-1} a_0)X^{k-1}$ and $Q(X) = 1 + c_1 X + \cdots + c_k X^k$.

Conversely, given such polynomials $P(X)$ and $Q(X)$, where $P(X)$ has degree less than k, and $Q(X)$ has degree k, there is a sequence $\{a_n\}_{n=0}^{\infty}$ whose generating function is $A(X) = P(X)/Q(X)$.

Moreover, the sequence $\{a_n\}_{n=0}^{\infty}$ satisfies a linear homogeneous recurrence relation with constant coefficients of degree k, where the coefficients of the recurrence relation are the coefficients of $Q(X)$.

In fact, if $Q(X) = b_0 + b_1 X + \cdots + b_k X^k$ where $b_0 \neq 0$ and $b_k \neq 0$, then

$$Q(X) = b_0 (1 + b_1/b_0 X + \cdots + b_k/b_0 X^k)$$
$$= b_0 (1 + c_1 X + \cdots + c_k X^k)$$

where $c_i = b_i/b_0$ for $i \geq 1$. Then

$$A(X) = \frac{P(X)}{Q(X)} = \frac{\frac{1}{b_0} P(X)}{1 + c_1 X + \cdots + c_k X^k},$$

and the coefficients of $A(X)$ are discovered by using partial fractions and the factors of $1 + c_1 X + \cdots + c_k X^k$. Then the recurrence relation satisfied by the coefficients of $A(X)$ is $a_n + c_1 a_{n-1} + \cdots + c_k a_{n-k} = 0$.

Much of the theory of recurrence relations, in particular linear recurrence relations with constant coefficients, can be developed using such generating function techniques. But of course, this requires a much more extensive knowledge of pairs of sequences and their generating functions than that summarized in the table of generating functions.

Exercises for Section 3.4

1. Solve the following recurrence relations by substitution.
   (a) $a_n = a_{n-1} + n$ where $a_0 = 2$.
   (b) $a_n = a_{n-1} + n^2$ where $a_0 = 7$.
   (c) $a_n = a_{n-1} + n^3$ where $a_0 = 5$.
   (d) $a_n = a_{n-1} + n(n - 1)$ where $a_0 = 1$.
   (e) $a_n = a_{n-1} + 1/n(n + 1)$ where $a_0 = 1$.
   (f) $a_n = a_{n-1} + 2n + 1$ where $a_0 = 1$.
   (g) $a_n = a_{n-1} + 3n^2 + 3n + 1$ where $a_0 = 1$.
   (h) $a_n = a_{n-1} + 3^n$ where $a_0 = 1$.
   (i) $a_n = a_{n-1} + n3^n$ where $a_0 = 1$.

2. Write a general expression for $A(X) = P(X)/Q(X)$ specifying the coefficients for $P(X)$ and $Q(X)$ where $A(X)$ generates the sequence $\{a_n\}_{n=0}^{\infty}$ and a_n satisfies the following recurrence relations:
   (a) $a_n + 5a_{n-1} + 3a_{n-2} = 0$ for $n \geq 2$, $a_0 = 1$, $a_1 = 2$;
   (b) $a_n + 7a_{n-1} + 8a_{n-2} = 0$ for $n \geq 2$, $a_0 = 1$, $a_1 = -2$;
   (c) $a_n - 5a_{n-1} + 8a_{n-2} - 4a_{n-3} = 0$ for $n \geq 3$,
        $a_0 = 1$, $a_1 = 0$, $a_2 = 1$; and
   (d) $a_n - 2a_{n-3} + a_{n-6} = 0$ for $n \geq 6$,
        $a_0 = 1$, $a_1 = a_2 = 3$, $a_3 = a_4 = a_5 = 2$.

3. Solve the following recurrence relations using generating functions.
   (a) $a_n - 6a_{n-1} = 0$ for $n \geq 1$ and $a_0 = 1$.
   (b) $a_n - 9a_{n-1} + 20a_{n-2} = 0$ for $n \geq 2$ and $a_0 = -3$, $a_1 = -10$.
   (c) $a_n - 5a_{n-1} + 6a_{n-2} = 0$ for $n \geq 2$, and $a_0 = 1$, $a_1 = -2$.
   (d) $a_n - 4a_{n-2} = 0$ for $n \geq 2$ and $a_0 = 0$, $a_1 = 1$.
   (e) $a_n - a_{n-1} - 9a_{n-2} + 9a_{n-3} = 0$ for $n \geq 3$ and $a_0 = 0$, $a_1 = 1$,
        $a_2 = 2$.
   (f) $a_n - 3a_{n-2} + 2a_{n-3} = 0$ for $n \geq 3$ and $a_0 = 1$, $a_1 = 0$, $a_2 = 0$.
   (g) $a_n - 2a_{n-3} + a_{n-6} = 0$ for $n \geq 6$ and $a_0 = 1$, $a_1 = 0 = a_2 = a_3 = $
        $a_4 = a_5$.

(h) $a_n + a_{n-1} - 16a_{n-2} + 20a_{n-3} = 0$ for $n \geq 3$ and $a_0 = 0$, $a_1 = 1$,
$\quad a_2 = -1$.

(i) $a_n - 10a_{n-1} + 33a_{n-2} - 36a_{n-3} = 0$ for $n \geq 3$ and $a_0 = 1$, $a_1 = 1$,
$\quad a_2 = -23$.

4. Find a general expression for a_n using generating functions.

(a) $a_n - 7a_{n-1} + 12a_{n-2} = 0$ for $n \geq 2$.

(b) $a_n - 4a_{n-1} - 12a_{n-2} = 0$ for $n \geq 2$.

(c) $a_n - 5a_{n-1} + 6a_{n-2} = 0$ for $n \geq 2$.

(d) $a_n - 3a_{n-1} + 3a_{n-2} - a_{n-3} = 0$ for $n \geq 3$.

(e) $a_n - 9a_{n-1} + 27a_{n-2} - 27a_{n-3} = 0$ for $n \geq 3$.

(f) $a_n - 7a_{n-1} + 16a_{n-2} - 12a_{n-3} = 0$ for $n \geq 3$.

5. If $A(X) = \sum_{n=0}^{\infty} a_n X^n$ generates the sequence $\{a_n\}_{n=0}^{\infty}$, and $a_n = f(n)$ is some function of n—like $f(n) = n^2 + 3n + 1$, what generates

(a) $b_n = f(n + 1)$?

(b) $b_n = f(n + 2)$?

(c) $b_n = f(n - 2)$ for $n \geq 2$ and $b_0 = b_1 = 0$?

6. One of the basic combinatorial procedures in computer science is sorting a list of items, and one of the best known sorting algorithms is called *bubble sort*, so named because small items move up the list the same way bubbles rise in a liquid. The bubble sort procedure is described as follows: Suppose that an n-tuple A of numbers are given, where $A(i)$ denotes the i-th entry of A. The n-entries of A are to be sorted into nondecreasing order; thus, the smallest entry is to be placed in $A(1)$ and the largest is to be placed in $A(n)$. The bubble sort procedure makes $n - 1$ passes over the n-tuple A, where a pass always starts at $A(n)$ and proceeds through the unsorted portion of A. Each pass consists of a sequence of steps, each of which compares $A(i)$ with $A(i + 1)$ and interchanges their values if they are in wrong relative order. Thus, if $A(i) > A(i + 1)$, then the entries are interchanged. The first pass starts with $i = n - 1$ and continues until $i = 1$. At the end of the first pass, the smallest entry of A has been "bubbled up" into the position $A(1)$ and need not be considered further. In the second pass, the value of i ranges from $n - 1$ to 2, this pass "bubbles" the smallest of $A(2),\ldots,A(n)$ into the second position. Now then $A(1)$ and $A(2)$ are in correct relative order. Finally, after the $(n - 1)$th pass, the values $A(1),A(2),\ldots,A(n - 1)$ are all in place, and consequently the largest entry of A has been moved to the n-th position.

Find a recurrence relation for the number of comparisons made in the n-1 passes and solve this relation by substitution.

7. Solve the following divide-and-conquer relations by substitution:

(a) $a_n = 7a_{n/3} + 5$
$\quad$ where $n = 3^k$ and $a_1 = 1$.

(b) $a_n = 2a_{n/4} + n$
   where $n = 4^k$ and $a_1 = 1$.

(c) $a_n = a_{n/2} + 2n - 1$
   where $n = 2^k$ and $a_1 = 1$.

8. Verify by mathematical induction that

$$a_n = A_1 n + A_2$$

is a solution to

$$a_n = d\, a_{n/d} + e$$

where

$$n = d^k.$$

9. Show that $C(\log_d (n) + 1)$ is the solution to the recurrence relation:

$$a_n = a_{n/d} + C$$

where $n = d^k$ and

$$a_1 = C.$$

10. Show that $a_n = e\, (C\, n^{\log_d C} - 1)/(C - 1)$ is the solution to the recurrence relation

$$a_n = C\, a_{n/d} + e \text{ for}$$
$$n = d^k, C \neq 1 \text{ and}$$
$$a_1 = e.$$

11. Show by substitution that $C(2n - 2, n - 1)(n - 1)!$ is the solution of the relation

$$a_n = (4n - 6)a_{n-1} \text{ for } n \geq 2 \text{ where}$$
$$a_1 = 1.$$

12. Give the generating function for the sequence $\{a_n\}_{n=0}^{\infty}$ if

(a) $a_n = 3^n$

(b) $a_n = \left(\frac{1}{3}\right)^n$

(c) $a_n = \left(-\frac{1}{3}\right)^n$

(d) $a_n = 3^n + 5^n$

(e) $a_n = 5 \left(\dfrac{1}{3}\right)^n - 7(5)^n + 3 \dbinom{n+3}{n}$

(f) $a_n = \dbinom{n+2}{n} 5^n$

(g) $a_n = 7(n+3)(n+2)(n+1)$

(h) $a_n = (n+1)(n)(n-1)$

(i) $a_n = n^2 3^n$

Selected Answers for Section 3.4

1.
(a) $a_n = \dfrac{n}{2}(n+1) + 2$

(b) $a_n = \dfrac{n(n+1)(2n+1)}{6} + 7$

(c) $a_n = \dfrac{n^2(n+1)^2}{4} + 5$

(e) $a_n = 1 + \displaystyle\sum_{k=1}^{\infty} \dfrac{1}{k(k+1)} = 1 + \displaystyle\sum_{k=1}^{n} \left(\dfrac{1}{k} - \dfrac{1}{k+1}\right)$

$= 1 + \dfrac{1}{1} - \dfrac{1}{2} + \dfrac{1}{2} - \dfrac{1}{3} + \cdots + \dfrac{1}{n} - \dfrac{1}{n+1}$

$= 1 + 1 - \dfrac{1}{n+1} = 2 - \dfrac{1}{n+1}$

$= \dfrac{2n+1}{n+1}$

(f) $a_n = (n+1)^2$

(g) $a_n = (n+1)^3$

(h) $a_n = \dfrac{3^{n+1} - 1}{2}$

2. (a) $A(X) = \dfrac{1 + 7X}{1 + 5X + 3X^2}$

3. (b) $a_n = 2 \cdot 5^n - 5 \cdot 4^n$

(c) $A(X) = \dfrac{1 - 7X}{1 - 5X + 6X^2} = \dfrac{5}{1 - 2X} - \dfrac{4}{1 - 3X}$; $a_n = 5(2^n) - 4(3^n)$

(d) $a_n = 0$ if n is even, $a_n = 2^{n-1}$ if n is odd

(e) $a_n = 1/12\{-3 + 4 \cdot 3^n - (-3)^n\}$

(f) $a_n = 8/9 - 6/9\,n + 1/9\,(-2)^n$

(h) $A(X) = \dfrac{X}{1 + X - 16X^2 + 20X^3} = \dfrac{-2/49}{1 - 2X} + \dfrac{7/49}{(1 - 2X)^2} - \dfrac{5/49}{1 + 5X}$,

$a_n = -2/49\,(2^n) + 7/49\,(n+1)2^n - 5/49\,(-5)^n$

(i) $A(X) = \dfrac{1 - 9X}{1 - 10X + 33X^2 - 36X^3} = \dfrac{1 - 9X}{(1 - 3X)^2(1 - 4X)}$

$\qquad = \dfrac{15}{1 - 3X} + \dfrac{6}{(1 - 3X)^2} - \dfrac{20}{1 - 4X}$,

$\qquad a_n = 15(3^n) + 6(n + 1)(3^n) - 20(4^n)$

11. Let

$$a_n = (4n - 6)a_{n-1} = (4n - 6)(4(n - 1) - 6)a_{n-2}$$
$$= (4n - 6)(4n - 10)a_{n-2} = (4n - 6)(4n - 10)(4(n - 2) - 6)a_{n-3}$$
$$= (4n - 6)(4n - 10)(4n - 14)a_{n-3}$$
$$= \ldots$$
$$= (4n - 6)(4n - 10)(4n - 14)\ldots(6)(2)a_1$$
$$= (4n - 6)(4n - 10)(4n - 14)\ldots(6)(2)$$

since $a_1 = 1$. We rewrite this as:

$$a_n = 2(2n - 3)(2)(2n - 5)(2)(2n - 7)\ldots(2)(3)(2)(1)$$
$$= 2^{n-1}[(1)(3)(5)\ldots(2n - 7)(2n - 5)(2n - 3)]$$

Multiply numerator and denominator by $(2)(4)(6)\ldots(2n - 6)$ $(2n - 4)(2n - 2)$ and we have

$$a_n = \dfrac{2^{n-1}(2n - 2)!}{(2)(4)\ldots(2n - 6)(2n - 4)(2n - 2)}$$
$$= \dfrac{2^{n-1}(2n - 2)!}{2^{n-1}(1)(2)\ldots(n - 3)(n - 2)(n - 1)}$$
$$= \dfrac{(2n - 2)!}{(n - 1)!} = \dfrac{(2n - 2)!(n - 1)!}{(n - 1)!(n - 1)!}$$
$$= C(2n - 2, n - 1)(n - 1)!$$

3.5   THE METHOD OF CHARACTERISTIC ROOTS

This new method is nothing more than a synthesis of all that we have learned from the method of generating functions. If we want to solve $a_n + c_1 a_{n-1} + \cdots + c_k a_{n-k} = 0$ where $c_k \neq 0$, then we can find $A(X) = P(X)/Q(X)$ where $Q(X)$ is a polynomial of degree k. Then the factors of $Q(X)$ completely determine the form of coefficients of $A(X)$.

But let us make one observation: In Example 3.4.4 the denominator $Q(X) = 1 - 7X + 10X^2$ and the general solution for a_n was $a_n = C_1 2^n + C_2 5^n$ because $Q(X)$ factors as $(1 - 2X)(1 - 5X)$. Note that the roots of $Q(X)$ were 1/2 and 1/5 while the solutions involve powers of their reciprocals. To avoid this reciprocal relationship, let us consider another polynomial where we replace X in $Q(X)$ by $1/t$ and multiply by t^2 to obtain the polynomial $C(t) = t^2 Q(1/t) = t^2 [1 - 7(1/t) + 10(1/t^2)] = t^2 - 7t + 10 = (t - 2)(t - 5)$. Now the roots of this polynomial, 2 and 5, are in direct relationship with the form of the solution for $a_n = C_1 2^n + C_2 5^n$.

In Example 3.4.5

$$A(X) = \frac{P(X)}{Q(X)} = \frac{P(X)}{1 - 9X + 26X^2 - 24X^3}$$

$$= \frac{P(X)}{(1 - 2X)(1 - 3X)(1 - 4X)}$$

and the form of the solution for that recurrence was $a_n = C_1 2^n + C_2 3^n + C_3 4^n$.

Note again that the roots of $Q(X)$ are 1/2, 1/3, and 1/4, but the roots of $C(t) = t^3 Q(1/t) = t^3 - 9t^2 + 26t - 24 = (t - 2)(t - 3)(t - 4)$ are 2, 3, 4, and these roots are in direct relationship to the form of the solution for a_n.

The polynomial $C(t)$ is called the **characteristic polynomial** of the recurrence relation. Note that if the recurrence relation is $a_n + c_1 a_{n-1} + \cdots + c_k a_{n-k} = 0$ for $n \geq k$, where $c_k \neq 0$, then the characteristic polynomial for this recurrence relation is $C(t) = t^k + c_1 t^{k-1} + \cdots + c_k$, and this, in turn, equals $t^k Q(1/t)$, where $Q(X) = 1 + c_1 X + \cdots + c_k X^k$. Then if $C(t)$ factors as $(t - \alpha_1)^{r_1} \cdots (t - \alpha_s)^{r_s}$ then in the expression $A(X) = P(X)/Q(X)$, the denominator $Q(X)$ factors as $(1 - \alpha_1 X)^{r_1} \cdots (1 - \alpha_s X)^{r_s}$.

Distinct Roots

If the characteristic polynomal has distinct roots $\alpha_1, \ldots, \alpha_k$, then the general form of the solutions for the homogeneous equation is $a_n = C_1 \alpha_1^n + \cdots + C_k \alpha_k^n$ where $C_1, C_2, \ldots C_k$ are constants which may be chosen to satisfy any initial conditions.

Example 3.5.1. To solve $a_n - 7a_{n-1} + 12a_{n-2} = 0$ for $n \geq 2$, the characteristic equation is $C(t) = t^2 - 7t + 12 = (t - 3)(t - 4)$. Thus, the general solution is $a_n = C_1 3^n + C_2 4^n$.

If the initial conditions are $a_0 = 2$, $a_1 = 5$, then we must solve the equations

$$C_1 + C_2 = 2 \quad \text{and} \quad 3C_1 + 4C_2 = 5$$

to find that $C_1 = 3$ and $C_2 = -1$, and the required solution is $a_n = (3)3^n - 4^n$.

Example 3.5.2. Solve $a_n - 5a_{n-1} + 6a_{n-2} = 0$ where $a_0 = 2$ and $a_1 = 5$.

Since the characteristic polynomial is $t^2 - 5t + 6 = (t - 3)(t - 2)$, $a_n = C_1 2^n + C_2 3^n$. From the initial conditions $a_0 = 2$ and $a_1 = 5$, we have the system of equations

$$C_1 + C_2 = 2$$
$$2C_1 + 3C_2 = 5.$$

Solving these equations, we find $C_1 = 1$ and $C_2 = 1$. Thus, $a_n = 2^n + 3^n$ for all integers $n \geq 0$.

Multiple Roots

In Example 3.4.7 we discovered by using generating functions that the general solution to $a_n - 6a_{n-1} + 12\,a_{n-2} - 8a_{n-3} = 0$ for $n \geq 3$ was of the form $a_n = C_1\,2^n + C_2\,(n + 1)\,2^n + C_3\,[(n + 2)\,(n + 1)\,2^n]/2$ because $Q(X) = 1 - 6X + 12X^2 - 8X^3 = (1 - 2X)^3$. But then this corresponds to the fact that the characteristic polynomial $C(t) = t^3 - 6t^2 + 12t - 8 = (t - 2)^3$ has 2 as a repeated root with multiplicity 3.

Let us make an observation here. If we rewrite

$$a_n = C_1 2^n + C_2(n + 1)2^n + C_3\frac{(n + 2)\,(n + 1)}{2}\,2^n \text{ as}$$

$$C_1 2^n + C_2 n 2^n + C_2 2^n + C_3\frac{n^2}{2}\,2^n + C_3\left(\frac{3n}{2}\right)2^n + C_3 2^n$$

and recombine, we have $a_n = (C_1 + C_2 + C_3)2^n + (C_2 + 3/2\,C_3)n2^n + 1/2C_3 n^2 2^n$. In other words, after simplification there are constants D_1, D_2, D_3 such that $a_n = D_1\,2^n + D_2 n2^n + D_3 n^2 2^n$. This type of result holds in general. Let us exhibit this in another example.

Example 3.5.3. Write the general form of the solutions to

(a) $a_n - 6a_{n-1} + 9a_{n-2} = 0$,
(b) $a_n - 3a_{n-1} + 3a_{n-2} - a_{n-3} = 0$, and
(c) $a_n - 9a_{n-2} + 27a_{n-2} - 27a_{n-3} = 0$.

Following the above idea since the characteristic polynomial in (a) is $t^2 - 6t + 9 = (t - 3)^2$ the general solution is the form $a_n = D_1 3^n + D_2 n 3^n$. Likewise the characteristic polynomial for (b) is $t^3 - 3t^2 + 3t - 1 = (t - 1)^3$ so that the general solution in (b) is $a_n = D_1 1^n + D_2 n 1^n + D_3 n^2 1^n = D_1 + D_2 n + D_3 n^2$.

In (c) the characteristic polynomial is $t^3 - 9t^2 + 27t - 27 = (t - 3)^3$ so that the general solution for (c) is $a_n = D_1 3^n + D_2 n 3^n + D_3 n^2 3^n$. We would expect this to generalize to cases where the characteristic polynomial has several multiple roots.

Example 3.5.4. If the characteristic polynomial of a linear homogeneous recurrence relation is $(t - 2)^2 (t - 3)^3$ then the general solution for a_n is $a_n = D_1 2^n + D_2 n 2^n + D_3 3^n + D_4 n 3^n + D_5 n^2 3^n$. (Of course since $(t - 2)^2 (t - 3)^3 = t^5 - 13t^4 + 67t^3 - 171t^2 + 216t - 108$ we see that the recurrence relation must have been $a_n - 13a_{n-1} + 67a_{n-2} - 171a_{n-3} + 216a_{n-4} - 108a_{n-5} = 0$, for $n \geq 5$.)

In general we have the following theorem.

Theorem 3.5.1. Let the distinct roots of the characteristic polynomial, $C(t) = t^k + c_1 t^{k-1} + \cdots + c_k$ of the linear homogeneous recurrence relation, $a_n + c_1 a_{n-1} + \cdots + c_k a_{n-k} = 0$, where $n \geq k$ and $c_k \neq 0$, be $\alpha_1, \alpha_2, \ldots, \alpha_s$ where $s \leq k$. Then there is a general solution for a_n which is of the form, $U_1(n) + U_2(n) + \cdots + U_s(n)$ where $U_i(n) =$

$$(D_{i_0} + D_{i_1} n + D_{i_2} n^2 + \cdots + D_{i_{m_i-1}} n^{m_i-1}) \alpha_i^n$$

and where m_i is the multiplicity of the root α_i.

Example 3.5.5. Suppose that the characteristic polynomial for a linear homogenous recurrence relation is $(t - 2)^3 (t - 3)^2 (t - 4)^3$. Then the general solution is $a_n = (D_1 + D_2 n + D_3 n^2) 2^n + (D_4 + D_5 n) 3^n + (D_6 + D_7 n + D_8 n^2) 4^n$.

Example 3.5.6. Solve the recurrence relation $a_n - 7a_{n-1} + 16a_{n-2} - 12a_{n-3} = 0$ for $n \geq 3$ with the initial conditions $a_0 = 1$, $a_1 = 4$, and $a_2 = 8$.

The characteristic polynomial is $t^3 - 7t^2 + 16t - 12 = (t - 2)^2 (t - 3)$. Thus, a_n may be written as $C_1 2^n + C_2 n 2^n + C_3 3^n$. But then the initial conditions give the system of equations

$$C_1 + C_3 = 1$$
$$2C_1 + 2C_2 + 3C_3 = 4$$
$$4C_1 + 8C_2 + 9C_3 = 8.$$

This system has the solution $C_1 = 5$, $C_2 = 3$, and $C_3 = -4$. Hence, the unique solution of the recurrence relation is

$$a_n = (5)(2^n) + (3)(n2^n) - (4)(3^n).$$

Remark. The methods we have described will work whether or not the roots of $Q(X)$, or the roots of $C(t)$, are real numbers. However if the roots are complex then a_n need not be real and in our discussion and definitions we have always discussed real-valued sequences. Hence we have restricted our attention in the examples to linear recurrence relations whose characteristic polynomials have had only real roots. We intend to do the same in the exercises.

Exercises for Section 3.5

1. Find and factor the characteristic polynomial for the recurrence relations in Exercise 3 of Section 3.4.

2. Do the same for Exercise 4 in Section 3.4.

3. Do the same for the recurrence relation $a_n - 5a_{n-1} + 8a_{n-2} - 4a_{n-3} = 0$ for $n \geq 3$.

4. Find the general form of a solution to the recurrence relations in (a)–(e) of Exercise 3 of Section 3.4.

5. Find the characteristic polynomial for the homogeneous recurrence relations whose general solution has the form
   (a) $a_n = B_1 + nB_2$,
   (b) $a_n = B_1 + nB_2 + n^2 B_3$,
   (c) $a_n = B_1 2^n + B_2 3^n$,
   (d) $a_n = B_1 2^n + B_2 n2^n$,
   (e) $a_n = B_1 2^n + B_2 n2^n + B_3 n^2 2^n$,
   (f) $a_n = B_1 2^n + B_2 n2^n + B_3 3^n + B_4 n3^n + B_5 6^n$.

6. Find C_1, C_2, C_3 if the recurrence relation $a_n + C_1 a_{n-1} + C_2 a_{n-2} + C_3 a_{n-3} = 0$ for $n \geq 3$ has a general solution of the form
   (a) $a_n = B_1 3^n + B_2 6^n$,
   (b) $a_n = B_1 3^n + B_2 n3^n$,
   (c) $a_n = B_1 3^n + B_2 n3^n + B_3 2^n$.

7. Solve the following recurrence relations using the characteristic roots
   (a) $a_n - 3a_{n-1} - 4a_{n-2} = 0$ for $n \geq 2$ and, $a_0 = a_1 = 1$;
   (b) $a_n - 4a_{n-1} - 12a_{n-2} = 0$ for $n \geq 2$ and, $a_0 = 4$, $a_1 = 16/3$;
   (c) $a_n - 4a_{n-1} + 4a_{n-2} = 0$ for $n \geq 2$ and, $a_0 = 5/2$, $a_1 = 8$;
   (d) $a_n + 7a_{n-1} + 8a_{n-2} = 0$ and, $a_0 = 2$, $a_1 = -7$;
   (e) $a_n + 5a_{n-1} + 5a_{n-2} = 0$ and, $a_0 = 0$, $a_1 = 2\sqrt{5}$;

(f) $a_n - 7a_{n-2} + 10a_{n-4} = 0$, $a_0 = 7$, $a_1 = 8\sqrt{2} + 5\sqrt{5}$, $a_2 = 41$,
    $a_3 = 16\sqrt{2} + 25\sqrt{5}$;
(g) $a_n - a_{n-1} - 6a_{n-2} = 0$ for $n \geq 2$ and, $a_0 = 12$, $a_1 = -1$;
(h) $a_n + 5a_{n-1} + 6a_{n-2} = 0$ and, $a_0 = -4$, $a_1 = -7$;
(i) $a_n - 6a_{n-1} + 8a_{n-2} = 0$ where $a_0 = 1$ and $a_1 = 0$;
(j) $2a_n - 7a_{n-1} + 3a_{n-2} = 0$ where $a_0 = 1$ and $a_1 = 1$;
(k) $a_n - 6a_{n-1} + 9a_{n-2} = 0$ where $a_0 = 1$ and $a_1 = 1$;
(l) $a_n - 5a_{n-1} + 6a_{n-2} = 0$ where $a_0 = 7$ and $a_1 = 1$;
(m) $a_n + a_{n-1} - 5a_{n-2} + 3a_{n-3} = 0$ where $a_0 = 0$, $a_1 = 1$, and $a_2 = 2$;
(n) $6a_n - 19a_{n-1} + 15a_{n-2} = 0$ where $a_0 = 0$ and $a_1 = -\dfrac{1}{6}$.

Selected Answers for Section 3.5

2. (d) $C(t) = t^3 - 3t^2 + 3t - 1 = (t-1)^3$
   (f) $C(t) = t^3 - 7t^2 + 16t - 12 = (t-2)^2 (t-3)$
3. $C(t) = t^3 - 5t^2 + 8t - 4 = (t-2)^2 (t-1)$
5. (a) $C(t) = (t-1)^2$
   (b) $C(t) = (t-1)^3$
   (c) $C(t) = (t-2)(t-3)$
   (d) $C(t) = (t-2)^2$
   (e) $C(t) = (t-2)^3$
   (f) $C(t) = (t-2)^2 (t-3)^2 (t-6)$
6. (a) $C_1 = -9$ $C_2 = 18$ $C_3 = 0$
   (b) $C_1 = -6$ $C_2 = 9$ $C_3 = 0$
   (c) $C_1 = -8$ $C_2 = 21$ $C_3 = -18$
7. (a) $a_n = \dfrac{2}{5} 4^n + \dfrac{3}{5} (-1)^n$

   (b) $a_n = \dfrac{5}{3} 6^n + \dfrac{7}{3} (-2)^n$

   (c) $a_n = \dfrac{5}{2} 2^n + \dfrac{3}{2} n\, 2^n$

   (d) $a_n = \left(\dfrac{-7+\sqrt{17}}{2}\right)^n + \left(\dfrac{-7-\sqrt{17}}{2}\right)^n$

   (e) $a_n = 2\left(\dfrac{-5+\sqrt{5}}{2}\right)^n - 2\left(\dfrac{-5-\sqrt{5}}{2}\right)^n$

   (f) $C(t) = (t-\sqrt{2})(t-\sqrt{2})(t-\sqrt{5})(t+\sqrt{5})$
       $a_n = 3(\sqrt{2})^n - 5(-\sqrt{2})^n + 7(\sqrt{5})^n + 2(-\sqrt{5})^n$

3.6 SOLUTIONS OF INHOMOGENEOUS LINEAR RECURRENCE RELATIONS

Let us turn our attention now to learn how to solve the inhomogeneous recurrence relation (IHR): $a_n + c_1 a_{n-1} + \cdots + c_k a_{n-k} = f(n)$ for $n \geq k$, where $c_k \neq 0$, and where $f(n)$ is some specified function of n. Call $f(n)$ the *forcing function* of (IHR).

We attempt to find a solution using generating functions. We follow the same procedure as in solving homogeneous recurrence relations (HR). Let $A(X) = \Sigma_{n=0}^{\infty} a_n X^n$, multiply each term in the IHR by X^n, sum from k to ∞, and replace the infinite sums by expressions from the table of equivalent expressions for $A(X)$. Let us present some examples for illustration.

Example 3.6.1. Find a solution to $a_n - a_{n-1} = 3(n-1)$ where $n \geq 1$ and where $a_0 = 2$.

Let $A(X) = \Sigma_{n=0}^{\infty} a_n X^n$. Multiply each term of the recurrence relation by X^n and sum from 1 to ∞. Then we have

$$\sum_{n=1}^{\infty} a_n X^n - \sum_{n=1}^{\infty} a_{n-1} X^n = 3 \sum_{n=1}^{\infty} (n-1) X^n.$$

Replace each infinite sum by an equivalent expression from the table, so that $A(X) - a_0 - XA(X) = 3X^2/(1-X)^2$.

But then $A(X)(1-X) = a_0 + 3X^2/(1-X)^2$ or $A(X) = a_0/(1-X) + 3X^2/(1-X)^3$. Using Table 3-1, the shifting property of generating functions, the inverse transformation, and the fact that $a_0 = 2$, we have $A(X) = 2 \Sigma_{n=0}^{\infty} X^n + 3/2 (\Sigma_{n=0}^{\infty} n(n-1)X^n)$ so that $a_n = 2 + 3/2 (n)(n-1)$ for $n \geq 0$.

Example 3.6.2. Find a general expression for a solution to the recurrence relation $a_n - 5a_{n-1} + 6a_{n-2} = n(n-1)$ for $n \geq 2$.

Let $A(X) = \Sigma_{n=0}^{\infty} a_n X^n$. Then multiply by X^n and sum from 2 to ∞ to get

$$\sum_{n=2}^{\infty} a_n X^n - 5 \sum_{n=2}^{\infty} a_{n-1} X^n + 6 \sum_{n=2}^{\infty} a_{n-2} X^n = \sum_{n=2}^{\infty} (n)(n-1) X^n.$$

Replacing the infinite sums by equivalent expressions, we have $(A(X) - a_0 - a_1 X) - 5X(A(X) - a_0) + 6X^2 A(X) = \Sigma_{n=2}^{\infty} (n)(n-1)X^n$. Using the shifting properties of generating functions and Table 3-1, we see that $\Sigma_{n=2}^{\infty} (n)(n-1) X^n = 2X^2/(1-X)^3$. Thus, $A(X)(1 - 5X + 6X^2) = a_0 +$

$(a_1 - 5a_0)X + (2X^2/(1 - X)^3)$ and

$$A(X) = \frac{a_0 + (a_1 - 5a_0)X}{(1 - 5X + 6X^2)} + \frac{2X^2}{(1 - X)^3(1 - 5X + 6X^2)}.$$

If we are given initial conditions for a_0 and a_1 then we can find the coefficients of $A(X)$. Suppose, for example, that $a_0 = 1$ and $a_1 = 5$, then

$$A(X) = \frac{1}{1 - 5X + 6X^2} + \frac{2X^2}{(1 - X)^3(1 - 5X + 6X^2)}$$

$$= \frac{1 - 3X + 5X^2 - X^3}{(1 - X)^3(1 - 2X)(1 - 3X)}.$$

But then by partial fractions,

$$A(X) = \frac{a_0 + (a_1 - 5a_0)X}{(1 - 5X + 6X^2)} + \frac{2X^2}{(1 - X)^3(1 - 5X + 6X^2)}.$$

for constants A, B, C, D, and E. Thus, $A(1 - X)^2 (1 - 2X) (1 - 3X) + B(1 - X) (1 - 2X) (1 - 3X) + C(1 - 2X) (1 - 3X) + D(1 - X)^3 (1 - 3X) + E(1 - X)^3 (1 - 2X) = 1 - 3X + 5X^2 - X^3$. Let $X = 1$, $1/2$, and $1/3$ and find $C = 1$, $D = -10$, and $E = 21/4$. Let $X = 0$ and we get the equation $A + B = 19/4$. Solve for the coefficient of X^4 and get $6A + 3D + 2E = 0$, but since $D = -10$ and $E = 21/4$, we find that $A = 39/12$, and then that $B = 3/2$. Then by the inverse transformation process, we have

$$a_n = \frac{39}{12} + \left(\frac{3}{2}\right)(n + 1) + \frac{(n + 2)(n + 1)}{2} - (10) 2^n + \left(\frac{21}{4}\right) 3^n$$

$$= \frac{23}{4} + 3n + \frac{n^2}{2} - 10(2^n) + \left(\frac{21}{4}\right) 3^n.$$

Now this process works in general for the IHR above. For if we let $A(X) = \sum_{n=0}^{\infty} a_n X^n$, multiply each term of the IHR by X^n, sum from k to ∞, and so forth, we obtain

$$A(X)(1 + c_1 X + c_2 X^2 + \cdots + c_k X^k) = \sum_{n=k}^{\infty} f(n)X^n + P(X),$$

where

$$P(X) = a_0 + (a_1 + c_1 a_0)X$$

$$+ \cdots + (a_{k-1} + c_1 a_{k-2} + \cdots + c_{k-1} a_0)X^{k-1}.$$

Then, for $Q(X) = 1 + c_1X + \cdots + c_kX^k$,

$$A(X) = \frac{\sum\limits_{n-k}^{\infty} f(n)X^n}{Q(X)} + \frac{P(X)}{Q(X)}.$$

Note that $P(X)/Q(X)$ is a solution of the HR: $a_n + c_1a_{n-1} + \cdots + c_ka_{n-k} = 0$ for $n \geq k$. But what does $(\sum_{n-k}^{\infty} f(n)X^n)/Q(X)$ represent? Perhaps a few more examples will give us a clue.

Example 3.6.3. Find a general expression for a solution to the recurrence relation $a_n - 5a_{n-1} + 6a_{n-2} = 4^n$ for $n \geq 2$.

Let $A(X) = \sum_{n-0}^{\infty} a_nX^n$. Then

$$A(X) = \frac{\sum\limits_{n-2}^{\infty} 4^nX^n}{1 - 5X + 6X^2} + \frac{a_0 + (a_1 - 5a_0)X}{1 - 5X + 6X^2}.$$

Now

$$\sum_{n-2}^{\infty} 4^nX^n = 4^2X^2 \sum_{n-2}^{\infty} 4^{n-2}X^{n-2},$$

which by a change of dummy variable becomes

$$4^2X^2 \sum_{n-0}^{\infty} 4^nX^n = 4^2 \frac{X^2}{1 - 4X}.$$

Therefore,

$$A(X) = \frac{4^2X^2}{(1 - 4X)(1 - 5X + 6X^2)} + \frac{a_0 + (a_1 - 5a_0)X}{1 - 5X + 6X^2}.$$

But since $1 - 5X + 6X^2 = (1 - 2X)(1 - 3X)$ we see that the homogeneous solutions have the form $C_12^n + C_23^n$. But likewise by partial fractions

$$\frac{4^2X^2}{(1 - 4X)(1 - 2X)(1 - 3X)} = \frac{C}{1 - 4X} + \frac{D}{1 - 2X} + \frac{E}{1 - 3X}$$

so that $4^2X^2/(1 - 4X)(1 - 2X)(1 - 3X)$ generates a sequence $\{b_n\}_{n-0}^{\infty}$ where $b_n = C4^n + D2^n + E3^n$ for some constants C, D, and E. Note that

$D2^n + E3^n$ also would solve the homogeneous recurrence relations $a_n - 5a_{n-1} + 6a_{n-2} = 0$, so the only new information gained is the part $C4^n$. When we compare this with the original function $f(n) = 4^n$, it seems that this function has almost reproduced itself. Thus, at least in this example, $(\Sigma_{n-k}^\infty f(n) X^n)/Q(X)$ has generated a sequence $b_n = Cf(n) + h(n)$ where $h(n)$ is a solution of the HR.

But this is not the whole picture yet. Let us subsitute $b_n = C4^n + D2^n + E3^n$ for a_n in the recurrence relation $a_n - 5a_{n-1} + 6a_{n-2} = 4^n$. After a moments reflection, we realize part of this is unnecessary since $D2^n + E3^n$ is a solution of the homogeneous relation so that substituting b_n is the same as substituting $C4^n$.

But then we have $4^n = C4^n - 5C4^{n-1} + 6C4^{n-2} = C4^{n-2}(4^2 - 5 \cdot 4 + 6)$ or $C = 8$. Therefore, we have concluded that, in this example $(\Sigma_{n-k}^\infty f(n)X^n)/Q(X)$ has generated a solution of the form $(8)4^n + h(n)$, where $(8)4^n$ is a particular solution of the inhomogeneous recurrence relation $a_n - 5a_{n-1} + 6a_{n-2} = 4^n$, and $h(n)$ is a solution of the homogeneous relation $a_n - 5a_{n-1} + 6a_{n-2} = 0$.

Example 3.6.4. Let us consider other examples of $f(n)$. For example, let us consider the case where we are to solve $a_n - 5a_{n-1} + 6a_{n-2} = 2^n$. Then, in this case, we obtain

$$A(X) = \frac{\sum_{n-2}^\infty 2^n X^n}{(1 - 3X)(1 - 2X)} + \frac{P(X)}{(1 - 3X)(1 - 2X)}.$$

Now

$$\sum_{n-2}^\infty 2^n X^n = 2^2 X^2 \sum_{r-0}^\infty 2^r X^r = \frac{2^2 X^2}{(1 - 2X)}.$$

Thus,

$$A(X) = \frac{2^2 X^2}{(1 - 3X)(1 - 2X)^2} + \frac{P(X)}{(1 - 3X)(1 - 2X)}.$$

Therefore, using a partial fraction decomposition we have

$$\frac{2^2 X^2}{(1 - 3X)(1 - 2X)^2} = \frac{A}{1 - 3X} + \frac{B}{1 - 2X} + \frac{C}{(1 - 2X)^2}.$$

Thus, $2^2 X^2/(1 - 3X)(1 - 2X)^2$ generates a sequence $\{b_n\}_{n-0}^\infty$ where $b_n = A3^n + B2^n + C(n + 1)2^n = A3^n + (B + C)2^n + Cn2^n$. Again note that

$A3^n + (B + C)2^n$ is a solution to the homogeneous relation, but $Cn2^n$ is a particular solution of the inhomogeneous relation for a specific choice of C. In fact by substituting $Cn2^n$ into the recurrence relation we obtain $(Cn2^n - 5C(n - 1)2^{n-1} + 6C(n - 2)2^{n-2} = 2^n$, or $C2^{n-2}[4n - 10(n - 1) + 6(n - 2)] = 2^n$. Thus $C = -2$, and $(-2)n2^n$ is a particular solution of the inhomogeneous relation $a_n - 5a_{n-1} + 6a_{n-2} = 2^n$.

Note that in Example 3.6.3, a particular solution was of the form $Cf(n)$, but in Example 3.6.4 a particular solution had the form $C nf(n)$. What is the difference? Upon reviewing the analysis of each of these cases, we observe that $f(n) = 4^n$ gave us the factor $(1 - 4X)$ in the denominator of the expression for $A(X)$, and in this case 4 was not a root of the characteristic polynomial so no higher power of $(1 - 4X)$ occurred in the denominator of $A(X)$. However, when $f(n) = 2^n$, then $(1 - 2X)^2$ occurs in the denominator of $A(X)$ because 2 was a root of the characteristic polynomial.

There are three clues here that seem to be consistent in both of the above examples:

(1) Any solution the IHR is the sum of a particular solution of the IHR and a solution of HR;

(2) the form of the particular solution is directly related to the function $f(n)$; and

(3) the form of a particular solution is affected by the roots of the characteristic polynomial $C(t)$.

Let us first discuss clue 1.

Theorem 3.6.1. Suppose that the IHR with constant coefficients is $a_n + c_1a_{n-1} + c_2a_{n-2} + \cdots + c_ka_{n-k} = f(n)$. Suppose, further, that HR $a_n + c_1a_{n-1} + c_2a_{n-2} + \cdots + c_ka_{n-k} = 0$ is the associated homogeneous relation. Then (1) if $\{a_n^{P_1}\}_{n=0}^{\infty}$ and $\{a_n^{P_2}\}_{n=0}^{\infty}$ are two solutions of IHR, then $\{a_n^{P_1} - a_n^{P_2}\}_{n=0}^{\infty}$ is a solution of HR.

(2) If $\{a_n^{P_1}\}_{n=0}^{\infty}$ is a particular solution of IHR and $\{a_n^H\}_{n=0}^{\infty}$ is a solution of HR, then $\{a_n^P + a_n^H\}_{n=0}^{\infty}$ is a solution of IHR.

(3) If $\{a_n^{H_1}\}_{n=0}^{\infty}, \{a_n^{H_2}\}_{n=0}^{\infty}, \ldots, \{a_n^{H_n}\}_{n=0}^{\infty}$ are different solutions to HR, then $\{C_1a_n^{H_1} + C_2a_n^{H_2} + \cdots C_ma_n^{H_m}\}_{n=0}^{\infty}$ is a solution of HR for any constants C_1, $C_2, \ldots, C_m$.

Proof. We only prove (2); the proofs for the others are similar. Since $a_n^P + c_1a_{n-1}^P + \cdots + c_ka_{n-k}^P = f(n)$ and $a_n^H + c_1a_{n-1}^H + \cdots + c_ka_{n-k}^H = 0$, we can add these two equations to get $(a_n^P + a_n^H) + c_1(a_{n-1}^P + a_n^H) + \cdots + c_k(a_n^P + a_n^H) = f(n)$, which is the desired conclusion.

The import of Theorem 3.6.1 is that the task of finding the complete solution for the IHR falls into two parts: *first, solve the HR in full generality, listing all possible solutions, and second, discover any particular solution at all of the IHR itself. The sum of these two parts provides a general solution to the IHR and if appropriate initial conditions are given, the arbitrary constants in the solution may be determined.* Let us apply this conclusion to another example.

Example 3.6.5. Find the complete solution to the IHR: $a_n - 7a_{n-1} + 10a_{n-2} = 4^n$ for $n \geq 2$.

We know that the solutions to the HR: $a_n - 7a_{n-1} + 10a_{n-2} = 0$ are of the form $a_n^H = C_1 2^n + C_2 5^n$ since $C(t) = t^2 - 7t + 10 = (t - 2)(t - 5)$ is the characteristic polynomial.

The main problem now is to determine a particular solution. We could use generating functions as we did in Example 3.6.3 and 3.6.4, but let us attempt to use the insights gained in that example. As is so often the case with mathematical problems, a good method is to *guess* the answer, or at least the general form of the answer, and subsequently to verify it, identifying the coefficients in the general expression.

In this case, we suggest that a particular solution will have the form $a_n^P = C4^n$. Let us determine if this is, in fact, the case, and, if so, what value of C will give a particular solution.

Substitute $C4^n$ for a_n into the inhomogeneous relation $a_n - 7a_{n-1} + 10a_{n-2} = 4^n$. Then we have $C4^n - 7C4^{n-1} + 10C4^{n-2} = 4^n$, or $C4^{n-2}(4^2 - 7 \cdot 4 + 10) = 4^n$. Thus, $C = -8$, $a_n^P = (-8)4^n$ is a particular solution, and $a_n = (-8)4^n + C_1 2^n + C_2 5^n$ is the complete solution.

If, in addition, we are given the initial conditions $a_0 = 8$ and $a_1 = 36$, then substituting $n = 0$ and $n = 1$ into the above expression for a_n, we have $8 = (-8) + C_1 + C_2$ and $36 = -32 + 2C_1 + 5C_2$. Simplying and solving, we find $C_1 = 4$ and $C_2 = 12$ as the solutions to these two equations in two unknowns. Thus, $a_n = (-8)4^n + (4)2^n + (12)5^n$ is the unique solution of $a_n - 7a_{n-1} + 10a_{n-2} = 4^n$ satisfying the initial conditions $a_0 = 8$ and $a_1 = 36$.

The Method of Undetermined Coefficients

The method of guessing the general form of a particular solution to an inhomogeneous recurrence relation and then determining the values of the coefficients in the general expression is called the method of **undetermined coefficients.**

In order for this method to be successful we require a little insight and more experience. We use generating functions to gain that experience and to verify what forms are good guesses for solutions. Once we have done this for several types of situations, we will not have to resort to generating functions anymore, rather we will make educated guesses based on our experience.

Inhomogeneous recurrence relations with certain types of known functions $f(n)$ are easier to solve than others, and the reason for that is that the generating functions for $\{f(n)\}_{n=0}^{\infty}$ are known. Basically, we shall consider only two cases: one where $f(n) = Da^n$ is an exponential function of n and the other where $f(n)$ is a polynomial in n. Of course, as our vocabulary of generating functions increases we can solve the IHR for other types of functions $f(n)$.

A trial solution for exponentials. Suppose that we are to solve the IHR where $f(n) = Da^n$, where D and a are constants. Let

$$A(X) = \sum_{n=0}^{\infty} a_n X^n.$$

Then we know that

$$A(X) = \frac{D\sum_{n=k}^{\infty} a^n X^n + P(X)}{Q(X)},$$

where

$$P(X) = a_0 + (a_1 + c_1 a_0)X + \cdots$$
$$+ (a_{k-1} + c_1 a_{k-2} + \cdots + c_{k-1} a_0)X^k$$

and

$$Q(X) = 1 + c_1 X + \cdots + c_k X^k.$$

Now

$$D \sum_{n=k}^{\infty} a^n X^n = Da^k X^k \sum_{n=k}^{\infty} a^{k-k} X^{n-k},$$

so that by change of dummy variable, letting $r = n - k$,

$$\sum_{n=k}^{\infty} a^{n-k} X^{n-k}$$

becomes

$$\sum_{r=0}^{\infty} a^r X^r = \frac{1}{1 - aX}.$$

Thus,

$$D \sum_{n=k}^{\infty} a^n X^n = \frac{Da^k X^k}{1 - aX},$$

and then by substituting this into the expression for $A(X)$, we have

$$A(X) = \frac{DX^k a^k + (1 - aX)P(X)}{(1 - aX)Q(X)}.$$

Note that the degree of the denominator is $k + 1$ and the degree of the numerator is at most k. Thus, we can apply the method of partial fractions to this quotient of polynomials.

But the question as to whether or not a is a root of the characteristic polynomial

$$C(t) = t^k + c_1 t^{k-1} + \cdots + c_k$$

immediately presents itself. Let us write

$$C(t) = (t - \alpha_1)^{r_1} (t - \alpha_2)^{r_2} \cdots (t - \alpha_s)^{r_s}$$

where $\alpha_1, \alpha_2, \ldots, \alpha_s$ are the distinct roots of $C(t)$. Thus,

$$Q(X) = (1 - \alpha_1 X)^{r_1} (1 - \alpha_2 X)^{r_2} \cdots (1 - \alpha_s X)^{r_s}.$$

The if a is not a root of $C(t)$, so that $a \neq \alpha_i$ for $i = 1, 2, \ldots, s$, then the partial fraction decomposition of $A(X)$ has the form

$$\frac{C}{1 - aX} + \frac{C_{11}}{(1 - \alpha_1 X)} + \cdots + \frac{C_{1r_1}}{(1 - \alpha_1 X)^{r_1}} + \frac{C_{21}}{(1 - \alpha_2 X)} + \cdots$$

$$+ \frac{C_{2r_2}}{(1 - \alpha_2 X)^{r_2}} + \cdots + \frac{C_{s_1}}{(1 - \alpha_s X)} + \cdots + \frac{C_{sr_s}}{(1 - \alpha_s X)^{r_s}},$$

for constants $C, C_{11}, C_{12}, \ldots, C_{sr_s}$. Note that all terms above except the first give rise to homogeneous solutions of the recurrence relation. Therefore, *in case a is not a root of the characteristic polynomial C(t)*, *we have a particular solution of the IHR of the form* $a_n^P = Ca^n$, and we must determine the value of C.

Example 3.6.6.   Find a particular solution to $a_n - 7a_{n-1} + 10a_{n-2} = 7 \cdot 3^n$ for $n \geq 2$.

We know from Example 3.4.5 that the characteristic polynomial $C(t) = t^2 - 7t + 10 = (t - 2)(t - 5)$ and that the homogeneous solutions have the form $a_n^H = C_1 2^n + C_2 5^n$. Since 3 is not a root of $C(t)$ we take a trial solution for $a_n^P = C3^n$ where the constant C is yet to be determined. Substituting $C3^n$ for a_n in the recurrence relation gives $C3^n - 7C3^{n-1} + 10C3^{n-2} = 7 \cdot 3^n$, or $C3^{n-2}(3^2 - 7 \cdot 3 + 10) = 7 \cdot 3^n$. This, in turn, implies that $C(-2) = 7 \cdot 3^2$ or $C = -63/2$. Thus, a particular solution is $a_n^P = (-63/2) 3^n$. Of course, the general solution to this relation is

$$a_n = \left(\frac{-63}{2} \right) 3^n + C_1 2^n + C_2 5^n.$$

Now if $\{a_n^{P_1}\}_{n=0}^{\infty}$ *is a particular solution to* $a_n + c_1 a_{n-1} + \cdots + c_k a_{n-k} = f_1(n)$ *and if* $\{a_n^{P_2}\}_{n=0}^{\infty}$ *is a particular solution to* $a_n + c_1 a_{n-1} + \cdots + c_k a_{n-k} = f_2(n)$, *then* $\{a_n^{P_1} + a_n^{P_2}\}_{n=0}^{\infty}$ *is a particular solution to* $a_n + c_1 a_{n-1} + \cdots + c_k a_{n-k} = f_1(n) + f_2(n)$.

Example 3.6.7.   Find a particular solution to $a_n - 7a_{n-1} + 10a_{n-2} = 7 \cdot 3^n + 4^n$.

To solve this, we use the above comments to resolve the problem into finding particular solutions to $a_n - 7a_{n-1} + 10a_{n-2} = 7 \cdot 3^n$ and $a_n - 7a_{n-1} + 10a_{n-2} = 4^n$. We know from Example 3.6.6 that $a_n^{P_1} = (-63/2) 3^n$ is a solution of the first relation. We know from example 3.6.5 that $a_n^{P_2} = (-8)4^n$ is a particular solution of the second relation. Therefore, $a_n^P = (-63/2)3^n + (-8)4^n$ is a particular solution to $a_n - 7a_{n-1} + 10a_{n-2} = (7)3^n + 4^n$.

Now, on the other hand, if a is a root of the characteristic polynomial $C(t)$, and if $f(n) = Da^n$, then the above argument needs some modification. Let us suppose $a = \alpha_i$ for some i and that the multiplicity of a as a root of $C(t)$ is m. Then when we express

$$A(X) = \frac{Da^k X^k + (1 - aX) P(X)}{(1 - aX) Q(X)}$$

as a sum of partial fractions, there will be one term of the form $C/(1 - aX)^{m+1}$. Hence a particular solution for the IHR has the form $a_n^P =$

$CC(n + m,n) \, a^n$. But

$$C(n + m,n) = \frac{(n + m)(n + m - 1) \dots (n + 1)}{m!},$$

and expanding this product will yield a polynomial in n of degree m. Thus, $CC(n + m,n) = P_0 + P_1 n + \cdots + P_m n^m$ and $a_n^P = CC(n + m,n)a^n = (P_0 + P_1 n + \cdots P_m n^m)a^n$. But because a is a root of $C(t)$ of multiplicity m, $(P_0 + P_1 n + \cdots + P_{m-1} n^{m-1})a^n$ is a solution of the HR. Thus, we can make a better choice for a particular solution of the IHR, namely, let $a_n^P = En^m a^n$, where E is some constant.

In summary, we have the following rules:

1. $a_n^P = Ca^a$ is *a particular solution* of IHR

$$a_n + C_1 a_{n-1} + \cdots + C_k a_{n-k} = Da^n$$

if a is not a root of the characteristic polynomial $C(t)$.

2. $a_n^P = Cn^m a^n$ *is a particular solution of IHR if a is a root of $C(t)$ of multiplicity m.*

Example 3.6.8. Find a particular solution of $a_n - 4a_{n-1} + 4a_{n-2} = 2^n$. Since the characteristic polynomial is $C(t) = t^2 = 4t + 4 = (t - 2)^2$, 2 is a root of multiplicity 2, thus a trial solution is $a_n^P = Cn^2 2^n$. Substituting into the recurrence relation, we obtain

$$Cn^2 2^n - 4C(n - 1)^2 \, 2^{n-1} + 4C(n - 2)^2 \, 2^{n-2} = 2^n,$$

or 　　　　　$C2^{n-2}[4n^2 - 8(n - 1)^2 + 4(n - 2)^2] = 2^n.$

Thus, $C[8] = 2^2$, or $C = 1/2$. Therefore, $a_n^P = n^2 2^n/2$ is a particular solution, and

$$a_n = \frac{n^2}{2} 2^n + C_1 2^n + C_2 n 2^n$$

is the general solution.

Trial solutions for products of polynomials and exponentials. Now let us suppose that

$$f(n) = (P_0 + P_1 n + \cdots + P_s n^s) \, a^n,$$

where P_i are constants. We desire the form of a particular solution in this case. Again to do this, we use generating functions to discover a candidate for a particular solution.

Let us do a simple example to illustrate what we can do in general.

Example 3.6.9.   Find the form of a particular solution to $a_n - 5a_{n-1} + 6a_{n-2} = n^2 4^n$ for $n \geq 2$. Let

$$A(X) = \sum_{n=0}^{\infty} a_n X^n.$$

Then

$$A(X) = \frac{\displaystyle\sum_{n=2}^{\infty} n^2 4^n X^n + a_0 + (a_1 - 5a_0) X}{1 - 5X + 6X^2}.$$

Now

$$\sum_{n=2}^{\infty} n^2 4^n X^n = 4^2 X^2 \sum_{n=2}^{\infty} n^2 4^{n-2} X^{n-2},$$

and we let $r = n - 2$, then we have

$$\sum_{n=2}^{\infty} n^2 4^n X^n = n^2 X^2 \sum_{r=0}^{\infty} (r + 2)^2 4^r X^r.$$

Let us write $(r + 2)^2 = r^2 + 4r + 4$ as $2C(r + 2,r) + C(r + 1,r) + C(r,r)$ so that

$$
\begin{aligned}
4^2 X^2 \sum_{r=0}^{\infty} (r + 2)^2 4^r X^r &= 4^2 X^2 \left[2 \sum_{r=0}^{\infty} C(r + 2,r) 4^r X^r \right. \\
&\quad \left. + \sum_{r=0}^{\infty} C(r + 1,r) 4^r X^r + \sum_{r=0}^{\infty} 4^r X^r \right] \\
&= 4^2 X^2 \left[\frac{2}{(1 - 4X)^3} + \frac{1}{(1 - 4X)^2} + \frac{1}{1 - 4X} \right] \\
&= \frac{4^2 X^2 [2 + (1 - 4X) + (1 - 4X)^2]}{(1 - 4X)^3}.
\end{aligned}
$$

Thus,

$$A(X) = \frac{4^2 X^2 [2 + (1 + 4X) + (1 - 4X)^2] + (1 - 4X)^3 P(X)}{(1 - 4X)^3 (1 - 5X + 6X^2)}.$$

where $P(X) = a_0 + (a_1 - 5a_0)X$. Thus,

$$A(X) = \frac{F(X)}{(1 - 4X)^3 (1 - 2X)(1 - 3X)}$$

where $F(X)$ is a polynomial of 4 or less. By partial fractions, we see that

$$A(X) = \frac{A}{(1 - 4X)^3} + \frac{B}{(1 - 4X)^2} + \frac{C}{(1 - 4X)}$$

$$+ \frac{D}{(1 - 2X)} + \frac{E}{(1 - 3X)}.$$

Now $D/(1 - 2X) + E/(1 - 3X)$ satisfies the homogeneous recurrence relation. The series

$$\frac{A}{(1 - 4X)^3} + \frac{B}{(1 - 4X)^2} + \frac{C}{1 - 4X}$$

$$= \sum_{n=0}^{\infty} [AC(n + 2,n) + BC(n + 1,n) + C] \, 4^n X^n$$

so that a particular solution has the form $[AC(n + 2,n) + BC(n + 1,n) + C] \, 4^n$. But after expanding the binomial coefficients, we see that the above solution takes the form $(P_0 + P_1 n + P_2 n^2) \, 4^n$. Thus, $f(n) = n^2 4^n$ determines a particular solution of the form a polynomial of degree 2 times 4^n.

With this example in mind, let us return to the general case. By imitating what we did in the above example, we can obtain the following conclusions:

3. $a_n^P = (A_0 + A_1 n + \cdots + A_s n^s)a^n$ is a particular solution of the IHR:

$$a_n + C_1 a_{n-1} + \cdots + C_k a_{n-k} = (P_0 + P_1 n + \cdots + P_s n^s)a^n$$

if a is not a root of the characteristic polynomial $C(t) = t^k + C_1 t^{k-1} + \cdots + C_k$.

4. $a_n^P = n^m (A_0 + A_1 n + \cdots + A_s n^s)a^n$ is a particular solution of the IHR if a is a root of $C(t)$ of multiplicity m.

Trial solutions for polynomials. The case where $f(n) = P_0 + P_1 n + \cdots + P_s n^s$ is just a special case of the above discussion where $a = 1$. Then we need be concerned only with whether or not 1 is a root of the characteristic polynomial $C(t)$.

Example 3.6.10. Find a particular solution of $a_n - 2a_{n-1} + a_{n-2} = 5 + 3n$. Since 1 is a root of $C(t) = t^2 - 2t + 1 = (t - 1)^2$ of multiplicity 2 we use $a_n^P = An^2 + Bn^3$ as a candidate for a solution and solve for A and B. Upon substitution, we have $[An^2 + Bn^3] - 2[A(n - 1)^2 + B(n - 1)^3] + [A(n - 2)^2 + B(n - 2)^3] = 5 + 3n$, and this simplifies to $(2A - 6B) + 6Bn = 5 + 3n$. In particular, this holds for all n, hence for $n = 0$ we must have $2A - 6B = 5$ and for $n = 1$, $2A - 6B + 6B = 5 + 3$ or $2A = 8$ or $A = 4$ and hence $B = \frac{1}{2}$. Thus, a particular solution is $a_n^P = 4n^2 + \frac{1}{2}n^3$, and the general solution is $a_n = 4n^2 + \frac{1}{2}n^3 + C_1 + nC_2$, where C_1 and C_2 are constants that can be determined by initial conditions.

Table 3-2 summarizes all we have said concerning the forms of particular solutions to the IHR. If $f(n)$ is the sum of different functions, we have noted that each function should be treated separately. If the function $f(n)$ includes a function that is a solution to the HR, then generally the form of the particular solution will include the product of powers of n with $f(n)$.

Table 3-2.

$f(n)$	Characteristic Polynomial $C(t)$	Form of Particular Solution a_n^P
Da^n	$C(a) \neq 0$	Aa^n
Da^n	a is a root of $C(t)$ of multiplicity m	$An^m a^n$
$Dn^s a^n$	$C(a) \neq 0$	$(A_0 + A_1 n + \cdots + A_s n^s)a^n$
$Dn^s a^n$	a is a root of $C(t)$ of multiplicity m	$n^m(A_0 + A_1 n + \cdots + A_s n^s)a^n$
Dn^s	$C(1) \neq 0$	$(A_0 + A_1 n + \cdots A_s n^s)$
Dn^s	1 is a root of $C(t)$ of multiplicity m	$n^m(A_0 + \cdots + A_s n^s)$

Remark: In each of the last four types of Table 3-2, the form of the particular solution is the same if Dn^s is replaced by $[P_0 + P_1 n + \cdots + P_s n^s]$.

Now let us apply what we have learned to several examples.

Example 3.6.11. Solve the recurrence relation $a_n - 6a_{n-1} + 8a_{n-2} = 9$ where $a_0 = 10$ and $a_1 = 25$.

First, we note that the characteristic polynomial $C(t) = t^2 - 6t + 8 = (t - 2)(t - 4)$. Therefore, the general solution of the homogeneous relation HR is: $a_n^H = C_1 4^n + C_2 2^n$. Since the forcing function has the form $D(1)^n$ and 1 is not a root of $C(t)$, we use $a_n^P = A(1)^n = A$ as a trial particular solution. But then after substituting $a_n^P = A$ into the IHR, we find $A = 3$. Thus, $a_n = C_1 4^n + C_2 2^n + 3$. Now using the initial conditions $a_0 = 10$ and

$a_1 = 25$, we have the system of equations $C_1 + C_2 + 3 = 10$ and $4C_1 + 2C_2 + 3 = 25$. But these equations simplify to $C_1 + C_2 = 7$ and $4C_1 + 2C_2 = 22$, which in turn have the solutions $C_1 = 4$ and $C_2 = 3$. Thus, $a_n = (4)(4^n) + (3)(2^n) + 3$.

Example 3.6.12.   Solve $a_n - 6a_{n-1} + 8a_{n-2} = 3^n$ where $a_0 = 3$ and $a_1 = 7$.

Here we try a particular solution of the form $A3^n$, and we discover $A = -9$ after substitution into the recurrence relation. Thus, $a_n = (-9)(3^n) + C_1 4^n + C_2 2^n$, and then the initial conditions give $C_1 = 5$ and $C_2 = 7$. Hence, $a_n = (-9)(3^n) + (5)(4^n) + (7)(2^n)$.

Example 3.6.13.   Solve $a_n - 6a_{n-1} + 8a_{n-2} = n4^n$ where $a_0 = 8$ and $a_1 = 22$.

Here the particular solution takes the form $n(A_0 + A_1 n)4^n$ since 4 is a root of characteristic polynomial of multiplicity 1. Substituting this expression into the recurrence relation gives

$$n(A_0 + A_1^n)4^n - 6(n-1)(A_0 + A_1(n-1))4^{n-1}$$
$$+ 8(n-2)(A_0 + A_1(n-2))4^{n-2} = n4^n$$

But by canceling the common term 4^{n-2}, we have

$$16n(A_0 + A_1 n) - 24(n-1)(A_0 + A_1(n-1))$$
$$+ 8(n-2)(A_0 + A_1(n-2)) = 16n$$

Now this is an expression that holds for all values of n, in particular for $n = 0$ we obtain the simplified equation $A_0 + A_1 = 0$, and then for $n = 1$ we obtain $A_0 + 3A_1 = 2$. These equations have the unique solution $A_0 = -1$ and $A_1 = 1$. Hence, $a_n^p = n(-1 + n)4^n = n(n-1)4^n$ is a particular solution.

But then $a_n = n(n-1)4^n + C_1 4^n + C_2 2^n$ is the general solution of the relation, and the initial conditions give $C_1 = 3$ and $C_2 = 5$. Hence, $a_n = n(n-1)4^n + (3)(4^n) + (5)(2^n)$ is the unique solution to the recurrence relation with the given initial conditions.

Solving Systems of Recurrence Relations

Let us illustrate how we might solve a system of recurrence relations by the following example.

Example 3.6.14. Suppose that A is the 2×2 matrix

$$\begin{bmatrix} 3 & -1 \\ 0 & 2 \end{bmatrix}.$$

For each integer $n \geq 1$, find an expression for A^n using recurrence relations. In particular, give the numerical entries of A^{100}.

First, we introduce the following notation. Let

$$A^n = \begin{bmatrix} a_n & b_n \\ c_n & d_n \end{bmatrix}$$

and observe that

$$A^n = A^{n-1} \begin{bmatrix} 3 & -1 \\ 0 & 2 \end{bmatrix}$$

for $n \geq 2$. (In fact, this equality holds for $n \geq 1$, since we could adopt the convention

$$A^0 = \begin{bmatrix} 1 & 0 \\ 0 & 1 \end{bmatrix}.)$$

Thus,

$$A^n = \begin{bmatrix} a_{n-1} & b_{n-1} \\ c_{n-1} & d_{n-1} \end{bmatrix} \begin{bmatrix} 3 & -1 \\ 0 & 2 \end{bmatrix}$$

$$= \begin{bmatrix} 3a_{n-1} & 2b_{n-1} - a_{n-1} \\ 3c_{n-1} & 2d_{n-1} - c_{n-1} \end{bmatrix}.$$

From the equality of matrices, we obtain four recurrence relations:

(1) $a_n = 3a_{n-1}$,   (2) $b_n = 2b_{n-1} - a_{n-1}$,
(3) $c_n = 3c_{n-1}$,   (4) $d_n = 2d_{n-1} - c_{n-1}$,

Where $a_1 = 3$, $b_1 = -1$, $c_1 = 0$, and $d_1 = 2$.

The recurrence relations (1) and (3) with their initial conditions imply that

$$a_n = 3^n \text{ for } n \geq 1 \qquad \text{and } c_n = 0 \text{ for } n \geq 1.$$

Thus, (2) and (4) become

(5) $b_n - 2b_{n-1} = -(3^{n-1})$ and
(6) $d_n - 2d_{n-1} = 0.$

But (6) has the unique solution $d_n = 2^n$ for $n \geq 1$, and we are left with only relation (5) to solve.

A particular solution of (5) has the form $B3^n$, and after substituting into (5), we find $B = -1$. Then, $b_n = A2^n - 3^n$ and the initial condition $b_1 = -1$ gives $A = 1$. Therefore, $b_n = 2^n - 3^n$ is the unique solution of (5).

Therefore,

$$A^n = \begin{bmatrix} 3^n & 2^n - 3^n \\ 0 & 2^n \end{bmatrix}$$

and

$$A^{100} = \begin{bmatrix} 3^{100} & 2^{100} - 3^{100} \\ 0 & 2^{100} \end{bmatrix}.$$

Solving Nonlinear Recurrence Relations

Sometimes nonlinear recurrence relations can be made into a linear recurrence by a substitution.

Example 3.6.15. Solve the recurrence relation $a_n^2 - 2a_{n-1}^2 = 1$ for $n \geq 1$ where $a_0 = 2$.

Let $b_n = a_n^2$. This substitution changes the relation $a_n^2 - 2a_{n-1}^2 = 1$ to a linear recurrence $b_n - 2b_n = 1$, and since $a_0 = 2$, we see that $b_0 = a_0^2 = 4$. By the technique of undetermined coefficients, we find that $b_n = (5)(2^n) - 1$, and hence that $a_n = \pm \sqrt{b_n}$. But in fact, since $a_0 = 2$, a_n cannot be $- \sqrt{b_n}$, so $a_n = \sqrt{(5)(2^n) - 1}$.

Likewise, divide-and-conquer recurrence relations can be solved using appropriate substitutions. Generally, a divide-and-conquer relation takes the form $a_n = Ca_{n/d} + f(n)$ where usually the values of n are determined by nonnegative integral powers of d. Now a_n is really a function of n, but since $n = d^k$, a_n can also be viewed as a function of k. When we change our perspective to viewing a_n as a function of k, we say that we make a "change of variables." Generally, what we do is this: let $b_k = a_n = a_{d^k}$ for $k \geq 0$. Then the relation $a_n = Ca_{n/d} + f(n)$ is transformed into the linear relation $b_k = Cb_{k-1} + f(d^k)$ for $k \geq 1$. But for $k = 0$, $d^0 = 1$, and any initial condition, $a_1 = A$ becomes $b_0 = A$.

Of course, after we solve the transformed relation for a general expression for b_k, we can then solve for a_n using the fact that $k = \log_d n$. Let us illustrate this technique in the following example.

Example 3.6.16. Solve the divide-and-conquer relation $a_n - 7a_{n/3} = 2n$ where $n = 3^k$ for $k \geq 1$ and $a_1 = 5/2$.

We employ the change of variables $b_k = a_n = a_{3^k}$. Then the transformed relation is

$$b_k - 7b_{k-1} = 2(3^k) \text{ for } k \geq 1 \text{ and } a_1 = 5/2 = b_0.$$

The linear relation has the characteristic polynomial $t - 7$ so that the homogeneous relation has a solution $b_k^H = B7^k$ for some constant B.

Moreover, a particular solution of the inhomogeneous relation takes the form $b_k^P = A3^k$. Substitution reveals $A = -3/2$ so that

$$b_k = -\frac{3}{2}(3^k) + B7^k,$$

but then the initial condition $b_0 = 5/2$ enables us to determine $B = 4$. Thus,

$$b_k = \frac{-3}{2}(3^k) + (4)(7^k).$$

Now let us give a solution for a_n in terms of n rather than k. Here we use the facts that $n = 3^k$ and $k = \log_3 n$. We observe then that

$$b_k = a_n = \left(-\frac{3}{2}\right)(n) + (4)(7^{\log_3 n}).$$

Moreover, from properties of logarithms, we know that

$$7^{\log_3 n} = n^{\log_3 7} \text{ so that } a_n = \left(-\frac{3}{2}\right)n + (4)(n^{\log_3 7}).$$

Exercises for Section 3.6

1. Find a particular solution to the following inhomogeneous recurrence relations using the method of undetermined coefficients.
   (a) $a_n - 3a_{n-1} = 3^n$.
   (b) $a_n - 3a_{n-1} = n + 2$.

(c) $a_n - 2a_{n-1} + a_{n-2} = 2^n$.
(d) $a_n - 2a_{n-1} + a_{n-2} = 4$.
(e) $a_n - 3a_{n-1} + 2a_{n-2} = 3^n$.
(f) $a_n - 3a_{n-1} + 2a_{n-2} = 3$.
(g) $a_n - 3a_{n-1} + 2a_{n-2} = 2^n$.
(h) $a_n - 3a_{n-1} + 2a_{n-2} = 5n + 3$.
(i) $a_n + 3a_{n-1} - 10a_{n-2} = n + 1$.
(j) $a_n + 3a_{n-1} - 10a_{n-2} = n^2 + n + 1$.
(k) $a_n + 3a_{n-1} - 10a_{n-2} = (-5)^n$.
(l) $a_n - 10a_{n-1} + 25a_{n-2} = 2^n$.
(m) $a_n + 6a_{n-1} + 12a_{n-2} + 8a_{n-3} = 3^n$.

2. Write the general form of a particular solution a_n^P (you need not solve for the constants) to the following recurrence relations.
   (a) $a_n - 2a_{n-1} = 3$.
   (b) $a_n - 2a_{n-1} = 2^n$.
   (c) $a_n - 2a_{n-1} = n2^n$.
   (d) $a_n - 7a_{n-1} + 12a_{n-2} = n$.
   (e) $a_n - 7a_{n-1} + 12a_{n-2} = 2^n$.
   (f) $a_n - 7a_{n-1} + 12a_{n-2} = 3^n$.
   (g) $a_n - 7a_{n-1} + 12a_{n-2} = 4^n$.
   (h) $a_n - 7a_{n-1} + 12a_{n-2} = n4^n$.

3. List the general solution (the general homogeneous solution plus the general form of a particular solution) of the recurrence relations in Exercise 2.

4. Suppose that the recurrence relation of degree k is

$$a_n + C_1 a_{n-1} + C_2 a_{n-2} + \cdots + C_k a_{n-k} = f(n),$$

for $n \geq k$, and that a_n^H denotes a solution of the associated homogeneous recurrence relation, a_n^P denotes a particular solution to the inhomogeneous relation. Moreover, $C(t)$ denotes the characteristic polynomial of the associated homogeneous recurrence relation.
   (a) Find $C_1, C_2, \ldots, C_k$ if $C(t) = (t-2)(t-3)(t-5)$.
   (b) Find $C_1, C_2, \ldots, C_k$ if $C(t) = (t-2)^2(t-3)(t-5)$.
   (c) List the general form of a_n^H for the case when $C(t) = (t-2)(t-4)(t-5)$.
   (d) List the general form of a_n^H for the case when $C(t) = (t-2)(t-4)^2(t-5)^3$.
   (e) List the general form of a_n^H for the case when $C(t) = (t-2)^5(t-4)^2(t-5)^3$.
   (f) List the general form of a_n^P when $f(n) = 3n^2 + 5n + 7$ and $C(t) = (t-2)^5(t-4)^2(t-5)^3$.
   (g) Same as (f) where $f(n) = 4^n$.

(h) Same as (f) where $f(n) = 5^n$.

(i) Same as (f) where $f(n) = 2^n$.

5. Solve the following recurrence relations using generating functions.

(a) $a_n - a_{n-1} = n$ for $n \geq 1$ and $a_0 = 0$.

(b) $a_n - a_{n-1} = 2(n-1)$ for $n \geq 1$ and $a_0 = 2$.

(c) $a_n - 2a_{n-1} = 4^{n-1}$ for $n \geq 1$ and $a_0 = 1, a_1 = 3$.

(d) $a_n - 2a_{n-1} + a_{n-2} = 2^{n-2}$ for $n \geq 2$ where $a_0 = 2$ and $a_1 = 1$.

(e) $a_n - 5a_{n-1} + 6a_{n-2} = 4^{n-2}$ for $n \geq 2$ and $a_0 = 1, a_1 = 5$.

(f) $a_n - 10a_{n-1} + 21a_{n-2} = 3^{n-2}$ for $n \geq 2$ and $a_0 = 1, a_1 = 10$.

6. Find the complete solution (homogeneous plus particular solutions) to $a_n - 10a_{n-1} + 25a_{n-2} = 2^n$ where $a_0 = 2/3$ and $a_1 = 3$. Use the result of Exercise 1 (1).

7. Find the complete solution to $a_n + 2a_{n-1} = n + 3$ for $n \geq 1$ and with $a_0 = 3$.

8. Suppose that a_n satisfies the relation $a_n - 9a_{n-1} + 26a_{n-2} - 24a_{n-3} = 5^n$ where $a_0 - 1, a_1 = 18, a_2 = 45$. Write the generating function

$$A(X) = \sum_{n-0}^{\infty} a_n X^n$$

as a quotient of 2 polynomials $P(X)/Q(X)$.

9. Solve the following recurrence relations for a particular solution.

(a) $a_n - 5a_{n-1} + 8a_{n-2} - 4a_{n-3} = n$.

(b) $a_n - 5a_{n-1} + 8a_{n-2} - 4a_{n-3} = 2^n$.

(c) $a_n - 5a_{n-1} + 8a_{n-2} - 4a_{n-3} = n2^n$.

(d) $a_n - 5a_{n-1} + 8a_{n-2} - 4a_{n-3} = 3^n$.

(e) $a_n - 5a_{n-1} + 8a_{n-2} - 4a_{n-3} = 1$.

10. Show that the divide-and-conquer relation

$$a_n = 2a_{n/2} + (n-1) \text{ for } n \geq 2; \text{ and } a_1 = 0$$

can be solved for $n = 2^k$ by using the change of variables $b_k = a_{2^k}$, solving the relation

$$b_k = 2b_{k-1} + 2^k - 1 \text{ for } k \geq 1$$

and $b_0 = 0$ by using the method of undetermined coefficients, and then finding a_{2^k} from b_k.

11. Choose an appropriate substitution to translate

$$a_n = 2a_{n/4} + n \text{ for } n = 4^k \geq 4 \text{ and } a_1 = 1$$

into a first order relation. Solve this relation by undetermined coefficients and then find a_{4^k}.

12. Solve the divide-and-conquer recurrence relations in Exercises 7, 8, 9, and 10 of Section 3.4 by the change of variables technique.

13. Solve the recurrence relation $a_n = 5(a_{n-1})^2$ for $n \geq 1$ and $a_0 = 1$. Here make the substitution $b_n = \log_2(a_n)$, solve the linear inhomogeneous recurrence relation for b_n, and then find a_n.

14. Solve the recurrence relation $a_n - 5a_{n/2} + 6a_{n/4} = n$ for $n = 2^k$, where $k \geq 0$, and where $a_1 = 1$ and $a_2 = 3$. Hint: make the substitution $b_k = a_{2^k}$.

15. Show that $a_n = A_1 \, C^{\log_d(n)} + A_2(n)$ is a solution to $a_n - Ca_{n/d} = en$, where c and e are constants and n is a power of d if C is not equal to d.

16. Find and solve a divide and conquer recurrence for the number of matches played in a tennis tournament with n players, where n is a power of 2.

17. In a large firm, every five salespeople report to a local manager, every five local managers report to a district manager, and so forth until finally five vice-presidents to the firm's president. If the firm has n salespeople, where n is a power of 5, find and solve the divide and conquer recurrence relations for:
    (a) the number of different managerial levels in the firm.
    (b) the number of managers (local managers up through the president) in the firm.

18. Solve the following recurrence relations by making an appropriate substitution to transform the relations into linear recurrences with constant coefficients.
    (a) $\sqrt{a_n} - \sqrt{a_{n-1}} - 2\sqrt{a_{n-2}} = 0$ where $a_0 = a_1 = 1$.
    (b) $na_n + na_{n-1} - a_{n-1} = 2^n$ where $a_0 = 10$.
    (c) $a_n^3 - 2a_{n-1} = 0$ where $a_0 = 8$. Hint: let $b_n = \log_2 a_n$.
    (d) $a_n - na_{n-1} = n!$ for $n \geq 1$ where $a_0 = 2$.
    (e) $a_n = \dfrac{\sqrt{a_{n-1}}}{a_{n-2}^2}$ where $a_0 = 1$ and $a_1 = 2$.
    (f) $a_n + 5na_{n-1} + 6n(n-1)a_{n-2} = 0$ where $a_0 = 6$ and $a_1 = 17$.
    (g) $a_n = (a_{n-1})^2 (a_{n-2})^3$ where $a_0 = 4$ and $a_1 = 4$.
    (h) $na_n - (n-2)a_{n-1} = 2n$ where $a_0 = 5$.
    (i) $na_n - (n+1)a_{n-1} = 2n$ where $a_0 = 1$.

19. Let $F = \begin{bmatrix} 1 & 1 \\ 1 & 0 \end{bmatrix}$. For $n \geq 0$, solve for the entries of F^n using recurrence relations. (Assume $F^0 = \begin{bmatrix} 1 & 0 \\ 0 & 1 \end{bmatrix}$.

20. Let $A = \begin{bmatrix} 3 & 2 \\ 1 & 4 \end{bmatrix}$. For $n \geq 0$ solve for the entries of A^n using recurrence relations.

21. Solve exercise 20 for the matrix $A = \begin{bmatrix} 4 & 6 \\ 1 & 5 \end{bmatrix}$.

22. Solve exercise 20 for the matrix $A = \begin{bmatrix} 2 & 0 \\ 1 & 3 \end{bmatrix}$.

23. Ackerman's function is a function defined on $N \times N$, where N is the set of nonnegative integers, as follows:

$$A(m,n) = \begin{cases} n + 1 & \text{if } m = 0 \\ A(m - 1,1) & \text{if } n = 0 \\ A(m - 1,A(m,n - 1)) & \text{otherwise.} \end{cases}$$

Then, for example, $A(1,1) = A(0,A(1,0))$
$$= A(0,A(0,1))$$
$$= A(0,2) = 3.$$

(a) Use the definition of Ackerman's function to show that $A(2,n) = A(2,n - 1) + 2$ for all integers $n \geq 1$ where $A(2,0) = 3$. But then by solving this recurrence relation, observe that

$$A(2,n) = 2n + 3 \text{ for } n \geq 0.$$

(b) As in (a), show that

$$A(3,n) = 2^{n+3} - 3 \text{ for } n \geq 0.$$

24. Solve the divide-and-conquer relations using a change of variables.
   (a) $a_n = 5a_{n/2} + 4$ where $a_1 = 0$ and $n = 2^k$ for $k \geq 0$.
   (b) $a_n - 2a_{n/3} = 4$ where $a_1 = 5$ and $n = 3^k$ for $k \geq 0$.
   (c) $a_n - 3a_{n/8} = 2n$ where $a_1 = 1$ and $n = 8^k$ for $k \geq 0$.
   (d) $a_n - 5a_{n/3} = n$ where $a_1 = 5/2$ and $n = 3^k$ for $k \geq 0$.
   (e) $a_n - 5a_{n/5} = n$ where $a_1 = 7$ and $n = 5^k$ for $k \geq 0$.

25. (a) Suppose that a person borrows \$40,000 at 13.9% interest. Suppose that the monthly payment is \$530, which includes payment on the principal and interest on the balance owed. At the end of seven years, the borrower must negotiate a new loan

for the balance owed. How much of the principal will be owed after seven years?

(b) If P is the original principal, p is the payment for each period, and r is the period rate of interest, derive a formula for B_n, the balance owed after n periods.

26. Find k where
   (a) $4^{\log_2 n} = n^k$
   (b) $27^{\log_3 n} = n^k$.

Review for Sections 2.8–3.6

1. Among all n-digit decimal numbers, how many of them contain the digits 2 and 5 but not the digits 0,1,8,9?

2. At a theater 20 men check their hats. In how many ways can their hats be returned so that
   (a) no man receives his own hat?
   (b) at least one of the men receives his own hat?
   (c) at least two of the men receive their own hats?

3. Find a recurrence relation for a_n, the number of ways a sequence of 1's and 3's can sum to n. For example, $a_4 = 3$ since 4 can be obtained with the following sequences: 1111 or 13 or 31.

4. Find a recurrence relation for the number of n-digit quinary sequences that have an even number of 0's. (Quinary sequences use only the digits 0,1,2,3, and 4.)

5. Write a general expression for the generating function $A(X)$ as the quotient of two polynomials $P(X)$ and $Q(X)$ by specifying the coefficients of $P(X)$ and $Q(X)$ where $A(X)$ generates the sequence a_n for $n \geq 0$ where $a_0 = 1$, $a_1 = 0$, $a_2 = 1$, and for $n \geq 3$, a_n satisfies the recurrence relation $a_n - 5a_{n-1} + 8a_{n-2} - 4a_{n-3} = 0$.

6. Write the general form of the solutions of the following:
   (a) $a_n - 9a_{n-1} + 14a_{n-2} = 0$.
   (b) $a_n - 6a_{n-1} + 9a_{n-2} = 0$.

7. Write the general form of a particular solution to the following:
   (a) $a_n - 9a_{n-1} + 14a_{n-2} = 5(3^n)$.
   (b) $a_n - 9a_{n-1} + 14a_{n-2} = 7^n$.
   (c) $a_n - 9a_{n-1} + 14a_{n-2} = 3n^2$.

8. Solve the divide and conquer relation $a_n = 3a_{n/5} + 3$ where $a_1 = 7$ and n is a power of 5.

9. Find a simple expression for the power series

$$\sum_{n=1}^{\infty} (n + 2)(n + 1)(n) X^n.$$

10. Find a simple expression for the sequence generated by

$$\frac{3}{(1-2X)} + \frac{5}{(1-X)^3} + \frac{X^2}{(1-3X)^3}.$$

Selected Answers for Section 3.6

1. (a) Try $a_n^p = An3^n$. Substitute to find $A = 1$ so $a_n^p = n3^n$.
    (b) Let $a_n^p = An + B$, get $-2A = 1, 3A - 2B = 2$, and solving, find $A = -1/2, B = -7/4$. Hence $a_n^p = (-n/2) - (7/4)$.
    (c) $a_n^p = 4 \cdot 2^n = 2^{n+2}$.
    (d) Try $a_n^p = An^2$ since 1 is a characteristic root.
    (f) Since 1 is a characteristic root, let $a_n^p = An$. Solving get $A = -3$. Thus, $a_n^p = -3n$.
    (j) Since $C(t) = (t + 5)(t - 2)$ and 1 is not a characteristic root, try $a_n^p = An^2 + Bn + C$. Substituting and equating coefficients, get $A = -1/6, 34A - 6B = 1, -37A + 17B - 6C = 1$. Solve for A, B, and C.
    (l) $a_n^p = A2^n$; $A = 4/9$; $a_n^p = 2^{n+2}/9$.
    (m) $a_n^p = (27/125)3^n$.

2. (a) $a_n^p = A$.
    (b) $a_n^p = An2^n$ since 2 is a characteristic root.
    (d) $a_n^p = An + B$.
    (e) $a_n^p = A2^n$.
    (f) $a_n^p = An3^n$ since 3 is a characteristic root of $C(t) = t^2 - 7t + 12 = (t - 3)(t - 4)$.
    (h) Let $a_n^p = n(An + B)4^n$.

4. (b) $C(t) = t^4 - 12t^3 + 51t^2 - 92t + 60$ implies that $C_1 = -12, C_2 = 51, C_3 = -92, C_4 = 60$.
    (c) $a_n^H = C_1 2^n + C_2 4^n + C_3 5^n$.
    (e) $a_n^H = C_1 2^n + C_2 n2^n + C_3 n^2 2^n + C_4 n^3 2^n + C_5 n^4 2^n + C_6 4^n + C_7 n4^n + C_8 5^n + C_9 n5^n + C_{10} n^2 5^n$.
    (g) $a_n^p = An^2 4^n$.

5. (a) $A(X) = \sum_{n=0}^{\infty} a_n X^n = \frac{a_0}{1 - X} + \frac{X}{(1 - X)^3}$
    $= \frac{X}{(1 - X)^3}$. Thus, $a_n = \frac{(n + 1)(n)}{2}$.
    (b) $A(X) = \frac{2}{1 - X} + \frac{2X^2}{(1 - X)^3}$; $a_n = 2 + (n)(n - 1)$.

(c) $A(X) = \dfrac{1}{1 - 2X} + \dfrac{X}{(1 - 2X)(1 - 4X)}$

$= \dfrac{1/2}{1 - 2X} + \dfrac{1/2}{1 - 4X}$; $a_n = \dfrac{1}{2}2^n + \dfrac{1}{2}4^n$.

(d) Observe that $A(X) = \dfrac{2 - 3X}{(1 - X)^2} + \dfrac{X^2}{(1 - 2X)(1 - X)^2}$

$= \dfrac{2 - 7X + 7X^2}{(1 - 2X)(1 - X)^2}$

$= \dfrac{3}{(1 - X)} - \dfrac{2}{(1 - X)^2} + \dfrac{1}{1 - 2X}$.

Thus, $a_n = 3 - 2(n + 1) + 2^n = 1 - 2n + 2^n$.

(f) $A(X) = \dfrac{1}{(1 - 3X)(1 - 7X)} + \dfrac{X^2}{(1 - 3X)^2(1 - 7X)}$

$= \dfrac{-35}{48}\left(\dfrac{1}{1 - 3X}\right) - \dfrac{1}{12}\dfrac{1}{(1 - 3X)^2} + \dfrac{29}{16}\dfrac{1}{(1 - 7X)}$

$a_n = \dfrac{-35}{48}3^n - \dfrac{1}{12}(n + 1)3^n + \dfrac{29}{16}7^n.$

6. $a_n = \dfrac{2}{9}5^n + \dfrac{1}{5}n5^n + \dfrac{2^{n+2}}{9}$.

7. $a_n = \dfrac{16}{9}(-2)^n + \dfrac{n}{3} + \dfrac{11}{9}$.

8. $A(X) = \dfrac{(1 + 9X - 91X^2)(1 - 5X) + X^3 5^3}{(1 - 5X)(1 - 9X + 26X^2 - 24X^3)}$.

4

Relations and Digraphs

4.1 RELATIONS AND DIRECTED GRAPHS

In Chapter 1 the concept of relation was introduced and used to illustrate how the language of set theory can be used to build a framework of precise definitions for more complex structures. The usefulness of relations goes beyond this, however. After sets, relations are probably the most basic and extensively used tools of mathematics. All functions are relations. The connectives $\in$, $\subseteq$, and $=$ of set theory are relations. Mathematical induction, on which virtually all of mathematics rests, is based on ordering relations. Applications of relations are found throughout computer science and engineering, including relations between the inputs and outputs of computer programs, relations between data attributes in databases, and relations between symbols in computer languages.

This chapter is devoted to a more detailed treatment of relations. We will review binary relations and special properties of binary relations, then look at some applications of binary and n-ary relations. We will begin by introducing a useful way of viewing binary relations as directed graphs.

Consider the diagram in Figure 4-1. This is a **directed graph** representing the kinship relation "is parent of" between eleven people. Each person is represented by a point, and an arrow is drawn from each parent to each of the respective children. Thus Terah has three children

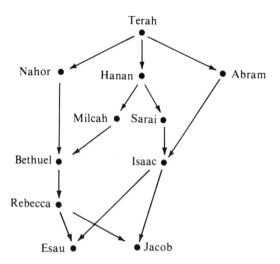

Figure 4-1. The relation "is parent of" on a set of people.

shown: Nahor, Hanan, and Abram. The binary relation represented by this directed graph is the set of pairs:

{(Terah, Hanan), (Terah, Nahor), (Terah, Abram), (Hanan, Milcah), (Hanan, Sarai), (Abram, Isaac), (Milcah, Bethuel), (Nahor, Bethuel), (Sarai, Isaac), (Bethuel, Rebecca), (Isaac, Esau), (Isaac, Jacob), (Rebecca, Esau), (Rebecca, Jacob)}

For many purposes, relations appear easier to understand when viewed as directed graphs than when viewed as sets of ordered pairs. It is probably easier to see, for example, that Isaac and Bethuel are cousins from the diagram than it is from the ordered pairs.

Definition 4.1.1.* A pair of sets $G = (V,E)$ is a directed graph (**digraph**) if $E \subseteq V \times V$. The elements of V are called **vertices** and the elements of E are called **edges.** An edge (x,y) is said to be *from* x to y, and is represented in a diagram by an arrow with the tail at x and the

---

*Unlike other areas of mathematics, such as geometry and algebra, where terminology has become fairly standard, the notation and definitions used in graph theory vary much from author to author. The terminology used here is widely accepted, but the reader should be prepared to encounter differences between this book and other books on the subject.

head at y. Such an edge is said to be **incident from** x, **incident to** y, and *incident on* both x and y. If there is an edge in E from x to y we say x is *adjacent to* y. The number of edges incident from a vertex is called the **out-degree** of the vertex and the number of edges incident to a vertex is called the **in-degree.** An edge from a vertex to itself is called a **loop,** and will ordinarily be permitted. A digraph with no loops is called **loop-free** or **simple.** (Though some authors prefer to require all digraphs to be loop-free, we will find loops useful in expressing certain binary relations.) Unless specified to the contrary, all directed graphs are presumed to be finite; that is, V is assumed to be a finite set.

Example 4.1.1.   For the graph shown in Figure 4-1, the edge (Terah, Abram) is from Terah to Abram. There are two edges incident on Abram. The edge (Terah, Abram) is incident to Abram and (Abram, Isaac) is incident from Abram. No vertex has in-degree or out-degree greater than two, except for Terah, which has out-degree three.

Note that for *any* digraph (V, E), E is a binary relation on V. Likewise, *any* binary relation $R \subseteq A \times B$ may also be viewed as a digraph $G = (A \cup B, R)$. In this sense the notion of binary relation on a set and the notion of digraph are equivalent.

In the remainder of this book, the terminology for digraphs and the terminology for binary relations will be used interchangeably. A relation will be treated as a digraph when concepts that are traditionally graph-theoretic, such as path properties, are involved. On the other hand, a digraph will be treated as a relation when properties that are tradition-ally phrased in terms of relations are involved, such as irreflexivity and transitivity.

Digraphs are a special case of a more general type of graph called a *directed multigraph*. When more than one edge is permitted incident from one vertex to another vertex, then the result is a directed multi-graph and then two or more edges incident from a vertex **x** to a vertex **y** are called *multiple edges*.

In addition to digraphs and directed multigraphs, there are also nondirected graphs and nondirected multigraphs where the direction of edges is not considered. Since the term **graph** applies to all of these, we must be careful when we use it that we are clear about the kind of graph we mean. Fortunately, there are a few contexts where, due to the similarity of these different kinds of graphs, it is not necessary to make such distinctions. For example, every form of graph may be viewed as a pair (V, E) of a set of vertices and a set of edges, so that the definition of **subgraph** given below is adequate for all kinds of graphs. That is why

Figure 4-2. A proper subgraph of the digraph in Figure 4-1.

the term graph is used instead of digraph. By contrast, in definitions and the statements of theorems that do not apply to all kinds of graphs we will be careful to specify which kind of graph we intend.

Definition 4.1.2.   A graph $G^1 = (V^1, E^1)$ is a **subgraph** of a graph $G = (V, E)$ if $V^1 \subseteq V$ and $E^1 \subseteq E \cap (V^1 \times V^1)$. G^1 is a **proper** subgraph of G if $G^1 \neq G$.

Example 4.1.2.   The digraph shown in Figure 4-2, comprised of the vertices {Sarai, Isaac, Abram} and the edges {(Sarai, Isaac), (Abram, Isaac)} is a proper subgraph of the digraph shown in Figure 4-1.

Frequently it happens that one wishes to ignore differences between graphs that have only to do with the *namings* of vertices in the vertex sets. For this purpose, the concept of **graph isomorphism** is introduced. Isomorphic means "having the same form."

Definition 4.1.3.   Two graphs $G_1 = (V_1, E_1)$ and $G_2 = (V_2, E_2)$ are **isomorphic** if there is a one-to-one onto function f: $V_1 \rightarrow V_2$ that preserves adjacency. By preserving adjacency, we mean for digraphs that for every pair of vertices v and w in V_1, (v,w) is in E_1 iff $(f(v), f(w))$ is in E_2. Another way of stating this is

$$E_2 = \{(f(v), f(w)) \mid (v,w) \in E_1\}.$$

In this case we call f a (directed graph) **isomorphism** from G_1 to G_2. An **invariant** of graphs (under isomorphism) is a function g on graphs such that $g(G_1) = g(G_2)$ whenever G_1 and G_2 are isomorphic.

Some examples of invariants of digraphs are the number of vertices, the number of edges, and the "degree spectrum," which is the collection of pairs (i,j), where i is the in-degree of a vertex and j is the out-degree of the same vertex, with one pair for each vertex of the graph. Note that graphs that are isomorphic have the same values of invariants, but that there is no known set of invariants that can guarantee that two graphs are isomorphic.

Example 4.1.3.   The digraphs in Figure 4-3 are isomorphic. They both have five vertices, eight edges, and degree spectrum (2,1), (2,1), (2,1), (2,1), (0,4). A digraph that is *not* isomorphic to either of the

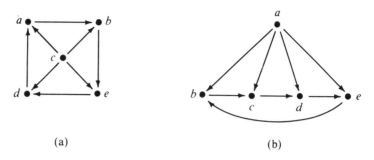

Figure 4-3. Isomorphic digraphs.

digraphs in Figure 4-3 is shown in Figure 4-4. Note that this graph, too, has five vertices, eight edges, and degree spectrum (2,1),(2,1), (2,1),(2,1),(0,4).

Exercises for Section 4.1

1. Draw the digraph of each of the following relations.
   (a) The relation "divides," defined by "a divides b iff there exists a positive integer c such that $a \cdot c = b$," on the integers $\{1,2,3,4,5,6,7,8\}$.
   (b) The relation $\subseteq$ on all the nonempty subsets of the set $\{0,1,2\}$.
   (c) The relation $\neq$ on the set $\{0,1,2\}$.

2. Specify the in-degree and out-degree of each vertex in the digraph in Figure 4-1.

3. (a) Give the edge sets for the digraphs (a) and (b) shown in Figure 4-3 as sets of ordered pairs.
   (b) Give a specific isomorphism from (a) to (b), described as a set of ordered pairs [vertex in (a), vertex in (b)].
   (c) How many isomorphisms are there between (a) and (b)?

4. Prove that the digraph shown in Figure 4-4 cannot be isomorphic to either of the ones shown in Figure 4-3.

Figure 4-4.

Figure 4-5.

5. (a) Draw all the subgraphs with four vertices and two edges of the digraph in Figure 4-5, *up to isomorphism.* That is, do not repeat graphs that are isomorphic.

   (b) Do the same for all subgraphs with four vertices and three edges.

   (c) Do the same for all subgraphs with four vertices and four edges.

   (d) Do the same for all subgraphs with four vertices and five edges.

6. Prove that each of the invariants cited in this section is truly an invariant:

   (a) the number of vertices;

   (b) the number of edges;

   (c) the degree spectrum.

7. Suppose G is an arbitrary digraph with n vertices. What is the largest possible number of isomorphisms between G and itself? (Choose G to maximize this number.)

8. Suppose G is an arbitrary digraph with n vertices. What is the largest possible number of distinct subgraphs with k vertices that G may have? (Treat isomorphic subgraphs as distinct. Choose G to maximize this number.)

Selected Answers for Section 4.1

1.

(a)

(b)

(c)

2. | vertex | in-degree | out-degree |
|---|---|---|
| Terah | 0 | 3 |
| Nahor | 1 | 1 |
| Hanan | 1 | 2 |
| Abram | 1 | 1 |
| Milcah | 1 | 1 |
| Sarai | 1 | 1 |
| Bethuel | 2 | 1 |
| Isaac | 2 | 2 |
| Rebecca | 1 | 2 |
| Esau | 2 | 0 |
| Jacob | 2 | 0 |

3. (a) $\{(a,b),(b,e),(e,d),(d,a),(c,a),(c,d),(c,b),(c,e)\}$
        $\{(a,b),(a,c),(a,d),(a,e),(b,c),(c,d),(d,e),(e,b)\}$

   (b,c) There are four possible isomorphisms. Vertex c in (a) must
         correspond to vertex a in (b), since these are the only vertices
         with out-degree four and in-degree zero. Once one of these
         other vertices in (a) is paired with a corresponding vertex in
         (b), the rest of the correspondences are fully determined by
         the definition of isomorphism. The isomorphisms are thus:

$$\{(c,a),(b,b),(e,c),(d,d),(a,e)\}$$
$$\{(c,a),(b,c),(e,d),(d,e),(a,b)\}$$
$$\{(c,a),(b,d),(e,e),(d,b),(a,c)\}$$
$$\{(c,a),(b,e),(e,b),(d,c),(a,d)\}$$

4. There are, of course, many proofs. They generally proceed by contradiction. Here is one:

The digraph of Figure 4-4 has a sequence of edges $(a,d),(d,a)$ that goes from vertex a to d, and back to a again. Suppose f is an isomorphism from this digraph to digraph (a) in Figure 4-3. Then $(f(a),f(d)),(f(d),f(a))$ must also be a sequence of edges leading from $f(a)$ to $f(d)$ and then back to $f(a)$. Since there is no such sequence of edges in digraph (a), no such isomorphism f can possibly exist.

5. (a)

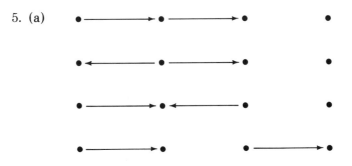

(d) Try taking away one edge in as many ways as possible:

6. Suppose that f is an isomorphism from G_1 to G_2.
   (a) It follows from the definition of one-to-one onto function that f gives a one-to-one correspondence between the vertices of G_1 and G_2.
   (b) There is a one-to-one correspondence between edges of G_1 and edges of G_2 given by: $\hat{f}((x,y)) = (f(x),f(y))$.
   (c) Suppose $v_1,\ldots,v_n$ is a list of the vertices of G_1, ordered in decreasing order of in-degree, and within vertices of equal in-degree, by increasing out-degree. For any v_i, $f(v_i)$ has the same in-degree in G_2 as v_i has in G_1, and likewise for out-degree. This is because every edge (v_i,v_j) in G_1 corresponds uniquely to an edge $(f(v_i),f(v_j))$ in G_2, and for every edge in G_2 there is

such a corresponding edge in G_1. Thus the degree spectrum (in-degree(v_1),out-degree(v_1)),. . .(in-degree(v_n),out-degree(v_n)) must be identical to the degree spectrum (in-degree($f(v_1)$),out-degree($f(v_1)$)),. . .(in-degree($f(v_n)$),out-degree($f(v_n)$)).

7. There are $n!$ isomorphisms between the complete digraph with loops on n vertices, $(\{1,. . .,n\},\{1,. . .,n\}\times\{1,. . .,n\})$, and itself.

8. The complete digraph on n vertices gives the largest number of subgraphs. There are $\binom{n}{k}$ distinct subsets of size k that may be formed from the n vertices. For each set of k vertices there are k^2 possible edges, and 2^{k^2} distinct subsets that may be formed from these edges. There are thus $\binom{n}{k} \cdot 2^{k^2}$ possible subgraphs with k vertices.

4.2   SPECIAL PROPERTIES OF BINARY RELATIONS

The following special properties, which may be possessed by a binary relation, occur often enough in mathematics that they have names:

1. **Transitivity** $\forall x,y,z,$ if $x\,R\,y$ and $y\,R\,z$, then $x\,R\,z$;

2. **Reflexivity** $\forall x,$ $x\,R\,x$;

3. **Irreflexivity** $\forall x,$ $x\,\not\!R\,x$;

4. **Symmetry** $\forall x,y$ if $x\,R\,y$, then $y\,R\,x$;

5. **Antisymmetry** $\forall x,y$ if $x\,R\,y$ and $y\,R\,x$, then $x = y$;

6. **Asymmetry** $\forall x,y$ if $x\,R\,y$, then $y\,\not\!R\,x$.

It is interesting to restate these properties in terms of digraphs. A digraph (relation) is **transitive** if for any three vertices x, y, and z, whenever there is an edge from x to y and an edge from y to z there is also an edge from x to z. Figure 4-6 illustrates a transitive relation on the set $\{u,v,w,x,y,z\}$. Note that x, y, and z in the definition of transitivity need not be distinct. For example, the digraph in Figure 4-7 is *not* a transitive

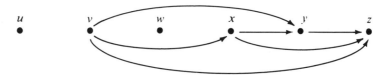

Figure 4-6. A transitive relation.

Figure 4-7. A relation that is not transitive.

Figure 4-8. A relexive relation.

Figure 4-9. An irreflexive relation.

Figure 4-10. A symmetric relation.

relation. [The edges (x,x), (y,y), and (z,z), which would be required by the definition, are missing.]

A digraph is **reflexive** if every vertex has an edge from the vertex to itself (sometimes called a self-loop). It is **irreflexive** if none of the vertices have self-loops. Figure 4-8 illustrates a reflexive relation on $\{x,y,z\}$ and Figure 4-9 illustrates a relation on $\{x,y,z\}$ that is irreflexive. (Note also that it is possible that a graph be neither reflexive nor irreflexive, as Figure 4-11 illustrates.)

A digraph is **symmetric** if for every edge in one direction between points there is also an edge in the opposite direction between the same two points. Figure 4-10 shows a symmetric relation on $\{x,y,z\}$. A digraph is **antisymmetric** if no two distinct points have an edge going between them in both directions. Figure 4-11 illustrates an antisymmetric relation. An **asymmetric** digraph is still further restricted. Self-loops are not even permitted. The digraph in Figure 4-11 is *not* asymmetric, since the self-loop at x violates the asymmetry property.

Properties (1)–(6) are used in various combinations as axioms to define special kinds of binary relations that are useful in mathematics. A binary relation that is transitive, reflexive, and symmetric is called an **equivalence** relation. A binary relation that is transitive, reflexive, and

Figure 4-11. An antisymmetric relation.

antisymmetric is called a **partial ordering** relation. Several of these properties were introduced in Chapter 1 and will be seen again more than once. In the next section, we shall consider an application of properties (1)–(6) to a relation that is neither a partial ordering nor an equivalence relation.

Exercises for Section 4.2

1. Give an example of a nonempty set and a relation on the set that satisfies each of the following combinations of properties; draw a digraph of the relation.
   (a) symmetric and transitive, but not reflexive.
   (b) symmetric and reflexive, but not transitive.
   (c) transitive and reflexive, but not symmetric.
   (d) transitive and reflexive, but not antisymmetric.
   (e) transitive and antisymmetric, but not reflexive.
   (f) antisymmetric and reflexive, but not transitive.

2. For each of the following digraphs, state which of the special properties (1–6) are satisified by the digraph's relation.

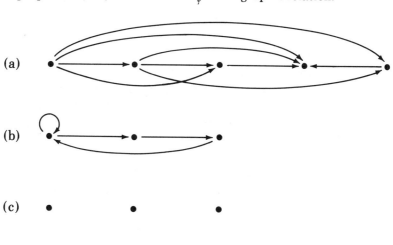

3. Prove or disprove each of the following:
   (a) Asymmetry implies antisymmetry.
   (b) Symmetry and transitivity together imply reflexivity.
   (c) Antisymmetry implies asymmetry.
   (d) Asymmetry and symmetry together imply transitivity.

4. Which of the properties (1–6) must apply to every subgraph of a digraph if it applies to the whole graph?

5. Draw six digraphs, each by making the minimum number of changes to the digraph shown below required to make it satisfy one of the properties (1–6).

6. Suppose that R is a symmetric and transitive relation defined on a set A. Consider the following argument. If $(a,b) \in R$ then symmetry implies that $(b,a) \in R$. However, transitivity then implies that $(a,a) \in R$. Thus, R is reflexive. Find the fallacy in this argument.

7. Consider the relation R defined on the set of positive integers by $(x,y) \in R$ if x divides y. Which of the 6 properties does R satisfy?

8. Prove that if R is a transitive and irreflexive relation on a set A, then R is antisymmetric and asymmetric.

9. Let R and S be relations on a set A. Prove or disprove the following:
   (a) If R and S are asymmetric, then $R \cup S$ and $R \cap S$ are asymmetric.
   (b) If R and S are antisymmetric, then $R \cup S$ and $R \cap S$ are antisymmetric.

10. State the negation of:
    (a) Relation R is both reflexive and transitive.
    (b) Relation R is either symmetric or transitive.

11. State the contrapositive of:
    (a) If a relation R is reflexive, symmetric, and transitive, then R is an equivalence relation.
    (b) If a relation R is reflexive, then $R = R^{-1}$.

12. Let A be the set of nonzero rational numbers. For $a,b \in$ A, define aRb if a/b is an integer. Prove that R is reflexive and transitive but not symmetric, antisymmetric, or asymmetric.

13. Let A be the set of all nonzero real numbers. For $a,b \in$ A, define $(a,b) \in R$ if a/b is a rational number. Prove that R is an equivalence relation on A.

14. Let A be the set of rational numbers. For $a,b \in A$, define $(a,b) \in R$ if $a-b$ is an integer. Prove that R is an equivalence relation on A.

15. Let D be the diagonal of A $\times$ A, that is, $D = \{(a,a) \mid a \in A\}$ is the identity relation on A. Prove that a relation R on A is:
    (a) reflexive iff $D \subseteq R$
    (b) irreflexive iff $D \cap R = \varnothing$, the empty set.

Selected Answers for Section 4.2

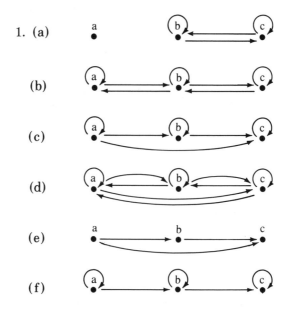

1. (a)

   (b)

   (c)

   (d)

   (e)

   (f)

2. (a) transitivity, irreflexivity, antisymmetry, asymmetry
   (b) antisymmetry
   (c) transitivity, irreflexivity, antisymmetry, asymmetry, symmetry
   (d) transitivity, reflexivity, symmetry, antisymmetry

3. (a) If x R y implies y Ɍ x, then (x R y and y R x) must be false. Thus, vacuously, (x R y and y R x) implies $x = y$. (Just as "if A then B" must always be true when A is false.)
   (b) Counterexample:

   (c) Counterexample:

(d) Vacuous, like (a), since a digraph that is symmetric and asymmetric must have no edges.

4.2.1 Big *O* Notation

There is a way of comparing the "sizes" of functions, commonly called "big oh notation," that has proven very useful in many areas of applied mathematics, especially in analysis of the running times of algorithms. This notation is particularly interesting because it expresses a relation between functions that is neither a partial ordering nor an equivalence relation, though it is sometimes mistakenly applied as if it were one or the other.

Definition 4.2.1. Let $g: N \to R$ be a function from the set of nonnegative integers into the real numbers. $O(g)$ denotes the collection of all functions $f: N \to R$ for which there exist constants c and k (possibly different for each f) such that for every $n \geq k, |f(n)| \leq c \cdot |g(n)|$. If f is in $O(g)$ we say that f is **of order** g.

It is worthwhile taking note that the definition of Big O can be simplified, by dropping the absolute values, when the functions involved are well behaved. This is expressed by the following lemma.

Lemma 4.2.1. If there exists a constant k_1 such that for every $n \geq k_1, f(n) \geq 0$ and $g(n) \geq 0$, then f is in $O(g)$ if and only if there exist constants c and k_2 such that for every $n \geq k_2, f(n) \leq c \cdot g(n)$.

Proof. Since $f(n)$ and $g(n)$ are both nonnegative for $n \geq k_1$, we have $|f(n)| = f(n)$ and $|g(n)| = g(n)$ for $n \geq \max(k_1, k_2)$. The lemma then follows immediately from Definition 4.2.1. $\square$

We shall use this lemma implicitly in the examples below, where all the functions are positive valued for sufficiently large arguments.

Example 4.2.1. Consider the functions $f(n) = 2^n$ and $g(n) = 3^n$. Since $2^n \leq 3^n$ for all $n \geq 0$, we know that f is in $O(g)$. (In this case we can choose $c = 1$ and $k = 0$ in the definition above.) On the other hand, g is not in $O(f)$. This can be shown by contradiction. Suppose g *is* in $O(f)$. Then there exist c and k such that for all (positive) $n \geq k, 3^n \leq c \cdot 2^n$, which implies

$$n \leq \frac{\log_e c}{\log_e \left(\frac{3}{2}\right)},$$

a contradiction.

Example 4.2.2. Consider the functions $f(n) = \log_2(n^x)$ and $g(n) = \log_e(n)$. Since $\log_2(n^x) = (x/\log_e(2)) \cdot \log_e(n)$, f is in $O(g)$. (This can be seen by choosing $c = x/\log_e(2)$ and $k = 1$.) Similarly, g is in $O(f)$. (This can be seen by choosing $c = \log_e(2)/x$ and $k = 1$.)

Example 4.2.3. Show that $5n^3 - 6n^2 + 4n - 2$ is in $O(n^3)$. Since $|5n^3 - 6n^2 + 4n - 2| \le 5n^3 + 6n^3 + 4n^3 + 2n^3 \le 17n^3$ for $n \ge 1$, we choose $c = 17$ and $k = 1$ and conclude $5n^3 - 6n^2 + 4n - 2 \in O(n^3)$.

Example 4.2.4. Consider the functions

$$f(n) = \begin{cases} 2^n \text{ if } n \text{ is an even integer} \\ n \text{ otherwise} \end{cases}$$

and

$$g(n) = \begin{cases} 2^n \text{ if } n \text{ is an odd integer} \\ n \text{ otherwise.} \end{cases}$$

These functions are pathological, as their definitions might lead one to suspect. Suppose that f is in $O(g)$. Then for some c and k it would be true that for all (positive) $n \ge k$, $f(n) \le c \cdot g(n)$. Taking n to be even, this would mean that $2^n \le c \cdot n$, which is a contradiction. Similar reasoning, considering the case when n is odd, shows that g cannot be in $O(f)$.

Note that in practice it is customary to extend this notation to formulas which define functions, so that the relationships described in the first two examples above would be expressed:

2^n is in $O(3^n)$, but 3^n is not in $O(2^n)$;

$\log_2(n^x)$ is in $O(\log_e(n))$, and $\log_e(n)$ is in $O(\log_2(n^x))$.

This avoids introducing special names, like f and g, for the functions being compared. It is also customary to write $O(f(n))$ instead of $O(f)$, and to use $O(f(n))$ where a value of a function in $O(f)$ is intended but the exact function may be unknown. For example, since $1^2 + 2^2 + \cdots + n^2 = (1/3)n(n + 1/2)(n + 1) = 1/3n^3 + 1/2n^2 + 1/6n$, we could write $1^2 + 2^2 + \cdots + n^2 = (1/3)n^3 + O(n^2)$. In this case $O(n^2)$ stands for $g(n) = 1/2n^2 + 1/6n$, which is a specific function g in $O(n^2)$. (The fact that g is in $O(n^2)$ is a consequence of a theorem that is given in the exercises for this section.)

Theorem 4.2.1. The relation $Q = \{(f,g) \mid f: N \to R, g: N \to R, f \text{ is in } O(g)\}$ is reflexive and transitive, but is not a partial ordering or an equivalence relation.

Proof. We must show four things: (1) Q is reflexive; (2) Q is transitive; (3) Q is not antisymmetric; and (4) Q is not symmetric.

1. Since $|f(n)| \leq 1 \cdot |f(n)|$ for all $n \geq 1$, we know that (f,f) is in Q for all $f: R \to R$, and so Q is reflexive.

2. Suppose (f,g) and (g,h) are in Q. Then there exist c_1, c_2, k_1, and k_2 such that for all $n \geq k_1, |f(n)| \leq c_1 \cdot |g(n)|$, and for all $n \geq k_2, |g(n)| \leq c_2 \cdot |h(n)|$. It follows that for all n greater than or equal to the maximum of k_1 and $k_2, |f(n)| \leq c_1 \cdot |g(n)| \leq c_1 \cdot c_2 \cdot |h(n)|$, and so Q is transitive.

3. Example 4.2.2 shows that it is possible to have (f,g) and (g,f) in Q and still have $f \neq g$.

4. Example 4.2.1 shows that it is possible to have (f,g) in Q without (g,f) being in Q. $\square$

Exercises for Section 4.2.1

1. Draw a digraph of the O notation relation Q as defined in Theorem 4.2.1 on the functions $\{f_1,f_2,f_3,f_4,f_5,f_6,f_7,f_8\}$ where

$$f_1(n) = 1 \qquad\qquad f_5(n) = n^2$$
$$f_2(n) = n \qquad\qquad f_6(n) = n^3$$
$$f_3(n) = \log_2(n) \qquad f_7(n) = n^2 \cdot \log_2(n)$$
$$f_4(n) = n \cdot \log_2(n) \qquad f_8(n) = 2^n$$

2. Prove the following theorem:
   If $p(n) = a_m n^m + \cdots + a_1 n + a_0$ then $p(n)$ is in $O(n^m)$.
   (Hint: Use $c = |a_m| + \cdots + |a_0|$ and $k = 1$.)

3. Prove or disprove each of the following:
   (a) If f is in $O(g)$ and c is a positive constant, then $c \cdot f$ is in $O(g)$.
   (b) If f_1 and f_2 are in $O(g)$ then $f_1 + f_2$ is in $O(g)$.
   (c) If f_1 is in $O(g_1)$ and f_2 is in $O(g_2)$ then $f_1 \cdot f_2$ is in $O(g_1 \cdot g_2)$.
   (d) If f_1 is in $O(g_1)$ and f_2 is in $O(g_2)$ then $f_1 + f_2$ is in $O(g_1 + g_2)$.

4. Big O notation is frequently used in the analysis of algorithms, where the functions involved are not always exactly known, but are known to take on only positive values and can be bounded by recurrence relations. The functions involved are ordinarily also known to be ***monotone increasing***—that is, $x \leq y$ implies $f(x) \leq f(y)$ for all x and y. Suppose a, b, and c are nonnegative integer constants. Prove that any positive valued monotone increasing function $T: N \to R$ that satisfies the recurrence relations

$$T(n) \leq b \text{ for } n = 1, \text{ and}$$
$$T(n) \leq a \cdot T(n/c) + b \cdot n \text{ for } n > 1$$

must be in

(a) $O(n)$ if $a < c$;
(b) $O(n \cdot \log(n))$ if $a = c$;
(c) $O(n^{\log_c a})$ if $a > c$.

Hint: The proof requires separate consideration of the three cases.

5. Suppose that $T : N \rightarrow R$ is a positive valued monotone increasing function that satisfies the recurrence relations

$$T(n) \leq b \text{ for } n = 1, \text{ and}$$
$$T(n) \leq a \cdot T(n/c) + b \text{ for } n > 1$$

Using big O notation, characterize T for each of the three cases:
(a) $a < c$
(b) $a = c$
(c) $a > c$

6. The simpler sorting algorithms are characterized by the fact that they require $f(n) \in O(n^2)$ comparisons to sort n items. On the other hand, most of the advanced sorting algorithms require $g(n) \in O(n \log_2 n)$ comparisons. It is instructive to compare n^2 and $n \log_2 n$. Construct a table that compares the values of n^2 and $n \log_2 n$ for the values $n = 10, 100, 1000,$ and 10000. Include the ratio $n^2 / n \log_2 n$ in your table.

7. Show that
(a) $n! \in O(n^n)$
(b) $\log_2 (n!) \in O(n \log_2 n)$
(c) $n \log_2 n \in O(\log_2 n!)$
(d) $2^n \in O(n!)$
(e) $n!$ is not in $O(2^n)$
(f) $(1 + 2 + 3 + \cdots + n) \in O(n^2)$
(g) $(1^2 + 3^2 + 5^2 + \cdots (2n - 1)^2) \in O(n^3)$

8. Prove the following for functions f, g, and h from N into the real numbers.
(a) $f \in O(g)$ iff $O(f) \subseteq O(g)$
(b) $f \in O(g)$ and $g \in O(f)$ iff the sets $O(f)$ and $O(g)$ are equal.
(c) If $f \in O(g)$ and $g \in O(h)$, then $f \in O(h)$.
(d) $O(f) = O(af)$ if a is a nonzero constant.

9. Prove the following. You may use the definition or exercises 2, 3, and 8.
(a) $n^k \in O(n^{k+1})$ where k is a positive integer.
(b) $3n + 2 \in O(n)$
(c) $9n^2 - 4n + 12 \in O(n^2)$
(d) $10n^2 + 5n - 6 \in O(n^3)$

(e) $3n + 2 \notin O(1)$

(f) $5(2^n) + n^2 \in O(2^n)$

(g) $10n^2 + 5n - 6 \notin O(n)$

(h) $O(\log_a n) = O(\log_b n)$ where a and b are integers greater than 1.

10. Give necessary and sufficient conditions for a function to be in $O(1)$.

11. Suppose that $f : N \to R$ is a function. We say that a number L is the limit of $f(n)$ as n approaches ∞ if for every number $\epsilon > 0$, there is a constant k such that for every $n \geq k, |f(n) - L| < \epsilon$. In this case we write $\lim_{n \to \infty} f(n) = L$.

   Use this definition of limit to prove the following:

   (a) $\lim\limits_{n \to \infty} 1/n = 0$

   (b) $\lim\limits_{n \to \infty} 1/n^2 = 0$

   (c) If $\lim_{n \to \infty} f(n) = A$ and $\lim_{n \to \infty} g(n) = B$ then $\lim f(n) + g(n) = A + B$, $\lim_{n \to \infty} f(n)g(n) = AB$, and $\lim_{n \to \infty} f(n)/g(n) = A/B$ (provided $B \neq 0$).

   (d) Suppose $\lim_{n \to \infty} f(n)/g(n) = L$. If $L \geq 0$, then $f \in O(g)$. If $L > 0$, then $f \in O(g)$ and $g \in O(f)$. If $L = 0$, then $f \in O(g)$ but $g \notin O(f)$.

12. Use exercise 11(d) to show the following:

   (a) $O(n + 1) = O(10n + 3) = O(n + \log_2 n) = O(n)$

   (b) $O(n^3 + n^2 + n + 3) = O(n^3 - 5) = O(n^3)$

   (c) $n^{1/2} \in O(n)$ but $n \notin O(n^{1/2})$

   (d) $O(\sqrt{n^2 + 1}) = O(n)$

   (e) $n^{10} \in O(n^{11} + n) = O(n^{11})$ but $n^{11} \notin O(n^{10})$

   (f) $2^n \in O(3^n)$ but $3^n \notin O(2^n)$

   (g) $2^n \in O(n2^n)$ but $n2^n \notin O(2^n)$

Answers for Section 4.2.1

1.

2. For $n > 1, |p(n)| = |a_m \cdot n^m + \cdots + a_1 \cdot n + a_0| \le |a_m| \cdot |n^m| + \cdots + |a_1| \cdot n + |a_0| \le (|a_m| + \cdots + |a_1| + |a_0|) n^m$.

3. (a) Suppose for all $n > k, |f(n)| < a \cdot |g(n)|$. Then $c \cdot |f(n)| \le c \cdot a \cdot |g(n)|$.

   (b) The crucial fact here is that $|x + y| \le |x| + |y|$. Suppose for all $n > k_1, |f_1(n)| < a_1 \cdot |g(n)|$ and for all $n > k_2, |f_2(n)| \le a_2 \cdot |g(n)|$. Then for all $n > \max (k_1, k_2), |f_1(n) + f_2(n)| \le |f_1(n)| + |f_2(n)| \le a_1 \cdot |g(n)| + a_2 \cdot |g(n)| \le (a_1 + a_2) \cdot |g(n)|$.

   (c) The crucial fact here is that $|x \cdot y| = |x| \cdot |y|$. Suppose for all $n > k_1, |f_1(n)| \le a_1 \cdot |g_1(n)|$ and for all $n > k_2, |f_2(n)| \le a_2 \cdot |g_2(n)|$. Then, for all $n \ge \max (k_1, k_2), |f_1(n) \cdot f_2(n)| = |f_1(n)| \cdot |f_2(n)| \le a_1 \cdot |g_1(n)| \cdot a_2 \cdot |g_2(n)| = (a_1 \cdot a_2) \cdot |g_1(n) \cdot g_2(n)|$.

   (d) This is true for positive valued functions, but is false if we consider negative valued functions. Consider $f(x) = x, g_1(x) = x^2$, and $g_2(x) = -x^2$. Clearly f is in $O(g_1)$ and in $O(g_2)$, due to the absolute values, but $O(g_1 + g_2) = O(0)$, and f cannot be in $O(0)$.

4. We know that T is positive valued, so we will use the formulation of Lemma 4.2.1. Since $n/c^i \le 1$ iff $\log_c n \le i$, we can expand this relation to

$$T(n) \le \sum_{i=0}^{\log_c(n)-1} (a/c)^i bn + a^{\log_c n} b.$$

If $0 \le a/c < 1$, the sum $\Sigma_{i=0}^{\infty} (a/c)^i$ is bounded by some constant k, so

$T(n) \le kbn + a^{\log_c n} b$. Since $a^{\log_c n} = n^{\log_c a}$, this means

$T(n) \le kbn + n^{\log_c a} b$. Since $\log_c a < 1$ for $a < c$, we have

$T(n) \le (k + 1) bn$.

If $a/c = 1$, we have $T(n) \le (\log_c n - 1) bn + n^{\log_c a} b = bn \log_c n$. If $a/c > 1$, we can use the solution $\Sigma_{i=0}^{k-1} x^i = (x^k - 1)/(x - 1)$, to obtain $T(n) \le bn((a/c)^{\log_c n} - 1)/(a/c - 1) + bn^{\log_c a}$, and from this, using $(a/c)^{\log_c n} = (n^{\log_c a})/n$, and some algebra, we can show that $T(n) \le k \cdot n^{\log_c a}$ for some constant k.

4.3   EQUIVALENCE RELATIONS

Equivalence relations, first defined in Chapter 1, are the primary tools employed in the process of **abstraction,** or selectively ignoring differences which are irrelevant to the purpose at hand. Within a given context, we say that two things are *equivalent* if the differences between them do not matter. For example, if a person wants to purchase something that costs one dollar, it would not matter whether he or she has a dollar coin or a dollar bill, or any particular dollar bill, since all of these would be accepted by merchants. That is, they are *equivalent* in purchasing power. On the other hand, there are other situations where these things are *not* equivalent. For example, a dollar coin weighs more than a dollar bill, so that if it were necessary to mail the dollar, one might not be willing to say the two were equivalent. At an extreme, a collector of rare bills would be very unlikely to say that all dollar bills are equivalent, since minor differences, such as serial numbers and errors in printing, may greatly affect the value of a bill as a collector's item. The meaning of equivalent thus depends on context and expresses the notion of being *the same in those respects relevant to the context.*

In terms of formal mathematics, a binary relation is an equivalence relation if it is reflexive, symmetric, and transitive. These properties express important aspects of being the same which are ordinarily taken for granted and are usually obvious for specific equivalence relations. For example, consider the relation "was born in the same month as." This relation is clearly reflexive, since each individual was born in the same month as himself. It is equally clearly symmetric—if individual A was born in the same month as individual B, then B was born in the same month as A. There is no question about transitivity either. Being told that A and B were born in the same month and B and C were born in the same month, no one is likely to deny that A and C must also have been born in the same month.

Another way of looking at equivalence relations is as ways of dividing things into classes. For the example above, the relation "was born in the same month as" partitions the set of all living human beings into twelve disjoint classes, corresponding to the twelve months of the year. Each of these equivalence classes consists of all the people who were born in a given month. Any time a set is partitioned into disjoint nonempty subsets an equivalence relation is involved. While sometimes it is more convenient to think in terms of relations and at other times it is more convenient to think in terms of partitions, the two notions are interchangeable.

Definition 4.3.1.   Given a set A, a **partition** of A is a collection P of disjoint subsets whose union is A. That is

1. for any $B \in P, B \subseteq A$;
2. for any $B, C \in P, B \cap C = \phi$, or $B = C$; and
3. for any $x \in A$ there exists $B \in P$ such that $x \in B$.

Definition 4.3.2.   Given any set A and any equivalence relation R on A, the **equivalence class** $[x]$ of each element x of A is defined $[x] =$

$\{y \in A \mid x\ R\ y\}$. Note that we can have $[x] = [y]$, even if $x \neq y$, provided $x\ R\ y$. That is, this notation does not give a unique "name" to each equivalence class.

Theorem 4.3.1. Given any set A and any equivalence relation R on A, $S = \{[x] \mid x \in A\}$ is a partition of A into disjoint nonempty subsets. Conversely, if P is a partition of A into nonempty disjoint subsets, then P is the set of equivalence classes for the equivalence relation E defined on A by $a\ E\ b$ iff a and b belong to the same subset of P.

Proof.

1. Clearly by the definition of $[x]$, $[x] \subseteq A$.
2. $[x] \cap [y] = \{z \in A \mid x\ R\ z$ and $y\ R\ z\}$. If this set is not empty, then for some $z \in A$, $x\ R\ z$ and $y\ R\ z$; but then, since R is transitive and symmetric, $x\ R\ y$, so that $[x] = [y]$.
3. For any $x \in A$, $[x] \in S$. □

Exercises for Section 4.3

1. Tell how many distinct equivalence classes there are for each of the following equivalence relations.
   (a) Two people are equivalent if they are born in the same week.
   (b) Two people are equivalent if they are born in the same year.
   (c) Two people are equivalent if they are of the same sex.

2. Suppose R is an arbitrary transitive reflexive relation on a set A. Prove that the relation E defined by $x\ E\ y$ iff $x\ R\ y$ and $y\ R\ x$ is an equivalence relation on A.

3. State definitions for equivalence relations that describe each of the following partitions.
   (a) the members of the Democratic Party; the members of the Republican Party; all the other people (three classes).
   (b) the negative integers; the nonnegative integers (two classes).
   (c) the sets $\{2i, 2i + 1\}$ for all $i \geq 0$ (infinitely many classes).
   (d) the even numbers; the odd numbers.

4. (a) Prove that isomorphism is an equivalence relation on digraphs.
   (b) How many equivalence classes are there for loop-free digraphs with three vertices?

5. How many different equivalence relations are there on a set with n elements for $n = 1,2,3,4$, and 5?

6. Prove that the relation $f\ E\ g$ iff f is in $O(g)$ and g is in $O(f)$ is an equivalence relation on functions from the real numbers to the real numbers.

7. Define the relation R on the set A of positive integers by $(a,b) \in R$ iff a/b can be expressed in the form 2^m, where m is an arbitrary integer.
   (a) Show that R is an equivalence relation.
   (b) Determine the equivalence classes under R.

8. Let A be the set of positive integers. Define R on A by $(a,b) \in R$ iff a divides b or b divides a. Show that R is reflexive and symmetric but not transitive.

9. A relation R on a set A is said to be *circular* if, for all $a,b,c \in A$, $(a,b) \in R$ and $(b,c) \in R$ imply that $(c,a) \in R$. Prove that R is reflexive and circular iff R is an equivalence relation.

10. Let R be a symmetric and transitive relation on a set A. Show that if for every $a \in A$ there exists $b \in A$ such that $(a,b) \in R$, then R is an equivalence relation. Show that the conclusion is false if symmetric is replaced by reflexive.

11. Let R be a binary relation on A. Let $S = \{(a,b) \mid (a,c) \in R$ and $(c,b) \in R$ for some $c \in A\}$. Show that if R is an equivalence relation on A, then S is also an equivalence relation.

12. Let R be a reflexive relation on a set A. Show that R is an equivalence relation iff (a,b) and (a,c) in R imply that $(b,c) \in R$.

13. Let R_1 and R_2 be two equivalence relations on a set A. Show that $R_1 \cap R_2$ is an equivalence relation on A, but that $R_1 \cup R_2$ need not be an equivalence relation.

14. (a) Let f be a function from a set A into A. Define a relation R on A by $(a,b) \in R$ iff $f(a) = f(b)$. Show that R is an equivalence relation.
   (b) One problem in computer assisted instruction is that of handling misspelled words or variant spellings. This is often done by a phonetic reduction routine, the first step of which is a function f that defines an equivalence relation on the set of letters of the alphabet as in part (a). For example, let f be the function from the alphabet into itself defined by the following scheme:

   $$f \downarrow \quad \begin{matrix} \text{ABCDEFGH I J KLMNOPQRSTUVWXYZ} \\ \text{ABCDABCHACCLMMABCRCDABH CAC} \end{matrix}$$

   That is, $f(A) = A$, $f(E) = A$, $f(F) = B$, etc ...
   Determine the equivalence classes determined by R where R is defined in part (a).

15. Use Theorem 4.3.1 to verify that the following relations are equivalence relations.
    (a) Let A be the set of all people in the world and let $(a,b) \in R$ iff a and b have the same father.
    (b) Let A be the set of positive integers and define R on A by $(a,b) \in R$ iff a and b have the same units digit in their decimal expansion.
    (c) Let A be the set of positive integers and define R on A by $(a,b) \in R$ iff a and b have the same remainder upon division by 7.
    (d) Let A be the set of positive integers and define R on A by $(a,b) \in R$ iff a and b have the same prime factors. (Thus, for example, $(6,12) \in R$.)
    (e) Let A be the set of positive integers and define R on A by $(a,b) \in R$ iff a and b have the same number of prime factors (counting multiplicities).
    (f) Let A be the set of all positive real numbers and define R on A by $(a,b) \in R$ iff $\lfloor a \rfloor = \lfloor b \rfloor$. (Recall $\lfloor a \rfloor$ is the greatest integer less than or equal to a.)
    (g) Let a be the set of all integers and define R on A by $(a,b) \in R$ iff $a^2 = b^2$.
    (h) Let A be the set of students in your discrete mathematics class and define R on A by $(a,b) \in R$ iff a and b sit in the same row in your classroom.
    (i) Let A be the set of United States citizens and define R on A by $(a,b) \in R$ iff a and b have the same zip code.
    (j) Let A be the set of all students at your university and define R on A by $(a,b) \in R$ iff a and b have the same grade point average.

16. Suppose that a relation R on a set A satisfies the following properties:
    (a) For each $a \in A$, there exists $b \in A$ such that $(a,b) \in R$.
    (b) If $(a,c) \in R$ and $(b,c) \in R$, then $(a,b) \in R$.

    Must R be an equivalence relation on A?

17. Show that the following relations on $R \times R$, where R is the set of real numbers, are equivalence relations and then give a geometric description of their equivalence classes:
    (a) $(a,b)\ R\ (c,d)$ iff $a^2 + b^2 = c^2 + d^2$.
    (b) $(a,b)\ S\ (c,d)$ iff $ab = cd$.
    (c) $(a,b)\ T\ (c,d)$ iff $a + b = c + d$.

18. Define the relation R on the set N of nonnegative integers as follows: $(a,b) \in R$ iff the sum of the (decimal) digits of a equals the sum of the digits of b. Show that R is an equivalence relation on N

and describe the equivalence class containing the number 202. List all the integers x in [202] where $0 < x < 100$.

19. Show that the relation R on the set Z of all integers is an equivalence relation, where R is defined as follows: $(a,b) \in R$ iff $a^2 - b^2$ is an integral multiple of 2. Find the equivalence classes determined by R.

20. Let $A = R \times R$, where R is the set of real numbers, and let $P = (x_0, y_0)$ be a given point in A. If a and b are in A, define aRb iff the distance from a to P is the same as the distance from b to P. Prove that R is an equivalence relation on A and that the equivalence classes are the circles in the plane having P as the center.

Selected Answers for Section 4.3

1. (a) 53, one class for each week
   (b) one class for each year in which a person was born, possibly an unbounded number
   (c) two, one for males and one for females

2. Transitivity: If $x \, E \, y$ and $y \, E \, z$, then $x \, R \, y$, $y \, R \, x$, $y \, R \, z$, and $z \, R \, y$. Since R is transitive, this means $x \, R \, z$ and $z \, R \, x$, which means $x \, E \, z$. Reflexivity: Since R is reflexive, $x \, R \, x$, and so also $x \, E \, x$. Symmetry: If $x \, E \, y$, then $x \, R \, y$ and $y \, R \, x$, which means $y \, E \, x$.

4.3.1   The Integers Modulo m

Since the most important use of equivalence relations is as a tool for abstraction (which means ignoring irrelevant details), real applications are ordinarily to structures that are rather complex. Such applications include the study of groups, including symmetries, and finite state machines. One application that is particularly useful in computer science, yet deals with familiar structures, is **modular arithmetic.** Due to the finite storage limitations and finite accuracy limitations of hardware arithmetic operations on computers, there is a frequent need for counting **modulo** some number m. A mundane example of modular counting can be seen in the common 12-hour clock, which counts seconds and minutes modulo 60 and hours modulo 12.

Definition 4.3.3.   Let m be any positive integer. The relation **congruence modulo** m [written $\equiv (\bmod \, m)$], is defined on the integers by $x \equiv y (\bmod \, m)$ iff $x = y + a \cdot m$ for some integer a.

Theorem 4.3.2.   For any positive integer m, the relation $\equiv (\bmod \, m)$ is an equivalence relation on the integers, and partitions the integers into m distinct equivalence classes: $[0], [1], \ldots, [m - 1]$.

Proof. That $\equiv$ (mod m) is an equivalence relation was proved in Example 1.3.7 of section 1.3.

If x is any integer, the division algorithm implies $x = mq + r$, where q and r are integers and $0 \le r < m$. Thus, $x \equiv r$ mod m and $[x] = [r]$. Thus, each equivalence class for this relation is one of the classes $[0], [1], \ldots, [m - 1]$. Moreover, if $[x] = [y]$ where $0 \le x \le y \le m - 1$, then $y = ma + x$ for some integer a. Therefore, $0 \le y - x = ma < m$ implies $a = 0$ and $x = y$. $\square$

The equivalence class $[r]$ is frequently called a *congruence class*, and the collection of congruence classes $[0], [1], \ldots, [m - 1]$ of integers with respect to the relation $\equiv$ (mod m) is customarily denoted by Z_m, for any positive integer m. That is, $Z_m = \{[0], [1], \ldots, [m - 1]\}$. Arithmetic on the integers can be extended to arithmetic on Z_m in a natural way:

$$[x] + [y] = [x + y];$$
$$-[x] = [-x];$$
$$[x] \cdot [y] = [x \cdot y].$$

Implicit in these definitions, of course, is the assumption that the operators so defined on Z_m are actually functions. This really should be *proven*.

Theorem 4.3.3. The operations $+$, $-$, and $\cdot$ on Z_m are well-defined functions.

Proof.

1. Suppose $x_1 \equiv x_2 (\text{mod } m)$ and $y_1 \equiv y_2 (\text{mod } m)$. We need to show that $x_1 + y_1 \equiv x_2 + y_2 (\text{mod } m)$. (This is the same as supposing that $[x_1] = [x_2]$ and $[y_1] = [y_2]$ and showing that $[x_1] + [y_1] = [x_2] + [y_2]$.) By the definition of $\equiv$ (mod m), we know that $x_1 = x_2 + a \cdot m$ for some a, and $y_1 = y_2 + b \cdot m$ for some b. It follows that $x_1 + y_1 = (x_2 + y_2) + (a + b) \cdot m$, so that $x_1 + y_1 \equiv x_2 + y_2 (\text{mod } m)$.
2. Suppose that $x_1 \equiv x_2 (\text{mod } m)$. Then $x_1 = x_2 + a \cdot m$ for some a, and $-x_1 = -x_2 + (-a) \cdot m$, so that $[-x_1] = [-x_2]$.
3. Suppose that $x_1 \equiv x_2 (\text{mod } m)$ and $y_1 \equiv y_2 (\text{mod } m)$. Then $x_1 = x_2 + a \cdot m$ for some a, and $y_1 = y_2 + b \cdot m$ for some b. It follows that $x_1 \cdot y_1 = x_2 \cdot y_2 + (x_2 \cdot b + y_2 \cdot a + a \cdot b \cdot m) \cdot m$, so that $x_1 \cdot y_1 \equiv x_2 \cdot y_2 (\text{mod } m)$. $\square$

Due to the way the operations of addition, multiplication, and subtraction are defined in Z_m, the usual laws of commutativity, associativity, and distributivity hold. That is,

1. $[x] + [y] = [y] + [x]$ (Addition is commutative.)
2. $[x] \cdot [y] = [y] \cdot [x]$ (Multiplication is commutative.)
3. $([x] + [y]) + [z] = [x] + ([y] + [z])$ (Addition is associative.)
4. $([x] \cdot [y]) \cdot [z] = [x] \cdot ([y] \cdot [z])$ (Multiplication is associative.)
5. $([x] + [y]) \cdot [z] = [x] \cdot [z] + [y] \cdot [z]$ (Multiplication distributes over addition.)

The proof of these assertions is left as an exercise.

The notation x **mod** m is ordinarily used to denote the smallest nonnegative integer y such that $x \equiv y \pmod m$. One way of looking at this is that x mod m is the *canonical representative* of the equivalence class of x. (Canonical means "according to the rule," which in this case is to select the smallest nonnegative member of the equivalence class.)

The reader should beware, however, of certain programming languages which use this notation in a different way when x or m is negative.

Definition 4.3.4.   Let x and y be integers. Recall that x **divides** y if there exists an integer z such that $x \cdot z = y$. The **greatest common divisor** of two positive integers is the largest positive integer that divides both of them. The notation gcd (x,y) denotes the greatest common divisor of x and y. Two integers are **relatively prime** if their greatest common divisor is 1.

Example 4.3.1.   The greatest common divisor of 237 and 204 is 3. Also, $\gcd(237,158) = 79$, and $\gcd(237,203) = 1$. Thus, 237 and 203 are relatively prime.

Theorem 4.3.4.   If x and m are relatively prime positive integers then, for every positive integer w, the equivalence classes $[w]$, $[w + x]$, $[w + 2 \cdot x], \ldots, [w + (m - 1) \cdot x]$ are all distinct.

In order to prove this theorem it is convenient to have a better characterization of the greatest common divisor, which we shall prove as a lemma.

Lemma 4.3.1.   Suppose x and m are positive integers and r is the smallest positive integer for which there exist integers c and d such that $r = c \cdot x + d \cdot m$. Then $r = \gcd(x,m)$.

Proof. We will first show that r divides x. Suppose $x = p \cdot r + q$, where $0 \le q < r$. (That is, q is the remainder when x is divided by r.) Then,

$$q = x - p \cdot r = x - p \cdot (c \cdot x + d \cdot m)$$
$$= (1 - p \cdot c) \cdot x + (-p \cdot d) \cdot m.$$

Since r is the smallest positive integer of this form, and $0 \le q < r$, it must be that $q = 0$, which is to say r divides x.

Interchanging x and m in the argument above gives a proof that r also divides m. To prove that r is the **largest positive integer that divides both x and m**, we will suppose there is another and show that it must be less than or equal to r.

Suppose s is a positive number that also divides x and m. Then, for some a and b, $x = a \cdot s$ and $m = b \cdot s$. Substituting $a \cdot s$ for x and $b \cdot s$ for m in $r = c \cdot x + d \cdot m$, we obtain $r = (c \cdot a + d \cdot b) \cdot s$. Because r and s are both positive, $c \cdot a + d \cdot b$ must also be positive. This means that $r \ge s$, since if $r < s$ then $c \cdot a + d \cdot b < 1$, which would be a contradiction.

Observe that the set S of all positive integers y such that $y = xs + mt$, for integers s and t, is a nonempty set since $x^2 + m^2$ is in S. Therefore, by the well ordering property of the positive integers, there exists a minimal element r in S. By Lemma 4.3.1, $r = \gcd(x, m)$.

Proof of Theorem 4.3.4.   Suppose $[w + x \cdot i] = [w + x \cdot j]$ and $0 \le j < i < m$. (We will show that this leads to a contradiction.) For some integer y, $w + x \cdot i = w + x \cdot j + y \cdot m$. Canceling out the w-terms and combining the x-terms, we obtain

(1) $x \cdot (i - j) = y \cdot m$. Since x and m are relatively prime, $\gcd(x, m) = 1$, and, by the preceding lemma, there exist c and d such that

(2) $c \cdot x + d \cdot m = 1$. From (1) we obtain

(3) $c \cdot x \cdot (i - j) - c \cdot y \cdot m = 0$. And from (2) we obtain

(4) $c \cdot x \cdot (i - j) + d \cdot m \cdot (i - j) = (i - j)$. Subtracting (3) from (4), we obtain

(5) $m \cdot (d \cdot (i - j) + c \cdot y) = i - j$.

This is a contradiction, since $0 < i - j < m$ implies $0 < d \cdot (i - j) + c \cdot y < 1$, and there is no integer between zero and one. $\square$

This theorem is the theoretical basis for a family of so-called "double hashing" algorithms for rapid average-case table searching. When searching a table of size m for a data item x, these techniques examine a sequence of locations $f(x) \bmod m$, $(f(x) + g(x)) \bmod m$,

$(f(x) + 2 \cdot g(x)) \bmod m \ldots (f(x) + (m - 1) \cdot g(x)) \bmod m$, called the "probe sequence" for x. (Here f and g are functions chosen for their ability to *randomize* the order of search.) For suitable f and g, and making suitable assumptions about the randomness of x and the number of occupied locations in the table, it is possible to prove that on the average fewer than 3 locations will need to be visited before finding either x or, if x is not in the table, an empty location. The importance of the theorem to this search strategy is that the sequence of locations searched can be relied upon to include every location of the table, provided $g(x)$ and m are relatively prime. In practice this is often achieved by choosing $g(x)$ to be odd and m to be a power of 2, or choosing m to be a prime number.

Exercises for Section 4.3.1

1. What positive integers less than 100 are equivalent to zero, modulo 10?

2. Consider the following attempts at defining division for Z_m. What is wrong with each?
   (a) $[x]/[y] = [x/y]$
   (b) $[x]/[y] = [\lfloor x/y \rfloor]$ (where $\lfloor x/y \rfloor$ is the greatest integer $\leq x/y$)
   (c) $[x]/[y] = [z]$ for some z such that $[y\,z] = [x]$

3. Verify that the commutative, associative, and distributive laws hold for addition and multiplication on Z_m.

4. Verify the following identities for Z_m.
   (a) $[0] + [x] = [x]$ ($[0]$ is an identity for addition.)
   (b) $[1] \cdot [x] = [x]$ ($[1]$ is an identity for multiplication.)
   (c) $[x] + ([-x]) = [0]$ ($-[x]$ is an additive inverse of $[x]$.)

5. Show that if $x = 2^k$ for some nonnegative k and y is odd, then x and y are relatively prime.

6. Suppose m were permitted to be *negative* in the definitions of this section. Which results would still hold? Which would not? How about $m = 0$?

7. Prove or disprove each of the following identities for the mod operator:
   (a) $(x \bmod m + y \bmod m) \bmod m = (x + y) \bmod m$.
   (b) $(-x) \bmod m = -(x \bmod m)$.

8. Define a relation R on the set of positive integers by $(a,b) \in R$ iff a and b are relatively prime. Which of the 6 basic properties of relations does R satisfy?

9. Define a relation R on the set Z of all integers by $(a,b) \in R$ iff $a^2 \equiv b^2 \bmod 7$. Show that R is an equivalence relation on Z. Determine the number of equivalence classes.

10. Find an integer x where $0 \le x \le 28$ such that
    (a) $[2][x] = [1]$ in Z_{29}
    (b) $3^{170} \equiv x \bmod 29$.

11. Prove that if p is a prime and $a^2 \equiv b^2 \bmod p$, then p divides $a + b$ or $a - b$.

12. (a) What is the last digit in the ordinary decimal expansion of 3^{400}? Hint: Note that $3^4 \equiv 1 \bmod 10$ so that $3^{4n} \equiv 1 \bmod 10$ for any integer n.
    (b) What is the last digit in the ordinary decimal expansion of 2^{400}?
    (c) What are the last 2 digits in the ordinary decimal expansion of 3^{400}? Hint: $3^{20} \equiv 1 \bmod 100$.

13. (a) Prove that any integer that is a square must have one of the following for its units digit: 0,1,4,5,6,9.
    (b) Prove that any fourth power must have one of 0,1,5,6 for its units digit.

14. Prove that 19 is not a divisor of $4n^2 + 4$ for any integer n.

15. Prove that if n is not a prime integer and $n > 4$, then $(n - 1)! \equiv 0 \bmod n$.

16. Determine whether or not the 328th day and the 104th day of the year fall on the same day of the week.

17. What is the remainder when 5^{1101} is divided by 6? When 3^{100} is divided by 8? When 5^{110} is divided by 4?

18. Suppose that x is an integer such that $x \equiv 1 \bmod 8$. Prove that $x^2 \equiv 1 \bmod 16$.

19. Show that the following statements are true:
    (a) $5 \equiv 21 \bmod 8$ (d) $117 \equiv 0 \bmod 39$
    (b) $-3 \equiv 7 \bmod 5$ (e) $8 \equiv -6 \bmod 7$
    (c) $57 \equiv 0 \bmod 19$ (f) $35 \equiv 17 \bmod 9$

20. Fill in the blanks at least 3 different ways:
    (a) _____ $\equiv 2 \bmod 7$ (b) $11 \equiv 3 \bmod$ _____
    (c) $6 \equiv -4 \bmod$ _____   (d) $5 \equiv$ _____ $\bmod 13$

21. Translate the following into statements about congruences:
    (a) The sum of two even integers is even.
    (b) 4 divides 36.
    (c) Any integer $n \ne 0$ divides itself.
    (d) The product of two odd integers is odd.

22. (Divisibility tests)
    (a) Use the congruence $10 \equiv 0 \bmod 2$ to obtain the following test for divisibility by 2: An integer x is divisible by 2 iff the units digit of x is divisible by 2.

(b) Use the congruence $10 \equiv 0 \bmod 2$ to obtain: An integer x is divisible by 5 iff the units digit of x is divisible by 5.

(c) Use the congruence $10^2 \equiv 0 \bmod 4$ to obtain: An integer x is divisible by 4 iff the number formed by the last 2 digits of x is divisible by 4.

(d) Generalize (c) to obtain divisibility tests for 8, 16, etc.

(e) Use the congruence $10 \equiv 1 \bmod 3$ to obtain: An integer x is divisible by 3 iff the sum of its digits is divisible by 3.

(f) Extend (e) to obtain a divisibility test for 9.

(g) Observe that $10^k \equiv -1 \bmod 11$ if k is odd. Then prove that an integer $x = a_n 10^n + \ldots + a_1 10 + a_0$ is divisible by 11 iff $a_0 - a_1 + a_2 - a_3 + \ldots + (-1)^n a_n$ is divisible by 11.

23. Using properties of congruences, solve the following congruences for x:

(a) $9546 \equiv x \bmod 9$

(b) $(16)\,(15)\,(22)\,(29)\,(31) \equiv x \bmod 7$

(c) $(1)\,(2)\,(3)\,(4)\,(5)\,(6) \equiv x \bmod 7$

(d) $(1)\,(2)\,(3)\,(4)\,(5)\,(6)\,(7)\,(8)\,(9)\,(10) \equiv x \bmod 11$

(e) $9546 \equiv x \bmod 11$

24. If a and m are relatively prime integers, prove that $ab \equiv ac \bmod m$ implies that $b \equiv c \bmod m$.

25. Prove that if $a \equiv b \bmod m$, where a and b are positive integers, then $\gcd(a,m) = \gcd(b,m)$.

26. (a) Prove that if x and m are relatively prime integers, where m is positive, then there is an integer c such that $xc \equiv \bmod m$.

(b) Give another proof of Theorem 4.3.4 using the operations of Z_m.

27. (a) Prove that two integers a and b are relatively prime iff there are integers x and y such that $1 = ax + by$.

(b) (*Euclid's Lemma*) If a, b, and c are integers such that a and c are relatively prime, and if c divides ab, then c divides b.

(c) If a prime p divides the product ab of two integers, then p divides a or p divides b.

(d) Suppose that a and m are integers where $m > 1$. Prove that the equation $ax \equiv 1 \bmod m$ has an integer solution iff a and m are relatively prime.

28. Show that some positive multiple of 63 has 452 as its final three digits.

29. (*Chinese Remainder Theorem*) If m and n are relatively prime integers greater than one, and a and b are arbitrary integers, then prove that there exists an integer x such that $x \equiv a \bmod m$ and $x \equiv b \bmod n$.

30. (a) Find the remainder when $1! + 2! + 3! + \ldots + 99! + 100!$ is divided by 12.
    (b) What is the remainder when $1^5 + 2^5 + 3^5 + \ldots + 100^5$ is divided by 4?

31. (*The Coconut Problem*). Three sailors are shipwrecked on a desert island and gather a pile of coconuts for food. They decide to retire for the night and divide the coconuts in the morning. One awakens and distrustfully decides to take his share during the night. He attempts to divide the pile of coconuts into three equal shares, but finds there is one coconut left over. He discards the extra one, takes his share to a hiding place, and goes back to sleep. Later, another of the sailors awakens and, upon dividing the pile of remaining coconuts into three equal shares, finds there is one left over. He also discards the extra one, hides his share, and goes back to sleep. Finally, the third sailor repeats the process once again. He finds, as the others did, that by discarding one coconut the remaining pile of coconuts can be divided into three equal shares. How many coconuts were in the original pile? Also solve the coconut problem for:
    (a) 4 sailors and 4 equal shares with a remainder of 1 after each division.
    (b) 5 sailors with a remainder of 1 after each division.

32. Describe all integer solutions to the equation $12x - 53y = 17$.

33. (a) When eggs in a basket are removed 2, 3, 4, 5, 6 at a time, there remain, respectively, 1, 2, 3, 4, 5 eggs. But when 7 are taken out at a time, none are left over. Find the smallest number of eggs that could have been in the basket.
    (b) Suppose that one egg remains in the basket when the eggs are removed 2, 3, 4, 5, or 6 at a time, but no eggs remain if 7 are removed at a time. Find the smallest number of eggs that could have been in the basket.

34. A band of 17 pirates stole a sack of gold coins. When they tried to divide the fortune into equal portions, 3 coins remained. In the ensuing brawl over who should get the extra coins, one pirate was killed. The wealth was redistributed, but this time an equal distribution left 10 coins. Again an argument developed in which another pirate was killed. But now the total fortune was evenly distributed among the survivors. What was the least number of coins that could have been stolen?

35. In a certain theater, the prices of admission are 2 dollars for men, one dollar for women, and 50 cents for children. How many of each attended if 100 people paid a total of 100 dollars?

36. How old was a man in 1900 if his age at death was 1/41 times the year of his birth?

37. About all we know of Diaphantus' life is his epitaph, from which his age at death is to be deduced: "Diaphantus spent one-sixth of his life in childhood, one-twelfth in youth, and another one-seventh in bachelorhood. A son was born five years after his marriage and died four years before his father at half his father's final age."

38. Find that integral solution to $13x + 21y = 1$ for which x is as small a positive integer as possible.

39. Three men possess a pile of money, their shares being 1/2, 1/3, and 1/6 of the pile. Each man takes some money from the pile until nothing is left. The first man then returns 1/2 of what he took, the second 1/3, and the third 1/6 of what he took. When the total so returned is divided equally among the 3 men it is found that each then possesses what he is entitled to. How much money was in the original pile, and how much did each man take from the pile?

Selected Answers for Section 4.3.1

1. 10, 20, 30, 40, 50, 60, 70, 80, 90

2. (a) Since x/y is not in general an integer, this is not always defined.
   (b) Take $m = 3$. Then $[2]/[2] = [1]$, but also $[3]/[2] = [1]$. It follows that $([3]/[2])[2] \neq [3]$, which is not what we expect from a division operator.
   (c) The problem is that in some cases no such z exists. If one does exist, however, we can show that $[x]/[y]$ is uniquely defined. In general, this definition has all the properties we expect of division if and only if m is prime.

5. Suppose x and y are as stated in the exercise. Suppose that $p = \gcd(x,y)$ and p is greater than 1. Then p must be a power of 2 greater than one, and $y = pq$ for some integer q greater than one. However, since p is even and the product of an even number with any other number is always even, y must be even. This is a contradiction.

4.4 ORDERING RELATIONS, LATTICES, AND ENUMERATIONS

A relation R on a set A is called a **partial order** on A when R is reflexive, antisymmetric, and transitive, and then the set A is called a **partially ordered set** or a **poset.** Frequently we will write $[A;R]$ to denote that A is partially ordered by the relation R.

Since the relation $\leq$ on the set of real numbers is the prototype of a partial order it is common to write $\leq$ to represent an arbitrary partial order on A and then, of course, the characteristic properties of a partial order can be described as follows:

1. $\forall\, a \in A, a \leq a$ (reflexivity)
2. $\forall\, a,b\ A$, if $a \leq b$ and $b \leq a$, then $a = b$ (antisymmetry)
3. $\forall\, a,b,c \in A$, if $a \leq b$ and $b \leq c$, then $a \leq c$ (transitivity).

Two elements a and b in A are said to be **comparable** under $\leq$ if either $a \leq b$ or $b \leq a$; otherwise they are **incomparable.** If every pair of elements of A are comparable, then we say that $[A;\leq]$ is **totally ordered** or that A is a **totally ordered set** or a **chain.** In this case, the relation $\leq$ is called a **total order.**

Example 4.4.1.

(a) Let U be an arbitrary set and let $A = \mathcal{P}\,(U)$ be the collection of all subsets of U. Then $[\mathcal{P}(U); \subseteq]$ is a poset but if U contains more than one element, then $\mathcal{P}(U)$ is not totally ordered under set inclusion. For if U contains the two distinct elements x and y, then $\mathcal{P}(U)$ contains two distinct elements $\{x\}$ and $\{y\}$, and these sets are incomparable under inclusion.

(b) If Z is the set of integers and $\leq$ is the usual ordering on Z, then not only is $[Z;\leq]$ partially ordered, but, more than that, it is totally ordered.

(c) Another familiar poset involves the set P of positive integers and the relation "divides" where we write $a \mid b$ iff a divides b or iff $b = ac$ for some integer c. Then $[P; \mid]$ is a partially ordered set. See Example 1.3.6 of Chapter 1 for verification. But $[P; \mid]$ is not totally ordered since 2 and 3 are incomparable, that is, 2 does not divide 3 and 3 does not divide 2.

(d) If n is a positive integer, let D_n denote the set of positive divisors of n. Then $[D_n; \mid]$ is a partially ordered set. For some values of n, $[D_n; \mid]$ is totally ordered but for other values of n $[D_n; \mid]$ is not totally ordered. For example, $[D_8; \mid]$ is totally ordered, but $[D_6; \mid]$ is not because 2 and 3 are incomparable. Of course, D_n is totally ordered under the different ordering relation $\leq$.

(e) If n is a positive integer, let I_n denote the set of integers $\{x \mid 1 \leq x \leq n\}$. Then $[I_n; \mid]$ is a poset but for $n \geq 3$ I_n is not totally ordered under $\mid$ because, for example, 2 and 3 are incomparable.

(f) The relation $<$ on Z is not a partial order because it is not reflexive.

Poset Diagrams

The diagrams we have described for digraphs can be used for partially ordered sets as well. Nevertheless, posets $[A;\leq]$ are traditionally represented in a more economical way by **poset** (or **Hasse**) **diagrams**. On a poset diagram there is a vertex for each element of A, but besides that, all loops are omitted eliminating explicit representation of the reflexive property. Moreover, an edge is not present in a poset diagram if it is implied by the transitivity of the relation. If we write $x < y$ to mean $x \leq y$ but $x \neq y$, then an edge connects a vertex x to a vertex y iff y **covers** x, that is iff there is no other element z such that $x < z$ and $z < y$.

Special Elements in Posets

Let $[A;\leq]$ be a poset and let B be a subset of A. Then

1. An element $b \in B$ is called the **least element** of B if $b \leq x$ for all $x \in B$. The set B can have at most one least element. For if b and b' were two least elements of B, then we would have $b \leq b'$ and $b' \leq b$. Hence, by antisymmetry $b = b'$.

    An element $b \in B$ is called the **greatest element** of B if $x \leq b$ for all $x \in B$. The set B can have at most one greatest element.
2. An element $b \in B$ is a minimal (maximal) element of B if $x < b$ ($x > b$) for no x in B. If the set B contains a least element b, then, of course, b is the only minimal element of B. However, if the set B contains a minimal element, it need not be the only minimal element of B.
3. An element $b \in A$ is called a **lower** (**upper**) bound of B if $b \leq x$ ($b \geq x$) for all $x \in B$.
4. If the set of lower bounds of B has a greatest element, then this element is called the **greatest lower bound** (or glb) of B; similarly, if the set of upper bounds of B has a least element, then this element is called the **least upper bound** (or lub) of B.

In a totally ordered set the concepts of minimal and least coincide, as do those of maximal and greatest. But in a partially ordered set (that is not totally ordered) these concepts are quite distinct. For example, the posets of Figure 4-12 (a),(b),(c),(d),(e), and (h) have a unique least element. The posets of (f) and (g) have several minimal elements, namely, 2 and 3 for (f) and 2, 3, and 5 for (g). The posets of (a),(b),(c),(d), and (g) have a unique greatest element while the poset $[I_{12}, |\,]$ of (e) has maximal elements 7, 8, 9, 10, 11, 12. Likewise, the poset of (f) has two maximal elements 12 and 18 while the poset of (h) has maximal elements 4, 6, and 9.

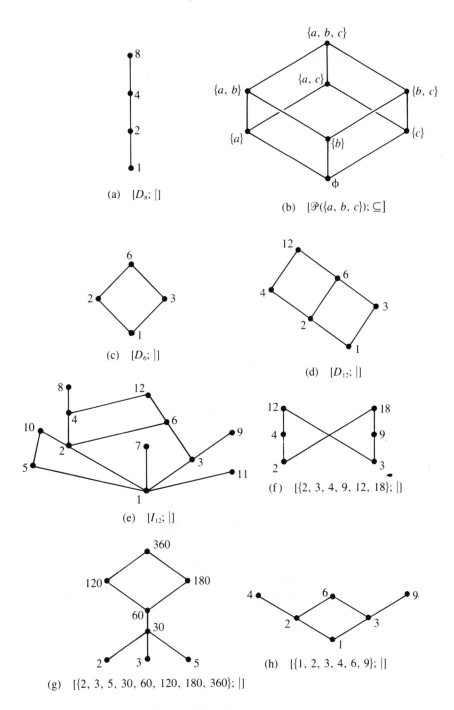

(a) $[D_8; |]$

(b) $[\mathscr{P}(\{a, b, c\}); \subseteq]$

(c) $[D_6; |]$

(d) $[D_{12}; |]$

(e) $[I_{12}; |]$

(f) $[\{2, 3, 4, 9, 12, 18\}; |]$

(g) $[\{2, 3, 5, 30, 60, 120, 180, 360\}; |]$

(h) $[\{1, 2, 3, 4, 6, 9\}; |]$

Figure 4-12. Poset diagrams.

Moreover the set $\{60,120,180,360\}$ of the poset of (g) has a minimal and least element (namely 60), whereas the subset $\{120,180,360\}$ has minimal elements (namely 120 and 180) but no least element. Note further that $\{120,180,360\}$ has a glb (namely 60) in contrast to $\{2,3,5,30\}$ which has no lower bounds at all and hence no glb. Moreover, the set $\{12,18\}$ in the poset of (f) has two lower bounds 2 and 3 but no glb. The elements 4 and 9 are not lower bounds of $\{12,18\}$.

Well-Ordered Sets

A total order $\leq$ on a set A is a **well order** if every nonempty subset B of A contains a least element. Moreover, $[A;\leq]$ is said to be **well ordered.**

Since A is a totally ordered set it follows that if a set B contains a minimal element, then B contains only one minimal element and moreover, this element is a least element of B.

Example 4.4.2.

(a) The poset $[N;\leq]$, where N is the set of nonnegative integers, is well ordered. In actual fact, the well-ordering property of $\leq$ is equivalent to the principle of mathematical induction, and is usually taken as an axiom.

(b) On the other hand, the poset $[Z;\leq]$ is not well ordered because the subset of negative integers does not contain a minimal element.

(c) The relation $\leq$ on the set Q of rational numbers is a total ordering but not a well ordering because some subsets of Q do not contain a minimal element. For example, the set P of positive rational numbers contains no minimal element for if $x \in P$, $x/2 < x$, and $x/2 \in P$.

(d) Any finite totally ordered set is well ordered.

Well-ordering relations are important in mathematics because they form the basis for proofs by generalized mathematical induction. If it is to be proven that a proposition $P(x)$ is true for all x in A, and there is a well ordering $\leq$ on A such that a_0 is the minimal element, then it is sufficient to prove two things:

1. $P(a_0)$ is true;
2. for any a in A, if $P(x)$ is true for all x in A such that $x < a$, then $P(a)$ is true.

Enumerations

Well-ordering relations are important in computing for yet another reason. To do computation on the elements of a set A, ordinarily it is necessary to enumerate them in some order $a_0, a_1, a_2, \ldots$. In other words, for the sets for which this is possible, we are transferring the well-ordered property of N to the set A. Let us illustrate.

Example 4.4.3. A well order on the set Z of integers can be constructed by listing the elements of N in ascending order and then pairing the elements of Z with those in N in a one-to-one correspondence. There are many such correspondences; we list one correspondence $f : N \to Z$ defined as follows:

$$
\begin{array}{ccccccccc}
N: & 0 & 1 & 2 & 3 & 4 & 5 & 6 & \ldots \\
   & \downarrow & \downarrow & \downarrow & \downarrow & \downarrow & \downarrow & \downarrow & \\
Z: & 0 & -1 & 1 & -2 & 2 & -3 & 3 & \ldots
\end{array}
$$

In other words, pair the even integers in N with the positive integers of Z and the odd integers of N with the negative integers of Z. Then we define a new ordering R of Z by aRb iff $f^{-1}(a) \le f^{-1}(b)$. Thus, $(-1)R(-3)$ because $1 < 5$.

The correspondence f above is a special case of an enumeration which we now define.

Definition 4.4.3. Let I be an "initial segment" of the nonnegative integers. That is, let $I = \{k \mid k \in N, \ k \le n\}$ for some constant n, or let $I = N$. A function $f : I \to S$ is an **enumeration** of S if f is onto; that is, for each $s \in S$ there exists an i such that $f(i) = s$. An enumeration has no repetitions if the function is one-to-one, that is, $f(i) = f(j)$ only if $i = j$.

Any set that has an enumeration is said to be **countable**. Sets that do not have an enumeration are said to be **uncountable,** or nondenumerable. The real numbers form a set which is totally ordered by $\le$ but uncountable.

For countable sets, the concepts of enumeration without repetition and well ordering are closely related, as can be seen by a simple construction: Suppose $f : I \to A$ is an enumeration of A. For each a in A, let $g(a)$ be the smallest integer in I such that $f(n) = a$. The relation $\le_f$ defined by $a \le_f b$ iff $g(a) \le g(b)$ is a well ordering of A for which each element $a \in A$ has only finitely many elements $b \in A$ such that $b \le_f$. Conversely, if R is a well ordering of a countable set A for which each $a \in A$ has only finitely many predecessors; that is, there are only finitely many elements $b \in A$ such that $(b, a) \in R$, then there is a unique enumeration without repetition f such that $\leq_f = R$. This enumeration is given by $f(0) = \min(A), f(i) = \min (A - \{f(j) \,|\, j < i\})$.

Lattices

We define a **join**-semilattice as a poset $[A;\leq]$ in which each pair of elements a and b of A have a least upper bound; we call this lub the *join* of a and b, and denote it by $a \vee b$.

Likewise we define a **meet**-semilattice as a poset in which each pair of elements a and b have a greatest lower bound; this glb is called the *meet* of a and b, and it is denoted by $a \wedge b$.

Thus, if $c = a \wedge b$, then c satisfies:

1. $c \leq a$ and $c \leq b$ (c is a lower bound of $\{a,b\}$),
2. If $d \leq a$ and $d \leq b$, then $d \leq c$ (c is the greatest lower bound of $\{a,b\}$).

Likewise if $c = a \vee b$, then c satisfies two similar properties by reversing the inequalities and changing the words *lower bound* to *upper bound*.

Definition 4.4.2. A **lattice** is a poset in which each pair of elements has a least upper bound and a greatest lower bound. In other words, a lattice is both a join-semilattice and a meet-semilattice.

Example 4.4.4.

(a) If U is any set, $[(\mathcal{P}(U); \subseteq]$ is a lattice in which the least upper bound of two subsets B and C of U is just $B \cup C$ and the greatest lower bound of $\{B,C\}$ is $B \cap C$. (Proof: Note that $B \cap C \subseteq B$ and $B \cap C \subseteq C$ so that $B \cap C$ is a lower bound of $\{B,C\}$. On the other hand, if $D \subseteq B$ and $D \subseteq C$, then $D \subseteq B \cap C$. Thus, $B \cap C$ is the greatest lower bound of $\{B,C\}$. Similarly, we can prove $B \cup C$ is the lub of $\{B,C\}$.)

(b) Any totally ordered set is a lattice in which $a \vee b$ is simply the greater and $a \wedge b$ is the lesser of a and b. For example, if R is the set of real numbers with the usual ordering $\leq$, then $a \vee b = \max\{a,b\}$ and $a \wedge b = \min\{a,b\}$.

(c) The poset $[P; \,|\,]$, where P is the set of positive integers, is a lattice in which $a \wedge b = \gcd(a,b)$ and $a \vee b = \text{lcm}(a,b)$ where gcd and lcm respectively stand for greatest common divisor and least common multiple. For instance, $6 \wedge 9 = 3$ and $6 \vee 9 = 18$.

(d) Of the posets in Figure 4-12, (a),(b),(c), and (d) are lattices, (e) and (h) are meet-semilattices, (g) is a join-semilattice, while (f) is neither. The poset (e) is not a join-semilattice since the elements 4 and 10 do not have a lub; (h) is not a join-semilattice since 4 and 6 have no lub; (g) is not a meet-semilattice since 2 and 3 have no glb.

Exercises for Section 4.4

1. Give definitions of functions that enumerate each of the following sets without repetitions:
    (a) the even nonnegative integers;
    (b) the nonnegative integers that are perfect squares;
    (c) the ordered pairs of nonnegative integers, $N \times N$.

2. Consider the relation Q defined on real functions of one variable by fQg iff $f: R \to R$, $g: R \to R$, and for all x, $f(x) \le g(x)$.
    (a) Prove that Q is a partial ordering.
    (b) Prove that it is not a total ordering.

3. Prove that if R is a partial ordering on a set S, then for $n \ge 2$, there cannot be a sequence $s_1, s_2, \ldots, s_n$ of distinct elements of S such that $s_1 \, R \, s_2 \, R \, s_3 \, R \, \ldots \, R \, s_n \, R \, s_1$.

4. Prove that the rational numbers are countable.

5. Consider the relation D on the integers defined by $(x,y) \in D$ iff there exists an integer a such that $x \cdot a = y$ (i.e., x divides y).
    (a) Prove that D is a partial ordering on the set A of integers > 1.
    (b) Prove that D is not a lattice ordering on A.
    (c) What are the minimal elements of the set A with respect to D?

6. Prove that any set that has an enumeration has an enumeration without repetitions.

7. Define a well ordering on the class of all (finite) digraphs with vertex sets chosen from some well ordered set A. Prove that it is a well ordering.

8. How many total orderings are there on a set with n elements?

9. Prove that if a set S has an enumeration, then so does every subset R of S.

10. For each of the following posets, draw a poset diagram and determine all maximal and minimal elements and greatest and least elements if they exist. Specify which posets are lattices.
    (a) $[D_{20}; |\,]$
    (b) $[D_{30}; |\,]$
    (c) $[I_{15}; |\,]$
    (d) $[A; \le]$ where $A = \{x \mid x$ is a real number and $0 < x \le 1\}$
    (e) $[A; |\,]$ where $A = \{2,3,4,6,8,24,48\}$

   (f) $[A; |]$ where $A = \{2,3,4,6,8,12,36,60\}$
   (g) $[A; |]$ where $A = \{2,3,4,6,12,18,24,36\}$

11. Given the subset B of a poset $[A; \leq]$, find, if they exist, all upper bounds of B, all lower bounds of B, the lub of B, and the glb of B. Determine whether or not the posets are meet- or join-semilattice.
   (a) $[\mathcal{P}(\{a,b,c\}); \subseteq]$ and $B = \mathcal{P}[\{a,b\})$
   (b) $[D_{30}; |]$ and $B = \{3,10,15\}$
   (c) $[I_{12}; |]$ and $B = \{2,7\}$
   (d) $[I_6; |]$ and $B = \{1,2,3,4,5\}$
   (e) $[D_{12} - \{1\}; |]$ and $B = \{2,6,10\}$
   (f) $[D_{12} - \{12\}; |]$ and $B = \{3,4\}$

12. Recall that if R is a relation on A, then R^{-1} is a relation on A defined by $R^{-1} = \{(a,b) \,|\, (b,a) \in R\}$.
   (a) Prove that if $[A; R]$ is a poset then $[A; R^{-1}]$ is also a poset (called the *dual* of $[A; R]$).
   (b) If $[A; R]$ is total ordered, then $[A; R^{-1}]$ is totally ordered.
   (c) If $[A; R]$ is lattice ordered, then $[A; R^{-1}]$ is lattice ordered.
   (d) Prove that R is a total order on A iff $A \times A = R \cup R^{-1} \cup \{(a,a) \,|\, a \in A\}$.

13. Define a well ordering on $N \times N$ using the well ordering on N, the set of nonnegative integers.

14. Define the relation C on $Z \times Z$ by $(a,b)C(c,d)$ iff $a \leq c$ and $b \leq d$.
   (a) Prove that C is a partial ordering but that C is not a total ordering.
   (b) Prove that C is a lattice ordering on $Z \times Z$.

15. Let $[A; \leq]$ be a poset and let B be a subset of A. Prove:
   (a) If b is a greatest element of B, then b is a maximal element of B.
   (b) If b is the greatest element of B, then b is a lub of B.
   (c) If $b \in B$ is an upper bound of B, then b is the greatest element of B.
   (d) If b and b' are greatest elements of B, then $b = b'$.

16. Let $[A; \leq]$ be poset where A is a finite set.
   (a) Prove that A contains at least one maximal element and at least one minimal element.
   (b) If, in addition, $[A; \leq]$ is a lattice, prove that A contains a least element and a greatest element.

17. Let R be a relation on a set A, and let B be a subset of A. Define the relation R' on B as follows: $R' = R \cap (B \times B)$. Then prove or disprove the following assertions.
   (a) If R is transitive on A, then R' is transitive on B.
   (b) If R is a partial ordering on A, then R' is a partial order on B.

(c) If R is a total ordering on A, then R' is a total ordering on B.

(d) If R is a well ordering on A, then R' is a well ordering on B.

18. Let $[A;\leq]$ be a lattice. Then prove $\forall a,b,c \in A$

(a) $a \wedge a = a$, $a \vee a = a$ (idempotent law).

(b) $a \wedge b = b \wedge a$, $a \vee b = b \vee a$ (commutative).

(c) $a \wedge (b \wedge c) = (a \wedge b) \wedge c$, $a \vee (b \vee c) = (a \vee b) \vee c$ (associative).

(d) $a \wedge (a \vee b) = a$, $a \vee (a \wedge b) = a$ (absorption).

(e) $a \leq b$ iff $a \wedge b = a$ iff $a \vee b = b$ (consistency).

19. **Algebraic definition of semilattice.** Suppose that A is a set and $\circ$ is a binary operation defined on A which is idempotent, commutative, and associative, that is, $a \circ a = a \; \forall \, a \in A$, $a \circ b = b \circ a \; \forall a,b \in A$, and $a \circ (b \circ c) = (a \circ b) \circ c \; \forall \, a,b,c \in A$.

(a) Prove the relation J is a partial ordering on A wh~ ~e J is defined by $(a,b) \in J$ iff $a \circ b = b$.

(b) Moreover, $[A;J]$ is a join-semilattice where the i~' ~ of a and b is $a \circ b$.

(c) Define the relation M on A defined b~   ~ ~M iff $a \circ b = a$ is

~J~ is a meet-semilattice

~...~ ~..~ ~meet of a and b is $a \circ b$.

20. Construct examples of the following:

(a) A nonempty totally ordered set in which some subsets do not contain a least element.

(b) A lattice $[L;\leq]$ where L is an infinite set but every totally ordered subset of L is finite.

(c) An infinite poset $[A;\leq]$ in which some subsets do not have a greatest element.

(d) An infinite poset $[A;\leq]$ with a subset B for which there is a glb but such that B does not contain a greatest element.

(e) An infinite poset $[A;\leq]$ with a subset B for which there is an upper bound for B but no lub.

21. (a) Determine necessary and sufficient conditions on n such that $[D_n;|\,]$ is totally ordered.

(b) What will be the nature of the poset diagrams for $[D_n;|\,]$ where $n = p_1^{k_1}p_2^{k_2}$ or where $n = p_1^{k_1}p_2^{k_2}p_3^{k_3}$ where p_1,p_2, and p_3 are distinct primes and k_1,k_2, and k_3 are positive integers?

22. Show that in the poset of real numbers with the usual order relation $\leq$, no element covers any other.

23. Draw as many as possible distinct poset diagrams of posets with one, two, or four elements. Label all greatest elements and least elements which appear.

24. The set of all n-digit binary sequences is denoted by B_n. Draw the poset diagram for $[B_4;\leq]$ where $(b_1,b_2,b_3,b_4) \leq (a_1,a_2,a_3,a_4)$ iff $b_i \leq a_i$

for each i. Determine the least and greatest elements of B_4. In $B_4 - \{(0,0,0,0),(1,1,1,1)\}$, find all maximal and minimal elements.

25. A poset $[A;\leq]$ may not be totally ordered but it may have totally ordered subsets. Any subset $\{a_1,a_2,\ldots,a_k\}$ of A such that $a_1 < a_2 < \cdots < a_k$ is a chain in $[A;\leq]$, and if there is no $b \in A$ such that $a_i < b < a_{i+1}$ then the chain is called a maximal chain with endpoints a_1 and a_k.

    (a) If $A = \mathcal{P}(U)$ is the set of all subsets of $U = \{a_1,a_2,\ldots,a_n\}$, determine all maximal chains in $[\mathcal{P}(U), \subseteq]$ with endpoints ϕ and U.

    (b) Determine all maximal chains with endpoints 1 and n in $[D_n; |\,]$ where $n = p_1^{k_1}p_2^{k_2}$ with p_1 and p_2 distinct prime integers and k_1 and k_2 are positive integers.

Selected Answers for Section 4.4

1. (a) $f(i) = 2i$
   (b) $f(i) = (i + 1)^2$
   (c) $f(i) = (i - r^2 - 1,r)$ for $r^2 < i \leq r^2 + r$
       $f(i) = (r, i - r^2 - r - 1)$ for $r^2 + r < i \leq (r + 1)^2$
   A complete solution to this exercise should also include a proof that each solution is correct.

3. Proof sketch: If such a sequence existed, then, by transitivity, we would have $s_1 R s_2, s_1 R s_3, \ldots, s_1 R s_n$, as well as $s_n R s_1$. Thus R could not be antisymmetric, since $s_1 \neq s_n$. (A better proof would prove that $s_1 R s_n$ by use of induction and the definition of transitivity.)

4. Proof sketch: To prove this, construct an enumeration of the rational numbers. (There is no need to avoid repetitions.) Every rational number can be expressed as a fraction of two integers, a/b. In exercise 1(c) we constructed an enumeration $f: I \to N \times N$, of the pairs of nonnegative integers. What we want to do here is to extend it to an enumeration of all pairs of integers, including negative integers. Let

$$\begin{aligned} g(4i) &= f(i) \\ g(4i + 1) &= (-a,b) \\ g(4i + 2) &= (a,-b) \\ g(4i + 3) &= (-a,-b) \end{aligned} \right\} \quad \text{where } f(i) = (a,b)$$

Now, verify that this is an enumeration of all the pairs of integers.

6. Proof sketch: Suppose $f: I \to D$ is onto D. Define $g(0) = f(0)$ and for $n \geq 0$ define $g(n + 1) = f(k)$, for the least k such that $f(k)$ is not in

$G(n) = \{g(i) \mid i \leq n\}$. Observe that the set $G(n)$ is always finite, so that there is such a k for every n. (A complete solution would also include a proof that g is an enumeration of D and has no repetitions.)

4.4.1 Application: Strings and Orderings on Strings

While the ordering relations $\leq$ and $<$ on the integers and real numbers are probably so familiar as to be taken for granted, there is another domain, of special importance in computer science, where the ordering relations are probably less familiar. This is the domain of character strings.

Definition 4.4.4. Let Σ be any finite set. A finite sequence of zero or more elements chosen from Σ is a **string** over Σ. In this context, Σ is called an **alphabet.** The length of a string w is denoted by $|w|$. The string of length 0 is denoted by Λ and called the **null string.** The set of all strings of length k is denoted by Σ^k. That is,

$\Sigma^0 = \{\Lambda\}$, and

$\Sigma^{k+1} = \{wa \mid w \in \Sigma^k \text{ and } a \in \Sigma\}$ for $k \geq 0$.

$\Sigma^* = \bigcup_{k \geq 0} \Sigma^k$, denoting the set of all strings over Σ.

$\Sigma^+ = \bigcup_{k > 0} \Sigma^k$, denoting the set of all nonnull strings over Σ.

Thus, for every $w \in \Sigma^k$, $|w| = k$. (Here wy denotes the **catenation** of strings w and y. If $w = w_1 \ldots w_n$ and $y = y_1 \ldots y_m$, $wy = w_1 \ldots w_n y_1 \ldots y_m$.)

Example 4.4.2. If $\Sigma = \{A,B,C,D,E,F,\ldots,X,Y,Z\}$, Σ^* includes all the words that can be written with capital letters and, in particular Σ^8 includes the string "ALPHABET."

There are many applications where an ordering relation on strings is needed. One ordering, commonly called "dictionary" or **lexicographic ordering,** is used to assist in searching for words in dictionaries and indices of books. This ordering is defined by *extending* a given total ordering $\leq_A$ on the alphabet A to a total ordering $\leq_L$ on A^* as follows: Let x and y be any two strings in A^*. Without loss of generality, suppose $|x| \leq |y|$. Let γ be the longest common prefix of x and y; that is, the longest string such that $\gamma w = x$ and $\gamma z = y$ for some w and z in A^*. There are three cases, exactly one of which must hold:

1. $w = z = \Lambda$ (x and y are identical);
2. $w = \Lambda$ and $z \neq \Lambda$ (x is a proper prefix of y); or
3. $w = a\alpha, z = b\beta$, $a \neq b, a, b \in A$, and $\alpha, \beta \in A^*$.

The relationship between x and y is defined in each case as follows:

1. $x \leq_L y$ and $y \leq_L x$;
2. $x \leq_L y$ and $y \not\leq_L x$;
3. if $a \leq_A b$ then $x \leq_L y$ and $y \not\leq_L x$, else $x \not\leq_L y$ and $y \leq_L x$.

Theorem 4.4.1.   Given any finite alphabet A and any total ordering $\leq_A$ on A, the lexicographic ordering $\leq_L$ defined by extending $\leq_A$ is a total ordering on A^*.

Proof.   It should be clear by examination of the three cases in the definition that $\leq_L$ is reflexive and antisymmetric. The proof will concentrate on showing that the relation is transitive.

The proof that $\leq_L$ is transitive is a "messy" proof by cases. It is unlikely that anyone enjoys this kind of proof, but the cases are forced upon us by the definition of $\leq_L$, which has three cases. We wish to prove that $x_1 \leq_L x_2$ and $x_2 \leq_L x_3$ implies $x_1 \leq_L x_3$. There are three ways that $x_1 \leq_L x_2$ can come about and three ways that $x_2 \leq_L x_3$ can come about. There are thus nine cases, which we can label by pairs (i,j), where (i) is the case by which $x_1 \leq_L x_2$ and (j) is the case by which $x_2 \leq_L x_3$. In order not to miss any cases, we label the sections of the proof according to this system. To keep the proof short, we combine cases whenever possible.

Case 1. $(1,1), (1,2), (1,3)$.   In all these cases $x_1 = x_2$. By substituting x_1 for x_2 in $x_2 \leq_L x_3$ we obtain $x_1 \leq_L x_3$, which is what we want to prove.

Case 2. $(2,1), (3,1)$.   In these cases $x_2 = x_3$. By substituting x_3 for x_2 in $x_1 \leq_L x_2$ we again obtain $x_1 \leq_L x_3$.

Case 3. $(2,2)$.   In this case $x_2 = x_1 z_1$ and $x_3 = x_2 z_2$. By substituting $x_1 z_1$ for x_2 in $x_3 = x_2 z_2$ we obtain $x_3 = x_1 z_1 z_2$, which satisfies case 2 of the definition of $\leq_L$. Thus $x_1 \leq_L x_3$.

Case 4. $(2,3)$.   In this case $x_2 = x_1 z = \gamma a \alpha$, $x_3 = \gamma b \beta$, and $a <_A b$. There are two subcases, depending on whether a falls in x_1 or in z. If a falls in x_1, then x_1 is divided into $x_1 = \gamma a \delta$ for some δ, and so case 3 of the definition gives $x_1 \leq_L x_3$. If a falls in z, then x_1 is a prefix of x_3 and $x_1 \leq_L x_3$ by case 2 of the definition of $\leq_L$.

Case 5. (3,2). In this case $x_1 = \gamma a\alpha$, $x_2 = \gamma b\beta$, $a < b$, and $x_3 = x_2 z$. Since $x_3 = \gamma b\beta z$ it follows that $x_1 \leq_L x_3$ by case 3 of the definition.

Case 6. (3,3). In this case $x_1 = \gamma_1 a_1 \alpha_1, x_2 = \gamma_1 b_1 \beta_1 = \gamma_2 a_2 \alpha_2$, $x_3 = \gamma_2 b_2 \beta_2$, $a_1 < b_1$, and $a_2 < b_2$. There are *three subcases*, depending on whether a_2 falls in γ_1, b_1, or β_1.

If a_2 occurs as part of the string γ_1, then $x_1 = \gamma_2 a_2 \alpha_3$ for some α_3, and so $x_1 \leq_L x_3$ by case 3. If a_2 falls in b_1, then $\gamma_1 = \gamma_2$, $a_1 \leq_A b_1 = a_2 <_A b_2$, and so $x_1 \leq_L x_3$ by case 3. Finally, if a_2 falls in β_1, then $x_3 = \gamma_1 b_1 \beta_3$ for some β_3, so that $x_1 \leq_L x_3$, again by case 3 of the definition.

We have covered every case and shown that in each case $x_1 \leq_L x_2$ and $x_2 \leq_L x_3$ implies $x_1 \leq_L x_3$, so that $\leq_L$ must be transitive. This completes the proof. □

Although the lexicographic ordering $\leq_L$ is a total ordering of A^*, it is *not a well ordering*. To see this, just consider the set $B = \{a^k b \mid k \geq 0\}$, which is a subset of $\{a,b\}^*$. B has no minimal element with respect to $\leq_L$, since

$$aab \leq_L ab, \ aaab \leq_L aab, \ aaaab \leq_L aaab, \ldots, a^{k+1}b \leq_L a^k b, \ldots.$$

This ordering does not correspond to any enumeration of A^*. It is therefore unsuitable as an ordering of A^* for some applications. When a well ordering of A^* is required, another extension of the alphabetic order on A is used. We shall call this new ordering on A^* the **enumeration ordering** and denote it by $\leq_E$ to distinguish it from the lexicographic ordering $\leq_L$ just defined. In enumeration order, strings of unequal ~~length~~ are ordered by *length,* and strings of equal length are ordered exactly as they are by $\leq_L$. For example, if $A = \{a,b\}$, the enumeration A^* we want is

$$\Lambda, a, b, aa, ab, ba, bb, aaa, aab, aba, abb, b \ldots.$$

Theorem 4.4.2. Enumeration ordering $\leq_E$ A^* (defined above) is a well ordering on A^* if $\leq A$ is a total ordering on A. Moreover, for each $w \in A^*$, there are only finitely many elements $w_i \in A^*$ such that $w_i \leq_E w$.

Proof. (This is left as an exercise for the reader.)

Exercises for Section 4.4.1

1. (a) Arrange the following strings into ascending order according to the definition of lexicographic ordering:

   ANIMAL, AND, AN, ANIMATION,
   BAND, CAN, BAN, CAR

(b) Arrange these same strings into ascending order according to the definition of enumeration ordering. (Assume the usual alphabetic ordering on the letters.)

2. (a) Suppose A has n elements. How many strings are there in A^k?
   (b) List all the elements of $\{a,b\}^3$.
   (c) List all the elements of $\{a,b,c\}^3 \cap \{b,c\}*$.
   (d) List all the elements of $\{a,b\}^2 \cdot \{c\}$, where $\cdot$ is the operation on sets of strings called "catenation", and defined by

$$A \cdot B = \{ab \mid a \in A \text{ and } b \in B\}.$$

3. For each of the following pairs of values for x and y, classify x and y according to the cases (1–3) of the definition of lexicographic ordering. Also, identify the longest common prefix of x and y, and state the relationship between x and y.
   (a) $x =$ "RED", $y =$ "RED"
   (b) $x =$ "REDOLENT", $y =$ "REDONE"
   (c) $x =$ "REDUCE", $y =$ "REDONE"
   (d) $x =$ "RED", $y =$ "REDONE"

4. Prove Theorem 4.4.2.
   Hint: The proof can be modeled on that of Theorem 4.4.1. To show (further) that every nonempty subset S of Σ^* has a minimal element, use induction on the length n of the shortest string(s) in S.

5. Extend the definition of enumeration ordering on strings to define a well ordering on the set of finite n-tuples of strings,

$$\{(w_1, w_2, \ldots, w_n) \mid w_i \in \Sigma^* \text{ and } n > 1\}.$$

Selected Answers for Section 4.4.1

1. (a) AN, AND, ANIMAL, ANIMATION, BAN, BAND, CAN, CAR
   (b) AN, AND, BAN, CAN, CAR, BAND, ANIMAL, ANIMATION

2. (a) n^k
   (b) aaa, aab, aba, abb, baa, bab, bba, bbb
   (c) bbb, bbc, bcb, bcc, cbb, cbc, ccb, ccc
   (d) aac, abc, bac, bbc

5. Define $(w_1, \ldots, w_n) \le (x_1, \ldots, x_m)$ iff $n < m$, or $n = m$ and there exists $1 \le k \le n$ such that for all $1 \le i \le k$, $w_i = x_i$, and, if $k < n$, $w_k \le x_k$ and $w_k \ne x_k$.

4.4.2 Application: Proving There Are Noncomputable Functions

Computer programming languages may be used to write algorithmic definitions of functions. It is an interesting question whether there are functions that *cannot* be defined in this fashion. To answer this question we need only a few properties of programming languages:

1. Each programming language makes use of an alphabet with a finite number of characters, say A.
2. Every legal program is a string of a finite length, in A^*.

Let us assume we are given a programming language that satisfies these two properties, and let g: $N \rightarrow A^*$ be the enumeration of A^* discussed after Definition 4.4.3 and in Theorem 4.4.2. Every program in this language is a string in A*. We would like to associate each program that computes a function with the function it computes, but some strings in A^* are not legal function definitions. Therefore let us define a function $f:N \rightarrow \{h:N \rightarrow N\}$ that gives a function for each string in A^*. Let $f(i)(x) = 0$ if $g(i)$ is not a syntactically legal program, or if $g(i)$ is a syntactically legal program, but when executed with input x this program outputs a value that is not a nonnegative integer, outputs more than one value, or simply does not terminate. Otherwise, let $f(i)(x)$ be the integer value output by program $g(i)$ on input x.

Let $d: N \rightarrow N$ be defined by $d(x) = f(x)(x) + 1$. Observe that if d were computed by any program $g(i)$ then $f(i)(i)$ would equal $d(i)$, but that would contradict the definition of d. This completes the proof of the following theorem.

Theorem 4.4.3.   For any programming language there is a mathematically definable function $d: N \rightarrow N$ that cannot be computed by any program that may be defined in the language.

The argument used in the proof of this theorem is a **diagonal** argument and is due to Cantor, who used it to show that the real numbers are not countable, and hence that there are irrational numbers. It is called diagonal because the definition of d pairs functions with inputs in the way they would be encountered if one were to lay all the possible functions and inputs in a square matrix and to run down the diagonal. This is illustrated in Figure 4-13. Such proofs are an important tool in computability theory and can be used to show that a number of problems of practical interest have no computable solution. Several more applications, including a proof that there are undefinable functions, are suggested in the exercises.

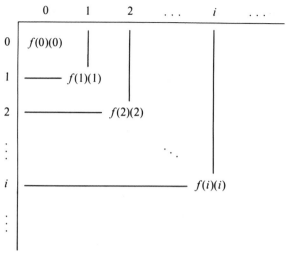

Figure 4-13. A diagonal construction.

Exercises for Section 4.4.2

1. Consider the programming language Ada (a registered trademark of the U.S. Department of Defense). In the proof of Theorem 4.4.3, suppose $g(i) =$

   "**function** h(x: integer) **is begin return** x + 1; **end** h;",

   where i is some specific integer. This string is a legal Ada function definition. What is the value of $f(i)(x)$? What is the value of $d(i)$?

2. Suppose $g(j) =$

   "**function** h(x: integer) **is begin loop null; end loop; end** h;",

   which is a legal Ada subprogram definition, but defines a computation that loops forever, without returning any value. What is the value of $f(j)(x)$? What is the value of $d(j)$?

3. Suppose $g(k) =$ "This is not a legal Ada function definition." What is the value of $f(j)(x)$? What is the value of $d(j)$?

4. Use a diagonal construction to show that the real numbers cannot be enumerated. Hint: Consider the real numbers between 0 and 1. Each of these real numbers may be represented by at most two infinite series $10^{-1}d_0 + 10^{-2}d_1 + \ldots + 10^{-i-1}d_i + \ldots$, where $0 \leq d_i \leq 9$.

5. Use a diagonal construction to show that the functions $f: N \to N$ cannot be enumerated.

6. Suppose we assume that all mathematical definitions can be expressed in writing using only a finite set of symbols. Then every mathematical definition consists of a finite string over some fixed finite alphabet. Use a diagonal construction to show that there must be functions $f: N \to N$ that cannot be defined mathematically.

Selected Answers for Section 4.4.2

1. $f(i)(x) = x + 1$
   $d(i) = i + 2$

4. Suppose $f: N \to [0,1]$ is an enumeration of the real numbers between 0 and 1. Let $dig(x,i)$ denote the ith digit in the decimal fraction expansion of x, as described in the hint. Define z to be the real number defined by letting the ith digit of z be $(dig(f(i), i) + 2) \bmod 10$. That is $z = dig(f(0),0) + 10 \cdot dig(f(1),1) + \cdots + 10^i \cdot ((dig(f(i),i) + 1) \bmod 10) + \cdots$ The existence of such a real number would be a contradiction, since if $f(i) = z$ for some i, we would have $dig(z,i) = (dig(z,i) + 1) \bmod 10$.

4.5 OPERATIONS ON RELATIONS

When one is working with relations, it is often useful to think of one relation as being derived from another. For example, the relation $\leq$ on integers may be viewed as a combination of the relations $<$ and $=$. Similarly, there is a clear connection between the human kinship relation "is a parent of," "is a child of," and "is a descendant of." In this section we will explore such connections and develop some algebra on relations. We will begin with some operations that apply to n-ary relations and end with some that apply only to binary relations.

Several operations on relations come directly from the standard operations on sets, without need for further definitions. Suppose $R_1 \subseteq A_1 \times \cdots \times A_n$. Then $\overline{R_1} = (A_1 \times \cdots \times A_n) - R_1$ is the **complement** of R_1. Suppose further that $R_2 \subseteq A_1 \times \cdots \times A_n$. The **union** $R_1 \cup R_2$ is also a relation, as are the **intersection** $R_1 \cap R_2$ and the **difference** $R_1 - R_2$.

Example 4.5.1. The relationships between the usual ordering relations $\leq$, $<$, and $=$ on the integers can be described in terms of union and complement. The relation $\leq$ is the union of the relations $<$ and $=$; the relation $<$ is the relation $\leq$ minus the relation $=$; the relation $=$ is the relation $\leq$ minus the relation $<$.

Another operation, called **projection**, is derived from the geometric concept with the same name. It produces a lower dimensional relation from one of higher dimension.

Definition 4.5.1   Let $R \subseteq A_1 \times \cdots \times A_n$ be an n-ary relation and let $s_1, s_2, \ldots, s_k$ be a subsequence of the component positions $1, \ldots, n$ of R. The **projection** of R with respect to $s_1, \ldots, s_k$ is the k-ary relation

$$\{(x_1, \ldots, x_k) \mid (x_1, \ldots, x_k) = (a_{s_1}, \ldots, a_{s_k}) \text{ for some } (a_1, \ldots, a_n) \in R\}.$$

Example 4.5.2.   If $R \subseteq \{a,b,c\}^3$ is the set of ordered triples $\{(a,a,a),$ $(a,b,c),$ $(b,b,c),$ $(a,a,c),$ $(b,a,c),$ $(b,c,c),$ $(a,c,c)\}$, the projection of R with respect to the first and third components is the binary relation $\{(a,a),$ $(a,c),$ $(b,c)\}$. In this case the projection of R with respect to the first component is the unary relation (i.e., the set) $\{a,b\}$.

Work on relational models for computer database systems has led to application of the union, intersection, complement and projection operations on n-ary relations, as well as a number of more esoteric operations. According to the relational model, a database is a collection of n-ary relations, corresponding to what are called "files" or "tables" in traditional data processing terminology. (The n-tuples of each relation correspond to what are called "records" or "rows.") One of the operations used in work on relational databases that is especially useful is the **join** (not to be confused with the join operation on lattices defined in Section 4.4). The join provides a means of combining two relations in a natural way in the event they have a component domain in common.

Definition 4.5.2.   Let $R_1 \subseteq A_1 \times A_2 \times \cdots \times A_n$ and $R_2 \subseteq B_1 \times B_2 \times \cdots \times B_m$ be relations and suppose $A_i = B_j$ for some i and j. The **join** of R_1 and R_2 **with respect to component i of R_1 and component** j of R_2 is the relation

$$\{(a_1, \ldots, a_n, b_1, \ldots, b_m) \mid (a_1, \ldots, a_n) \in R_1, (b_1, \ldots, b_m) \in R_2, \text{ and } a_i = b_j\}.$$

Example 4.5.3.   Suppose $R_1 = \{(a,b), (a,c), (b,a)\} \subseteq \{a,b,c\}^2$ and $R_2 = \{(a,b,x), (c,a,y), (a,a,x), (a,c,x)\} \subseteq \{a,b,c\}^2 \times \{x,y\}$. Then the join of R_1 and R_2 with respect to the first component of R_1 and the second component of R_2 is the relation

$$\{(a,b,c,a,y), (a,b,a,a,x), (a,c,c,a,y), (a,c,a,a,x), (b,a,a,b,x)\}.$$

Example 4.5.4.   A manufacturer who wished to store information concerned with ordering parts might do so as three relations:

SUPPLIERS $\subseteq$ SUPPLIER-NUMBERS $\times$ SUPPLIER-NAMES $\times$ STATUS-CODES $\times$ ADDRESSES
PARTS $\subseteq$ PART-NUMBERS $\times$ PART-NAMES $\times$ COLORS $\times$ WEIGHTS
ORDERS $\subseteq$ SUPPLIER-NUMBERS $\times$ PART-NUMBERS $\times$ QUANTITIES

The interpretation of a quadruple in *SUPPLIERS* is that it gives the unique identification number of a supplier, the supplier's name, a status code indicating reliability of the supplier, and the supplier's address. The interpretation of a quadruple in *PARTS* is that it gives the unique identification number of a part, the name of the part, its color, and its weight. The interpretation of a triple in *ORDERS* is that it gives the number of a supplier, the number of a part ordered from the supplier, and the quantity of this part ordered, but not yet delivered. This scheme is probably much simpler than a real database would be for such an application, but it will be adequate to demonstrate the utility of some of the relational operations.

Suppose the manufacturer is interested in knowing from which suppliers with status code 3 he has ordered the part with number k, and how many parts each has yet to deliver. Pairs (s,q) such that s is one of these suppliers and q is the quantity of parts outstanding on order could be obtained from the relations *SUPPLIERS* and *ORDERS* by means of a series of projections and joins.

Since $\{k\}$ is a unary relation on the domain *PART-NUMBERS*, we can take the join of *ORDERS* and $\{k\}$ with respect to *PART-NUMBERS*, obtaining a relation which we shall call $R1$. The members of $R1$ are all the quadruples (x,k,z,k) such that (x,k,z) is a triple in *ORDERS*. To eliminate undesired information, we take the projection of $R1$ with respect to the first and third components, yielding a binary relation in *SUPPLIER-NUMBERS* $\times$ *QUANTITIES* which we shall call $R2$. $R2$ consists of all pairs (s,q) such that s is the number of a supplier of the part with number k and q is the quantity of the part outstanding on order. Similarly, since $\{3\}$ is a unary relation on the domain *STATUS-CODES*, we can take the join of *SUPPLIERS* and $\{3\}$ with respect to *STATUS-CODES*, calling the result $R3$. $R3$ consists of all the quintuples $(x,y,3,z,3)$ such that $(x,y,3,z)$ is in *SUPPLIERS*. Again, taking the projection of $R3$ with respect to the first component we obtain $R4$, the set of numbers of all suppliers with status code 3. Relations $R2$ and $R4$ share component domain *SUPPLIER-NUMBER*, so that we can take the join of $R2$ and $R4$ with respect to this domain, obtaining a new relation $R5$. The elements of $R5$ are the triples (s,q,s) such that s is the number of a supplier of the part with number k with status 3 and q is the quantity of the part outstanding on order. All that remains is to eliminate the redundant supplier number. This can be done by projecting $R5$ with respect to the first and second components, resulting in the desired set of pairs.

Some further operations are defined only on binary relations. Three of these are inverse, composition, and transitive closure. We review the definitions of inverse and composition given in Chapter 1.

Definition 4.5.3. Suppose $R \subseteq A \times B$. The **inverse** of R, denoted by R^{-1}, is the relation $\{(y,x) \mid (x,y) \in R\}$.

Figure 4-14. A relation and its inverse.

Example 4.5.5. The inverse R^{-1} of the relation $R = \{(x,y), (y,z),$ $(z,y), (z,x)\}$ is the relation $\{(y,x), (z,y), (y,z), (x,z)\}$. These relations are shown as digraphs in Figure 4-14. Notice that the digraph of the inverse of a relation has exactly the edges of the digraph of the original relation, but the directions of the edges are reversed.

Example 4.5.6. As another example, consider the usual, ordering relation $\leq$ on the integers. The inverse of this relation is the relation $\geq$. Observe also that the inverse of a relation is the relation itself iff the relation is symmetric. In particular, the inverse of the relation $=$ on the integers is itself.

Definition 4.5.4. Suppose $R_1 \subseteq A \times B$ and $R_2 \subseteq B \times C$. The **composition** of R_1 and R_2, denoted by $R_1 \cdot R_2$ is the relation $\{(x,z) \mid (x,y) \in R_1 \text{ and } (y,z) \in R_2\}$.

Example 4.5.7. The composition $R_1 \cdot R_2$ of the relations $R_1 = \{(a,a),$ $(a,b), (c,b)\}$ and $R_2 = \{(a,a), (b,c), (bd)\}$ is the relation $\{(a,a), (a,c), (a,d),$ $(c,c), (c,d)\}$. These relations are shown in Figure 4-15.

Composition of relations is associative, that is, if R, S, and T are relations, then (R S) T = R (S T).

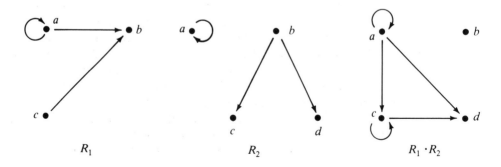

Figure 4-15. The composition of binary relations.

The notation R^k is used for the iterated composition of R with itself. That is, $R^1 = R$, and $R^{k+1} = R^k \cdot R$, for $k \geq 1$. Taken to its limit, this construction leads to another important operation on binary relations—the transitive closure. Observe that $R^{k+n} = R^k R^n$.

Definition 4.5.5. Suppose $R \subseteq A \times A$. The **transitive closure** of R, denoted by R^+ is $R \cup R^2 \cup R^3 \cdots = \bigcup_{k \geq 1} R^k$. The **transitive reflexive** closure of R, denoted by R^*, is $R^+ \cup \{(a,a) \mid a \in A\}$.

Example 4.5.8. The transitive closure R^+ of the relation $R = \{(a,b), (b,c), (c,d)\}$ is the relation $\{(a,b), (a,c), (a,d), (b,c), (b,d), (c,d)\}$. This is illustrated by the digraphs in Figure 4-16.

As another example of composition and transitive closure, consider the relation "is a parent of." The composition of this relation with itself is the relation "is a grandparent of," and the transitive closure of this relation is the relation "is an ancestor of."

Theorem 4.5.1. R^+ is the smallest relation containing R that is transitive.

Proof. First we shall show that R^+ is transitive. Suppose $x\, R^+ y$ and $y\, R^+ z$. Then, since $R^+ = \bigcup_{k \geq 1} R^k$, $x\, R^i y$ and $y\, R^j z$ for some $i, j \geq 1$. Thus $x\, R^{i+j} z$, and so $x\, R^+ z$.

Now suppose $R \subseteq Q$ and Q is transitive. Suppose $R^k \subseteq Q$. Then, since $R \subseteq Q$ and Q is transitive, $R^{k+1} \subseteq Q$. Thus, by induction on k, $R^k \subseteq Q$ for every $k \geq 1$, and so $R^+ \subseteq Q$. $\square$

Similarly, the transitive reflexive closure of a binary relation is the smallest relation that contains it and is both transitive and reflexive. These are only two of a whole realm of possible closures. For example, the symmetry property says that if R includes a pair (x,y) then R must also include (y,x). The **symmetric** closure of a relation R is thus the set $R \cup R^{-1}$, which is the smallest symmetric relation that includes R. In general, if P is a property such that P can be made true for any set by adding

Figure 4-16. A relation and its transitive closure.

certain elements to the set, we can call P a **closure property** and define the P-closure of a set to be the smallest set that contains it and satisfies property P.

Note that Σ^+, the set of all nonnull character strings over alphabet Σ, is another example of such a closure. In this case the property with respect to which the closure is taken is: If x and y are elements of Σ^+ then xy is an element of Σ^+.

Exercises for Section 4.5

1. Consider the relation $R = \{(a,b), (b,c), (b,d), (d,a), (c,c)\}$.
   (a) Draw a digraph for the relation R.
   (b) Draw a digraph for the complement of R, $\overline{R}$.
   (c) Draw a digraph for the inverse of R, R^{-1}.
   (d) Draw a digraph for the intersection of R and the inverse of R, $R \cap R^{-1}$.

2. Answer the questions asked in exercise (1) for the relation $Q = \{(a,b),(b,c),(b,d),(d,d),(c,c),(a,c)\}$.

3. With R and Q defined in exercises 1 and 2, determine the composition $R \cdot Q$. Draw a digraph. What is the difference of R and Q, R-Q? Draw a digraph. Give a set of n-tuples for the join of R and Q with respect to the first component of R and the second component of Q.

4. Let $P = \{(x,y,x \cdot y) \,|\, x$ and y are intergers$\}$ and $Q = \{(x,x,z) \,|\, x$ and z are integers$\}$.
   (a) What is $P \cap Q$?
   (b) What is the projection of $P \cap Q$ with respect to the first and third components?
   (c) Let R be the join of P and $\{3,5\}$ with respect to the first component of P. Describe R.
   (d) Let T be the join of R and $\{7\}$ with respect to the second component of R. Describe T.
   (e) What is the projection of T with respect to the third component?

5. Let $P = \{(x,y,x \cdot y) \,|\, x$ and y are integers$\}$ and, $S = \}(x,y,x + y) \,|\, x$ and y are integers$\}$.
   (a) What is $P \cap S$?
   (b) What is the join of P and S with respect to the third component of each?
   (c) What is the projection of this join, with respect to the third component?

6. Let $R = \{(x,y) \,|\, x = y \cdot z$ for some z greater than one, and x,y,z are positive integers$\}$.
   (a) What is $R \cap R^{-1}$? Is R symmetric? Reflexive?

(b) Prove that $R \supseteq R \cdot R$. What is the transitive closure of R?

(c) What is the projection of R with respect to the first component?

7. Give the transitive closure, the transitive reflexive closure, and the symmetric closure, represented as digraphs, for each of the following relations:

(a) $x \, R \, y$ iff x is an integral multiple of y, on the set $\{2,3,4,5,6\}$.

(b) $x \, R \, y$ iff $x = y + 1$, on the set $\{0,1,2,3,4,5\}$.

8. Give the closures of the set $\{1\}$ with respect to each of the following properties:

(a) If x is an element of S then $x + 1$ is an element of S.

(b) If x is an element of S then $x + 1$ and $-x$ are elements of S.

(c) If x is an element of S then $x + 2$ is an element of S.

(d) If x is an element of S then $x \cdot 2$ is an element of S.

9. Answer the same questions as in (8) for the set $\{0\}$.

10. Prove composition of relations is associative. That is if R, S, T are relations on A, then $(R \cdot S) \cdot T = R \cdot (S \cdot T)$.

11. Prove by mathematical induction that if R is a relation on A, then

(a) $R^m \cdot R^n = R^{m+n}$

(b) $(R^m)^n = R^{mn}$

12. Give an example of a relation R and a positive k such that $R^{k+1} \cdot R^{-1} \neq R^k$.

13. (a) Show that the transitive closure of a symmetric relation is symmetric.

(b) Is the transitive closure of an antisymmetric relation always antisymmetric?

(c) Show that the transitive closure of a reflexive and symmetric relation is an equivalence relation.

14. Let R_1 and R_2 be arbitrary binary relations on a set A. Prove or disprove the following assertions.

(a) If R_1 and R_2 are reflexive, then $R_1 \cdot R_2$ is reflexive.

(b) If R_1 and R_2 are irreflexive, then $R_1 \cdot R_2$ is irreflexive.

(c) If R_1 and R_2 are symmetric, then $R_1 \cdot R_2$ is symmetric.

(d) If R_1 and R_2 are antisymmetric, then $R_1 \cdot R_2$ is antisymmetric.

(e) If R_1 and R_2 are transitive, then $R_1 \cdot R_2$ is transitive.

15. Let R be a binary relation on a set A where A has n elements. Prove that the transitive closure of $R = \bigcup_{i=1}^{n} R^i$.

16. Prove that if R is a transitive relation on a set A, then for each positive integer n, $R^n \subseteq R$.

17. Let R be a relation on a set A. Prove:

(a) If R is reflexive, then $R \subseteq R^2$.

(b) R is transitive iff $R^2 \subseteq R$.

18. Suppose that R and S are relations on a set A, where $R \subseteq S$ and S is transitive. Prove that $R^n \subseteq S$ for each positive integer n.

19. Suppose that R and S are symmetric relations on a set A. Prove:
    (a) If $(x,y) \in S \cdot R$, then $(y,x) \in R \cdot S$.
    (b) If $R \cdot S \subseteq S \cdot R$, then $R \cdot S = S \cdot R$.
    (c) $R \cdot S$ is symmetric iff $R \cdot S = S \cdot R$.
    (d) R^n is symmetric for each positive integer n.

20. Suppose R and S are relations on a set A. Prove or disprove:
    (a) If R and S are reflexive, then so is $R \cdot S$.
    (b) If R and S are both reflexive and symmetric, then $R \cdot S$ is reflexive and symmetric iff $R \cdot S = S \cdot R$.
    (c) If R and S are transitive, then $R \cdot S$ is transitive iff $R \cdot S = S \cdot R$.
    (d) If R and S are equivalence relations on A, then $R \cdot S$ is an equivalence relation iff $R \cdot S = S \cdot R$.

21. Suppose R and S are relations on a set A.
    (a) Prove that $(R \cdot S)^{-1} = S^{-1} \cdot R^{-1}$.
    (b) Is it true that $(R \cup S)^{-1} = R^{-1} \cup S^{-1}$?
    (c) Is is true that $(R \cap S)^{-1} = R^{-1} \cap S^{-1}$?
    (d) If R is an equivalence relation on A, is it true that R^{-1} is an equivalence relation on A?
    (e) If R satisfies any of the six properties of relations defined in Section 4.2, determine whether or not R^{-1} satisfies the same properties.
    (f) Suppose that $R \subseteq S$. Show that $R^{-1} \subseteq S^{-1}$.

22. Assume that R is a reflexive relation on a set A.
    (a) Show that $R \cdot R^{-1}$ is reflexive and symmetric.
    (b) Prove or disprove $R \cdot R^{-1}$ is transitive.
    (c) Prove or disprove that the transitive closure $R \cdot R^{-1}$ is an equivalence relation.

23. Let R and S be relations from A to B and let T and W be relations from B to C. Prove:
    (a) $R \cdot (T \cup W) = (R \cdot T) \cup (R \cdot W)$
    (b) $R \cdot (T \cap W) = (R \cdot T) \cap (R \cdot W)$
    (c) If $R \subseteq S$, then $R \cdot T \subseteq S \cdot T$.
    (d) If V is a relation from D to A, and if $R \subseteq S$, prove that $V \cdot R \subseteq V \cdot S$.

24. Let R and S be relations on a set A. Prove that $(R \cap S)^n \subseteq R^n \cap S^n$ for each integer $n \geq 1$.

25. If R and S are equivalence relations on a set A, prove that $(R \cup S)^+$ is an equivalence relation containing both R and S. Moreover, if T is any equivalence relation on A such that $R \cup S \subseteq T$, prove that $(R \cup S)^+ \subseteq T$.

26. Prove that if the relation R is a partial order (total order) on a set A, then R^{-1} is a partial order (total order) on A.

27. If R is the relation on the set Z of integers defined by aRb iff $b = 2a$, and if $S = \{(c,d) \in Z \times Z \mid d = 3c\}$, find $R \cdot S$.

28. Prove by induction that if R is a reflexive and transitive relation on a set A, then $R^n = R$ for each positive integer n.

29. If R is a symmetric relation defined on a set A, then $R \cup R^2$ is symmetric.

30. Find the transitive closure of R if
    (a) $R = \{(a,b),(b,c),(c,d),(d,e)\}$
    (b) $R = \{(a,a),(a,b),(b,c),(b,d),(d,c),(d,d)\}$
    (c) $R = \{(a,b),(b,c),(c,d),(d,e),(e,a)\}$
    (d) $R = \{(a,b),(a,c),(c,c),(c,d),(d,c),(c,e),(e,f),(f,d)\}$

Selected Answers for Section 4.5

1. (b)   (d)

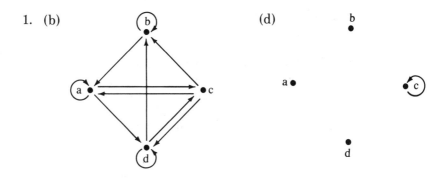

4. (a) $\{(x,x,x^2) \mid x$ is an integer$\}$
    (b) $\{(x,x^2) \mid x$ is an integer$\}$
    (c) $\{(3,y,3 \cdot y,3) \mid y$ is an integer$\} \cup \{(5,y,5 \cdot y,5) \mid y$ is an integer$\}$
    (d) $\{(3,7,21,3,7), (5,7,35,5,7)\}$
    (e) $\{21,35\}$

6. (a) The intersection is empty. The relation R is not symmetric, nor is it reflexive.
    (b) If $x = y \cdot a$ and $y = z \cdot b$ then $x = z \cdot (b \cdot a)$. If a and b are greater than one, then so is $b \cdot a$. The relation R is its own transitive closure, since it is already transitive.

(c) The projection of R with respect to the first component is the set of integers ≥ 2.

8. (a) the positive integers
   (b) all the integers
   (c) the odd positive integers
   (d) the integer powers of two

4.6   PATHS AND CLOSURES

Many applications of directed graphs involve questions of connectivity. Two vertices of a graph are **connected** if there is a path that goes from one vertex to the other along some of the edges of the graph. Different authors disagree about details, such as what consitutes a path, however. Usually a path is defined as a sequence of edges, though occasionally it may be defined as a sequence of vertices. Whether repetition of edges or vertices is permitted is a common source of disagreement, as well as other details, such as whether there is always a path of length zero from each vertex to itself. These differences are not due to whimsy. They correspond to real differences in what "connected" may mean as it is applied to the different real-life structures that graphs may be called upon to model. The definitions that follow have been chosen for their generality of application, but the reader should be prepared to pay attention for possible discrepancies in the meanings of terms as they may be used between one book and another.

Definition 4.6.1. A **directed path** in a digraph $A = (V,E)$ is a sequence of zero or more edges $e_1, \ldots, e_n$ in E such that for each $2 \leq i \leq n$, e_{i-1} is to the vertex that e_i is from; that is, e_i may be written as (v_{i-1}, v_i) for each $1 \leq i \leq n$. Such a path is said to be *from* v_0 to v_n, and its length is n. In this case, v_0 and v_n are called the **endpoints** of the path. A **nondirected path** in G is a sequence of zero or more edges $e_1, \ldots, e_n$ in E for which there is a sequence of vertices $v_0, \ldots, v_n$ such that $e_i = (v_{i-1}, v_i)$ or $e_i = (v_i, v_{i-1})$ for each $1 \leq i \leq n$. A path is **simple** if all edges and vertices on the path are distinct, except that v_0 and v_n may be equal. A path of length ≥ 1 with no repeated edges and whose endpoints are equal is a **circuit**. A simple circuit is called a **cycle**.

Note that the definitions of simple, circuit, and cycle apply equally to directed and nondirected paths. It is also important to notice that a path of length zero is permitted, but it does not have a unique pair of endpoints. Such a path has no edges and can be viewed as being from a vertex to itself. When we wish to exclude paths of length zero we will use the term **nontrivial path.** Ordinarily the term path will be qualified as directed or nondirected, simple or not. When it is not qualified the kind of path may be inferred from the context, or does not matter.

A path $e_1, \ldots, e_n$ is said to **traverse** a vertex x if one (or more) of the e_i's is to or from x and x is not serving as one of the endpoints of the path, or more precisely, if $e_i = (x,y)$, then $2 \le i \le n$, or if $e_i = (y,x)$, then $1 \le i \le n - 1$.

Example 4.6.1.   The digraph shown in Figure 4-17 includes some of each kind of path and cycle.

There are two simple directed paths from a to d. They are (a,b), (b,c), (c,d), and (a,c), (c,d). In addition to all the simple directed paths, there are several more simple nondirected paths from a to d, including (a,b), (b,d). There are a number of nontrivial directed cycles, including (a,b), (b,c), (c,d), (d,e), (e,a) and a larger number of nondirected cycles, including all the directed cycles as well as cycles such as $(a,b),(b,c),(c,a)$.

We shall now explore some ways in which the concept of path is related to other concepts we have discussed previously, starting with the composition of relations. (Recall that $E^n = E \cdot E \cdot \ldots \cdot E$, the composition of n copies of E.)

Theorem 4.6.1.   If $A = (V,E)$ is a digraph, then for $n \ge 1$, $(x,y) \in E^n$ iff there is a directed path of length n from x to y in A.

Proof.   The proof is by induction on n. For $n = 1$, $E^n = E$. The definition of path guarantees that $(x,y) \in E$ iff there is a path of length 1 from x to y, since a path of length 1 is an edge of A. For $n > 1$, we assume the theorem is true for $n - 1$, and break the proof into two parts:

(if) Suppose $(v_0,v_1), \ldots, (v_{n-1},v_n)$ is a directed path from v_0 to v_n. Then $(v_0,v_{n-1}) \in E^{n-1}$, by induction. By definition $E^n = E^{n-1} \cdot E$, and $(v_{n-1},v_n) \in E$, so that $(v_0, v_n) \in E^n$.

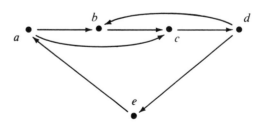

Figure 4-17. A directed graph with paths.

(only if) Suppose $(v_0, v_n) \in E^n$. Then, since $E^n = E^{n-1} \cdot E$, there exists some v_{n-1} such that $(v_0, v_{n-1}) \in E^{n-1}$ and $(v_{n-1}, v_n) \in E$. By the inductive hypothesis, there is a directed path $(v_0, v_1), \ldots, (v_{n-2}, v_{n-1})$ of length $n - 1$ from v_0 to v_{n-1}. Adding (v_{n-1}, v_n) to this path gives the path of length n desired. $\square$

Paths and composition are also related to the property of transitivity. If $A = (V, E)$ is a digraph then E is transitive iff every directed path in A has a "short-cut." That is, if there is a nontrivial directed path from a vertex x to a vertex y there must also be an edge directly from x to y. This reasoning leads to a useful corollary.

Corollary 4.6.1. If $A = (V, E)$ is a digraph then for any two vertices x and y in V, $(x, y) \in E^+$ iff there is a nontrivial directed path from x to y in A.

Proof. This is a consequence of the previous theorem. If there is a directed path from x to y in A of some length $n \geq 1$, then $(x, y) \in E^n$, so that $(x, y) \in E^+$. Conversely, if $(x, y) \in E^+$, then $(x, y) \in E^n$ for some $n \geq 1$, and so the previous theorem guarantees that there is a directed path of length n from x to y. $\square$

Example 4.6.2. Let $A = \{a, b, c, d, e\}$ and let $R = \{(a,a), (a,b), (b,c), (c,d), (c,e), (d,e)\}$. We compute the transitive closure of R in two different ways.

Method 1. First, we use Definition 4.5.5. Since $R^+ = \bigcup_{k \geq 1} R^k$ we need only compute R^k, for each k, and take their union. By definition of composition, we get

$$R^2 = \{(a,a), (a,b), (a,c), (b,e), (b,d), (c,e)\}$$
$$R^3 = \{(a,a), (a,b), (a,c), (a,d), (a,e), (b,e)\}$$
$$R^4 = \{(a,a), (a,b), (a,c), (a,d), (a,e)\}$$
$$R^5 = \{(a,a), (a,b), (a,c), (a,d), (a,e)\}$$

Notice that $R^4 = R^5$ so it will follow that $R^4 = R^5 = R^6 = R^k$ for all $k \geq 4$. Then R^+ is just the union of these sets, so that

$$R^+ = \{(a,a), (a,b), (a,c), (a,d), (a,e), (b,c), (b,d), (b,e), (c,d), (c,e), (d,e)\}$$

Method 2. We apply Corollary 4.6.1, consult the graph of R, and compute paths of all possible lengths from each vertex. Then, for example, we know $(a,d) \in R^+$ because there is a path of length 3 from a to d, namely, the path $(a,b), (b,c), (c,d)$. Similarly, $(a,b) \in R^+$ because there is a path of length 1 and length 2 from a to b. Likewise, $(a,e) \in R^+$

because there is a path of length 3 from a to e (and a path of length 4 as well).

We can continue this process, but where does it end? Note that there is a path of length 20 from a to c by going around loop (a,a) 18 times, then along (a,b), and finally along (b,c). But note that since there are only 5 vertices, any path of length greater than 5 between vertices x and y must have repeated a vertex and so must include a circuit. Therefore, such a path from x to y could be shortened to one of length less than or equal to 5. In other words, $R^+ = R \cup R^2 \cup R^3 \cup R^4 \cup R^5$, and we need not consider paths of length greater than 5. Thus, we can again conclude that R^+ is the same set as above.

In general, if R is a relation on a set of n vertices, we can extend the reasoning in the above example and conclude that any path can be shortened to a path of length at most n. Therefore, in general, $R^+ = R \cup R^2 \cup \ldots \cup R^n$, if R is a relation on a set with n elements.

Based on the definitions of paths it is possible to define exactly what is meant by a graph being connected. There are three kinds of connectivity relevant to digraphs that we shall discuss now.

Definition 4.6.2. A pair of vertices in a digraph are **weakly connected** if there is a nondirected path between them. They are **unilaterally connected** if there is a directed path between them. They are **strongly connected** if there is a directed path from x to y *and* a directed path from y to x. A graph is (weakly, unilaterally, or strongly) connected if every pair of vertices in the graph is (weakly, unilaterally, or strongly) connected. A subgraph A^1 of a graph A is a (weakly, unilaterally, or strongly) **connected component** if it is a maximal (weakly, unilaterally, or strongly) connected subgraph; that is, there is no (weakly, unilaterally, or strongly) connected subgraph of A that properly contains A^1.

Example 4.6.3. The graph shown in Figure 4-18 illustrates all of the connectivity relations.

The entire graph, comprising vertices $\{a,b,c,d,e,f,g\}$ and their incident edges, is not even weakly connected. The subgraph comprising vertices $\{a,b,c\}$ and the edges $\{(a,b), (c,b)\}$ is weakly connected, but not connected by either of the stronger definitions, since there is no directed path between a and c. The subgraph comprising vertices $\{d,e,f,g\}$ and their incident edges is unilaterally connected, but not strongly connected, since there is no directed path from e to d. The only nontrivial strongly connected subgraph is comprised of vertices $\{e,f,g\}$ and the edges $\{(g,e),$

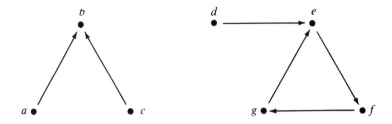

Figure 4-18. A graph illustrating connectivities.

(e,f), (f,g)}. This graph has two weakly connected components, which are subgraphs with vertex sets $\{a,b,c\}$ and $\{d,e,f,g\}$. There are a number of unilaterally connected components. Two of them are the subgraphs $(\{a,b\}, \{(a,b)\})$, and $(\{b,c\}, \{(c,b)\})$. The strongly connected components are the individual vertices a, b, c, and d, and the subgraph $(\{e,f,g\}, \{(e,f),$ $(f,g), (g,e)\})$.

Note that when the term "connected" is used without further qualification it will mean weakly connected.

Exercises for Section 4.6

1. Consider the digraph in Figure 4-16.
   (a) Find all of the simple directed paths from a to d.
   (b) Find all of the simple nondirected paths from a to d.
   (c) Find all of the directed cycles that start at d.
   (d) Find all of the nondirected cycles that start at d.

2. Consider the digraph in Figure 4-17.
   (a) Find all the weakly connected components.
   (b) Find all the unilaterally connected components.
   (c) Find all the strongly connected components.

3. Draw a digraph with 5 vertices that has 4 strongly connected components, 2 weakly connected components, and 3 unilaterally connected components, exactly.

4. Draw a digraph with 4 vertices and one strongly connected component.

5. Draw a digraph with 5 vertices whose longest simple directed path has length 3, whose longest simple undirected path has length 5, whose longest directed cycle has length 2, and whose longest undirected cycle has length 5.

6. What is the longest length possible for a simple directed path in a digraph with n vertices? How about the longest cycle?

7. Define the relation C on the vertices of a digraph such that $x \, C \, y$ iff there is a nondirected path from x to y. Prove or disprove:
   (a) C is an equivalence relation.
   (b) the equivalence classes of C are each weakly connected components of the digraph.

8. Give the definition of an equivalence relation on the vertices of a digraph such that the equivalence classes together with the edges between these vertices are the strongly connected components.

9. Define the relation D on the vertices of a digraph such that $x \, D \, y$ iff there is a directed path from x to y. Prove or disprove:
   (a) D is not an equivalence relation on the vertices of a digraph.
   (b) D is a partial order iff the digraph has no cycles of length greater than one.

10. Let $A = (V, E)$ be a digraph. Define $A^1 = (V, E^1)$ where $(x, y) \in E^1$ iff $(x, y) \in E$ or $(y, x) \in E$. Prove or disprove:
   (a) E^1 is an equivalence relation.
   (b) A^1 is unilaterally connected iff A is weakly connected.

11. Let $A = (V, E)$ be a digraph. Define $A^+ = (V, E^+)$. Prove or disprove that A^+ is strongly connected iff A is unilaterally connected.

12. Prove that the following definitions of E^n are equivalent:
   (a) $E^1 = E$ and $E^n = E^{n-1} \cdot E$ for $n > 1$;
   (b) $E^1 = E$ and $E^n = E \cdot E^{n-1}$ for $n > 1$.

13. Prove that the following definitions of E^+ are equivalent:
   (a) $E^+ = \bigcup_{i=1}^{\infty} E^i$;
   (b) $C^1 = E$, $C^{n+1} = C^n \cdot C^n$, and $E^+ = \bigcup_{i=1}^{\infty} C^i$.

14. Let R be a relation on a set A and let $S = R^2$. Prove that $(x, y) \in S^+$ iff there is a directed path in R from x to y of even length.

15. Prove by induction: If P is a path from a vertex x to a vertex y, then P contains a simple path.

16. Prove that a digraph G is unilaterally connected iff there is a directed path in G containing all the vertices of G.

17. Prove that a digraph G is strongly connected iff there is a closed directed path containing every vertex in G. (A path (v_0, v_1), $(v_1, v_2), \ldots, (v_{n-1}, v_n)$ is *closed* if $v_0 = v_n$.)

Selected Answers for Section 4.6

1. (a) $(a,b), (b,c), (c,d)$
        $(a,c), (c,d)$

(b) the above, plus
$(a,b),(d,b)$
$(a,c),(b,c),(d,b)$
$(e,a),(d,e)$

(c) $(d,b),(b,c),(c,d)$
$(d,e),(e,a),(a,c),(c,d)$
$(d,e),(e,a),(a,b),(b,c),(c,d)$

(d) the above, plus
$(d,b),(a,b),(e,a),(d,e)$
$(d,b),(a,b),(a,c),(c,d)$
$(d,b),(b,c),(a,c),(e,a),(d,e)$
$(c,d),(b,c),(a,b),(e,a),(d,e)$
$(c,d),(b,c),(d,b)$
$(c,d),(a,c),(e,a),(d,e)$
$(c,d),(a,c),(a,b),(d,b)$
$(d,e),(e,a),(a,b),(d,b)$
$(d,e),(e,a),(a,c),(b,c),(d,b)$

2. (a) There are two weakly connected components. They are the subgraph with vertices $\{a,b,c\}$ and all incident edges, and the subgraph with vertices $\{d,e,f,g\}$ and all incident edges.

(b) There are four unilaterally connected components. They are the subgraph with vertices $\{a,b\}$ and edge $\{(a,b)\}$, the subgraph with vertices $\{b,c\}$ and edge $\{(c,b)\}$, the subgraph with vertices $\{d,e\}$ and edge $\{(d,e)\}$, and the subgraph with vertices $\{d,e,f,g\}$ and edges $\{(d,e),(e,f),(f,g),(g,e)\}$.

(c) There are five strongly connected components. They are: $(\{a\},\varnothing);(\{b\},\varnothing);(\{c\},\varnothing);(\{d\},\varnothing);(\{e,f,g\},\{(e,f),(f,g),(g,e)\}$.

3.

4.

5.

6. Since only the starting vertex may be repeated on a simple path, each edge except for the last must go to a new vertex. If there are only n vertices in the graph, then a simple path may not have length greater than n. The same is true of a cycle, which is a simple path.

8. Define $v \ E \ w$ iff there is a directed path from v to w and there is a directed path from w to v. Of course it needs to be verified that this

is an equivalence relation, and that the equivalence classes are the strongly connected components.

12. Let $E(a,n)$ denote E^n as defined in (a), and let $E(b,n)$ denote E^n as defined in (b). We wish to prove that $E(a,n) = E(b,n)$ for all $n > 0$. The proof is by induction on n. For $n = 1$, we know that $E(a,n) = E = E(b,n)$. For $n = 2$, we know that $E(a,n) = E \cdot E = E(b,n)$. For $n > 2$, let us assume that $E(a,n - 1) = E(b,n - 1)$. $E(a,n) = E(a,n - 1) \cdot E$, and, by induction, $E(a,n - 1) = E(b,n - 1)$. Thus $E(a,n) = E(b,n - 1) \cdot E$. We know that $n - 1 > 1$, and so $E(a,n) = E(b,n - 1) \cdot E = E \cdot E(b,n - 2) \cdot E$. Thus, since $E(b,n - 2) = E(a,n - 2)$, by induction, we have $E(a,n) = E \cdot E(b,n - 2) \cdot E = E \cdot E(a,n - 2) \cdot E = E \cdot E(a,n - 1) = E \cdot E(b,n - 1) = E(b,n)$.

4.7 DIRECTED GRAPHS AND ADJACENCY MATRICES

We have seen that digraphs and binary relations are closely related to each other. They are both related just as closely to another important mathematical structure: matrices. We will now therefore review the definition of matrix, to see how a matrix can represent a binary relation and how this matrix representation can be useful in extracting information about a digraph.

Definition 4.7.1. Let S be any set and m,n be any positive integers. An $m \times n$ **matrix** over S is a two-dimensional rectangular array of elements of S with m rows and n columns. The elements are doubly indexed, with the first index indicating the row number and the second index indicating the column number, as shown in Figure 4-19.

Matrices are interesting in their own right and are the subject of a rich mathematical theory. However, most of this theory is beyond the scope of

$$
A = \begin{array}{c} \\ \text{Row 1} \\ \\ \text{Row 2} \\ \\ \vdots \\ \\ \text{Row } m \end{array}
\begin{array}{cccc}
\text{Column 1} & \text{Column 2} & \ldots & \text{Column } n \\
\left[\begin{array}{cccc}
A(1, 1) & A(1, 2) & \ldots & A(1, n) \\
A(2, 1) & A(2, 2) & \ldots & A(2, n) \\
\vdots & \vdots & & \vdots \\
A(m, 1) & A(m, 2) & \ldots & A(m, n)
\end{array}\right]
\end{array}
$$

Figure 4-19

the present book. We are presently only interested in matrices for one special application—the representation of digraphs or binary relations. For this purpose we will consider **Boolean matrices,** which are matrices over the set $\{0,1\}$. There is a natural one-to-one correspondence between the binary relations and the square Boolean matrices, as demonstrated by the following definition.

Definition 4.7.2.   Let E be any binary relation on a finite set $V = \{v_1,\ldots,v_n\}$. The **adjacency matrix** of E is the $n \times n$ Boolean matrix A defined by $A(i,j) = 1$ iff $(v_i,v_j) \in E$.

Note that every $n \times n$ Boolean matrix is the adjacency matrix of a unique binary relation on V. *Note also that the interpretation of adjacency matrices depends on the presumed ordering of the set V.*

Example 4.7.1.   The relation $\le$ on the set $\{0,1,2,3,4\}$ is represented by the adjacency matrix

$$\begin{bmatrix} 1 & 1 & 1 & 1 & 1 \\ 0 & 1 & 1 & 1 & 1 \\ 0 & 0 & 1 & 1 & 1 \\ 0 & 0 & 0 & 1 & 1 \\ 0 & 0 & 0 & 0 & 1 \end{bmatrix}$$

Notice that all the diagonal entries in the matrix above are 1. This is because $\le$ is reflexive, and will be true for any reflexive relation.

Example 4.7.2.   The digraph in Figure 4-20 is represented by the adjacency matrix,

$$\begin{bmatrix} 0 & 1 & 0 & 0 & 0 \\ 0 & 0 & 1 & 0 & 0 \\ 0 & 0 & 0 & 1 & 1 \\ 1 & 1 & 0 & 0 & 0 \\ 0 & 0 & 0 & 0 & 0 \end{bmatrix}$$

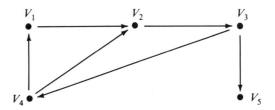

Figure 4-20

It is often very convenient to use the Boolean matrix representation of binary relations when computations involving relations are performed. By using this representation, and a generalized way of combining two scalar operators to get a matrix operator (defined below), it is possible to describe in simple terms algorithms for extracting much useful information from a binary relation.

In the following definition we define an **operator on operators,** that takes two arbitrary scalar operators $\oplus$ and $\otimes$ as arguments and yields a matrix operator $\oplus . \otimes$ as result. (The notation used for this operator here is chosen because it is the notation used in the programming language APL for a similar operator.)

Definition 4.7.3. Let S be any set and let $\oplus$ and $\otimes$ be any two binary operators defined on the elements of S. Assume that $\oplus$ is associative. The **inner product** of $\oplus$ and $\otimes$, denoted by $\oplus . \otimes$, is defined for $n \times n$ matrices over S by $A \oplus . \otimes B = D$ such that

$$D(i, j) = (A(i,1) \otimes B(1, j)) \oplus \ldots \oplus (A(i,n) \otimes B(n, j)).$$

By extension, when we wish to iteratively apply such an inner product to a single matrix, we write $(\oplus . \otimes)^k A$ to denote the matrix A in the case that $k = 1$ and for $k > 1$ to denote the matrix $((\oplus . \otimes)^{k-1} A \oplus . \otimes A)$.

For any single scalar operator $\oplus$ we will also write $A \oplus B$ for matrices A and B to denote the matrix E such that

$$E(i, j) = A(i, j) \oplus B(i, j).$$

Note that in the case that $\oplus$ and $\otimes$ are the usual addition and multiplication operators on the integers or the real numbers, the inner product defined above is the usual definition of matrix product.

Example 4.7.3. Let S be the set $\{0,1\}$ and let $\oplus$ and $\otimes$ be the operators OR and AND defined by the table:

x	y	x OR y	x AND y
0	0	0	0
0	1	1	0
1	0	1	0
1	1	1	1

Let A be the matrix in Example 4.7.1 and B be the matrix in Example 4.7.2. The inner product A OR.AND B is the matrix

$$\begin{bmatrix} 1 & 1 & 1 & 1 & 1 \\ 1 & 1 & 1 & 1 & 1 \\ 1 & 1 & 0 & 1 & 1 \\ 1 & 1 & 0 & 0 & 0 \\ 0 & 0 & 0 & 0 & 0 \end{bmatrix}$$

We will show how two representative elements of this matrix are obtained. The entry in row 2, column 2 is obtained from the second row of A and the second column of B, and is (0 AND 1) OR (1 AND 0) OR (1 AND 0) OR (1 AND 1) OR (1 AND 0) = 1. The entry in row 2, column 3 is obtained from the second row of A and the third column of B, and is (0 AND 1) OR (0 AND 1) OR (1 AND 1) OR (0 AND 0) OR (0 AND 0) = 1.

Definition 4.7.4. When the two operations $\oplus$ and $\otimes$ are the particular operations of OR and AND, respectively, then we shall refer to their inner product as the **Boolean product**.

The following theorem expresses an important computational relationship between binary relations and these operations on Boolean matrices.

Theorem 4.7.1. Let R_A and R_B be binary relations on a set $V = \{v_1, \ldots, v_n\}$, represented by adjacency matrices A and B respectively. Then the Boolean product A OR.AND B is the adjacency matrix of the relation $R_A \cdot R_B$, and the matrix $(\text{OR.AND})^n A$ is the adjacency matrix of the relation R_A^n. Here $(\text{OR.AND})^2 A$ means A OR.AND A.

Proof. Recall that $R_A \cdot R_B = R_C$ where $R_C = \{(v_i,v_k) \mid (v_i,v_j) \in R_A$ and $(v_j,v_k) \in R_B$ for some $j\}$. Thus if C is the adjacency matrix of R_C then

$C(i,k) = 1$ iff for some j, $A(v_i,v_j) = 1$ and $A(v_j,v_k) = 1$. This is exactly the same as saying $C(i,k) = (A(i,1)$ AND $B(1,j))$ OR ... OR $(A(i,n)$ AND $B(n,j)$, which is $(A$ OR.AND $B)(i,j)$. The theorem is thus a direct consequence of the definitions. □

Corollary 4.7.1. Let A be the adjacency matrix of any binary relation R on a set $V = \{v_1,\ldots,v_n\}$. Then the adjacency matrix of the transitive closure R^+ is given by A OR $(OR.AND)^2 A$ OR ... OR $(OR.AND)^n A$.

Proof. This follows from the preceding theorem and the fact that $R^+ = R \cup R^2 \cup \cdots \cup R^n$. □

Corollary 4.7.2. Let A be the adjacency matrix of any finite binary relation R. The adjacency matrix of the transitive reflexive closure of R, R^*, is given by $(I$ OR A OR $(OR.AND)^2 A$ OR ... OR $(OR.AND)^n A)$, where I is the identity matrix

$$
I = \begin{bmatrix}
1 & 0 & \cdot & \cdot & \cdot & 0 \\
0 & 1 & & & & 0 \\
\cdot & & \cdot & & & \cdot \\
\cdot & & & \cdot & & \cdot \\
\cdot & & & & \cdot & \cdot \\
0 & 0 & \cdot & \cdot & \cdot & 1
\end{bmatrix}.
$$

Proof. This is left to the reader as an exercise. □

Example 4.7.4. Let us apply Corollary 4.7.1 to the relation R of Example 4.6.2. In other words, $A = \{a,b,c,d\}$ and $R = \{(a,a)(a,b),(b,c),(c,d),(c,e),(d,e)\}$. Let A_R^k denote the adjacency matrix of the relation R^k. By Theorem 4.7.1, A_R^2 is the Boolean product of A_R and A_R, that is, $A_R^2 = A_R$ OR.AND A_R. Likewise, for $k > 1$, A_R^k is the Boolean product of A_R^{k-1} and A_R.

We see that

$$
A_R = \begin{bmatrix}
1 & 1 & 0 & 0 & 0 \\
0 & 0 & 1 & 0 & 0 \\
0 & 0 & 0 & 1 & 1 \\
0 & 0 & 0 & 0 & 1 \\
0 & 0 & 0 & 0 & 0
\end{bmatrix}
\qquad
A_R^2 = \begin{bmatrix}
1 & 1 & 1 & 0 & 0 \\
0 & 0 & 0 & 1 & 1 \\
0 & 0 & 0 & 0 & 1 \\
0 & 0 & 0 & 0 & 0 \\
0 & 0 & 0 & 0 & 0
\end{bmatrix}
$$

$$A_R^3 = \begin{bmatrix} 1 & 1 & 1 & 1 & 1 \\ 0 & 0 & 0 & 0 & 1 \\ 0 & 0 & 0 & 0 & 0 \\ 0 & 0 & 0 & 0 & 0 \\ 0 & 0 & 0 & 0 & 0 \end{bmatrix} \qquad A_R^4 = A_R^5 = \begin{bmatrix} 1 & 1 & 1 & 1 & 1 \\ 0 & 0 & 0 & 0 & 0 \\ 0 & 0 & 0 & 0 & 0 \\ 0 & 0 & 0 & 0 & 0 \\ 0 & 0 & 0 & 0 & 0 \end{bmatrix}.$$

Of course, we can compute the separate matrices by appeal to the graph and observing paths of different lengths. For instance, the second row of A_R^2 means that there is a path of length 2 from b to d and one of length 2 from b to e. The third row of A_R^2 reflects the fact that there is a path of length 2 from c to e.

Now if $\vee$ stands for the Boolean operation OR, then the adjacency matrix of the transitive closure R^+ is:

$$A_R{}^+ = A_R \vee A_R^2 \vee A_R^3 \vee A_R^4 \vee A_R^5 = \begin{bmatrix} 1 & 1 & 1 & 1 & 1 \\ 0 & 0 & 1 & 1 & 1 \\ 0 & 0 & 0 & 1 & 1 \\ 0 & 0 & 0 & 0 & 1 \\ 0 & 0 & 0 & 0 & 0 \end{bmatrix}$$

Making use of operations other than "AND" and "OR", and integer matrices, it is possible to extract other useful information from an adjacency matrix. Among such useful information is the number of distinct paths of a given length from one vertex to another in a digraph, the length of the shortest path between two vertices, and the length of the longest path between two vertices.

Theorem 4.7.2. Suppose $G = (V, E)$ is a directed graph and A is its adjacency matrix. Let $\oplus$ and $\otimes$ denote the operations

$$x \oplus y = \begin{cases} x \text{ if } x > y \\ y \text{ otherwise,} \end{cases}$$

$$x \otimes y = \begin{cases} x + y \text{ if } x > 0 \text{ and } y > 0 \\ 0 \text{ otherwise,} \end{cases}$$

and

$$L^k = \begin{cases} A \text{ for } k = 1 \\ L^{k-1} \oplus (L^{k-1} \oplus \cdot \otimes A) \text{ for } k > 1. \end{cases}$$

Then $L^k(i,j)$ is the length of the longest nontrivial directed path from v_i to v_j that has length $\leq k$, unless $L^k(i,j) = 0$, in which case no such path exists.

Proof.   The proof is by induction on k. For $k = 1$ we have $L^k = A$. Since $A(i,j) = 1$ iff there is a directed path of length 1 from v_i to v_j, the conclusion follows directly from the definition of adjacency matrix. For $k > 1$, we can assume the theorem holds for smaller values of k. $L^k(i,j) = L^{k-1}(i,k) \oplus (L^{k-1}(i,1) \otimes A(1,j)) \oplus \cdots \oplus (L^{k-1}(i,n) \otimes A(n,j))$. In other words, $L^k(i,j)$ is the maximum of $L^k(i,j)$ and all of $L^{k-1}(i,t) \otimes A(t,j)$ for $1 \leq t \leq n$. We will consider two cases:

Case 1.   Suppose there is no nontrivial directed path from v_i to v_j of length $\leq k$. We need to show that $L^{k-1}(i, j)$ and all of the $L^{k-1}(i,t) \otimes A(t,j)$ are zero in this case. We will argue that if one of these is nonzero there must be a nontrivial directed path from v_i to v_j of length $\leq k$. If $L^{k-1}(i, j)$ is nonzero, then by induction there is a nontrivial path of length $\leq k - 1$ from v_i to v_j. If $L^{k-1}(i,t) \otimes A(t, j)$ is nonzero for some t, then by induction there is a directed path from v_i to v_t and an edge from v_t to v_j that can be combined to form a directed path of length $\leq k$ from v_i to v_j. (This is shown in Figure 4-21.)

Case 2.   Suppose there is a nontrivial directed path from v_i to v_j of length $\leq k$. Choose one such path that has maximum length. Let l be the length of this path and (v_t, v_j) be the last edge on it. By definition of adjacency matrix, $A(t,j) = 1$. We break this into subcases, depending on whether l is greater than 1.

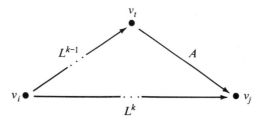

Figure 4-21

(a) If $l = 1$, since we are considering the case where $k > 1$, we know that l is also the length of the longest path of length $\leq k - 1$, and so by induction $L^{k-1}(i, j) = 1$. This means that $L^k(i, j)$ is *at least* 1, but we are not done, since it might be that one of the $L^{k-1}(i,t)\ A(t, j)$ is greater. To see that this cannot be, suppose that $L^{k-1}(i,t) \otimes A(t, j)$ is nonzero. Then, as was already seen in Case 1, there must be a directed path from v_i to v_j of length >1 and $\leq k$, which would be longer than l, a contradiction (see Figure 4-21). Thus $L^k(i, j) = 1$.

(b) If $l > 1$, there is a nontrivial directed path from v_i to v_t of length $l - 1$. There can be no directed path from v_i to v_t longer than $l - 1$ and shorter than k, since otherwise such a path could be joined with (v_t, v_j) to obtain a path longer than l from v_i to v_j, contradicting the definition of l. Thus, by induction, $L^{k-1}(i,t) = l - 1$. It follows immediately from the definition of $\otimes$ that $L^{k-1}(i,t) \otimes A(t, j) = l$. Once again, we have only shown that $L^k(i, j)$ is *at least* l. Suppose that $L^k(i, j)$ is greater than l. This can only be if either $L^{k-1}(i,j)$ is greater than l or some term $L^{k-1}(i,t) \otimes A(t, j)$ is greater than l. $L^{k-1}(i,j)$ cannot be greater than l, since this would contradict the induction hypothesis. On the other hand, if $L^{k-1}(i,t) \otimes A(t,j)$ is greater than l, $L^{k-1}(i,t)$ must be greater than $l - 1$. By induction, there must be a path of length $L^{k-1}(i,t)$ from v_i to v_t and this must not be longer than $k - 1$. Such a path could be joined with (v_t, v_j) to obtain a path of length greater than l and less than or equal to k from v_i to v_j, a contradiction of the definition of l (see Figure 4-21). This concludes the proof that $L^k(i, j) = l$. $\square$

Note that this theorem deals with paths which in our definition need not be simple and so may include repeated vertices and edges. Finding the length of the longest *simple* path between two vertices is a *much* harder problem.

Adjacency matrices provide a simple and very elegant means of representing digraphs and for describing algorithms on digraphs. In the next section we will look at one such algorithm, due to S. Warshall, for computing the transitive closure of a relation. Before moving on, however, we must point out that adjacency matrices are only one of several convenient representations of digraphs, and are not the best representation for some purposes. They are especially inefficient for graphs with many vertices and very few edges. Regrettably, however, a complete study of graph representations and algorithms is beyond the scope of this book.

Exercises for Section 4.7

1. (a) Give the adjacency matrix of the digraph $G = (\{a,b,c,d\},R)$, where $R = \{(a,b),(b,c),(d,c),(d,a)\}$.

(b) Give the Boolean matrix representation of the transitive closure, R^+.

(c) Give the Boolean matrix representation of the transitive reflexive closure, R^*.

(d) Give the matrix L^3 defined in Theorem 4.7.2 for this digraph.

2. (a) Give the adjacency matrix of the digraph shown in Figure 4-16.

(b) Give the Boolean matrix representation of the transitive closure of the relation represented by this digraph.

(c) Give the Boolean matrix representation of the transitive reflexive closure of this relation.

(d) Give the matrix L^2 defined in Theorem 4.7.2 for this digraph.

3. Prove Corollary 4.7.2.

4. Give alternative definitions for $\oplus$ and $\otimes$ so that substituting them for the definitions in Theorem 4.7.2 make $L^k(i,j)$ the length of the *shortest* nontrivial directed path from v_i to v_j of length $\leq k$.

5. Suppose that G is a directed graph, $v_1, v_2, \ldots, v_n$ are the vertices of G, and A is the adjacency matrix of G. Let A^k denote the inner product of A^{k-1} and A where $k \geq 1$ and $\oplus$ and $\otimes$ denote the usual addition and multiplication operations.

(a) Prove that the i–j th entry of A^k, which we denote by $A^k(i, j)$, is the number of distinct nontrivial directed paths from v_i to v_j of length exactly k.

(b) Then prove that $M = A + A^2 + \ldots + A^k$ is a matrix whose i–j th entry $M(i, j)$ is the total number of distinct directed nontrivial paths from v_i to v_j of length at most k.

(c) For the graph G in exercise 1(a) compute A^2, $A + A^2$, and $A + A^2 + A^3 + A^4$ and interpret the entries of each of the matrices.

6. Suppose that R and S are relations defined on a set A. Suppose also that A_R and A_S are the adjacency matrices of R and S respectively. Prove that

(a) $A_R \vee A_S$ is the adjacency matrix of $R \cup S$, where $\vee$ denotes the binary operation OR.

(b) $A_R \wedge A_S$ is the adjacency matrix of $R \cap S$, where $\wedge$ is the binary operation AND.

(c) The transpose of A_R is the adjacency matrix of R^{-1}. (The transpose of a matrix M is a matrix M^T where $M^T(i, j) = M(j, i)$.)

7. In the manner of Corollary 4.7.1, describe how to obtain the adjacency matrix of the transitive reflexive closure of a relation R defined on a set A.

8. For each of the 6 properties of relations defined in Section 4.2, give an interpretation of their definition in terms of adjacency matrices.

Selected Answers for Section 4.7

1. (a) $\begin{bmatrix} 0 & 1 & 0 & 0 \\ 0 & 0 & 1 & 0 \\ 0 & 0 & 0 & 0 \\ 1 & 0 & 1 & 0 \end{bmatrix}$

(b) $\begin{bmatrix} 0 & 1 & 1 & 0 \\ 0 & 0 & 1 & 0 \\ 0 & 0 & 0 & 0 \\ 1 & 1 & 1 & 0 \end{bmatrix}$

(c) $\begin{bmatrix} 1 & 1 & 1 & 0 \\ 0 & 1 & 1 & 0 \\ 0 & 0 & 1 & 0 \\ 1 & 1 & 1 & 1 \end{bmatrix}$

(d) $\begin{bmatrix} 0 & 1 & 2 & 0 \\ 0 & 0 & 1 & 0 \\ 0 & 0 & 0 & 0 \\ 1 & 2 & 3 & 0 \end{bmatrix}$

4.7.1   Warshall's Algorithm

This section, and the rest of this chapter, deals with applications of the theory of digraphs and relations to the study of algorithms. Several aspects of algorithms lend themselves to mathematical analysis. The first and most basic aspect is the *correctness* of an algorithm. An algorithm is considered to be correct if, when performed on input that satisfies its input requirements, the algorithm terminates and yields output that satisfies its output requirements. Correctness is frequently broken into two parts: *termination* and *partial correctness*. Termination means that the algorithm halts on every input that meets its input requirements. Partial correctness means that if ever the algorithm halts, and if the input met the input requirements, the output must meet all the output requirements. Precisely specifying the input and output requirements of an algorithm, and then verifying its correctness, are very challenging but very important applications of the techniques of mathematical definition and proof.

After correctness, the next most basic issue in the study of algorithms is *efficiency*. The most widely accepted approach to studying the efficiency of algorithms is in terms of complexity measures, or functions that relate the resource requirements of an algorithm, such as time or memory, to the size of the input. Because the details of how a computer algorithm is implemented, such as the computer on which it is executed, may affect its exact resource requirements, and because exact figures are very hard to obtain, it is customary to derive formulas that describe the rate of growth of the resources required by an algorithm approximately, within the tolerance of a constant factor. Big-O notation is frequently used in this context.

For example, if we assert that the running time t of an algorithm is $O(n^2)$, it may be that this time is expressible in the form

$$t = a_0 + a_1 n + a_2 n^2,$$

and then such things as program details and execution speeds affect only the coefficients a_0, a_1, and a_2, but do not alter the *order* of the running time. (Note, however, that the analysis of any algorithm generally assumes that certain primitive operations can be performed with constant cost. Any implementation that violates this assumption may increase the order of the running time. In this way, the choice of data structures may affect the order of the running time of an algorithm.)

We emphasize that the order of the running time indicates the rate of increase of running time with input size. Thus, for example, if the running times of two algorithms are, respectively, $O(n^2)$ and $O(n^3)$, then we know that for large values of n, a two-fold increase in input size will increase the running time of the first algorithm by a factor of 4 and the second by a factor of 8.

Essentially, then, it is the time complexity of an algorithm that determines how large a problem an algorithm can solve. For example, if some graph-theoretic algorithm has a running time of order 2^n, where n is the number of vertices, then an increase of 10 in the number of vertices—say from 10 to 20, or from 20 to 30—will increase the running time by a constant multiple of 2^{10}, or by approximately 1000. Also, for such an algorithm, even a tenfold increase in computer speed adds only three to the size of problem that can be solved in a given time, since $2^{3.3}$ is approximately 10.

Therefore, the order of the running time of an algorithm gives us an estimate of its practical feasibility. Nevertheless, we should not forget that the order of the complexity function of an algorithm is a measure of the *asymptotic* performance of the algorithm, as the size of the input goes to infinity. It is possible, for small inputs, that an algorithm of higher order actually can be more efficient than one of lower order. There is a tendency, for instance, to assume that an algorithm with running time of order n^2 is better than one with running time of order n^3. Indeed, that assumption is correct for large values of n, but in some practical situations the latter *may* have a better performance for small values of n.

An algorithm is said to be **polynomial-bounded** if its running time is bounded by a function of order n^k, where n is the input size and k is a constant. Accordingly, an algorithm is regarded as being relatively *fast* or *efficient* if it is polynomial-bounded, and *inefficient* otherwise. Therefore, a problem is said to be *easy* if some polynomial-bounded algorithm has been found to solve it.

Associating polynomial-boundedness with computational efficiency is theoretically justified in that, above a certain input size, a polynomial-bounded algorithm will always have a smaller running time than a nonpolynomial-bounded algorithm. (Of course, for very small input sizes the nonpolynomial algorithm could have a better performance.)

We shall show in the remainder of Chapter 4 that there are several graph-theoretic problems that have efficient algorithms for their solution. (Kruskal's algorithm, discussed in Chapter 5, is also an efficient algorithm.) On the other hand, there are a number of important graph problems for which no efficient algorithms have ever been found. Among these more difficult problems, we have:

1. *The Subgraph Isomorphism Problem.* Given two graphs G and H, does G contain a subgraph isomorphic to H?
2. *The Planar Subgraph Problem.* Given a graph $G = (V,E)$ and a positive integer $k \leq |E|$, is there a subset E' of the edges of G with $|E'| \geq k$ such that $G' = (V,E')$ is planar? (Planarity is discussed in Chapter 5.)
3. *The Hamiltonian Cycle Problem.* Does a given graph contain a Hamiltonian cycle? (Hamiltonian cycles are discussed in Chapter 5.)
4. *The Chromatic Number Problem.* Given a graph G and a positive integer k, is it possible to color the vertices of G with k colors in such a way that no two adjacent vertices are painted the same color? (Chromatic numbers are discussed in Chapter 5.)

All the problems listed above belong to a larger class of problems called **NP-complete** problems, a class of problems introduced by Stephen Cook of the University of Toronto in 1972. This class of NP-complete problems is now known to contain literally hundreds of different problems notorious for their computational intractability.

NP-complete problems, which occur in such areas as computer science, mathematics, operations research, and economics, are in some sense the hardest problems that can be solved in polynomial time by algorithms that are allowed to "guess" and then verify that their guesses are correct. These problems have two important properties: First, all are equivalent in the sense that all or none can be solved by efficient algorithms. (More precisely, each NP-complete problem can be transformed into any other

NP-complete problem by a polynomial-bounded transformation; clearly if a problem is easily transformable into an easy problem, then it is also an easy problem. Thus, the NP-complete problems are all easy or none of them are easy.) Second, the running times of all methods currently known for finding general solutions for any of the NP-complete problems can always blow up exponentially in a manner similar to the behavior of 2^n. (Even for relatively small values of n, 2^n is a very large number. For example, when $n = 70$, a computer that can perform 10^6 operations per second would require 300,000 centuries to perform 2^{70} operations.)

Since many of the NP-complete problems have been studied intensively for decades, and no efficient algorithms have been found for any of them, it seems likely that no such algorithms exist. Indeed, many mathematicians strongly suspect that the inability to find an efficient solution procedure is inherent in the nature of NP-complete problems: they believe that no such procedure can exist.

NP-complete problems may be rightfully considered "hard" problems, since no efficient algorithm is known for solving any of them. The reader should be aware, however, that there exists an infinite hierarchy of classes of problems of all degrees of "hardness", including many that can be proven to be much harder than the NP-complete problems. The study of such complexity classes, and the classification of problems according to complexity of the algorithms that can solve them, is an important branch of computer science known as **computational complexity** theory.

The algorithm described in this section is one for which the issues of termination and complexity are rather simple. The outer loop is always performed n times, where n is the dimension of the input matrix. Each iteration of this loop looks at each pair (i,j) of vertices, and performs a constant-cost operation on the pair. The total running time of the algorithm can thus be estimated to be $O(n^3)$, and its total memory requirement to be $O(n^2)$, for the matrix. The issue of partial correctness, however, is less simple, and it is to that issue that we will pay the most attention.

The following algorithm, for computing the transitive closure of a relation, is due to S. Warshall (1962). Besides being a way to calculate the transitive closure of a relation this algorithm can be generalized to solve a number of other related problems.

Algorithm 4.7.1 Computing the Transitive Closure

Input: The adjacency matrix M of a digraph (V,E), where $V = \{v_1, \ldots, v_n\}$.

Output: A new adjacency matrix M, which is the adjacency matrix of (V,E^+).

Method: For each k from 1 up to n (sequentially) do the following:

For each pair (i,j) such that $1 \le i,j \le n$ (in any order) do the following:

$(*)$ If $M(i,k) = 1$, $M(k,j) = 1$, and $M(i,j) = 0$ then change $M(i,j)$ to 1.

The basic idea behind this algorithm is shown in Figure 4-22. The reasoning that goes with the figure is that if there is a path from vertex v_i to vertex v_j that traverses only vertices in $\{v_1,\ldots,v_k\}$ it must fall into one of two cases. It may be that the path traverses only vertices in $\{v_1,\ldots,v_{k-1}\}$. Otherwise, it traverses only vertices in $\{v_1,\ldots,v_{k-1}\}$ to get to v_k for the first time, may visit v_k several more times via subpaths that traverse only vertices in $\{v_1,\ldots,v_{k-1}\}$, and finally reaches v_j via a subpath that traverses only vertices in $\{v_1,\ldots,v_{k-1}\}$. By incrementing k successively from 0 through n, we eventually consider all paths between v_i and v_j.

Another way of looking at it is that for each vertex v_k and all the paths through it, the main loop of the algorithm considers all of the possible two-step paths from v_i to v_j that go into and out of v_k, and, if any are found, the algorithm builds a bypass (v_i,v_j), provided such a bypass doesn't already exist. Ultimately, each vertex is bypassed, which means that for each path in the original graph there will be a direct connection (edge) in the result.

To see that this algorithm works, it is useful to view the progress of the algorithm as a sequence of n stages, as k successively takes on the values $1,\ldots,n$. Let M_0 denote the initial matrix M. Let M_k denote the matrix at the end of the kth stage. Note that no entry of M is ever set to 0, so that

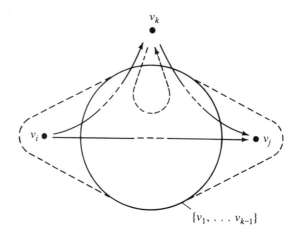

Figure 4-22. Warshall's algorithm.

$M_k(i,j) = 1$ implies that $M_{k+1}(i,j) = 1$. What we wish to prove is that M_n is the adjacency matrix of (V,E^+). This will follow from two separate theorems.

Theorem 4.7.3 $M_k(i,j) = 1$ if there is a nontrivial directed path from v_1 to v_j that traverses only vertices in $\{v_1,\ldots,v_k\}$.

In order to prove this theorem we will need a lemma.

Lemma 4.7.1   If there is a directed path in a digraph from vertex a to vertex b and S is the set of vertices traversed by this path, then for any vertex c in S there exist directed paths from a to c and from c to b such that each traverses only vertices in $S - \{c\}$.

Proof.   The proof of Lemma 4.7.1 is by induction on the size of S. If $S = \varnothing$, the lemma is trivially true, since there is no vertex c in S. Suppose S is nonempty and c is an element of S. Let $e_1,\ldots,e_k$ be a directed path from a to b that traverses exactly the vertices in S. By the definition of traverse, there is at least one edge e_i in the path that is incident to c. Let e_k be the first edge on the path that is incident to c and let e_j be the last edge that is incident from c. We know that $e_i = (x,c)$ and $e_j = (c,y)$ for some x and y, from the definition of path. Thus $e_1,\ldots,e_i$ is a path from a to c and $e_j,\ldots,e_k$ is a path from c to b. Neither of these traverses c, but both are subpaths of $e_1,\ldots,e_n$ so that they traverse only vertices in $S - \{c\}$. $\square$

Now that the lemma is proven we can proceed with the proof of Theorem 4.7.3.

Proof.   The proof is by induction on k. For $k = 0$, we know that $M_0(i,j) = 1$ iff (v_i,v_j) is in E. To prove the theorem for $k > 0$ we assume the theorem holds for smaller k. Suppose there is a nontrivial directed path from v_i to v_j that traverses only vertices in $\{v_1,\ldots,v_k\}$. By the preceding lemma, either there is a nontrivial directed path from v_i to v_j that traverses only vertices in $\{v_1,\ldots,v_{k-1}\}$ or else there are nontrivial paths from v_i to v_k and from v_k to v_j using only vertices in $\{v_1,\ldots,v_{k-1}\}$. In the first case, by the inductive hypothesis, $M_{k-1}(i,j) = 1$, and in the second case $M_k(i,j)$ is set to 1 by the algorithm. $\square$

Taking $k = n$, we obtain an immediate corollary, which is what we really wanted to show. (Note that this is typical of induction proofs—it is often necessary to prove something more specific than what is ultimately desired in order to get a "handle" on the induction.)

Corollary 4.7.3.   If there is a nontrivial directed path from v_i to v_j in (V, E), then $M_n(i,j) = 1$ where $n = |V|$.

This is not enough. We also need to know that $M(i,j) = 1$ at the end of the algorithm *only* if there is a nontrivial directed path from v_i to v_j. The next theorem proves that.

Algorithm 4.7.1 does not specify precisely in what order the pairs (i,j) are to be considered, but in any execution of the algorithm we assume that some order is chosen. Thus the step (*) is performed repeatedly, for different values of i, j and k in some order. By *time t* we will mean that step (*) has been performed t times. Thus time 0 is before the step (*) has been performed at all, time 1 is just after it has been performed once, and so forth. This will enable us to prove the following theorem by induction.

Theorem 4.7.4.   If $M(i,j) = 1$ at any time during the execution of the algorithm, there is a nontrivial directed path from v_i to v_j in (V, E).

Proof.   Suppose $M(i,j) = 1$. The proof is by induction on the time t at which entry $M(i,j)$ is first set to 1. (As observed earlier, the algorithm can never change $M(i,j)$ to 0, so that this can happen at most once for each pair (i,j).) For $t = 0$, we have the original adjacency matrix, so that if $M(i,j) = 1$ at time $t = 0$ it must be true that (v_i, v_j) is an edge in E. For $t > 0$, suppose entry $M(i,j)$ is changed from 0 to 1 at time t. Then at time $t - 1$ it must be true for the current value of k that $M(i,k) = 1$ and $M(k,j) = 1$. By induction, there must be nontrivial directed paths from v_i to v_k and from v_k to v_j. Joining these at v_k we obtain a nontrivial directed path from v_i to v_j. $\square$

Application of Warshall's Algorithm.   To apply Algorithm 4.7.1, we start with M_0, the adjacency matrix of the relation R, and then successively construct the matrices $M_1, M_2, \ldots, M_n$, where n is the number of vertices for the relation R. Moreover, for each $k \geq 1$, we can construct M_k in terms of the previously constructed M_{k-1}. Condition (*) of Algorithm 4.7.1 tells us how to obtain $M_k(i,j)$, the (i,j) entry of M_k, from certain entries of M_{k-1}. In particular, if $M_{k-1}(i,j) = 1$, then $M_k(i,j) = 1$ also. In other words, every entry of M_{k-1} that is a 1 remains a 1 in M_k. Moreover, if $M_{k-1}(i,j) = 0$, then we get a new 1 in position (i,j) of M_k only if there were ones in positions (i,k) and (k,j) of M_{k-1}. To put it another way, $M_k(i,j) = 1$ if $M_{k-1}(i,k) = 1$ and $M_{k-1}(k,j) = 1$. Thus, if $M_{k-1}(i,j) = 0$, we need only examine column k and row k of M_{k-1}, and then if there is a 1 in position i of column k and a 1 in position j of row k, a 1 will be entered in position (i,j) of M_k.

All of this is described succinctly in the following expression:

$$M_k(i,j) = M_{k-1}(i,j) \vee (M_{k-1}(i,k) \wedge M_{k-1}(k,j)).$$

Thus, we may construct M_k from M_{k-1} by employing the following procedures:

Step 1. First transfer to M_k all 1's in M_{k-1}.

Step 2. Record all positions $p_1, p_2, \ldots$ in column k of M_{k-1}, where the entry is 1, and the positions $q_1 q_2, \ldots$ in row k of M_{k-1}, where the entry is 1.

Step 3. Put a 1 in each position (p_s, q_t) of M_k (provided a 1 is not already there from a previous step).

Example 4.7.5. We find the transitive closure of the relation R discussed previously in Examples 4.6.2 and 4.7.4. This time we will apply Warshall's algorithm.

First, we let

$$M_0 = \begin{bmatrix} 1 & 1 & 0 & 0 & 0 \\ 0 & 0 & 1 & 0 & 0 \\ 0 & 0 & 0 & 1 & 1 \\ 0 & 0 & 0 & 0 & 1 \\ 0 & 0 & 0 & 0 & 0 \end{bmatrix}$$

Next we find M_1. To see if there are any new 1's, we observe that M_0 has 1's in position 1 of column 1 and positions 1 and 2 of row 1. Thus, $M_1(1,1) = 1 = M_1(1,2)$, but since the (1,1) and (1,2) entries of M_0 were already transferred to M_1, we introduce no new 1's. Thus, $M_1 = M_0$.

Now we compute M_2 so that in this computation we let $k = 2$. In column 2 of M_1, there is a 1 in position 1 and there is a 1 in position 3 of row 2. Thus, $M_2(1,3) = 1$. This is the only new 1 to be added to M_1. Hence,

$$M_2 = \begin{bmatrix} 1 & 1 & 1 & 0 & 0 \\ 0 & 0 & 1 & 0 & 0 \\ 0 & 0 & 0 & 1 & 1 \\ 0 & 0 & 0 & 0 & 1 \\ 0 & 0 & 0 & 0 & 0 \end{bmatrix}$$

Proceed next to compute M_3. We observe that positions 1 and 2 of column 3 have 1's, while positions 4 and 5 of row 3 have 1's. Thus, $M_3(1,4) =$

$M_3(1,5) = M_3(2,4) = M_3(2,5) = 1$. Therefore,

$$M_3 = \begin{bmatrix} 1 & 1 & 1 & 1 & 1 \\ 0 & 0 & 1 & 1 & 1 \\ 0 & 0 & 0 & 1 & 1 \\ 0 & 0 & 0 & 0 & 1 \\ 0 & 0 & 0 & 0 & 0 \end{bmatrix}$$

Next observe that M_3 has a 1 in positions 1, 2, and 3 of column 4 while row 4 has a 1 in position 5. Thus, $1 = M_4(1,5) = M_4(2,5) = M_4(3,5)$. None of these require changes. Therefore, $M_3 = M_4$.

Finally, M_4 has a 1 in several positions of column 5 but no 1's in row 5. Thus, no new 1's need be added to M_4. Therefore, $M_5 = M_4 = M_3$ is the adjacency matrix of R^+. Of course, it is no surprise that we obtained the same result as in Example 4.7.4.

Corollary 4.7.4. Warshall's algorithm computes the adjacency matrix of the transitive closure (V,E^+) of digraph (V,E).

Exercises for Section 4.7.1

1. Using Warshall's algorithm, compute the adjacency matrix of the transitive closure of the digraph in Figure 4-16.

2. Using Warshall's algorithm, compute the adjacency matrix of the transitive closure of the digraph $G = (\{a,b,c,d,e\}, \{(a,b), (b,c),(c,d),(d,e),(e,d)\}$.

3. Using the basic idea of Warshall's algorithm, devise an algorithm to compute a matrix P such that $P(i,j)$ is the *number* of distinct directed paths (including paths with cycles) from v_i to v_j in a digraph (V,E). Verify that your algorithm is correct. Note that any vertex v_k with a loop or which lies on a directed cycle has infinitely many paths from v_k to v_k. In order to represent this situation, extend the integers to include ∞ and define $x + \infty = \infty$ for each $x \in Z$.

4. Using the basic idea of Warshall's algorithm, devise an algorithm to compute a matrix D such that $D(i,j)$ is the *length* of the *shortest* directed path from v_i to v_j in a digraph (V,E). Verify that your algorithm is correct.

5. Similarly, devise an algorithm to compute a matrix L such that $L(i,j)$ is the length of the *longest* directed path from v_i to v_j in a digraph (V,R). Verify that your algorithm is correct. (Hint: If there is a path with a cycle there is one of infinite length.)

6. In each of the following, let R be a relation on $A = \{a,b,c,d\}$ whose

adjacency matrix is given. Compute the adjacency matrix of R^+ using Warshall's algorithm.

(a) $\begin{bmatrix} 0 & 1 & 0 & 0 \\ 1 & 0 & 1 & 0 \\ 0 & 0 & 0 & 1 \\ 0 & 0 & 0 & 0 \end{bmatrix}$
(b) $\begin{bmatrix} 1 & 0 & 0 & 1 \\ 1 & 1 & 0 & 0 \\ 0 & 0 & 0 & 0 \\ 0 & 0 & 1 & 0 \end{bmatrix}$

(c) $\begin{bmatrix} 1 & 1 & 0 & 0 \\ 1 & 0 & 0 & 0 \\ 0 & 0 & 0 & 0 \\ 0 & 0 & 1 & 0 \end{bmatrix}$
(d) $\begin{bmatrix} 1 & 0 & 0 & 1 \\ 0 & 0 & 1 & 1 \\ 0 & 0 & 1 & 1 \\ 1 & 0 & 0 & 1 \end{bmatrix}$

(e) $\begin{bmatrix} 0 & 1 & 0 & 0 \\ 0 & 0 & 0 & 1 \\ 0 & 1 & 0 & 0 \\ 0 & 0 & 1 & 0 \end{bmatrix}$
(f) $\begin{bmatrix} 1 & 0 & 0 & 1 \\ 0 & 1 & 1 & 0 \\ 0 & 1 & 1 & 0 \\ 1 & 0 & 0 & 1 \end{bmatrix}$

Selected Answers Solutions for Section 4.7.1

1. We will consider the vertices $a, \ldots, e$ to be numbered $1, \ldots, 5$. The original adjacency matrix is

$$
\begin{array}{c c}
 & \begin{array}{c c c c c} a & b & c & d & e \end{array} \\
\begin{array}{c} a \\ b \\ c \\ d \\ e \end{array} &
\begin{bmatrix} 0 & 1 & 1 & 0 & 0 \\ 0 & 0 & 1 & 0 & 0 \\ 0 & 0 & 0 & 1 & 0 \\ 0 & 1 & 0 & 0 & 1 \\ 1 & 0 & 0 & 0 & 0 \end{bmatrix}
\end{array}
$$

During the first iteration, which looks at paths through a, edges are added from e to b and from e to c, resulting in the matrix shown below.

$$
\begin{bmatrix} 0 & 1 & 1 & 0 & 0 \\ 0 & 0 & 1 & 0 & 0 \\ 0 & 0 & 0 & 1 & 0 \\ 0 & 1 & 0 & 0 & 1 \\ 1 & 1 & 1 & 0 & 0 \end{bmatrix}
$$

During the second iteration, which looks at paths through b, an edge is added from d to c, resulting in the matrix shown below.

$$
\begin{bmatrix} 0 & 1 & 1 & 0 & 0 \\ 0 & 0 & 1 & 0 & 0 \\ 0 & 0 & 0 & 1 & 0 \\ 0 & 1 & 1 & 0 & 1 \\ 1 & 1 & 1 & 0 & 0 \end{bmatrix}
$$

During the third iteration, which looks at paths through c, edges are added from a,b,e, and d, to d, resulting in the matrix below, which has the column for d now completely filled.

$$\begin{bmatrix} 0 & 1 & 1 & 1 & 0 \\ 0 & 0 & 1 & 1 & 0 \\ 0 & 0 & 0 & 1 & 0 \\ 0 & 1 & 1 & 1 & 1 \\ 1 & 1 & 1 & 1 & 0 \end{bmatrix}$$

During the fourth iteration, since every vertex now has an edge from it to d, and d has an edge to c and an edge to e, all possible remaining edges are added to c and e, resulting in the columns for c and e filling up. The resulting matrix is shown below.

$$\begin{bmatrix} 0 & 1 & 1 & 1 & 1 \\ 0 & 1 & 1 & 1 & 1 \\ 0 & 1 & 1 & 1 & 1 \\ 0 & 1 & 1 & 1 & 1 \\ 1 & 1 & 1 & 1 & 1 \end{bmatrix}$$

During the fifth and final iteration, since every vertex now has an edge to e, and e has an edge to every vertex, all remaining possible edges are added, resulting in a matrix full of 1's (the adjacency matrix of the complete digraph on 5 vertices).

3. Let $P_n(i,j)$ be the number of directed paths from v_i to v_j that traverse only vertices $\{v_1, v_2, \ldots, v_k\}$. Then $P_0(i,j) = M(i,j)$ if $M(i,j) = 0$ or if $i \neq j$, and $P_0(i,i) = \infty$ if $M(i,i) = 1$. For $k \geq 1$, $P_k(i,j) = P_{k-1}(i,j) + P_{k-1}(i,k)\,(P_{k-1}(k,k) + 1)\,(P_{k-1}(k,j))$. These equations can be proved using induction and case analysis.

   We must be more careful in using these equations to derive an algorithm than we were with Warshall's algorithm, since we do not want to count any path twice. To avoid this, we will use two matrices, M and P, so that the values of P_{k-1} can be held fixed until all the values of P_k have been computed.

Input: the adjacency matrix M of a digraph $G = (V,E)$, where $V = \{v_1, v_2, \ldots, v_n\}$.

Output: A new matrix M such that $M(i,j)$ gives the number of distinct paths from v_i to v_j in the digraph represented by the input matrix.

Method: For each k from 1 to n sequentially, if $M(k,k) = 1$, change $M(k,k)$ to ∞.

For each k from 1 to n sequentially, do the following:

Copy the values of the matrix M into the matrix P.

For each pair (i,j) change $M(i,j)$ to $P(i,j) + P(i,k)(P(k,k) + 1)P(k,j)$.

4.8   APPLICATION: SORTING AND SEARCHING

Suppose $\leq$ is a total ordering relation on a set D. A sequence $A = \langle a_1, \ldots, a_n \rangle$ of elements chosen from D is said to be **sorted** if $a_i \leq a_{i+1}$ for every i such that $1 \leq i \leq n$. Sorted sequences are encountered frequently in computer algorithms, sometimes in the guises of lists, arrays, files, or tables. One important property of sorted lists is that each item is greater than or equal to all the items preceding it and less than or equal to all the items following it in the list. The following lemma expresses this formally.

Lemma 4.8.1.   If $\langle a_1, \ldots, a_n \rangle$ is a sorted sequence with respect to a total order on D and $1 \leq i \leq n$, then

(i) $a_i \leq a_j$ for every j such that $i \leq j \leq n$, and
(ii) $a_j \leq a_i$ for every j such that $1 \leq j \leq i$;

The proof follows immediately from the transitive property of the ordering relation.

One operation that is frequently performed on sorted sequences in computer applications is **searching** a sequence $A = \langle a_1, \ldots, a_n \rangle$ for some item x. The objective is to find an i such that $1 \leq i \leq n$ and $a_i = x$, provided such an i exists. A classic solution to this problem, called **binary search,** repeatedly divides the sequence to be searched as nearly in half as is possible, narrowing the search to one of the two halves at each stage, until x is found or the search is reduced to the null sequence.

Algorithm 4.8.1 Binary Search.

Input: A sorted sequence of n elements $\langle a_1, \ldots, a_n \rangle$ drawn from a set with a total ordering relation, denoted by $\leq$, and another element x drawn from the same set.

Output: The index j of an element a_j such that $a_j = x$, if such a j exists, otherwise zero.

Method:

1. Let $i = 1$, $l(1) = 1$, and $u(1) = n$.
2. If $l(i) \geq u(i)$ go to step 5.
3. Let $m(i) = \lfloor (l(i) + u(i))/2 \rfloor$.*

---

*$\lfloor x \rfloor$ denotes the greatest integer $\leq x$.

Let

$$l(i + 1) = \begin{cases} l(i) & \text{if } x < a_{m(i)} \\ m(i) & \text{if } x = a_{m(i)} \, , \\ m(i) + 1 & \text{if } x > a_{m(i)} \end{cases}$$

and

$$u(i + 1) = \begin{cases} m(i) - 1 & \text{if } x < a_{m(i)} \\ m(i) & \text{if } x = a_{m(i)} \, , \\ u(i) & \text{if } x > a_{m(i)} \end{cases}$$

4. Increase i by one and go to step 2.
5. Output $l(i)$ if $l(i) = u(i)$ and $x = a_{l(i)}$; otherwise output 0.

Example 4.8.1. On the sequence $\langle 1,3,3,4,7,12,13,15 \rangle$ searching for 12, the binary search algorithm would go through step 3 three times, computing the following values:

i	$l(i)$	$m(i)$	$u(i)$	$\langle a_{l(i)}, \ldots, a_{u(i)} \rangle$
1	1	4	8	$\langle 1,3,3,4,7,12,13,15 \rangle$
2	5	6	8	$\langle 7,12,13,15 \rangle$
3	6		6	$\langle 12 \rangle$

The output would be 6, as desired.

As with any algorithm, there are two essential things that need to be proved about Algorithm 4.8.1. The first is termination—that it halts after a finite number of steps, for every legal input. The second is partial correctness—that when it terminates the output is correct. In addition, it would be nice to know the time complexity, or number of steps the algorithm may take before termination, expressed as a function of the length of the input sequence.

Two important properties of Algorithm 4.8.1 are stated in the following theorem. This theorem gives what is often called an *invariant assertion*. That is, it states a fact about the variables and parameters of the algorithm that is true *every* time a certain point in the algorithm is reached. In this case, the assertion is strong enough that we will be able to use it to prove the correctness of the algorithm.

Theorem 4.8.1. Every time Step 2 of Algorithm 4.8.1 is reached, the following are true:

(a) x is in the subsequence $\langle a_{l(i)}, \ldots, a_{u(i)} \rangle$ iff x is in the sequence $\langle a_1, \ldots, a_n \rangle$;

(b) $u(i) - l(i) < n/2^{i-1}$;

(c) the total number of times Step 2 has been reached is exactly i.

As part of the proof of this theorem, we first prove the following lemma, which expresses an important consequence of a sequence being sorted according to some total ordering.

Lemma 4.8.2. Let $A = \langle a_1, \ldots, a_n \rangle$ be a sorted sequence with respect to a total order $\leq$ on D, let $1 \leq m \leq n$, and let $x = a_i$ for some i $(1 \leq i \leq n)$. Then exactly one of the following cases holds:

(a) $x < a_m$ and $i < m$, or

(b) $x = a_m$, or

(c) $x > a_m$ and $i > m$.

Proof. It is a consequence of $\leq$ being a total ordering that exactly one of $x < a_m$, $x = a_m$, and $x > a_m$ must hold. What remains to be shown therefore is that in case (a) $i < m$ and in case (b) $i > m$. By the preceding lemma, a_m is the minimum of $\langle a_m, \ldots, a_n \rangle$, so in case (c), if $i \geq m$, $x < a_m \leq a_i$, which would be a contradiction. Thus $i < m$ in case (a). Similarly, a_m is the maximum of $\langle a_1, \ldots, a_m \rangle$, so that in case (b) if $i \leq m$ then $a_i \leq a_m < x$, a contradiction. Thus $i > m$ in case (b). $\square$

Proof of Theorem 4.8.1. The proof is by induction on the number of times Step 2 has been reached previously. If this is the first time, $i = 1$, $l(i) = 1$, and $u(i) = n$, so that (a) is satisfied trivially and, since $u(1) - l(1) = n - 1 < n/2^0$, so is (b). Part (c) is satisfied, since this is the first time Step 2 is reached and $i = 1$. Suppose Step 2 is reached and it is not the first time. Fix i to denote the value of i at this time. Due to the structure of the algorithm, it must be that the previous three steps were 2,3,4. By induction, the theorem held the last time Step 2 was reached. At that time the value of i was one less than it is now, since the only change to i is made in Step 4, where it is increased by one. The fact that Step 3 was performed after Step 2, rather than Step 5, permits us to conclude that $l(i-1) < u(i-1)$. Looking at what was done in Step 3, we see that one of the following three cases must now hold:

(1) $x < a_{m(i-1)}$, $l(i) = l(i-1)$, and $u(i) = m(i-1) - 1$;

(2) $x = a_{m(i-1)}$, $l(i) = m(i-1) = u(i)$;

(3) $x > a_{m(i-1)}$, $l(i) = m(i-1) + 1$, and $u(i) = u(i-1)$.

In case (1), we know by Lemma 4.8.2 that x cannot be in $\langle a_{m(i-1)}, \ldots, a_{u(i-1)} \rangle$, so that it must be in $\langle a_{l(i-1)}, \ldots, a_{m(i-1)-1} \rangle$ if it is in $\langle a_1, \ldots, a_n \rangle$ at all. Thus (a) is proven for case (1). Since $u(i) = \lfloor (l(i-1) + u(i-1))/2 \rfloor - 1$, $u(i) - l(i) \leq (l(i-1) + u(i-1))/2 - l(i) - 1 = (u(i-1) - l(i-1))/2 - 1$, and, by induction, this is less than $n/2^{i-1}$. Part (c) follows by induction, since at the last time Step 2 was reached the value of i was one less than it is now. This concludes the proof for case (1). In case (2), part (a) of the theorem holds trivially, since $x = a_{l(i)} = a_{u(i)}$, and therefore x must be in the full sequence $\langle a_1, \ldots, a_n \rangle$. Equally trivially, part (b) must hold in this case, since $u(i) - l(i) \le 0$. The argument that part (c) holds is the same as in case (1). Finally, there is case (3), which is so much like case (1) that the completion of the proof is left to the reader. $\square$

We will begin the analysis of this algorithm by verifying termination. We observe that the order in which the steps are to be executed is $1, 2, 3, 4, 2, 3, 4, 2, 3, 4, \ldots, 2, 5$. That is, for the algorithm not to terminate, Step 2 must be executed infinitely often. By the preceding theorem, this would imply that when we reach Step 2 for the $i = (\log_2(n) + 1)th$ time, we would have $u(i) - l(i) < n/2^{\log 2n} = 1$. If this were true, however, the next thing to do would be to go to Step 5 and halt. It is thus impossible for Algorithm 4.8.1 to take more than $\log_2(n) + 1$ iterations.

Notice that we have not only proven termination. We have also shown that the time complexity of this algorithm is $0(\log n)$.

The next thing to prove about this algorithm is its partial correctness. This follows from part (a) of the preceding theorem. Suppose the algorithm terminates. That means the last two steps were 2,5. At step 2 the theorem held. The fact that Step 5 was performed next tells us that $l(i) \ge u(i)$. There are two possible cases. Either $l(i) = u(i)$, in which case the theorem says that x is in $\langle a_1, \ldots, a_n \rangle$ iff $x = a_{l(i)}$, or $l(i) > u(i)$, in which case the theorem says that x is not in $\langle a_1, \ldots, a_n \rangle$ at all. Looking at Step 5, we see that in the former case the output is $l(i)$ if $x = a_{l(i)}$ and zero otherwise, and in the latter case algorithm outputs 0. In both cases the output is correct.

The binary search algorithm permits searching a sorted list far more quickly than an unsorted list may be searched by any algorithm. Largely because this and a number of other algorithms for efficiently performing useful operations on lists require that the list be sorted, the problem of rearranging the elements of an unsorted list into sorted order is a very important one. The problem of sorting a list has been extensively studied, and a number of rather complex solutions have evolved. One of the simplest approaches to solving this problem, not very efficient but easily understood, is described by the following abstract algorithm.

Algorithm 4.8.2   Interchange Sort.

Input: A sequence A of n elements $\langle a_1, a_2, \ldots, a_n \rangle$ drawn from a set with a total ordering relation, denoted by $\le$.

Output: The elements of A arranged into a sorted sequence $\langle a_{\pi(1)}, a_{\pi(2)}, \ldots, a_{\pi(n)} \rangle$.

Method:

1. Search the current arrangement of the sequence for a pair (a_i, a_j) that is out of order.
2. If no such pair is found, halt.

3. If a pair is found that is out of order, interchange the positions of the two elements and go back to step 1.

It is worthwhile to take notice that this algorithm, like Algorithm 4.7 but unlike Algorithm 4.8.1, is *nondeterministic*. That is, it does not specify exactly what steps are to be performed and in what order. In Algorithm 4.7 the nondeterminism was in the order in which the pairs (i,j) were considered. In the present algorithm the nondeterminism is in the choice of a particular pair that is out of order, to be interchanged. This is in contrast to Algorithm 4.8.1, which specifies uniquely what to do at each step, and is therefore called *deterministic*.

When analyzing an algorithm it is ordinarily desirable to describe it in the simplest, most abstract form, so that whatever results are obtained can be applied to the widest possible range of implementations. Stating an algorithm in a nondeterministic form is a useful technique for achieving this objective. In the case of Algorithm 4.8.2, there are a number of distinct deterministic sorting algorithms in use that may be viewed as refinements of this algorithm, including algorithms that have become rather well known under the names "Bubble Sort", "Successive Minima" and "Shell Sort". Any results we can prove about this abstract nondeterministic algorithm will also apply to all of these deterministic versions of it.

Largely because of their generality, nondeterministic algorithms tend to pose special problems for analysis. For example, though it is probably obvious that Algorithm 4.8.2 cannot terminate unless the list has been arranged into the desired order, it is less obvious that the algorithm must eventually terminate. Assuming the algorithm does eventually terminate, it is still less obvious how many interchanges may be performed before this happens. We could solve this problem by making the algorithm deterministic, by specifying a particular order in which pairs are considered, and then apply the technique of invariant assertions, as we did for Algorithm 4.8.1. It will be more instructive, however, to analyze this algorithm in its most general, nondeterministic form. In doing so we shall see how the theory of ordering relations can be put to good use.

To simplify the discussion of these problems, suppose that the list A to be sorted has no duplications. (It will be seen, when all is done, that this restriction can be lifted.)

Each permutation of the n elements of A can be described abstractly by a sequence consisting of the numbers $1,\ldots,n$ arranged in some order. We will follow the convention that the number i will stand for the *ith* largest element of A. Thus the desired final sorted order will be represented by $\langle 1,2,\ldots,n \rangle$. The initial order of A will be described by $\langle p_1, p_2, \ldots, p_n \rangle$, where p_i is the position that a_i is supposed to be in after the sequence is sorted. For example, if $A = \langle 64, 10, 2 \rangle$ initially, the desired sorted sequence would be $B = \langle 2, 10, 64 \rangle$. We would represent A

by $\langle 3, 2, 1 \rangle$ and B by $\langle 1, 2, 3 \rangle$. As a further example, if A were $\langle 6789, 9000, 345, 547, 100001 \rangle$, we would represent A by $\langle 3, 4, 1, 2, 5 \rangle$, and the desired sorted sequence, $\langle 345, 547, 6789, 9000, 100001 \rangle$, by $\langle 1, 2, 3, 4, 5 \rangle$. In this abstract form, the problem of sorting can be viewed as taking some permutation $\langle p_1, \ldots, p_n \rangle$ of the sorted order and *undoing* it to get back to the sorted order, $\langle 1, 2, \ldots, n \rangle$.

Let $P_n = \{\langle x_1, \ldots, x_n \rangle \mid 1 \le x_i \le n \text{ and } (x_i = x_j \text{ only if } i = j)\}$, that is, all the n-tuples describing permutations of n items. Define the relation R on P_n so that x R y iff y may be obtained from x by interchanging one pair of elements that are out of order in y; that is:

$$R = \{(x,y) \mid x = \langle x_1, \ldots, x_n \rangle, y = \langle y_1, \ldots, y_n \rangle,$$
$$\exists i < j, (x_i = y_j, y_i = x_j, x_i > x_j),$$
$$\forall k \ne i, j, (x_k = y_k)\}.$$

The importance of R is that if x is the permutation of a set S achieved at some stage of the Algorithm 4.8.2 and y is the permutation achieved after step 3 is performed for the next time, then x R y. To obtain some insight into the nature of R, consider the representation of R as a directed graph, shown in Figure 4-23 for $S = \langle 64, 20, 10 \rangle$. In this diagram an arrow is drawn from x to y for each ordered pair (x,y) in R. If Figure 4-22 seems familiar, it may be because it is similar to the lattice diagrams of Section 4.4. In fact, for any n, the transitive reflexive closure R^* of R is a partial ordering of P_n, and the transitive closure R^+ is irreflexive. Proving the latter will be the main effort of the remainder of this section. First, however, consider the relevance to this algorithm of R^+ being irreflexive.

From the definitions of transitive reflexive closure and of R, it follows that if x and y are any two permutations obtained by the algorithm at two stages for the same input sequence $\langle a_1, a_2, \ldots, a_n \rangle$, and x is equal to or occurs before y, then x R^* y. Suppose now that for some input sequence the algorithm never terminates. By the pigeonhole principle, since there are only finitely many permutations of n items, some permutation is

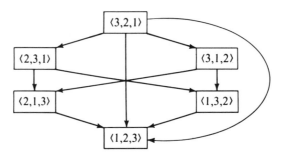

Figure 4-23. The relation R on P_3.

eventually repeated. That is, for some x, $x\,R^+\,x$. This cannot happen if R^+ is irreflexive, so that proving R^+ is irreflexive will prove that the algorithm must terminate.

We will prove R^+ is irreflexive by relating it to the known total ordering of the positive integers. Define the *degree of disorder* of each permutation $x = \langle x_1, \ldots, x_n \rangle$ to be $\delta(x) = \Sigma_{1 \le i \le j \le n}\, \sigma_x(i,j)$, where

$$\sigma_x(i,j) = 1 \text{ iff } x_i > x_j \qquad \text{and} \qquad \sigma_x(i,j) = 0 \text{ otherwise.}$$

The idea is that $\delta(x)$ measures the number of pairs that are out of order in x.

Example 4.8.2. Let us apply Algorithm 4.8.2 to a sequence of numbers and observe what happens to the degree of disorder after each recorded change. We sort the sequence $\langle 60,22,28,8,14,81,3,69 \rangle$.

Sequence	Permutation	Degree of disorder
$\langle 60,22,28,8,14,81,3,69 \rangle$	$\langle 6,4,5,2,3,8,1,7 \rangle$	15
$\langle 22,60,28,8,14,81,3,69 \rangle$	$\langle 4,6,5,2,3,8,1,7 \rangle$	14
$\langle 22,28,8,14,60,3,69,81 \rangle$	$\langle 4,5,2,3,6,1,7,8 \rangle$	9
$\langle 22,8,14,28,3,60,69,81 \rangle$	$\langle 4,2,3,5,1,6,7,8 \rangle$	6
$\langle 8,22,14,3,28,60,69,81 \rangle$	$\langle 2,4,3,1,5,6,7,8 \rangle$	4
$\langle 8,22,3,14,28,60,69,81 \rangle$	$\langle 2,4,1,3,5,6,7,8 \rangle$	2
$\langle 8,3,14,22,28,60,69,81 \rangle$	$\langle 2,1,3,4,5,6,7,8 \rangle$	1
$\langle 3,8,14,22,28,60,69,81 \rangle$	$\langle 1,2,3,4,5,6,7,8 \rangle$	0

In the above chart there was one interchange in going from the sequence in line one to the sequence in line two. However, there were several interchanges in going from line two to line three; namely, 60 and 28, then 60 and 8, 60 and 14, 81 and 3, and finally 81 and 69. These five interchanges led to the reduction of the degree of disorder by five.

Lemma 4.8.3. If $x\,R\,y$ then $\delta(x) > \delta(y)$.

Proof. Suppose $x\,R\,y$. Recall that $x\,R\,y$ iff x and y are identical except for the interchange of one pair. That is, there exist $p < q$ such that $x_p = y_q, y_p = x_q, x_p > x_q$ and for all $k \ne p,q$, $x_k = y_k$. The relationship between σ_x and σ_y can be tabulated completely:

$$\sigma_x(p,q) = 1 \qquad \text{and} \qquad \sigma_y(p,q) = 0;$$
$$\sigma_x(p,k) = \sigma_y(q,k), \sigma_x(k,p) = \sigma_y(k,q), \sigma_x(q,k) = \sigma_y(p,k), \text{and}$$
$$\sigma_x(k,q) = \sigma_y(k,p) \text{ for all } k \ne p,q;$$
$$\sigma_x(i,j) = \sigma_y(i,j) \text{ for all } i,j \ne p,q.$$

Applying these relationships to $\delta(x) = \Sigma_{1 \le i \le j \le n} \sigma_x(i,j)$, we obtain

$$\delta(x) = 1 + \sum_{1 \le k < p} (\sigma_x(k,p) + \sigma_x(k,q)) + \sum_{p < k < q} (\sigma_x(p,k) + \sigma_x(k,q))$$

$$\sum_{q < k \le n} (\sigma_x(p,k) + \sigma_x(q,k)) + \sum_{\substack{1 \le i,j \le n, \\ i,j \ne p,q}} \sigma_x(i,j)$$

$$= 1 + \sum_{1 \le k < p} (\sigma_y(k,q) + \sigma_y(k,p))$$

$$+ \sum_{p < k < q} (\sigma_y(q,k) + \sigma_y(k,p))$$

$$+ \sum_{q < k \le n} (\sigma_y(q,k) + \sigma_y(p,k)) + \sum_{\substack{1 \le i,j \le n \\ i,j \ne p,q}} \sigma_y(i,j).$$

By breaking $\delta(y)$ down similarly and subtracting we obtain

$$\delta(x) - \delta(y) = 1 + \sum_{p < k < q} (\sigma_y(q,k) + \sigma_y(k,p) - \sigma_y(p,k) - \sigma_y(k,q)).$$

All that is lacking to conclude $\delta(x) > \delta(y)$ is to show that

$$\sigma_y(q,k) + \sigma_y(k,p) \ge \sigma_y(p,k) + \sigma_y(k,q) \text{ for all } p < k < q.$$

This can be seen by examining the cases, which we do now.

  (i) if $y_k < y_p < y_q$ then $\sigma_y(q,k) + \sigma_y(k,p) = 1 + 0 = \sigma_y(p,k) + \sigma_y(k,q) = 1 + 0$;

 (ii) if $y_p < y_k < y_q$ then $\sigma_y(q,k) + \sigma_y(k,p) = 1 + 1 > \sigma_y(p,k) + \sigma_y(k,q) = 0 + 0$;

(iii) if $y_p < y_q < y_k$ then $\sigma_y(q,k) + \sigma_y(k,p) = 0 + 1 = \sigma_y(p,k) + \sigma_y(k,q) = 0 + 1$. $\square$

Lemma 4.8.4. If $x R^+ y$ then $\delta(x) > \delta(y)$.

Proof. Suppose $x R^+ y$. By definition of transitive closure, $x R^k y$ for some $k \ge 1$. If $k = 1$, $\delta(x) > \delta(y)$, by the preceding lemma. If $k > 1$, there must be some z such that $x R^{k-1} z R y$. By induction and Lemma 4.8.3, it follows that $\delta(x) > \delta(z) > \delta(y)$. $\square$

Theorem 4.8.2. The transitive closure of R, R^+, is irreflexive.

Proof. If $x R^+ x$, then the preceding lemma gives $\delta(x) > \delta(x)$, which is an obvious contradiction. $\square$

So far, we have shown that Algorithm 4.8.2 must always terminate. The intuitive notion of relative degrees of disorder, corresponding to distances on the diagram in Figure 4-22, was the basis of this proof. Essentially, what Lemma 4.8.3 shows is that each step of the algorithm decreases the degree of disorder, and therefore only a finite number of steps may be performed before the sequence is completely in order. We will now proceed with a more careful analysis, to see just how many steps may be taken in this process.

Lemma 4.8.3 gives a quick upper bound on the number of steps.

$$\delta(x) = \sum_{1 \le i \le j \le n} \sigma_x(i,j) \le \sum_{i=1}^{n-1} \sum_{j=i+1}^{n} 1 = \sum_{i=1}^{n-1} (n-i) = \frac{n(n-1)}{2}.$$

Thus, since each step of the algorithm reduces $\delta(x)$ by at least one, it can never take more than $n(n-1)/2$ steps on any input sequence of length n.

Of course, we have not shown that this number of steps is *required* for any input sequence, though there are several widely taught sorting algorithms that do perform this number of interchanges for some inputs. One of these is the "Bubble Sort," which makes repeated bottom-to-top passes over the sequence to be sorted, comparing adjacent pairs of elements.

Though Algorithm 4.8.2 has provided an interesting example of the technique of proving termination by means of an ordering relation, it is not typical of the algorithms in use where long sequences must be sorted. A more typical method of sorting is based on the concept of *merging* two sorted sequences into one sorted sequence. This operation is also of interest in itself, because it lies at the heart of a number of other fundamental algorithms, including the usual method of updating sequential files in data processing applications. For variety, and because it is natural, we shall present an algorithm for this operation as a recursive function definition.

Algorithm 4.8.3   Merging Two Sorted Sequences.

Input: Two sequences, $A = \langle a_1, \ldots, a_n \rangle$ and $B = \langle b_1, \ldots, b_m \rangle$, sorted according to some total ordering, $\le$.

Output: A single sentence, merge (A,B), which is a sorted permutation of the sequence $\langle a_1, \ldots, a_n, b_1, \ldots, b_m \rangle$.

Method:

$$\text{merge } (A,B) = \begin{cases} A \text{ if } |B| = 0; \\ B \text{ if } |A| = 0; \\ \langle a_1 \rangle \cdot \text{merge } (\langle a_2, \ldots, a_n \rangle, B) \text{ if } a_1 \le b_1; \text{ and} \\ \langle b_1 \rangle \cdot \text{merge } (A, \langle b_2, \ldots, b_m \rangle) \text{ if } a_1 \not\le b_1. \end{cases}$$

Here the symbol "$\cdot$" stands for catenation of sequences. That is,

$$\langle a_1, \ldots, a_n \rangle \cdot \langle b_1, \ldots, b_n \rangle = \langle a_1, \ldots, a_n, b_1, \ldots, b_n \rangle.$$

Example 4.8.3. If $A = \langle 1,2,5,12,12,16 \rangle$ and $B = \langle 2,3,7,13,21 \rangle$ then merge $(A,B) = \langle 1,2,2,3,5,7,12,12,13,16,21 \rangle$. The important properties of the merge of two sequences are stated in the following lemma.

Lemma 4.8.5. Let $C = $ merge (A,B), where A and B are sorted sequences with respect to a total order on S. Then C is also a sorted sequence and a permutation of $A \cdot B$.

Proof. If $|A| = 0$ or $|B| = 0$ the lemma is trivially true, since $C = A = A \cdot B$ or $C = B = A \cdot B$. Supposing $|A| > 0$ and $|B| > 0$, there are two cases left: either $a_1 \leq b_1$ or $b_1 < a_1$. (One must hold, since $\leq$ is assumed to be total.) In the first case, $C = \langle a_1 \rangle \cdot$ merge $(\langle a_2,\ldots,a_n \rangle, B)$. We argue that this is clearly a permutation of $A \cdot B$, since we may assume by induction (on n and m) that $\langle c_2,\ldots,c_{n+m} \rangle = $ merge$(\langle a_2,\ldots,a_n \rangle, B)$ is a sorted permutation of $\langle a_2,\ldots,a_n \rangle \cdot B$. To show that C is sorted, we rely on induction to conclude that $\langle c_2,\ldots,c_{n+m} \rangle$ is sorted and reason that $a_1 \leq c_2$ (as follows): By Lemma 4.8.1, c_2 is the minimum of $\langle a_2,\ldots,a_n \rangle \cdot B$, a_2 is the minimum of $\langle a_2,\ldots,a_n \rangle$ and b_1 is the minimum of B. It follows that $c_2 = a_2$ or $c_2 = b_1$. We know $a_1 \leq a_2$ since A is sorted, and that $a_1 \leq b_1$ by the assumption of the case being considered. Thus $a_1 \leq c_2$ and C is sorted.

There is one case left, when $b_1 < a_1$. In this case $C = \langle b_1 \rangle \cdot$ merge$(A, \langle b_2,\ldots,b_m \rangle)$. The reasoning that C is a sorted permutation of $A \cdot B$ is analogous to the previous case, and is left to the reader. $\square$

Merging is the basic concept behind an important family of sorting algorithms. These algorithms sort a large list by breaking the unsorted list into small sorted lists and then merging these sorted lists together until they have all been combined back into a single sorted list.

Example 4.8.4. The sequence $\langle 5,4,1,2,6,3,2,3 \rangle$ might be broken into $\langle 5 \rangle \langle 4 \rangle \langle 1,2 \rangle \langle 6 \rangle \langle 3 \rangle \langle 2,3 \rangle$. The sequence of merges below might then be performed, resulting in the desired sorted list.

$$\langle 5 \rangle \langle 4 \rangle \rightarrow \langle 4,5 \rangle$$
$$\langle 1,2 \rangle \langle 6 \rangle \rightarrow \langle 1,2,6 \rangle$$
$$\langle 3 \rangle \langle 2,3 \rangle \rightarrow \langle 2,3,3 \rangle$$
$$\langle 4,5 \rangle \langle 1,2,6 \rangle \rightarrow \langle 1,2,4,5,6 \rangle$$
$$\langle 1,2,4,5,6 \rangle \langle 2,3,3 \rangle \rightarrow \langle 1,2,2,3,3,4,5,6 \rangle$$

There are a number of different algorithms for sorting by merges, differing according to how the original sequence is broken up and the rule used to determine the pairs of sequences that are merged together. One of these is given in the algorithm below.

Algorithm 4.8.4 Merge Sort.

Input: A sequence S of n elements $\langle a_1, a_2, \ldots, a_n \rangle$ drawn from a set with a total ordering relation, denoted by $\leq$.

Output: The elements of S arranged into a nondecreasing sequence

$$\langle a_{\pi(1)}, a_{\pi(2)}, \ldots, a_{\pi(n)} \rangle \text{ where } \langle \pi(1), \pi(2), \ldots, \pi(n) \rangle$$

is a permutation of $1, 2, \ldots, n$.

Method:

1. Break S into a collection of up to m sorted sequences $S_1^{(1)}, \ldots, S_m^{(1)}$.
2. Repeat the following until only one list remains: Suppose this is the beginning of the ith stage and the sequences so far are $S_1^{(i)}, \ldots, S_k^{(i)}$. Let $S_j^{(i+1)} = \text{merge}(S_{2 \cdot j - 1}^{(i)}, S_{2 \cdot j}^{(i)})$ for $i = 1, \ldots, \lfloor k/2 \rfloor$ and $S_{\lfloor k/2 \rfloor}^{(i+1)} = S_k^{(i)}$ in the case that k is odd.

The objective behind this algorithm is to build up large lists with few merges. It can be shown that by the end of the ith stage there is no list of length less than 2^i.

Theorem 4.8.3. By the end of the ith stage ($i \geq 1$), the sequences produced by Algorithm 4.8.2 all have lengths $\geq 2^i$, except for (possibly) the last one.

Proof. Initially, no sequence may have length less than 1. Thus, as a basis for induction, after the first stage every list has length at least 2, except (possibly) the last. By induction, at the end of stage $i - 1$ every list has length $\geq 2^i$, except (possibly) the last, $S_k^{(i)}$.

Every new list $S_j^{(i+1)}$ produced by the end of stage i comes from two lists $S_{2j-1}^{(i)}$ and $S_{2j}^{(i)}$, each of length $\geq 2^i$, except for the last one, which may be $S_k^{(i)}$ or a list created by merging $S_k^{(i)}$ with another list. It follows that all but the last of the $S_j^{(i+1)}$ have length $\geq 2^{i+1}$. $\square$

Using this theorem it is possible to obtain an upper bound on the number of stages required to sort a list of length n. Since step 2 of Algorithm 4.8.4 terminates when one list is left, suppose the end of stage i is the first time the number of sequences is reduced to one. Then there are two lists at the beginning of stage i. This means either $i = 1$ or the two lists were produced at the end of stage $i - 1$. In the latter case, we know that one of the two lists is of size $\geq 2^{i-1}$. Thus $n \geq 2^{i-1} + 1$, which implies $i \leq \log_2(n - 1) + 1$.

Analysis of the complexity of merging would show that merging two sequences of combined length ℓ takes $0(\ell)$ time. Thus each stage of Algorithm 4.8.4 takes $0(n)$ time. Since the algorithm goes through at most

$\log_2(n - 1) + 1$ stages, the total time complexity of the algorithm is thus $0(n \cdot \log_2 n)$.

There is no need to show separately that this algorithm must terminate, for we have already shown a stronger result by demonstrating a specific upper bound on the number of stages it may go through. Similarly, the partial correctness of this algorithm, that when it terminates S is a sorted sequence, follows directly from the fact that S is initially broken into sorted sequences and, by Lemma 4.8.5, the merge operation preserves the property of being sorted.

Exercises for Section 4.8

1. Give the sequence of values for i, $l(i)$, $m(i)$, and $u(i)$ computed by Algorithm 4.8.1 with inputs $x = 24$ and $A = \langle 1,8,9,15,24, 32,34,35,37,43,95,99 \rangle$.

2. Give the digraph of the relation R on P_4 defined in this section.

3. Give the degrees of disorder of the following permutations.
   (a) $\langle 1,2,3,4 \rangle$
   (b) $\langle 4,3,2,1 \rangle$
   (c) $\langle 1,3,2,4 \rangle$
   (d) $\langle 3,1,4,2 \rangle$

4. Give the sequence merge(A,B) where $A = \langle 1,24,95,100,101 \rangle$ and $B = \langle 1,24,93,97,101,102 \rangle$. Show how it is defined in terms of the merges of shorter sequences.

5. Suppose the sequence $S = \langle 10,8,9,7,5,6,4,2,3,1 \rangle$ is given as input to Algorithm 4.8.3, and at Step 1 it is broken into the collection of sequences $\langle 10 \rangle$, $\langle 8,9 \rangle$, $\langle 7 \rangle$, $\langle 5,6 \rangle$, $\langle 4 \rangle$, $\langle 2,3 \rangle$, $\langle 1 \rangle$. Give the collection of sequences as it would be at the end of each succeeding stage, until the algorithm terminates.

6. Complete the proof of Theorem 4.8.1.

7. Derive and prove a formula for the total number of (recursive) evaluations of the function *merge* to evaluate *merge*(A,B), where A and B are sequences of length n and m, respectively.

8. Which properties of a total ordering are actually used (explicitly or implicitly) in the proof of Lemma 4.8.1? Prove, by counterexample, that this lemma fails if any one of these properties is not satisfied.

9. Which properties of a total ordering are actually used (explicitly or implicitly) in the proof of Lemma 4.8.2? Prove, by counterexample, that this lemma fails if any one of them is not satisfied.

10. Show that any permutation of the sequence $\langle 1,2,\ldots,n \rangle$ can be obtained by a series of at most $n - 1$ pairwise interchanges. Invent an algorithm that takes as input two sequences, $A = \langle a_1,\ldots,a_n \rangle$

and $B = \langle b_1, \ldots, b_n \rangle$ and rearranges the sequence A according to the permutation described by B. (Assume that B is a permutation of $\langle 1, 2, \ldots, n \rangle$. The desired output is $\langle a_{b_1}, a_{b_2}, \ldots, a_{b_n} \rangle$. Use only pairwise interchanges.)

11. An important variation on merging is taking the *union* of two sorted sequences, each of which has no repetitions. This is defined by

$$
\text{union}(A,B) = \begin{cases}
A & \text{if } |B| = 0; \\
B & \text{if } |A| = 0; \\
\langle a_1 \rangle \cdot \text{union}(\langle a_2, \ldots, a_n \rangle, B) & \text{if } a_1 < b_1; \\
\langle a_1 \rangle \cdot \text{union}(\langle a_2, \ldots, a_n \rangle, \langle b_2, \ldots, b_m \rangle) & \text{if } a_1 = b_1; \\
\langle b_1 \rangle \cdot \text{union}(A, \langle b_2, \ldots, b_m \rangle) & \text{if } b_1 < a_1.
\end{cases}
$$

12. Observe that the degree of disorder of any permutation is an odd or an even integer. Then prove that if a pair of elements are interchanged in any permutation, then the degree of disorder is either increased or decreased by an odd integer. In fact, if $i < j$ and x_i and x_j are interchanged, prove that $\delta(x)$ is decreased by an odd integer if $x_i > x_j$ and increased if $x_i < x_j$.

Selected Answers for Section 4.8

1.

i	$l(i)$	$m(i)$	$u(i)$
1	1	6	12
2	1	3	5
3	4	4	5
4	5		5

3. (a) 0
   (b) 6 (The pairs are (4,3), (4,2), (4,1), (3,2), (3,1), (2,1).)
   (c) 1 (The only pair out of order is (3,2).)
   (d) 3 (The only pairs out of order are (3,1), (3,2), (4,2).)

4. $\langle 1,1,24,24,93,95,97,101,101,102 \rangle$
   This is obtained as
   $\langle 1 \rangle$ merge($\langle 24,95,100,101 \rangle$, $\langle 1,24,93,97,101,102 \rangle$) =
   $\langle 1 \rangle \langle 1 \rangle$ merge($\langle 24,95,100,101 \rangle$, $\langle 24,93,97,101,102 \rangle$) =
   $(1) \langle 1 \rangle \langle 24 \rangle$ merge($\langle 95,100,101 \rangle$, $\langle 24,93,97,101,102 \rangle$) = ...
   $\langle 1 \rangle \langle 1 \rangle \langle 24 \rangle \langle 24 \rangle \langle 93 \rangle \langle 97 \rangle \langle 101 \rangle \langle 101 \rangle \langle 101 \rangle \langle 102 \rangle$ merge($\langle \rangle$, $\langle 102 \rangle$).

5. $\langle 8,9,10 \rangle \langle 5,6,7 \rangle \langle 2,3,4 \rangle \langle 1 \rangle$
   $\langle 5,6,7,8,9,10 \rangle \langle 1,2,3,4 \rangle$
   $\langle 1,2,3,4,5,6,7,8,9,10 \rangle$

11. The main difference between this and merging is the treatment of the case $a_1 = b_1$. Prove the following lemma.

Lemma. Let $A = \langle a_1, \ldots, a_n \rangle$ and $B = \langle b_1, \ldots, b_m \rangle$ be sorted sequences without repetitions, and $C = \text{union}(A,B) = \langle c_1, \ldots, c_\ell \rangle$. Then C is a sorted sequence without repetitions and

$$\{c_i \mid 1 \le i \le \ell\} = \{a_i \mid 1 \le i \le n\} \cup \{b_i \mid 1 \le i \le m\}.$$

4.9   APPLICATION: TOPOLOGICAL SORTING

There are many situations where a natural partial ordering on a set exists and it is desired to extend this partial ordering to a total ordering; that is, to enumerate the elements of the set in some order consistent with the given partial ordering.

One example of this arises in compiling programs in the programming language Ada.† In Ada, a program may be broken into many independent compilation units. Each unit includes a list specifying certain other units which must be compiled before it. From this information an Ada compiler must discover an order in which these units may be compiled that is consistent with these specifications, provided such an order exists. This is a problem that fits naturally into the terminology of binary relations and digraphs. Let U be the set of compilation units. Let R be the binary relation on compilation units defined by $u_1 \, R \, u_2$ iff it is specified that u_1 must be compiled before u_2. The compiler's problem is thus:

1. Determine whether R^* is a partial ordering on U [i.e., whether the digraph (U,R) contains any nontrivial directed cycles]. If not, report that there is no legal order of compilation.

2. If R^* is a partial ordering, construct a sequence $\langle u_1, \ldots, u_n \rangle$ such that $U = \{u_1, \ldots, u_n\}$ and for each (u_j, u_k) in R^*, $j \le k$.

We shall call a sequence such as the one described above a **topological enumeration** of U with respect to R, and call the process of discovering one **topological sorting**. There are several natural algorithms for solving this problem. We shall consider one of them that is fairly simple to state and prove, though not necessarily the most efficient.

Algorithm 4.9.1   Topological sort.

Input: A digraph $G = (V,E)$, with n vertices.

Output: A topological enumeration $S_n = \langle s_1, \ldots, s_n \rangle$ of V with respect to E, provided E^* is a partial ordering on V.

†Ada is a registered trademark of the U.S. Department of Defense (AJPO).

Method:

1. Let $U_0 = V$, $S_0 = \langle \ \rangle$ (the sequence of length zero), and $T_0(v) = \{u \mid (u,v) \in E$ and $u \neq v\}$.

2. Repeat the following for $i = 1, \ldots, n$:

   (a) Choose s_i from U_{i-1} such that $T_{i-1}(s_i) = \phi$, provided such an s_i exists. Otherwise, halt and output a message that E^* is not antisymmetric.

   (b) Let $U_i = U_{i-1} - \{s_i\}$, $S_i = S_{i-1} \cdot \langle s_i \rangle$, and $T_i(v) = T_{i-1}(v) - \{s_i\}$ for all $v \in V$.

3. If not already halted, output S_n.

Example 4.9.1. Consider the digraph $G = (\{a,b,c,d,e\},E)$ where $E = \{(a,b), (a,c), (a,e), (b,d), (b,e), (c,d), (d,e)\}$. (This is shown in figure 4-24.) The topological sorting algorithm described above would compute the following sequence of sets on input G:

i	U_i	S_i	$T_i(a)$	$T_i(b)$	$T_i(c)$	$T_i(d)$	$T_i(e)$
0	$\{a,b,c,d,e\}$	$\langle \ \rangle$	ϕ	$\{a\}$	$\{a\}$	$\{b,c\}$	$\{a,b,d\}$
1	$\{b,c,d,e\}$	$\langle a \rangle$	ϕ	ϕ	ϕ	$\{b,c\}$	$\{b,d\}$
2	$\{c,d,e\}$	$\langle a,b \rangle$	ϕ	ϕ	ϕ	$\{c\}$	$\{d\}$
3	$\{d,e\}$	$\langle a,b,c \rangle$	ϕ	ϕ	ϕ	ϕ	$\{d\}$
4	$\{e\}$	$\langle a,b,c,d \rangle$	ϕ	ϕ	ϕ	ϕ	ϕ
5	ϕ	$\langle a,b,c,d,e \rangle$	ϕ	ϕ	ϕ	ϕ	ϕ

To prove that the S_n computed by this algorithm is actually a topological enumeration of V with respect to E, provided E^* is a partial ordering on V, we actually need to prove a stronger theorem, one which characterizes the state of completion at the end of each stage i.

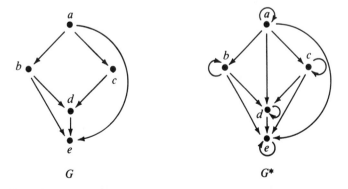

Figure 4-24. A digraph to be sorted and its transitive reflexive closure.

Theorem 4.9.1. For each $i = 0,1,\ldots,n$ the sets computed by the topological sort algorithm satisfy the following:

1. $S_i = \langle s_1,\ldots,s_i \rangle$ is a sequence of i distinct vertices in V and U_i consists of all the vertices in V that are not in S_i;
2. $T_{i-1}(s_i) = \phi$;
3. $T_i(v) = \{u \mid (u,v) \in E, u \neq v, \text{ and } u \notin \{s_1,\ldots,s_i\}\}$ for all $v \in V$;
4. If $(s_j,s_k) \in E^*$ and $1 \leq j,k \leq i$ then $j \leq k$.

Proof. The proof is by induction on i. For $i = 0$, $S_i = \phi$, and $U_i = V$, so that (1)–(4) are satisfied, trivially. For $i > 0$ we assume that the theorem is true for smaller values of i. Each part of the theorem is considered separately:

(1) By induction, $S_{i-1} = \langle s_1,\ldots,s_{i-1} \rangle$ is a sequence of $i - 1$ distinct vertices in V and U_{i-1} consists of all the vertices in V that are not in S_i. By step 2(b) of the algorithm, $S_i = \langle s_1,\ldots,s_i \rangle$ and $U_i = U_{i-1} - \{s_i\}$, where s_i is in U_{i-1} and therefore distinct from $s_1,\ldots,s_{i-1}$. It follows that $s_1,\ldots,s_i$ are distinct and U_i consists of all the vertices in V that are not in S_i.

(2) If $T_{i-1}(s_i)$ were not empty, s_i would not have been chosen in step 2(a) of the algorithm. [The algorithm would have terminated early if *all* the $T_{i-1}(v)$ were nonempty.]

(3) By induction, $T_{i-1}(v) = \{u \mid (u,v) \in E, u \neq v, \text{ and } u \notin \{s_1,\ldots,s_i\}\}$. By step 2(b) of the algorithm, $T_i(v) = T_{i-1}(v) - \{s_i\} = \{u \mid (u,v) \in E, u \neq v, \text{ and } u \notin \{s_1,\ldots,s_i\}\}$.

(4) By induction, if $(s_j,s_k) \in E^*$ and $1 \leq j,k \leq i - 1$ then $j \leq k$. We therefore only have to consider the case when one or both j and k is equal to i. Suppose $(s_j,s_k) \in E^*$ and $1 \leq j,k \leq i$. If $k = i$, there is no problem, since this implies $j \leq k$. The only remaining case to prove is thus for $j = i$. Suppose part (4) of the theorem is false for some $k < i$; that is, $(s_i,s_k) \in E^*$ and $k < i$. Choose k to be the *smallest* k for which this happens. This means that for some v in V, $(s_i,v) \in E^*$ and $(v,s_k) \in E$. [Note that it might be that $v = s_i$ and $(s_i,s_k) \in E$.] We have already shown that $T_{i-1}(s_i) = \phi$, and by induction we know that $T_{i-1}(s_i) = \{u \mid (u,s_i) \in E, u \neq s_i \text{ and } u \notin \{s_1,\ldots,s_{i-1}\}\}$. It follows that $v = s_t$ for some $t \leq i$. If $t = i$ then $v = s_i$ and $(s_i,s_k) \in E$. Otherwise, by induction, since $(s_t,s_k) \in E^*$ and t and k are both less than i, we know that $t \leq k$. Since k was chosen to be the *smallest* k for which $(s_i,s_k) \in E^*$ and $k < i$, we know that $t = k$. We have shown so far that (s_i,s_k) must be in E. By induction, $T_{k-1}(s_k) = \phi$ and $T_{k-1}(s_k) = \{u \mid (u,s_k) \in E, u \neq s_k, \text{ and } u \notin \{s_1,\ldots,s_{k-1}\}\}$. This is a contradiction, since $(s_i,s_k) \in E$ and $s_i \notin \{s_1,\ldots,s_k\}$. It follows that part (4) of the theorem must be true. $\square$

Parts (1) and (4) of the theorem immediately yield the following corollary, for the case when $i = n$.

Corollary 4.9.1. The sequence S_n produced by the topological sort algorithm is a topological enumeration of V with respect to E, provided the algorithm does not halt before the stage $i = n$.

What remains to be proven is that the algorithm does not halt before stage $i = n$ unless E^* is not antisymmetric. We will argue that this is so without stating it as a formal theorem. Suppose the algorithm does halt in step 2(a) for some $i \leq n$. This means that $T_{i-1}(v) = \phi$ for every v in U_{i-1}. By the preceding theorem, $T_{i-1}(v) = \{u \mid (u,v) \in E, u \neq v,$ and $u \in U_{i-1}\}$. In other words, the vertices in U_{i-1} and the edges in the $T_{i-1}(v)$'s form a subgraph of G that has no self-loops and in which every vertex has nonzero in-degree. We shall prove that such a subgraph must have a nontrivial cycle and therefore that E^* cannot be antisymmetric. It will follow that the algorithm never halts before reaching step 5 unless E^* is not a partial ordering on V.

Lemma 4.9.1. If $G = (V,E)$ is a digraph the following statements are equivalent:

(1) G has a nonempty subgraph without self-loops in which every vertex has nonzero in-degree;
(2) G contains a nontrivial directed cycle;
(3) E^* is not antisymmetric.

Proof. $(1 \rightarrow 2)$ We prove the contrapositive, that is, we prove that if G contains no nontrivial directed cycles, then in each nonempty subgraph of G without loops there is a vertex with zero in-degree.

Let H be any nonempty subgraph of G without loops. Since G contains no directed cycles neither does H. Let P denote a directed path $(v_1,v_2),\ldots,(v_{p-1},v_p)$ in H of maximal length. We assert that the in-degree of v_1 is zero. (Actually, the out-degree of v_p is also zero, but we leave that result as an exercise.)

If the in-degree of v_1 is not zero, then there exists a vertex w in H such that the edge (w,v_1) is in H. Two cases present themselves:

Case 1. Suppose $w \neq v_i$ for each i where $1 \leq i \leq p$. Then there is a directed path P': $(w,v_1),(v_1,v_2),\ldots,(v_{p-1},v_p)$, whose edges contain all the edges of P and the edge (w,v_1). But this contradicts that P was a path of maximal length in H.

Case 2. Suppose $w = v_i$ for some i. Then in H, there is a directed circuit C: $(v_1,v_2),\ldots,(v_i,v_1)$. This again is a contradiction, since H has no loops and no nontrivial directed cycles.

The contradiction in the above two cases leads to the conclusion that there is no edge (w,v_1) in H. In other words, the in-degree of v_1 is zero.

$(2 \rightarrow 3)$ Suppose G satisfies (2). Let $(v_1,v_2), \ldots ,(v_k,v_1)$ be such a cycle. Since it is nontrivial, we can assume $v_1 \neq v_k$. Then (v_1,v_k) and (v_k,v_1) are both in E^+, by Corollary 4.6.1, which means that E^+ (and hence E^*) is not antisymmetric.

$(3 \rightarrow 2)$ Suppose E^* is not antisymmetric. Then there exist $v \neq w$ such that (v,w) and (w,v) are both in E^*. By Corollary 4.6.1, there must be nontrivial directed paths from v to w, and from w to v. If one of these paths includes a cycle, we are done. Otherwise, putting them together we get a nontrivial directed cycle.

$(2 \rightarrow 1)$ Suppose G includes a nontrivial directed cycle $(v_1,v_2), \ldots ,$ (v_k,v_1). Then the subgraph $(\{v_1,v_2,\ldots ,v_k\},\{(v_1,v_2),\ldots ,(v_k,v_1)\})$ satisfies (1). $\square$

An interesting corollary of the proof of the topological sorting algorithm is that any finite antisymmetric binary relation on a countable set may be extended to a well ordering. Note that this is false for infinite relations as can be seen from the lexicographic ordering relation defined in section 4.4.1.

Corollary 4.9.2.   Let R be a finite antisymmetric relation on a finite or countably infinite set. Then R^* is a partial ordering and there is a well ordering W such that $R \subseteq W$.

This fact is crucial in some proofs of lower bounds on the algorithmic complexity of problems involving searching and sorting, one of which is considered in the exercises.

Exercises for Section 4.9

1. Find a well ordering of the set $\{a,b,c,d,e,f,g\}$ that extends the relation $\{(a,a),(a,g),(b,a),(c,a),(d,b),(d,c),(d,f),(e,d),(e,g),(e,f)\}$.

2. Show that there can be no well ordering of the set $\{a,b,c,d,e,f\}$ that extends the relation $\{(a,b),(a,d),(a,c),(b,d),(b,c),(c,e),(d,f),(e,f),(f,a)\}$.

3. Perform Algorithm 4.9.1 on the digraph $G = (\{a,b,c,d,e\},E)$, where $E = \{(a,d),(b,c),(b,d),(b,e),(e,a),(e,d)\}$. Show all steps, as in Example 4.9.1. Draw this digraph and the digraph of its transitive reflexive closure.

The following two exercises form the heart of a proof that any algorithm that can choose the kth largest element out of an input sequence by means of pairwise comparisons must use at least $n - 1$ comparisons.

4. Prove the following lemma:

Let R be an antisymmetric relation on a set $S = \{s_1,\ldots ,s_n\}$ and suppose that in every total ordering of S that extends R the element

s_e is the kth largest. Then for every $i \neq e, 1 \leq i \leq n$, either $s_e\, R^+ s_i$ or $s_i\, R^+ s_e$. (Hint: Suppose that some element s_x is not related to s_e by R^+. Show that then there exist more than one total ordering of S that extend R and that in one of them s_e is not the kth largest.)

5. Prove the following lemma:

    Let $R, S,$ and s_e be as in the preceding exercise. Then R contains at least $n - 1$ ordered pairs (s_i,s_j) such that $s_i \neq s_j$. [Hint: Set up a one-to-one correspondence $f: S - \{e\} \to R$ between a subset of the ordered pairs in R and the elements $s_i, i \neq e$. The preceding exercise may be viewed as saying that there is at least one directed path in (S,R) from s_i to s_e or from s_e to s_i. Taking $f(s_i)$ to be any one of these edges on such a path gives a function $f: S - \{e\} \to R$, but not necessarily a one-to-one function. The problem is to define f so that it is one-to-one.]

6. Consider a path P: $(v_1,v_2),\dots,(v_{p-1},v_p)$ of maximal length in a directed graph G, where G contains no directed cycles. Prove that v_p has out-degree zero.

Selected Answers for Section 4.9

1. e,d,c,b,a,g,f

2. This relation contains several cycles. One is $(a,d),(d,f),(f,a)$. It therefore cannot be extended to a well-ordering.

REVIEW FOR CHAPTER FOUR

1. Define the relation R on the set of real numbers by $(x,y) \in R$ iff $x^2 + y^2 = 1$. Of the six special properties of relations studied in Section 4.2,
   (a) determine those properties that R satisfies.
   (b) determine those properties R does not satisfy.

2. For the following two digraphs, either prove that they are not isomorphic or exhibit an isomorphism by listing corresponding pairs of vertices and also corresponding pairs of edges.

3. For the set $A = \{a,b,c,d\}$, determine
   (a) the number of different equivalence relations on A. (Show your analysis.)
   (b) the number of nonisomorphic digraphs of equivalence relations on A.

4. Let R be the relation defined on the set A of nonzero real numbers by $(a,b) \in R$ iff $ab > 0$.
   (a) Show that R is an equivalence relation on A.
   (b) Determine the equivalence class of 1.
   (c) Determine the equivalence class of -1.
   (d) What partition of A is determined by R?

5. Complete the following definitions:
   (a) An *equivalence relation* on a set A is a relation that is _____

   _____ .

   (b) A *partial order* on a set A is a relation that is _____

   _____ .

   (c) A total order on A is a partial order such that _____

   _____ .

   (d) A total order on A is a well ordering if _____

   _____ .

   (e) If R is a partial order on A and $a \in A$, then a is a *minimal* element of A (with respect to the relation R) if _____

   _____ .

6. Exhibit the digraphs of all possible partial orders on $A = \{a,b,c\}$ such that a is a minimal element.

7. Show that if R is a reflexive and symmetric relation on a set A, then so is the relation R^{-1}.

8. Given the relations $R = \{(a,a),(a,b),(b,c),(c,b)\}$ and $S = \{(a,b), (c,b),(c,c)\}$ defined on the set $A = \{a,b,c\}$, list the elements of the following:
   (a) $R - S$
   (b) $R \cdot S$
   (c) R^{-1}
   (d) $R \cap S$
   (e) The transitive closure of R.

9. Let M_R and M_S be the adjacency matrices for the relations R and S defined in problem 8 above.
   (a) List the entries of M_R and M_S.
   (b) Compute M_R OR.AND M_S.
   (c) Carry out Warshall's algorithm for M_R, listing each intermediate matrix.

10. Disprove the following by giving a counterexample.
    (a) If a relation R on a set A is not symmetric, then R is antisymmetric.
    (b) The lexicographic ordering on the set of strings on an alphabet A is a well ordering.
    (c) Any symmetric and transitive relation R on a set A is an equivalence relation.

11. Using the topological sorting algorithm on the following directed graph, list a topological sorting of the vertices.

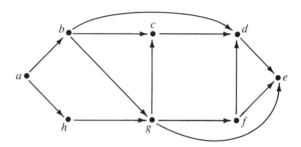

12. Compute the adjacency matrix of the transitive closure of the relation R whose adjacency matrix is

$$\begin{bmatrix} 1 & 1 & 0 & 0 & 0 \\ 1 & 1 & 0 & 0 & 0 \\ 0 & 0 & 1 & 1 & 0 \\ 0 & 0 & 1 & 1 & 1 \\ 0 & 0 & 0 & 0 & 1 \end{bmatrix}$$

13. Find an integer x where $0 \le x \le 12$ and $5^{243} \equiv x \bmod 13$.

14. Let $A = \{2,5,8,10,20,25,50,60,100\}$ and define the relation R on A by $(a,b) \in R$ iff a divides b. Then R is a partial order on A. Find the minimal and maximal elements of A under the relation R.

15. Given the words: MY, ME, MINE, MAMA, MOO, MOST, MARE, MAID, MADE, MEAL, MAIL, MART, MARRY, REST, RAIL, RED, REASON,
    (a) Arrange them into ascending order according to the definition of lexicographic ordering.
    (b) Arrange them into ascending order according to the definition of enumeration ordering.

16. Give the degree of disorder of the following permutations:
    (a) $\langle 2,3,4,5,1,6 \rangle$
    (b) $\langle 3,2,1,6,5,4 \rangle$

17. Sort the sequence $\langle 3,2,1,5,4 \rangle$ using no more than four interchanges.

18. Use the binary search algorithm to search for 15 in the sequence $\langle 1,2,2,4,5,7,8,10,10,12,14,15,16,16,18,20,21 \rangle$.

19. Prove by mathematical induction:
    (a) If R is a symmetric relation defined on a set A, then R^n is symmetric for each positive integer n.
    (b) If R is a reflexive and transitive relation defined on a set A, then $R = R^n$ for each positive integer n.

20. Show that $3n^3 + 6n^2 - 4n + 7$ is in $O(n^3)$.

21. Prove that an integer $N = a_n 10^n + a_{n-1} 10^{n-1} + \cdots + a_1 10 + a_0$, where each a_i is an integer such that $0 \le a_i \le 9$, is divisible by 25 iff $a_1 10 + a_0$ is divisible by 25.

5

Graphs

5.1 BASIC CONCEPTS

Directed graphs, and a number of related concepts, were introduced earlier as a way of viewing relations on sets. This chapter deals with more general kinds of graphs, including digraphs, nondirected graphs, and multigraphs. We begin by extending the terminology introduced for directed graphs to nondirected graphs. We discuss a number of important cases of directed and nondirected graphs, as well as a few important properties that may be possessed by graphs, such as planarity and colorability.

Definition 5.1.1. A graph G is a pair of sets (V,E), where V is a set of vertices and E is a set of edges. If G is a **directed graph** (digraph) the elements of E are ordered pairs of vertices. In this case an edge (u,v) is said to be from u to v, and to join u to v. If G is a nondirected graph the elements of E are unordered pairs (sets) of vertices. In this case an edge $\{u,v\}$ is said to **join** u and v or to be **between** u and v. An edge that is between a vertex and itself is called a self-loop (loop for short). A graph with no loops is said to be **simple** or **loop-free**. Strictly speaking, a loop $\{v,v\}$ between v and v is not a pair, but we shall allow this slight abuse of terminology. If G is a graph, $V(G)$ and $E(G)$ denote its sets of vertices and edges, respectively. Ordinarily $V(G)$ is assumed to be a finite set, in which case $E(G)$ must also be finite, and we say that G is finite. If G is finite, $|V(G)|$ denotes the number of vertices in G, and is called the **order** of G. Similarly, if G is finite, $|E(G)|$ denotes the number of edges in G, and is called the **size** of G. If one allows more than one edge to join a pair of vertices, the result is then called a *multigraph*.

One of the appealing features of the study of graphs and multigraphs lies in the geometric or pictorial aspect of the subject. Graphs may be expressed by diagrams, in which each vertex is represented by a point in the plane and each edge by a curve joining the points. *In such diagrams the curve representing an edge should not pass through any points that represent vertices of the graph other than the two endpoints of the curve.* Such diagrams should be familiar from Chapter 4, where they are used to represent directed graphs.

It is convenient to refer to a diagram of a graph G as the graph itself since the sets V and E are clearly discernible from the diagram.

Example 5.1.1. The diagrams of Figure 5-1(a) and (b) illustrate two nondirected graphs. The graph G, shown in Figure 5-1(a), is not simple since there is a loop incident on vertex c. By contrast, the graph G' shown in Figure 5-1(b) is simple. On the other hand, the diagram G'' in Figure 5-1(c) represents a multigraph since there are three edges between the vertices b and c. From the diagram, it is clear that $V(G) = \{a,b,c,d\}$ and $E(G) = \{\{a,b\},\{a,c\},\{b,c\},\{c,c\},\{a,d\},\{c,d\}\}$. Moreover, $V(G) = V(G')$ but $E(G') = E(G) - \{\{c,c\}\}$. The edges of the multigraph G'' do not, in fact, form a set in the strictest sense of the word, so to list the edges of G'' we need to indicate the multiplicity of edges between b and c somehow. For instance, we could list $E(G'') = \{\{a,b\},\{a,c\},3\{b,c\},\{a,d\},\{c,d\}\}$, where $3\{b,c\}$ indicates that there are three edges between b and c. By no means is this the only way to indicate a multigraph; we discuss other ways in Section 5.9.

The graph G has order 4 and size 6 while G' has order 4 and size 5, and, of course, it would be appropriate to say that the multigraph G'' has order 4 and size 7.

Observe that nondirected graphs may be viewed as symmetric directed graphs, in which for every edge (u,v) between two vertices in one direction there is also an edge (v,u) between the same vertices in the other direction. Thus, the nondirected graph in Figure 5-1(a) could be viewed as being "the same" as the symmetric directed graph shown in Figure 5-1 (d).

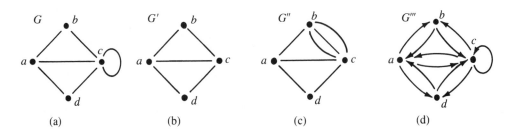

Figure 5-1. A nonsimple graph, a simple graph, a multigraph, and a symmetric directed graph.

Following this convention, it is possible to translate the entire theory of nondirected graphs into the language of directed graphs. However, though it is always possible, this is not always desirable. For certain concepts (such as colorability, to be discussed in a later section), direction is irrelevant. Such concepts are traditionally discussed in terms of nondirected graphs. In situations where the direction of edges is not relevant, the language of nondirected graphs is clearer and simpler, and so it will be used in this chapter.

Sometimes it may be useful to view a directed graph as a nondirected graph, within a context where we wish to ignore the direction of edges. In this case we shall speak of the *underlying nondirected graph* of the digraph, which has the same set of vertices and has a nondirected edge $\{u,v\}$ between two vertices iff the digraph has an edge (u,v) or (v,u). Similarly, we shall use the term *underlying simple graph* to denote the simple graph obtained by deleting all loops from a graph.

Therefore, the graph of Figure 5-1(a) is the underlying nondirected graph of the directed graph in Figure 5-1(d). Likewise, the graph in Figure 5-1(b) is the underlying simple graph of the graph in Figure 5-1(a).

The terminology of Chapter 4 for directed graphs is retained and extended in the present chapter:

Definition 5.1.2. In a directed graph an edge (u,v) is said to be **incident from** u, and to be **incident to** v. Within a particular graph, the number of edges incident to a vertex is called the **in-degree** of the vertex and the number of edges incident from it is called its **out-degree**. The in-degree of a vertex v in a graph G is denoted by $\text{degree}_G^+(v)$ and the out-degree by $\text{degree}_G^-(v)$. In the case of a nondirected graph, an unordered pair $\{u,v\}$ is an edge **incident on** u and v. The **degree** of a vertex is determined by counting each loop incident on it twice and each other edge once. The degree of a vertex v in a graph G may be denoted by $\text{degree}_G(v)$ or by $\deg_G(v)$. (The subscript G may be omitted, if it is clear from context.) A vertex of degree zero is called an **isolated vertex**.

In either case, if there is an edge incident from u to v or incident on u and v, then u and v are said to be **adjacent** or to be **neighbors.**

The minimum of all the degrees of the vertices of a graph G is denoted by $\delta(G)$, and the maximum of all the degrees of the vertices of G is denoted by $\Delta(G)$. If $\delta(G) = \Delta(G) = k$, that is, if each vertex of G has degree k, then G is said to be *k-regular* or *regular of degree k*. Usually, a 3-regular graph is called a *cubic* graph.

If $v_1, v_2, \ldots, v_n$ are the vertices of G, then the sequence $(d_1, d_2, \ldots, d_n)$, where $d_i = \text{degree}(v_i)$, is the *degree sequence* of G. Usually, we order the vertices so that the degree sequence is monotone increasing, that is, so that $\delta(G) = d_1 \le d_2 \le \ldots \le d_n = \Delta(G)$.

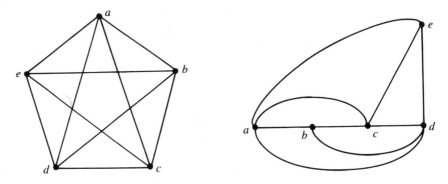

Figure 5-2.

Example 5.1.2. The vertex c of the graph G in Figure 5-1(a) has degree 5 while the degree of c in G' is 3. The degree sequence of G is $(2,2,3,5)$ while the degree sequence of G' is $(2,2,3,3)$.

Example 5.1.3. How one draws a diagram to represent a graph is basically immaterial. For instance, the two diagrams in Figure 5-2 look considerably different, yet they represent the same graph since each diagram conveys the same information, namely, that the graph has five vertices a, b, c, d, and e, and each vertex is adjacent to every other vertex. Moreover, note that this graph is 4-regular since each vertex has degree 4.

The basic relationship of degrees of vertices to the number of edges is described in the following theorem, frequently referred to as "The First Theorem of Graph Theory" or as "The Sum of Degrees Theorem." Both the theorem and its corollary are valid for multigraphs as well as for graphs.

Theorem 5.1.1. If $V = \{v_1,\ldots,v_n\}$ is the vertex set of a nondirected graph G, then

$$\sum_{i=1}^{n} \deg(v_i) = 2|E|.$$

If G is a directed graph, then

$$\sum_{i=1}^{n} \deg^+(v_i) = \sum_{i=1}^{n} \deg^-(v_i) = |E|.$$

Proof. The proof is easy, since when the degrees are summed, each edge contributes a count of one to the degree of each of the two vertices on which the edge is incident. ☐

The theorem has an interesting corollary:

Corollary 5.1.1.   In any nondirected graph there is an even number of vertices of odd degree.

Proof.   Let W be the set of vertices of odd degree and let U be the set of vertices of even degree. Then

$$\sum_{v \in V(G)} \deg(v) = \sum_{v \in W} \deg(v) + \sum_{v \in U} \deg(v) = 2|E|.$$

Certainly, $\Sigma_{v \in U} \deg(v)$ is even; hence $\Sigma_{v \in W} \deg(v)$ is even, implying that $|W|$ is even and thereby proving the corollary. $\square$

Corollary 5.1.2.   If $k = \delta(G)$ is the minimum degree of all the vertices of a nondirected graph G, then

$$k|V| \leq \sum_{v \in V(G)} \deg(v) = 2|E|.$$

In particular, if G is a k-regular graph, then

$$k|V| = \sum_{v \in V(G)} \deg(v) = 2|E|.$$

Example 5.1.4.

(a) Is there a graph with degree sequence $(1,3,3,3,5,6,6)$?
No, because the existence of such a graph would violate the conclusion of Corollary 5.1.1.

(b) Is there a simple graph with degree sequence $(1,1,3,3,3,4,6,7)$?
Assume there is such a graph. Then the vertex of degree 7 is adjacent to all other vertices, so in particular it must be adjacent to both vertices of degree 1. Hence, the vertex v of degree 6 cannot be adjacent to either of the two vertices of degree 1. But this leaves only six vertices (including v itself) to which the vertex v is adjacent. Since it is assumed that the graph is simple, v cannot be adjacent to itself and, therefore, there can be only five vertices adjacent to v. But then v cannot have degree 6. This contradiction shows that there is no simple graph with the given degree sequence.

(c) Is there a nonsimple graph G with degree sequence $(1,1,3, 3,3,4,6,7)$?
Yes, the diagram in Figure 5-3 exhibits such a graph.

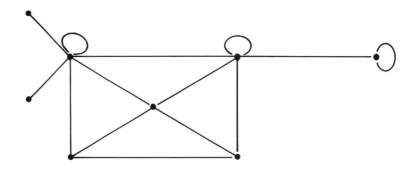

Figure 5-3.

Definition 5.1.3. In a nondirected graph G a sequence P of zero or more edges of the form $\{v_0,v_1\}, \{v_1,v_2\},\ldots,\{v_{n-1},v_n\}$ (sometimes abbreviated as $v_0 - v_1 - \ldots - v_n$) is called a path from v_0 to v_n; v_0 is the **initial** vertex and v_n is the **terminal** vertex, and they both are called endpoints of the path P. Moreover, it is convenient to designate P as a $v_0 - v_n$ path.

In the definition of path, vertices and edges may be repeated. In fact, if $v_0 = v_n$, then P is called a **closed path**. But, on the other hand, if $v_0 \neq v_n$, then P is an **open path**.

In general, we should observe that the path P is a graph itself where $V(P) = \{v_0,v_1,\ldots,v_n\} \subseteq V(G)$ and $E(P) = \{\{v_0,v_1\},\{v_1,v_2\},\ldots,\{v_{n-1},v_n\}\} \subseteq E(G)$. Moreover, $1 \leq |V(P)| \leq n + 1$ and $0 \leq |E(P)| \leq n$, where $|V(P)|$ may be less than $n + 1$ if there are repeated vertices and $|E(P)| < n$ if there are repeated edges.

Of course, P may have no edges at all, in which case, the length of P is zero, P is called a *trivial path*, and $V(P)$ is just the singleton set $\{v_0\}$.

A path P is **simple** if all edges and vertices on the path are distinct except possibly the endpoints. Thus, an open simple path of length n has $n + 1$ distinct vertices and n distinct edges, while a closed simple path of length n has n distinct vertices and n distinct edges. The trivial path is taken to be a simple closed path of length zero.

A path of length ≥ 1 with no repeated edges and whose endpoints are equal is called a **circuit**. A circuit may have repeated vertices other than the endpoints; a **cycle** is a circuit with no other repeated vertices except its endpoints. Thus, a cycle is a simple circuit, and, in particular, a loop is a cycle of length 1. Of course, in a graph a cycle that is not a loop must have length at least 3, but there may be cycles of length 2 in a multigraph. Two paths in a graph are said to be *edge-disjoint* if they share no common edges; they are *vertex-disjoint* if they share no common vertices.

Example 5.1.5. In Figure 5-1(a) the path $\{c,c\}$ is a cycle of length 1, while the sequence of edges $\{a,b\},\{b,c\},\{c,a\}$ and $\{a,d\},\{d,c\},\{c,a\}$ form cycles

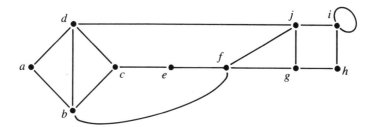

Figure 5-4.

of length 3. The path $\{a,b\},\{b,c\},\{c,d\},\{d,a\}$ is a cycle of length 4. Moreover, $\{a,b\},\{b,c\},\{c,c\},\{c,a\}$ is a circuit of length 4; it is not a cycle because the sequence of vertices a-b-c-c-a includes more than one repeated vertex. Similarly the sequence of edges $\{a,b\},\{b,c\},\{c,a\},\{a,d\},\{d,c\},\{c,a\}$ forms a closed path of length 6, but this path is not a circuit because the edge $\{c,a\}$ is repeated twice.

Example 5.1.6. For the graph in Figure 5-4, we have the following facts:

Path	Length	Simple path?	Closed path?	Circuit?	Cycle?
a-d-c-e-f-j-d-a	7	no	yes	no	no
b-c-e-f-g-j-f-b	7	no	yes	yes	no
a-b-a	2	no	yes	no	no
a-d-c-b-a	4	yes	yes	yes	yes
i-i	1	yes	yes	yes	yes
a	0	yes	yes	no	no
e-f-g-j-f-b	5	no	no	no	no
d-b-c-d	3	yes	yes	yes	yes

By definition, a simple path is certainly a path and although the converse statement need not be true, we have the following result.

Theorem 5.1.2. In a graph G, every u–v path contains a simple u–v path.

Proof. If a path is a closed path, then it certainly contains the trivial path. Assume, then, that P is an open u–v path. We complete the proof by induction on the length n of P. If P has length one, then P is itself a simple path. Suppose that all open u–v paths of length k, where $1 \leq k \leq n$, contains a simple u–v path. Now suppose that P is the open u–v path $\{v_0,v_1\},\ldots,\{v_n,v_{n+1}\}$ where $u = v_0$ and $v = v_{n+1}$. Of course, it may be that P has repeated vertices, but if not, then P is a simple u–v path and we're done. If, on the other hand, there are repeated vertices in P, let i and j be distinct positive integers where $i < j$ and $v_i = v_j$. If the closed path $v_i - v_j$ is removed from P, an open path P' is obtained having length $\leq n$ since at least the edge $\{v_i, v_{i+1}\}$ was deleted from P. Thus, by the inductive hypothesis, P' contains a simple u–v path and, thus, so does P. $\square$

Finally, we mention that there are a number of applications where it is natural or necessary to apply "weights," "capacities," or "labels" to the edges or vertices of a graph. These are merely values that are associated with the elements of a graph by a function. We shall refer to these all as **labelings.**

Definition 5.1.4.   An **edge labeling** of a graph G is a function f: $E(G) \to D$, where D is some domain of labels. A **vertex labeling** of G is a function f: $V(G) \to D$.

The degree function which assigns to each vertex in a nondirected graph its degree is an example of a vertex labeling. Labelings will be seen again in the context of minimal spanning trees in Section 5.4, in the context of binary trees in Section 5.6, in the context of coloring in Section 5.11, and in the context of network flows in Chapter 7.

Exercises for Section 5.1

1. Draw a picture of each of the following graphs and state whether it is directed or nondirected and whether it is simple.
   (a) $G_1 = (V_1, E_1)$, where $V_1 = \{a,b,c,d,e\}$ and
       $E_1 = \{\{a,b\},\{b,c\},\{a,c\},\{a,d\},\{d,e\}\}$.
   (b) $G_2 = (V_2, E_2)$, where $V_2 = \{a,b,c,d,e\}$ and
       $E_2 = \{(a,b),(a,c),(a,d),(a,e),(e,c),(c,a)\}$.
   (c) $G_3 = (V_3, E_3)$, where $V_3 = \{a,b,c,d,e\}$ and
       $E_3 = \{(a,a),(a,b),(b,c),(c,d),(e,d),(d,e)\}$.
   (d) Determine the in-degree and out-degree of each vertex in the graphs G_2 and G_3.

2. For each of the directed graphs given in Exercise 1 give the underlying nondirected graph. For each of the nondirected graphs in Exercise 1 give the translation to an equivalent (symmetric) directed graph. Give the sets of vertices and edges, as well as drawing the graphs.

3. A sequence $d = (d_1, d_2, \dots, d_n)$ is **graphic** if there is a simple nondirected graph with degree sequence d. Show that the following sequences are not graphic.
   (a) $(2,3,3,4,4,5)$
   (b) $(2,3,4,4,5)$

(c) (1,3,3,3)
(d) (2,3,3,4,5,6,7)
(e) (1,3,3,4,5,6,6)
(f) (2,2,4)
(g) (1,2,2,3,4,5)
(h) Any sequence $(d_1, d_2 \ldots, d_n)$ where all the d_i's are distinct and $n > 1$.
(i) (1,1,3,3,3,3,4,5,8,9)

4. Suppose that G is a nondirected graph with 12 edges. Suppose that G has 6 vertices of degree 3 and the rest have degrees less than 3. Determine the minimum number of vertices G can have.

5. Let G be a nondirected graph of order 9 such that each vertex has degree 5 or 6. Prove that at least 5 vertices have degree 6 or at least 6 vertices have degree 5.

6. (a) Suppose that we know the degrees of the vertices of a nondirected graph G. Is it possible to determine the order and size of G? Explain.
   (b) Suppose that we know the order and size of a nondirected graph G. Is it possible to determine the degrees of the vertices of G? Explain.

7. Give a simplest possible example of a nonnull nondirected graph:
   (a) having no vertices of odd degree;
   (b) having no vertices of even degree;
   (c) having exactly one vertex of odd degree;
   (d) having exactly one vertex of even degree;
   (e) having exactly two vertices of odd degree;
   (f) having exactly two vertices of even degree.

8. Suppose you are married and you and your spouse attended a party with three other married couples. Several handshakes took place. No one shook hands with himself nor with one's own spouse, and no one shook hands with the same person more than once. After all the handshaking was completed, suppose that you asked each person, including your spouse, how many hands that person had shaken. Each person gave a different answer.
   (a) How many hands did you shake?
   (b) How many hands did your spouse shake?
   (c) What is the answer to (a) and (b) if there were a total of 5 couples?

9. Let $\delta(G)$ and ΔG denote the minimum and maximum degrees of all the vertices of G, respectively. Show that for a nondirected graph G, $\delta(G) \leq 2 \cdot |E|/|V| \leq \Delta(G)$.

10. How many vertices will the following graphs have if they contain:
    (a) 16 edges and all vertices of degree 2;
    (b) 21 edges, 3 vertices of degree 4, and the other vertices of degree 3;
    (c) 24 edges and all vertices of the same degree.

11. For each of the following questions, describe how the problem may be viewed in terms of a graph model, and then answer the question:
    (a) Must the number of people at a party who do *not* know an odd number of other people be even?
    (b) Must the number of people ever born who had (or have) an odd number of brothers and sisters be even?
    (c) Must the number of families in Florida with an odd number of children be even?

12. What is the largest possible number of vertices in a graph with 35 edges and all vertices of degree at least 3?

13. Show that in a graph G there is a vertex of degree at least $\lceil 2|E|/|V| \rceil$ and a vertex of degree at most $\lfloor 2|E|/|V| \rfloor$.

14. (a) Let G be a graph with n vertices and m edges such that the vertices have degree k or $k + 1$. Prove that if G has N_k vertices of degree k and N_{k+1} vertices of degree $k + 1$, then $N_k = (k + 1)n - 2m$.
    (b) Suppose all vertices of a graph G have degree k, where k is an odd number. Show that the number of edges in G is a multiple of k.

15. For any simple graph G, prove that the number of edges of G is less than or equal to $1/2n(n - 1)$, where n is the number of vertices of G.

16. Which of the following are true for every graph G? For those statements that are not true for every graph give a counter example.
    (a) There are an odd number of vertices of even degree.
    (b) There are an even number of vertices of odd degree.
    (c) There are an odd number of vertices of odd degree.
    (d) There are an even number of vertices of even degree.

17. Applications of Graph Theory to Organic Chemistry: Molecules are made up of a number of atoms that are chained together by chemical bonds. (We can model this by a graph. The atoms will be vertices and the bonds will be the edges of the graph.) A hydrocarbon molecule contains only carbon and hydrogen atoms. Moreover, each hydrogen atom is bonded to a single carbon atom and each carbon atom may be bonded to 2, 3, or 4 atoms, which can be either

carbon or hydrogen atoms. For example, an ethane molecule has 2 carbon atoms and 6 hydrogen atoms, thus, ethane is represented by the molecular formula C_2H_6 and the molecule can be represented by the following graph:

We know that in any such graph the degree of any carbon atom must be 4 and the degree of any hydrogen atom must be 1.

A. Find the number of bonds in
   (a) the cyclobutane molecule C_4H_8
   (b) a cyclohexane molecule C_6H_{12}
   (c) a hydrocarbon molecule C_nH_{2n}
   (d) a decane molecule $C_{10}H_{22}$.
   (e) a hydrocarbon molecule C_nH_{2n+2}.

B. Can there exist (at least theoretically) a hydrocarbon molecule with the following molecular formulas?
   (a) C_3H_5
   (b) C_5H_{10}
   (c) C_5H_{11}
   (d) $C_{20}H_{39}$

18. Prove that the number of simple paths of length 2 in a graph G is $d_1^2 + d_2^2 + \ldots + d_n^2$, where $(d_1, d_2, \ldots, d_n)$ is the degree sequence for G. Hint: for each vertex v in G, count the number of paths of length 2 with v as midpoint.

19. How many different simple graphs are there with the given vertex set $\{v_1, v_2, \ldots, v_n\}$?

*20. (Havel-Hakimi) Prove that there exists a simple graph with degree sequence $(d_1, d_2, \ldots, d_n)$ where $d_1 \le d_2 \le \ldots \le d_n$ if and only if there exists one with degree sequence $(d_1^1, d_2^1, \ldots, d_{n-1}^1)$, where

$$d_k^1 = \begin{cases} d_k & \text{for } k = 1, 2, \ldots, n - d_n - 1, \\ d_k - 1 & \text{for } k = n - d_n, \ldots, n - 1. \end{cases}$$

Thus $(1,1,1,2,2,2,3,3,3,3,5)$ is graphic if and only if $(1,1,1,2,2,1,2,2,2,2)$ is graphic.

Apply the Havel-Hakimi result to determine if the following sequences are graphic:
(a) (1,1,1,1,2,2,2,2,2,2)
(b) (1,1,1,2,2,2,3,3,4,7)
(c) (0,1,2,3,4,4)
(d) (1,1,2,2,2,2,3,3)
(e) (1,3,3,4,5,5,5,5,5)
(f) (1,2,3,4,4,5,6,7)

Selected Answers for Section 5.1

3. (a) Sum of degrees is odd, or there is an odd number of vertices of odd degree.
   (b) A simple graph of order 5 cannot have a vertex of degree 5.
   (c) Suppose that G is a graph with this degree sequence. Each vertex of degree 3 has an edge leading to each other vertex. Let a, b, c, d be the vertices where $\deg(a) = 1$. Since b, c, d have degree 3, there must be an edge joining b to a, one joining c to a, and one joining d to a. Hence a has degree 3 or more. Contradiction.
   (d) A simple graph of order 7 cannot have a vertex of degree 7.
   (e) Suppose that a graph G exists with the given degree sequence. Let v be the vertex of degree 1, $G - v$ has 6 vertices and at least 1 vertex of degree 6. This is impossible.

5. Use the Pigeon Hole Principle.

8. Hint: Represent this situation by a graph where the vertices correspond to people and each edge represents a handshake. Now consider the degrees of this graph. The degree of any vertex is at most 6. The 7 different responses must have been 0, 1, 2, 3, 4, 5, and 6.

10. (a) 16
    (b) 13
    (c) Solve $k \mid V \mid = 48$ for all divisors k of 48.

12. $2 \mid E \mid \geq 3 \mid V \mid$ or $2/3 \mid E \mid = (2/3) \, 35 \geq \mid V \mid$ implies $\mid V \mid \leq 23$.

15. The number of edges $\leq C(n,2)$ since each edge connects 2 vertices.

5.2   ISOMORPHISMS AND SUBGRAPHS

The concepts of isomorphism and subgraph, defined previously in Chapter 4, extend naturally to nondirected graphs.

Definition 5.2.1. Two graphs G and G' are **isomorphic** if there is a function $f\colon V(G) \to V(G')$ from the vertices of G to the vertices of G' such that

(i) f is one-to-one,
(ii) f is onto, and
(iii) for each pair of vertices u and v of G, $\{u,v\} \in E(G)$ iff $\{f(u),f(v)\} \in E(G')$.

Any function f with the above three properties is called an **isomorphism** from G to G'. The condition (iii) says that vertices u and v are adjacent in G iff $f(u)$ and $f(v)$ are adjacent in G'. In other words, we say that the function f *preserves adjacency.*

If the graphs G and G' are isomorphic and f is an isomorphism of G to G', then intuitively the only difference between the graphs is the names of the vertices. Indeed, if we were to change the names of the vertices of G' from $f(v)$ to v for each $v \in V(G)$, then G' with the newly named vertices would be identical to the graph G, for then they both would have the same lists of vertices and edges.

Of course, if G and G' are isomorphic graphs the isomorphism f is by no means unique, there may be several isomorphisms from G to G'. But if such an isomorphism f exists, then there are several conclusions we can make, namely,

1. $|V(G)| = |V(G')|$
2. $|E(G)| = |E(G')|$
3. If $v \in V(G)$, then $\deg_G(v) = \deg_G(f(v))$, and, thus, the degree sequences of G and G' are the same.
4. If $\{v,v\}$ is a loop in G, then $\{f(v),f(v)\}$ is a loop in G', and more generally, if $v_0 - v_1 - v_2 - \ldots - v_{k-1} - v_k = v_0$ is a cycle of length k in G, then $f(v_0) - f(v_1) - f(v_2) - \ldots - f(v_{k-1}) - f(v_k)$ is a cycle of length k in G'. In particular, the cycle vectors of G and G' are equal, where the **cycle vector** of G is by definition the vector $(c_1,c_2,\ldots c_n)$ where c_i is the number of cycles in G of length i. Of course, $c_1 = 0$ for simple graphs and c_2 is nonzero only for multigraphs.

Discovering Isomorphisms

The problem of determining whether or not two graphs are isomorphic is known as the *isomorphism problem,* and for arbitrary graphs, the only known algorithms which guarantee a correct answer to the isomorphism problem usually require approximately 2^n operations where n is the number of vertices. For instance, if G and G' are two graphs with the same number of vertices, we might attempt to answer the isomorphism question by exhaustively searching the entire list of one-to-one onto maps $f:V(G) \rightarrow V(G')$. But if $n = |V(G)|$, there are $n!$ such one-to-one onto maps f, and, of course, $n!$ is quite large even for small values of n.

The isomorphism problem itself is particularly difficult, but once we have a one-to-one onto map $f:V(G) \rightarrow V(G')$ the process of checking whether or not this map is, in fact, an isomorphism is fairly easy. We simply employ the adjacency matrix as a bookkeeping device for recording all adjacencies.

If $v_1, v_2, \ldots, v_n$ are the vertices of G, then the **adjacency matrix** for this ordering of the vertices of G is the $n \times n$ matrix A, where the ij^{th} entry $A(i,j)$ of A is 1 iff the edge $\{v_i, v_j\}$ is an edge of G; otherwise, $A(i,j) = 0$. Thus, A is a symmetric matrix each of whose entries is either 0 or 1. Moreover, 1 will appear on the ith position of the diagonal of A iff there is a loop at v_i. Of course, if we change the ordering of the vertices of G, then the entries of A will be rearranged.

Then we have the following fact, the proof of which we leave to the reader.

> Suppose that G and G' are two graphs and that $f:V(G) \rightarrow V(G')$ is a one-to-one onto function. Let A be the adjacency matrix for the vertex ordering $v_1, v_2, \ldots, v_n$ of the vertices of G. Let A' be the adjacency matrix for the vertex ordering $f(v_1), f(v_2), \ldots, f(v_n)$. Then f is an isomorphism from $V(G)$ to $V(G')$ iff the adjacency matrices A and A' are equal.

Of course, if the adjacency matrices A and A' are not equal, then all that proves is that the function f itself is not an isomorphism; it may still be the case that graphs G and G' are isomorphic under some other function.

The next question is this: What can we do to help locate a potential isomorphism? Or in other words, can we refine the brute force method of considering all possible one-to-one onto maps so that we consider only those maps from a smaller list.

For one thing, we can attempt to use the fact that an isomorphism must map a vertex of G of degree d to a vertex of G' of degree d. For many examples, this will shorten the search for isomorphisms considerably.

Example 5.2.1. The graphs G and G' of Figure 5-5 are isomorphic.

First we observe that if G and G' are isomorphic, then the vertices b, d, and e must be mapped to the vertices b', d', and e' respectively since these

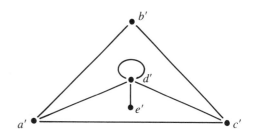

Figure 5-5.

are the unique vertices of degree 2, 5, and 1. Now instead of 5! maps, we have only 2 maps to consider. In this case, both of these maps are isomorphisms. For instance, we can verify that the map f which maps a to c', b to b', c to a', d to d', and e to e' is an isomorphism because the adjacency matrix for G for the ordering a,b,c,d,e and the adjacency matrix for G' for the ordering $f(a) = c'$, $f(b) = b'$, $f(c) = a'$, $f(d) = d'$, $f(e) = e'$ is the matrix

$$\begin{bmatrix} 0 & 1 & 1 & 1 & 0 \\ 1 & 0 & 1 & 0 & 0 \\ 1 & 1 & 0 & 1 & 0 \\ 1 & 0 & 1 & 1 & 1 \\ 0 & 0 & 0 & 1 & 0 \end{bmatrix}$$

It is routine to check that the map f' which maps a to a', b to b', c to c', d to d', and e to e' is also an isomorphism.

Thus, in some cases at least, the degree sequence of a graph can be used to shorten the search for isomorphisms. In fact, if the simple graphs G and G' each have V_i vertices of degree i, then there are $(V_0!)(V_1!) \ldots (V_{n-1}!)$ one-to-one onto maps from $V(G)$ to $V(G')$ that will map the V_i vertices of degree i to vertices of degree i. With any luck, this number will be more manageable than $n!$.

Determining When Graphs Are Not Isomorphic

Next we address this question: How do we show that graphs are not isomorphic, short of exhaustive search?

One way to show that two graphs are not isomorphic is to find some property that isomorphic graphs must share, but which the two graphs G and G' do *not* share. Thus, for example, if G and G' have different numbers of vertices, then the graphs are not isomorphic. Likewise, if they have different degree sequences, they are not isomorphic.

But unfortunately, two graphs may have the same degree sequence (and hence the same number of vertices and edges) and still not be isomorphic. In this case, we must dig deeper for reasons why they are not isomorphic. Here's where the rub comes; there is generally no set procedure that works and we must resort to a whole grab-bag of tricks. Nevertheless, one approach seems to work fairly often. Let us attempt to describe that approach in general terms.

The basic idea is that we attempt to classify the vertices into classes according to some property preserved by isomorphism. Then if the two graphs are isomorphic the vertices of a given class in one graph must correspond to the vertices of the same class in the other graph. But, on the other hand, if the vertices in these classes do not correspond, then we have to conclude that the graphs are not isomorphic. For example, if we classify into a single class the vertices of degree two in each graph, these classes form a graph in their own right and we could ask if these subgraphs are isomorphic. To clarify what we mean we need to discuss the concept of subgraph.

Definition 5.2.2.   If G and H are graphs then H is a **subgraph** of G iff $V(H)$ is a subset of $V(G)$ and $E(H)$ is a subset of $E(G)$. A subgraph H of G is called a *spanning subgraph* of G iff $V(H) = V(G)$. If W is any subset of $V(G)$, then the *subgraph induced by* W is the subgraph H of G obtained by taking $V(H) = W$ and $E(H)$ to be those edges of G that join pairs of vertices in W.

Example 5.2.2.   Consider the graphs shown in Figure 5-6. The graph G' shown in (b) is a subgraph of the graph G shown in (a), with $V(G') = \{v_1, v_2, v_4, v_5\}$. The graph G'' shown in (c) is a spanning subgraph of G, while the graph G''' in (d) is the subgraph induced by the set $W = \{v_1, v_2, v_4, v_5\}$.

The graph G'''' shown in (e) is not a subgraph of G because the edge $\{v_1, v_5\}$ was not in $E(G)$.

If e is an edge of a given graph G, we use the notation $G - e$ to denote the graph obtained from G by deleting the edge e; in other words, $E(G - e)$ is the set of all edges in G except e. More generally, $G - \{e_1, \ldots, e_k\}$ stands for the graph obtained from G by deleting the edges $e_1, \ldots, e_k$. Similarly, if v is a vertex of G, we use the notation $G - v$ to denote the graph obtained by removing the vertex v from G, *together with all edges incident on v*; more generally, we write $G - \{v_1, \ldots, v_k\}$ for the graph obtained by deleting the vertices $v_1, \ldots, v_k$ and all edges incident on any of them.

A simple nondirected graph with n mutually adjacent vertices is called a *complete graph on n vertices*, and may be represented by the symbol

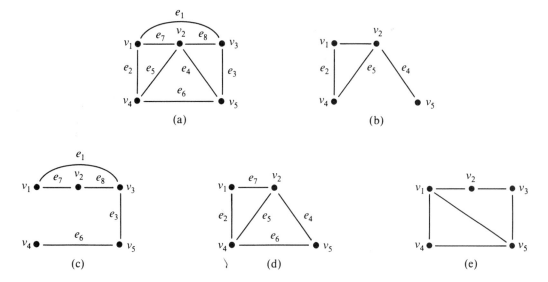

Figure 5-6. Graphs G, G', G'', G''', and G''''.

K_n. A complete graph on n vertices has $n \cdot (n-1)/2$ edges, and each of its vertices has degree $n-1$. A complete subgraph on two vertices, K_2, is just two vertices joined by an edge. Any graph may be viewed as made up of "building blocks" which are complete subgraphs. For example, both graphs in Figure 5-5 consist of two K_3 graphs joined at a common edge, a K_2 graph, and a loop.

Example 5.2.3. We show that the two graphs in Figure 5-7 are not isomorphic.

Both graphs have 8 vertices and 11 edges; both have 3 vertices of degree 2, 4 vertices of degree 3, and one vertex of degree 4. However, there the similarity ends. If the two graphs were isomorphic then the respective

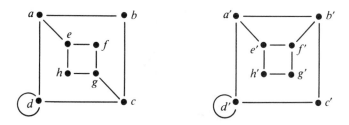

Figure 5-7. Two nonisomorphic graphs.

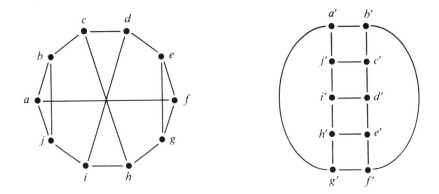

Figure 5-8. Two nonisomorphic graphs.

subgraphs induced by the vertices of degree 2 would be isomorphic. In the first graph no pair of vertices of degree 2 are adjacent, whereas g' and h' are vertices of degree 2 in the second graph that are adjacent. (There is also a difference in the structure of the subgraphs induced by the vertices of degree 3 in each graph.) A third reason these graphs are not isomorphic is that in the first graph the vertex of degree 4 is adjacent to two vertices of degree 3, whereas the vertex of degree 4 in the second graph is adjacent to a vertex of degree 2.

Example 5.2.4.   The graphs in Figure 5-8 are not isomorphic.

These two graphs have 10 vertices and 15 edges, and all vertices are of degree 3. The first graph has a subgraph of 3 adjacent vertices (the triangle $a–b–j–a$, for example), but the second contains no cycles of length 3.

Example 5.2.5.   The graphs in Figure 5-9 are isomorphic.

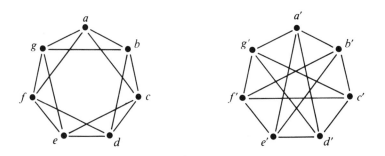

Figure 5-9. Two isomorphic graphs.

Both of these graphs have 7 vertices and 14 edges. Every vertex in each graph has degree 4. We construct an isomorphism. Starting with vertex a in the left graph, we can match a to any vertex of the right graph, since each graph is symmetric. Let us match a to a'. Now the set of neighbors of a must be matched to the set of neighbors of a'. It might seem conceivable that g could be matched with g', b with b', f with e', and c with d'. However, g and b are adjacent in the left graph and g' and h' are not adjacent in the right graph. Thus, we must try another approach. Let us consider the subgraph of neighbors of a and the subgraph of neighbors of a'. (See Figure 5-10).

Clearly f must be matched with either g' or b'. Say f is matched to g', then match g to d', b to e', and c to b'. Now there remain only two unmatched vertices in each graph: d and e, and c' and f'. Vertex g is adjacent to e but not d, whereas d' (matched with g) is adjacent to c' but not to f'. Hence we must match e to c' and d to f'. We conclude that if there is an isomorphism between these graphs then the matching we have obtained,

$$a \rightarrow a'$$
$$b \rightarrow e'$$
$$c \rightarrow b'$$
$$d \rightarrow f'$$
$$e \rightarrow c'$$
$$f \rightarrow g'$$
$$g \rightarrow d'$$

must be such an isomorphism. Checking that edges match, we see that in fact it is an isomorphism.

There is a simpler way to verify that these graphs are isomorphic. It involves the use of complements of graphs.

Definition 5.2.3.   If H is a subgraph of G, the **complement of H in** G, denoted by $\overline{H}(G)$, is the subgraph $G - E(H)$; that is, the edges of H are

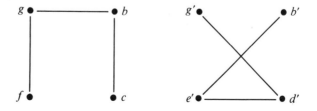

Figure 5-10. Neighbors of a and a'.

Figure 5-11. A graph and its complement.

deleted from those of G. If H is a simple graph with n vertices the **complement** $\overline{H}$ of H is the complement of H in K_n.

It follows from this definition that $V(\overline{H}) = V(H)$ and any two vertices are adjacent in $\overline{H}$ if and only if they are *not* adjacent in H. Note that the degree of a vertex in $\overline{H}$ plus its degree in H is $n - 1$, where $n = |V(H)|$.

Example 5.2.6. A graph and its complement are shown in Figure 5-11(a) and (b).

Now two simple graphs are isomorphic iff their complements are isomorphic. (This follows directly from the definitions of isomorphism and complement.) The complements of the graphs shown in Figure 5-9 are shown in Figure 5-12.

Clearly these complementary graphs are nothing more than cycles of length 7, and hence are isomorphic. In general, if a graph has more pairs of vertices joined by edges than not, its complement will have fewer edges and thus probably will be simpler to analyze.

Definition 5.2.4. Let G and G' be two graphs. The **intersection** of G and G', written $G \cap G'$, is the graph whose vertex set is $V(G) \cap V(G')$

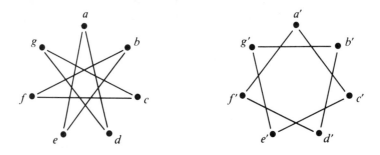

Figure 5-12. The complements of the graphs in Figure 5-9.

and whose edge set is $E(G) \cap E(G')$. Similarly, the **union** of G and G' is the graph with vertex set $V(G) \cup V(G')$ and edge set $E(G) \cup E(G')$.

In general, if G is a simple graph with n vertices, then $G \cup \overline{G}$ is a complete graph on n vertices.

Special Graphs

There are a number of interesting special classes of graphs, sufficiently important to have names. One example is the complete graphs, discussed previously. Every complete graph with n vertices is isomorphic to every other complete graph with n vertices. We now introduce a few more such special classes of graphs, which, like the complete graphs, each form a class of isomorphic graphs.

Definition 5.2.5. A **cycle graph** of order n is a connected graph whose edges form a cycle of length n. Cycle graphs are denoted by C_n. A **wheel** of order n is a graph obtained by joining a single new vertex (the "hub") to each vertex of a cycle graph of order $n - 1$. Wheels of order n are denoted by W_n. A **path graph** of order n is obtained by removing an edge from a C_n graph. Path graphs of order n are denoted by P_n. A **null graph** of order n is a graph with n vertices and no edges. Null graphs of order n are denoted by N_n. (Note that this is in contrast to the empty graph, which has no vertices and no edges.)

Example 5.2.7. Graphs of classes K_5, C_5, W_5, P_5, and N_5 are shown in Figure 5-13(a) through (e), respectively. Note that N_5 is the complement of K_5.

Definition 5.2.6. A **bipartite graph** is a nondirected graph whose set of vertices can be partitioned into two sets M and N in such a way that each edge joins a vertex in M to a vertex in N. A **complete bipartite graph** is a bipartite graph in which every vertex of M is adjacent to every vertex of N. The complete bipartite graphs that may be partitioned into sets M and N as above such that $|M| = m$ and $|N| = n$ are denoted by $K_{m,n}$

(a)   (b)   (c)   (d)   (e)

Figure 5-13. K_5, C_5, W_5, P_5, and N_5.

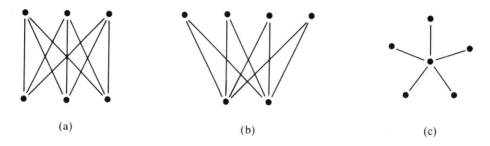

(a)   (b)   (c)

Figure 5-14. Complete bipartite graphs.

(where we normally order m and n such that $m \le n$). Any graph that is $K_{1,n}$ is called a **star graph.**

Example 5.2.8. Graphs that are $K_{3,3}$, $K_{2,4}$ and $K_{1,5}$ are shown in (a), (b), and (c) of Figure 5-14, respectively.

The next figure, Figure 5-15, shows five other especially interesting graphs, the graphs of the edges of the five platonic solids: tetrahedron, octahedron, cube, icosahedron, and dodecahedron.

Exercises for Section 5.2

1. Determine all nonisomorphic simple nondirected graphs of order 3. Do the same for those of order 4. (Hint: there are 4 of order 3 and 11 of order 4.)

2. Which of the following pairs of nondirected graphs in Figure 5-16 are isomorphic? Justify your answer carefully.

Figure 5-15.

Figure 5-16.

Figure 5-16. continued

Figure 5-16. continued

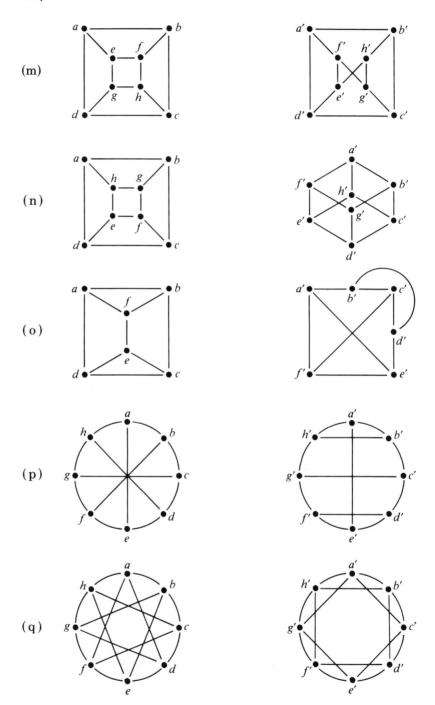

Figure 5-16. continued

(r)

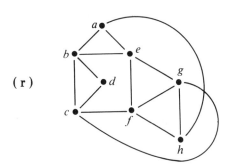

Figure 5-16. continued

3. (a) Give an example of two nondirected graphs with 4 vertices and 2 edges that are not isomorphic. Verify that they are not isomorphic.

   (b) Let G, G', and G'' be any 3 nondirected simple graphs of order 4 and size 2. Prove that at least two of these graphs are isomorphic.

4. The 3 graphs illustrated in Figure 5-17 are isomorphic. The vertices of the first graph are labeled a, b, c, d, e, f, g, h, i, and j. Complete the labelling of the vertices of the second graph as $a', \ldots, j'$ in such a way that the map $f(x) = x'$ for each letter x in $\{a,b,c,d,e,f,g,h,i,j\}$ is an isomorphism. Also label the vertices of the third graph as $a'', b'', \ldots, j''$ in such a way that the map $g(x) = x''$ is an isomorphism. The graph is known as the Petersen graph.

5. Prove that if G is a simple nondirected graph with 6 vertices, then G or the complement of G contains 3 mutually adjacent vertices.

6. The two graphs G_1 and G_2 shown in Figure 5-18 have 8 vertices and 8 edges. Moreover, they have the same degree sequence

(a)   (b)   (c)

Figure 5-17

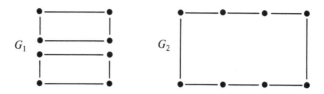

Figure 5-18

(2,2,2,2,2,2,2,2). Verify that they are not isomorphic. Is it true that every graph G with 8 vertices and 8 edges and degree sequence (2,2,2,2,2,2,2,2) has to be isomorphic to one of G_1 and G_2? Explain.

7. Give an example of 2 nonisomorphic graphs of order 6 and size 6 with degree sequence (2,2,2,2,2,2).

8. Let G be a simple graph all of whose vertices have degree 3 and $|E| = 2|V| - 3$. What can be said about G?

9. Prove that a graph G is isomorphic to a graph H if there is an ordering of the vertices of G and H such that the resulting adjacency matrices are equal.

10. Draw the graph of $K_{2,5}$.

11. (a) Show that two simple graphs are isomorphic if and only if their complements are isomorphic.

    (b) If a simple graph with n vertices is isomorphic to its complement, how many vertices does it have?

    (c) Can a simple graph with 7 vertices be isomorphic to its complement?

12. Let C_n be the cycle graph with n vertices. Prove that C_5 is the only cycle graph isomorphic to its complement.

13. Note that the Petersen graph may be obtained by taking an outer cycle graph with 5 vertices, 5 "spokes" incident to the vertices of this C_5, and an inner cycle graph on 5 vertices attached by joining its vertices to every *second* spoke. M. E. Watkins has defined the *generalized Petersen graph* $P(n,k)$, which consists of an outer n-cycle, n spokes incident to the vertices of this n-cycle, and an inner n-cycle attached by joining its vertices to every k-th spoke. Thus, $P(5,2)$ is the Petersen graph.
    (a) Draw a diagram of $P(7,2)$, $P(9,2)$, $P(7,3)$, and $P(9,4)$.
    (b) Prove that $P(n,k)$ is isomorphic to $P(n,n-k)$.

14. Let $\mathcal{G}_n$ be the class of all simple graphs with n vertices and where each vertex has degree 3, that is $\mathcal{G}_n$ is the class of all *cubic graphs*

on n vertices. Then $\mathscr{G}_n$ can be partitioned into equivalence classes where we define two graphs as equivalent if they are isomorphic.

(a) Show that if $G = (V,E)$ is in $\mathscr{G}_n$, then n is a multiple of 2 and $|E|$ is a multiple of 3.

(b) Show that there is only one equivalence class for $\mathscr{G}_4$, namely the class determined by K_4.

(c) Show that $K_{3,3}$ and the graph H:

are nonisomorphic graphs in $\mathscr{G}_6$.

(d) Show that every graph in $\mathscr{G}_6$ is isomorphic to $K_{3,3}$ or to H.

(e) Show that none of the 6 graphs illustrated below are isomorphic. Note that all 6 graphs are in $\mathscr{G}_8$.

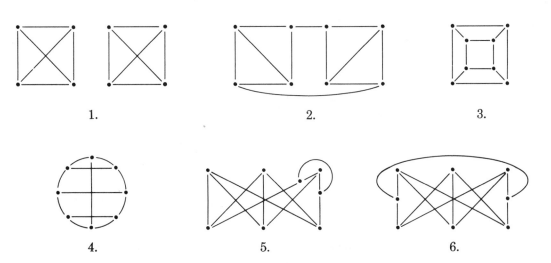

1.   2.   3.

4.   5.   6.

*(f) Show that every graph in $\mathscr{G}_8$ is isomorphic to one of the above 6 graphs listed in (e).

(g) Exhibit at least 2 nonisomorphic graphs in $\mathscr{G}_{10}$.

15. The length of the longest simple path between two distinct vertices in a graph G is called the *diameter* of G. The length of the shortest cycle in G is the *girth* of G while the length of the longest cycle is

the *circumference* of G. Find the diameter, girth, and circumference of the following graphs.

(a) C_5 (b) K_5 (c) $K_{3,3}$.

(d) The Petersen graph of Figure 5-17

16. Prove that every circuit contains a cycle.

17. State the contrapositive and the converse of the following statements:

    (a) If G and H are isomorphic graphs, then G and H have the same number of vertices.

    (b) If G and H are isomorphic graphs, then G and H have the same degree sequence.

    Which of these statements are true?

18. Determine the number of edges in (a) K_n (b) $K_{m,n}$ (c) C_n (d) P_n. Also determine the number of vertices in $K_{m,n}$.

19. Show that a simple graph with $|V| = n$ is not bipartite if $|E| > \lfloor n^2/4 \rfloor$.

20. (a) Suppose that G is a bipartite graph where $V(G) = M \cup N$ and each edge of G joins a vertex in M to a vertex in N. Prove that the sum of the degrees of the vertices in M is equal to the sum of the degrees of the vertices in N.

    (b) Are there bipartite graphs with the following degree sequences?

    (i) (3,3,3,5,6,6,6,6,6,6).

    (ii) (3,5,5,5,5,5,5,5,5,5).

    (iii) (4,5,5,5,5,5,5,5,7,10,10,10).

21. If a graph G has n vertices, all but one of odd degree, how many vertices of odd degree are there in the complement?

22. If $v_1, v_2, \ldots, v_p$ are the vertices of a graph G and A is the adjacency matrix of G, prove the following:

    (a) $A^n(i,j)$ = the number of different paths from v_i to v_j of length n in G.

    (b) $A^2(i,i)$ = degree of v_i.

    (c) $(1/6)\text{Tr}(A^3)$ = the number of 3-cycles in G. Here $\text{Tr}(B)$ for a matrix B means the trace of B which is defined as the sum of the diagonal entries of B.

23. If $\{v_1, v_2, \ldots, v_n\} = V(G)$ and $\{e_1, e_2, \ldots, e_m\} = E(G)$, then define the incidence matrix of the graph G as the $n \times m$ matrix B where $B(i,j) = 1$ iff vertex v_i is incident with edge e_j and $B(i,j) = 0$ otherwise. Prove the following facts about incidence matrices:

    (a) No two columns of B are equal.

    (b) The sum of the entries in any column is 2.

    (c) If G and G' are two graphs, then G and G' are isomorphic iff their incidence matrices are equal for some ordering of their vertices and edges.

24. Show that a graph G is bipartite iff the vertices of G can be labeled in such a way that its adjacency matrix A can be represented in the form

$$\begin{bmatrix} 0 & B \\ B^t & 0 \end{bmatrix}$$

where B^t means the transpose of the matrix B and 0 is a matrix whose only entries are zero.

25. Determine the graph G with adjacency matrix A such that

$$A^2 = \begin{bmatrix} 4 & 1 & 1 & 4 & 1 & 1 \\ 1 & 3 & 2 & 1 & 2 & 2 \\ 1 & 2 & 3 & 1 & 2 & 2 \\ 4 & 1 & 1 & 4 & 1 & 1 \\ 1 & 2 & 2 & 1 & 3 & 2 \\ 1 & 2 & 2 & 1 & 2 & 3 \end{bmatrix} \quad \text{and} \quad A^3 = \begin{bmatrix} 4 & 9 & 9 & 4 & 9 & 9 \\ 9 & 4 & 5 & 9 & 4 & 4 \\ 9 & 5 & 4 & 9 & 4 & 4 \\ 4 & 9 & 9 & 4 & 9 & 9 \\ 9 & 4 & 4 & 9 & 4 & 5 \\ 9 & 4 & 4 & 9 & 5 & 4 \end{bmatrix}$$

26. Prove that if $d(G) \geq 3$, where $d(G)$ is the diameter of the simple graph G, then $d(\overline{G}) \leq 3$, where $\overline{G}$ is the complement of G. Then if G is isomorphic to $\overline{G}$, $d(G) = 2$ or 3.

27. A *tripartite* graph is a graph where the set of vertices may be partitioned into three subsets so that no edge has incident vertices in the same subset.
    (a) Give an example of a tripartite graph.
        In general, a graph G is *k-partite* if it is possible to partition the vertex set of G into k disjoint subsets $V_1, V_2, \ldots, V_k$ such that each edge has one endpoint in one subset V_i and the other endpoint in some V_j where $i \neq j$. A *complete k-partite graph* is a simple k-partite graph with the additional property that each vertex in V_i is adjacent to each vertex in V_j where $i \neq j$, for all i and j. If $n_i = |V_i|$, then the complete k-partite graph is denoted by $K_{n_1, n_2, \ldots, n_k}$.
    (b) Draw an example of $K_{2,2,3}$.
    (c) Draw an example of $K_{2,2,2,2}$.
    (d) Determine the size of $K_{n_1, n_2, \ldots, n_k}$.

Selected Answers for Section 5.2

2. (a) Nonisomorphic; vertices of degree 3 are adjacent in one graph, nonadjacent in the other.

(b) Isomorphic; remove vertices of degree 2 and compare the remaining graphs.

(c) Nonisomorphic; different number of vertices.

(d) Nonisomorphic; vertices of degree 3 are adjacent in one graph and nonadjacent in the other.

(e) Nonisomorphic; consider their degree sequences.

(f) Nonisomorphic; consider their complements.

(i) Isomorphic; remove the vertices of degree 5 and their incident edges from each graph, observe an isomorphism for the remaining graphs and extend.

(k) Nonisomorphic; consider the neighbors of the vertex of degree 5 in each graph.

3. Hint: If a graph has order 4 and size 2, then the 2 edges may be adjacent or nonadjacent.

8. Note that $|V| = 6$: Conclude that G is isomorphic to $K_{3,3}$ or the graph

13. (a)

P(7,2)

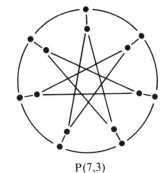

P(7,3)

5.3   TREES AND THEIR PROPERTIES

In this section we will study a very important special kind of graph known as a *tree*.

Definition 5.3.1. A **tree** is a simple graph G such that there is a unique simple nondirected path between each pair of vertices of G. A **rooted tree** is a tree in which there is one designated vertex, called a *root*. A rooted tree is a **directed tree** if there is a root from which there is a *directed* path to each vertex. In this case there is exactly one such

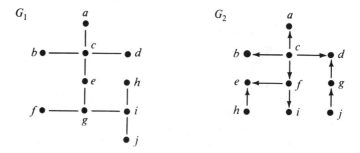

Figure 5-19. Two kinds of nondirected trees.

root, and we shall designate it as *the root*. The **level** of a vertex v in a rooted tree is the length of the path to v from the root. A tree T with only one vertex is called a *trivial tree;* otherwise T is a *nontrivial tree*.

Note that the first part of this definition applies to nondirected graphs as well as to digraphs. A tree may be either a digraph or a nondirected graph. Note that in a tree any vertex may be designated as a root.

Example 5.3.1. Two trees, G_1 and G_2, are shown in Figure 5-19. $G_1 = (V, E_1)$ and $G_2 = (V, E_2)$, where

$$V = \{a,b,c,d,e,f,g,h,i,j\},$$
$$E_1 = \{\{a,c\},\{b,c\},\{c,d\},\{c,e\},\{e,g\},\{f,g\},\{g,i\},\{h,i\},\{i,j\}\}, \text{ and}$$
$$E_2 = \{(c,a),(c,b),(c,d),(c,f),(f,e),(f,i),(g,d),(h,e),(j,g)\}.$$

Neither of these trees is a directed tree. If vertex c is designated as the root of each tree, vertex j is at level 4 in G_1 and at level 3 in G_2.

Example 5.3.2. A directed tree T is shown in Figure 5-20. $T = (V, E)$, were $V = \{a,b,c,d,e,f,g,h\}$ and $E = \{(a,b),(a,c),(a,d),(b,e),$

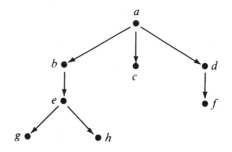

Figure 5-20. A directed tree.

$(d,f),(e,g),(e,h)$}. The root of T is the vertex a and the vertices at level 2 are e and f. Directed trees are conventionally drawn with the root at the top and all edges going from the top of the page toward the bottom, so that the direction of edges is sometimes not explicitly shown.

Trees arise in many practical applications; frequently they occur in situations where many elements are to be organized into some sort of hierarchy that expresses what is more important, what must be done first, or what is more desirable. For instance, a tree can be used to show the order in which tasks are to be completed in the assembly of some product; the root can represent the finished product, and tasks that can be done concurrently appear on the same level, whereas if task A must be completed before task B can start, then A would have to lie on a higher level than B.

Example 5.3.3. Frequently a computer scientist describes algebraic formulas as tree structures. For example, $a + b$ can be diagrammed as

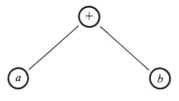

Figure 5-21

If b were the expression $(c \times d)$ we may write

$$a + (c \times d)$$

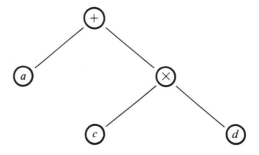

Figure 5-22

The expression $(a + 5) \times [(3b + c)/(d + 2)]$ can be pictured as

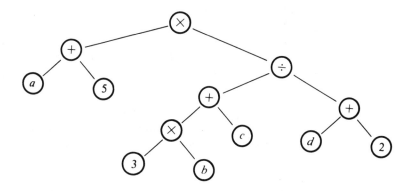

Figure 5-23

If we now read the vertices of this tree once each starting from the top and proceeding counterclockwise, we write

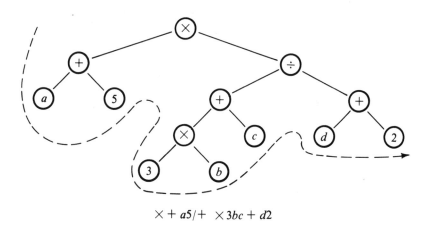

$$\times + a5/+ \times 3bc + d2$$

Figure 5-24

which is the same algebraic formula written in the **operator prefix notation** invented by Jan Lukasiewicz (1878–1956). The advantage of this notation is that it is unambiguous despite the fact that it does not employ parentheses.

Any algebraic expression can be written in operator prefix notation with the operator written before its arguments (or its operands). Analo-

gously, one can place the operator after its arguments, and this is called
operator postfix notation. Thus, the algebraic expression depicted in
Figure 5-21 can be written in postfix notation as $ab+$, the expression of
Figure 5-22 becomes $acd \times +$, and the postfix notation for the expression
of Figures 5-23 and 5-24 is a $5 + 3\,b \times c + d\,2 + \div \times$. (Prefix and postfix
notation are nothing more than the preorder or postorder tree traversals
described in Section 5.6.)

Now let us prepare to give a characterization of trees. To do this we
need to discuss the concept of connectivity. Two vertices a and b of a
graph are said to be **connected** iff there is a nondirected path from a to b
in G, and then the graph G is **connected** iff each pair of its vertices is
connected.

In general, if we define the relation R on the vertices of a graph G by aRb
iff a and b are connected, then R is an equivalence relation. Consequently,
the vertices of G can be partitioned into disjoint nonempty sets $V_1, V_2, \ldots, V_k$
and the subgraphs $H_1, H_2, \ldots, H_k$ of G induced by $V_1, V_2, \ldots,$ and V_k,
respectively, are called the **connected components** of G or simply the
components of G. Usually we denote the number of components of G by
$C(G)$ and, of course, G is connected iff $C(G) = 1$. Equivalently, a component
of a graph G is a connected subgraph of G not properly contained in any
other connected subgraph of G; that is, a component of G is a subgraph of G
that is maximal with respect to the property of being connected. In other
words, a connected subgraph H of a graph G is a component of G if for each
connected subgraph F of G where $H \subseteq F \subset G$, $V(H) \subseteq V(F)$, and $E(H) \subseteq$
$E(F)$, then it follows that $H = F$.

Definition 5.3.2. If a graph G is connected and e is an edge such
that $G - e$ is not connected, then e is said to be a **bridge** or a **cut edge**.
If v is a vertex of G such that $G - v$ is not connected, then v is a **cut
vertex**.

Example 5.3.4. Let G be the graph depicted in Figure 5-25.
This graph G has 3 components; the vertices a and d are connected as

Figure 5-25. A graph with 3 components.

are i and g and j and k, but i and k are not connected. Moreover, c is a cut vertex of the first component.

Now let us present a few interesting properties of trees.

Theorem 5.3.1.   A simple nondirected graph G is a tree iff G is connected and contains no cycles.

Proof.   Suppose that G is a tree. Since each pair of vertices are joined by a path, G is connected. If G contains a cycle containing distinct vertices u and v, then u and v are joined by at least two simple paths, the one along one portion of the cycle and the other path completing the cycle. This contradicts the hypothesis that there is a *unique* simple path between u and v, and thus a tree has no cycles.

Conversely, suppose that G is connected and contains no cycles. Let a and b be any pair of vertices of G. If there are 2 different simple paths, P_1 and P_2, from a to b, then we can find a cycle in G as follows. Since P_1 and P_2 are different paths there must be a vertex v_1 (possibly $v_1 = a$) on both paths such that the vertex following v_1 on P_1 is not the same as the vertex following v_1 on P_2. Since P_1 and P_2 terminate at b, there is a first vertex after v_1, call it v_2, which P_1 and P_2 have in common (possibly $v_2 = b$). Thus, the part of P_1 from v_1 to v_2 together with that part of P_2 from v_1 to v_2 form a cycle in G. This contradicts the assumption that G has no cycles. Therefore, G has exactly one path joining a and b.   □

Theorem 5.3.2.   In every nontrivial tree there is at least one vertex of degree 1.
(We are excluding the trivial tree with only one vertex.)

Proof.   Start at any vertex v_1. If $\deg(v_1) \neq 1$, move along any edge to a vertex v_2 incident with v_1. If $\deg(v_2) \neq 1$, continue to a vertex v_3 along a different edge. We can continue this process to produce a path $v_1 - v_2 - v_3 - v_4\ldots$. (Here we mean that there is an edge from v_1 to v_2, one from v_2 to v_3 and so on.) None of the v_i's is repeated in this path since then we would have a circuit—which a tree may not have. Since the number of vertices in the graph is finite, this path must end somewhere. Where it ends must be a vertex of degree 1 since we can enter this vertex but cannot leave it.   □

Theorem 5.3.3.   A tree with n vertices has exactly $n - 1$ edges.

Proof.   We employ mathematical induction on the number of vertices. If $n = 1$, there are no edges. Hence, the result is trivial for $n = 1$. Assume, then, for $n \geq 1$ that all trees with n vertices have exactly $n - 1$

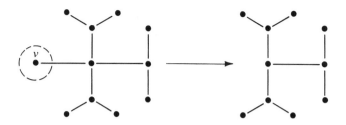

Figure 5-26

edges. Then consider an arbitrary tree T with $n + 1$ vertices. By the previous theorem, there is a vertex v in T of degree 1. Let us "prune" this tree by removing this vertex and its associated edge e from T that is, consider $T' = T - v$. Let us illustrate by the picture in Figure 5-26.

Note that T' has n vertices and one fewer edge than T. But more than that, T' is connected since for any pair of vertices a and b in T' there is a unique simple path from a to b in T. Moreover, this path has not been affected by the removal of the vertex v and the edge e. Likewise, there are no cycles in T' since there were none in T. Thus, T' is a tree and the inductive hypothesis implies that T' has $n - 1$ edges. But then T must have n edges as T has one more edge than T'. $\square$

We can improve upon Theorem 5.3.2.

Corollary 5.3.1.   If G is a nontrivial tree then G contains at least 2 vertices of degree 1.

Proof.   Let n = the number of vertices of G. By the sum of degrees formula,

$$\sum_{i=1}^{n} \deg(v_i) = 2\,|E| = 2(n - 1) = (2n - 2).$$

Now if there is only one vertex, say v_1, of degree 1, then

$$\deg(v_i) \geq 2 \quad \text{for} \quad i = 2, \ldots, n$$

and $\displaystyle\sum_{i=1}^{n} \deg(v_i) = 1 + \sum_{i=2}^{n} \deg(v_i) \geq 1 + 2n - 2 = 2n - 1.$

But then

$$2n - 2 \geq 2n - 1 \quad \text{or} \quad -2 \geq -1, \text{ a contradiction.} \quad \square$$

Figure 5-27. The trees on 6 vertices.

Example 5.3.5. There are 6 nonisomorphic trees with 6 vertices. They are shown in Figure 5-27.

The following fact about trees will also prove useful.

Theorem 5.3.4. If 2 nonadjacent vertices of a tree T are connected by adding an edge, then the resulting graph will contain a cycle.

Proof. If T has n vertices then T has $n - 1$ edges and then if an additional edge is added to the edges of T the resulting graph G has n vertices and n edges. Hence G cannot be a tree by Theorem 5.3.3. However, the addition of an edge has not affected the connectivity. Hence G must have a cycle. $\square$

Example 5.3.6. Adding any of the dotted lines to the tree in Figure 5-28 will create a cycle.

We have already given one characterization of trees in addition to the definition. There are several other characterizations; we list one more at this point.

Theorem 5.3.5. A graph G is a tree if and only if G has no cycles and $|E| = |V| - 1$.

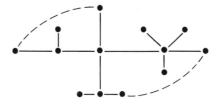

Figure 5-28

Proof. In Theorem 5.3.3 we have already proved one half of the theorem. To prove the other half we need only show that if G has no cycles and $|E| = |V| - 1$, then G is connected. Denote by $G_1, G_2, \ldots, G_k$ the components of G, where $k \geq 1$. Let $|V_i| = $ the number of vertices of G_i. Now each G_i is a tree, for G_i is connected and G_i contains no cycles since G does not. Thus, G_i has $|V_i| - 1$ edges. Hence G has $(|V_1| - 1) + (|V_2| - 1) + \cdots + (|V_k| - 1) = |V_1| + |V_2| + \cdots + |V_k| - k = |V| - k$ edges. By hypothesis, G has $|V| - 1$ edges. Thus, $k = 1$, and G is connected. $\square$

Exercises for Section 5.3

1. Tell how many different (pairwise nonisomorphic) trees there are of order
   (a) 2
   (b) 3
   (c) 4
   (d) 5

2. (a) Prove that if G is a tree, then the sum of degrees equals $2|V| - 2$.
   (b) Draw all pairwise nonisomorphic trees with $|V| = 7$ and at least one vertex of degree ≥ 4. Hint: Determine the degree sequence first.
   (c) Is there a tree with $|V| = 5$ and 2 vertices of degree 3?

3. Give an example of a graph G such that for each pair of distinct vertices a and b there are exactly 2 paths from a to b.

4. List all pairwise nonisomorphic trees with degree sequence
   (a) (1,1,1,1,2,2,3,3)
   (b) (1,1,1,1,2,2,2,3,3)
   (c) (1,1,1,1,1,1,2,3,5)

5. Show that there is no tree with degree sequence
   (a) (1,1,2,2,3,3)
   (b) (1,1,2,2,2,2,3,3,3,3,3,3)
   (c) (1,1,1,1,1,1,2,3,7)

6. Determine the different possible degree sequences of a tree
   (a) with 17 vertices and exactly one vertex of degree 5, one of degree 4, and two vertices of degree 3.
   (b) with 12 vertices and at least 3 vertices of degree 3.

7. An edge e in a connected graph G is a *cut edge* if $G - e$ is not connected. Prove that a connected graph is a tree if and only if each edge is a cut edge.

8. Prove that a simple graph G is a tree iff G contains no cycles and the addition of any new edge forms a cycle. Hint: Consider the case where G contains at least 2 components.

9. Let G be a graph with k components, where each component is a tree. Obtain a formula for $|E|$ in terms of $|V|$ and k.

10. Show that a graph G is a tree iff G is connected and contains no circuits.

11. (a) Suppose $n \geq 2$ and $d_1, d_2, \ldots, d_n, d_{n+1}$ are $n + 1$ positive integers such that their sum equals $2n$. Use the pigeonhole principle to prove that there exists an index i such that $d_i = 1$ and there is an index j such that $d_j > 1$.

    (b) Use part (a) and mathematical induction to prove that if n is an integer ≥ 2 and $d_1, d_2, \ldots, d_n$ are positive integers such that $\Sigma_{i-1}^n d_i = 2n - 2$, then there is a tree T_n with n vertices whose degrees are $d_1, d_2, \ldots, d_n$. Hint: think about how you would construct a tree with degree sequence $(1,1,1,2,2,3)$ from one with degree sequence $(1,1,2,2,2)$.

12. Characterize all connected graphs having the same number of vertices as edges (that is, what must such a graph look like and explain why).

13. Let T be a nontrivial tree with vertices $v_1, v_2, \ldots, v_n$. Let N_1 be the number of vertices of T with degree equal to 1. Prove that $N_1 = 1 + (1/2) \Sigma_{i-1}^n |\deg(v_i) - 2|$ where $|x|$ means absolute value of x.

14. (a) Show that a vertex v in a tree T is a cut vertex of T iff $\deg(v) > 1$.

    (b) Show that a connected graph with $|E| = |V| - 1 \geq 2$ contains at least 2 vertices that are not cut vertices.

    (c) Show that a connected graph with $|E| = |V| - 1 \geq 2$ contains at least one cut vertex.

15. (a) Suppose that a tree T has N_1 vertices of degree 1, 2 vertices of degree 2, 4 vertices of degree 3 and 3 vertices of degree 4. Find N_1.

    (b) Suppose that a tree T has N_1 vertices of degree 1, N_2 vertices of degree 2, N_3 vertices of degree 3, $\ldots, N_k$ vertices of degree k. Find N_1 in terms of $N_2, N_3, \ldots,$ and N_k.

16. Characterize all trees with exactly 2 vertices of degree 1.

17. Write the expression

$$\{[(a + b) \times c] \times (d + e)\} - [f - (g \times h)]$$

as a tree and then express the result in operator prefix notation.

18. Let $T = (V, E)$ be a directed graph that is a tree and v_0 be a vertex in V. Suppose $v_1, \ldots, v_k$ are the vertices adjacent to v_0. Let $S_i = \{u \mid u$ is connected to v_i by a path that does not traverse $v_0\}$, $E_i = E (S_i \times S_i)$, and $T_i = (S_i, E_i)$, for $i = 1, \ldots, k$.
    (a) Show that the subgraphs T_i are disjoint.
    (b) Show that each T_i is a tree.
    (c) Show that every edge of T not incident to v_0 is in one of the T_i's.

19. Suppose that G is a simple graph with $|V| = n$. Prove that G must be connected if $|E| > (1/2) (n - 1) (n - 2)$.

20. A *forest* is a simple graph with no circuits. Show that the connected components of a forest are trees.

21. Show that the removal of an edge from a tree results in a forest with exactly 2 connected components.

22. Given a forest G with k connected components, how many new edges must be added to it to obtain a tree?

23. Prove that any 2 simple connected graphs with n vertices, all of degree 2, are isomorphic.

24. (a) If G is a simple graph with $C(G)$ connected components, show that if e is any edge of G, then $G - e$ has at most $C(G) + 1$ components. In particular, if G is a forest, than $G - e$ has exactly $C(G) + 1$ components. See Exercise 21.
    (b) For a simple graph G with $C(G)$ connected components, prove by induction on the number of edges of G that $|V(G)| \leq C(G) + |E(G)|$.

25. Draw all pairwise nonisomorphic forests with
    (a) 2 or more connected components and 5 vertices in all.
    (b) 3 or 4 connected components and $|V| = 6$.

26. (a) Suppose that G is a forest with k components and $|V| = 17$ and $|E| = 8$. Determine k. Hint: See Exercise 9.
    (b) If G is a forest with 7 connected components and $|V| = 33$, find $|E|$.

27. Show that a graph with no vertices of odd degree and no vertices of degree 0 must contain a circuit.

28. If a graph G is not connected, prove that its complement $\overline{G}$ is connected. Is the converse true?

29. In a connected graph G, prove that any two paths of maximal length have a common vertex.

30. Prove by induction on the number of vertices that the vertices of a tree can be colored with at most two colors so that no two vertices of the same color are adjacent.

31. Suppose that G_1 and G_2 are isomorphic graphs. Prove that if G_1 is connected, then G_2 is connected.

Selected Answers for Section 5.3

1. (a) 1   (b) 1   (c) 2   (d) 3

3. Any cycle graph C_n.

9. If $G_1, \ldots, G_k$ are the components of G, then $|E(G_i)| = |V(G_i)| - 1$ for each i. Then $|E(G)| = \Sigma_{i=1}^{k} |E(G_i)| = \Sigma_{i=1}^{k} (|V(G_i)| - 1) = |V(G)| - k$.

15. (a) Recall that for a tree $|E| = |V| - 1$ and by the sum of degrees formula

$$N_1 + 2 \cdot 2 + 4 \cdot 3 + 3 \cdot 4 = 2|E| = 2(|V| - 1)$$
$$= 2(N_1 + 2 + 4 + 3 - 1)$$
$$= 2N_1 + 16. \text{ Thus, } N_1 = 12.$$

   (b) Similar to (a); observe that

$$N_1 - 2 = N_3 + 2N_4 + 3N_5 + \ldots + (k - 2)N_k.$$

17.

$$-\times\times + abc + de - f \times gh$$

22. Label the components $G_1, G_2, \ldots, G_k$. Fix roots v_i in each G_i. Join one edge between v_1 and v_2, one between v_2 and v_3, etc. There are $k - 1$ such adjunctions. Also we could apply Exercise 9.

5.4   SPANNING TREES

Definition 5.4.1.   A subgraph H of a graph G is called a **spanning tree** of G if

(a) H is a tree, and
(b) H contains all the vertices of G.

A spanning tree that is a directed tree is called a **directed spanning tree** of G.

Spanning trees play an important role in many computer algorithms that work on graphs. Some of the consequences of the definition are explored in the exercises for this section.

Example 5.4.1.   Consider the digraph $G = (V,E)$ where $V = \{a,b,c,d,e\}$ and $E = \{(a,c),(b,a),(b,b),(b,c),(c,d),(c,e),(d,c),(d,d),(e,b)\}$ (shown in Figure 5-29). There are many directed spanning trees of G, one of which is $T = (V,\{(b,a),(a,c),(c,d),(c,e)\})$, (shown in Figure 5-30).

Example 5.4.2.   Consider the graph G in Figure 5-31(a). Suppose that this graph represents a communication network in which the vertices correspond to stations and the edges correspond to communication links. What is the largest number of edges that can be deleted while still allowing the stations to communicate with each other?

First observe that cycles are not necessary, since cycles give two ways in which stations can communicate and all that is needed is one way to communicate. For example, the cycle d-c-e-d gives two ways for d and e to communicate. Namely, along the path d-c-e or directly from d to e. If the edge $\{d,e\}$ is deleted, d and e can still communicate via c. Likewise, one edge of each cycle in G can be deleted. The edges left are the fewest necessary to maintain communication between all stations. One way of accomplishing this is shown in Figure 5-31(b). The result is a spanning tree for the graph of Figure 5-31(a). Of course, by deleting another

Figure 5-29

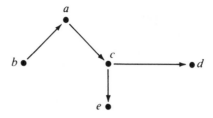

Figure 5-30. A spanning tree *T* of the graph in Figure 5-29.

sequence of edges to eliminate cycles, we may obtain other spanning trees for G. The graph G has 15 edges and the spanning tree for G has 10 edges so 5 edges have to be deleted.

In general, if G is a connected graph with n vertices and m edges, a spanning tree of G must have $n - 1$ edges by Theorem 5.3.3. Hence, the number of edges that must be removed before a spanning tree is obtained must be $m - (n - 1) = m - n + 1$. This number is frequently called the *circuit rank* of G.

The idea illustrated in the above example is the essence of the following theorem.

Theorem 5.4.1.   A nondirected graph G is connected if and only if G contains a spanning tree. Indeed, if we successively delete edges of cycles until no further cycles remain, then the result is a spanning tree of G.

Proof.   If G has a spanning tree T, there is a path between any pair of vertices in G along the tree T. Thus G is connected.

Conversely, we prove that a connected graph G has a spanning tree by mathematical induction on the number k of cycles in G. If $k = 0$, then G is connected with no cycles and hence G is already a tree. Suppose that all connected graphs with fewer than k cycles have a spanning tree. Now suppose that G is a connected graph with k cycles. Remove an edge e from

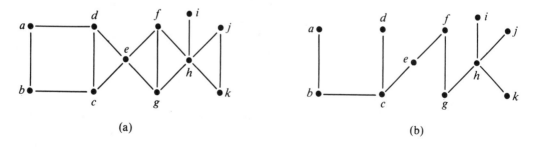

(a)   (b)

Figure 5-31

one of the cycles. Then $G - e$ is still connected and has a spanning tree by the inductive hypothesis because $G - e$ has fewer cycles than G. But since $G - e$ has all the vertices of G, the spanning tree for $G - e$ is also one for G. The result follows by mathematical induction. □

Before describing other methods for obtaining spanning trees for connected graphs, let us introduce the following terminology.

Definition 5.4.2. Let T be a rooted tree with designated root v_0. Suppose that u and v are vertices in T and that $v_0-v_1-\cdots-v_n$ is a simple path in T. Then

(a) v_{n-1} is the **parent** of v_n.
(b) $v_0,v_1,\ldots,v_{n-1}$ are the **ancestors** of v_n.
(c) v_n is a **child** of v_{n-1}.
(d) If u is an ancestor of v, then v is a **descendant** of u.
(e) If u has no children, then u is a **leaf** of T.
(f) If v is not a leaf of T, then v is an **internal vertex** of T.
(g) The subgraph of T consisting of v and all its descendants, with v designated as a root, is the **subtree** of T **rooted at v**.

Example 5.4.3. If, in the tree of Figure 5-19, a is designated as the root, then b,d,f,h, and j are leaves of T; i is the parent of h and j; f,h,i, and j are the descendants of g; a and c are the ancestors of e; and the children of c are b,d, and e. Moreover, the vertices a,c,e,g, and i are internal vertices to the tree rooted at a.

If, on the other hand, we let c be the root, then a becomes a leaf, and c is the parent of a.

Breadth-First Search and Depth-First Search

An algorithm based on the proof of the Theorem 5.4.1 could be designed to produce a spanning tree for a connected graph. Recall from the proof that all that one need do is destroy cycles in the graph by removing an edge from a cycle until no cycles remain. Unfortunately, such an algorithm is not very efficient because of the time-consuming process of finding cycles. But, on the other hand, we can define other rather efficient algorithms for finding a spanning tree of a connected graph. These algorithms are known as **breadth-first search** (abbreviated BFS) and **depth-first search** (abbreviated DFS).

The idea of breadth-first search is to visit all vertices sequentially on a given level before going onto the next level. Depth-first search, on the other hand, proceeds successively to higher levels at the first opportunity.

Example 5.4.4. Let us illustrate breadth-first search on the graph of Figure 5-31(a).

We select the ordering of the vertices $abcdefghijk$. Then we select a as the first vertex chosen in the spanning tree T and designate it as the root of T. Thus, at this stage, T consists of the single vertex a. Add to T all edges $\{a,x\}$, as x runs in order from b to k, that do not produce a cycle in T. Thus, we add $\{a,b\}$ and $\{a,d\}$. These edges are now called *tree edges* for the breadth-first search tree.

Now repeat the process for all vertices on level one from the root by examining each vertex in the designated order. Thus, since b and d are at level one, we first examine b.

For b, we include the edge $\{b,c\}$ as a tree edge. Then for d, we reject the edge $\{d,c\}$ since its inclusion would produce a cycle in T. But we include $\{d,e\}$.

Next, we consider the vertices at level two. Reject the edge $\{c,e\}$; include $\{e,f\}$ and $\{e,g\}$.

Then repeat the procedure again for vertices on level three. Reject $\{f,g\}$, but include $\{f,h\}$. At g, reject $\{f,g\}$ and $\{g,h\}$.

On level four, include $\{h,i\}$, $\{h,j\}$, and $\{h,k\}$.

Next, we attempt to apply the procedure on level five at i, j, and k, but no edge can be added at these vertices so the procedure ends. The spanning tree T therefore includes the vertices a,b,c,d,e,f,g,h,i,j, and k, and the edges $\{a,b\}$, $\{a,d\}$, $\{b,c\}$, $\{d,e\}$, $\{e,f\}$, $\{e,g\}$, $\{f,h\}$, $\{h,i\}$, $\{h,j\}$, and $\{h,k\}$.

The edges that were rejected in the breadth-first search are called *cross edges*. Clearly, breadth-first search partitions the edges of the graph G into the two sets of tree edges and cross edges.

Now let us give a formal description of breadth-first search.

Algorithm 5.4.1. Breadth-First Search for a Spanning Tree.

Input: A connected graph G with vertices labeled $v_1, v_2, \ldots, v_n$.

Output: A spanning tree T for G.

Method:

1. (Start) Let v_1 be the root of T. Form the set $V = \{v_1\}$.
2. (Add new edges.) Consider the vertices of V in order consistent with the original labeling. Then for each vertex $x \in V$, add the edge $\{x, v_k\}$ to T where k is the minimum index such that adding the edge $\{x, v_k\}$ to T does not produce a cycle. If no edge can be added, then stop; T is a spanning tree for G. After all the vertices of V have been considered in order, go to Step 3.
3. (Update V.) Replace V by all the children v in T of the vertices x of

V where the edges $\{x,v\}$ were added in Step 2. Go back and repeat Step 2 for the new set V.

An alternative to breadth-first search is depth-first search.

Example 5.4.5.   Let us illustrate how to find a depth-first search spanning tree for the graph G in Figure 5-31(a).

As in breadth-first search, choose an ordering of the vertices, say, *abcdefghijk*. Select a as the root of T. The vertex a is said to be *visited*. Next, we select the edge $\{a,x\}$ where x is the first label in the designated order that does not form a cycle with those edges already chosen in T. In this case, we add the edge $\{a,b\}$. The edge $\{a,b\}$ is now said to be *examined* and becomes a tree edge. In this context, a is the *parent* of b and b is the *child* of a.

In general, while we are at some vertex x, two situations arise:

Situation 1.   If there are some unexamined edges incident on x, then we consider the edge $\{x,y\}$, where y is the first vertex in the designated ordering on the vertices for which $\{x,y\}$ is unexamined.

In this situation, two cases present themselves:

Case 1.   If y has not been previously visited, visit y, select $\{x,y\}$ as a tree edge, and continue the search from y. In this case, x is the parent of y.

Case 2.   If y has been visited previously, then reject the edge $\{x,y\}$, consider it examined, and proceed to select another unexamined edge $\{x,z\}$ incident on x where z is the first vertex for which $\{x,z\}$ is an unexamined edge. Each such rejected edge in the context of depth-first search is called a *back edge*.

In the example at hand, we would select the edge $\{b,c\}$, continue the search at c, and select $\{c,d\}$. Then we would continue the search at d and first reject the edge $\{d,a\}$ and then select the edge $\{d,e\}$. At e, we reject the edge $\{e,c\}$, select $\{e,f\}$. Continuing in this manner, we select $\{f,g\}$, reject $\{g,e\}$ select $\{g,h\}$, reject $\{h,f\}$, and select $\{h,i\}$.

At this point, we are presented with a second general situation.

Situation 2.   If all the edges incident on x have already been examined, then we return to the parent of x and continue the search from the parent of x. The vertex x is now said to be *completely scanned*. Moreover, the process of returning to the parent of x is called *backtracking*.

Thus, in the example that we are considering, since there are no unexamined edges at i, we must backtrack to h and continue the search from h. Then we select $\{h,j\}$ and $\{j,k\}$ and, finally, reject $\{k,h\}$.

Actually, we are through, because there are no more unexamined edges. But if we had limited vision (as a computer may have), we may be aware only that there are no unexamined edges at k. Therefore, we backtrack, according to Situation 2, to j. But then we must backtrack to h, etc. Eventually we must backtrack all the way back to the root a.

In general, depth-first search terminates when the search returns to the root and all vertices have been visited.

As before, let us give a formal description of the depth-first search algorithm.

Algorithm 5.4.2.   Depth-First Search for a Spanning Tree.

Input: A connected graph G with vertices labeled $v_1, v_2, \ldots, v_n$.
Output: A spanning tree T for G.
Method:

1. (Visit a vertex.) Let v_1 be the root of T, and set $L = v_1$. (The name L stands for the vertex last visited.)
2. (Find an unexamined edge and an unvisited vertex adjacent to L.) For all vertices adjacent to L, choose the edge $\{L, v_k\}$, where k is the minimum index such that adding $\{L, v_k\}$ to T does not create a cycle. If no such edge exists, go to Step 3; otherwise, add edge $\{L, v_k\}$ to T and set $L = v_k$; repeat Step 2 at the new value for L.
3. (Backtrack or terminate.) If x is the parent of L in T, set $L = x$ and apply Step 2 at the new value of L. If, on the other hand, L has no parent in T (so that $L = v_1$) then the depth-first search terminates and T is a spanning tree for G.

Minimal Spanning Trees

The application of spanning trees are many and varied, and in order to gain some appreciation for this fact, we will describe what is sometimes called the *connector problem*. Suppose that we have a collection of n cities, and that we wish to construct a utility, communication, or transportation network connecting all of the cities. Assume that we know the cost of building the links between each pair of cities and that, in addition, we wish to construct the network as cheaply as possible.

The desired network can be represented by a graph by regarding each city as a vertex and by placing an edge between vertices if a link runs between the two corresponding cities. Moreover, given the cost of constructing a link between cities v_i and v_j, we can assign the weight c_{ij} to the edge $\{v_i,v_j\}$. The problem, then, is to design such a network so as to minimize the total cost of construction. If M is the graph of a network of minimal cost, it is essential that M be connected for all of the cities are to be connected by links. Moreover, it is also necessary that there be no circuits in the graph M, for otherwise we can remove an edge from a circuit and thereby reduce the total cost by the cost of construction of that edge. Hence, a graph of minimal cost must be a spanning tree of the graph of the n vertices.

Thus, the problem of building a network at minimal cost can now be stated in general terms. Let G be the graph of all possible links between the cities with the nonnegative cost of construction $C(e)$ assigned to each edge e in G. Then if H is any subgraph of G with edges $e_1,\ldots,e_m$ the total cost of constructing the network H is $C(H) = \Sigma_{i=1}^m C(e_i)$. A spanning tree T where $C(T)$ is minimal is called a *minimal spanning tree* of G.

It should be clear at this point that finding a solution to the connector problem is equivalent to finding a minimal spanning tree for a connected graph G where each edge of G is labelled with a nonnegative cost.

Now let us describe an algorithm that will, in fact, construct a minimal spanning tree. The algorithm is known as *Kruskal's Algorithm* (after the mathematician J. B. Kruskal, Jr.).

Algorithm 5.4.3.   Kruskal's Algorithm for Finding a Minimal Spanning Tree.

Input: A connected graph G with nonnegative values assigned to each edge.

Output: A minimal spanning tree for G.

Method:

1. Select any edge of minimal value that is not a loop. This is the first edge of T. (If there is more than one edge of minimal value, arbitrarily choose one of these edges.)
2. Select any remaining edge of G having minimal value that does not form a circuit with the edges already included in T.
3. Continue Step 2 until T contains n-1 edges, where n is the number of vertices of G.

We leave it to the reader to verify that this algorithm does, if fact, produce a spanning tree. Let us call any tree obtained by this process an economy tree. The point of the next theorem is that an economy tree is a minimal spanning tree and thus solves the connector problem.

Suppose that a problem calls for finding an optimal solution (either maximum or minimal). Suppose, further, that an algorithm is designed to make the optimal choice from the available data at each stage of the process. Any algorithm based on such an approach is called a **greedy algorithm.** A greedy algorithm is usually the first heuristic algorithm one may try to implement and it does lead to optimal solutions sometimes, but not always. Kruskal's algorithm is an example of a greedy algorithm that does, in fact, lead to an optimal solution.

Theorem 5.4.2. Let G be a connected graph where the edges of G are labelled by nonnegative numbers. Let T be an economy tree of G obtained from Kruskal's Algorithm. Then T is a minimal spanning tree.

Proof. As before, for each edge e of G, let $C(e)$ denote the value assigned to the edge by the labelling.

If G has n vertices, an economy tree T must have $n - 1$ edges. Let the edges $e_1, e_2, \ldots, e_{n-1}$ be chosen as in Kruskal's Algorithm. Then $C(T) = \sum_{i=1}^{n-1} C(e_i)$. Let T_0 be a minimal spanning tree of G. We show that $C(T_0) = C(T)$, and thus conclude that T is also minimal spanning tree.

If T and T_0 are not the same let e_i be the first edge of T not in T_0. Add the edge e_i to T_0 to obtain the graph G_0. Suppose $e_i = \{a, b\}$. Then a path P from a to b exists in T_0 and so P together with e_i produces a circuit C in G_0 by Theorem 5.3.4. Since T contains no circuits, there must be an edge e_0 in C that is not in T. The graph $T_1 = G_0 - e_0$ is also a spanning tree of G since T_1 has $n - 1$ edges. Moreover,

$$C(T_1) = C(T_0) + C(e_i) - C(e_0)$$

However, we know that $C(T_0) \leq C(T_1)$ since T_0 was a minimal spanning tree of G. Thus,

$$C(T_1) - C(T_0) = C(e_i) - C(e_0) \geq 0$$

implies that

$$C(e_i) \geq C(e_0).$$

However, since T was constructed by Kruskal's algorithm, e_i is an edge of smallest value that can be added to the edges $e_1, e_2, \ldots, e_{i-1}$ without producing a circuit. Also, if e_0 is added to the edges $e_1, e_2, \ldots, e_{i-1}$, no circuit is produced because the graph thus formed is a subgraph of the tree T_0. Therefore, $C(e_i) = C(e_0)$, so that $C(T_1) = C(T_0)$.

We have constructed from T_0 a new minimal spanning tree T_1 such that the number of edges common to T_1 and T exceeds the number of edges common to T_0 and T by one edge, namely e_i.

Repeat this procedure, to construct another minimal spanning tree T_2

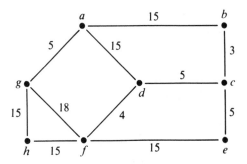

Figure 5-32

with one more edge in common with T than was in common between T_1 and T.

By continuing this procedure, we finally arrive at a minimal spanning tree with *all* edges in common with T, and thus we conclude that T is itself a minimal spanning tree. $\square$

The following example illustrates the use of Kruskal's algorithm.

Example 5.4.6. Determine a railway network of minimal cost for the cities in Figure 5-32.

We collect lengths of edges into a table:

Edge	Cost
$\{b,c\}$	3
$\{d,f\}$	4
$\{a,g\}$	5
$\{c,d\}$	5
$\{c,e\}$	5
$\{a,b\}$	15
$\{a,d\}$	15
$\{f,h\}$	15
$\{g,h\}$	15
$\{e,f\}$	15
$\{f,g\}$	18

1. Choose the edges $\{b,c\}$, $\{d,f\}$, $\{a,g\}$, $\{c,d\}$, $\{c,e\}$.
2. Then we have options: we may choose only one of $\{a,b\}$ and $\{a,d\}$ for the selection of both creates a circuit. Suppose that we choose $\{a,b\}$.
3. Likewise we may choose only one of $\{g,h\}$ and $\{f,h\}$. Suppose we choose $\{f,h\}$.

4. We then have a spanning tree as illustrated in Figure 5-33.

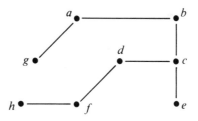

Figure 5-33

The minimal cost for construction of this tree is

$$3 + 4 + 5 + 5 + 5 + 15 + 15 = 52.$$

Exercises for Section 5.4

1. Consider the graph G in Figure 5-34. Which of the graphs a–f in Figure 5-35 are spanning trees of G and why?

Figure 5-34

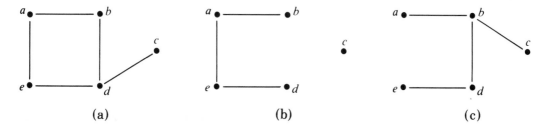

| (a) | (b) | (c) |

Figure 5-35

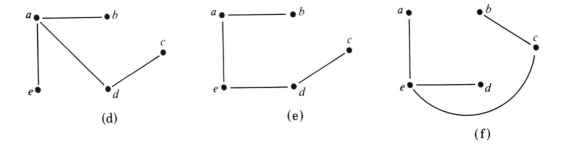

(d)   (e)

(f)

Figure 5-35 (Continued)

2. Prove that a graph G is a tree if and only if G is connected and $|V| - 1 = |E|$.

3. Prove that if G is a connected graph then $|E| \geq |V| - 1$.

4. Prove that a connected graph G is a tree if and only if G has fewer edges than vertices.

5. Suppose that G is a connected graph. Prove that any circuit and the complement of any spanning tree have an edge in common. Is the complement of a spanning tree also a spanning tree?

6. Find spanning trees for each of the graphs in Figure 5-36 by using all 3 algorithms, BFS, DFS, and destruction of cycles with the vertex orderings $abcdefg$ and $gfedcba$.

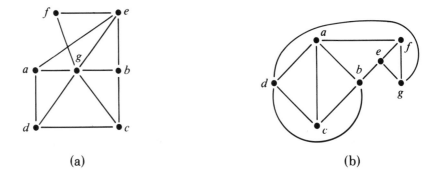

(a)   (b)

Figure 5-36

7. Determine the different nonisomorphic spanning trees for the graph in Figure 5-37.

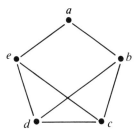

Figure 5-37

8. Define the circuit rank of a disconnected graph to be the sum of the circuit ranks of all its connected components. Derive a formula for the circuit rank of G involving $|E|, |V|$, and $C(G)$.

9. Let $v_0\text{-}v_1\text{-}\ldots\text{-}v_m$ be a path of maximal length in a connected graph G. Prove that $G\text{-}v_0$ is connected. Conclude that any connected graph contains a vertex that is not a cut vertex.

10. If the intersection H of a collection of subtrees of a tree T is nonempty, then H is a subtree of T.

11. Let H be a subgraph of a connected graph G. Show that H is a subgraph of some spanning tree T of G iff H contains no cycle.

12. Describe how to obtain all possible spanning trees for the following graphs and list the total number of different spanning trees:

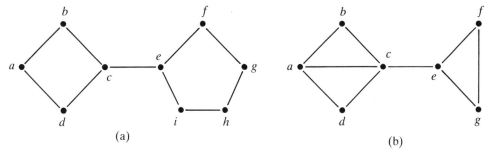

Figure 5-38

(a) C_n, the n-cycle graph.
(b) The graph G in Figure 5-38(a).
(c) The graph H in Figure 5-38(b).
(d) The star graph $K_{1,n}$.
(e) The tetrahedron W_4.
(f) The wheel W_5 of order 5.

13. Find a minimal spanning tree for each of the graphs in Figure 5-39.

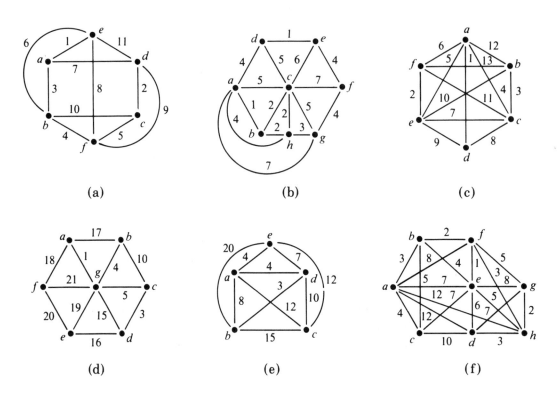

Figure 5-39

14. A company wishes to build an intercommunication system connecting its 7 branches. The distances are given in the following table.

	a	b	c	d	e	f	g
a	0	20	42	31	28	29	33
b		0	25	35	29	24	31
c			0	41	33	22	38
d				0	34	36	40
e					0	41	32
f						0	25
g							0

For example, the distance from a to f is 29. Suppose that the cost of construction of lines between 2 branches is some constant k times the distance between them.

(a) Find the cheapest way to build the system.

(b) Find the total cost.

15. (a) Define what is meant by a *maximal* spanning tree.

    (b) Then modify Kruskal's algorithm so that one has a greedy algorithm that finds a maximal spanning tree.

    (c) Find a maximal spanning tree for the graph in exercise 1(a).

16. Let G be a connected graph such that each edge e has a positive cost $C(e)$. If no two edges have the same cost, prove that G has a *unique* minimal spanning tree.

17. (Another method for finding a minimal spanning tree.) This method is based on the fact that it is foolish to use a costly edge unless it is needed to insure the connectedness of the graph. Thus, let us delete one by one those costliest edges whose deletion does not disconnect the graph.

    (a) Explain why this process gives a spanning tree for any connected graph G.

    (b) Modify the proof of Theorem 5.4.2 to show that this process produces a minimal spanning tree.

    (c) Find a minimal spanning tree for the graph in Figure 5-40 using the process described in (a).

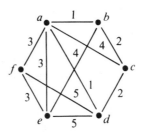

Figure 5-40

18. (a) Prove that any edge of a connected simple graph G is an edge of some spanning tree of G.

    (b) Prove that for any two edges of a simple connected graph G, there is a spanning tree containing the two edges. Hint: Kruskal.

19. Disprove the following:
    (a) If a connected simple graph G with more than 2 edges has a unique minimal spanning tree, then the costs of all the edges are distinct.
    (b) Given any 3 edges e_1, e_2, e_3 of a simple connected graph G, there is a spanning tree of G containing e_1, e_2, and e_3.
    (c) Any two spanning trees of a simple connected graph G have a common edge.
    (d) If G is a connected simple graph and T is a spanning tree for G, then there is an ordering on the vertices of G so that BFS produces T as a BFS spanning tree.
    (e) Different orderings on the vertices of a connected graph G always produce different DFS spanning trees.

20. Prove that if a simple connected graph G has only one spanning tree, then G itself is a tree.

21. (*Prim's algorithm for a minimal spanning tree.*) Let G be a connected graph with nonnegative values assigned to each edge. First, let T be the tree consisting of any vertex v_1 of G. Among all the edges not in T that are incident on a vertex in T and do not form a circuit when added to T, select one of minimal cost and add it to T. The process terminates after we have selected $n - 1$ edges where $n = |V(G)|$.

    Apply Prim's algorithm to obtain minimal spanning trees for each of the graphs in Figure 5-39 and 5-40.

22. Let T_1 and T_2 be spanning trees of a simple connected graph G and let e be an edge of T_1 not in T_2.
    (a) Show that there is a spanning tree T_3 of G containing e and all but one edge of T_2.
    (b) Show that T_1 can be transformed into T_2 through a sequence of trees, each arising from the previous one by removing one edge and adding another.

23. Compare breadth-first search and depth-first search spanning trees for an n-cycle.

24. An edge $e = \{a, b\}$ of a graph G is *contracted* if it is deleted and the endpoints a and b are identified. When an edge is contracted the resulting multigraph is denoted by $G \cdot e$. If $e = \{a, b\}$ is not a loop, observe that
    (a) $|V(G \cdot e)| = |V(G)| - 1$
    (b) $|E(G \cdot e)| = |E(G)| - 1$
    (c) $C(G)$ = the number of connected components of $G = C(G \cdot e)$ = the number of connected components of $G \cdot e$.

        Let $T(G)$ = the number of spanning trees of a connected graph G. Then for an edge e of G that is not a loop, prove:

(d) $T(G - e)$ is the number of spanning trees of G that do not contain the edge e.

(e) There is a one-to-one correspondence between the set of spanning trees of G containing the edge e and the set of spanning trees of $G \cdot e$.

(f) Use (d) and (e) to conclude $T(G) = T(G - e) + T(G \cdot e)$.

25. Show that a tree is a bipartite graph.

26. Suppose that v is a vertex of degree 1 in a connected simple graph G and suppose that the edge e is incident on v. Explain why each spanning tree of G must contain the edge e.

27. Let $v_1, v_2, \ldots, v_n$ be n vertices and let $d_1, d_2, \ldots, d_n$ be positive integers such that $\sum_{i=1}^{n} d_i = 2n - 2$. We know (see Exercise 11 of Section 5.3) that there is a tree T with vertices $v_1, v_2, \ldots, v_n$ where $\deg(v_i) = d_i$.

(a) Prove by induction on n that the number of such trees is
$$[(n - 2)!]/[(d_1 - 1)! \ldots (d_n - 1)!]$$

(b) Use (a) to prove that the number of spanning trees of K_n is n^{n-2}. (This is a classical result due to A. Cayley.)

28. Let T be a tree with vertices $v_1, v_2, \ldots, v_n$. With T we associate the sequence $(t_1, t_2, \ldots, t_{n-2}, t_{n-1})$ obtained as follows: Delete the vertex v_{s_1} of degree 1 with the least index and let t_1 be the index of its neighbor. Let v_{s_2} be the vertex of degree one in $T - v_{s_1}$ of least index and let t_2 be the index of its neighbor in $T - v_{s_1}$. Delete v_{s_2} and so on, repeating this process until a tree with only one vertex remains. This produces a sequence $(t_1, t_2, \ldots, t_{n-2}, t_{n-1})$ of $n - 1$ integers called the *Prüfer code of T*.

(a) Note that $t_{n-1} = n$. Observe that the index of a vertex v_i occurs $\deg(v_i) - 1$ times in $(t_1, t_2, \ldots, t_{n-1})$.

(b) In particular, the vertices of T of degree 1 are precisely those vertices whose indices do not appear in the Prüfer code of T.

(c) Show that $(4,3,5,3,4,5,8)$ is the Prüfer code of the following tree:

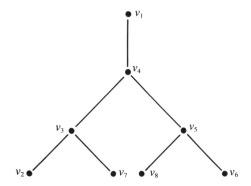

(d) Obtain the Prüfer code for each of the following trees:

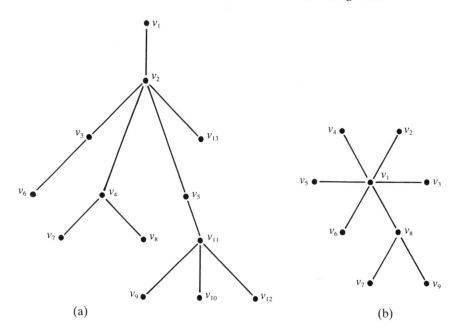

(a)   (b)

(e) Observe that the edges of T are just the edges $\{v_{s_i}, v_{t_i}\}$ where $s_1, s_2, \ldots, s_{n-2}$ are the indices of the vertices removed and t_i is the index of the neighbor of v_{s_i}.

(f) Observe that $s_1 = \min \{k \mid 1 \le k \le n$ and k does not occur in $(t_1, t_2, \ldots, t_{n-2})\}$, $s_2 = \min \{k \mid k \notin \{s_1, t_2, \ldots, t_{n-1}\}\}$, and in general s_i can be defined recursively by $s_i = \min \{k \mid k \notin \{s_1, s_2, \ldots, s_{i-1}, t_i, t_{i+1}, \ldots, t_{n-1}\}\}$. Thus a Prüfer code $(t_1, t_2, \ldots, t_{n-1})$ where $1 \le t_i \le n$ and $t_{n-1} = n$ determines the indices of the removed vertices v_{s_i}. For example, $(3,4,4,5,5,6,6,8,10)$ determines $s_1 = 1$, $s_2 = 2$, $s_3 = 3$, $s_4 = 4$, $s_5 = 7$, $s_6 = 5$, $s_7 = 9$, $s_8 = 6$, $s_9 = 8$.

Then if we connect the vertices v_{s_i} to v_{t_i} we obtain the tree

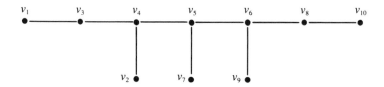

Note that the Prüfer code of this tree is $(3,4,4,5,6,6,8, 10)$.

(g) Given the following sequences, use the process described in (f) to determine the indices of the removed vertices. Then obtain the tree for which the given sequence is its Prüfer code.

    (i) (1,4,4,5,6)

    (ii) (1,4,4,6,6)

    (iii) (1,4,4,5,6,5,7,7,10)

(h) The number of sequences $(t_1, t_2, \ldots, t_{n-1})$ where $1 \le t_i \le n$ and $t_{n-1} = n$ is obviously n^{n-2}. Each of these is the Prüfer code of a spanning tree of K_n. Moreover, there is a one-to-one correspondence between such sequences and spanning trees of K_n. Thus, we obtain another proof of Cayley's formula that there are n^{n-2} spanning trees of K_n.

Selected Answers for Section 5.4

1. Only (c), (e), and (f).

3. A connected graph has a spanning tree with $|V| - 1$ edges.

4. Apply Exercise 3 and Exercise 2.

5. Recall how to obtain a spanning tree by removing edges from circuits.

8. If $G_1, \ldots, G_k$ are the components of G, and if $m_i = |E(G)|$ and $n_i = |V(G_i)|$ then

$$\sum_{i=1}^{k} (m_i - n_i + 1) \text{ circuit rank of } G$$

$$= \sum_{i=1}^{k} m_i - \sum_{i=1}^{k} n_i + k$$

$$= m - n + C(G)$$

$$= |E(G)| - |V(G)| + C(G).$$

13.

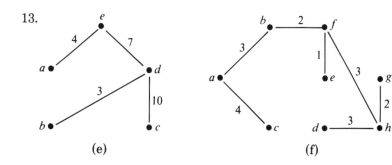

(e)   (f)

17. (c) Delete $\{d,e\}$, $\{d,f\}$, $\{b,e\}$, $\{a,c\}$, and one of $\{a,f\}$, $\{e,f\}$, and $\{a,e\}$.

20. Use induction on the number of edges. The result is obvious for a graph with 0 edges. Suppose for any (connected) graph G with k edges that if G has only one spanning tree, then G is already a tree. Consider a graph H with only one spanning tree and having $k + 1$ edges. If H is not a tree, H contains a circuit (because H must be connected). Remove one edge e_1 from the circuit, $H_1 = H - e_1$ is still connected and any spanning tree for H_1 is also one for H. By the inductive hypothesis H_1 is a tree, and hence is the unique spanning tree of H. Replace e_1 and remove another edge e_2 of the circuit in H. Get a new spanning tree H_2 for H. This contradiction proves the $(k + 1)$th case, and the result is proved by mathematical induction.

5.5   DIRECTED TREES

The question now arises: Under what conditions does a digraph have a *directed* spanning tree? The answer requires us to give a characterization of directed trees and of quasi-strongly connected graphs.

Definition 5.5.1.   Two vertices u and v of a directed graph G are said to be **quasi-strongly connected** if there is a vertex w from which there is a directed path to u and a directed path to v.

Of course, if there is a directed path P from u to v then certainly u and v are quasi-strongly connected, because we may take w to be u itself, and then there is the trivial path with no edges from u to u and the path P from u to v.

The graph G is said to be *quasi-strongly connected* if each pair of vertices of G is quasi-strongly connected.

It should be clear that if a directed graph G is quasi-strongly connected, then the underlying nondirected graph will be connected.

The digraph in Figure 5-41(a) is quasi-strongly connected, but the digraph in Figure 5-41(b) is not.

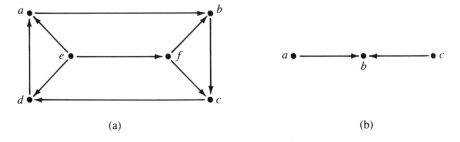

(a)   (b)

Figure 5-41

In the graph of Figure 5-41(a) the vertex e is special in the sense that there is a directed path from e to every other vertex. By no means is this accidental as shown by the following theorem.

Theorem 5.5.1. Let G be a digraph. Then the following are equivalent:

1. G is quasi-strongly connected.
2. There is a vertex r in G such that there is a directed path from r to all the remaining vertices of G.

Proof. Clearly (2) implies (1). On the other hand, suppose that G is quasi-strongly connected and consider its vertices $v_1, v_2, \ldots, v_n$. There is a vertex w_2 from which there is a path to v_1 and a path to v_2. Likewise, there is a vertex w_3 from which there is a path to w_2 and to v_3, and so on until finally we conclude that there is a vertex w_n from which there is a path to w_{n-1} and a path to v_n. Clearly, there is a directed path from w_n to each vertex v_i of G since w_n is connected to $v_1, v_2, \ldots, v_{n-1}$ through w_{n-1}. Thus, (1) implies (2). $\square$

We present in the next theorem a number of equivalent characterizations of a directed tree. These are both analogous to and dependent upon the results in Section 5.3 characterizing nondirected trees.

Theorem 5.5.2. Let G be a digraph with $n > 1$ vertices. Then the following statements are equivalent:

1. G is a directed tree.
2. There is a vertex r in G such that there is a unique directed path from r to every other vertex of G.
3. G is quasi-strongly connected and $G - e$ is not quasi-strongly connected for each edge e of G.
4. G is quasi-strongly connected and contains a vertex r such that the in-degree of r is zero (that is, $\deg^+(r) = 0$) and $\deg^+(v) = 1$ for each vertex $v \neq r$.
5. G has no circuits (that is, G has no directed circuits and no nondirected circuits) and has a vertex r such that $\deg^+(r) = 0$ and $\deg^+(v) = 1$ for each vertex $v \neq r$.
6. G is quasi-strongly connected without circuits.
7. There is a vertex r such that $\deg^+(r) = 0$ and $\deg^+(v) = 1$ for each vertex $v \neq r$, and the underlying nondirected graph of G is a tree.

Proof. $(1 \rightarrow 2)$ The root r in G satisfies the property.

$(2 \rightarrow 3)$ By Theorem 5.5.1, G is quasi-strongly connected. Suppose quasi-strong connectivity is not destroyed when some edge (u,v) is removed. Then there is a vertex w such that there are two directed paths, one to u and one to v, neither of which uses the edge (u,v). Thus, in G there are two distinct directed paths from w to v and hence two distinct directed paths from the root r of G to v. This contradicts (2).

We leave the proof of $(3 \rightarrow 4)$ to the reader.

$(4 \rightarrow 5)$ The sum of the in-degrees is equal to $n - 1$. Therefore, by Theorem 5.1.1, G has $n - 1$ edges. Clearly, the underlying nondirected graph G^* is connected. Therefore, Theorem 5.3.5 implies that G^* is a tree and hence has no cycles. Thus, G has no cycles and likewise no circuits.

$(5 \rightarrow 6)$ By Theorem 5.1.1, the graph G has $n - 1$ edges since the sum of the in-degrees is $n - 1$. Since G has no circuits the underlying nondirected graph G^* must be a tree. Thus, there is a unique simple nondirected path from the vertex r to every other vertex of G. Now for any path P from r to another vertex v we assert that P is, in fact, a directed path from r to v, for otherwise at least one of the vertices in the path would have in-degree greater than 1, contrary to statement 5. Thus, G is quasi-strongly connected by Theorem 5.5.1.

The ideas of the proof of $(6 \rightarrow 7)$ are contained in the proof of $(4 \rightarrow 5)$. Moreover, the proof $(7 \rightarrow 1)$ is contained in $(5 \rightarrow 6)$, for since the underlying graph G^* is a tree there is a unique simple path from r to any other vertex of G. As above, we can observe that any such path is, in fact, directed. □

We know that a nondirected graph G has a spanning tree if and only if G is connected. The corresponding theorem in the case of directed graphs is the following.

Theorem 5.5.3.   A digraph G has a directed spanning tree if and only if G is quasi-strongly connected.

Proof.   If G has a directed spanning tree T then obviously the root of T satisfies the conditions of Theorem 5.5.1 so that G is quasi-strongly connected.

Conversely if G is quasi-strongly connected and is not a directed tree, then by statement 3 of Theorem 5.5.2, there are edges whose removal from G will not destroy the quasi-strongly connected property of G. Therefore, if we remove successively all these edges from G, then the resulting graph is a directed spanning tree. □

Now let us discuss some related combinatorial facts about directed trees, and also study two special classes of directed trees.

Definition 5.5.1.   A directed **forest** is a collection of directed trees. The **height of a vertex** v in a directed forest is the length of the longest directed path from v to a leaf. The **height of a (nonempty) tree** is the height of its root. We adopt the convention that -1 is the height of the empty tree.

The **level** of a vertex v in a forest is the length of the path to v from the root of the tree to which it belongs. A directed tree T is said to have **degree** k if k is the maximum of the out-degrees of all the vertices in T.

Example 5.5.1. A forest is shown in Figure 5-42. This forest consists of three trees, T_1, T_2, and T_3.

$$T_1 = (\{a,b,c\},E_1), \quad \text{where} \quad E_1 = \{(a,b),(a,c)\}$$
$$T_2 = (\{d\},\phi), \quad \text{and}$$
$$T_3 = (\{e,f,g,h,i,j,k\},E_3), \quad \text{where}$$
$$E_3 = \{(e,f),(e,g),(g,h),(h,i),(h,j),(h,k)\}.$$

The root of T_1 is a, the root of T_2 is d, and the root of T_3 is e. Vertex g is the parent of h and a child of e. Vertex g is the proper ancestor of h, i, j, and k, which is the same thing as saying that h, i, j, and k are proper descendants of g. The leaves of this forest are b, c, d, f, i, j, and k. The vertices with height 1 are a and h, and the only vertex with height 2 is g. T_1 has height 1, T_2 has height 0, and T_3 has height 3. The vertices at level 0 are a, d, and e. The vertices at level 1 are b, c, f, and g. Vertex h is the only vertex at level 2. T_1 has degree 2, T_2 has degree 0, and T_3 has degree 3.

The combinatorial relationships between the degree of a tree, its height, the number of vertices in it, and several other parameters, such as

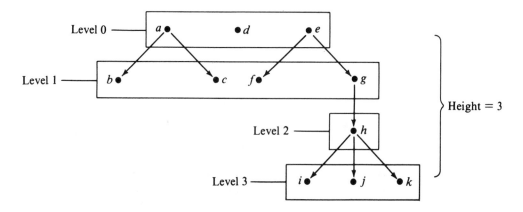

Figure 5-42. A forest.

the number of leaves, are of great interest in computer applications, because they influence the time and storage costs of algorithms. Some of these relationships will be explored in the next few theorems, and others will be derived in the exercises.

Combinatorial facts about trees are most naturally derived by means of recurrence relations. For example, suppose one wishes to know the maximum number of vertices $L(\ell,k)$ in the ℓth level of a (nonempty) directed tree of degree k. Since for $\ell = 0$ there can be only one vertex, which is the root, it follows that

$$L(0,k) = 1, \text{ for all } k.$$

For $\ell > 0$, each vertex at level ℓ is the child of a vertex of level $\ell - 1$. There are up to $L(\ell - 1,k)$ vertices at level $\ell - 1$, and each of them has up to k children. It follows that

$$L(\ell,k) = L(\ell - 1,k) \cdot k.$$

The solution to this recurrence relation is $L(\ell,k) = k^\ell$, which can be easily verified by induction: $k^0 = 1$; $k^{\ell-1} \cdot k = k^\ell$. This reasoning proves a theorem, which we shall now state.

Theorem 5.5.4.   There are between 0 and k^ℓ vertices at level ℓ in any directed tree of degree k.

It is also interesting to know the maximum number of vertices in the entire tree, for a tree of height h and degree k. In such a tree the last nonempty level is level h, so that, if each level is as full as possible, the number of vertices in the tree is

$$\sum_{\ell=0}^{h} L(\ell,k) = \sum_{\ell=0}^{h} k^\ell = \frac{k^{h+1} - 1}{k - 1}.$$

This constitutes half of the proof of a theorem, which is given below. (The rest is left as an exercise.)

Theorem 5.5.5.   There are between $h + k$ and $\dfrac{k^{h+1} - 1}{k - 1}$ vertices in a directed tree of height h and degree k.

Two special kinds of directed trees are of interest because they are "compact." That is, they are no taller than they need to be. Because of this, it is possible to get a tighter lower bound on the number of vertices, in terms of the height and the degree of a tree.

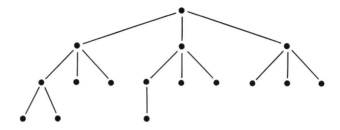

Figure 5-43. A complete tree of degree 3.

Definition 5.5.2. **A complete (directed) tree of degree k** has the maximum possible number (exactly k^ℓ) of vertices in each level, except possibly the last.

Example 5.5.2. The tree in Figure 5-43 is a complete tree of degree 3.

Corollary 5.5.1. There are between

$$\frac{k^h + k - 2}{k - 1} \quad \text{and} \quad \frac{k^{h+1} - 1}{k - 1}$$

vertices in a complete directed tree of degree k and height h.

Proof. We already know that

$$\frac{k^{h+1} - 1}{k - 1}$$

is an upper bound on the number of vertices, from Theorem 5.5.5. Since every level is full except for the hth level, and that must have at least one vertex, the number of vertices in the entire tree is at least

$$1 + \sum_{\ell=0}^{h-1} L(\ell,k) = 1 + \sum_{\ell=0}^{h-1} k^\ell = 1 + \frac{(k^h - 1)}{k - 1} = \frac{k^h + k - 2}{k - 1}. \quad \square$$

Definition 5.5.3. **A B-tree of order k,** (or a k-way B-tree) is a directed tree such that:

1. all the leaves are at the same level;
2. every internal vertex, except possibly the root, has at least $\lceil k/2 \rceil$ · children; (here $\lceil x \rceil$ means the least integer $\geq x$.)

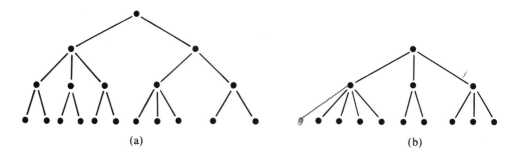

Figure 5-44. A *B*-tree and a non-*B*-tree.

3. the root is a leaf or has at least two children; and
4. no vertex has more than k children.

Example 5.5.3. Of the two trees shown in Figure 5-44, only (a) is a B-tree. The order of tree (a) is 3 or 4. Tree (b) cannot be a B-tree. Its order must be at least 5, since one vertex has five children, but another vertex has only two children, which is less than $\lceil 5/2 \rceil$.

B-trees have important applications in implementing indexed sequential files. They are used for directories of files stored in disk systems. A large value for k is desirable, since the cost to access one vertex is high, but does not increase significantly with k.

Theorem 5.5.6. A B-tree of order k and height h has at least $2 \cdot \lceil k/2 \rceil^{h-1}$ leaves, for $h \geq 1$.

Proof. The proof is by induction on h, and is left to the reader as an exercise. $\square$

Exercises for Section 5.5

1.(a) Draw a directed tree of height 4 with exactly one vertex at level one, exactly two vertices at level 2, exactly three vertices at level 3, and exactly four vertices at level 4.

  (b) How many directed trees are there that fit the description above, up to isomorphism (that is, count isomorphic trees as being the same).

2. Prove that every vertex in a directed tree different from the root has a unique parent.

3. Derive formulae for the minimum and maximum possible heights of a directed tree of degree k with n vertices.

4. Derive formulae for the minimum and maximum possible heights of a directed tree of degree k with n *leaves*.

5. Derive formulae for the minimum and the maximum possible k for

which there exists a complete directed tree of degree k with n vertices and height h.

6. Complete the proof of Theorem 5.5.5.

7. Derive a formula for the maximum possible height of a complete directed tree of degree k with n vertices.

8. Derive a formula for the maximum possible height of a B-tree of order k with n leaves.

9. Prove Theorem 5.5.6.

10. Derive a formula for the maximum number of internal vertices in a B-tree with n leaves.

11. A sentence consists of a number of syntactic entities (such as noun phrases, verb phrases, or prepositional phrases) which are concatenated with each other in accordance with certain grammatical rules. The process of *parsing* or resolving a sentence into its syntactic components leads naturally to a tree. For example, think of the entire sentence as the root of the tree, the syntactic categories as the internal vertices, and the words of the sentence as the vertices of degree one. Using this procedure, give a tree depiction of the following sentence:

The tall boy smiled at the blond girl.

12. A directed true T of degree k is said to be *regular* if every internal vertex has exactly k descendants. In other words, T is regular if every vertex has 0 or k out-degree. For example,

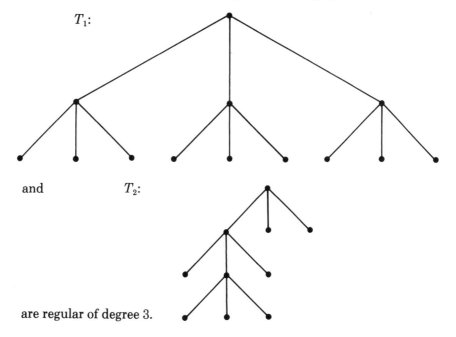

are regular of degree 3.

The tree T_1 is *full* in the sense that each level has the maximum number of vertices. The tree T_2 is *thin* in the sense that each level has the minimum number of internal vertices (either 0 or 1).

Note that any full regular tree of degree k is a complete tree of degree k, but the converse need not be true.

(a) Draw a thin regular tree of degree 3 and height 4. Note that there must be 9 leaves.

(b) Draw a regular tree of degree 4, height 4, and having 13 leaves.

(c) Draw a full regular tree of degree 2 and height 3.

(d) Show that for a regular tree T of degree k, if i is the number of internal vertices and ℓ is the number of leaves, then $(k-1)i = \ell - 1$. In particular, $i = \ell - 1$ if T is regular of degree 2.

(e) Consider the problem of connecting 28 lamps to a single electrical outlet by using extension cords each of which has 4 outlets. Although there are many ways to connect the lamps, how many extension cords are always needed? Hint: view this problem as a tree.

(f) How many matches must be played to determine a tennis champion if a single elemination tournament is played with 64 players?

(g) Suppose that a certain computer has an instruction which computes the sum of 3 numbers. How many times will the instruction have to be executed to find the sum of 27 numbers?

(h) Derive a formula for the maximum and minimum number of leaves in a regular tree of degree k and height h.

(i) Show that a regular tree of degree 2 has an odd number of vertices.

13. Define a tree T to be a 2-3 tree if each internal vertex of T has 2 or 3 descendants and all leaves are at the same level.

(a) Draw two different examples of a 2-3 tree with 10 leaves and height 3.

(b) A 2-3 tree is a B-tree of order k. Find the possible values for k.

(c) Derive a formula for the maximum and minimum number of vertices in a 2-3 tree of height h.

(d) Derive a formula for the maximum and minimum number of leaves in a 2-3 tree of height h.

14. Suppose T is a tree with the property that the root has at least 2 descendants, all leaves occur at the same level, m = minimum number of descendants of internal vertices different from the root, and M = maximum number of descendants for *all* internal vertices (including the root). Prove that if $M \leq 2m$, then T is a B-tree of order k for any integer k such that $M \leq k \leq 2m$.

Selected Answers for Section 5.5

7. By Corollary 5.5.1, if a complete directed tree of degree k has height h and n vertices, we know that $k^{h-1} \leq n$. From this we obtain $h \leq \log_k(n) + 1$.

8. By Theorem 5.5.6, if a B-tree of order k has height h and n vertices, then $n \geq 2\lceil k/2 \rceil^{\,h-1}$. It follows that $h \leq \log_{\lceil k/2 \rceil}(n/2) + 1$.

9. Let n_i be the smallest possible number of vertices on level i of a B-tree of order k and height $\geq i$. From the definition of B-tree, we obtain the recurrence:

$$n_0 = 1,$$
$$n_1 = 2,$$
$$n_i = n_{i-1}\lceil k/2 \rceil, \text{ for } i \geq 2.$$

Solving this we obtain $n_i = 2\lceil k/2 \rceil^{\,i-1}$.

12. (d) There are ki descendants of the i internal vertices. But $ki = $ the number of internal vertices plus the number of leaves minus 1, since the root is not a descendant of any vertex. Thus,

$$ki = i + \ell - 1 \quad \text{or} \quad (k-1)i = \ell - 1.$$

(g) View the problem of finding the sum of the numbers as a tree where there are 27 leaves representing the numbers and internal vertices for each time the instruction is executed.

Thus we have $(3-1)i = 27 - 1$ or $i = 13$ internal vertices. Therefore, the addition instruction must be executed 13 times.

5.6 BINARY TREES

Definition 5.6.1. A **binary tree** is a directed tree $T = (V,E)$, together with an edge-labeling $f: E \rightarrow \{0,1\}$ such that every vertex has at most one edge incident from it labeled with 0 and at most one edge incident from it labeled with 1. Each edge (u,v) labeled with 0 is called a **left edge;** in this case u is called the **parent** of v and v is called the **left child** of u. Each edge (u,v) labeled with 1 is called a **right edge;** in this case u is also called the parent of v, but v is called the **right child** of u. The subtrees of which the left and right children of a vertex u are the roots are called the **left** and **right subtrees** of u, respectively. We may represent a binary tree by a triple (V,E,f).

It is implicit in this definition that every vertex in a binary tree has a unique parent, a unique left child, and a unique right child, if it has any at

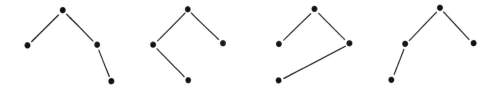

Figure 5-45. Four distinct binary trees.

all. That each vertex has a unique parent (if any) follows from the definition of tree, where it is required that there be a unique path from the root to each vertex. (If any vertex v were to have two parents there would be a path to each of them from the root, and extending these paths to v would yield two paths to v.) That each vertex has a unique left child and a unique right child (if any) follows from the labeling of the edges of the tree with 0's and 1's. (At most one edge from the parent can have a 0 label and at most one edge can have a 1.) It is also implicit in the definition that every vertex other than the root has a parent. This is so because every vertex v in a tree must have a path to it from the root. The last vertex before v on such a path must be the parent of v.

The left-right orientation on the vertices of a binary tree requires that we view as distinct binary trees those that would otherwise be the same as directed trees of degree at most 2. (See Figure 5-45 for an illustration.)

Example 5.6.1   Details such as edge labels and the direction of edges are usually represented only implicitly in drawings of binary trees. The convention is that for each vertex v the root of v's left subtree lies below v and to its left on the page, whereas the root of v's right subtree lies below v and to its right on the page. Figure 5-46 shows an example of a binary tree drawn with and without edge labels and directed edges.

There are a few special kinds of binary trees that are important in computer applications; one of these is the complete binary tree.

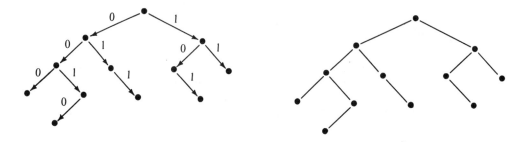

Figure 5-46. Conventions for drawing binary trees.

Definition 5.6.2.   Let T be a binary tree. Every vertex v in T has a unique **level-order index,** defined as follows: If v is the root, let index $(v) = 1$; if v is the left child of a vertex u, let index $(v) = [$index $(u) \cdot] \, 2$; otherwise, if v is the right child of some vertex u, let index $(v) = 1 + [$index $(u)] \cdot 2$.

Note that the index of a child is obtained by doubling the index of its parent and adding the label on the edge that goes from the parent to the child. Another way of looking at this vertex numbering scheme is based on the fact that each vertex corresponds to a unique string w in $\{0,1\}^*$, determined by the sequence of labels on the path to it from the root. The index of the vertex corresponding to w is the integer represented by $1w$, viewed as a binary number. Thus, the level of any vertex v can be computed as $\lfloor \log_2(\text{index } (v)) \rfloor$.

Example 5.6.2.   The binary tree shown in Figure 5-46 is repeated in Figure 5-47, this time with the level-order index written at the location of each vertex. Left children all have even indices, and right children have odd indices. The string corresponding to vertex number 18 is 0010. Putting a 1 in front of this, we get 10010, which is the base two representation of 18.

Definition 5.6.3.   A **complete binary tree** is a binary tree for which the level-order indices of the vertices form a complete interval $1, \ldots, n$ of the integers. That is, if such a tree has n vertices there is a vertex in the tree with index i for every i from 1 to n.

Example 5.6.3.   Of the two binary trees shown in Figure 5-48, only (a) is complete. In particular, (b) has ten vertices but has no vertex with index 10.

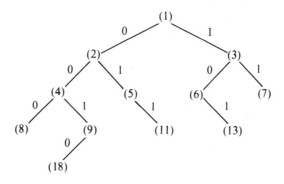

Figure 5-47. A binary tree, with level-order indices.

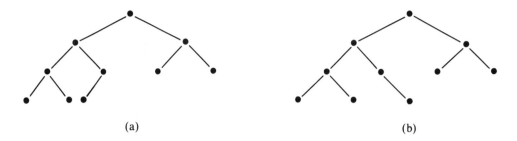

(a)   (b)

Figure 5-48. Complete and noncomplete binary trees.

It is probably apparent by now that each complete binary tree corresponds to a special labeling of a complete directed tree of degree 2. This is formally a consequence of the following lemma, which shows that every level of a complete binary tree has exactly 2^ℓ vertices, except for possibly the last.

Lemma 5.6.1. In a complete binary tree with n vertices the indices of the vertices in the ℓth level comprise the complete interval 2^ℓ through $2^{\ell+1} - 1$, or from 2^ℓ through n if n is less than $2^{\ell+1} - 1$.

Proof.   The proof is by induction on ℓ. For $\ell = 0$ and $n = 0$ the lemma holds vacuously. For $\ell = 0$ and $n > 0$ there is exactly one vertex with index $2^0 = 2^{0+1} - 1 = 1$, and that is the root, which is also the only vertex at level 0. For larger values of ℓ, we assume the lemma holds for $\ell - 1$, by induction. If $n < 2^\ell$, the lemma holds vacuously. Otherwise, we invoke the definition of level. The vertices in level ℓ are exactly those at distance ℓ from the root. The vertices in level $\ell - 1$ are exactly those at distance $\ell - 1$ from the root. It follows that the vertices at level ℓ are precisely the children of the vertices at level $\ell - 1$, which the inductive hypothesis asserts are those with indices $2^{\ell-1}$ through $2^\ell - 1$. By the definition of level-order index, the children have indices in the range 2^ℓ through $2^{\ell+1} - 1$. This complete interval, or the initial segment of it up through n, must be in T, by the definition of complete binary tree.   □

Because of the natural way their vertices correspond to an initial segment of the positive integers, complete binary trees can be represented very efficiently on computers. They are applied in a number of excellent algorithms, including "Heap Sort," priority queue implementation, and algorithms for the efficient ordering of data in hash tables.

Definition 5.6.4.   A **height balanced** binary tree is a binary tree such that the heights of the left and right subtrees of every vertex differ by at most one.

Height balanced trees are important in computer science because the height of a height balanced tree is always $O(\log n)$ with respect to the number n of vertices in the tree. Thus, a number of algorithms for frequently performed operations can be implemented to perform in $O(\log n)$ time on tables that are organized as height balanced trees.

Example 5.6.4. The tree shown in Figures 5-46 and 5-47 is a height-balanced binary tree of height 4. Deleting vertex (13) would result in a tree that would no longer be height balanced, since it would have a left subtree of height 3 and a right subtree of height 1. Deleting vertex (8) would also result in a tree that would no longer be height balanced, since vertex (4) would have a left subtree of height -1 and a right subtree of height 1.

Note that every complete binary tree is also a height-balanced binary tree. Height-balanced binary trees are of interest because they are more general than complete binary trees but it is still possible to obtain a nontrivial lower bound on the number of vertices in a tree of a given height.

Theorem 5.6.1. There are at least $(1/\sqrt{5})\ [(1 + \sqrt{5})/2]^{h+3} - 2$ vertices in any height-balanced binary tree with height h.

Proof. This bound can be obtained from a recurrence. Let $V(h)$ denote the least achievable number of vertices in a (nonempty) height balanced binary tree of height h. Then clearly $V(0) = 1$ and $V(1) = 2$. For $h > 1$, we observe that there must be a root and two subtrees, possibly empty. The height of one subtree must be $h - 1$. The height of the other may be $h - 1$ or $h - 2$. It is clear that it is not possible to construct a height-balanced subtree of height $h - 1$ with fewer vertices than are required to construct a height-balanced subtree of height $h - 2$. A height-balanced tree of height h with the fewest possible vertices thus consists of a root, one height-balanced subtree of height $h - 1$ and one height-balanced subtree of height $h - 2$. The total number of vertices in such a tree is $V(h) = 1 + V(h - 1) + V(h - 2)$.

This recurrence should be very familiar, since it is nearly the same as the recurrence for the Fibonacci numbers, which was solved in Chapter 3. Using similar techniques, we obtain the solution:

$$V(h) = \left(\frac{5 + 2\sqrt{5}}{5}\right)\phi^h + \left(\frac{5 - 2\sqrt{5}}{5}\right)(1 - \phi)^h - 1$$

where

$$\phi = \frac{1 + \sqrt{5}}{2} \text{ and } 1 - \phi = \frac{1 - \sqrt{5}}{2}.$$

We also obtained a solution of this recurrence relation in Example 1.10.11 of Chapter 1, namely that

$$V(h) = \frac{1}{\sqrt{5}} (\phi)^{h+3} - \frac{1}{\sqrt{5}} (1 - \phi)^{h+3} - 1$$

$$= F_{h+2} - 1, \text{ where } F_n \text{ is the } n^{\text{th}} \text{ Fibonacci number.}$$

But since it is immediate that

$$\frac{5 + 2\sqrt{5}}{5} = \frac{1}{\sqrt{5}} \phi^3 \quad \text{and} \quad \frac{5 - 2\sqrt{5}}{5} = \frac{-1}{\sqrt{5}} (1 - \phi)^3$$

we see that these two apparently different solutions are the same.

It is interesting to compare this bound with the lower bound of 2^h vertices for complete binary trees of height h. Since ϕ is between 1.61803 and 1.61804

$$-1 < (1 - \phi)^{h+3} < 1, \text{ and, in particular,}$$

$$\frac{-1}{\sqrt{5}} (1 - \phi)^{h+3} > -1 \text{ so that}$$

$$V(h) > \frac{1}{\sqrt{5}} (\phi)^{h+3} - 2.$$

Thus, $V(h)$ is greater than $(1.89)(1.61)^h - 2$; that is, we have an exponential lower bound on the number of vertices in a height balanced tree comparable to the bound for a complete binary tree.   □

Definition 5.6.5.   A **binary search tree** is a binary tree with a vertex labeling $l: V \rightarrow A$, where $A = \{a_1, a_2, \ldots, a_n\}$ is a totally ordered set with $a_1 < a_2 < \cdots < a_n$, and where the labeling l satisfies the properties:

(i) For each vertex u in the left subtree of a vertex v, $l(u) < l(v)$
(ii) For each vertex u in the right subtree of a vertex v, $l(u) > l(v)$.

Example 5.6.5.   Given a sequence of numbers 17,23,4,7,9,19, 45,6,2,37,99 let us build a binary search tree for the set A obtained by sorting the numbers in their proper order. First label the root with 17; then since 23 is larger than 17, make a right child for the root and label it 23. Next since 4 is less than the root, label a left child of the root with the label 4. Continue to the next number 7 in the list. Since 7 is less than the root 17, 7 will appear as the label of some vertex in the left subtree of the root, yet 7 is greater than 4, so label with 7 the right child of the vertex 4.

Next use the label 9. Again 9 must be the label of a vertex in the left subtree of the root since $9 < 17$. Moreover, 9 is greater than 4, so 9 must be a label in the right subtree of 4. Likewise $9 > 7$, so 9 is a label for the right child of 7. Next, 19 is greater than 17, so 19 will label some vertex in the right subtree of 17; but 19 is less than 23, the right child of 17, so label the left child of 23 with the label 19. Next consider 45; $45 > 17$ so 45 is a label for a vertex in the right subtree of 17; $45 > 23$, so label the right child of 23 with 45. Continuing as above, we see that 6 must be the label for the left child of the vertex 7, 2 is the left child of 4, 37 and 99 are respectively the left and right children of 45. Thus, we have the following binary tree:

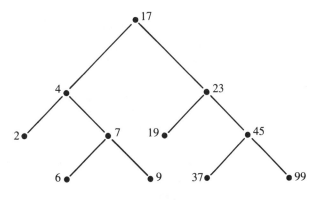

Figure 5-49

One nice feature of the binary search tree is that now we can perform an inorder traversal of the tree (see Exercise 7 in this section), and this traversal will give a rearrangement of the original sequence into its proper order.

Also, if T' is a subtree of the binary search tree T with root u, then the minimum and maximum labels on the vertices of T' are easily located. To find the minimum label, start at the root u and visit successively left children until a leaf is reached. The label on this leaf is the minimum label on the subtree T'. Likewise, we can find the maximum label on the vertices of T' by successively visiting right children until we arrive at a leaf.

But that is not all. Binary search trees have a dynamic feature as well; additional numbers appended to the original sequence can be incorporated into the binary search tree just as if these had been in the sequence all along. For instance, if 1,5,11,39,84,87 are appended to the original sequence, then 1 becomes a left child of 2, 5 is a left child of 6, 11 is a right child of 9, 39 is a right child of 37, 84 is a left child of 99, and 87 becomes a right child of 84.

Vertices can be deleted in such a way that the resulting tree is still a binary search tree, but we leave the discovery of such a procedure as an exercise.

Exercises for Section 5.6

1. Draw the binary tree whose level order indices are

$$\{1,2,4,5,8,10,11,20\}.$$

2. Show that if $H(n)$ is the maximum possible height of a *complete* binary tree with n vertices then H is in $O(\log(n))$.

3. Show that if $H(n)$ is the maximum possible height of a *height-balanced* binary tree with n vertices then it is in $O(\log(n))$.

4. Another kind of balanced binary tree is called an α-balanced binary tree. Let T be a binary tree and ℓ and r be the number of vertices in T's left and right subtrees, respectively. Define the *balance* of T to be the ratio $(1 + \ell)/(2 + \ell + r)$, and say that T is α-balanced if the balance of T is between α and $1 - \alpha$. Derive a lower bound on the number of vertices in an α-balanced binary tree of height h.

5. Draw a height balanced binary tree of height 4 with the minimum possible number of vertices.

6. The terminology for binary trees borrows heavily from human kinship relations. Are the parent-child relationships between human beings really binary trees? Justify your answer.

7. (Tree Traversal Algorithms.) A *traversal* of a tree is a process that enumerates each of the vertices in the tree exactly once. When a vertex is encountered in the order of enumeration specified by a particular process, we say that we *visit* the given vertex. We describe here three principal ways that may be used to traverse a binary tree; each scheme will be defined by specifying the order for processing the 3 entities: the root (N), the left subtree (L), and the right subtree (R). There are several names for these traversals; we choose the names that indicate the order of occurrence of the root.

   1. Preorder Traversal (abbreviated NLR).

      (a) For any subtree, first visit the root.
      (b) Then perform preorder traversal on the entire left subtree from that root (if a left subtree exists.)
      (c) Perform preorder traversal on the root's right subtree.

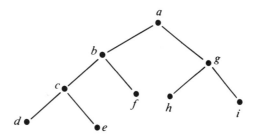

Figure 5-50

2. Inorder Traversal (abbreviated LNR).

   (a) First, perform inorder traversal on the root's left subtree (if it exists).

   (b) Then visit the root.

   (c) Perform inorder traversal on the root's right subtree.

3. Postorder Traversal (abbreviated LRN).

   (a) Perform postorder traversal on the root's left subtree.

   (b) Perform postorder on the root's right subtree.

   (c) Visit the root.

   For example, the path traversals in the tree shown in Figure 5-50 are given by:

   Preorder: $a - b - c - d - e - f - g - h - i$

   Inorder: $d - c - e - b - f - a - h - g - i$

   Postorder: $d - e - c - f - b - h - i - g - a$

   Show the sequential orders in which the vertices of the following trees are visited in a preorder, an inorder, and a postorder traversal.

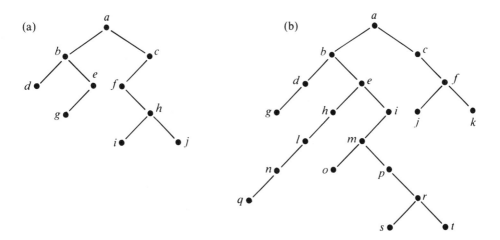

8. Give an example of a binary tree complete in the sense of Definition 5.5.2 but which is not a complete binary tree in the sense of Definition 5.6.3.

9. Determine whether or not the following trees are height balanced. If not, explain why not.

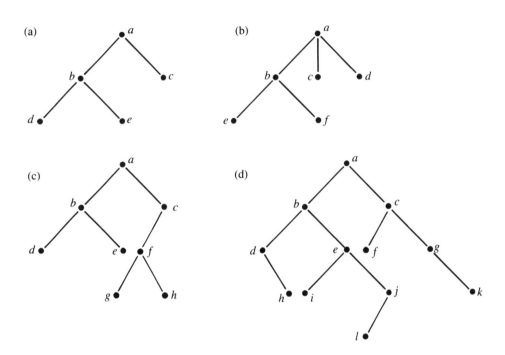

10. Draw all regular binary trees
    (a) with exactly 7 vertices.
    (b) with exactly 9 vertices.
    (A regular binary tree is a binary tree where each vertex has 0 or 2 children.)

11. Draw all distinct binary trees
    (a) with 3 vertices.
    (b) with 4 vertices.

12. Draw binary search trees for the following lists:
    (a) 18,44,2,5,73,45,14,6,8,10,20,11
    (b) 2,1,5,6,8,9,7,3,4
    (c) Carol, Bob, Dumpty, Ace, Ellen, King, Humpty, Joe, Myrtie
    (d) Ask, Art, Ate, Able, Alto, Also, Avid
    (Hint: on (c) and (d) use the usual dictionary ordering.)

13. An *ordered tree* is a directed tree for which the children of each vertex are ordered sequentially. Draw an ordered tree of degree 3 with
    (a) 7 vertices.
    (b) 10 vertices and height 4.
    Explain how to view a binary tree as an ordered tree.

14. Determine the height of the binary tree whose largest level-order index is
    (a) 2^4
    (b) $2^5 + 17$
    (c) $2^5 + 31$
    (d) $2^5 + 33$
    (e) $4^5 + 7$
    (f) 2^k for k a positive integer
    (g) n for n a positive integer.

15. Given the level-order index of a vertex v in a binary tree, determine the level-order index of v's left child, right child, and parent if the level-order index of v is
    (a) 29
    (b) 410
    (c) n, where n is a positive integer.

16. Design an algorithm to delete a vertex from a binary search tree so that the result is still a binary search tree. Hint: consider the three cases where v is a leaf, has one child, or 2 children.

17. Determine the conditions that will cause each internal vertex of a binary search tree to have exactly one child.

Selected Answers for Section 5.6

1.

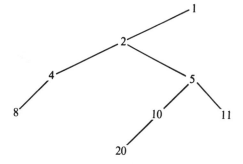

Figure 5-51

3. By Theorem 5.6.2, if a height balanced binary tree has n vertices and height h,

$$n \geq \frac{1}{\sqrt{5}}\left(1 + \frac{\sqrt{5}}{2}\right)^h - 2. \quad \text{That is,}$$

$$h \leq \log_{\frac{(1 + \sqrt{5})}{2}}(\sqrt{5}(n + 2)).$$

To see that h (and hence $H(n)$) is in $O(\log(n))$, observe that $\sqrt{5}(n + 2) \leq 7n$ for $n \geq 1$, and $\log_x(7n) = \log_x 7 + \log_x n \leq 2 \log_x n$, for $n \geq 7$. Thus, $h < (2) \log_{1.6}(n)$ for $n \geq 7$.

6. First, they are not binary, since a person may have more than two children. Second, they need not be trees, since a person may have a common ancestor on two sides of the family: for example,

Figure 5-52

Of course, human kinship terms *do* form binary trees when viewed linguistically, even in the above case, by ignoring crossbreeding.

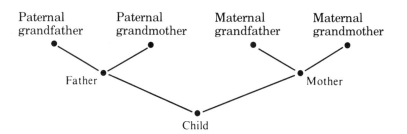

Figure 5-53

7. (a) Preorder: $a - b - d - e - g - c - f - h - i - j$
       Inorder: $d - b - g - e - a - f - i - h - j - c$
       Postorder: $d - g - e - b - i - j - h - f - c - a$

REVIEW FOR SECTIONS 5.1–5.6

1. For the graph below find the spanning tree (draw the tree and label each vertex with an integer from 1 to 8 to indicate the order of visitation) obtained by:

   (a) BFS
   (b) DFS
   (c) Kruskal's algorithm.

   (In each problem use the vertex ordering *abcdefgh*.)

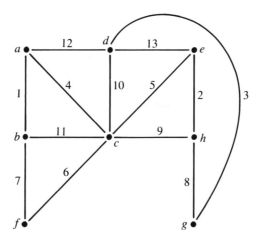

2. For the tree below list the vertices in the required order:

   (a) Preorder: _____
   (b) Postorder: _____
   (c) Inorder: _____

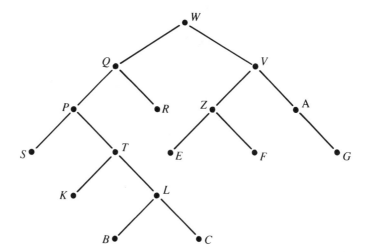

3. Draw, if possible (if not, explain why not), and label the binary tree with the level-order indices 1,2,3,4,5,6,7,8,9,10,11,13,18,26,52.

4. Draw all nonisomorphic, nondirected trees with degree sequence
   (a) (1,1,1,1,2,3,3)
   (b) (1,1,1,1,1,1,1,1,3,4,5).

5. Draw all nonisomorphic, nondirected forests on 4 vertices.

6. Fill in the blanks:
   (a) If G is a simple connected graph with 70 vertices, then the number of edges of G is between _____ and _____. Explain.
   (b) Suppose that N_i denotes the number of vertices of degree i in a tree T. If $N_2 = 5$, $N_3 = 1$, $N_4 = 7$, $N_5 = 2$, then $N_1 =$ _____. Explain.
   (c) The maximum number of vertices in a B-tree of order 4 and height 5 is _____. Explain.
   (d) The minimum number of vertices in a B-tree of order 5 and height 5 is _____. Explain.
   (e) Suppose that G is a graph such that each vertex has degree 4 and $|E| = 4|V| - 36$. Then $|V| =$ _____ and $|E| =$ _____.
   (f) The number of connected components of a forest A is _____ if $|E| = 24$ and $|V| = 32$.
   (g) (1,3,3,4,5,5) is/is not the degree sequence of a simple graph because _____.
   (h) If (1,3,3,4,4,5) is the degree sequence of a graph G, then $|E| =$ _____.
   (i) The height of a binary tree is _____ if the largest level-order index is $4^{12} + 5$. Explain.
   (j) If the level-order index of a vertex v in a binary tree is 223, then the level-order index of v's left child is _____ and v's right child is _____, and of v's parent is _____.
   (k) For the tree T depicted below, height $T =$ _____; height of the vertex e is _____; the level of vertex n is _____ and the degree of the tree T is _____.

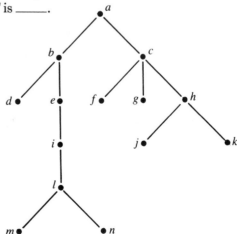

(1) K_n has _____ edges and $K_{m,n}$ has _____ edges.

(m) The wheel graph W_n on n vertices has _____ edges.

(n) P_n, the path graph on n vertices has _____ edges.

(o) If the simple graph G has n vertices, then $\overline{G}$, the complement of G, has _____ edges.

7. Prove or disprove.

(a) The following two graphs are isomorphic.

 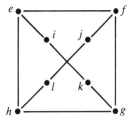

(b) Every connected simple graph with $|E| = n$ and $|V| = n + 1$ where $n \geq 2$ contains a cut point.

(c) If G is a simple graph with two connected components, $|E(G)| \geq |V(G)| - 2$.

(d) Given any two edges e_1 and e_2 of a simple connected graph G, there is a spanning tree containing the two edges e_1 and e_2. Hint: Kruskal.

(e) A "full" binary tree has an odd number of vertices. (Definition: A full binary tree is one where the last level has the maximum number of vertices possible.)

(f) A simple graph with 5 edges and 6 vertices is connected.

(g) A simple graph with degree sequence (3,3,3,3,3,3,3,3) is connected.

(h) Every connected simple graph with $n \geq 2$ vertices contains at least 2 vertices that are not cut vertices.

(i) Suppose that G is a connected, simple graph and e is an edge of G. If e is on no cycle of G, then e is a cut edge of G.

(j) Suppose that G is a connected simple graph and e is a cut edge of G. Then for each spanning tree T of G, e is an edge of T.

(k) Each connected, simple graph G with n vertices and m edges where $3 \leq n \leq m$ contains a circuit.

(1) If $k - 1$ edges are added to a forest with k connected components, then the resulting graph is a tree.

(m) If a simple graph contains a circuit C which contains all the edges and vertices of G, then $|E(G)| = |V(G)|$.

8. True or False.

____ (a) A simple graph contains no cycles.

____ (b) A complete binary tree is a B-tree of order 2.

___ (c) A tree has no cycles.

___ (d) A tree always contains a cut point.

___ (e) A complete directed tree of degree 2 is a complete binary tree.

___ (f) Every circuit in a graph is also a cycle.

___ (g) A connected, simple graph with $n > 3$ vertices contains at least 2 cut vertices.

___ (h) The sequence (1,3,3,3,5,5) is graphic.

___ (i) The following is a valid argument: If a graph T is connected and contains no circuits, then T is a tree. If T is a tree, then $|E(T)| = |V(T)| - 1$. $|E(T)| \neq |V(T)| - 1$. Therefore, T is not connected or T contains a circuit.

___ (j) The following is a valid argument: If graphs G and H are isomorphic, then G and H have the same degree sequence. The graphs G and H have the same degree sequence. Therefore, the graphs G and H are isomorphic.

___ (k) The following is a valid argument: If the graph T is a directed tree, then T is quasi-strongly connected. The graph T is not a directed tree. Therefore, T is not quasi-strongly connected.

___ (l) The sequence (1,1,2,2,3,3,3,3) is the degree sequence of a tree.

___ (m) The two graphs below are isomorphic.

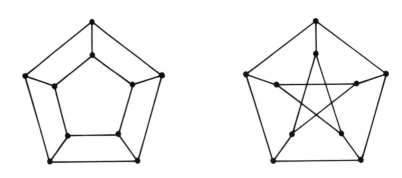

___ (n) The following graph is quasi-strongly connected:

5.7   PLANAR GRAPHS

When drawing a graph on a piece of paper, we often find it convenient (or even necessary) to permit edges to intersect at points other than at vertices of the graph. These points of intersection are called **crossovers** and the intersecting edges (or crossing edges) are said to *cross over* each other. For example, the graph of Figure 5-54 (a) exhibits three crossovers: $\{b,e\}$ crosses over $\{a,d\}$ and $\{a,c\}$, and $\{b,d\}$ crosses over $\{a,c\}$. A graph G is said to be **planar** if it can be drawn on a plane without any crossovers; otherwise G is said to be **nonplanar.** Note that if a graph G has been drawn with crossing edges, this does not mean that G is nonplanar—there may be another way to draw the graph without crossovers. For example, the graph in Figure 5-54 (a), can be redrawn in Figure 5-54 (b) without crossovers. Accordingly we say that a planar graph is a **plane graph** if it is already drawn in the plane so that no two edges cross over. Therefore, the graph in Figure 5-54 (b) is a plane graph while its depiction in (a) is not.

Example 5.7.1. Suppose we have three houses and three utility outlets (electricity, gas, and water) situated so that each utility outlet is connected to each house. Is it possible to connect each utility to each of the three houses without lines or mains crossing?

We can represent this situation by a graph whose vertices correspond to the houses and the utilities, and where an edge joins two vertices iff one vertex denotes a house and the other a utility. The resulting graph is the complete bipartite graph $K_{3,3}$. The 3 houses-3 utilities problem can then be rephrased in terms of graph theory: Is $K_{3,3}$ a planar graph?

Before we answer this question let us try to find a systematic way to draw a graph in the plane without edges crossing. We want to be able to conclude that a graph is nonplanar if our construction fails. The method

(a)   (b)

Figure 5-54

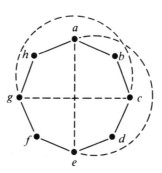

Figure 5-55

we will use involves two simple ideas:

1. If we have drawn a cycle in the plane, then any edge not on the cycle must be either *inside* the cycle, *outside* the cycle, or the edge must cross over one of the edges of the cycle.
2. The roles of being inside or outside the cycle are symmetric that is, the graph can be redrawn so that edges and vertices formerly outside the cycle are now inside the cycle and vice versa.

Figure 5-55 indicates various possible configurations for the edges $\{a,e\}$ and $\{c,g\}$ relative to the cycle: $a - b - c - d - e - f - g - h - a$.

Now let us use these ideas to show that $K_{3,3}$ is nonplanar and hence that the 3 houses-3 utilities problem has a negative answer. With $K_{3,3}$ labeled as indicated in Figure 5-56, we draw the cycle of $K_{3,3}$: $a - d - c - e - b - f - a$. Since the roles of inside and outside the cycle are equivalent, we can assume that the edge $\{a,e\}$ is inside the cycle. The edge $\{b,d\}$ then must be drawn outside the cycle since otherwise $\{a,e\}$ and $\{b,d\}$ would cross over. But note there is no place to draw the edge $\{c,f\}$, either inside

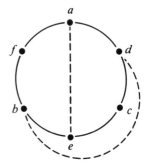

Figure 5-56

or outside the cycle, without crossing over either $\{b,d\}$ or $\{a,e\}$. Therefore, $K_{3,3}$ is not planar.

Questions of planarity arise in applications of graph theory to computer science especially in designing and building electrical circuitry. A printed circuit board is a planar network, and so minimizing the amount of nonplanarity is a key design criterion.

Furthermore, since flowcharts are prepared in such a way as to allow statements executed sequentially to be placed in reasonable proximity, it is desirable that there be as few overlapping or crossing flowlines as possible. Thus planarity or at least near planarity becomes an objective here also.

A plane graph G can be thought of as dividing the plane into **regions** or **faces**. Intuitively the regions are the connected portions of the plane remaining after all the curves and points of the plane corresponding, respectively, to edges and vertices of G have been deleted. Each plane graph G determines a region of infinite area called the *exterior region* of G. The vertices and edges of G incident with a region r make up the boundary of the region r. If G is connected, then the boundary of a region r is a closed path in which each cut-edge is traversed twice. When the boundary contains no cut edges of G, then the boundary of r is a cycle of G. In either case, the degree of r is the length of its boundary. Of course, a cycle of G need not be the boundary of a region.

For example, in the graph shown in Figure 5-57(a), $a - b - c - f - g - h - a$ is a cycle but there are only 4 regions determined by this graph, namely the 3 regions with boundaries $a - b - g - h - a$, $b - c - f - g - b$, $c - d - e - f - c$, and the exterior region whose boundary is $a - b - c - d - e - f - g - h - a$. In Figure 5-57(b), the exterior region has degree 10 because of the 2 cut edges t,u and w,x; the boundary of the exterior region is the closed path $t - u - v - w - x - y - z - x - u - t$.

Heretofore, we have not formally used the concept of a multigraph. We referred to its definition only in passing in sections 4.1 and 5.1. In the definition given below, the dual of a graph need not itself be a graph in

(a)   (b)

Figure 5-57

the strictest sense of the word. In fact, the dual of a 3-cycle will be a multigraph with 2 vertices and 3 multiple edges joining the 2 vertices. Thus, the reader should review the earlier definitions. The primary technical fact that we need is that the "First Theorem of Graph Theory" holds for multigraphs as well as for graphs.

Definition 5.7.1. Given a plane graph G, we can define another multigraph G^* as follows: Corresponding to each region r of G there is a vertex r^* of G^*, and corresponding to each edge e of G there is an edge e^* of G^*; two vertices r^* and s^* are joined by the edge e^* in G^* iff their corresponding regions r and s are separated by the edge e in G. In particular, a loop is added at a vertex r^* of G^* for each cut-edge of G that belongs to the boundary of the region r. The multigraph G^* is called the **dual** of G.

There is a natural way to draw G^* in the plane: Place r^* in the corresponding region r of G (think of r^* as the capital of the region r), and then draw each e^* in such a way that it crosses the corresponding edge e of G exactly once (and crosses no other edge of G). This procedure is illustrated in Figure 5-58, where the dual edges are indicated by dashed lines and the dual vertices by asterisks. Note that if e is a loop of G, the e^* is a cut edge of G^* and conversely.

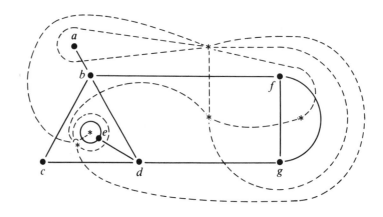

Figure 5-58

Let $|E^*|$, $|R^*|$, $|V^*|$, and $|E|$, $|R|$, $|V|$ denote the number of edges, regions, and vertices of G^* and G respectively. The the following relations are direct consequences of the definition of G^*:

For all plane graphs G,
  i) $|E^*| = |E|$
  ii) $|V^*| = |R|$, and

iii) $\deg_G{}^*(r^*) = \deg_G(r)$ for each region r of G. Moreover, if G is connected, it can be shown that

iv) $|R^*| = |V|$.

In (iii) we mean that the degree of the vertex r^* in G^* is the same as the degree of the corresponding region r determined by G.

Theorem 5.7.1. If G is a plane graph, then the sum of the degrees of the regions determined by G is $2|E|$, where $|E|$ is the number of edges of G.

Proof. Let us use the notation $\Sigma_{r \in R(G)}\mathrm{degree}(r)$ for the sum of the degrees of all the regions determined by G. Then if G^* is the dual of G, let $\Sigma_{r^* \in V(G^*)}\mathrm{degree}(r^*)$ denote the sum of the degrees of the vertices of G^*. Then

$$\sum_{r \in R(G)} \mathrm{degree}(r) = \sum_{r^* \in V(G^*)} \mathrm{degree}(r^*)$$

by (iv) above. But we already know that the sum of the degrees of all vertices in any graph is twice the number of edges. In particular, we know that

$$\sum_{r^* \in V(G^*)} \mathrm{degree}(r^*) = 2|E^*|.$$

But $|E| = |E^*|$ by (i). Therefore the theorem is proved. $\square$

Exercises for Section 5.7

1. A plane graph G is *self-dual* if it is isomorphic to its dual.
   For $n = 2,3,4,5$, find a self-dual graph on n vertices.

2. Show that K_5 is nonplanar by the technique used in Example 5.7.1.

3. Draw the dual graph for each of the following graphs:

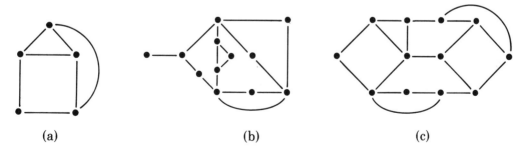

(a)   (b)   (c)

Figure 5-59

4. Show that the following graphs are planar:

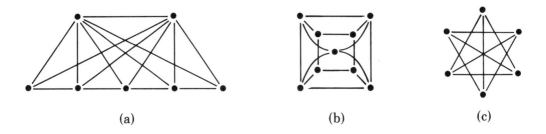

(a)   (b)   (c)

Figure 5-60

5. Suppose that the 3 houses-3 utilities problem was instead the 5 houses and 2 utilities problem. What would the solution be?
6. Which of the following graphs are planar?

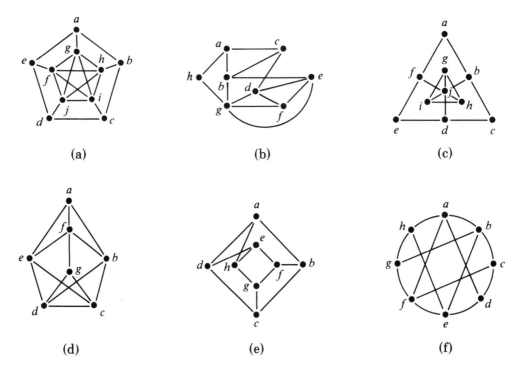

(a)   (b)   (c)

(d)   (e)   (f)

Figure 5-61

7. (a) Show that $K_5 - e$ is planar for any edge e of K_5.

   (b) Show that $K_{3,3} - e$ is planar for any edge e of $K_{3,3}$.

8. A graph G is *critical planar* if G is nonplanar but any subgraph obtained by removing a vertex is planar.

   (a) Which of the following graphs are critical planar?

       (i) $K_{3,3}$

       (ii) K_5

       (iii) K_6

       (iv) $K_{4,3}$

   (b) Show that critical planar graphs must be connected and cannot have a vertex whose removal disconnects the graph.

9. Show that the following graphs are self-dual:

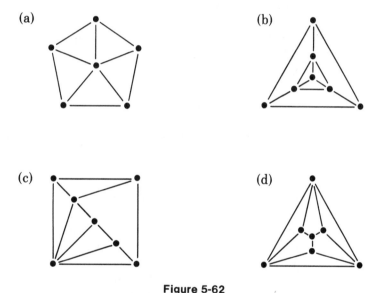

Figure 5-62

10. Show that the complete tripartite graph $K_{1,2,3}$ is nonplanar.

11. Show that the wheel graph W_n on n vertices is isomorphic to its dual.

12. Show that any graph with 4 or fewer vertices is planar.

13. Let G be a connected planar graph and let G^* be its dual. If T is a spanning tree of G, show that those edges of G^* that do not correspond to edges of T form a spanning tree for G^*. Conclude that G and G^* have the same number of spanning trees.

Selected Answers for Section 5.7

1. (b)

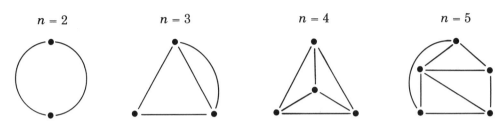

Figure 5-63

5.8   EULER'S FORMULA

If G is a connected planar graph, then any drawing of G in the plane as a plane graph will always form $|R| = |E| - |V| + 2$ regions, including the exterior region, where $|R|$, $|E|$, and $|V|$ denote, respectively, the number of regions, edges, and vertices of G. This remarkable formula was discovered by Euler in 1752.

Theorem 5.8.1 (Euler's Formula).   If G is a connected plane graph, then $|V| - |E| + |R| = 2$.

Proof.   We prove this by first observing the result for a tree. By convention, a tree determines only one region. We know already that the number of edges of a tree is one less than the number of vertices. Thus, for a tree the formula $|V| - |E| + |R| = 2$ holds. Moreover, we note that a connected plane graph G with only 1 region must be a tree since otherwise there would be a circuit in G, and the existence of a circuit implies an internal region and an external region.

We prove the general result by induction on the number k of regions determined by G. We have proved the result for $k = 1$. Assume the result for $k \geq 1$ and suppose then that G is a connected plane graph that determines $k + 1$ regions. Delete an edge common to the boundary of two separate regions. The resulting graph G^1 has the same number of vertices, one fewer edge, but also one fewer region since two previous regions have been consolidated by the removal of the edge. Thus if $|E^1|$, $|V^1|$, and $|R^1|$ are, respectively, the numbers of edges, vertices, and regions for G^1, $|E^1|$, $= |E| - 1$, $|R^1| = |R| - 1$, $|V^1| = |V|$. But then $|V| - |E| + |R| = |V^1| - |E^1| + |R^1|$. By the inductive hypothesis, $|V^1| - |E^1| + |R^1| = 2$. Therefore, $|V| - |E| + |R| = 2$ and the theorem is proved by mathematical induction. $\square$

The above theorem allows the graph to have loops (though the following corollary does not). We recall that the degree of a region is the number of edges in the closed path that forms the boundary. Then the above theorem actually holds for graphs where regions may have degree 1. *However, we shall assume throughout this section that the graph is simple and* $|\mathrm{E}| > 1$; *thus we are assuming that the degree of each region is greater than or equal to 3.* We call a connected plane graph **polyhedral** if degree $(r) \geq 3$ for each region $r \in R(G)$; and if, in addition, degree $(v) \geq 3$ for each vertex $v \in V(G)$. In particular, we observe the following fact: In a plane graph G, if the deree of each region is $\geq k$, then $|R| \leq 2|E|$. In particular, we have $3|R| \leq 2|E|$. The proof of this fact is easy—just observe that $\Sigma_{r \in R(G)}$ degree $(r) = 2|E| \geq k|R|$ since degree $(r) \geq k$ for each region $r \in R(G)$.

Corollary 5.8.1. In a connected plane (simple) graph G, with $|E| > 1$,

(1) $|E| \leq 3|V| - 6$ and
(2) there is a vertex v of G such that degree $(v) \leq 5$.

Proof. By Euler's formula $|R| + |V| = |E| + 2$, and since G is simple $3|R| \leq 2|E|$ or $|R| \leq 2/3|E|$. Hence, $2/3|E| + |V| \geq |R| + |V| = |E| + 2$. Thus, $|V| - 2 \geq 1/3|E|$ or $3|V| - 6 \geq |E|$.

As for (2) if each vertex has degree ≥ 6, then since $\Sigma_{v \in V(G)}$ degree $(v) = 2|E|$, it follows that $6|V| \leq 2|E|$ or $|V| \leq 1/3|E|$. Likewise $|R| \leq 2/3|E|$. But then since $|R| + |V| = |E| + 2$, we have $2/3|E| + 1/3|E| \geq |R| + |V| = |E| + 2$ or $|E| \geq |E| + 2$ or $0 \geq 2$, an obvious contradiction. $\square$

The simple fact in (2) is very useful in connection with the four-color problem, which we will discuss in the Section 5.12.

Theorem 5.8.2. A complete graph K_n is planar iff $n \leq 4$.

Proof. It is easy to see that K_n is planar for $n = 1,2,3,4$. Thus, we have only to show that K_n is nonplanar if $n \geq 5$, and for this it suffices to show that K_5 is nonplanar. We prove this by an indirect argument. If K_5 were planar, then $|R| = |E| - |V| + 2 = 10 - 5 + 2 = 7$. But since K_n is simple and loop free, we would also have $3|R| \leq 2|E|$ which in this case would imply that $3 \cdot 7 = 21 \leq 2 \cdot 10 = 20$, an obvious contradiction. (Note we could have obtained a contradiction also by appealing to the inequality $|E| \leq 3|V| - 6$.) $\square$

Theorem 5.8.3. A complete bipartite graph $K_{m,n}$ is planar iff $m \leq 2$ or $n \leq 2$.

Proof.   It is easy to see that $K_{m,n}$ is planar if $m \leq 2$ or $n \leq 2$. Now let $m \geq 3$ and $n \geq 3$. To prove that $K_{m,n}$ is nonplanar it suffices to prove that $K_{3,3}$ is nonplanar. (We did this in section 5.7, but let us give another proof based on Euler's Formula.)

Since $K_{3,3}$ has six vertices and nine edges, if $K_{3,3}$ were planar, Euler's formula would give that $|R| = |E| - |V| + 2 = 9 - 6 + 2 = 5$. Since $K_{3,3}$ is bipartite there can be no cycles of odd length. Hence each cycle has length ≥ 4 and thus the degree of each region would have to be greater than or equal to 4. But then we would have to have $4|R| \leq 2|E|$ or $20 = 4 \cdot 5 \leq 2 \cdot 9 = 18$, a contradiction. $\square$

Euler's formula and the method of proof by contradiction will solve the following problem.

Example 5.8.1.   Prove that there does not exist a polyhedral graph with exactly seven edges.

If there were such a polyhedral graph with $|E| = 7$, then $3|R| \leq 2|E| = 14$ since each region has degree ≥ 3. Moreover, each vertex has degree ≥ 3, so that $3|V| \leq 2|E| = 14$. Thus, $|R| \leq 4$ and $|V| \leq 4$. By Euler's formula, $|R| + |V| = |E| + 2 = 9$ and then $8 \geq |R| + |V| = |E| + 2 = 9$, a contradiction. We conclude that a polyhedral graph with 7 edges does not exist.

Exercises for Section 5.8

1. Prove that there is no polyhedral graph with exactly 30 edges and 11 regions.

2. Prove that for any polyhedral graph
   (a) $|V| \geq 2 + |R|/2$.
   (b) $|R| \geq 2 + |V|/2$.
   (c) $3|R| - 6 \geq |E|$.

3. Using the results of Problem 2 together with $3|R| \leq 2|E|$ and $3|V| \leq 2|E|$ prove that $|V| \geq 4$, $|R| \geq 4$, and $|E| \geq 6$ for any polyhedral graph.

4. (a) If G is a polyhedral graph with 12 vertices and 30 edges, prove that the degree of each region is 3.
   (b) If G is a connected plane graph with 6 vertices and 12 edges, prove that the degree of each region is 3.

5. Show that a plane connected graph with less than 30 edges has a vertex of degree ≤ 4.

6. Suppose that G is a connected plane graph with less than 12 regions and such that each vertex of G has degree ≥ 3. Then prove that G has a region of degree ≤ 4.

7. Show that if G is a polyhedral graph, then there is a region of degree ≤ 5.

8. Give a direct proof that a plane-connected graph with each region of degree ≥ 5 and each vertex of degree ≥ 3 must have at least 30 edges.

9. Prove that a connected plane graph with 7 vertices and degree $(v) = 4$ for each vertex v of G must have 8 regions of degree 3 and one region of degree 4.

10. Let G be a connected plane graph with $|V| \geq 3$. Let $\delta(G)$ and $\Delta(G)$ denote, respectively the minimum and maximum of the degrees of all the vertices of G. Suppose that G has exactly V_k vertices of degree k.
    (a) Show that $5V_1 + 4V_2 + 3V_3 + 2V_4 + V_5 \geq V_7 + 2V_8 + \ldots + (n-6)V_n + 12$ when $n = \Delta(G)$.
    (b) Use the result in (a) to prove the existence of a vertex in G of degree ≤ 5.
    (c) Observe that equality holds in (a) iff the degree of each region is 3.
    (d) Suppose that $\delta(G) = 5$. Prove that there are at least 12 vertices of degree 5.
    (e) Suppose that $\delta(G) \geq 3$ and $|V| \geq 4$. Prove that there are at least 4 vertices of degree less than or equal to 5.

11. Suppose that a plane graph G is not connected but instead consists of several components, that is, disjoint connected subgraphs.
    (a) Find the appropriate modification of Euler's formula for a plane graph with C components.
    (b) Show that Corollary 5.8.1 is true even for plane graphs that are not connected.

12. Draw 2 polyhedral graphs with 6 vertices and 10 edges.

13. Give an example of a connected plane graph such that:
    (a) $|E| = 3|V| - 6$.
    (b) $|E| < 3|V| - 6$.

14. *Prove or disprove:* If G is a connected graph such that $|E| = 3|V| - 6$, then G is planar. Explain.

15. Show that part (b) of the corollary to Euler's formula is false if we do not assume G is simple.

16. Show that K_n is planar for $1 \leq n \leq 4$.

17. Show that $K_{m,n}$ is planar for $1 \leq m \leq 2$ or $1 \leq n \leq 2$.

18. Show that if G is a simple planar graph with $|V| \geq 11$, then the complement of G is nonplanar.

19. Let G be a plane graph such that each vertex has degree 3. Prove that a dual of G will have an odd number of regions of finite area.

20. If G is a connected plane graph with all cycles of length at least r, show that the inequality $|E| \le 3|V| - 6$ can be strengthened to $|E| \le [r/(r-2)] (|V| - 2)$.

21. The crossing number $c(G)$ of a graph G is the minimum number of pairs of crossing edges among all the depictions of G in the plane. For example, if G is planar, then $c(G) = 0$.
    (a) Determine $c(G)$ for the following graphs:
        (1) $K_{3,3}$.
        (2) K_5.
    (b) Determine that $c(K_6) = 3$. (Hint. Introduce new vertices at each cross-over and use the corollary to Euler's formula.)

22. Show that if a plane graph is self-dual (see Exercise 1 of Section 5.7 for the definition), then $|E| = 2|V| - 2$.

23. Sometimes we are able to say that a simple graph is nonplanar simply because there are too many edges. Explain.

24. Give a proof of Euler's formula by using induction on the number of edges.

25. Suppose that G is a connected planar graph. Determine $|V|$ if
    (a) G has 35 regions each of degree 6.
    (b) G has 14 regions each of degree 4.
    (c) G has $|R|$ regions each of degree k.

26. A planar graph G is called *maximal planar* if, for each pair of nonadjacent vertices u and v of G, the addition of the edge $\{u,v\}$ destroys planarity.

    Prove that if G is a maximal planar graph and $|V| \ge 3$, then $|E| = 3|V| - 6$ and the degree of each region is 3.

Selected Answers for Section 5.8

2. (a) $|V| = 2 + (|E| - |R|) \ge 2 + |R|/2$ since $|E| \ge 3/2|R|$ implies that $|E| - |R| \ge |R|/2$.

6. Suppose degree $(r) \ge 5$ for each region r. Then $2|E| \ge 5|R|$ or $5/2|R| \le |E|, 2|E| \ge 3|V|$ or $|V| \le 2/3|E|$ and $-|V| \ge -2/3|E|$. Then, $|R| = |E| - |V| + 2 \ge |E| - 2/3|E| + 2 = |E|/3 + 2 \ge 5/6|R| + 2$. But then $|R|/6 \ge 2$ or $|R| \ge 12$, a contradiction.

7. Observe that in the dual G^* there is a vertex r^* of degree ≤ 5. Hence the region r has degree ≤ 5.

10. $|V| = \Sigma_{i=1} n_i$ and $2|E| = \Sigma_{v \in V(G)} \deg(v) = n_1 + 2n_2 + 3n_3 + \cdots$. Also $3|R| \le 2|E|$ and equality holds iff each region has degree 3. $2 = |R| + |V| - E$ by Euler's formula, so $|V| - 1/3|E| = 2/3|E| + |V| - |E| \ge 2$ or $6|V| - 2|E| \ge 12$. Write $6|V| = 6n_1 + 6n_2 + 6n_3 + \cdots - 2|E| = -n_1 - 2n_2 - 3n_3 \ldots$ and sum.

11. (a) Observe that if the exterior region is not counted then for each component of G, we have $|R_i| = |E_i| - |V_i| + 1$. Then summing over all components and adding 1 for the exterior region gives $|R| = |E| - |V| + C + 1$.

18. The number of edges of G plus the number of edges of $\overline{G}$ is $1/2$ $|V|(|V| - 1)$. Thus, G or $\overline{G}$ has at least $1/4 |V|(|V| - 1)$ edges. But since $1/4 |V|(|V| - 1) > 3|V| - 6$ for $|V| \geq 11$ (verify !), G cannot have that many edges or otherwise the corollary to Euler's formula would be violated. But then $\overline{G}$ violates the corollary and hence is nonplanar.

5.9   MULTIGRAPHS AND EULER CIRCUITS

The earliest recorded use of the concept of graph in mathematics is by the Swiss mathematician Leonhard Euler (1707–1782), who in 1736 settled a famous unsolved problem of his day known as the Problem of the Königsberg Bridges. The East Prussian city of Königsberg (now Kaliningrad) was located on the banks of the Pregel River. Included in the city were two islands, which were linked to each other and to the banks of the river by seven bridges, as shown in Figure 5-64.

The problem was to begin at any of the four land areas, denoted by the letters a, b, c, and d, to walk across a route that crossed each bridge exactly once, and to return to the starting point. In proving that this particular problem is unsolvable, Euler replaced each of the areas a, b, c, and d by a vertex and each bridge by an edge joining the corresponding vertices, thereby producing the "graph" shown in Figure 5-65.

Of course, to express this kind of structure, the more general notion of multigraph is required.

There are several ways of formally expressing the existence of multiple edges between a single pair of points. One way is by labeling the edges with numbers, called **multiplicities.** According to this convention, the

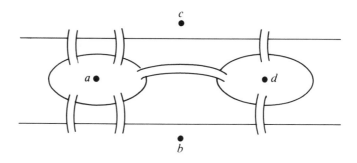

Figure 5-64. The Königsberg Bridges.

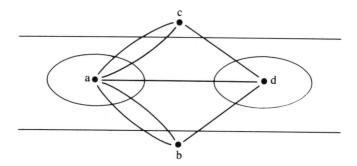

Figure 5-65. Euler's "graph."

Königsberg bridges could be represented by the labeled graph shown in Figure 5-66. The edges {a,c} and {a,b} are labeled with multiplicity 2, indicating that there are two bridges between each of these pairs of vertices, whereas the other edges are each labeled with multiplicity 1, since each corresponds to a single bridge.

This convention for representing multiple edges corresponds to a natural extension of the notion of adjacency matrix, which we shall call a *multiplicity matrix*. The multiplicity matrix of the graph shown in Figure 5-66 is:

$$
\begin{array}{c c}
 & \begin{array}{cccc} a & b & c & d \end{array} \\
\begin{array}{c} a \\ b \\ c \\ d \end{array} &
\left[\begin{array}{cccc}
0 & 2 & 2 & 1 \\
2 & 0 & 0 & 1 \\
2 & 0 & 0 & 1 \\
1 & 1 & 1 & 0
\end{array} \right]
\end{array}
$$

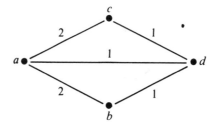

Figure 5-66. The graph of Königsberg Bridges.

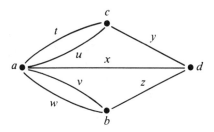

Figure 5-67. The multigraph of the Königsberg Bridges.

While this representation is adequate for this problem, and convenient for computer manipulation, it is not adequate for all applications, since it does not provide for the possibility of there being multiple *distinct* edges between a pair of vertices. We could put labels $\{t,u,v,w,x,y,z\}$ on the Königsberg bridges as shown in Figure 5-67.

Now let us show how Euler proved that the problem of Königsberg Bridges is unsolvable.

Definition 5.9.1. An **Euler path** in a multigraph is a path that includes each edge of the multigraph exactly once and intersects each vertex of the multigraph at least once. A multigraph is said to be *traversable* if it has an Euler path. An **Euler circuit** is an Euler path whose endpoints are identical. (That is, if an Euler path is a sequence of edges $e_1, e_2, \ldots, e_k$ corresponding to the sequence of pairs of vertices $(x_1,x_2),(x_2,x_3), \ldots, (x_{k-1},x_k)$, then the e_i's are all distinct, and $x_1 = x_k$.) A multigraph is said to be an **Eulerian multigraph** if it has an Euler circuit.

We will now prove the main theorem characterizing nondirected multigraphs that have Euler paths.

Theorem 5.9.1. A nondirected multigraph has an Euler path iff it is connected and has 0 or exactly 2 vertices of odd degree. In the latter case, the two vertices of odd degree are the endpoints of every Euler path in the multigraph.

Proof. (only if) Let multigraph G have an Euler path. It is clear that G must be connected. Moreover, every time the Euler path meets a vertex it traverses two edges which are incident on the vertex and which have not been traced before. Except for the two endpoints of the path, the degree of all other vertices must therefore be even. If the endpoints are distinct, their degrees are odd. If the two endpoints coincide, their degrees are even and the path becomes an Euler circuit. (if) Let us construct an Euler path by starting at one of the vertices of odd degree and traversing each edge of G exactly once. If there are no vertices of odd degree we will start at an arbitrary vertex. For every vertex of even degree the path will enter the vertex and leave the vertex by tracing an edge that was not traced before. Thus the construction will terminate at a vertex with an odd degree, or return to the vertex where it started. This tracing will produce an Euler path if all edges in G are traced exactly once this way.

If not all edges in G are traced, we will remove those edges that have been traced and obtain the subgraph G' induced by the remaining edges. The degrees of all vertices in this subgraph must be even and at least one vertex must intersect with the path, since G is connected. Starting from one of these vertices, we can now construct a new path, which in this case will be a cycle, since all degrees are now even. This path will be joined into the previous path. The argument can be repeated until a path that traverses all edges in G is obtained.   □

The proofs of the following corollaries follow from the preceding theorem.

Corollary 5.9.1.   A nondirected multigraph has an Euler circuit iff it is connected and all of its vertices are of even degree.

Corollary 5.9.2.   A directed multigraph G has an Euler path iff it is unilaterally connected and the in-degree of each vertex is equal to its out-degree, with the possible exception of two vertices, for which it may be that the in-degree of one is one larger than its out-degree and the in-degree of the other is one less than its out-degree.

Corollary 5.9.3.   A directed multigraph G has an Euler circuit iff G is unilaterally connected and the in-degree of every vertex in G is equal to its out-degree.

Example 5.9.1.   The problem of drawing a multigraph on paper with a pencil without lifting the pencil or repeating any lines is clearly a problem of finding an Euler path in the multigraph. A multigraph can be drawn in this way iff it has an Euler path. For example, the multigraph in Figure 5-68 (a) can be drawn in this fashion with each edge being traced exactly once, while the directed multigraph in Figure 5-68 (b) cannot.

It is interesting to observe that Theorem 5.9.1 gives the basis of an efficient algorithm for determining whether a nondirected multigraph has an Euler circuit. Assume that G is represented by an $n \times n$ adjacency matrix A. To check that all vertices of G are of even degree we add all the

(a)

(b)

Figure 5-68

1's in each row and check whether the sum is even. For each row, this takes O(n) steps, and since there are n rows the whole process can be performed in O(n^2) steps. To check that G is connected, we can compute the transitive reflexive closure A^* of A, which takes O(n^3) steps using Warshall's algorithm. (This can be done in O(n^2) steps by methods that are not covered in this book.)

Application to Computer Science

De Bruijn Sequences. Let $\Sigma = \{0,1,\dots,n-1\}$ be an alphabet of n symbols. Clearly there are n^k different sequences of length k over these symbols. A de Bruijn sequence, known also as a maximum-length shift register sequence, is a circular array $a_0 a_1 \dots a_{L-1}$ over Σ such that for every sequence α in Σ^k there exists a unique j such that

$$a_j a_{j+1} \cdots a_{j+k-1} = \alpha,$$

where the computation of the indices is modulo L. Clearly if the sequence satisfies this condition, then $L = n^k$. The most important case in computer science is the binary alphabet, where $n = 2$, and thus $\Sigma = \{0,1\}$. Binary de Bruijn sequences are very useful in coding theory and are implemented by shift registers. To show that it is possible to arrange 2^k binary digits in a circular array such that the 2^k sequences of k consecutive digits in the arrangement are all distinct, we construct a directed graph with 2^{k-1} vertices which are labeled with the 2^{k-1} ($k-1$)-digit binary numbers. From the vertex $b_1 b_2 \cdots b_{k-1}$ there is an edge to the vertex $b_2 b_3 \cdots b_{k-1} 0$ which is labeled $b_1 b_2 \cdots b_{k-1} 0$ and an edge to the vertex labeled $b_2 b_3 \cdots b_{k-1} 1$ which is labeled $b_1 b_2 \cdots b_{k-1} 1$. According to Corollary 5.9.3 the graph has an Euler circuit, which corresponds to a circular arrangement of the 2^k binary digits.

These graphs are known as de Bruijn diagrams or Good's diagrams, or shift register state diagrams, and are denoted by $G_{n,k}(V,E)$ where

$$V = \Sigma^{k-1}$$
$$E = \{[(b_1,\ldots,b_{k-1}),(b_2,\ldots,b_{k-1},b_k)] \mid b_1,b_2,\ldots,b_{k-1},b_k \in \Sigma\}$$

and each edge $[(b_1,\ldots,b_{k-1}),(b_2,\ldots,b_{k-1},b_k)]$ is labeled $b_1,\ldots,b_{k-1},b_k$. Figure 5-69 illustrates $G_{2,4}(V,E)$.

For example, consider the directed Euler circuit of $G_{2,4}(V,E)$ consisting of the sequence of edges with the following labels:

0000, 0001, 0010, 0101, 1010, 0100, 1001, 0011,

0111, 1111, 1110, 1101, 1011, 0110, 1100, 1000

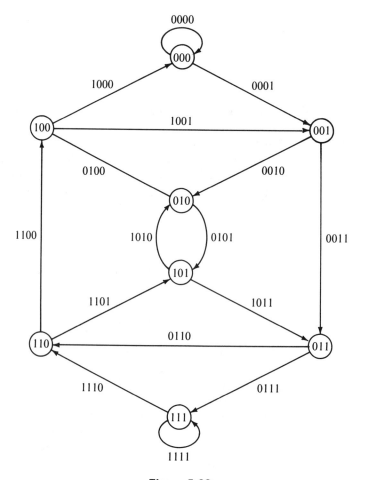

Figure 5-69

and the sequence of 16 binary digits is

$$0000101001111011,$$

where the circular arrangement is obtained by closing the two ends of the sequence.

One application of this concept is in the generation of unique codes. For example, the 16 binary digits in the above sequence generated by use of the Euler circuit on $G_{2,4}(V,E)$ create 16 unique code words that differ by only one digit. This means that the sequence of 16 binary digits given above, when implemented on a rotating drum, will generate 16 different positions of the drum and by that produce 16 distinct codes. In general it is possible to arrange 2^n binary digits in a circular array such that 2^n sequences of n consecutive digits in the arrangement are all distinct.

Exercises for Section 5.9

1. Show that a directed graph that contains an Euler circuit is strongly connected.

2. Prove that a connected graph has a Euler circuit iff it can be decomposed to a set of elementary cycles that have no edge in common.

3. Find an Euler circuit in Figure 5-70.

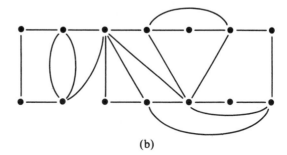

(a)   (b)

Figure 5-70

4. In present day Kaliningrad (Königsberg) two additional bridges have been constructed. One bridge is between regions b and c and the other between b and d. Is it now possible to construct a route over all bridges of Kaliningrad without recrossing any of them?

5. *Prove or Disprove:* A graph which possesses an Euler circuit may have an edge whose removal disconnects the graph. (Such an edge is called a "bridge.")

6. Give an **example** of a graph with ten edges that has a bridge as well as an Euler path.

7. In the definition of Euler circuit discuss the requirement that the Euler circuit intersects with every vertex at least once.

8. Is it possible for a knight to move on an 8×8 chessboard so that it makes every possible move exactly once? A move between two squares on the chessboard is complete when it is made in either direction.

9. Build a 27-digit circular ternary (0, 1, or 2) sequence in which every 3-digit subsequence appears exactly once.

10. Let $L(G)$ of a graph G be another graph which has a vertex for each edge in G and two of these vertices are adjacent iff the corresponding edges in G have a common end vertex. Show that $L(G)$ has an Euler circuit if G has an Euler circuit.

11. Find a graph G which has no Euler circuit but for which $L(G)$ has one.

12. Prove that if a connected graph has $2n$ vertices of odd degree then
    (a) n Euler paths are required to contain each edge exactly once.
    (b) There exists a set of n such paths.

13. Prove that for positive integers $\delta + 1$ and k there exists a de Bruijn sequence.

14. Prove that for positive integers $\delta + 1$ and n, $G_{\delta+1,n}(V,E)$ has a directed Euler circuit.

15. Which of the multigraphs in Figure 5-71 have Euler paths, circuits, or neither?

(a)

(b)

(c)

(d)

Figure 5-71

5.10   HAMILTONIAN GRAPHS

Suppose that a traveling salesman's territory includes several cities with roads connecting certain pairs of these cities. Suppose additionally that the salesman's job requires that he visit each city personally. Is it possible for him to schedule a round trip by car enabling him to visit each specified city exactly once?

We can represent the salient features of this problem by a graph G whose vertices correspond to the cities in the salesman's territory, and such that two vertices are joined by an edge iff there is a road directly connecting the two cities (that is, the road does not pass through any other city in the territory). The solution of this problem depends on whether G has a cycle containing every vertex of G.

Thus, we see that this problem suggests the concept of Hamiltonian graphs. A graph G is said to be **Hamiltonian** if there exists a cycle containing every vertex of G. Such a cycle is referred to as a **Hamiltonian cycle.** Thus, a Hamiltonian graph is a graph containing a Hamiltonian cycle. We define a **Hamiltonian path** as a simple path that contains all vertices of G but where the end points may be distinct. Since *a graph is Hamiltonian iff its underlying simple graph is Hamiltonian we limit our discussion to simple graphs.*

Whereas the Euler circuit is a circuit that traverses each edge exactly once and, therefore, traverses each vertex at least once, a Hamiltonian cycle traverses each vertex exactly once (and hence may miss some edges altogether.) Thus, there is a striking similarity between the concepts of Eulerian graph and Hamiltonian graph, and therefore one might expect an elegant characterization of Hamiltonian graphs as in the case of Eulerian graphs. Such is not the case, and, in fact, the development of such a characterization is a major unsolved problem in graph theory.

To be sure, a Hamiltonian cycle always provides a Hamiltonian path, upon deletion of any edge. On the other hand, a Hamiltonian path may not lead to a Hamiltonian cycle (it depends on whether or not the end points of the path happen to be joined by an edge in the graph).

The name "Hamiltonian" is derived from the Irish mathematician Sir William Rowan Hamilton, who invented a game in 1857 consisting of a solid regular dodecahedron, twenty pegs, one inserted at each corner of the dodecahedron, and a supply of string. Each corner was marked with the name of an important city of the time, and the aim of the game was to find a round trip route along the edges of the dodecahedron that passed through each city exactly once. In order for the players to recall which cities in a route had aleady been visited, the string was used to connect the appropriate pegs in order. Later on Hamilton introduced a graphical version of the game where the object was to find a Hamiltonian cycle on the graph of the dodecahedron. Try your hand at the game.

Generally, it has been assumed that Hamilton's game represented the first interest in Hamiltonian graphs, but in fact, the English mathematician Thomas P. Kirkman posed a problem about Hamiltonian graphs in a paper submitted to the Royal Society in 1855, two years prior to the appearance of Hamilton's game.

Clearly the graph illustrated in Figure 5-72 is Hamiltonian for we can find a Hamiltonian cycle by inspection [following the numbering and omitting the edge $\{v_4,v_8\}$].

But, of course, there are some graphs that are not Hamiltonian and they need not contain a Hamiltonian path. The problem of proving that no Hamiltonian cycle (or path) exists in a given graph can be very difficult—frequently requiring the analysis of several cases.

Let us focus our attention for the moment on showing that a Hamiltonian cycle or path does not exist, for the nonexistence problem requires the type of systematic logical analysis that is the essense of most applied graph theory.

To prove the nonexistence of a Hamiltonian path or cycle, we must begin building parts of a Hamiltonian path and show that the construction must always fail, that is, we cannot visit all vertices without visiting some vertices at least twice. The following examples demonstrate how such contradictions can be obtained. But first let us state some basic rules that must be followed in building Hamiltonian paths. The idea underlying these rules is that a Hamiltonian cycle must contain exactly two edges incident at each vertex and a Hamiltonian path must contain at least one of the edges.

Some Basic Rules for Constructing Hamiltonian Paths and Cycles

Rule 1. If G has n vertices, then a Hamiltonian path must contain exactly $n - 1$ edges, and a Hamiltonian cycle must contain exactly n edges.

Rule 2. If a vertex v in G has degree k, then a Hamiltonian path must contain at least one edge incident on v and at most two edges incident on v. A Hamiltonian cycle will, of course, contain exactly two edges incident on v. In particular, both edges incident on a vertex of degree two will be contained in every Hamiltonian cycle. In sum: there cannot be three or more edges incident with one vertex in a Hamiltonian cycle.

Rule 3. No cycle that does not contain all the vertices of G can be formed when building a Hamiltonian path or cycle.

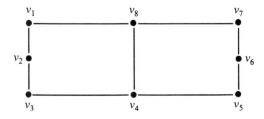

Figure 5-72

Rule 4. Once the Hamiltonian cycle we are building has passed through a vertex v, then all other unused edges incident on v can be deleted because only two edges incident on v can be included in a Hamiltonian cycle.

Example 5.10.1. The path through the vertices of G_1 (in Figure 5-73) in the order of appearance in the English alphabet forms a Hamiltonian path. However, G_1 has no Hamiltonian cycle since if so, any Hamiltonian cycle must contain the edges $\{a,b\}$, $\{a,e\}$, $\{c,d\}$, $\{d,e\}$, $\{f,g\}$, and $\{e,g\}$. But then there would be three edges of the cycle incident on the vertex e.

Example 5.10.2. Likewise the graph G_2 (in Figure 5-73) has neither a Hamiltonian path nor a cycle for the following reason. Note that the vertex l has degree 5 so that at least three edges incident on l cannot be included in any Hamiltonian path. The same is true for the vertices h and j. There are 13 vertices of degree 3 and, in particular, b, d, f, and n are such that at least one of the three edges incident on each of these vertices cannot be included in a Hamiltonian path. Thus, at least $9 + 4 = 13$ of the 27 edges of G_2 cannot be included in any Hamiltonian path. Hence there are not enough edges to form a Hamiltonian path on the 16 vertices of G_2. Thus, G_2 has no Hamiltonian path.

The appeal to the symmetry of the graph often saves some effort. See Figure 5-74.

Example 5.10.3. If a Hamiltonian cycle exists for G_3 then the cycle must include the edges $\{a,d\}$, $\{d,g\}$, $\{b,e\}$, $\{e,h\}$, $\{c,f\}$, and $\{f,i\}$ by Rule 2. Next consider the vertex b. Since the graph is symmetric with respect to the edges $\{a,b\}$ and $\{b,c\}$, it does not matter which of these two edges we choose as the other edge incident on b to be in the cycle. Suppose we choose the edge $\{a,b\}$ (if we obtain a contradiction using $\{a,b\}$, then we

Figure 5-73

G_3

G_4

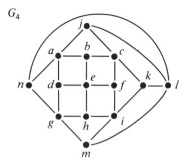

Figure 5-74

would also obtain a contradiction with $\{b,c\}$). Now by Rule 4, we can delete the other edge $\{b,c\}$. Deleting $\{b,c\}$ reduces the degree of c to 2 so then a Hamiltonian cycle must include the edge $\{a,c\}$ also. But then there would be three edges incident at a. Therefore, G_3 has no Hamiltonian cycle. There is, however, a Hamiltonian path, namely, the path that traverses the vertices in the following order: a-d-g-h-e-b-c-f-i.

Example 5.10.4. The graph G_4 in Figure 5-74 has vertical and horizontal symmetry (the vertex l is off to one side but its adjacencies are symmetrical). There are no vertices of degree 2 in this graph so we seek a vertex so that once two edges are chosen at the vertex, then the use of Rules 2 and 4 will force the successive deletion and inclusion of many edges. Vertex e is such a vertex. We can use two edges incident on e, 180 degrees apart, or use two edges incident on e that form a 90 degree angle. We examine both cases.

Case 1. Suppose that we consider the situation where edges from opposite sides of e are in a proposed Hamiltonian cycle. We choose the edges $\{d,e\}$ and $\{e,f\}$ as part of our Hamiltonian cycle (the choice of edges $\{b,e\}$ and $\{e,h\}$ would give the same conclusion by symmetry). Then by Rule 4, we can delete $\{b,e\}$ and $\{e,h\}$. Then at b and h we must use both remaining edges incident on b and h respectively. Thus, we must choose the edges $\{g,h\}$, $\{h,i\}$, $\{a,b\}$ and $\{b,c\}$. Now at d we may choose either $\{a,d\}$ or $\{d,g\}$. The two cases are symmetrical with respect to the edges chosen for the cycle thus far. Therefore, without loss of generality, we choose $\{a,d\}$ and, consequently, delete $\{d,g\}$. Now at f, we cannot use $\{c,f\}$ or else the subcycle a-b-c-f-e-d-a would result, so we chose $\{f,i\}$. But then by Rule 4 the other edges incident on a and i, respectively, must be deleted. Delete $\{a,n\}$, $\{a,j\}$, $\{i,k\}$, and $\{i,m\}$. We now have arrived at a situation which is contrary to properties of Hamiltonian cycles. Vertices j and k currently have degree 2, but adding their two remaining edges forces three edges to be incident on c (the same discrepancy occurs at g). We conclude that there is no Hamiltonian cycle in Case 1.

Case 2.   Suppose now that we include two edges incident on e that form a 90 degree angle. By symmetry we may choose any pair of such edges. Suppose that we choose $\{b,e\}$ and $\{e,f\}$. Then by Rule 4 we may delete $\{d,e\}$ and $\{e,h\}$. Then at d we must choose $\{a,d\}$ and $\{d,g\}$, and at h we must choose $\{g,h\}$ and $\{h,i\}$. If at f we choose $\{c,f\}$ and at b we choose $\{b,c\}$ then we have a subcycle: b-c-f-e-b. If instead we choose $\{f,i\}$ and $\{a,b\}$ we get a subcycle: a-b-e-f-i-h-g-d-a. Thus, we conclude that at f and at b we must choose $\{c,f\}$ and $\{a,b\}$ or $\{f,i\}$ and $\{b,c\}$. By symmetry either choice is equivalent. Let us choose $\{c,f\}$ and $\{a,b\}$. Then having used two edges at c, and at g, we delete the other edges $\{a,n\}$, $\{a,j\}$, $\{g,n\}$ and $\{g,m\}$. This leaves only one edge incident on n. There can be no Hamiltonian cycle with that property.

Therefore, having obtained a contradiction in both cases we conclude that G_4 has no Hamiltonian cycle.

Example 5.10.5.   There is no Hamiltonian cycle for the graph G_5. (See Figure 5-75.)

Each vertex of G_5 has degree either 3 or 4. Moreover, every edge connects a vertex of degree 3 with one of degree 4. No other type edge exists in this graph. If we had a Hamiltonian cycle it would visit each vertex of the graph passing alternately through a 3-vertex and a 4-vertex. This cycle would establish a one-to-one correspondence between the set of 3-vertices and the set of 4-vertices. Then it would follow that if this graph had a Hamiltonian cycle there would be exactly as many 3-vertices as 4-vertices. But by inspection we see that there are eight 3-vertices and only six 4-vertices. Therefore, there can be no Hamiltonian cycle for G_5.

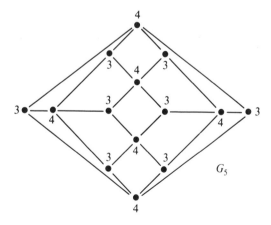

Figure 5-75

We can also conclude that there is no Hamiltonian path for G_5 by similar reasoning, for if there were such a path the number of 3-vertices and the number of 4-vertices could only differ by one. Since this is not the case there is no Hamiltonian path.

The above graph is a special case of a *bipartite* graph, one whose vertex set can be partitioned into two sets in such a way that each edge joins a vertex of the first set to a vertex of the second set. If we think of coloring the vertices of the first set one color, say red, and coloring the vertices of the second set blue, then every edge goes between vertices of different colors. By reasoning as we did in the above example we obtain the following simple fact: *If a connected bipartite graph G has a Hamiltonian cycle then the numbers of red and blue vertices must be equal; if G has a Hamiltonian path, then the numbers of red and blue vertices can differ by at most one.*

The next theorem gives a condition that must prevail for any *plane* Hamiltonian graph. To understand the statement of this theorem let us review some terminology. Let G be a plane Hamiltonian graph with n vertices. Moreover, suppose that C is a fixed Hamiltonian cycle in G. With respect to this cycle, a *diagonal* is an edge of G that does not lie on C. Let $r_i (i = 3,4,\ldots,n)$ denote the number of regions of G in the interior of C whose boundary contains exactly i edges (that is, r_i is the number of regions of degree i in the interior of C). Similarly, let r_i^1 denote the number of regions of degree i in the exterior of C. To illustrate these definitions, let G be the following graph (Figure 5-76) with Hamiltonian cycle C: $v_1 - v_2 - v_3 - v_4 - v_5 - v_6 - v_7 - v_8 - v_9 - v_1$.

Thus, $r_3 = 3$, $r_3^1 = 2$, $r_4 = 2$, $r_4^1 = 1$, $r_5 = 0$, and $r_5^1 = 1$. Moreover, $\{v_1,v_3\}$, $\{v_4,v_9\}$, $\{v_5,v_7\}$, and $\{v_7,v_9\}$, are diagonals in the interior of C, while $\{v_1,v_4\}$ is a diagonal in the exterior of C.

Theorem 5.10.1 (Grinberg).   Let G be a simple plane graph with n vertices. Suppose that C is a Hamiltonian cycle in G. Then with respect

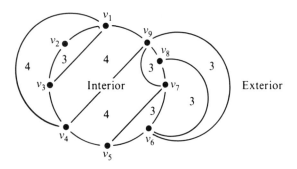

Figure 5-76

to this cycle C,

$$\sum_{i=3}^{n} (i - 2)(r_i - r_i^1) = 0.$$

Proof. First consider the interior of C. Suppose that exactly d diagonals occur there. Since G is a plane graph, none of its edges intersect. Thus a diagonal splits the region through which it passes into two parts. Thinking of putting in the diagonals one at a time, we see that the insertion of a diagonal increases by one the number of regions inside the cycle. Consequently d diagonals divide the interior of C into $d + 1$ regions. Therefore,

$$\sum_{i=3}^{n} r_i = d + 1 \qquad \text{and} \qquad d = \sum_{i=3}^{n} r_i - 1.$$

Let N denote the sum of the degrees of the regions interior to C. Then $N = \Sigma_{i=3}^{n} i r_i$. However, N counts each diagonal twice (since each diagonal bounds two of the regions interior to C) and each edge of C once (since each bounds only one region interior to C). Thus,

$$N = \sum_{i=3}^{n} i r_i = 2d + n.$$

Substituting for d, we have

$$\sum_{i=3}^{n} i r_i = 2 \sum_{i=3}^{n} r_i - 2 + n$$

so that

$$\sum_{i=3}^{n} (i - 2)\, r_i = n - 2.$$

By considering the exterior of C we conclude in a similar fashion that

$$\sum_{i=3}^{n} (i - 2)\, r_i^1 = n - 2.$$

Therefore, combining the two results gives

$$\sum_{i=3}^{n} (i - 2)(r_i - r_i^1) = 0. \quad \square$$

Example 5.10.6. Show that the graph G_6 (Figure 5-77) has no Hamiltonian cycle.

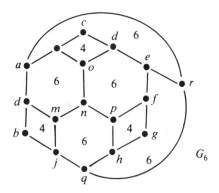

Figure 5-77

We have indicated the degree of each region in the plane depiction of G_6. There are three regions of degree 4 and six regions of degree 6. Thus, by Grinberg's theorem if a Hamiltonian cycle existed, then we would have $r_4 + r_4^1 = 3, r_6 + r_6^1 = 6$, and $2(r_4 - r_4^1) + 4(r_6 - r_6^1) = 0$, or $(r_4 - r_4^1) = -2(r_6 - r_6^1)$. But then $r_4 - r_4^1$ must be an even integer. However, since $r_4 + r_4^1 = 3$, the only possibilities for r_4 and r_4^1 are 0 and 3, and 1 and 2. Neither of the possibilities is such that their difference is even. Therefore, the assumption that there was a Hamiltonian cycle for G_6 led to a contradiction, and G_6 has no Hamiltonian cycle.

Example 5.10.7.   The graph G_7 (in Figure 5-78) does possess Hamiltonian cycles. Show, however, that any such cycle containing one of the edges e, e^1, must avoid the other.

There are five regions of degree 4 and two regions of degree 5. Thus, for any Hamiltonian cycle of G_7 we must have $2(r_4 - r_4^1) + 3(r_5 - r_5^1) = 0$ or $2(r_4 - r_4^1) = -3(r_5 - r_5^1)$ so that 3 divides $r_4 - r_4^1$. But then since $r_4 + r_4^1 = 5$, the only possible values of r_4 and r_4^1 are 4 and 1, making $r_4 - r_4^1$ either 3 or -3.

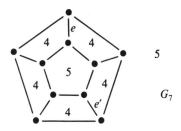

Figure 5-78

Now each of the edges e and e^1 separates a pair of regions of degree 4. Thus, a Hamiltonian cycle would have one of e's quadrilaterals inside and the other outside. Similarly a Hamiltonian cycle containing the edge e^1 would split e^1's quadrilaterals. If both e and e^1 belong to a Hamiltonian cycle then there would be at least two regions of degree 4 on the inside and at least two on the outside. This makes impossible the four-one split guaranteed by Grinberg's theorem.

We have been discussing ways to show that certain graphs are not Hamiltonian. Now let us reverse our point of view and mention one sufficient condition for the existence of a Hamiltonian cycle. The result that we state was proved by Dirac in 1952; a proof can be found in several books on graph theory—for example, *Graph Theory With Applications* by J. A. Bondy and U. S. R. Murty [4]. There are also several other similar results that are known.

Dirac's Theorem.   A simple graph with n vertices (n ⩾ 3) in which each vertex has degree at least $n/2$ has a Hamiltonian cycle.

Corollary 5.10.1.   If G is a complete simple graph on n-vertices ($n \geq 3$), then G has a Hamiltonian cycle.

The corollary is true even for directed graphs.

Exercises for Section 5.10

1. Find a Hamiltonian cycle in each of the following graphs:

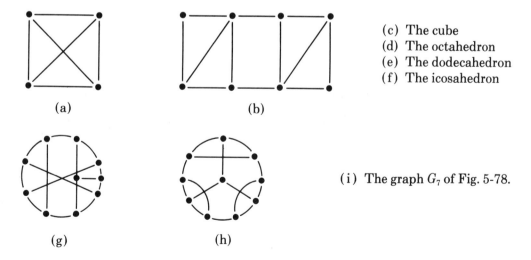

(c) The cube
(d) The octahedron
(e) The dodecahedron
(f) The icosahedron

(i) The graph G_7 of Fig. 5-78.

Figure 5-79

2. Prove that there is no Hamiltonian cycle in each of the following graphs:

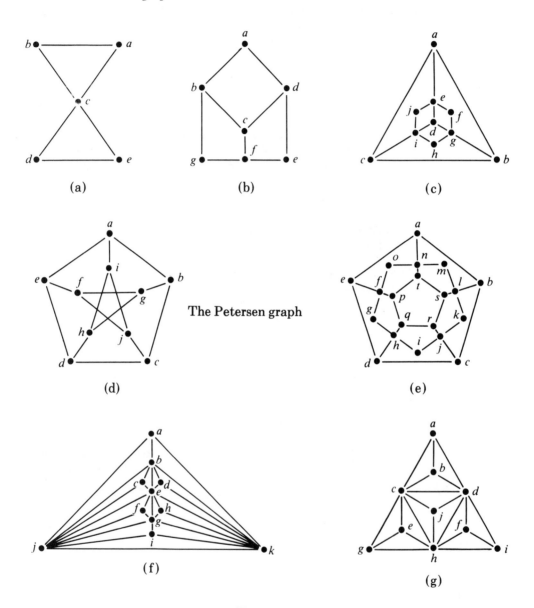

(a)   (b)   (c)

The Petersen graph

(d)   (e)

(f)

(g)

Figure 5-80

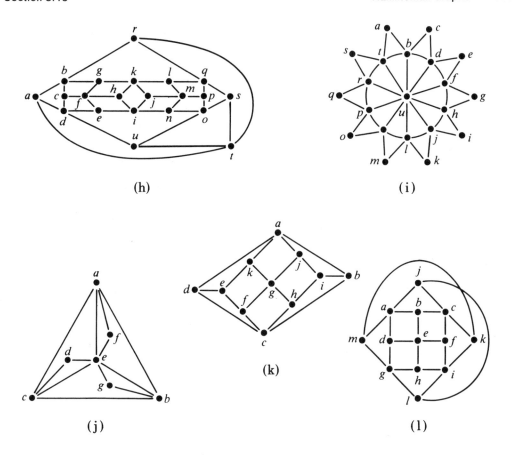

Figure 5-80. continued

3. Prove that there are no Hamiltonian paths in each of the following graphs:
   (a) See Graph (e) in Exercise 2.   (b) See Graph (h) in Exercise 2.

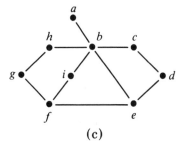

(c)

Figure 5-81

4.(a) Show that any Hamiltonian cycle in the graph H_1 in Fig. 5-82 which contains the edge $\{c,d\}$ must also contain the edge $\{g,h\}$.

(b) Show that any Hamiltonian cycle in the graph H_2 must contain exactly two of the edges $\{a,h\},\{c,d\},\{i,j\}$. Show that any Hamiltonian cycle that contains both edges $\{d,e\}$ and $\{e,j\}$ cannot also contain $\{a,h\}$.

(c) Using the results of Example 5.10.7., show that no Hamiltonian cycle in H_3 can contain both the edges $\{a,f\}$ and $\{h,n\}$.

(d) Using (c), show that every Hamiltonian cycle in graph H_4 must contain the edge e.

(e) Show that the pentagon P must lie outside any Hamiltonian cycle in the graph H_5, and that a Hamiltonian cycle in H_5 must contain exactly four of the edges of P.

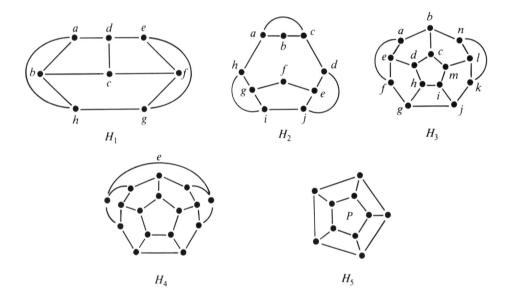

Figure 5-82

5. Use Grinberg's theorem to show that there are no planar Hamiltonian graphs with

(a) regions of degree 5, 8, and 9 with exactly one region of degree 9.

(b) regions of degree 5, 8, 9, and 11 with exactly one region with degree 9.

(c) regions of degree 4 and 5 and only one region of degree 4.

(d) regions of degree 4, 5, and 8 and only one region of degree 4.

6. If possible, use Grinberg's theorem to show that the following graphs have no Hamiltonian cycle:
   (a) graph (c) in Exercise 2.
   (b) graph (k) in Exercise 2.
   (c) graph (b) in Exercise 2.

(a)

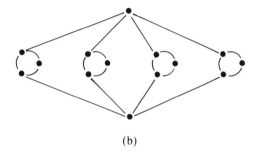

(b)

Figure 5-83

7. How many different Hamiltonian cycles are there in K_n, a complete graph on n vertices?

8. The Knight's Tour Puzzle. Consider the standard 8×8 chessboard, with squares colored alternately white and black. Suppose we place a knight on one of the 64 squares. According to the rules of chess, a knight moves by proceeding two squares vertically or horizontally from its starting square, followed by moving one square in a perpendicular direction. The Knight's Tour Puzzle asks the question: Following these rules, is it possible for the knight to tour the chessboard, visiting every square once and only once, and then return to its original square?
   (a) The question has an affirmative answer. Can you find such a tour?
   (b) Formulate the question in terms of graphs.
   (c) Investigate the Knight's Tour Puzzle for a 4×4 chessboard.
   (d) Investigate the Knight's Tour Puzzle for a 4×5 chessboard.

9. Suppose that a classroom has 25 students seated in desks in a square 5×5 array. The teacher wants to alter the seating by having every student move to an adjacent seat (just ahead, just behind, or on the left, or on the right). Show by a parity (even or odd) argument that such a move is impossible.

10. Suppose a set I of k vertices in a graph G is chosen so that no pair of vertices in I are adjacent. Then for each v in I, degree $(v) - 2$ of the edges incident on v will not be used in a Hamiltonian cycle. Summing over all vertices in I, we have $E^1 = \Sigma_{v \in I}[\text{degree }(v) - 2] =$

$\Sigma_{v \in I}$degree $(v) - 2k$ edges that cannot be used in a Hamiltonian cycle.

(a) Let V and E be the number of vertices and edges in G, respectively. Show that if $E - E^1 < V$, then G can have no Hamiltonian cycle.

(b) Why is part (a) valid only when I is a set of nonadjacent vertices?

(c) With a suitably chosen set I, use part (a) to show that the following graphs have no Hamiltonian cycles:
  (1) the graph in Exercise 2(i).
  (2) the graph G_2 in Example 5.10.2.
  (3) the graph G_4 in Example 5.10.4.

11. Prove that a directed Hamiltonian cycle of $G_{\delta+1,k}(V,E)$ corresponds to a directed Euler cycle of $G_{\delta+1,k-1}(V,E)$. Is it true that $G_{\delta+1,k}(V,E)$ always has a directed Hamiltonian cycle?

12. Characterize the class of graphs in which an Euler path is also a Hamiltonian path.

Selected Answers for Section 5.10

2. (b) By considering the vertices of degree 2 we see that the edges $\{a,b\}$, $\{a,d\}$, $\{b,g\}$, $\{g,f\}$, $\{e,f\}$, and $\{d,e\}$ must be included in a Hamiltonian cycle. But this forms a proper subcycle of G.

   (e) Consider the vertices o, g, i, k, and m. Similar to the proof of (b).

   (g) To reach a and g we must go $c - a - b - d$ or $c - b - a - d$; similarly for i,f and g,e. But pasting these subpaths together leaves no way to visit j.

   (h) Observe that there are 2 regions of degree 6 and 15 regions of degree 4. If a Hamiltonian cycle exists, then Grinberg's Theorem would give $2(r_4 - r_4^1) + (r_6 - r_6^1) = 0$ and thus that $2(r_6 - r_6^1) = r_4 - r_4^1$. But $r_4 - r_4^1 =$ even integer and $r_4 + r_4^1 = 15$ imply that $2r_4$ is odd, a contradiction.

   (l) Argument is similar to Example 5.10.4.

3. (a) A Hamiltonian path for this graph must contain 19 edges since there are 20 vertices. There are 30 edges in G, so to show that no Hamiltonian path exists we must eliminate 12 or more edges. Consider first the vertices f, h, j, l, and n. A Hamiltonian path can contain at most 2 edges incident with each of these vertices. Thus, we must eliminate a total of 10 of the edges incident with these vertices. Moreover, there must be at least one edge to the outer pentagon $a - b - c - d - e - a$ and at least one edge leading to the inner pentagon $p - q - r - s - t - p$ from the middle subgraph $f - g - h - i - j - k - l -$

$m - n - o - f$. Then observe that we must eliminate at least one of the edges from each of the outer and inner pentagons, giving a total of at least 12 edges eliminated.

4. (a) Suppose $\{g,h\}$ is not used but $\{e,d\}$ is used. Then edges $\{a,h\}$, $\{b,h\}$, $\{g,f\}$ and $\{e,g\}$ must be included in a Hamiltonian cycle. But then $\{a,b\}$ and $\{e,f\}$ must be deleted lest a triangle subcycle be formed. Thus, $\{b,c\}$ and $\{c,f\}$ must be included. But then there are three edges of the cycle incident on c. Contradiction.

   (b) If two of $\{a,h\}$, $\{c,d\}$, and $\{i,j\}$ are not used (suppose $\{c,d\}$ and $\{i,j\}$ are not used), then $\{d,e\}$, $\{e,j\}$, and $\{d,j\}$ must be used, giving a triangle subcycle; impossible. Thus, at least two of $\{a,h\}$, $\{c,d\}$, and $\{i,j\}$ must be used. However, all three cannot be used, for in that case, (a) if $\{d,j\}$ is used, then $\{d,e\}$ and $\{e,j\}$ are not used, meaning two of the three edges at e are not used; impossible. Thus (b) $\{d,j\}$ cannot be used, and then $\{d,e\}$ and $\{e,j\}$ are both included. Similarly, $\{g,h\}$,$\{g,i\}$ and $\{a,b\}$,$\{b,c\}$ are used, forming a subcycle with nine edges. Therefore, exactly two of the edges $\{a,h\}$, $\{c,d\}$, and $\{i,j\}$ can be used.

   If a Hamiltonian cycle contains both $\{d,e\}$ and $\{e,j\}$ then $\{d,j\}$ cannot be used, forcing $\{i,j\}$,$\{c,d\}$ to be used. But by the first part, exactly two of the three vertices must be used, eliminating $\{a,h\}$.

   (e) The Grinberg theorem gives $2(r_4 - r_4^1) + 3(r_5 - r_5^1) = 0$; r_4 and r_4^1 cannot be equal since $r_4 + r_4^1 = 5$. Thus, $r_4 - r_4^1 \neq 0$ and $r_5 - r_5^1 \neq 0$. Since $r_5 + r_5^1 = 2$, $r_5 - r_5^1 = \pm 2$. Thus, both pentagons must lie on the same side of any Hamiltonian cycle. The infinite region lies on the exterior of any Hamiltonian cycle, so then does P. Therefore, $r_5 - r_5^1 = -2$, giving $2(r_4 - r_4^1) - 6 = 0$ or $r_4 - r_4^1 = 3$. Since $r_4 + r_4^1 = 5$, it follows that $r_4 = 4$ and $r_4^1 = 1$. Since the edges of P all separate P from some region of degree 4, and $r_4 = 4$ we see that exactly four of the edges of P must be included in any Hamiltonian cycle.

5. (a) $3(r_5 - r_5^1) + 6(r_8 - r_8^1) + 7(r_9 - r_9^1) = 0$ and only one region of degree 9 implies that $(r_9 - r_9^1) = \pm 1$. Thus, $3(r_5 - r_5^1) + 6(r_8 - r_8^1) = \pm 7$ implies 3 divides ± 7. Contradiction.

6. (a) Observe that $r_3^1 = 1$ and $r_3 - r_3^1 + 2(r_4 - r_4^1) + 3(r_5 - r_5^1) = 0$ gives that $2(r_4 - r_4^1) + 3(r_5 - r_5^1) = 1$. Moreover any Hamiltonian cycle will separate the three regions of degree 4 into either 2 inside and 1 outside or 2 outside and 1 inside. Thus, $r_4 - r_4^1 = \pm 1$. If $r_4 - r_4^1 = 1$, then $3(r_5 - r_5^1) = -1$ and 3 will divide 1. Contradiction. If $r_4 - r_4^1 = -1$, then $3(r_5 - r_5^1) = 3$ implies $r_5 - r_5^1 = 1$, but this fact and $r_5 + r_5^1 = 3$ gives that $r_5 = 2$, no contradiction. Thus Grinberg's Theorem gives no conclusion.

(c) $2(r_4 - r_4^1) + 4(r_6 - r_6^1) = 0$ implies 2 divides $r_4 - r_4^1$. But since the external region has degree 6, $r_6 - r_4^1 = -1$ and $r_4 - r_4^1 = 2$. But then $r_4 + r_4^1 = 3$ implies $r_4 = 5/2$, a contradiction.

7. $(n - 1)!/2$

9. The number of students moving to the left = the number moving to the right; so the total number of students moving to the left or right is even. Similarly the total moving forward or backward is even. Hence, the total number of students must be even, but the total is 25.

10. (a) At most $E - E^1$ edges can be used in a Hamiltonian cycle, but a Hamiltonian cycle has V edges.

(b) The sum E^1 counts some edges twice if I is not a set of nonadjacent vertices, and hence E^1 would not be the correct bound on the number of edges that cannot be used.

(c) (1) Let $I = \{a,e,g,i,c\}$
   (2) $I = \{b,d,f,l,j,h,n\}$.
   (3) $I = \{a,c,e,g,i,l\}$.

5.11   CHROMATIC NUMBERS

The Scheduling Problem

Suppose that the state legislature has a list of 21 standing committees. Each committee is supposed to meet one hour each week. What is wanted is a weekly schedule of committee meeting times that uses as few different hours in the week as possible so as to maximize the time available for other legislative activities. The one constraint is that no legislator should be scheduled to be in two different committee meetings at the same time. The question is: What is the minimum number of hours needed for such a schedule?

First, we model this problem with a "committee" graph G_0 that has a vertex for each committee and has an edge between vertices corresponding to committees with a common member. But then we need to introduce a new graph theoretic concept.

By a **vertex coloring** of a graph G, we mean the assignment of colors (which are simply the elements of some set) to the vertices of G, one color to each vertex, so that adjacent vertices are assigned different colors. (This is nothing more than a special kind of labeling as described in Section 5.1.) An *n-coloring* of G is a coloring of G using n colors. If G has an n-coloring, then G is said to be *n-colorable*.

Figure 5-84 shows a 4-coloring as well as a 3-coloring of the graph G.

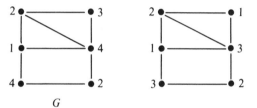

Figure 5-84

The question is: What is the minimum number of colors required? We define the chromatic number of a graph G to be the minimum number n for which there exists an n-coloring of the vertices of G. We denote the chromatic number of G by $\chi(G)$, and if $\chi(G) = k$ we say that G is k-chromatic.

In investigating the chromatic number of a graph, *we shall restrict ourselves to simple graphs*. This is reasonable since if there is a loop, then no coloring of G is possible.

Example 5.11.1. We show that $\chi(G) = 4$ for the graph of G of Figure 5-85.

Clearly the triangle abc requires three colors; assign the colors 1, 2, and 3 to a, b, and c respectively. Then since d is adjacent to a and c, d must be assigned a color different from the colors for a and c, color d the color 2. But then e must be assigned a color different from 2 since e is adjacent to d. Likewise e must be assigned a color different from 1 or 3 because e is adjacent to a and to c. Hence a fourth color must be assigned to e. Thus, the 4-coloring exhibited indicates $\chi(G) \leq 4$. But, at the same time, we have argued that $\chi(G)$ cannot be less than 4. Hence $\chi(G) = 4$.

We now return to our scheduling problem and the resulting graph G_0.

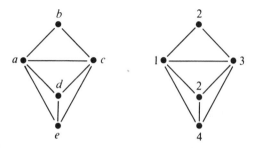

Figure 5-85

Theorem 5.11.1. The minimum number of hours for the schedule of committee meetings in our scheduling problem is $\chi(G_0)$.

Proof. Suppose $\chi(G_0) = k$ and suppose that the colors used in coloring G_0 are $1,2,\ldots,k$. First we assert that all committees can be scheduled in k one-hour time periods. In order to see this, consider all those vertices colored 1, say, and the committees corresponding to these vertices. Since no two vertices colored 1 are adjacent, no two such committees contain the same member. Hence, all these committees can be scheduled to meet at the same time. Thus, all committees corresponding to same-colored vertices can meet at the same time. Therefore, all committees can be scheduled to meet during k time periods.

Next we show that all committees cannot be scheduled in less than k hours. We prove this by contradiction. Suppose that we can schedule the committees in m one-hour time periods, where $m < k$. We can then give G_0 an m-coloring by coloring with the same color all vertices which correspond to committees meeting at the same time. To see that this is, in fact, a legitimate m-coloring of G_0, consider two adjacent vertices. These vertices correspond to two committees containing one or more common members. Hence, these committees meet at different times, and thus the vertices are colored differently. However, an m-coloring of G_0 gives a contradiction since we have $\chi(G_0) = k$. $\square$

Theorem 5.11.1 would completely solve our scheduling problem except for one unfortunate fact: Ordinarily it is extremely difficult to determine the chromatic number of a graph. For graphs with a small number of vertices, it is often not too difficult to guess the chromatic number. But to verify rigorously that the chromatic number is a given integer k, we must also show that the graph cannot be colored with $k-1$ colors. The goal is to show that any $(k-1)$-coloring that we might construct for the graph must force two adjacent vertices to have the same color.

To assist in this process we list a few rules that may be helpful; these are by no means all the rules that could have been listed. We leave the verification of some of the rules to the reader.

Rule 1. $\chi(G) \leq |V|$, where $|V|$ is the number of vertices of G.

Rule 2. A triangle always requires three colors, that is, $\chi(K_3) = 3$; more generally, $\chi(K_n) = n$, where K_n is the complete graph on n vertices.

Rule 3. If some subgraph of G requires k colors then $\chi(G) \geq k$.

Rule 4. If degree$(v) = d$, then at most d colors are required to color the vertices adjacent to v.

Rule 5. $\chi(G) = \text{maximum}\{\chi(C) \mid C \text{ is a connected component of } G\}$.

By studying the chromatic number of arbitrary graphs, it is often useful to restrict oneself to graphs which are critical in some sense. For example, in studying k-chromatic graphs in general, we often restrict our attention to graphs that are k-chromatic—but only just so in the sense that although G requires k colors, any proper subgraph of G can be colored with fewer than k colors.

More precisely we define a graph G to be k-**critical** if $\chi(G) = k$ and $\chi(G - v) < \chi(G)$ for each vertex v of G. It is easy to see that a k-chromatic graph has a k-critical subgraph.

The following properties of k-critical graphs were proven by G. A. Dirac.

Theorem 5.11.2.   Let G be a k-critical graph. Then

   (i) G is connected.
  (ii) The degree of each vertex of G is at least $k - 1$, that is, $\delta(G) \geq k - 1$.
 (iii) G cannot be expressed in the form $G_1 \cup G_2$, where G_1 and G_2 are graphs which intersect in a complete graph. In particular, G contains no cut vertices.

Proof.   (i) If G is not connected, let C be any component of G with $\chi(C) = k$, and let v be any vertex of G which is not in C. But then $\chi(G - v) \geq k$, contradicting the fact that G is k-critical.

(ii) Since G is k-critical, we have $\chi(G - v) \leq k - 1$ for each vertex v of G. If degree$(v) < k - 1$, then the neighbors of v will be colored with at most $k - 2$ colors. But then it follows that any $(k - 1)$-coloring of $G - v$ can be extended to a $(k - 1)$-coloring of G by coloring v by the color different from the colors on the neighbors of v, contradicting the fact that $\chi(G) = k$.

(iii) If $G = G_1 \cup G_2$ and $G_1 \cap G_2 = K_r$, then $\chi(G_1) \leq k - 1$ and $\chi(G_2) \leq k - 1$, since G_1 and G_2 are subgraphs of G, and G is k-critical. But by relabeling the colors, we can color the vertices of $G_1 \cap G_2$ the same way in both graphs. These two colorings can then be combined to give a $(k - 1)$-coloring of G, contradicting the fact that $\chi(G) = k$. By considering the case $r = 1$, it follows immediately that G can contain no cut vertex.   □

Rule 6.   Every k-chromatic graph has at least k vertices v such that degree$(v) \geq k - 1$.

Proof.   Let G be a k-chromatic graph and let H be a k-critical subgraph of G. By Theorem 5.11.2, each vertex of H has degree at least

$k - 1$ in H, and hence also in G. Since H is k-chromatic, H clearly has at least k vertices.   □

The next rule follows immediately from Rule 6.

Rule 7.   For any graph G, $\chi(G) \leq 1 + \Delta(G)$, where $\Delta(G)$ is the largest degree of any vertex of G.

Rule 8.   When building a k-coloring of a graph G, we may delete all vertices of degree less than k (along with their incident edges). In general, when attempting to build a k-coloring of a graph, it is desirable to start by k-coloring a complete subgraph of k vertices and then successively finding vertices adjacent to $k - 1$ different colors, thereby forcing the color choice of such vertices.

Rule 9.   These are equivalent:

(i) A graph G is 2-colorable.
(ii) G is bipartite.
(iii) Every cycle of G has even length.

Rule 10.   If $\delta(G)$ is the minimum degree of any vertex of G, then $\chi(G) \geq |V| / |V| - \delta(G)$ where $|V|$ is the number of vertices of G.

This rule follows easily from the observation that for any vertex v there are at least $\delta(G)$ neighbors of v that are colored some color different from that of v. Hence there are at most $|V| - \delta(G)$ vertices colored the same color as v. If we place vertices in the same class iff they are colored the same color, there will be $\chi(G)$ classes, and each class will contain at most $|V| - \delta(G)$ vertices. Thus, $[|V| - \delta(G)] \chi(G) \geq |V|$ since every vertex is in some class.

Example 5.11.2.   Find the chromatic number of the "wheel" graph of Figure 5-86.

Figure 5-86

Since degree $(f) = 5$ and all other vertices have degree 3, we see that $\Delta(G) = 5$ and hence that $\chi(G) \le 1 + \Delta(G) = 6$. Since there is a triangle subgraph of G, $\chi(G) \ge 3$.

We can see in at least three different ways that $\chi(G) \le 4$. First, if $\chi(G) \ge 5$, then Rule 6 would imply that there would be at least five vertices of degree greater than or equal to 4. But since this is not the case, $\chi(G) \le 4$.

Alternately, we could see Rule 8 to build a 4-coloring as follows. Delete all vertices (and incident edges) of degree less than 4. In particular, all vertices except f will be deleted. But since the remaining graph is 4-colorable we conclude that G is 4-colorable.

But while it is readily apparent that $3 \le \chi(G) \le 4$, we do not yet know exactly what the chromatic number is. Let us attempt to build a 3-coloring of G. We start by coloring the triangle a,b,f with the colors 1,2,3 respectively. Now since c is adjacent to vertices b and f of colors 2 and 3, respectively, c is forced to be colored 1, and then d is forced to be 2. However, now the adjacent vertices a and e cannot both have color 1. Thus, the graph cannot be 3-colored. On the other hand, using a fourth color for e yields a 4-coloring of G. Therefore $\chi(G) = 4$.

Exercises for Section 5.11

1. What is the chromatic number of (a) a cycle? (b) a tree?

2. What does Rule 7 indicate about the chromatic number of (a) $K_{3,3}$, (b) $K_{4,4}$, and (c) $K_{n,n}$? Determine each chromatic number.

3. A mathematics department plans to offer seven graduate courses next semester, namely combinatorics (C), group theory (G), field theory (F), numerical analysis (N), topology (T), applied mathematics (A), and real analysis (R). The mathematics graduate students and the courses they plan to take are:

Abe: C,F,T	George: A,N
Bob: C,G,R	Herman: F,G
Carol: G,N	Ingrid: C,T
DeWitt: C,F	Jim: C,R,T
Elaine: F,N	Ken: A,R
Fred: C,G	Linda: A,T

   How many time periods are needed for these 7 courses?

4. Determine the chromatic numbers of each of the following graphs. (Give a careful argument to show that fewer colors will not suffice.):

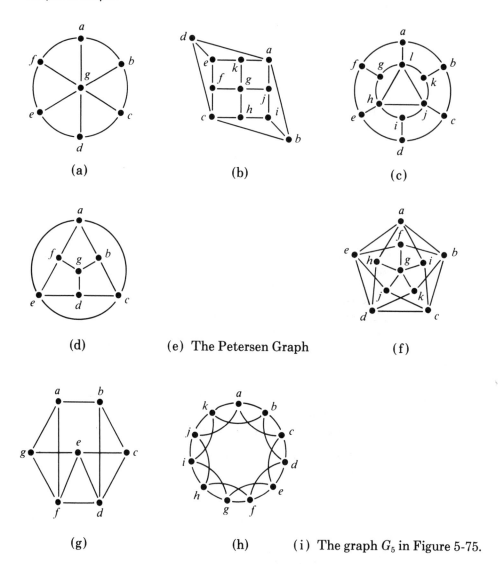

(a)

(b)

(c)

(d)

(e) The Petersen Graph

(f)

(g)

(h)

(i) The graph G_5 in Figure 5-75.

Figure 5-87

5. Instead of coloring vertices, we can color edges so that edges with
   common end points are colored different colors. The edge chromatic number of G is the minimum number of colors to color all
   the edges of G. If $\Delta(G)$ is the largest degree of the vertices of G,
   prove that $\Delta(G)$ is less than or equal to the edge chromatic number
   of G.

6. Find the edge chromatic number for the following graphs:

(a)

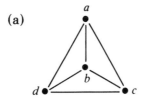

Figure 5-88

(b) $K_{3,3}$
(c) The Petersen Graph
(d) K_n

7. Show that the graphs of Exercise 4(f) and 4(g) are k-critical for some integer k.

8. Give a proof for Rule 9.

9. We give an algorithm by D. J. A. Welsh and M. B. Powell to color a graph G. First order the vertices according to decreasing degrees $d_1 \geq d_2 \geq \cdots \geq d_k$. (Such an ordering may not be unique.) Then use the first color to color the first vertex and to color, in sequential order, each vertex which is not adjacent to a previously colored vertex. Repeat the process using the second color and the remaining unpainted vertices. Continue the process with the third color, and so on until all vertices are colored. For example, in the graph in Figure 5-89, we order the vertices according to decreasing degrees: e,c, g,a,b,d,f,h. Use the first color to color e and a. Use the second color on vertices c, d, and h. Use the third color to color vertices g, b, and f. Thus, G is 3-colorable. Note that G is not 2-colorable since there is a triangle subgraph. Hence $\chi(G) = 3$. Use the Welsh-

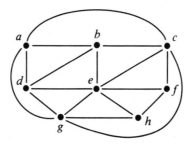

Figure 5-89

Powell algorithm to determine an upper bound to the chromatic number of the following graphs:

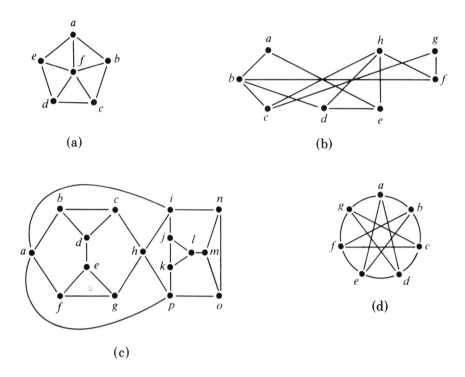

(a)

(b)

(c)

(d)

Figure 5-90

10. A local restaurant has 8 different banquet rooms. Each banquet requires some subset of these 8 rooms. Suppose that there are 12 evening banquets that are to be scheduled in a given 7-day period. Two banquets that are scheduled the same evening must use different banquet rooms. Model and restate this scheduling problem as a graph-coloring problem.

11. The organizers of a peace conference have rooms available in six local hotels. There are n participants in the conference and, because of political conflicts, various pairs of participants must be put in different hotels. The organizers wonder whether six hotels will suffice to separate all conflicts. Model this conflict problem with a graph and restate the problem in terms of vertex coloring.

12. In a round-robin tournament where each pair of n contestants plays each other, a major problem is scheduling play over a

minimal number of days (each contestant plays at most one match a day).

(a) Restate this problem as an edge-coloring problem (see Exercise 5).

(b) Solve this problem for $n = 6$.

13. Give a proof for Rule 7.

14. The greedy algorithm for vertex coloring: Order the vertices in some order, say as $v_1, v_2, \ldots, v_n$, and then color them one by one: color v_1 the color 1, then color v_2 the color 1 if v_1 and v_2 are *not* adjacent, color v_2 the color 2 otherwise. Continue this process giving each vertex the smallest numbered color it can have at that stage. This so-called *greedy algorithm* produces a coloring, but this coloring may (and usually does) use many more colors than necessary. Figure 5-91 shows a bipartite (hence, 2-colorable) graph for which the greedy algorithm wastes 4 colors.

(a) Given a graph G, show that its vertices can be ordered in such a way that the greedy algorithm uses exactly $\chi(G)$ colors. (Thus, it is not surprising that it pays to investigate the number of colors needed by the greedy algorithm in various orders of the vertices.)

(b) For each $k \geq 3$, find a bipartite graph with vertices $v_1, v_2, \ldots, v_n$ for which the greedy algorithm uses k colors. Show that this cannot be done if $n = 2k - 3$. Can it be done if $n = 2k - 2$?

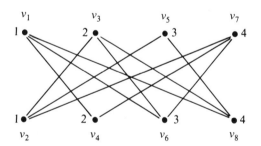

Figure 5-91

15. Let $d_1 \geq d_2 \geq d_3 \geq \ldots d_n$ be the degree sequence of G. Show that in an order $v_1, v_2, \ldots, v_n$ where $\deg(v_i) = d_i$, then the greedy algorithm uses at most max min $\{d_i + 1, i\}$ colors, and so if k is the maximal integer for which $k \leq d_k + 1$, then $\chi(G) \leq k$. (This is the basis for the Welsh-Powell algorithm.)

16. Show that any graph G has at least $C(\chi(G),2)$ edges.

17. Describe all graphs with n vertices and chromatic number n.

18. Analyze the possibilities for the chromatic number of wheel graph W_n. (See Definition 5.2.5.)

19. Prove that a wheel graph W_n, where n is odd, is a 4-critical graph.

20. Prove that any k-chromatic graph contains a k-critical subgraph.

21. (a) Show that the cycles of odd length n, where $n \geq 3$, are the only graphs which are 3-critical. Hint: Apply rule 9.
    (b) Show that K_1 is the only 1-critical graph and K_2 is the only 2-critical graph.

22. Disprove by giving a counter example: Connected nonplanar graphs with 6 vertices are not 4-colorable.

Selected Answers for Section 5.11

1. (a) The chromatic number of a cycle is either 2 or 3, depending on whether its length is even or odd.

2. (a) 4,   (b) 5,   (c) $n + 1$

3. Four time periods.

4. (a) 3
   (c) 2
   (d) 4
   (e) 3
   (f) 4
   (g) 4
   (h) 4
   (i) 2

6. (a) 3
   (b) 3
   (c) 4

8. (i) → (ii). Suppose that G is 2-colorable. Let M and N be the set of vertices colored the first color and second color respectively.
   (ii) → (iii). Suppose G is bipartite and suppose M and N form a bipartite partition of the vertices of G. If a cycle begins at a vertex $v \in M$, then it will go to a vertex in N, and back to a vertex in M then to N and so on. Hence when the cycle returns to v it must be of even length.

9. (a) Order the vertices as f,a,b,c,d,e. Color f color 1, color a, c, and e color 2, then use color 3 to color b and d. Thus $\chi(G) \leq 3$.
   (b) Order the vertices as h,a,d,f,b,c,e,g. Color h, b, and g color 1, color a and d the color 2, then use 3 to color f, c, and e. Thus $\chi(G) \leq 3$.

    (c) $\chi(G) = 4$.
    (d) $\chi(G) = 4$.

10. Let vertex = banquet, let edges join banquet vertices with a common room, and let color = day of the week. The question is: Is the graph 7-colorable?

5.12 THE FOUR-COLOR PROBLEM

Although the chromatic number of an arbitrary graph cannot be estimated at all accurately, the opposite situation holds for planar graphs. In fact, it is easy to prove that $\chi(G) \leq 5$ for planar graphs. Interest in the chromatic numbers for planar graphs came originally from problems in map coloring. One wants to color the regions of a map in such a way that no two adjacent regions (that is, regions sharing some common boundary) are of the same color. Many mathematicians thought that each map, no matter how complicated, would require no more than four colors. Whether or not this is true became known as the four-color problem. However, the problem of coloring the *regions* of a planar graph is the same as that of coloring the *vertices* of the dual graph. Hence, the original four-color problem can be reformulated in terms of chromatic numbers: Is $\chi(G) \leq 4$ for any planar graph G?

In 1976 Appel and Haken answered the four-color problem affirmatively by dividing the problem into nearly two thousand cases and then writing computer programs to analyze the various cases. The final solution required more than 1,200 hours of computer calculations.

Even though the solution of the four-color problem must be classified as a monumental achievement, some mathematicians have been dissatisfied with (and even skeptical of) the proof. Thus, this question remains: Does there exist a purely mathematical proof, unaided by computers?

Recent and past attempts at a completely mathematical proof have met with failure. The most famous failure occurred in 1879 when A. B. Kempe published a paper that purported to solve the four-color problem. For approximately ten years the problem was considered settled, but in 1890 P. J. Heawood pointed out an error in Kempe's argument. Nevertheless, Heawood was able to show, using Kempe's ideas, that every planar graph is 5-colorable. We give Heawood's proof of the following theorem.

Theorem 5.12.1. Every simple planar graph is 5-colorable.

Proof. We use induction on the number of vertices of the graph, and assume the theorem to be true for all planar graphs with at most n vertices.

Let G be a planar graph with $n + 1$ vertices. By the corollary to Euler's

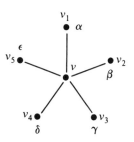

Figure 5-92

formula, G contains a vertex v whose degree is at most 5. The graph $G - v$ is a planar graph with n vertices, and so can be colored with five colors, by the inductive hypothesis. Our aim is to show how this coloring of the vertices of $G - v$ can be modified to give a coloring of the vertices of G. We may assume that v has exactly five neighbors, and that they are differently colored, since otherwise there would be at most four colors adjacent to v, leaving a spare color which would be used to color v; this would complete the coloring of the vertices of G. So the situation is now as in Figure 5-92, with the vertices $v_1, \ldots, v_5$ colored $\alpha, \beta, \gamma, \delta, \epsilon$, respectively. If λ and μ are any two colors, we define $H(\lambda, \mu)$ to be the two-colored subgraph of G induced by all those vertices colored λ or μ. We shall first consider $H(\alpha, \gamma)$; there are two possibilities:

(1) If v_1 and v_3 lie in different components of $H(\lambda, \gamma)$ (see Figure 5-93), then we can interchange the colors α and γ of all the vertices in the component of $H(\alpha, \gamma)$ containing v_1. The result of this recoloring is that v_1 and v_3 both have color γ, enabling v to be colored α. This completes the proof in this case.

(2) If v_1 and v_3 lie in the same component of $H(\alpha, \gamma)$ (see Figure 5-94), then there is a circuit C of the form $v \to v_1 \to \cdots \to v_3 \to v$, the path between v_1 and v_3 lying entirely in $H(\alpha, \gamma)$. Since v_2 lies inside

Figure 5-93

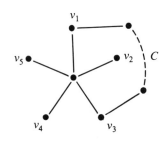

Figure 5-94

C and v_4 lies outside C, there cannot be a two-colored path from v_2 to v_4 lying entirely in $H(\beta,\delta)$. We can therefore interchange the colors of all the vertices in the component of $H(\beta,\delta)$ containing v_2. The vertices v_2 and v_4 are both now colored δ, enabling v to be colored β. This completes the proof. $\square$

The argument used in the proof of 5-color theorem (namely that of looking at a two-colored subgraph $H(\alpha,\gamma)$ and interchanging the colors) is often called a *Kempe-chain argument,* since it was initiated by A. B. Kempe in his abortive attack on the four-color problem.

Exercises for Section 5.12

1. (a) Use a Kempe-chain argument to show that a planar graph G with less than 30 edges is 4-colorable. (Hint: See exercise 5 of section 5.8.)

   (b) Explain why the regions of a planar graph with less than 30 edges can be colored with at most 4 colors.

2. Prove that every planar graph with less than 12 vertices has a vertex of degree ≤ 4. Then prove that every such graph is 4-colorable.

3. Show that a simple connected planar graph with 17 edges and 10 vertices cannot be colored with 2 colors.

4. Show that a simple connected planar graph with 8 vertices and 13 edges cannot be 2-colored.

5. Show that the regions of a simple planar graph G can be 2-colored iff each vertex of G has even degree.

6. Show that a simple connected graph with 7 vertices each of degree 4 is nonplanar. (Hint: Use Rule 9 of section 5.11 and exercise 5.)

7. If $\overline{G}$ is the complement of G, then show that:
   (a) $\chi(G) + \chi(\overline{G}) \leq |V| + 1$, where $|V|$ is the number of vertices of G.
   (b) $\chi(G)\,\chi(\overline{G}) \geq |V|$.
   (c) $\left(\dfrac{|V| + 1}{2}\right)^2 \geq \chi(G) + \chi(\overline{G}) \geq 2\sqrt{|V|}$.
   (d) $\chi(G) \geq n/n - d$ where d is minimum degree of the vertices of G.

8. A variation of the coloring problem is to color the edges of a graph so that all the edges incident on one vertex are colored distinctly. Let the line graph $L(G)$ of a graph G have a vertex for each edge of G and suppose that two of these vertices are adjacent iff the corresponding edges in G have a common endpoint. Show that a

graph G can be edge colored with k colors iff the vertices of $L(G)$ can be colored with k colors.

9. Prove that the regions of a plane graph can be 4-colored if G has a Hamiltonian cycle.

10. Suppose that G is a plane graph where each region has degree 3. Show that G is 3-colored unless $G = K_4$.

Selected Answers for Section 5.12

1. (a) There is a vertex of degree ≤ 4 by Exercise 5 of section 5.8. Follow the ideas of the proof of Theorem 5.12.1.
   (b) Take duals.

2. Apply Exercise 6 of section 5.8 to the dual of G.

3. Let G be a plane graph with 17 edges and 10 vertices. Suppose that G can be 2-colored. Then by Rule 9 of section 5.11 each cycle in G has even length. By Euler's Formula, the number of regions for G is $|E| - |V| + 2 = 9$. The problem of coloring the vertices of G is equivalent to the problem of coloring the regions of the dual of G. The dual G^* contains 9 vertices, 17 edges, and the degree of each vertex of G^* must be even. Now since G is simple each region of G has degree ≥ 3. Thus, each vertex of G^* has degree ≥ 3 and since the vertices of G^* are even, in fact, their degrees are ≥ 4. Thus in G^*, $2|E^*| \geq 4|V^*|$ or $34 \geq 36$. This contradiction shows that G cannot be 2-colored.

4. First we prove by Euler's Formula that there is a circuit of length 3. The number of regions is $|R| = |E| - |V| + 2 = 7$. If all circuits have length ≥ 4, then $2|E| \geq 4|R|$, or $|E| \geq 2|R| = 14$. But $|E| = 13$. Thus, there is a circuit of length ≤ 3. Since G is a simple graph, there are no cycles of length 1, and since G is a graph (and not a multigraph) there are no cycles of length 2. Hence there is a cycle of length exactly 3. These 3 vertices require exactly 3 colors.

5. Take the dual of G. Apply Rule 9 to G^*. Interpret this result for G.

6. If such a graph is planar, use the sum of degrees formula to conclude $|E| = 14$ and Euler's Formula to get $|R| = 9$. The degree of each region is ≥ 3 and the sum of these 9 number is 28. Hence there are 8 regions of degree 3 and 1 of degree 4. Any region of degree 3 will require 3 colors. Hence the regions of G are not 2-colorable. But this observation and the hypothesis that each vertex has even degree violates the conclusion of exercise 5.

7. (b) Place all vertices of G with the same color in the same class. Thus, there are $\chi(G)$ classes and each vertex of G is in some

class. Let k be the size of the largest class. Then all vertices in this largest class are nonadjacent since they all have the same color. Thus, in $\overline{G}$, these k vertices are adjacent, that is, there is the complete subgraph K_k inside of $\overline{G}$. Therefore, $\chi(\overline{G}) \geq k$. But then $|V| = n \leq \chi(G)k \leq \chi(G)\,\chi(\overline{G})$.

(c) Let $k = \chi(G), \overline{k} = \chi(\overline{G})$. Then from the fact that $(k - \overline{k})^2 \geq 0$ we have $(k + \overline{k})^2 \geq 4k\overline{k}$. Now $k\overline{k} \geq n$ by 7(b). Thus, $(k + \overline{k})^2 \geq 4n$ or $(k + \overline{k}) \geq 2\sqrt{n}$. By 7(a), $k + \overline{k} \leq n + 1$ so $(n + 1)^2 \geq (k + \overline{k})^2 \geq 4k\overline{k}$ or $[(n + 1)/2]^2 \geq k\overline{k}$.

10. Let C be a Hamiltonian cycle for the plane graph G. The cycle C separates the plane into the interior of C and the exterior of C. The regions determined by G that are inside C can be colored alternately, say, red and blue while those outside C can be colored alternately green and yellow.

REVIEW FOR SECTIONS 5.7–5.12

1. Is it possible to have a soccer league with 15 teams in which each team plays exactly 9 games? Model the problem with a graph or multigraph and justify your answer with a result from graph theory.

2. Fill in the blanks.
   (a) Suppose that a graph G has 2 vertices of degree 4, 4 vertices of degree 3, and 2 vertices of degree 5. Then G has _____ edges.
   (b) If a planar connected graph G has 5 regions of degree 4, 11 regions of degree 5, 2 regions of degree 7, 2 regions of degree 8, and 1 region of degree 9, then G has _____ edges and _____ vertices.
   (c) If a planar graph G has 15 vertices, 19 edges, and 8 regions, then G has _____ connected components.

3. For the following graph H,

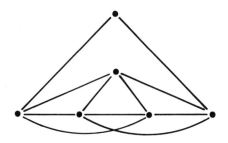

    (a) Use Euler's formula and its corollaries to prove that H is not planar.

    (b) Determine the chromatic number of H.

4. Let G be a simple graph with 116 vertices and chromatic number 5.

    (a) Prove that in any 5-coloring of G, some color is used at least 24 times.

    (b) Prove that the complement $\overline{G}$ of G contains a subgraph isomorphic to K_{24}, the complete graph on 24 vertices.

5. Prove that if G is any connected graph for which each vertex has degree at least six, then G is not planar.

6. Prove that the graph G obtained by removing 5 edges from the complete graph K_n is not planar.

7. Determine whether or not the following graph has
    (a) a Euler circuit.
    (b) a Euler path.
    (c) a Hamiltonian cycle.
    (d) a Hamiltonian path.
    (e) chromatic number 3.

8. (a) In the graph G below find a Hamiltonian cycle or show carefully that none exists.

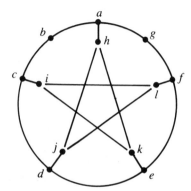

(b) Also determine $\chi(G)$.

9. Use Grinberg's theorem to determine whether or not there is a Hamiltonian cycle in the following graphs:

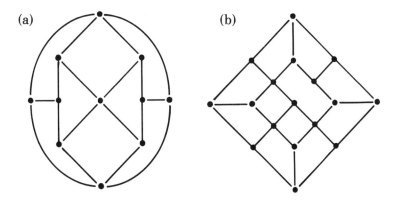

(a)   (b)

10. Prove by induction: For all integers $n \geq 2$, the directed complete simple graph on n vertices has a directed Hamiltonian path.

11. There are 30 different one-hour final examinations on one day. For this day we want a schedule of examinations that uses as few different hours as possible. The schedule must be arranged so that no student can be in different examinations at the same time. Explain how to model this problem as a graph coloring problem.

12. (a) How many different Hamiltonian cycles are there in K_{17}, the complete (simple nondirected) graph on 17 vertices?
    (b) Show that there are 8 edge-disjoint Hamiltonian cycles in K_{17}.
    (c) If 17 mathematics professors dine together at a circular table during a conference and if each night each professor sits next to a pair of different professors, how many days can the conference last? Explain how to model this problem as a problem about graphs.
    (d) Show that if n is an odd integer, then the edges of K_n can be partitioned into $(1/2)(n-1)$ edge-disjoint Hamiltonian cycles.

13. (a) Show that a group of $n \geq 3$ people can be seated around a table such that everyone will have two of his friends at his two sides if everyone knows at least half of the people in the group.
    (b) Show that the complete bipartite graph $K_{m,m}$ has a Hamiltonian cycle if $m > 1$.

14. Suppose that G is a planar graph with n vertices each of which has degree at most 5 and at least one vertex of degree 4. Use induction to prove that G is 4-colorable.

15. A coloring of a simple graph G partitions the set of vertices into disjoint subsets of vertices of the same color.
    (a) Let k be a maximal set of vertices of one color in G. Show that for the complement $\overline{G}$ the chromatic number $\chi(\overline{G}) \geq |k|$.
    (b) If G has n vertices, $\chi(G) \chi(\overline{G}) \geq n$.

16. Draw the dual of the graph in Exercise 9(a).

17. Determine the chromatic number of
    (a) a bipartite graph,
    (b) a wheel with 8 vertices W_8,
    (c) a wheel with 13 vertices W_{13},
    (d) a complete graph K_n,
    (e) a cycle graph C_n where n is odd,
    (f) a cycle graph C_n where n is even,
    (g) $K_n - e$, where e is an edge in K_n.

18. Determine what values of n are such that K_n has a Euler circuit.

19. A knight has made n moves on an 8×8 chessboard and has returned to the square from which it started. Prove that n must be an even integer. Hint: use Rule 9 of Section 5.11.

20. Show that if a simple planar connected graph has no cycles of length 3, then $|E| \leq 2|V| - 4$.

6

Boolean Algebras

6.1 INTRODUCTION

Boolean algebras are named after the English mathematician George Boole (1815–1864) who in 1854 published the magnum opus "An Investigation of the Laws of Thought." The mathematical methods, as tools to the study of logic, have made algebras one of the more interesting classes of mathematical structures when the application to the area of computer sciences is concerned. The algebras are of special significance to computer scientists because of their direct applicability to switching theory and the logical design of digital computers.

In essence a lattice that contains the elements 0 and 1, and which is both distributive and complemented is called a Boolean algebra. Since in a distributive complemented lattice the complement of every element is unique, complementation can be regarded as a bona fide operation over the domain of such a lattice.

The importance of Boolean algebras stems from the fact that many algebraic systems in both pure and applied mathematics are isomorphic to them.

Of special interest to computer scientists is the "smallest " Boolean algebra whose domain is the set of elements {0,1}. Boolean expressions generated by n Boolean variables over this specific Boolean algebra are realized by combinational networks and are used extensively in computer design. A **switching** algebra is developed in this chapter for the analysis and synthesis of such networks.

Since digital computers are built predominantly out of binary components—that is, components which assume only two possible distinct positions, various functional units in a digital computer can be viewed as combinational (or switching) networks.

The main reasons for employing such binary devices are their speed of operation, cheaper cost of manufacturing, and higher reliability compared with nonbinary devices.

The chapter begins with a detailed discussion of Boolean algebras and Boolean functions. We then develop the theory of switching mechanisms, and discuss the simplification of Boolean functions. Two specific applications of computer design are illustrated—the initial design of the arithmetic logic unit (ALU) of a digital computer and the use of multiplexers in logic design.

6.2   BOOLEAN ALGEBRAS

A **Boolean algebra** is a distributive, complemented lattice having at least two distinct elements as well as a zero element 0 and a one element 1. Namely, the Boolean algebra can be represented by the system

$$\mathcal{B} = \langle B, +, \cdot, {}^{-}, 0, 1 \rangle$$

where B is a set, $+$ and $\cdot$ are binary operations, and $^{-}$ is a unary operation (complementation) such that the following axioms hold:

1. There exist at least two elements $a, b \in B$ such that $a \neq b$.
2. $\forall\, a, b \in B$,
   (a) $a + b \in B$,
   (b) $a \cdot b \in B$.
3. $\forall\, a, b \in B$,
   (a) $a + b = b + a$,⎫
   (b) $a \cdot b = b \cdot a$.⎭   commutativity laws
4. (a) $\exists\, 0 \in B$ such that $a + 0 = a$, $\forall a \in B$.   existence of zero
   (b) $\exists\, 1 \in B$ such that $a \cdot 1 = a$, $\forall a \in B$.   existence of unit
5. $\forall a, b, c \in B$,
   (a) $a + (b \cdot c) = (a + b) \cdot (a + c)$,⎫
   (b) $a \cdot (b + c) = (a \cdot b) + (a \cdot c)$,⎭   distributivity laws
6. $\forall a \in B$, $\exists\, \bar{a} \in B$ (complement of a) such that
   (a) $a + \bar{a} = 1$ and
   (b) $a \cdot \bar{a} = 0$.   existence of complements

The associative laws

$$a + (b + c) = (a + b) + c \qquad \text{and} \qquad a \cdot (b \cdot c) = (a \cdot b) \cdot c$$

for all $a, b, c \in B$ can be derived from the above postulates. (See Theorem 6.2.9.)

When parentheses are not used, it will be implied that $\cdot$ operations are performed before $+$ operations. Also, we write ab for $a \cdot b$.

The reader may have observed a similarity between the above axioms and those of ordinary algebra. It should be noted, however, that the distributive law over addition

$$a + (b \cdot c) = (a + b) \cdot (a + c)$$

does not hold for ordinary algebra.

The simplest example of a Boolean algebra consists of only two elements, 0 and 1, defined to satisfy

$$1 + 1 = 1 \cdot 1 = 1 + 0 = 0 + 1 = 1$$
$$0 + 0 = 0 \cdot 0 = 1 \cdot 0 = 0 \cdot 1 = 0$$
$$\overline{1} = 0$$
$$\overline{0} = 1.$$

Clearly, all of the axioms of a Boolean algebra are satisfied.

It is easy to verify that the set of postulates defining the Boolean algebra above is **consistent** and **independent.** That is, none of the postulates in the set may contradict any other postulate in the set, and none of the postulates can be proved from the other postulates in the set.

Some comment should be made concerning our notation and terminology. We have adopted the notation frequently used by computer designers. The binary operation $+$ in our definition of Boolean algebra satisfies the properties:

$$0 + 0 = 0, \quad \text{and} \quad 0 + 1 = 1 + 0 = 1 + 1 = 1$$

Mathematicians, however, commonly use the symbol $+$ in Boolean algebra in the following sense:

$$0 + 0 = 0, 1 + 1 = 0 \quad \text{and} \quad 0 + 1 = 1 = 1 + 0 = 1.$$

Normally, computer designers use the symbol $\oplus$ to denote this latter binary operation of addition modulo 2, so that for them, $1 \oplus 1 = 0$, and, in general, $x_1 \oplus x_2 = x_1\overline{x}_2 + \overline{x}_1 x_2$ for each $x_1, x_2 \in B$.

Moreover, mathematicians usually use $\vee$ and $\wedge$ in place of our $+$ and $\cdot$ operations. The principal reason computer designers used the symbols $+$ and $\cdot$ was that in the past commercial line printers did not have the symbols $\vee$ and $\wedge$. It is frequently the case that the terms *sum, join,* and *disjunction* are used interchangeably in a Boolean algebra as are *product, meet,* and *conjunction.*

A partial order $\leq$ can be defined on any Boolean algebra as follows:

$$x \leq y \quad \text{iff } x \cdot y = x \quad \text{and} \quad x + y = y.$$

While in fact the entire system—the set B, called the *domain* of the Boolean algebra, the operations $+$, $\cdot$, and $-$, and the zero and unit—make up the Boolean algebra, we will frequently suppress some information and say, for example, that $\langle B, +, \cdot, - \rangle$ or that the set B is a Boolean algebra.

It is easy to see, from the above axioms, that they are arranged in pairs and that either axiom can be obtained from the other by interchanging the operations of $+$'s and $\cdot$'s, and the elements 0 and 1. This is the *principle of duality*. For example,

$$a + (b \cdot c) = (a + b) \cdot (a + c)$$
$$\downarrow \qquad\qquad \downarrow \qquad\qquad \downarrow$$
$$a \cdot (b + c) = (a \cdot b) + (a \cdot c).$$

Every theorem that can be proved for Boolean algebra has a dual which is also true.

We shall now prove some theorems necessary for the convenient manipulation of Boolean algebra. The basic tool of proof of these theorems is the use of the axioms as well as principles of substitution and duality.

Theorem 6.2.1. $\forall a \in B, a + a = a.$

Proof.

$$
\begin{array}{ll}
a + a = (a + a) \cdot 1 & \text{Axiom 4(b)} \\
\quad = (a + a) \cdot (a + \bar{a}) & \text{Axiom 6(a)} \\
\quad = a + a\bar{a} & \text{Axiom 5(a)} \\
\quad = a + 0 & \text{Axiom 6(b)} \\
\quad = a & \text{Axiom 4(a).} \quad \square
\end{array}
$$

Theorem 6.2.2. $\forall a \in B, a \cdot a = a.$

Proof.   (a) By the principle of duality

$$(b) \quad aa = aa + 0 \qquad \text{Axiom 4(a)}$$
$$= aa + a\bar{a} \qquad \text{Axiom 6(b)}$$
$$= a(a + \bar{a}) \qquad \text{Axiom 5(b)}$$
$$= a \cdot 1 \qquad \text{Axiom 6(a)}$$
$$= a \qquad \text{Axiom 4(b).} \quad \square$$

As shown here, the complete proof is obtained by the use of the dual axioms to those used in the proof of Theorem 6.2.1. Theorems 6.2.1 and 6.2.2 are known as the **idempotent laws** for Boolean algebra.

Theorem 6.2.3. The elements 0 and 1 are unique.

Proof. Suppose that there are two zero elements, 0_1 and 0_2. For each $a_1, \in B$ and $a_2 \in B$ we have, by Axiom 4(a)

$$a_1 + 0_1 = a_1 \qquad \text{and} \qquad a_2 + 0_2 = a_2.$$

Let $a_1 = 0_2$ and $a_2 = 0_1$. Thus,

$$0_2 + 0_1 = 0_2 \qquad \text{and} \qquad 0_1 + 0_2 = 0_1.$$

But by Axiom 2(a) $0_2 + 0_1 = 0_1 + 0_2$, and thus $0_1 = 0_2$.

By the use of the principle of duality, the reader can easily show that the element 1 is also unique.

As a matter of fact, we shall state the dual results of the theorem without a proof throughout the rest of this section. The proof is by the principle of duality.

Theorem 6.2.4. $\forall a \in B, a + 1 = 1$ and $a \cdot 0 = 0$.

Proof.

$$a + 1 = (a + 1) \cdot 1 \qquad \text{Axiom 4(b)}$$
$$= (a + 1) \cdot (a + \bar{a}) \qquad \text{Axiom 6(a)}$$
$$= a + 1 \cdot \bar{a} \qquad \text{Axiom 5(a)}$$
$$= a + \bar{a} \qquad \text{Axioms 4(b) and 2(b)}$$
$$= 1 \qquad \text{Axiom 6(a)}$$
$$a \cdot 0 = 0 \qquad \text{Principle of duality.} \quad \square$$

Theorem 6.2.5. The elements 0 and 1 are distinct and $\bar{1} = 0; \bar{0} = 1$.

Proof.   Let $a \in B$; namely,

$$a \cdot 1 = a \qquad \text{Axiom 4(b)}$$

and

$$a \cdot 0 = 0 \qquad \text{Theorem 6.2.4.}$$

Suppose $0 = 1$. Hence the above is satisfied only if $a = 0$. But we know that there are at least two elements in B and thus $0 \neq 1$. Clearly,

$$\overline{0} = \overline{0} + 0 = 1 \qquad \text{Axioms 4(a) and 6(a)}$$

and

$$\overline{1} = \overline{1} \cdot 1 = 0 \qquad \text{Axioms 4(b) and 6(b).} \quad \square$$

Theorem 6.2.6. $\forall a \in B$ there exists a unique complement $\bar{a}$.

Proof.   Suppose a has two complements, $\bar{a}_1$ and $\bar{a}_2$. Thus by Axiom 6 (a,b)

$$a + \bar{a}_1 = 1, a + \bar{a}_2 = 1$$
$$a \cdot \bar{a}_1 = 0, a \cdot \bar{a}_2 = 0.$$

Hence,

$$
\begin{aligned}
\bar{a}_1 &= 1 \cdot \bar{a}_1 & &\text{Axioms 2(b) and 4(b)} \\
&= (a + \bar{a}_2) \cdot \bar{a}_1 & &\text{Axiom 6(a)} \\
&= a\bar{a}_1 + \bar{a}_2\bar{a}_1 & &\text{Axiom 5(b)} \\
&= 0 + \bar{a}_2\bar{a}_1 & &\text{Axiom 6(b)} \\
&= a\bar{a}_2 + \bar{a}_1\bar{a}_2 & &\text{Axioms 6(b) and 2(b)} \\
&= (a + \bar{a}_1) \cdot \bar{a}_2 & &\text{Axiom 5(b)} \\
&= 1 \cdot \bar{a}_2 & &\text{Axiom 6(a)} \\
&= \bar{a}_2 & &\text{Axiom 4(b).} \quad \square
\end{aligned}
$$

Theorem 6.2.7. (Absorption laws.)   Let $a,b \in B$. Then

$$a + a \cdot b = a \qquad \text{and} \qquad a \cdot (a + b) = a.$$

Proof.

$$a + a \cdot b = a \cdot 1 + a \cdot b \qquad \text{Axiom 4(b)}$$
$$= a(1 + b) \qquad \text{Axiom 5(b)}$$
$$= a \cdot 1 \qquad \text{Theorem 6.2.4}$$
$$= a \qquad \text{Axiom 4(b)}$$
$$a \cdot (a + b) = a \qquad \text{Principle of duality.} \quad \Box$$

Theorem 6.2.8. (Involution law.) $\forall a \in B, \bar{\bar{a}} = a.$

Proof.   Since $\bar{\bar{a}} = \overline{(\bar{a})}$ we are searching for a complement of $\bar{a}$. However,

$$\bar{a} + a = 1 \qquad \text{and} \qquad a\bar{a} = 0.$$

So a is one complement of $\bar{a}$. By Theorem 6.2.6 the complement is unique and thus $\bar{\bar{a}} = a.$ $\Box$

Theorem 6.2.9.   A Boolean algebra is associative under the operations of $+$ and $\cdot$; namely, for all $a, b,$ and c in B,

$$a + (b + c) = (a + b) + c$$

and

$$a \cdot (b \cdot c) = (a \cdot b) \cdot c.$$

Proof.   Let

$$\alpha = [(a + b) + c] \cdot [a + (b + c)]$$
$$= [(a + b) + c] \cdot a + [(a + b) + c] \cdot (b + c)$$
$$= [(a + b) \cdot a + c \cdot a] + [(a + b) + c] \cdot (b + c)$$
$$= a + [(a + b) + c] \cdot (b + c)$$
$$= a + \{[(a + b) + c] \cdot b + [(a + b) + c] \cdot c\}$$
$$= a + (b + c).$$

But also

$$\alpha = (a + b)[a + (b + c)] + c[a + (b + c)]$$
$$= (a + b)[a + (b + c)] + c$$
$$= \{a[a + (b + c)] + b[a + (b + c)] + c\}$$
$$= (a + b) + c.$$

Thus $\qquad\qquad a + (b + c) = (a + b) + c.$

Similarly, $\qquad\qquad a \cdot (b \cdot c) = (a \cdot b) \cdot c$

by the principle of duality. $\quad\square$

Theorem 6.2.10 (DeMorgan's laws). For any $\forall a,b \in B$, $\overline{a + b} = \bar{a} \cdot \bar{b}$ and $\overline{a \cdot b} = \bar{a} + \bar{b}.$

Proof. The method of proof here is to show that

$$(a + b) + \bar{a}\bar{b} = 1 \qquad \text{and} \qquad (a + b) \cdot \bar{a}\bar{b} = 0.$$

This shows that $(a + b)$ and $\bar{a}\bar{b}$ are complements and by the Theorem 6.2.6 we establish DeMorgan's laws.

$$
\begin{aligned}
(a + b) + \bar{a}\bar{b} &= [(a + b) + \bar{a}] \cdot [(a + b) + \bar{b}] \\
&= [\bar{a} + (a + b)] \cdot [a + (b + \bar{b})] \\
&= [(\bar{a} + a) + b] \cdot [a + (b + \bar{b})] \\
&= (1 + b) \cdot (a + 1) \\
&= 1 \cdot 1 \\
&= 1.
\end{aligned}
\qquad (6.2.1)
$$

$$
\begin{aligned}
(a + b) \cdot \bar{a}\bar{b} &= a(\bar{a}\bar{b}) + b(\bar{b}\bar{a}) \\
&= (a\bar{a})\bar{b} + (b\bar{b})\bar{a} \\
&= 0 + 0 \\
&= 0.
\end{aligned}
\qquad (6.2.2)
$$

Clearly $\overline{a \cdot b} = \bar{a} + \bar{b}$ by the principle of duality.

Theorem 6.2.11. $\forall a,b \in B, a + \bar{a}b = a + b$ and $a(\bar{a} + b) = ab.$

Proof. $a + \bar{a}b = (a + \bar{a})(a + b) = 1 \cdot (a + b) = a + b.$ $a(\bar{a} + b) = ab$ by the principle of duality. $\quad\square$

Before concluding this section, we would like to make several observations. First, if $\langle B, +, \cdot, \bar{\ }, 0, 1 \rangle$ is a Boolean algebra and Q is a subset of B which is closed under the operations $+$, $\cdot$, and $\bar{\ }$, and $0, 1 \in Q$, then $\langle Q, +, \cdot, \bar{\ }, 0, 1 \rangle$ is a subalgebra of $\langle B, +, \cdot, \bar{\ }, 0, 1 \rangle$ called a **Boolean subalgebra.** A Boolean subalgebra is a Boolean algebra.

Second, we can observe that set theory is an example of Boolean algebra. Let A be any set and let $\bar{\ }$ denote the operation of set complementation relative to A. Thus $\overline{C}$ contains those elements found in the universal set but not in set C. Then $\langle \mathcal{P}(A), \cup, \cap, \overline{}, \phi, A \rangle$ is a Boolean algebra where $\mathcal{P}(A)$ is the power set of A. This is an example of a Boolean set algebra. The power set is not really necessary; it can be replaced by any collection of sets which is closed under the set operations of union, intersection, and complementation relative to some universal set.

Another example of a Boolean algebra is the set of all functions from a set U to a two-element set, say $\{0, 1\}$, where if f and g are two such functions, then $f + g$ is the function defined by $(f + g)(u) = \max \{f(u), g(u)\}$ for any $u \in U$. Likewise, $f \cdot g$ and $\overline{f}$ are defined by $(f \cdot g)(u) = \min \{f(u), g(u)\}$ and $\overline{f}(u) = 1 - f(u)$, for each $u \in U$. It need not be a tedious job to check that all the axioms of a Boolean algebra are satisfied, for each function f corresponds uniquely to a subset S_f of U, where $S_f = \{u \in U \mid f(u) = 1\}$. Then it is easy to see that $f + g, f \cdot g$, and $\overline{f}$ correspond respectively to $S_f \cup S_g, S_f \cap S_g$ and the complement of S_f in U. Because we know that the power set of U forms a Boolean algebra we know that the set of functions defined above is also a Boolean algebra. (It is a fact that the set of all functions from a set U into any Boolean algebra B is also a Boolean algebra, but we shall not verify that fact.)

The Boolean set algebra $\mathcal{P}(U)$ and the Boolean algebra of all functions from U into $\{0, 1\}$ are essentially the same in the sense of the following definition.

Definition 6.2.1. If $\langle A, +, \cdot, -, 0, 1 \rangle$ and $\langle B, \vee, \wedge, \tilde{}, 0', 1' \rangle$ are two Boolean algebras, a function $h: A \to B$ is called a *Boolean algebra homomorphism* if h preserves the two binary operations and the unary operation in the following sense: for all $a, b \in A$

(1) $h(a + b) = h(a) \vee h(b)$
(2) $h(a \cdot b) = h(a) \wedge h(b)$
(3) $h(\overline{a}) = \tilde{h}(a)$

A Boolean homomorphism $h: A \to B$ is a Boolean *isomorphism* if h is one-to-one onto B. If such an isomorphism exists, then the two Boolean algebras are said to be *isomorphic*.

If $\langle B, +, \cdot, -, 0, 1 \rangle$ is a Boolean algebra then the cartesian product B^n of n copies of B can be made into a Boolean algebra by defining

$$(b_1, b_2, \ldots, b_n) + (c_1, c_2, \ldots, c_n) = (b_1 + c_1, b_2 + c_2, \ldots, b_n + c_n),$$
$$(b_1, b_2, \ldots, b_n)(c_1, c_2, \ldots, c_n) = (b_1 c_1, b_2 c_2, \ldots, b_n c_n)$$

and

$$\overline{(b_1, b_2, \ldots, b_n)} = (\overline{b_1}, \overline{b_2}, \ldots, \overline{b_n}).$$

Call this Boolean algebra the *direct sum* of n copies of B.

It is interesting to note that there is a natural isomorphism from the Boolean set algebra $\langle \mathcal{P}(U), \cup, \cap, \bar{\ }, \phi, U \rangle$, where $U = \{u_1, u_2, \ldots, u_n\}$, and the direct sum of n copies of the Boolean algebra B_2 containing only two elements $\{0, 1\}$. This isomorphism is given by

$$h : \mathcal{P}(U) \rightarrow B_2^n$$

where

$$H(S) = (b_1, b_2, \ldots, b_n)$$

with

$$b_i = \begin{cases} 1 & \text{if } u_i \in S \\ 0 & \text{otherwise} \end{cases}$$

for $i = 1, 2, \ldots, n$. For example when $n = 4$, $h(\{u_1, u_2, u_4\}) = (1, 0, 1, 1)$. The proof that h is an isomorphism is left to the reader.

Exercises for Section 6.2

1. Let l_1 and l_2 be elements in a poset L. Prove that if l_1 and l_2 have a glb (lub) then this glb (lub) is unique.

2. Show that if a poset L has a least element (greatest element), then this least (greatest) element is unique.

3. Let $L' = \{1,2,3,4,6,8,12,24\}$ be a poset with respect to the partial ordering $|$ ("is a divisor of"). Determine the glb and lub of every pair of elements. Does the poset have a least element? A greatest element?

4. Prove that $\langle \mathbb{N}; \text{lcm}, \text{gcd} \rangle$ is a lattice under $/$, where the relation $/$ on a set of positive integers $\mathbb{N}$ represents that n_1/n_2 iff n_1 is a divisor of n_2. (lcm = least common multiple; gcd = greatest common divisor).

5. Prove that if l_1 and l_2 are elements of a lattice $\langle L; \vee, \wedge \rangle$, then

$$(l_1 \vee l_2 = l_1) \leftrightarrow (l_1 \wedge l_2 = l_2) \leftrightarrow (l_2 \leq l_1).$$

6. Let L be a poset with a least element and a greatest element. Show that L forms a lattice if for any $x_1, x_2, y_1, y_2 \in L$, where $x_i \leq y_j$, $(i, j \in \{1,2\})$, there is an element $z \in L$ such that $x_i \leq z \leq y_j$ $(i,j \in \{1,2\})$.

7. A lattice $\langle L; \vee, \wedge \rangle$ is modular if for all $l_1, l_2, l_3 \in L$,

$$(l_2 \leq l_1) \rightarrow (l_2 \vee (l_1 \wedge l_3) = l_1 \wedge (l_2 \vee l_3)).$$

Show that in a modular lattice

$$(l_2 \geq l_1) \rightarrow (l_2 \wedge (l_1 \vee l_3) = l_1 \vee (l_2 \wedge l_3)).$$

8. Let $E(Q)$ be the set of all equivalence relations on a set Q. For any $q_1, q_2 \in E(Q)$, define $q_1 \leq q_2$ appropriately, and show that $\langle E(Q); \vee, \wedge \rangle$ is a lattice under the defined operation $\leq$.

9. Show that every subsystem of a Boolean algebra is a Boolean algebra.

10. Show that in any Boolean algebra if $a \cdot x = 0$ and $a + x = 1$, then $x = \bar{a}$.

11. Show that in any Boolean algebra the following four equations are mutually equivalent:

$$a \cdot b = a, \, a + b = b, \, \bar{a} + b = 1, \, a \cdot \bar{b} = 0.$$

12. Show that a Boolean algebra B satisfies the *modular law:*

$$a + (bc) = (a + b)c \text{ for all } a, b, c, \in B \text{ where } a \leq c.$$

(Hint: see exercise 5 in section 1.4.)

13. Determine the number of elements in the Boolean algebra of all functions from a set A containing n elements to the set $\{0,1\}$.

14. Draw a diagram of the partial ordering relations between all elements of the Boolean algebra of all functions from the set $\{a,b,c,d\}$ to $\{0,1\}$.

15. (a) Let $D_{110} = \{1,2,5,10,11,22,55,110\}$ be the set of positive divisors of 110. Show that $\langle D_{110}, lcm, gcd, \bar{\ }, 1, 110 \rangle$ is a Boolean algebra where $\bar{x} = 110/x$ for any $x \in D_{110}$. The zero element is 1 and the unit is 110. What does $a \leq b$ mean in this algebra?

(b) Show that if $D_{18} = \{1,2,3,6,9,18\}$ is the set of positive divisors of 18 then $\langle D_{18}, lcm, gcd, \bar{\ }, 1, 18 \rangle$ is not a Boolean algebra for any definition of the operation $\bar{\ }$.

(c) The essential difference between the above two examples is that 110 is square-free and 18 is not. A positive integer n is square-free if m^2 divides n for positive integer m implies $m = 1$. Show that if D_n is the set of all positive divisors of a square-free integer n, then $\langle D_n, lcm, gcd, \bar{\ }, 1, n \rangle$ is a Boolean algebra where $\bar{x} = n/x$ for any $x \in D_n$.

16. Show that in any Boolean algebra B, the following hold for $a,b,c \in B$:
    (a) $a + b = a \oplus b \oplus ab$,
    (b) $a\overline{b} + b\overline{c} = a\overline{c}$ if $c \leq b \leq a$,
    (c) $a \oplus b \leq c$ if $a \leq c$ and $b \leq c$,
    (d) $a + c = b + c$ iff $a \oplus b \leq c$,
    (e) $ac = bc$ iff $c \leq a \oplus b$.

Selected Answers for Section 6.2

9. Let $\langle \tilde{B}; \vee, \wedge, ^- \rangle$ be a subsystem of the Boolean algebra $\langle B; \vee, \wedge, ^- \rangle$ and thus the commutative, associative, and distributive laws are preserved in $\langle \tilde{B}; \vee, \wedge, ^- \rangle$. If $x \in \tilde{B}$, then $\overline{x} \in \tilde{B}$; hence $x \vee \overline{x}, x \wedge \overline{x} \in \tilde{B}$. Thus, the elements 0 and 1 are included in $\tilde{B}$ and $\langle \tilde{B}; \vee, \wedge, ^- \rangle$ satisfy the identity and complement laws.

6.3 BOOLEAN FUNCTIONS

Let $\mathcal{B} = \langle B, +, \cdot, ^-, 0, 1 \rangle$ be a Boolean algebra. An element $a \in B$ is called an **atom** if $a \neq 0$ and for every $x \in B$,

$$x \cdot a = a$$

or
$$x \cdot a = 0.$$

It is easy to show that if B is a *finite* Boolean algebra, and if R is the set of all atoms in B, then B is isomorphic to

$$\langle \mathcal{P}(R), \cup, \cap, ' \rangle$$

where $'$ denotes set complementation. (See exercises 14–19 at the end of this section.)

An immediate corollary of the above is that the cardinality of B is 2 to the power of the cardinality of R. Namely, the cardinality of the domain of every finite Boolean algebra is a power of 2. Also, Boolean algebras whose domains have the same cardinality must be isomorphic (details are left as an exercise to the reader).

It is clear from the above discussion that the "smallest" Boolean algebra is

$$\mathcal{B}_2 = \langle B_2, +, \cdot, ^-, 0, 1 \rangle$$

whose domain is $\{0,1\}$, with the operations given in Table 6-1.

Table 6-1. Operations in B_2

x	$\bar{x}$	$+$	0	1	$\cdot$	0	1
0	1	0	0	1	0	0	0
1	0	1	1	1	1	0	1

In any algebraic system one may define functions mapping the algebra into itself. For Boolean algebras in general, and for the finite Boolean algebra B_2 in particular, we define Boolean expressions generated by elements of the n-tuple $\vec{x} = (x_1, x_2, \ldots, x_n)$ over B, recursively as

1. Any element of B and any of the Boolean variables in $\vec{x}$ are Boolean expressions generated by elements of $\vec{x}$ over B.
2. If β_1 and β_2 are expressions generated by elements of x over B, so are

$$\beta_1, \bar{\beta}_1, \beta_2, \bar{\beta}_2, \beta_1 + \beta_2, \beta_1 \cdot \beta_2.$$

For example, $0 \cdot \bar{1}$, $\overline{0 \cdot x_1 + \bar{x}_2 x_3}$, $x_1 + \bar{x}_1 x_2$ are Boolean expressions generated by elements of $\vec{x}$ over the Boolean algebra B_2.

If the elements of $\vec{x}$ are interpreted as Boolean variables that can assume only values in B, then the Boolean expressions represent elements in B. Thus, these expressions can be interpreted as functions of the form

$$f: B^n \rightarrow B$$

where $f(\vec{x})$, for any argument $\vec{x}$, can be determined using elements of $\vec{x}$ and the operations $+$, $\cdot$, $\bar{}$. We refer to these functions as Boolean functions of n variables over B.

For example, $x_1 + \bar{x}_2$ determines the function $f(x_1, x_2) = x_1 + \bar{x}_2$; thus, $f(0,0) = 1$, $f(0,1) = 0$, $f(1,0) = 1$, and $f(1,1) = 1$.

It is clear that different Boolean expressions may determine the same Boolean functions. For example, $x_1 \cdot (x_2 + x_3)$ and $(x_1 \cdot x_2) + (x_1 \cdot x_3)$ always determine the same Boolean functions. DeMorgan's Law, the Absorption Laws, the Distributive Laws, and other identities for Boolean algebras bring out forcibly the redundancy of Boolean expressions. One of the main objectives of this section will be to eliminate the ambiguity which would otherwise result, by developing a systematic process that will reduce every Boolean expression $f(x_1, x_2, \ldots, x_n)$ to a simple **canonical form** such that two Boolean expressions represent the same Boolean function if and only if their canonical forms are identical.

Furthermore, given any two Boolean functions f_1 and f_2 over the same n-tuple $\vec{x} \in B$, new Boolean functions can be determined through the use of the following three Boolean operations.

$$g(x_1,\ldots,x_n) = \bar{f}_1(x_1,\ldots,x_n),$$
$$h(x_1,\ldots,x_n) = f_1(x_1,\ldots,x_n) + f_2(x_1,\ldots,x_n),$$
$$k(x_1,\ldots,x_n) = f_1(x_1,\ldots,x_n) \cdot f_2(x_1,\ldots,x_n).$$

Iteration of this process a finite number of times will result in the development of a complete class of Boolean functions over n variables. Since there are 2^n elements in $\{0,1\}^n$ and 2 elements in $\{0,1\}$, there are 2^{2^n} Boolean functions of n variables over B_2.

Let F_n be the set of all Boolean functions of n variables over B_2. The system

$$\mathscr{F}_n = \langle F_n, +, \cdot, \bar{}, 0, 1 \rangle$$

is a Boolean algebra of the 2^{2^n} functions of n variables over B_2. This algebra is called the **free Boolean algebra** on n generators over B_2.

Definition 6.3.1.   A **literal** x^* is defined to be a Boolean variable x or its complement, $\bar{x}$.

Definition 6.3.2.   A Boolean expression generated by $x_1,\ldots,x_n$ over B, which has the form of a conjunction (product) of n distinct literals is called a **minterm**. It is clear that there are 2^n minterms generated by n variables in B_2.

Example 6.3.1.   The four minterms generated by the two variables, x_1 and x_2, in B_2 are

$$x_1 x_2, \ \bar{x}_1 x_2, \ x_1 \bar{x}_2, \text{ and } \bar{x}_1 \bar{x}_2.$$

Similarly, a Boolean expression of the form of a disjunction (sum) of n distinct literals is called a **maxterm** generated by $x_1,\ldots,x_n$.

The next theorem gives the canonical form for any Boolean function of n variables, but before stating and proving it, we would like to denote a minterm by

$$m_{j_1 \ldots j_n}$$

where

$$j_i = \begin{cases} 0 & \text{if} \quad x_i^* = \bar{x}_i \\ 1 & \text{if} \quad x_i^* = x_i, \quad \text{for} \quad i = 1, 2, \ldots, n. \end{cases}$$

Similarly, a maxterm is denoted by

$$M_{j_1 \ldots j_n}$$

where

$$j_i = \begin{cases} 0 & \text{if} \quad x_i^* = x_i \\ 1 & \text{if} \quad x_i^* = \bar{x}_i, \quad \text{for} \quad i = 1,2,\ldots,n. \end{cases}$$

Theorem 6.3.1. Every Boolean expression $f(x_1,\ldots,x_n)$ over B can be written in the forms

$$f(x_1,\ldots,x_n) = \sum_{k=00\ldots0}^{11\ldots1} \alpha_k m_k \qquad \text{(disjunctive normal form)},$$

$$f(x_1,\ldots,x_n) = \prod_{k=00\ldots0}^{11\ldots1} (\beta_k + M_k) \qquad \text{(conjunctive normal form)},$$

where k assumes all 2^n possible configurations $j_1 j_2 \ldots j_n$, such that $j_i \in \{0,1\}$, and where

$$\alpha_{j_1 \ldots j_n} = \beta_{j_1 \ldots j_n} = f(j_1,\ldots,j_n).$$

Proof. Using the definitions of $m_{j_1 \ldots j_n}$ and $M_{j_1 \ldots j_n}$, it is clear that $m_{j_1 \ldots j_n} = 1$ and $M_{j_1 \ldots j_n} = 0$ iff $x_i = j_i \; \forall i, i = 1,2,\ldots,n$. Hence

$$f(x_1,\ldots,x_n) = \alpha_{j_1 \ldots j_n} \cdot m_{j_1 \ldots j_n}$$

if $x_1 x_2 \cdots x_n = j_1 j_2 \cdots j_n$, and 0 otherwise; also

$$f(x_1,\ldots,x_n) = \beta_{j_1 \ldots j_n} + M_{j_1 \ldots j_n}$$

if $x_1 x_2 \cdots x_n = j_1 j_2 \cdots j_n$, and 1 otherwise. $\square$

It is left as an exercise to the reader to show that the expanded normal forms of Theorem 6.3.1 are unique.

Based on the previous discussion, we may conclude that two Boolean expressions represent the same Boolean function iff they have the same canonical forms.

The disjunctive form is a sum of minterms and each minterm is a product of literals. Accordingly, this canonical form is sometimes referred to as the *minterm form* or as a *sum-of-products form*. Likewise the conjunctive normal form is also called the *maxterm* form or a *product-of-sums* form.

Example 6.3.2. The representation of $f(x_1,x_2) = (x_1 + x_2)\,(\overline{x}_1 + \overline{x}_2)$ in disjunctive normal form is

$$f(x_1,x_2) = (0 \cdot \overline{x}_1 \cdot \overline{x}_2) + (1 \cdot \overline{x}_1 \cdot x_2) + (1 \cdot x_1 \cdot \overline{x}_2) + (0 \cdot x_1 \cdot x_2)$$
$$= \overline{x}_1 x_2 + x_1 \overline{x}_2.$$

Example 6.3.3. For the Boolean expression $f(x_1,x_2,x_3) = \overline{(\overline{x}_1 x_2)} \cdot (x_1 + x_3)$ we write the table of functional values for all possible values of the variables x_1,x_2,x_3.

x_1	x_2	x_3	f
0	0	0	0
0	0	1	1
0	1	0	0
0	1	1	0
1	0	0	1
1	0	1	1
1	1	0	1
1	1	1	1

Then the disjunctive normal form of f is:

$$f(x_1,x_2,x_3) = 0 \cdot (\overline{x}_1\overline{x}_2\overline{x}_3) + 1 \cdot (\overline{x}_1\overline{x}_2 x_3) + 0 \cdot (\overline{x}_1 x_2\overline{x}_3) + 0 \cdot (\overline{x}_1 x_2 x_3)$$
$$+ 1 \cdot (x_1\overline{x}_2\overline{x}_3) + 1 \cdot (x_1\overline{x}_2 x_3) + 1 \cdot (x_1 x_2 x_3)$$
$$= \overline{x}_1\overline{x}_2 x_3 + x_1\overline{x}_2\overline{x}_3 + x_1\overline{x}_2 x_3 + x_1 x_2\overline{x}_3 + x_1 x_2 x_3.$$

Note that the disjunctive normal form is nothing more than the sum of those five minterms which correspond to the combination of values of $x_1,x_2 x_3$ for which f takes on the value 1.

Likewise, the conjunctive normal form of f is the product of those three maxterms that correspond to those values of x_1,x_2,x_3 for which f is equal to 0. This is because if $\beta = 1$ then $\beta + M_{j_1 \ldots j_n} = 1$ and this term is effectively eliminated from the conjunctive normal form. Therefore, the conjunctive normal form of f is:

$$f(x_1,x_2,x_3) = (\overline{x}_1 + \overline{x}_2 + \overline{x}_3)(\overline{x}_1 + x_2 + \overline{x}_3)(x_1 + \overline{x}_2 + \overline{x}_3).$$

We have used the operations of $+$, $\cdot$, and $\overline{}$ on elements of B. But other operations can be defined as well. A set of operations $Q = \{\tau_1, \ldots, \tau_r\}$ on B is called **functionally complete** if for every Boolean function f there exists a form F_Q constructed from $x_1, \ldots, x_n, \tau_1, \ldots, \tau_r$ such that F_Q denotes f. The usual way to test if a given set of operations is functionally complete is to attempt to generate the sets $\{+, ^-\}$ or $\{\cdot, ^-\}$ from a given set, since it is well known that these two are functionally complete (by Theorem 6.3.1 and DeMorgan's laws).

Example 6.3.4. (a) The sheffer stroke function

$$f(x_1,x_2) = x_1 \,|\, x_2 = \overline{x_1 x_2}$$

is functionally complete since

$$x_1 \,|\, x_1 = \overline{x_1 x_1} = \overline{x}_1$$

and

$$(x_1 \,|\, x_1) \,|\, (x_2 \,|\, x_2) = \overline{x}_1 \,|\, \overline{x}_2 = \overline{\overline{x}_1 \overline{x}_2} = x_1 + x_2.$$

(b) The dagger function

$$f(x_1,x_2) = x_1 \downarrow x_2 = \overline{x_1 + x_2}$$

is functionally complete since

$$x_1 \downarrow x_1 = \overline{x_1 + x_1} = \overline{x}_1$$

and

$$(x_1 \downarrow x_1) \downarrow (x_2 \downarrow x_2) = \overline{x}_1 \downarrow \overline{x}_2 = \overline{\overline{x}_1 + \overline{x}_2} = x_1 x_2.$$

Exercises for Section 6.3

1. The following is a Boolean expression generated by x,y over the Boolean algebra $\langle B; -, \vee, \wedge, ^- \rangle$, where $B = \{0, \alpha, \beta, 1\}$: $f(x,y) = (\overline{x} \wedge \overline{y}) \vee (x \wedge (\alpha \vee y))$. Tabulate the values of $f(x,y)$ for all arguments $(x,y) \in B^2$.

2. The minterm normal form of a Boolean expression generated by x,y over the Boolean algebra described in Exercise 1 above is given by

$$f(x,y) = (\alpha \wedge \overline{x} \wedge \overline{y}) \vee (\beta \wedge \overline{x} \wedge y) \vee (0 \wedge x \wedge \overline{y}) \vee (1 \wedge x \wedge y).$$

Give the maxterm normal form of f.

3. Enumerate all 16 Boolean functions of one variable over the Boolean algebra described in Exercise 1 above.

4. Which of the following statements are always true? Justify your answer.
   (a) If $x(y + \bar{z}) = x(y + \bar{w})$, then $z = w$.
   (b) If $z = w$, then $x(y + \bar{z}) = x(y + \bar{w})$.

5. Obtain the sum-of-products canonical forms of the following Boolean expressions:
   (a) $x_1 \oplus x_2$

   (b) $\overline{(x_1 \oplus x_2)} \oplus (\bar{x}_1 \oplus x_3)$

   (c) $x_1\bar{x}_2 + x_3$

   (d) $\bar{x}_1 + [(x_2 + \bar{x}_3)(\overline{x_2 x_3})] \, (x_1 + x_2\bar{x}_3)$

6. If $\alpha(x_1,x_2,\ldots,x_n)$ is the dual of $\beta(x_1,x_2,\ldots,x_n)$, then show that $\bar{\beta}(x_1,x_2,\ldots,x_n) = \alpha(\bar{x}_1,\bar{x}_2,\ldots,\bar{x}_n)$.

7. Let B be a Boolean algebra with 2^n elements. Show that the number of sub-Boolean algebras of B is equal to the number of partitions of a set with n elements.

8. Write the dual of each Boolean equation:
   (a) $x + \bar{x}y = x + y$
   (b) $(x \cdot 1)(0 + \bar{x}) = 0$

9. Let $f(x,y,z) = x\bar{y} + xy\bar{z} + \bar{x}y\bar{z}$. Show that
   (a) $f(x,y,z) + x\bar{z} = f(x,y,z)$
   (b) $f(x,y,z) + x \neq f(x,y,z)$
   (c) $f(x,y,z) + \bar{z} \neq f(x,y,z)$

10. Let B be the set of all functions from $\{a,b,c\}$ to $\{0,1\}$. Determine the set of atoms of B. For each function f, determine the subset of atoms $a \leq f$.

11. Prove that in any Boolean algebra if $a \geq z$ and $a \neq z$, then $a\bar{z} \geq c$ for some atom c.

12. (a) Show that $\{2,5,11\}$ are the only atoms in the Boolean algebra D_{110} described in exercise 15 of section 6.2.
    (b) Find all atoms in D_n when n is a square-free integer.
    (c) Determine the number of Boolean subalgebras of D_{110}.

13. Determine that the set of atoms in the Boolean algebra of all subsets of a set A is just the singleton subsets of A.

14. Prove that if a_1 and a_2 are two atoms in a Boolean algebra such that $a_1 a_2 \neq 0$, then $a_1 = a_2$.

15. Prove that if $a_1 + a_2 + \cdots + a_n \geq b$, where $b,a_1,a_2,\ldots,a_n$ are all atoms of a Boolean algebra B, then $a_i = b$ for some i.

16. Prove that in every finite Boolean algebra, the sum of all atoms is 1.

17. Suppose that b is a nonzero element of a finite Boolean algebra B. Suppose that a is an atom of B. Then precisely one of the following hold:

$$a \leq b \quad \text{or} \quad a \leq \overline{b}.$$

18. Prove that every nonzero element c of a finite Boolean algebra B is the sum of precisely all the atoms a such that $a \leq c$. Moreover, no other sum of atoms is equal to c.

19. Let B be a finite Boolean algebra and let R be the set of atoms of B. Define $h : B \to \mathcal{P}(R)$ as follows:

$$h(x) = \begin{cases} \phi & \text{if } x = 0 \\ \{a \in R \,|\, a \leq x\} & \text{if } x \neq 0. \end{cases}$$

Show that h is a Boolean algebra isomorphism.

20. Let B be a Boolean algebra. An element $a \in B$ is said to be *minimal* if $a \neq 0$ and if, for every $x \in B$, $x \leq a$ implies $x = a$ or $x = 0$. Show that a is an atom iff a is minimal.

Selected Answers for Section 6.3

5.   (a) $x_1 \oplus x_2 = x_1\overline{x}_2 + \overline{x}_1 x_2$
    (c) $x_1\overline{x}_2 x_3 = x_1\overline{x}_2(x_3 + \overline{x}_3) + (x_1 + \overline{x}_1)(x_2 + \overline{x}_2)x_3 =$

$$= x_1\overline{x}_2 x_3 + x_1\overline{x}_2\overline{x}_3 + x_1 x_2 x_3 + \overline{x}_1 x_2 x_3 + \overline{x}_1\overline{x}_2 x_3$$

9.   (c) $f(x,y,z) + \overline{z} = x\overline{y} + xy\overline{z} + \overline{x}y\overline{z} + \overline{z} = x\overline{y} + \overline{z} \neq f(x,y,z)$

6.4   SWITCHING MECHANISMS

In the first three sections, we discussed Boolean algebras in general, and specifically the Boolean algebra of all subsets of a given set and the Boolean algebra of all functions into a two-element set. Boolean algebras are very important in theoretical considerations, but by far the most important application lies in the realm of electrical engineering and computer design. Such an application is not surprising as digital signals, mechanical switches, diodes, magnetic dipoles, and transistors are all two-state devices. These two states may be realized as current or no current, magnetized or not magnetized, high potential or low potential, and closed or open. It is easy to see that one may have a one-to-one correspondence between Boolean variables and digital signals where 0 and 1 represent the two states.

Accordingly, various functional units in a digital computer can be viewed as *switching mechanisms* (*combinational circuits* or *logic networks*) which accept a collection of inputs and generate a collection of outputs. Each input and output is "binary" in the sense that it is capable of assuming only two distinct values, which are designated 0 and 1 for convenience. More formally, a switching mechanism with n inputs and m outputs is a realization of a function $f : B_2^n \to B_2^m$ where if $(z_1, z_2, \ldots, z_m) = f(x_1, x_2, \ldots, x_n)$, then $x_1, x_2, \ldots, x_n$ are the n inputs and $z_1, z_2, \ldots, z_m$ are the m outputs. The function f is called a *switching function,* and usually f (and, hence, the switching mechanism realizing f) is specified by a *truth table* where each row of the table lists the output $(z_1, z_2, \ldots, z_m)$ for one of the 2^n inputs $(x_1, x_2, \ldots, x_n)$. Two switching mechanisms are said to be *equivalent* if they have the same truth table, that is, if they are the realizations of the same switching function. A *gate* is a switching mechanism with only one output; thus, a gate realizes a Boolean function $f : B_2^n \to B_2$.

In this section we develop an assortment of switching mechanization techniques using some specific design examples.

The first step in the design of a switching mechanism is to define the problem concisely. This is done by translating the general description of the problem into either logic equation or truth table form. As a specific example consider a common output device used to display decimal numbers. This device is known as the seven-segment display (SSD) shown in Figure 6-1(a). The seven segments are labeled with standard letters from a through g. The 10 displays, representing decimal digits 0 through 9, are shown in Figure 6-1(b).

Let us examine the problem of designing a logic system that will turn on the correct segments in response to binary coded decimal (BCD) input. The four input wires indicate the number to be displayed by the binary pattern of 1's and 0's on them. The seven outputs of the logic system must turn on in the proper pattern to display the desired digit.

From the word description and the pictured displays in Figure 6-1(b) we can tabulate the desired outputs for each valid combination of inputs. Each row on the table represents a different number displayed, and each column represents an input or output signal. Since the input codes are BCD representations of the digit displayed, the four input columns are generated by writing in the BCD codes for each of the decimal digits. The output columns are then filled in by inspecting the displays and filling in

(a)

Input codes　0　1　2　3　4　5　6　7　8　9

(b)

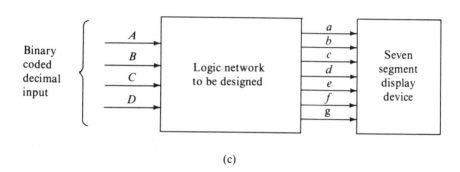

(c)

Decimal displayed	Inputs				Outputs						
	A	B	C	D	a	b	c	d	e	f	g
0	0	0	0	0	1	1	1	1	1	1	0
1	0	0	0	1	0	1	1	0	0	0	0
2	0	0	1	0	1	1	0	1	1	0	1
3	0	0	1	1	1	1	1	1	0	0	1
4	0	1	0	0	0	1	1	0	0	1	1
5	0	1	0	1	1	0	1	1	0	1	1
6	0	1	1	0	0	0	1	1	1	1	1
7	0	1	1	1	1	1	1	0	0	0	0
8	1	0	0	0	1	1	1	1	1	1	1
9	1	0	0	1	1	1	1	0	0	1	1

(d)

Figure 6-1. (a) Segment identification, (b) desired displays, (c) logic problem, (d) truth table.

a 1 if the segment is shown or a 0 if it is not. For example, the 7 requires only segments a, b, and c, so we write 1110000 in row 7. We can thus fill out the entire table from the word and picture description of the problem.

We have thus translated the problem statement into the concise, standardized language of the truth table from the generalized requirement of driving a seven-segment display.

Actually the truth table of Figure 6-1 is incomplete. Since there are 16 possible combinations of four inputs, we could have a more complete truth table by generating all 16 rows and writing X's in the output columns in the last six rows to indicate that we "don't care" what the outputs are for these rows. Their omission from the table, however, indicates the same thing as "don't care." The truth table is easy to generate and very effective for showing several switching functions at once.

Another very useful way of expressing logic functions is the logic equation. A logic equation expresses only one function, so that, for example, a separate logic equation could be written for each output function (a–g) in the truth table of Figure 6-1. Since logic equations are very similar in form to discrete gating structures, they are very useful when the switching is to be mechanized with logic gates. The logical symbols $-$, $+$, and $\cdot$, respectively, are used to indicate "not," "or," and "and." Looking at Figure 6-1 we see that segment d of our seven-segment display must be on when displaying the digits 0, or 2, or 3, or 5, or 6, or 8. Using the $+$ symbol in place of or, we can write the logic equation for the variable d as

$$d = 0 + 2 + 3 + 5 + 6 + 8.$$

This is an equation in terms of the displayed digit, but we really want an equation in terms of the input variables, A, B, C, and D. The input code for 0 is "not-A and not-B and not-C and not-D"; we can therefore write an equation:

$$\text{digit } 0 = \overline{A} \cdot \overline{B} \cdot \overline{C} \cdot \overline{D}.$$

Likewise

$$\text{digit } 2 = \overline{A} \cdot \overline{B} \cdot C \cdot \overline{D},$$
$$\text{digit } 3 = \overline{A} \cdot \overline{B} \cdot C \cdot D,$$
$$\text{digit } 5 = \overline{A} \cdot B \cdot \overline{C} \cdot D,$$
$$\text{digit } 6 = \overline{A} \cdot B \cdot C \cdot \overline{D},$$

and

$$\text{digit } 8 = A \cdot \overline{B} \cdot \overline{C} \cdot \overline{D}.$$

We can now substitute these expressions for the digits in terms of inputs into the original equation for d as follows:

$$d = \overline{A} \cdot \overline{B} \cdot \overline{C} \cdot \overline{D} + \overline{A} \cdot \overline{B} \cdot C \cdot \overline{D} + \overline{A} \cdot \overline{B} \cdot C \cdot D + \overline{A} \cdot B \cdot \overline{C} \cdot D$$
$$+ \overline{A} \cdot B \cdot C \cdot \overline{D} + A \cdot \overline{B} \cdot \overline{C} \cdot \overline{D}.$$

We thus have a logic equation for output d as a function of the inputs A, B, C, and D. We can write similar equations for each of the other outputs as follows:

$$a = \overline{A} \cdot \overline{B} \cdot \overline{C} \cdot \overline{D} + \overline{A} \cdot \overline{B} \cdot C \cdot \overline{D} + \overline{A} \cdot \overline{B} \cdot C \cdot D + \overline{A} \cdot B \cdot \overline{C} \cdot D$$
$$+ \overline{A} \cdot B \cdot C \cdot D + A \cdot \overline{B} \cdot \overline{C} \cdot \overline{D} + A \cdot \overline{B} \cdot \overline{C} \cdot D$$

$$b = \overline{A} \cdot \overline{B} \cdot \overline{C} \cdot \overline{D} + \overline{A} \cdot \overline{B} \cdot \overline{C} \cdot D + \overline{A} \cdot \overline{B} \cdot C \cdot \overline{D} + \overline{A} \cdot \overline{B} \cdot C \cdot D$$
$$+ \overline{A} \cdot B \cdot \overline{C} \cdot \overline{D} + \overline{A} \cdot B \cdot C \cdot D + A \cdot \overline{B} \cdot \overline{C} \cdot \overline{D}$$
$$+ \overline{A} \cdot \overline{B} \cdot \overline{C} \cdot D$$

$$c = \overline{A} \cdot \overline{B} \cdot \overline{C} \cdot \overline{D} + \overline{A} \cdot \overline{B} \cdot \overline{C} \cdot D + \overline{A} \cdot \overline{B} \cdot C \cdot D + \overline{A} \cdot B \cdot \overline{C} \cdot \overline{D}$$
$$+ \overline{A} \cdot B \cdot \overline{C} \cdot D + \overline{A} \cdot B \cdot C \cdot \overline{D} + \overline{A} \cdot B \cdot C \cdot D$$
$$+ A \cdot \overline{B} \cdot \overline{C} \cdot \overline{D} + A \cdot \overline{B} \cdot \overline{C} \cdot D$$

$$e = \overline{A} \cdot \overline{B} \cdot \overline{C} \cdot \overline{D} + \overline{A} \cdot \overline{B} \cdot C \cdot \overline{D} + \overline{A} \cdot B \cdot C \cdot \overline{D} + A \cdot \overline{B} \cdot \overline{C} \cdot \overline{D}$$

$$f = \overline{A} \cdot \overline{B} \cdot \overline{C} \cdot \overline{D} + \overline{A} \cdot B \cdot \overline{C} \cdot \overline{D} + \overline{A} \cdot B \cdot \overline{C} \cdot D + \overline{A} \cdot B \cdot C \cdot \overline{D}$$
$$+ A \cdot \overline{B} \cdot \overline{C} \cdot \overline{D} + A \cdot \overline{B} \cdot \overline{C} \cdot D$$

$$g = \overline{A} \cdot \overline{B} \cdot \overline{C} \cdot \overline{D} + \overline{A} \cdot \overline{B} \cdot C \cdot D + \overline{A} \cdot B \cdot \overline{C} \cdot \overline{D} + \overline{A} \cdot B \cdot \overline{C} \cdot D$$
$$+ \overline{A} \cdot B \cdot C \cdot \overline{D} + A \cdot \overline{B} \cdot \overline{C} \cdot \overline{D}.$$

These equations for the seven-segment display are presented in disjunctive normal form. There are $2^4 = 16$ minterms of four variables. For example:

$$m_0 = \overline{A} \cdot \overline{B} \cdot \overline{C} \cdot \overline{D},$$
$$m_1 = \overline{A} \cdot \overline{B} \cdot \overline{C} \cdot D,$$
$$m_2 = \overline{A} \cdot \overline{B} \cdot C \cdot \overline{D},$$
$$m_5 = \overline{A} \cdot B \cdot \overline{C} \cdot D,$$
$$m_{14} = A \cdot B \cdot C \cdot \overline{D},$$
$$m_{15} = A \cdot B \cdot C \cdot D.$$

These numbered minterms (note that the numbers correspond to the binary weighted value) are simply a useful shorthand for writing functions in canonical form. For **example,**

$$a = m_0 + m_2 + m_3 + m_5 + m_7 + m_8 + m_9$$
$$= \Sigma m \ (0,2,3,5,7,8,9).$$

Since the minterm numbers directly correspond to rows on the truth table and displayed digits, this form of writing the equation is not only shorter but also easier to write directly from Figure 6-1.

The other canonical form, the conjunctive form, uses the maxterms shown below:

$$M_0 \ = \overline{A} + \overline{B} + \overline{C} + \overline{D},$$
$$M_1 \ = \overline{A} + \overline{B} + \overline{C} + D,$$
$$\vdots$$
$$M_{15} = A + B + C + D.$$

Clearly, each maxterm is true for all but one combination of the variables, and any function can be written as a product of maxterms. For example, $a = M_9 \cdot M_{11} \cdot M_{14} = \Pi \ M(9,11,14)$.

It is customary to classify the gate complexity of an integrated circuit (IC) in one of the three following categories.

1. A SSI (small scale integration) device has a complexity of less than 10 gates. These are ICs that contain several gates or flip-flops in one package.

2. A MSI (medium scale integration) device has a complexity of 10 to 100 gates. These are ICs that provide elementary logic functions such as registers, counters, and decoders.

3. A LSI (large scale integration) device has a complexity of 100 to 10,000 gates. Examples of LSI ICs are large memories, microprocessors, and calculator chips.

4. A VLSI (very large scale integration) device has a complexity of more than 10,000 gates. Right now a piece of silicon about half a centimeter square could contain over 100,000 gates. LSI is now giving way to VLSI in all phases of design including larger memories and microcomputing devices.

Though MSI, LSI, and VLSI have made it possible to mechanize logic functions much more economically than the method of connecting discrete gates, much logic must still be done at the gate level—even in the most sophisticated designs. Though the discrete gate part of the system may perform the minority of the logic "work" in the system, it represents a large part of the computer scientist's detail design and checkout work. *It is therefore important that computer science students fully master the techniques of gate minimization.*

Logic gates perform functions that can be described both by switching equations and by truth tables.

The most common pictorial representation of logic circuits is the **block diagram,** in which the logic elements are represented by standard symbols. The standard logic symbols by IEEE Standard No. 91 (ANSI Y 32.14, 1973) are shown in Figures 6-2 to 6-7.

The reader will note that there are two types of symbols, the **uniform-shape** symbols and the **distinctive-shape** symbols. The uniform-shape symbols are those established by the International Electrotechnical Commission (IEC Publication 117-15) and are widely used in Europe. The IEEE has included these symbols in its standard, but the distinctive-shape symbols remain the standard of preference in the United States and have found wide acceptance in other parts of the world. The distinctive-shape symbols will be used here.

Input		Output
A	B	F
0	0	0
0	1	0
1	0	0
1	1	1

Figure 6-2. AND function.

Input		Output
A	B	F
0	0	0
0	1	1
1	0	1
1	1	1

Figure 6-3. INCLUSIVE OR function.

Input		Output
A	B	F
0	0	1
0	1	1
1	0	1
1	1	0

Figure 6-4. NAND function.

Input		Output
A	B	F
0	0	1
0	1	0
1	0	0
1	1	0

Figure 6-5. NOR function.

Input	Output
A	F
0	1
1	0

Figure 6-6. Inverter.

Input		Output
A	B	F
0	0	0
0	1	1
1	0	1
1	1	0

Figure 6-7. EXCLUSIVE-OR.

Figure 6-2 represents the AND function. The AND output is high iff all the inputs are high (1).

The symbol shown in Figure 6-3 represents the INCLUSIVE OR function. The OR output is low (0) iff all inputs are low.

The NAND symbol characterizes a function whose output is low (0) iff all inputs are high (1) (Figure 6-4).

The NOR symbol shown in Figure 6-5 characterizes a function whose output is high iff all inputs are low.

The INVERTER, shown in Figure 6-6, is a device that provides the complement.

EXCLUSIVE-OR, illustrated in Figure 6-7, characterizes an even-odd recognizer.

Though all the gates are shown with two inputs, they can have as many inputs as desired. Package pin limitations make it fairly standard to package together *four two-input gates, three three-input gates, two four-input gates, or one eight-input gate per package.*

The one-to-one relationship between the gate structure and the logic equation makes it easy to see why logic equations are so useful in gating design.

Exercises for Section 6.4

1. A majority function is a digital circuit whose output is 1 iff the majority of the inputs are 1. The output is 0 otherwise. Obtain the truth table of a three-input majority function and show that the circuit of a majority function can be obtained with 4 NAND gates.

2. Two digital functions, f_1 and f_2, are used as control mechanisms:

$$f_1 = xyT_1 + \overline{x}\,\overline{y}T_2,$$
$$f_2 = xT_1 + \overline{y}T_2.$$

Under what conditions of input variables x and y and timing variables T_1 and T_2, will the two digital functions be 1 at the same time?

3. Design a combinational circuit that accepts a 3-bit number and generates an output binary number equal to the square of the input number.

4. Two 2-bit numbers $A = a_1 a_0$ and $B = b_1 b_0$ are to be compared by a four-variable function $f(a_1, a_0, b_1, b_0)$. The function f is to have the value 1 whenever $\alpha(A) \leq \alpha(B)$, where $\alpha(X) = x_1 \times 2^1 + x_0 \times 2^0$ for a 2-bit number. Assume that the variables A and B are such that $|\alpha(A) - \alpha(B)| \leq 2$. Design a combinational system to implement f using as few gates as possible.

5. A number code where consecutive numbers are represented by binary patterns that differ in one bit position only is called a Gray code. A truth table for a 3-bit Gray-code to binary-code converter is shown.

3-bit Gray Code Inputs			Binary Code Outputs		
a	b	c	f_1	f_2	f_3
0	0	0	0	0	0
0	0	1	0	0	1
0	1	1	0	1	0
0	1	0	0	1	1
1	1	0	1	0	0
1	1	1	1	0	1
1	0	1	1	1	0
1	0	0	1	1	1

Implement the three functions f_1, f_2, f_3 using only NAND gates.

Selected Answers for Section 6.4.

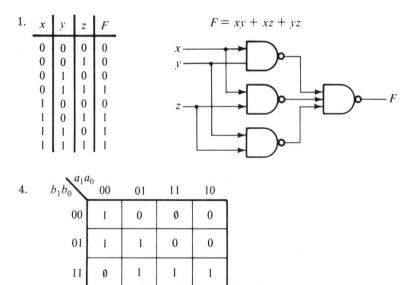

1.

x	y	z	F
0	0	0	0
0	0	1	0
0	1	0	0
0	1	1	1
1	0	0	0
1	0	1	1
1	1	0	1
1	1	1	1

$$F = xy + xz + yz$$

4.

$$f_{\min} = \bar{a}_1 \bar{a}_0 + \bar{b}_1 \bar{b}_0 + \bar{a}_0 \bar{b}_1 + \bar{a}_1 \bar{b}_0 + \bar{a}_1 \bar{b}_1$$
where x means 0 or 1.

6.5 MINIMIZATION OF BOOLEAN FUNCTIONS

Our aim in minimizing a switching function f is to find an expression g which is equivalent to f and which minimizes some cost criteria. The most popular criteria to determine minimal cost is to find an expression with a minimal number of terms in a sum-of-product expression, provided there is no other such expression with the same number of terms and with fewer literals. (A literal is a variable in complemented or uncomplemented form.)

Consider the minimization of the function:

$$f(x_1, x_2, x_3) = \bar{x}_1 x_2 \bar{x}_3 + \bar{x}_1 \bar{x}_2 \bar{x}_3 + x_1 \bar{x}_2 \bar{x}_3 + x_1 x_2 x_3 + x_1 \bar{x}_2 x_3.$$

The combination of the first and second product terms yields

$$\bar{x}_1 x_2 \bar{x}_3 + \bar{x}_1 \bar{x}_2 \bar{x}_3 = \bar{x}_1 \bar{x}_3 (x_2 + \bar{x}_2) = \bar{x}_1 \bar{x}_3.$$

Similarly, the combination of the second and the third terms yields

$$\bar{x}_1 \bar{x}_2 \bar{x}_3 + x_1 \bar{x}_2 \bar{x}_3 = \bar{x}_2 \bar{x}_3 (x_1 + \bar{x}_1) = \bar{x}_2 \bar{x}_3,$$

and the combination of the fourth and fifth terms yields

$$x_1 x_2 x_3 + x_1 \bar{x}_2 x_3 = x_1 x_3 (x_2 + \bar{x}_2) = x_1 x_3.$$

Thus the reduced expression is

$$f(x_1, x_2, x_3) = \bar{x}_1 \bar{x}_3 + \bar{x}_2 \bar{x}_3 + x_1 x_3.$$

This expression is in **irredundant** (or irreducible) form, in the sense that any attempt to reduce it any further by eliminating any term or any literal in any term will yield an expression which is not equivalent to f.

The above reduction procedure is not unique. In fact, if we combine the first and second terms of f, the third and fifth, and fourth and fifth, we obtain the expression

$$f(x_1, x_2, x_3) = \bar{x}_1 \bar{x}_3 + x_1 \bar{x}_2 + x_1 x_3,$$

which represents a different irredundant expression of f.

The algebraic procedure of combining various terms and applying to them the rules of Boolean algebra is quite tedious. The map technique presented here provides a systematic method for minimization of switching functions.

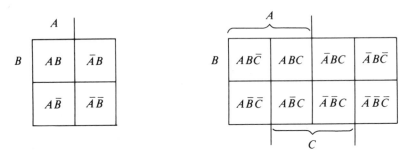

Figure 6-8. Veitch diagrams.

The Veitch diagram, developed by E. W. Veitch in 1952, is a refinement of the Venn diagram in that circles are replaced by squares and arranged in the form of a matrix. Figure 6-8 illustrates a Veitch diagram for two and three variables.

By the Veitch diagram we represent graphically the various combinations in such a manner that minimization is simplified. An inspection of the various cells in the matrices reveals that there is only a one-variable change between any two adjacent cells. Each cell is identified by a minterm.

Clearly minterms that can be combined are adjacent. Figure 6-10(a) shows a Veitch diagram for functions of four variables. The numbers inside the squares indicate the minterm number represented by that square. The brackets labeled A, B, C, and D indicate the *regions where the indicated variables are true;* therefore $C = 0$ in all the squares in the top half and $B = 1$ in the squares in the bottom half, for example.

Any function of four variables can be represented by simply filling in 1's and 0's to indicate the function, as with a truth table. For example, Figure 6-10(b) represents the function

$$f = \overline{A} \cdot \overline{B} \cdot \overline{C} \cdot D + A \cdot B \cdot C \cdot D.$$

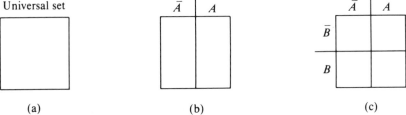

Figure 6-9. Development of Karnaugh maps by Venn diagram approach.

An improvement to the above idea was represented by M. Karnaugh in 1953, who rearranged the alphabetical assignments on the map.

Since there must be a square for each minterm, we can begin with a universal set and divide it into 2^n squares by using the Veitch diagram approach as seen in Figure 6-9.

The decimal representation of the minterms is given in Figure 6-10(a).

The Karnaugh map technique is thought to be the most valuable tool available for dealing with Boolean functions. It provides instant recognition of basic patterns, can be used to obtain all possible combinations and minimal terms, and is easily applied to all varieties of complex problems.

Minimization with the map is accomplished through recognition of basic patterns. The appearance of 1's in adjacent cells immediately identifies the presence of a redundant variable.

Figure 6-11 illustrates the grouping of one, two, and four cells on a four-variable map. It takes all four variables to define a single cell of a four-variable map: the grouping of two cells eliminates one variable, the grouping of four cells eliminates two variables, and the grouping of eight cells eliminates three variables.

Any given grouping of 1's in the Karnaugh map is identified by a product term. That is, the single cell in Figure 6-11(a) is defined as $ABCD$. The grouping of two cells, as in Figure 6-11(b), is identified by $AB\overline{D}$. Minimization involves the gathering of the various groups in the most efficient manner, where the variables are arranged in a "ring" pattern of symmetry, so that these squares would be adjacent if the map were inscribed on a torus (a doughnut-shaped form). If you have difficulty visualizing the map on a torus, just remember that squares in the same row or column, but on opposite edges of the map, may be combined.

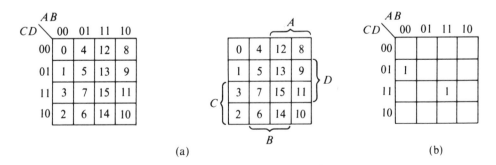

Figure 6-10. Mapping a 4-variable function: (a) minterm numbers, (b) $f = \overline{A} \cdot \overline{B} \cdot \overline{C} \cdot D + A \cdot B \cdot C \cdot D$.

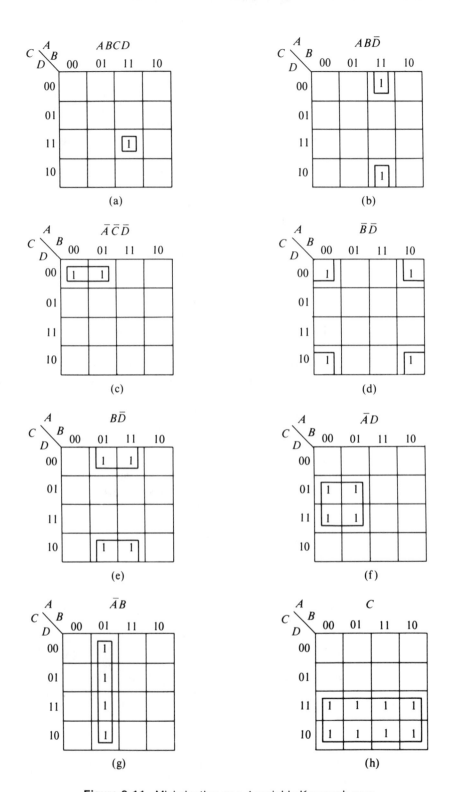

Figure 6-11. Minimization on a 4-variable Karnaugh map.

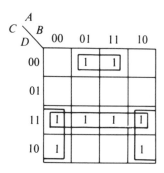

Figure 6-12.

The function $f(A,B,C,D) = \Sigma\, m(2,3,4,7,10,11,12,15)$ is minimized on the Karnaugh map shown in Figure 6-12.

$$f = \Sigma\, m(2,3,4,7,10,11,12,15) = B\overline{C}\overline{D} + \overline{B}C + CD.$$

Quite often some of the possible combinations of input values never occur. In this case we "don't care" what the function does if these input combinations appear. Diagramming makes it easy to take advantage of these "don't care" conditions by letting the "don't care" minterms be 1 or 0, depending on which value results in a simpler expression.

Figure 6-13 shows an example of the use of "don't cares" to simplify the seven-segment display functions for segments a to g previously referred to. Since minterms 10 through 15 will never occur, we put 0's on the diagram in those positions. We then put 1's on the diagram for the appropriate minterms. The minimized functions of the seven-segment display are shown in Figure 6-13.

Five and six-variable maps are shown below in Figure 6-14 and Figure 6-15, respectively.

The Karnaugh map for five-variables has two four-variable maps placed side by side. They are identical in $BCDE$, but one corresponds to $A = 1$, the other to $A = 0$. The standard four-variable adjacencies apply in each map. In addition, squares in the same relative position on the two maps, e.g., 4 and 20, are also logically adjacent. Similar arguments apply to the six-variable map.

Exercises for Section 6.5

1. Simplify the following Boolean expression.

$$a c \overline{d} e + a c d + \overline{e}\,\overline{h} + a c \overline{f}\overline{g}\overline{h} + a c d \overline{e}.$$

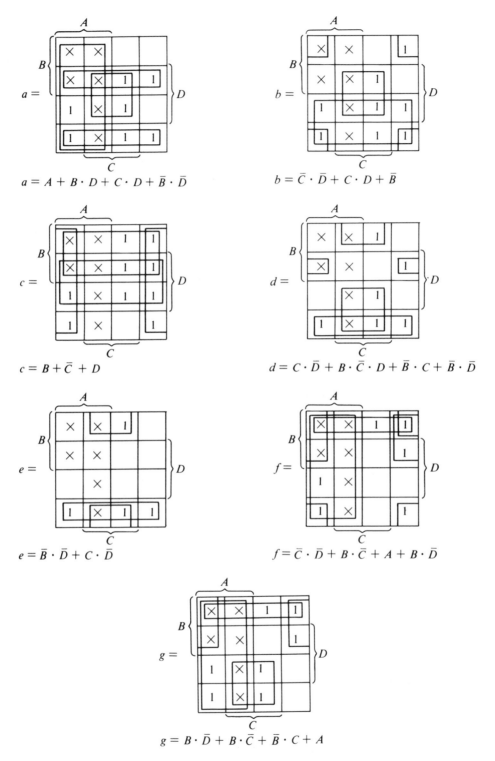

Figure 6-13. Seven-segment display driver minimization.

	BC	$A = 0$		
DE	00	01	11	10
00	0	4	12	8
01	1	5	13	9
11	3	7	15	11
10	2	6	14	10

	$A = 1$			BC
00	01	11	10	DE
16	20	28	24	00
17	21	29	25	01
19	23	31	27	11
18	22	30	26	10

Figure 6-14. Five-variable Karnaugh map.

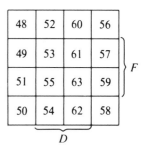

Figure 6-15. Six-variable Karnaugh map.

2. Reduce the following expression to a minimal sum-of-products expression.

$$f = (x \oplus yz) + \overline{(\overline{x}\,\overline{y} \oplus w)} + \overline{x}yw.$$

3. Minimize the following expressions using a truth table or map technique.

(a) $f = AB\overline{C}D + ABC\overline{D} + B\overline{C}D + \overline{A}BC\overline{D}.$
(b) $f = \overline{A}\,\overline{B}CD + \overline{A}C\overline{D} + ABC\overline{D} + \overline{A}BCD.$
(c) $f = AB\overline{C} + \overline{A}B\overline{C} + A\overline{B}\,\overline{C} + \overline{A}\,\overline{B}\,\overline{C}.$
(d) $f = B\overline{C}D + A\overline{C}\,\overline{D} + AB\overline{C}D + A\overline{B}\,\overline{C}D.$
(e) $f = (A + B + \overline{C})(\overline{A} + B + \overline{C}).$

4. Prove or disprove the following equalities using a truth table or map technique.

(a) $\overline{A}\,\overline{B}C + A\overline{B}\,\overline{C} + \overline{A}\,\overline{B}\,\overline{C} + A\overline{B}C = (\overline{B} + C)(\overline{B} + \overline{C}).$
(b) $A\overline{C} + A\overline{B} + AC\overline{D} = (A + B)(A + \overline{B})(\overline{B} + \overline{C} + \overline{D}).$
(c) $ABC + (\overline{A} + \overline{B})D = (AB + D)(\overline{A} + \overline{B} + C).$
(d) $A\overline{B}\,\overline{D} + BC + CD = \overline{A}\,\overline{B}\,\overline{D} + \overline{C}D + B\overline{C}.$

5. To design product-of-sums (POS) forms, select sets of the 0's of the function. Realize each set as a sum term, with variables being the complements of those that would be used if this same set were being realized as a product to produce 1's.

Obtain a minimal POS realization of $f(A,B,C,D) = \Sigma\, m(0,2,10,11,12,14) = \Pi\, M(1,3,4,5,6,7,8,9,13,15).$

6. Minimize the following switching functions.

(a) $\Sigma\, m(1,2,3,13,15).$
(b) $\Sigma\, m(0,2,10,11,12,14).$
(c) $\Sigma\, m(0,2,8,12,13).$
(d) $\Sigma\, m(1,5,6,7,11,12,13,15).$
(e) $\Sigma\, m(0,1,4,5,6,11,12,14,16,20,22,28,30,31).$
(f) $\Sigma\, m(6,7,10,14,19,27,37,42,43,45,46).$

7. Simplify each of the following expressions by using the rules of Boolean algebra.

(a) $xy\overline{z} + \overline{x}y\overline{z} + x\overline{y}(z + w) + (\overline{z} + w).$
(b) $x + \overline{y}z + \overline{w}\,(x + \overline{y}z).$
(c) $x\overline{y}(z + \overline{w}) + x\overline{y}(z + \overline{w}) + xy\overline{z}.$
(d) $\overline{x}\,\overline{y}z + \overline{x}y\overline{w} + \overline{x}\,\overline{y}w + y\overline{z}\,\overline{w}.$

8. Minimize the following functions, using the map technique.
   (a) $f = \Sigma\, m(0,5,10,15) + \Sigma_\phi(1,7,11,13)$ where Σ_ϕ denotes don't care minterms.
   (b) $f = \Sigma\, m(0,1,4,5,8,9) + \Sigma_\phi(7,10,12,13)$.
   (c) $f = \Sigma\, m(1,3,4,6,9,11) + \Sigma_\phi(2,5,7)$.

Selected Answers for Section 6.5

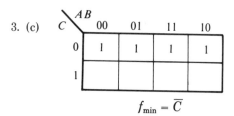

3. (c)

		AB			
C		00	01	11	10
0		1	1	1	1
1					

$$f_{min} = \overline{C}$$

6. (c)

		xy			
wz		00	01	11	10
00		1		1	1
01				1	
11					
10		1			

$$f_{min} = \overline{x}\,\overline{y}\,\overline{z} + xy\overline{w} + x\overline{w}\,\overline{z}$$
$$\text{or } f_{min} = \overline{x}\,\overline{y}\,\overline{z} + xy\overline{w} + \overline{y}\,w\overline{z}$$

6.6   APPLICATIONS TO DIGITAL COMPUTER DESIGN

Initial Design of the Arithmetic Logic Unit of a Digital Computer

An arithmetic logic unit (ALU) of a digital computer can be partitioned into stages, one for each pair of bits of the input operands. For operands with m bits, the ALU consists of m identical stages, where each stage receives as inputs the bits of inputs A and B which are designated by subscript numbers from 1 (low order bit) to m. Figure 6-16 shows the block diagram of an ALU stage k.

The carries are connected in a chain through the ALU stages, where C_{k-1} is the input carry to stage k and C_k is the output carry of stage k. The function selection lines are identical to all stages of the ALU and are designated as selectors of the arithmetic or logic micro-operation to be

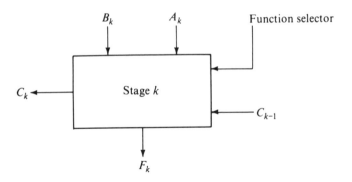

Figure 6-16. kth stage of an ALU.

performed by the ALU. Terminals F_1 to F_m generate the required output function of the ALU. In many cases, a 4-bit ALU is enclosed within one integrated circuit (IC) package. Such a package will contain four stages with four inputs for A, four inputs for B, and four outputs for F. The number of lines for the function selector determines the number of operations that the ALU can perform. A m-bit ALU can be constructed from 4-bit ALUs by cascading several packages. The output carry from one IC package must be connected to the input carry of the package with the next higher-order bits.

The internal construction of the ALU depends on the micro-operations that it implements. In any case, it always needs *full-adders* to perform the arithmetic operations. Additional gates are sometimes included for logic micro-operations. In order to minimize the number of terminals for the function selection, IC ALUs use k selection lines to specify 2^k micro-operations.

Half-Adder. The most basic digital arithmetic function is the addition of two binary digits. A combinational circuit that performs the arithmetic addition of two bits is called a **half-adder.** One that performs the addition of three bits (two significant bits and a previous carry) is called a **full-adder.** The name of the former stems from the fact that two half-adders can be employed to implement a full-adder.

The input variables of a half-adder are called the *augend* and *addend* bits. The output variables are called the *sum* and *carry*. It is necessary to specify two output variables because the sum of $1 + 1$ is binary 10, which has two digits. We assign symbols A_i and B_i to the two input variables, and F_i (for the sum function) and C_i (for carry) to the two output variables. The truth table for the half-adder is shown in Fig. 6-17(a). The C_i output is 0 unless both inputs are 1.

The F_i output represents the least significant bit of the sum. The

A_i	B_i	C_i	F_i
0	0	0	0
0	1	0	1
1	0	0	1
1	1	1	0

(a)   Truth table   (b)   Logic diagram

Figure 6-17. Half-adder.

Boolean functions for the two outputs can be obtained directly from

$$F_i = \overline{A}_i B_i + A_i \overline{B}_i = A_i \oplus B_i,$$
$$C_i = A_i B_i.$$

The logic diagram is shown in Figure 6-17(b). It consists of an exclusive-OR gate and an AND gate.

Full-Adder.   A full-adder is a combinational circuit that forms the arithmetic sum of three input bits. It consists of three inputs and two outputs. Two of the input variables, denoted by A_i and B_i, represent the two significant bits to be added.

The third input, C_{i-1}, represents the carry from the previous lower significant position.

Two outputs are necessary because the arithmetic sum of three binary digits ranges in value from 0 to 3, and the decimal numbers 2 and 3 need two binary digits. The two outputs are designated by the symbols F_i (for sum) and C_i (for carry). The binary variable C_i gives the output carry. The truth table of the full-adder is shown in Table 6-2. The eight rows under the input variables designate all possible combinations of 1's and 0's that these variables may have. The 1's and 0's for the output variables are determined from the arithmetic sum of the input bits. When all input bits are 0's the output is 0. The F_i output is equal to 1 when only one input is equal to 1. The C_i output has a carry of 1 if two or three inputs are equal to 1.

Thus the two functions for the full-adder are

$$F_i = A_i \oplus B_i \oplus C_{i-1}$$

and

$$C_i = A_i B_i + A_i C_{i-1} + B_i C_{i-1}.$$

Table 6-2. Truth Table for
Full-Adder

Inputs	Outputs
$A_iB_iC_{i-1}$	C_iF_i
0 0 0	0 0
0 0 1	0 1
0 1 0	0 1
0 1 1	1 0
1 0 0	0 1
1 0 1	1 0
1 1 0	1 0
1 1 1	1 1

Clearly, we can also write

$$C_i = A_iB_i + (\overline{A}_iB_i + A_i\overline{B}_i)\,C_{i-1} = A_iB_i + (A_i \oplus B_i)\,C_{i-1}.$$

Two-level realizations of these two functions require a total of nine gates with 25 inputs plus one inverter to generate $\overline{C}_{i-1}$. A NAND circuit for this form is shown in Figure 6-18.

Since the full adder is of basic importance in digital computers, a great deal of effort has gone into the problem of producing the most economical realization. The form leading to best economy is a function of the technology used.

Since today full-adders are realized as IC, as pointed out before, the forms of interest are those suitable for medium scale integration (MSI) form.

Two of these realizations are shown below in Figures 6-19 and 6-20.

Although the realization of Figure 6-20 appears to be more expensive, this form is better suited for IC realization and is used in full-adder chips such as SN7480 and SN7482.

A popular realization of a full-adder is by utilizing two half-adders. Using this design we will also add control bits so that the circuit will perform more operations than just addition. The circuit of the full-adder with the three control bits S_0, S_1, and S_2, as well as the mode bit M is shown in Figure 6-21.

Control line S_0 controls input A_i. Lines S_1 and S_2 control input B_i. The mode line M controls the input carry C_{i-1}. When $S_0 S_1 S_2 = 101$ and $M = 1$, the terminals marked x, y, and z have the values of A_i, B_i, and C_i, respectively.

Control lines $S_2 S_1 S_0$ may have eight possible bit combinations and each combination provides a different function for F_i and C_i. The mode

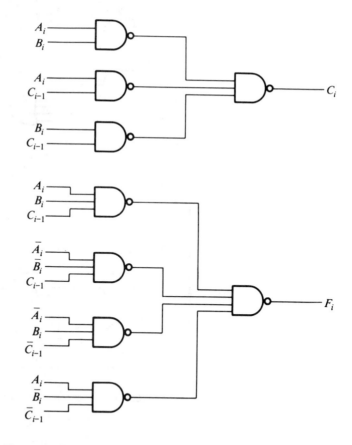

Figure 6-18. Two-level realization of a full-adder.

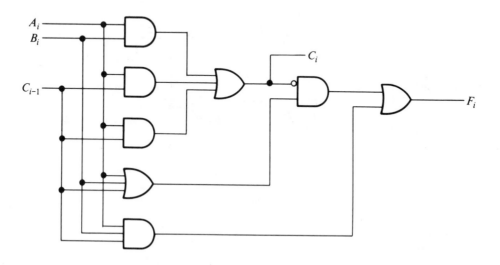

Figure 6-19. Full-adder realization with 8 gates and 19 inputs.

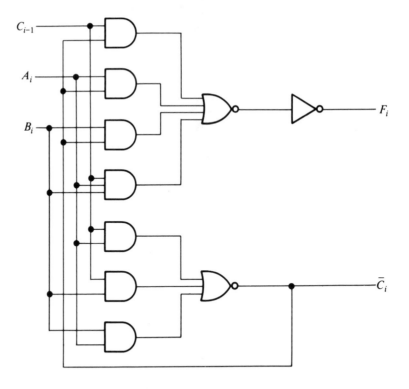

Figure 6-20. Full-adder realization with 10 gates and 22 inputs.

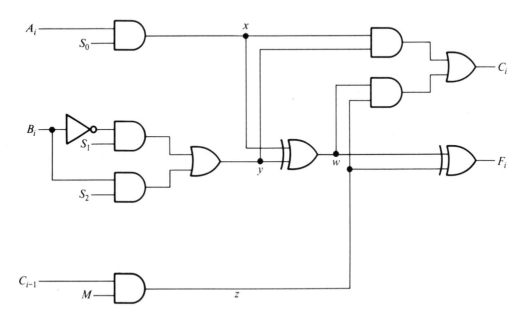

Figure 6-21. Controlled full-adder.

618

Table 6-3. Effect of Control lines on Full-Adder.

S_0	x	S_1	S_2	y	M	z
0	0	0	0	0	0	0
1	A_i	0	1	B_i	1	C_{i-1}
		1	0	$\overline{B_i}$		
		1	1	1		

control M enables the input carry C_{i-1} and is used to *differentiate* between an *arithmetic and logic function*. When $M = 1$, input carry C_{i-1} propagates through the gate making $z = C_{i-1}$. This allows the propagation of the carry through all the ALU stages for an arithmetic micro-operation. When $M = 0$, the input carry is inhibited, making $z = 0$. This is a necessary condition for a *logic micro-operation*. Hence, the ALU can provide up to eight arithmetic operations and eight logic operations.

Table 6-3 shows how the control lines control inputs A_i, B_i, and C_i. The value of x may be 0 or A_i depending on whether S_0 is 0 or 1. Control lines S_1 and S_2 control the value of y which may be 0, 1, B_i or $\overline{B_i}$. Terminal z may be equal to 0 or C_{i-1} depending on whether M is 0 or 1.

Table 6-4 shows a list of the Boolean functions for each of the eight positive combinations of the control lines: w represents $x \oplus y$; when $M = 0$, output F_i is the same as w since $F_i = w \oplus 0 = w$; and output carry $C_i = xy$ since $z = 0$. However, this signal is not allowed to propagate to the z terminal of the next higher stage when $M = 0$. The Boolean functions listed under F_i (with $M = 0$) provide the eight logic functions of the ALU.

The arithmetic operations of the ALU are generated when $M = 1$. The Boolean functions of the eight arithmetic operations in the one-stage ALU are listed in Table 6-4.

Table 6-4. Boolean functions for one stage of the ALU.

$S_0 S_1 S_2$	x	y	w	F_i $M = 0$	F_i $M = 1$	C_i for $M = 1$
0 0 0	0	0	0	0	C_{i-1}	0
0 0 1	0	B_i	B_i	B_i	$B_i \oplus C_{i-1}$	$B_i C_{i-1}$
0 1 0	0	$\overline{B_i}$	$\overline{B_i}$	$\overline{B_i}$	$\overline{B_i} \oplus C_{i-1}$	$\overline{B_i} C_{i-1}$
0 1 1	0	1	1	1	$\overline{C_{i-1}}$	C_{i-1}
1 0 0	A_i	0	A_i	A_i	$A_i \oplus C_{i-1}$	$A_i C_{i-1}$
1 0 1	A_i	B_i	$A_i \oplus B_i$	$A_i \oplus B_i$	$A_i \oplus B_i \oplus C_{i-1}$	$A_i B_i + A_i C_{i-1} + B_i C_{i-1}$
1 1 0	A_i	$\overline{B_i}$	$A_i \oplus \overline{B_i}$	$\overline{A_i \oplus B_i}$	$A_i \oplus \overline{B_i} \oplus C_{i-1}$	$A_i \overline{B_i} + A_i C_{i-1} + \overline{B_i} C_{i-1}$
1 1 1	A_i	1	$\overline{A_i}$	$\overline{A_i}$	$\overline{A_i} \oplus C_{i-1}$	$A_i + C_{i-1}$

Table 6-5. Logic Micro-Operations in the ALU.

M	$S_0 S_1 S_2$	Micro-Operation	Description
0	0 0 0	$F = 0$	Clear all bits
0	0 0 1	$F = B$	Transfer B
0	0 1 0	$F = \overline{B}$	Complement B
0	0 1 1	$F = 1$	Set all bits
0	1 0 0	$F = A$	Transfer A
0	1 0 1	$F = A \oplus B$	Exclusive-OR
0	1 1 0	$F = \overline{A \oplus B}$	Exclusive-NOR
0	1 1 1	$F = \overline{A}$	Complement A

The ALU is constructed by connecting m identical stages in cascade. The logic micro-operations performed by the ALU are listed in Table 6-5. These are the F functions for $M = 0$.

Note that there are 16 possible micro-operations for two logic operands and only 8 of them are available in the ALU. In fact, the two important logic operations AND and OR are not generated in this ALU. By providing a fourth control line it is possible to include these functions.

The arithmetic operations are derived from Table 6-3 and the conditions of Table 6-4.

In each case, a parallel binary adder composed of full-adder circuits is used, but some of the input lines are either missing or complemented. Thus, in row 001, input A is missing because all the x inputs of the full-adders change to zero by selection line S_0. The output function for this condition is $F = B$ when $C_1 = 0$ and $F = B + 1$ when $C_0 = 1$. In row 010, input A is changed to zero and all B inputs are complemented so $F = \overline{B}$ when $C_0 = 0$ and $F = \overline{B} + 1$ when $C_0 = 1$. In row 110, all bits of input B are complemented so that F generates the arithmetic operation of A plus the 1's complement of B. The Boolean function for row 111 represents one stage of a decrement micro-operation.

The input carry C_0 that enters the first low-order stage of the ALU is employed for adding 1 to the sum in four micro-operations. Hence, arithmetic micro-operations require five control lines. M must always be 1. The three control lines specify an operation and input carry C_0 must be set to 0 or 1 for a particular micro-operation. Some of the arithmetic functions generate the same operation as the logic functions when $C_0 = 0$. Others have no useful application.

The arithmetic micro-operations in the ALU are shown in Table 6-6.

It should be noted that by making the input carry $C_0 = 1$, a one is added to $A + \overline{B}$ when $M = 1$ and $S_0 S_1 S_2 = 110$ and the result is $A + \overline{B} + 1$ which is equal to A plus the 2's complement of B. This is equivalent to a subtraction operation, since the output logic function for a full subtractor

Table 6-6. Useful Arithmetic Micro-Operations in the ALU

M	$S_0S_1S_2$	C_0	Micro-operation	Description
1	0 0 1	1	$F = B + 1$	Increment B
1	0 1 0	1	$F = \bar{B} + 1$	2's complement B
1	1 0 0	1	$F = A + 1$	Increment A
1	1 0 1	0	$F = A + B$	Add A and B
1	1 1 0	0	$F = A + \bar{B}$	A plus 1's complement of B
1	1 1 0	1	$F = A + \bar{B} + 1$	A plus 2's complement of B
1	1 1 1	0	$F = A - 1$	Decrement A

is exactly the same as the output function of a full-adder, and the next borrow function resembles the function for the carry in the full-adder except that the minuend is complemented. Thus by creating the 2's complement of the subtrahend and adding it to the minuend we achieve the operation of arithmetic subtraction.

In conclusion, we have used a simple full-adder stage to achieve a variety of arithmetic and logic operations in the digital computer.

Multiplexers

A multiplexer is a digital system that receives binary information from 2^n lines and transmits information on a single output line. The one input line being selected is determined from the bit combination of n selection lines. It is analogous to a mechanical or electrical switch, such as the selector switch of a stereo amplifer, which selects the input that will drive the speakers. The input can come from either phono, tape, AM, FM, or AUX lines worked by the position of the switch. An example of a 4×1 multiplexer is shown in Figure 6-22. The four input lines are applied to the circuit but only one input line has a path to the output at any given time.

The selection lines W_1 and W_2 determine which input is selected to have a direct path to the output. A multiplexer is also known as a **data selector** since it selects one of multiple input data lines and steers the information to the output line. The size of the multiplexer is specified by the number of its inputs, 2^n. It is also implied that it has n selection lines and one output line.

Clearly the function of the multiplexer is in gating of data that may come from a number of different sources. The logic function describing the 4×1 multiplexer is

$$F = \overline{W}_1\,\overline{W}_2\,I_0 + \overline{W}_1\,W_2\,I_1 + W_1\,\overline{W}_2\,I_2 + W_1\,W_2\,I_3.$$

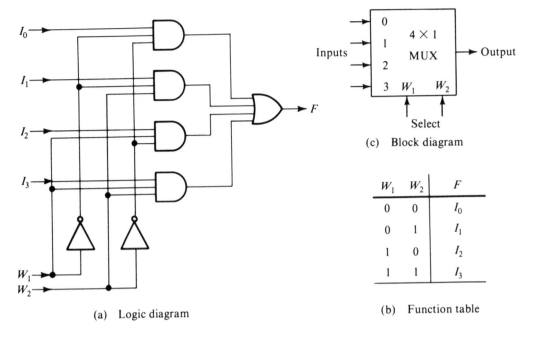

(a) Logic diagram

(b) Function table

Figure 6-22. 4 by 1 multiplexer.

In general for a $2^n \times 1$ multiplexer

$$F = \sum_{i=0}^{2^n-1} m_i\, I_i$$

where m_i represents the minterm i of the selection variables, and I_i is the ith input line.

Thus when $n = 3$, the multiplexer logic function will be represented by F, where

$$\begin{aligned}
F = \sum_{i=0}^{7} m_i\, I_i = \; & \overline{W}_1\,\overline{W}_2\,\overline{W}_3\,I_0 + \overline{W}_1\,\overline{W}_2\,W_3\,I_1 + \overline{W}_1\,W_2\,\overline{W}_3\,I_2 \\
& + \overline{W}_1\,W_2\,W_3\,I_3 + W_1\,\overline{W}_2\,\overline{W}_3\,I_4 \\
& + W_1\,\overline{W}_2\,W_3\,I_5 + W_1\,W_2\,\overline{W}_3\,I_6 \\
& + W_1\,W_2\,W_3\,I_7.
\end{aligned}$$

Now, if one wants to load a 16-bit data register from one to four distinct sources, this can be accomplished with 16 four-input multiplexers that come in eight IC packages.

x_1	x_2	x_3	F
0	0	0	0
0	0	1	0
0	1	0	0
0	1	1	1
1	0	0	1
1	0	1	1
1	1	0	1
1	1	1	0

x_1	x_2	F
0	0	0
0	1	x_3
1	0	1
1	1	$\bar{x}_3$

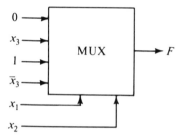

Figure 6-23. Multiplexer implementation of a logic function.

Multiplexers are also very useful as basic elements for implementing logic functions. Consider a function F defined by the truth table of Figure 6-23. It can be represented as shown by factoring out the variables x_1 and x_2. Note that for each valuation of x_1 and x_2, the function F corresponds to one of four terms: 0, 1, x_3 or $\bar{x}_3$. This suggests the possibility of using a four-input multiplexer circuit, where x_1 and x_2 are the two select inputs that choose one of the four data inputs. Then, if the data inputs are connected to 0, 1, x_3, or $\bar{x}_3$, as required by the truth table, the output of the multiplexer will correspond to the function F. The approach is completely general. Any function of three variables can be realized with a single four-input multiplexer. Similarly, any function of four variables can be implemented with an eight-input multiplexer, etc.

Using multiplexers in this fashion is a straightforward approach, which often reduces the total number of ICs needed to realize a given function. If the function of Figure 6-23 is constructed with AND, OR, and NOT gates, its minimal form is

$$F = x_1\,\bar{x}_2 + x_1\,\bar{x}_3 + \bar{x}_1\,x_2\,x_3$$

which implies a network of three AND gates and one OR gate. Thus parts of more than one IC are needed for this implementation. In general, the multiplexer approach is more attractive for functions that do not yield simple sum-of-products expressions. Of course, the relative merits of the two approaches should be judged by the number of ICs needed to implement a given function.

Example 6.6.1 Let $F(x_1,x_2,x_3) = \Sigma\, m\, (1,3,5,6)$.

We will implement F with a 4 by 1 multiplexer. First, we express F in its sum of minterms form. The next step is to connect the last $n - 1$ variables (x_2, x_3) to the selection lines with x_2 connected to the high-order selection line and x_3 to the lowest-order selection line W_2. The first variable x_1 will be connected to the input lines in both complemented $(\overline{x}_1)$ and uncomplemented form (x_1) as needed.

From the truth table of F shown in Figure 6-24 it is clear that x_1 is 1 for minterms 5 and 6 and 0 for minterms 1 and 3, and thus I_2 should be connected to x_1 and I_3 to $\overline{x}_1$, where $I_0 = 0$ and $I_1 = 1$.

It is not necessary to choose the leftmost variable in the ordered sequence of a variable list for the inputs to the multiplexer. In fact, we can choose any one of the variables for the inputs of the multiplexer, provided we modify the multiplexer implementation table. Suppose we want to implement the same function with a multiplexer, but using variables x_1 and x_2 for selection lines W_1 and W_2 and variable x_3 for the inputs of the multiplexer. Variable x_3 is complemented in the even-numbered minterms and uncomplemented in the odd-numbered minterms, since it is the last variable in the sequence of listed variables.

This implementation is shown in Figure 6-25.

Multiplexer ICs may have an enable input to control the operation of the unit. When the enable input is in a given binary state, the outputs are disabled, and when it is in the other state (the enable state), the circuit functions as a normal multiplexer. The enable input (sometimes called strobe) can be used to expand two or more multiplexer ICs to a digital multiplexer with a larger number of inputs.

In some cases two or more multiplexers are enclosed within one IC package. The selection and enable inputs in multiple-unit ICs may be common to all multiplexers, as illustrated in Figure 6-26 in which we show a quadruple two-line to one-line multiplexer IC, similar to IC package 74157. This unit has four multiplexers, each capable of selecting one of two input channels; namely, F_i can be selected to be equal to either I_i or J_i, $1 \le i \le 4$. The enable E disables the multiplexers in state 1 and enables them in state 0.

It is very important for the computer science student to be familiar

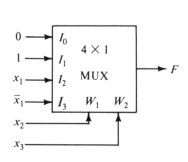

Minterm	x_1	x_2	x_3	F
0	0	0	0	0
1	0	0	1	1
2	0	1	0	0
3	0	1	1	1
4	1	0	0	0
5	1	0	1	1
6	1	1	0	1
7	1	1	1	0

(a) Multiplexer Implementation   (b) Truth table

Figure 6-24. Implementing $F(x_1, x_2, x_3) = \Sigma m(1, 3, 5, 6)$ with a 4 by 1 multiplexer.

with the various digital functions encountered in computer hardware design. With the advent of MSI, LSI, and VLSI functions, computer architecture has taken on a new dimension, giving the designer the ability to create systems that were previously uneconomical or impractical. For this reason, the material introduced throughout this chapter is being illustrated with two key applications that explain the typical characteristic of the subject matter covered in Chapter 6. This material represents an essential prerequisite to the understanding of the internal organization of digital computers and the design of computer architecture.

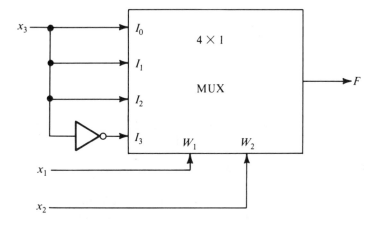

Figure 6-25. Another implementation of Figure 6-24(a).

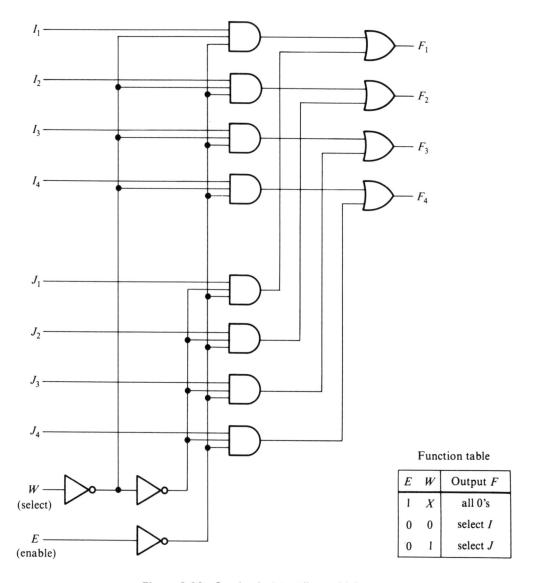

Figure 6-26. Quadruple 2-to-1 line multiplexers.

Exercises for Section 6.6

1. Show that a full-adder circuit consists of a three-input exclusive OR and a three-input majority function.

2. Obtain the simplified Boolean function of the full-adder in sum-of-products form and draw the logic diagram using NAND gates.

3. There are 16 logic functions for two Boolean variables. Table 6-6 lists 8 of these functions. List the remaining 8 functions.

4. The OR and AND logic functions can be included in the ALU by modifying the output circuit and using a fourth selection line S_3 as shown in the Figure 6-27. Using the two Boolean identities $A + B = A \oplus B + AB$ and $AB = 0 + AB$, determine the values of $S_3 S_2 S_1 S_0$ for M for the OR and AND logic functions.

5. For the S_3 of Exercise 4 show that when $S_3 = 0$, none of the other operations of the ALU are altered. What other logic operations can be implemented with this modification and what are their selection values?

(Original output)

(Modified output)

Figure 6-27.

6. Let us denote $A_i B_i$ by G_i and $A_i \oplus B_i$ by P_i. The output carry of the full adder can now be expressed as $C_i = G_i + P_i C_{i-1}$. G_i is called a carry-generate and produces an output when both A_i and B_i are 1's, irrespective of the input carry. P_i is called a carry-propagate because it is the term associated with the propagation of the carry from C_{i-1} to C_i. Any output carry may be expressed as

$$C_k = G_k + P_k C_0$$

where

$$G_k = G_{k-1} + P_{k-1} G_{k-2} + \cdots + P_{k-1} P_{k-2} \cdots P_2 G_1$$
$$P_k = P_{k-1} P_{k-2} \cdots P_2 P_1.$$

This technique for reducing the carry propagation time in the parallel-adder part of the ALU is known as the *carry look-ahead technique*.

(a) Show that the technique reduces the carry propagation time.

(b) Show that P_i in a full-adder can be expressed by the Boolean function $A_i + B_i$.

7. Draw the logic diagram of a 4-bit adder with look-ahead carry. List

the Boolean functions and draw the logic diagram for outputs G_5 and P_5.

8. Realize the following two functions using multiplexers and as little external gating as possible.
   (a) $f_1(x,y,z,w) = \Sigma m(2,4,6,10,12,15)$
   (b) $f_2(x,y,z,w) = \Sigma m(3,5,9,11,13,14,15)$.

9. Given the function $f(x,y,z,w) = \Sigma m(1,2,4,5,6,7,8,9,10,11,14,15)$.
   (a) Realize this function using an eight-input multiplexer.
   (b) Realize this function using four-input multiplexers plus external gates as required.

Selected Answers for Section 6.6

1. $F = xy + xz + yz$ is identical to the carry. Hence,

5. When $S_3 = 0$ then $S_3A_iB_i = 0$ and $F_i = G_i$. For $M = 0$

S_3	S_2	S_1	S_0	F_i	$F_i + A_i B_i$	$=$	G_i
1	0	0	0	0	A_iB_i		$A_i B_i$ (AND)
1	0	0	1	B_i	$B_i + A_i B_i$		B_i
1	0	1	0	$\overline{B_i}$	$\overline{B_i} + A_i B_i$		$A_i + \overline{B_i}$
1	0	1	1	1	$1 + A_i B_i$		1
1	1	0	0	A	$A_i + A_i B_i$		A_i
1	1	0	1	$A_i \oplus B_i$	$A_i \oplus B_i + A_iB_i$		$A_i + B_i$ (OR)
1	1	1	0	$A_i \oplus \overline{B_i}$	$A_i \oplus \overline{B_i} + A_iB_i$		$\overline{A_i \oplus B_i}$ (EQ)
1	1	1	1	$\overline{A_i}$	$\overline{A_i} + A_i B_i$		$\overline{A_i} + B_i$

REVIEW FOR CHAPTER 6

1. The following is the truth table for several functions of 3 input variables. Without formally deriving any Boolean expressions, deduce the values for the functions f_1, f_2, f_3, f_4 in terms of the variables x_1, x_2, x_3.

x_1	x_2	x_3	f_1	f_2	f_3	f_4
0	0	0	0	1	0	1
0	0	1	0	1	0	1
0	1	0	0	1	0	0
0	1	1	0	1	0	0
1	0	0	0	1	1	1
1	0	1	0	1	1	1
1	1	0	0	1	1	0
1	1	1	0	1	1	0

2. Simplify the following Boolean expressions using laws of Boolean algebra:
   - (a) $xy + x\overline{y}$
   - (b) $(x + y)(x + \overline{y})$
   - (c) $xz + xy\overline{z}$
   - (d) $\overline{(x + y)} + \overline{(\overline{x} + \overline{y})}$
   - (e) $x + \overline{x}y + \overline{x}\,\overline{y}$

3. Prove the following using the laws of Boolean algebra:
   - (a) $xy + x\overline{y} = x$
   - (b) $(x + y)(x + \overline{y}) = x$
   - (c) $x(\overline{x} + y) = xy$
   - (d) $xyz + \overline{x}y + xy\overline{z} = y$
   - (e) $y(w\overline{z} + wz) + xy = y(w + x)$
   - (f) $\overline{x}\,\overline{y}z + \overline{x}yz + x\overline{y}z + xyz = z$

4. Give the canonical sum of products (or disjunctive normal form) for the following functions:
   - (a) $f(x,y,z) = (x + \overline{y})(z + \overline{x})$
   - (b) $f(x,y,z)$ where $f = 1$ iff an odd number of input variables equals 1.
   - (c) $f(x,y,z) = M_2 M_3 M_5$, where M_i are maxterms.

5. Using Karnaugh maps, minimize the function $f(a,b,c,d) = \Sigma\, m(0,1,2,3,4,5,8,10,12,13,14)$.

6. Minimize the function $f(a,b,c,d)$ whose Karnaugh map with "don't-cares" is the following:

ab \ cd	00	01	11	10
00	1	0	0	1
01	1	1	0	0
11	X	1	X	0
10	1	X	0	0

7. Implement the function $f(a,b,c,d) = \Sigma\, m(4,10,13,15)$ using an 8×1 multiplexer.

8. (a) Construct a 32×1 multiplexer using five 8×1 multiplexers.
   (b) Construct a 16×1 multiplexer using two 8×1 multiplexers, an enable input, and an OR gate.

9. Design a full-adder using half-adders and an OR gate.

10. Using full-adders only design a circuit that will output the sum of the switching variables x_1, x_2, x_3, and x_4.

11. (a) Show that the set $\{F,G\}$ is functionally complete where $F(x,y) = x \longrightarrow y = \overline{x} + y$ is the implication function and $G(x) = \overline{x}$ is the negation function.
    (b) Show that $\{H,K\}$ is not functionally complete where $H(x,y) = x \longleftrightarrow y$ is the function that outputs 1 iff x and y have the same value and $K(x,y) = x + y$ is the inclusive OR function.

12. Draw the logic diagram of the function $f(A,B,C,D) = A\overline{B}CD + A\overline{C} + BCD$ using only NAND gates.

7

Network Flows

7.1 GRAPHS AS MODELS OF FLOW OF COMMODITIES

Many systems involve the communication, transmission, transportation, or flow of commodities. The commodity may be something tangible such as oil drums, automobiles, manufactured goods, or money; or it may be something intangible such as information, disease, or heredity. Thus, systems of telephones and lines; railway networks; or interconnections of factories, warehouses, and retail outlets all involve the flow of commodities through a network.

The modeling of such physical systems is frequently accomplished by directed graphs, with the edges of the graph representing roads, telephone lines, railway tracks, airline routes, oil pipelines, or power lines—in general, channels through which the commodity flows. The vertices of the graph can represent highway intersections, telephone outlets, railway stations, airline terminals, oil reservoirs, or power relay stations—in general, points where the flow of the commodity originates, is relayed, or terminates.

The structural information conveyed by a graph is basically information indicating which vertices are adjacent and which vertices are connected by directed paths. Two models of physical systems may be structurally similar—may even be isomorphic as graphs—but have significantly different physical interpretations. For instance, the interconnections of an electric network and a telephone system can be specified by a graph, but edges of the electric network model might be interpreted by parameters such as resistance or voltage whereas the edges of the telephone network model might be interpreted by parameters such as cost per unit length or maximum number of calls per unit of time. We must account for these parameters as part of the model, and we do so by either edge or vertex labels. Thus, for each edge of a graph we can associate a number of parameters that represent natural limitations and capabilities of the edges. (We could do the same for the vertices of the graph, but in our discussion in this chapter we shall restrict our attention to edge labelings.)

Example 7.1.1. A Traffic System. Let each vertex of a graph represent a city. Two vertices are connected by an edge if there is a highway between the corresponding cities that doesn't pass through a third city. One edge label could denote the distance between the two cities. A second label could indicate the speed limit, while a third label could represent the maximum number of cars that can travel along the highway per unit of time.

Example 7.1.2. An Economic Model. Suppose that we have a system of production centers, warehouses, and markets connected by a set of highways, railways, and waterways. This system can be modeled by a graph where edges represent transportation channels and vertices represent the production centers, warehouses, and markets. Typical edge labels could include the maximum volume per unit time, the cost of transportation per unit of commodity, or the number of available transport vehicles.

Example 7.1.3. A Dynamic Model. We may incorporate a dynamic feature into either the traffic model or the economic model. For instance, we may have several cities located on a road map, and edge capacities may represent the maximum number of units shipped per day. Moreover, we may also know the number of days it takes to ship the commodity between cities. If we wish to find out the number of units of commodity shipped from city A to city B, say, over a period of 3 days, we can add the dimension of time to the network by replacing each city vertex v by four vertices v_0, v_1, v_2, v_3, where v_i represents city v on day i. (Of course, v_0 represents city v on the starting day.) Then for each road (u,v) which takes k days to traverse, we can make the edges (u_0,v_k), (u_1,v_{k+1}), $\cdots$ (u_{3-k},v_3), each with the same capacity as (u,v). Finally, the edges (v_i,v_{i+1}) can be assigned an infinitely large capacity to allow commodity to be stored temporarily at any vertex.

Let us be more precise in the following example.

Example 7.1.4. Suppose that oil is being transported from refineries at a and b to markets at g, h, and i. Suppose, moreover, that there are four relay stations or reservoirs denoted by vertices c, d, e, and f. Each week there are 20,000 and 60,000 barrels of oil available to a and b, while 20,000, 30,000, and 20,000 barrels are needed respectively by g, h, and i.

(a)

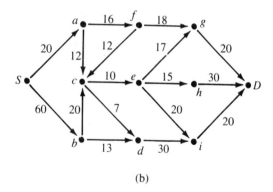
(b)

Figure 7-1.

The pipelines connecting the refineries, reservoirs, and markets are denoted by edges in the graph of Figure 7-1(a). Each edge is labeled with a number indicating the maximum amount of oil in thousands of barrels that can be transported via that pipeline per week.

In cases where there are multiple sources (such as a and b) and multiple terminals (such as g, h, and i) we can artificially combine them to a single source vertex and a single terminal vertex by adding two extra vertices as we did in Figure 7-1(b). These two vertices, S and D, represent the total supply and demand. Each edge (S,x) is labeled with a number representing the total amount of oil that refinery x can supply each week. Likewise each edge of the form (x,D) is labeled with a number representing the total amount of oil demanded by market x each week. The edge (h,D), for instance, is labeled 30 because market h requires 30,000 barrels of oil each week; the edge (S,a) is labeled 20 because it can supply 20,000 barrels of oil each week.

With these preliminary examples we see the essential characteristics of a (transport) network as defined below.

Definition 7.1.1. A directed graph G that is (weakly) connected and contains no loops is called a **(transport) network** if and only if

   (i) there are two distinguished vertices S and D of G, called the **source** and **sink** of G, respectively; and

   (ii) there is a nonnegative real valued function k defined on the edges of G.

The function k is called the **capacity function** of G and if e is any edge of G, the value $k(e)$ is called the **capacity of** e. We think of $k(e)$ as representing the maximum rate at which a commodity can be transported along e. The vertices, distinct from the source S and the sink D, are called **intermediate** vertices.

Frequently it is the case that in a transport network the source S has in-degree 0 and the sink D has out-degree 0, but this is not always the case.

We generally will say that (G,k) is a transport network, and by this we indicate that G is the directed graph and k is the capacity function defined on the edges of G.

If $G = (V(G),E(G))$ is a directed graph with vertex set $V(G)$ and edge set $E(G)$, then for any $v \in V(G)$, let $A(v)$ denote the set of all $y \in V(G)$ such that y is incident from v. Thus $A(v) = \{y \in V(G) \mid (v,y) \in E(G)\}$ could be called the set of vertices of G that "follow v" or are "after v." Similarly, let $B(v) = \{y \in V(G) \mid (y,v) \in E(G)\}$. Then $B(v)$ is the set of vertices that are "before v."

For example, in the graph of Figure 7-1(b), $A(c) = \{d,e\}$ and $B(c) = \{a,b,f\}$, while $A(e) = \{g,h,i\}$ and $B(e) = \{c\}$. Note that $A(D) = \phi$ and $B(S) = \phi$, the empty set.

Exercises for Section 7.1

1. The accompanying graph (Figure 7-2) represents a pipeline system in which oil for 3 cities, j, k, and m is delivered from 3 refineries, a, b, and c. Vertices d, e, f, g, h, and i represent intermediate pumping stations. The capacities of the pipelines are shown on the edges.
   (a) Model this system as a transport network.
   (b) Add to your model the additional facts that refineries a, b, and c can produce at most 2, 10, and 5 units respectively.
   (c) Incorporate in your model the facts that cities j, k, and m require 4, 3, and 7 units respectively.

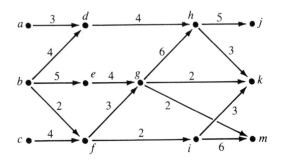

Figure 7-2.

2. It is possible to go from city A to city C either directly or by going through city B. During the period 4:00 to 5:00 p.m., the average trip times are 30 minutes from A to B, 15 minutes from B to C, and 30 minutes from A to C. The maximum capacities of the routes are 4000 vehicles on the A to B route, 3000 vehicles on the B to C route, and 6000 vehicles on the A to C route. Represent the flow of traffic from A to C from 4:00 to 5:00 p.m. as a transport network. (Hint: let vertices represent cities at a particular time.)

3. Suppose 3 applicants A_1, A_2, A_3 apply for 4 jobs J_1, J_2, J_3, J_4, where A_1 is qualified for J_1 and J_3; A_2 is qualified for J_2 and J_4, and A_3 is qualified for J_1 and J_4. Represent this information in a transport network.

4. Suppose there are 5 girls G_1, G_2, ..., G_5 and 5 boys B_1, B_2, ..., B_5. Suppose, further, that G_1 dates B_1, B_2, and B_4; G_2 dates B_2 and B_3; G_3 dates B_2; G_4 dates B_3 and B_5; and G_5 dates B_3, B_4, and B_5.
   (a) Represent this information in a transport network.
   (b) Is it possible for all 5 girls to go to a party with a boy she usually dates?

5. Suppose 5 persons A, B, C, D, E belong to 4 committees C_1, C_2, C_3, C_4 where $C_1 = \{B,C\}$, $C_2 = \{A,C,D\}$, $C_3 = \{C,D,E\}$, and $C_4 = \{A,D\}$.
   (a) Represent this information in a transport network.
   (b) Is it possible to select 4 chairpersons for the committees where no person chairs more than one committee?

6. A manufacturing firm that produces a variety of articles can forecast the demand for each product over a certain period of time. The task of producing the various articles can be assigned to different workers, who take varying amounts of time to produce the articles. Let n be the number of articles and d_i be the demand for each article for $i = 1, 2, ...,n$; let m be the number of workers and t_{ij} the time required by the jth worker to produce the ith article. Assume also that the jth worker can produce for at most c_j units of time.
   (a) At first ignore the time t_{ij} required for producing article i by worker j and design a transport network for the case $n = 3$ and $m = 2$.
   (b) Describe how to design a transport network in general.
   (c) Discuss how the time t_{ij} could be incorporated into your transport network.

7. From a group of n cities, C_1, C_2, ..., C_n buses run to a single destination C_{n+1}. If there is a road from C_i to C_j, let t_{ij} be the time required to go from C_i to C_j by this road, and let b_{ij} be the maximum number of buses which can use the road from C_i to C_j per unit of time ($b_{ij} = 0$ if there is no road from C_i to C_j). Let b_{ii} be the maximum number of buses that can be stationed at C_i, and let A_i be the

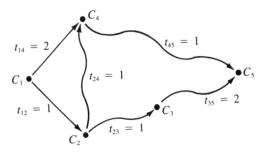

Figure 7-3.

number of buses stationed at C_i originally (at time $t = 0$). The question is: How should the buses be scheduled in order to have as many buses as possible arrive at C_{n+1} in a given time interval θ?

Draw a transport network with vertices $C_i(t) = (C_i, t)$, edges $(C_i(t), C_i(t + t_{ij}))$ of capacity b_{ij} and edges $(C_i(t), C_i(t + 1))$ of capacity b_{ii} for $i = 1, 2, 3, 4, 5$ and $t = 0, 1, 2, 3, 4$ where the 5 cities and times of travel between them are labeled on the road map shown in Figure 7-3.

7.2   FLOWS

A practical question to ask about a system such as the one pictured in Figure 7-1(b) is: Is it possible to organize the shipment of oil so that the demands of markets g, h, and i are met? The total amount of oil available to a is 20,000 barrels per week. (This is indicated in the graph of Figure 7-1(b) that a has only one incoming edge $(B(a) = \{S\})$ and the capacity of this edge is 20.) Since $A(a) = \{c, f\}$, c and f are the only recipients of a's oil. Thus, a could send all 20,000 barrels to c and 0 barrels to f, except for the fact that the edge (a, c) has a 12,000 barrel capacity. Thus, a can send a maximum of 12,000 barrels to c. But likewise a can send a maximum of 16,000 barrels to f. Hence, a could send 16,000 barrels to f and 4,000 to c; or 8,000 to f and 12,000 to c; or 10,000 to f and 10,000 to c; and so on.

Although we have not begun to consider how b could distribute its oil, we already see a large number of potential plans. Let us begin to organize our thoughts concerning this problem, and as we proceed we will begin to see the basic ideas of the concept of **flow**.

Let us label edges with two numbers, first the capacity of the edge and second the amount of flow through that edge. We will make two rules:

(1) The amount of oil transported along any edge cannot exceed the capacity of that edge.

(2) Except at the source S and the sink D, the amount of oil flowing into a vertex v must be equal to the amount of oil flowing out of the vertex. (Thus, no oil gets "stored" at an intermediate vertex.)

Of the possible ways refinery a could distribute its oil, suppose we arbitrarily decide to send 16,000 barrels to f and 4,000 barrels to c. We could send 20,000 to c from b, but since $A(c) = \{e,d\}$ and the capacities of edges (c,e) and (c,d) are 10 and 7 respectively, we see that c can send out a maximum of 17,000 barrels. Thus, since no oil can be stored at c, c can receive at most 17,000 barrels. Since c has been chosen to receive 4,000 barrels from a, c can receive at most 13,000 barrels from b. Let us make the choice to send 13,000 barrels to c from b. Likewise we choose to send 13,000 from b to d. Thus, there must be only 26,000 sent from S to b.

We continue making choices subject to the above two constraints as indicated in the diagram shown in Figure 7-4.

We see that, for this choice of flow, market i received its demand but g and h did not. The following is a summary of the schedule we have developed:

a sends 16,000 to f	g receives 16,000
4,000 to c	h receives 10,000
b sends 13,000 to c	
13,000 to d	i receives 20,000
Total 46,000	Total 46,000

Thus, even though a and b are capable of producing a total of 80,000 barrels of oil and the total demand is only 70,000, the capacity restrictions have affected the flow to the extent that some markets did not receive their demand. We will have to wait until later to see whether or not this flow is the best possible schedule.

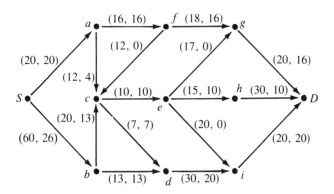

Figure 7-4.

Definition 7.2.1.   Let (G,k) be a transport network with source S and sink D. Assume the capacity function k is defined on the edges of G. A **flow** in G is a nonnegative real-valued function F defined on the edges of G such that

(i) $0 \le F(e) \le k(e)$ for each edge $e \in E(G)$.

(ii) If x is any vertex of G, different from the source or the sink, then the sum of all values $F(x,y)$ such that $y \in A(x)$ must equal the sum of all values $F(z,x)$ such that $z \in B(x)$, and

(iii) $F(e) = 0$ for any edge e incident to the source S or incident from the sink D.

Condition (i) is called the **capacity constraint,** and it insures that the amount of flow along an edge does not exceed the capacity for that edge. Condition (ii) requires that the sum of all incoming flows at x is equal to the sum of all outgoing flows at x. In symbols, (ii) becomes

$$\text{(ii')} \quad \sum_{y \in A(x)} F(x,y) = \sum_{z \in B(x)} F(z,x)$$

$$\text{for every } x \in V(G) - \{S,D\}. \tag{7.2.1}$$

Thus, $\Sigma_{y \in A(x)} F(x,y)$ is the flow out of x and $\Sigma_{z \in B(x)} F(z,x)$ is the flow into x. Moreover, the **net flow out of a vertex** x is defined to be $\Sigma_{y \in A(x)} F(x,y) - \Sigma_{z \in B(x)} F(z,x)$, while the **net flow into** x is, of course, $\Sigma_{z \in B(x)} F(z,x) - \Sigma_{y \in A(x)} F(x,y)$. Thus, conditions (ii) and (ii') assure that for an intermediate vertex x the net flow out of x equals the net flow into x, and this common value is zero. In other words, no flow is either created or destroyed at an intermediate vertex.   *conservation*

Frequently, condition (ii) or (ii') is referred to as the **conversation equation.**

Condition (iii) insures that the flow moves from source S to sink D and not in the reverse direction.

Note that every transport network has at least one flow, namely, the zero flow, where $F(e) = 0$ for each edge $e \in E(G)$. Another example of a flow is depicted in Figure 7-5. We still follow the convention of labeling each edge e as $(k(e),F(e))$ where $k(e)$ is the capacity of e and $F(e)$ is the flow along e.

Example 7.2.1.   Consider the example of a network flow shown in Figure 7-5.

Note the zero flow into the source S along the edge (b,S). While it is true that for each intermediate vertex x, the flow into x equals the flow out of x, we only illustrate this fact for the vertex e. The flow into e is

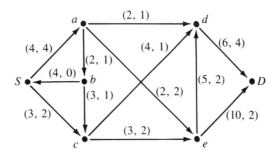

Figure 7-5.

$F(a,e) + F(c,e) = 2 + 2 = 4$, while the flow out of e is $F(e,D) + F(e,d) = 2 + 2 = 4$.

The flow along edges (S,a) and (a,e) is actually equal to the capacity. Such edges are said to be **saturated**; all other edges of Figure 7-5 are **unsaturated.** If an edge e is unsaturated, then we define the **slack** of e in a flow F to be $s(e) = k(e) - F(e)$. Therefore, the slack of edge (e,D) in Figure 7-5 is 8 while the slack of edge (d,D) is 2.

Note further that in this example the flow out of the source S is equal to $F(S,a) + F(S,c) = 4 + 2 = 6$, and this, in turn, is equal to the flow into the sink D. That this is always the case will be proved later, but first we need some additional notation and terminology.

In general, if X and Y are any subsets of $V(G)$, we shall write (X,Y) for the set of all edges which go from a vertex $x \in X$ to a vertex $y \in Y$. In particular, we shall write (x,Y) for $(\{x\},Y)$ and (Y,x) for $(Y,\{x\})$. Let us note that $(V(G),x)$ consists of all edges incident to x, and since these edges are all incident from a vertex $y \in B(x)$, we have $(V(G),x) = (B(x),x)$. Similarly $(x,V(G)) = (x,A(x))$.

Now suppose g is any function from $E(G)$ into the real numbers (g might be, for example, a capacity function or a flow function). We shall write $g(X,Y)$ for the sum of all values $g(x,y)$ where $x \in X$, $y \in Y$, and $(x,y) \in E(G)$. Thus, $g(X,Y) = \Sigma_{(x,y)\in(X,Y)}g(x,y)$ is the sum of all values $g(e)$ where e is an edge from an $x \in X$ to a $y \in Y$. We adopt the convention that $g(X,Y) = 0$ if (X,Y) is the empty set. Likewise, $g(X,\phi)$ will mean a sum of no values of the function if ϕ denotes the empty set. Thus, $g(X,\phi) = 0 = g(\phi,Y)$. For example, in Figure 7-5 let $X = \{S,c,e\}$ and $Y = \{a,b,d,D\}$, then $(X,Y) = \{(S,a),(c,d),(e,d),(e,D)\}$. Note that while there are edges (a,e), (b,S), and (b,c) in the graph G, nevertheless, the directions of these edges are such that these edges are not in (X,Y) but instead are in (Y,X). Moreover, if g is the indicated flow function, then $g(X,Y) = g(S,a) + g(c,d) + g(e,d) + g(e,D) = 4 + 1 + 2 + 2 = 9$. If $X = \{a\}$ and $Y = \{D\}$, then $g(X,Y) = 0$ since there are no edges in (X,Y) in this case. Likewise, $g(Y,X) = 0$ for $X = \{S,a,b,c,d,e\}$ and $Y = \{D\}$, while $g(X,Y) = 6$.

In particular, for $X \subseteq V(G), g(X,V(G))$ is the sum of all values $g(x,y)$, where $x \in X$ and y is any vertex in $V(G)$ for which (x,y) is an edge in $E(G)$. Thus, $g(X,V(G)) = \Sigma_{x \in X}g(x,V(G)) = \Sigma_{x \in X}g(x,A(x))$. Moreover, $g(V(G),X) = \Sigma_{x \in X}g(V(G),x) = \Sigma_{x \in X}g(B(x),x)$. Thus, if g is a flow function, then $g(V(G),x)$ is the flow into x and $g(x,V(G))$ is the flow out of x.

In general if X, Y, and Z are subsets of $V(G)$, then $g(X,Y \cup Z) = g(X,Y) + g(X,Z) - g(X,Y \cap Z)$ by the principle of inclusion and exclusion. In particular, if Y and Z are disjoint sets, then $g(X,Y \cup Z) = g(X,Y) + g(X,Z)$.

Theorem 7.2.1. Let S and D be the source and sink, respectively, of a network (G,k). Let F be a flow in G. Then the flow out of $S = F(S,V(G)) = F(V(G),D) = $ the flow into D.

Proof. We first observe that

$$F(V(G),V(G)) = \sum_{x \in V(G)} F(x,V(G))$$

$$= \sum_{x \in V(G)} F(V(G),x) \qquad (7.2.2)$$

However, by condition (ii) of Definition 7.2.1, $F(x,V(G)) = F(V(G),x)$ for each $x \in V(G) \backslash \{S,D\}$. Thus, the above sum becomes

$$F(S,V(G)) + F(D,V(G)) = F(V(G),S) + F(V(G),D) \qquad (7.2.3)$$

But, in addition, we know that $F(V(G),S) = 0 = F(D,V(G))$ since there is no flow into S and no flow out of D. Thus, equation (7.2.3) becomes $F(S,V(G)) = F(V(G),D)$ and the theorem is proved. □

Definition 7.2.2. Let F be a flow defined on the transport network (G,k). Let $a_1, a_2, \ldots, a_m$ denote all the vertices that are incident from the source S, and let $b_1, b_2, \ldots, b_t$ denote all the vertices incident to the sink D. Thus, $A(S) = \{a_1, a_2, a_3, \ldots, a_m\}$ and $B(D) = \{b_1, b_2, b_3, \ldots, b_t\}$. The quantity

$$|F| = F(S,a_1) + F(S,a_2) + \cdots + F(S,a_m)$$

$$= F(S,V(G))$$

$$= F(b_1,D) + F(b_2,D) + \cdots + F(b_t,D)$$

$$= F(V(G),D)$$

is called the **value** of the flow F.

The flow in Figure 7-5 has value 6, while the flow in Figure 7-4 has value 46.

Definition 7.2.3. A flow F in a network (G,k) is called a **maximal flow** if $|F| \geq |F'|$ for every flow F' in (G,k).

Thus, a flow F is maximal if the value of F is the largest possible value of any flow. In other words, a flow F is maximal if there is no flow F^* such that $|F| < |F^*|$.

We should hasten to remark that, at this stage, it is not clear that a given network (G,k) must have a maximal flow at all. It is at least conceivable that the set of values of flows could be an infinite set of numbers that does not contain a maximal number in that set. Our objective will be to show that there always is a maximal flow, and then our major concern will be to determine the value of a maximal flow in a given network and to discuss a method for constructing such a flow.

Example 7.2.2. Let us explain why the flow in Figure 7-6 is maximal.

The value of any flow defined on the network is such that

$$|F| = F(S,a) + F(S,b)$$

Since $F(S,b) \leq k(S,b) \leq 2$ and $F(S,a) = F(a,d) + F(a,c) \leq k(a,d) + k(a,c) = 1 + 2 = 3, |F| = F(S,a) + F(S,b) \leq 5$. Then, without question, it follows that the indicated flow is a maximal flow.

Not every example will be this easy to analyze. We will have to develop more sophisticated techniques for determining maximal flows.

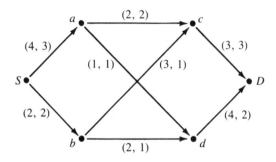

Figure 7-6.

In summary, Definition 7.2.1 and Theorem 7.2.1 combine to produce the following fact:

$$F(x,V(G)) - F(V(G),x) = \begin{cases} |F| & \text{if } x = S \\ -|F| & \text{if } x = D \\ 0 & \text{if } x \neq S,D \end{cases}$$

Exercises for Section 7.2

1. Determine which of the graphs in Figure 7-7 are transport networks. Explain your answer.

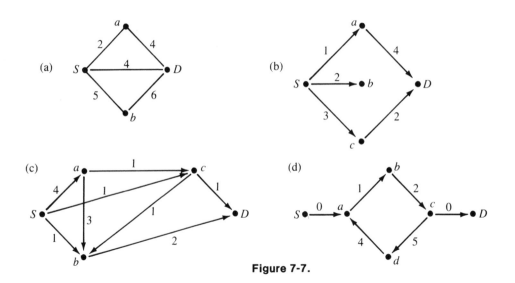

(a)

(b)

(c)

(d)

Figure 7-7.

2. Consider the flow F in the transport network shown in Figure 7-8. In the graph we continue to label edges as (m,n) to indicate that the capacity is m and the flow is n.
   (a) Find the capacity of the edges (S,a) and (a,c).
   (b) Find $k(a,D)$ and $k(b,d)$.
   (c) Find $F(S,a)$, $F(a,c)$, $F(a,D)$, and $F(b,d)$.
   (d) Determine which edges are saturated.
   (e) Determine the slack of edges (a,D) and (S,b).
   (f) Find $|F|$.
   (g) Determine $A(a)$, $A(b)$, and $A(d)$.
   (h) Determine $B(D)$.

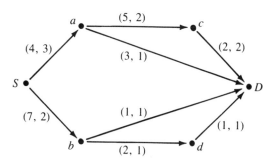

Figure 7-8.

3. Using the capacities in the network of Figure 7-8, determine which of the following are flows. Explain.

(a)

(b)

(c)

(d)

(e)

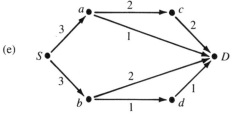

4. Consider the network with capacities as in Figure 7-8.
   (a) Find two different maximal flows.
   (b) Find the value of each of these flows.

5. Seven kinds of military equipment are to be flown to a battlefield by five cargo planes. There are 4 units of each kind of equipment and the five planes can carry 9, 8, 6, 4, and 3 units, respectively. Show that the equipment cannot be loaded in such a way that no 2 units of the same kind are on the same plane. Hint: draw a graph with 14 vertices: 5 vertices representing the planes, 7 vertices representing the kinds of equipment, and 2 vertices S and D representing total number of units of equipment and total loading capacity.

6. For the network and flow F in Figure 7-5, compute $F(X, Y)$ where
   (a) $X = \{a,b,c\}$ $Y = \{d,e\}$
   (b) $X = \{S,c\}$ $Y = \{a,b\}$
   (c) $X = \{S,c\}$ $Y = \{b\}$
   (d) $X = \{b\}$ $Y = \{S,c\}$
   (e) $X = \{a\}$ $Y = \{b\}$
   (f) $X = \{S,a,b,c\}$ $Y = \{D\}$
   (g) $X = \{a,d,e\}$ $Y = \{D\}$
   (h) $X = \{d,e\}$ $Y = \{a,b\}$
   (i) $X = \{S,b,c\}$ $Y = \{a,d,e,D\}$

7. In the following networks with indicated flows, determine the value of the flow and a way to increase the value of the flow if possible. (If it is not possible to increase the value, explain why not.)

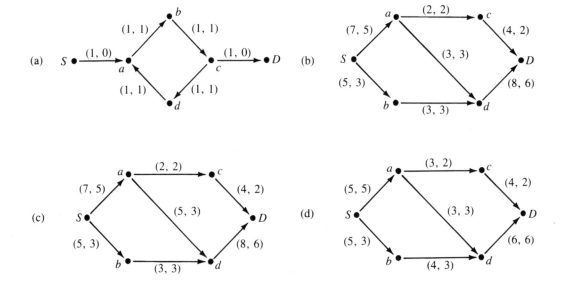

8. In the following transport networks, fill in the missing edge flows so that the result is a flow in the given network. Determine the value of each of the flows.

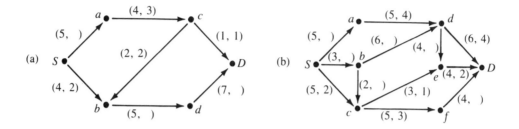

Selected Answers for Section 7.2

5. Label the planes $P_1, P_2, P_3, P_4,$ and P_5 whose capacities are 9, 8, 6, 4, and 3 respectively. Then note that no plane can have more than 7 units of equipment since no 2 units of the same kind are to be in the same plane. Therefore even though P_1 and P_2 can carry 9 and 8 units, the maximum they can carry, in fact, is 7 units. Thus, the 5 planes can carry at most $7 + 7 + 6 + 4 + 3 = 27$ units of equipment. Since there are a total of $4 \times 7 = 28$ units, no such loading is possible.

7. (a) 0
   (b) 8, maximal flow

7.3 MAXIMAL FLOWS AND MINIMAL CUTS

We saw in Sections 7.1 and 7.2 that the solution of many practical problems requires finding a maximal flow for a network. Sometimes, as in Example 7.2.2, a maximal flow can be found quite easily by inspection. Other problems are more difficult.

Example 7.3.1.   Consider the transport network in Figure 7-9(a).

Obviously there are certain limiting values on the possibilities of the value of a flow. For instance, the total capacity of edges incident from the source S limits the value of any flow. Let T_1 be the total capacity of edges from the source and T_2 be the total capacity of edges into the sink. Then,

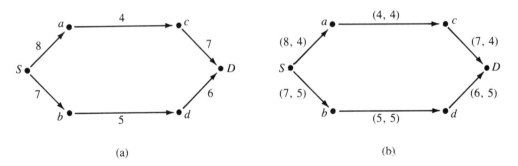

Figure 7-9.

in Figure 7-9(a),

$$T_1 = k(S,a) + k(S,b)$$
$$= 8 + 7 = 15$$
$$T_2 = k(c,D) + k(d,D)$$
$$= 7 + 6 = 13.$$

Is is patent that the value of any flow cannot exceed the smaller of these two values. Thus, $|F| \leq 13$ for any flow F. There are, however, other limitations on the values of a flow. For example, it may be that some edges are unreliable, and if "enough" of these edges fail, then we will not be able to transmit the desired flow. Consider, for example, the edges (a,c) and (b,d) in Figure 7-9. If these edges were removed from the network no amount of flow from S could reach the sink. Thus, we conclude that the capacities of these edges likewise limit the value of a flow; therefore, $|F| \leq k(a,c) + k(b,d) = 4 + 5 = 9$. If we find a flow of value 9, then we know that that flow would be a maximal flow. The flow in Figure 7-9(b) is, therefore, maximal.

Example 7.3.2.   Consider the example shown in Figure 7-10.

The flow out of the source and the flow into the sink are nowhere near capacity. Nevertheless, the "bottleneck" edge (c,d) is saturated, and therefore the flow illustrated is a maximal flow. Of course, the examples of flows shown in Figure 7-11 are also maximal flows in the same transport network.

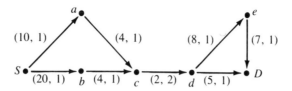

Figure 7-10.

These examples suggest the following definition.

Definition 7.3.1. Suppose that (G,k) is a transport network with source S and sink D. Suppose that X is a set of vertices such that $S \in X$, but $D \notin X$. Let $\overline{X}$ denote the complement of X in $V(G)$. Then the set $(X,\overline{X})$ of all edges from a vertex in X to a vertex in $\overline{X}$ is called an *S-D* **cut.**

In Figure 7-9, the sets $X_1 = \{S\}$, $X_2 = \{S,a,b\}$, and $X_3 = \{S,a,b,c,d\}$ are examples of sets such that $(X_i,\overline{X_i})$ are *S-D* cuts.

The concept of a cut is central to the study of network flow. For example, if a north-south river is viewed as a screen line, the set X can be chosen as the set of vertices east of the river, and then the cut $(X,\overline{X})$ is the set of all westbound roads on bridges crossing the river. Then we can calculate the total traffic from east to west by just counting the traffic across these roads in $(X,\overline{X})$. We are led, therefore, to the notion of the capacity of a cut.

Definition 7.3.2. If C is any set of edges in a transport network (G,k), then the **capacity** of C is the sum of the capacities of the edges of C. Thus, the capacity $k(C)$ is defined by:

$$k(C) = \sum_{e \in C} k(e).$$

In particular, we are interested in the capacities of *S-D* cuts $(X,\overline{X})$. We write $k(X,\overline{X})$ for the sum of all capacities of edges from X to $\overline{X}$. We

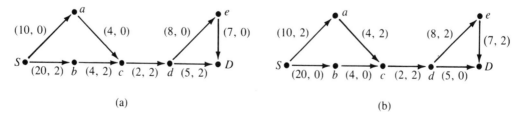

(a)   (b)

Figure 7-11.

caution that there may be edges from $\overline{X}$ to X but they do not enter into the computation of $k(X,\overline{X})$.

Specifically, the cuts with smallest capacities are critical. Thus, we call an S-D cut $(X,\overline{X})$ a **minimal cut** if there is no S-D cut $(Y,\overline{Y})$ such that $k(Y,\overline{Y}) < k(X,\overline{X})$.

In the following example of a transport network, there are 4 vertices other than the source and sink. Therefore, there are $2^4 = 16$ S-D cuts $(X,\overline{X})$ since any subset of $\{a,b,c,d\}$ together with S forms a possible choice for X.

Example 7.3.3. Refer to the transport network shown in Figure 7-12 and the list of S-D cuts in Table 7-1.

Table 7-1. Possible S-D Cuts.

X	$\overline{X}$	Capacity of $(X,\overline{X})$
$\{S\}$	$\{a,b,c,d,D\}$	5
$\{S,a\}$	$\{b,c,d,D\}$	11
$\{S,b\}$	$\{a,c,d,D\}$	10
$\{S,c\}$	$\{a,b,d,D\}$	16
$\{S,d\}$	$\{a,b,c,D\}$	19
$\{S,a,b\}$	$\{c,d,D\}$	12
$\{S,a,c\}$	$\{b,d,D\}$	22
$\{S,a,d\}$	$\{b,c,D\}$	20
$\{S,b,c\}$	$\{a,d,D\}$	15
$\{S,b,d\}$	$\{a,c,D\}$	17
$\{S,c,d\}$	$\{a,b,D\}$	21
$\{S,a,b,c\}$	$\{d,D\}$	17
$\{S,a,b,d\}$	$\{c,D\}$	14
$\{S,a,c,d\}$	$\{b,D\}$	22
$\{S,b,c,d\}$	$\{a,D\}$	13
$\{S,a,b,c,d\}$	$\{D\}$	10

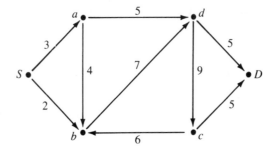

Figure 7-12.

We emphasize that if $X = \{S,a,b\}$, $k(X,\overline{X}) = 12 = k(a,d) + k(b,d) = 5 + 7$. Note that the capacity of the edge (c,b) does not enter into the computation of the capacity of X because the edge (c,b) is in the wrong direction, that is, (c,b) is an edge from $\overline{X}$ to X and not from X to $\overline{X}$. Likewise the capacity of $(Y,\overline{Y})$, where $Y = \{S,b\}$, is $k(b,d) + k(S,a) = 7 + 3 = 10$ since $(Y,\overline{Y}) = \{(b,d),(S,a)\}$. The edges (c,b) and (a,b) do not contribute any value to the capacity of $(Y,\overline{Y})$.

We have seen in Example 7.3.1 that the value of any flow is limited by the capacities of the cuts. That this is true in general is part of the content of the following theorem.

Theorem 7.3.1.   If F is a flow in a transport network (G,k) and if $(X,\overline{X})$ is any S-D cut, then

(a) $|F| = F(X,\overline{X}) - F(\overline{X},X)$, and consequently
(b) $|F| \le k(X,\overline{X})$.

Proof.   If $(X,\overline{X})$ is an S-D cut, then

$$F(X,V(G)) - F(V(G),X) = \sum_{x \in X} F(x,V(G)) - F(V(G),x),$$

$$= F(S,V(G)) - F(V(G),S)$$

$$+ \sum_{x \in X \setminus S} F(x,V(G)) - F(V(G),x)$$

But for $x \notin \{S,D\}$

$$F(x,V(G)) - F(V(G),x) = 0$$

so that the above equation simplifies to

$$F(S,V(G)) - F(V(G),S) = |F|.$$

Thus, $F(X,V(G)) - F(V(G),X) = |F|$.

Moreover, $F(X,V(G)) = F(X,X \cup \overline{X}) = F(X,X) + F(X,\overline{X})$ since $X \cap \overline{X} = 0$. Similarly $F(V(G),X) = F(X \cup \overline{X},X) = F(X,X) + F(\overline{X},X)$. Therefore, $F(X,V(G)) - F(V(G),X)$ simplifies to $F(X,\overline{X}) - F(\overline{X},X)$.

If we define the **net flow across the cut** $(X,\overline{X})$ as $F(X,\overline{X}) - F(\overline{X},X)$, then we see that the value of any flow is equal to the net flow across any S-D cut $(X,\overline{X})$. We may think of $F(X,\overline{X})$ as the flow out of X and $F(\overline{X},X)$ as the flow into X.

Now $|F| = F(X,\overline{X}) - F(\overline{X},X) \le F(X,\overline{X})$ since $F(y,x) \ge 0$ for any $y \in$

$\overline{X}$ and $x \in X$. But also, since $F(x,y) \le k(x,y)$ by the capacity constraint in the definition of a flow, we see that $F(X,\overline{X}) \le k(X,\overline{X})$ so that, in general, the value of any flow is less than or equal to the capacity of any cut.   □

In summary, we have established that the value of a flow F can be computed in several apparently different ways:

(1) $|F|$ = the total flow out of the source S,
(2) $|F|$ = the total flow into the sink D,
(3) $|F|$ = the net flow across any S-D cut.

In truth, methods (1) and (2) are nothing but special cases of (3), for in (1) the total flow out of S equals the net flow across the cut $(X,\overline{X})$, where $X = \{S\}$. Moreover, the total flow into D equals the net flow across the cut $(Y,\overline{Y})$, where $\overline{Y} = \{D\}$.

From this theorem we can readily see that if F^* is a maximal flow and $(M,\overline{M})$ is a minimal cut, then $|F^*| \le k(M,\overline{M})$.

Corollary 7.3.1. Let F be a flow and $(X,\overline{X})$ be an S-D cut such that $|F| = k(X,\overline{X})$. Then F is a maximal flow and $(X,\overline{X})$ is a minimal cut.

Proof. Let F^* be any flow of larger value and $(Y,\overline{Y})$ be any cut of smaller capacity. Then $|F| \le |F^*| \le k(Y,\overline{Y}) \le k(X,\overline{X})$. But the assumption that $|F| = k(X,\overline{X})$ makes all inequalities in fact equalities. Thus, $|F| = |F^*|$ for any flow F^* such that $|F| \le |F^*|$. Hence F is a maximal flow. Moreover, $k(Y,\overline{Y}) = k(X,X)$ for any cut with capacity less than or equal to $k(X,\overline{X})$. Therefore, $(X,\overline{X})$ is a minimal cut.   □

Corollary 7.3.2. Suppose that F is a flow in a network and suppose that $(X,\overline{X})$ is an S-D cut. Then the value of F equals the capacity of $(X,\overline{X})$ if and only if

(i) $F(e) = k(e)$ for each edge $e \in (X,\overline{X})$, (that is, $F(X,\overline{X}) = k(X,\overline{X})$), and
(ii) $F(e') = 0$ for each edge $e' \in (\overline{X},X)$ (or in other words, $F(\overline{X},X) = 0$).

(If these conditions prevail, then F is a maximal flow and $(X,\overline{X})$ is a minimal cut.)

Proof.   Suppose conditions (i) and (ii) hold. Then

$$|F| = \text{net flow across } (X,\overline{X})$$

$$= \sum_{e \in (X,\overline{X})} F(e) - \sum_{e' \in (\overline{X},X)} F(e')$$

$$= \sum_{e \in (X,\overline{X})} F(e) = \sum_{e \in (X,\overline{X})} k(e)$$

$$= k(X,\overline{X})$$

since the edges in $(\overline{X},X)$ do not contribute to the computation of the capacity of $(X,\overline{X})$.

Conversely, if $|F| = k(X,\overline{X})$, then $k(X,\overline{X}) = |F| = \Sigma_{e \in (X,\overline{X})} F(e) - \Sigma_{e' \in (\overline{X},X)} F(e') \le \Sigma_{e \in (X,\overline{X})} k(e) - \Sigma_{e' \in (\overline{X},X)} F(e') = k(X,\overline{X}) - \Sigma_{e' \in (\overline{X},X)} F(e')$ since $F(e) \le k(e)$ for each edge of $(\overline{X},X)$. But then subtracting $k(X,\overline{X})$ from the above inequality gives $0 \ge \Sigma_{e' \in (\overline{X},X)} F(e')$. But since $F(e') \ge 0$ for each edge e' by definition of a flow, we conclude that $\Sigma_{e' \in (\overline{X},X)} F(e') = 0$ and $F(e') = 0$ for each edge $e' \in (\overline{X},X)$. But then $k(X,\overline{X}) = \Sigma_{e \in (X,\overline{X})} F(e) = \Sigma_{e \in (X,\overline{X})} k(e)$ implies $F(e) = k(e)$ for each edge $e \in (X,\overline{X})$. $\square$

Example 7.3.4.   The flow in Figure 7-13 has value 5, and since in Example 7.3.3 we observed that the minimal cut had capacity 5 we conclude that this flow is a maximal flow by Corollary 7.3.1.

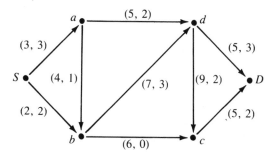

Figure 7-13.

Example 7.3.5.   Consider the flow shown in Figure 7-14.

This flow is maximal since for $X = \{S,a,b\}$, the cut $(X,\overline{X})$ satisfies the conditions of Corollary 7.3.2.

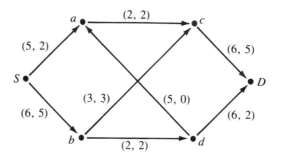

Figure 7-14.

Exercises for Section 7.3

1. Figure 7-15 shows a transport network with the edges labeled with their capacities. Find all S-D cuts and their capacities. What is the minimum cut capacity?

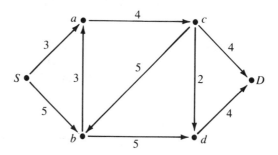

Figure 7-15.

2. Suppose that (G,k) is a transport network with source S and sink D. Show that if there is no directed path from S to D, then the value of a maximal flow in G and the capacity of a minimal cut in G are both zero.

3. Let (G,k) be the network with source S and sink D illustrated in Figure 7-16, where the label on each edge is the capacity.
   (a) Give an example of a flow F in G such that $|F| > 5$.
   (b) Determine $|F|$.
   (c) Let $X = \{S,a,d\}$. List the edges of $(X,\overline{X})$, and find the capacity of $(X,\overline{X})$.

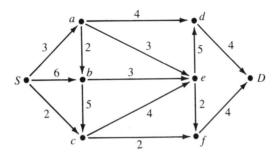

Figure 7-16.

4. In the network shown in Figure 7-17, find at least 7 cuts of capacity 6.

5. (a) Consider 4 people—Allen, Baker, Case, and Dunn—and 4 jobs—j_1, j_2, j_3, and j_4. Allen has the skill to perform jobs j_1, j_2, and j_3. Baker can do jobs j_1 and j_2. Case can do jobs j_2 and j_3, while Dunn can only do job j_4. We wish to either assign the people to the jobs so that all jobs can be done simultaneously or state that the assignment is impossible. It is assumed that each person can do only one job at a time. Formulate this problem as a flow problem. Determine what the value of the flow must be for all 4 people to be assigned. Determine a maximal flow.

   (b) Suppose that 4 people, 2 carpenters, a plumber, and one person who is qualified as a carpenter and as a plumber, apply for 4 job openings, one in carpentry and three in plumbing.
   (1) Explain why the 4 jobs will not all be filled.
   (2) Formulate this as a flow problem as in (a) and find a maximal flow.

Figure 7-17.

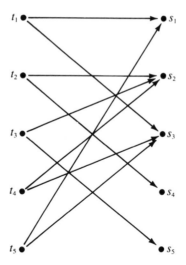

Figure 7-18.

(c) The vertices in Figure 7-18 represent teachers $t_1, t_2, t_3, t_4,$ and t_5, and students $s_1, s_2, s_3, s_4,$ and s_5 at an academy. The edge (t_i, s_j) indicates that the student s_j would like to be taught by teacher t_i.

   (1) Draw a transport network that will answer whether or not the 5 teachers can be assigned to the 5 students, 1 teacher per student, so that each student is taught by a teacher that he or she wants.

   (2) Let S be the source and D the sink. State what $k(S, t_1) = 1$ $k(s_2, D) = 1$ and $k(t_1, s_2) = 1$ means.

   (3) Find a maximal flow in this network.

   (4) Use the maximal flow to arrange the required teacher-student assignment.

6. A set of edges in a directed graph G is called an (edge) **disconnecting set** if its removal from G breaks all directed paths from at least one vertex of G. We say that a disconnecting set C is an a–b disconnecting set if every directed path from a to b is broken by the removal of C.

   (a) Prove every S-D cut is an S-D disconnecting set.

   (b) Show that the set $Y = \{(S,a), (S,b), (a,c)\}$ for the directed graph in Figure 7-9 is an S-D disconnecting set but is not an S-D cut.

7. A set of edges C in a directed graph G is an **(edge) cut-set** if its removal from G breaks all directed paths from at least one vertex of G to at least one other vertex of G and no proper subset of C

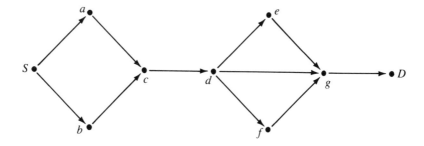

Figure 7-19.

breaks all directed paths between the same pair of vertices. Thus, a set C of edges in G is a cut-set if C is a disconnecting set and no proper subset of C is a disconnecting set. We say that C is an a-b cut-set if every directed path from a to b has been broken by the removal of C. In Figure 7-19 the sets $\{(S,a),(S,b)\}$, $\{(S,b),(a,c)\}$, $\{(c,d)\}$, $\{(d,e),(d,g),(f,g)\}$, and $\{(g,D)\}$ all form S-D edge cut-sets.

(a) Prove that the class of S-D edge cut-sets in a transport network is contained in the class of cuts.

(b) In the transport network of Figure 7-19 find an S-D cut that is not an S-D edge cut-set.

(c) Prove that a minimal S-D cut is necessarily an S-D edge cut-set.

8. Suppose that E is the set of all edge S-D cut-sets in a network (G,k) where S and D are the source and sink. Suppose, moreover, that C is the set of S-D cuts. We know that $E \subseteq C$ and thus, $m_1 = \min\{k(Y)\,|\,Y$ is an edge S-D cut-set$\}$ = the minimum of all capacities of edge S-D cut-sets $\geq \min\{k(X,\overline{X})\,|\,(X,\overline{X})$ is an S-D cut$\}$ = m_2 because the minimum over a larger set is always smaller than the minimum over a subset of that set. Prove that, in fact, $m_1 = m_2$.

9. For the flow F of Example 7.3.5 and $X = \{S,d\}$, compute $F(X,\overline{X})$ and $F(\overline{X},X)$. Do the same for $X = \{S,a,c\}$.

10. (a) If $k(e) = 1$ for each edge e in a transport network (G,k), find another description for the capacity of a cut $(X,\overline{X})$.

(b) If $k(e) = 1$ for each edge e in (G,k), give another description of the value of a flow and of a maximal flow.

11. For each of the following examples in Figure 7-20, determine (a) $|F|$, (b) $F(X,\overline{X})$, and (c) $k(X,\overline{X})$ where $X = \{S,a\}$.

In each of the examples, it is possible to increase the flow. Can you see how to increase the flow? If so, determine the value of a maximal flow.

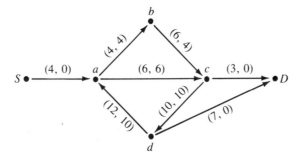

Figure 7-20.

12. Disprove the following:
    (a) If all the edges of a transport network have the same capacity, then all S-D cuts have the same capacity.
    (b) If F is a flow such that $|F| = 0$, then F is the zero flow.
    (c) If all edges in a transport network from the source S are saturated, then all edges into the sink D are saturated.
    (d) If $(X,\overline{X})$ is a minimal S-D cut in a transport network, then there are no edges in G with capacity smaller than $C = \min\{k(e) \mid e \in (X,\overline{X})\}$.
    (e) If F is a maximal flow, then there are no edges with larger flow than $\max\{F(e) \mid e$ is an edge from $S\}$.
    (f) In the transport network shown in Figure 7-21, with indicated capacities, $(X,\overline{X})$ is a minimal cut where $X = \{S,a,b,c,d\}$.

13. A set of words is to be transmitted as messages. We want to investigate the possibility of representing each word by one of the letters of that word chosen such that the words can be represented uniquely. If such a representation is possible, we can transmit a single letter instead of a complete word for a message we want to send. For the following sets of words, design a network model and determine whether or not such a representation is possible.
    (a) $\{bcd,aef,abef,abdf,abc\}$
    (b) $\{ace,bc,dab,df,fe\}$
    (c) $\{abc,bde,ac,bc,c\}$

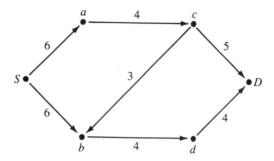

Figure 7-21.

14. (a) Let π_1 and π_2 be two partitions of a set with n elements, both of which contain exactly r disjoint nonempty subsets. State a necessary and sufficient condition for the possibility of selecting r of the n elements such that the r disjoint subsets in π_1 as well as the r disjoint subsets of π_2 are represented.

(b) The integers $2,3,\ldots,20,21$ are partitioned into 4 disjoint subsets according to their remainders upon division by 4. They are also partitioned into 4 disjoint subsets according to the number of prime factors they contain (counting multiplicity), namely $\{\{2,3,5,7,11,13,17,19\}, \{4,6,9,10,14,15,21\}, \{8,12,18,20\}, \{16\}\}$. Is it possible to select 4 integers such that there is a representative for each possible remainder modulo 4 and a representative for each possible number of prime factors?

15. A company advertises for a carpenter, a mechanic, a plumber, and an electrician. The company interviews four applicants. The first is a plumber and a carpenter, the second a mechanic and an electrician, the third a mechanic, and the fourth a carpenter and an electrician. Can the jobs be filled by these four applicants?

16. At a party, there are 6 girls and 6 boys. Girl G_1 know boys B_1, B_2, B_6; girl G_2 knows B_2 and B_5; girl G_3 knows B_2, B_3, B_4; girl G_4 knows B_1, B_3, B_5; while girls G_5 and G_6 know B_1, B_6 and B_2, B_3, B_4, B_5 respectively. During a single dance, can each girl dance with a boy she knows?

17. Let E be a set and $A_1, A_2, \ldots, A_n$ are n subsets of E (these sets need not be disjoint). If we can choose n distinct elements $a_1, a_2, \ldots, a_n$ where $a_i \in A_i$ for each i, then we say that we have a **system of distinct representatives**. For instance, the sets A_i may be different clubs, and it may be that all of the members want to send precisely one representative from each club to a convention. The question is: When is this possible? In the following, use network

flows to determine whether or not the indicated subsets of $E = \{a,b,c,d,e\}$ have a system of distinct representatives. If so, list such a system.

(a) $A_1 = \{a,b\}$, $A_2 = \{a,c,d\}$, $A_3 = \{b,c\}$, $A_4 = \{c,d\}$

(b) $A_1 = \{a,d\}$, $A_2 = \{a,b,c\}$, $A_3 = \{c,d,e\}$, $A_4 = \{c,d,e\}$

(c) $A_1 = \{a,d\}$, $A_2 = \{a,b,c,d,e\}$, $A_3 = \{d\}$, $A_4 = \{a,d\}$

(d) $A_1 = \{a,b,c\}$, $A_2 = \{b,d,e\}$, $A_3 = \{a,c\}$, $A_4 = \{b,c\}$, $A_5 = \{c\}$

(e) $A_1 = \{b,d,e\}$, $A_2 = \{a,e\}$, $A_3 = \{c,d\}$, $A_4 = \{c,d\}$

Selected Answers for Section 7.3

1. The 16 cuts have capacities 8,8,8,9,9,10,11,11,12,13,14,16, 17,18,19,21.

3. (a)

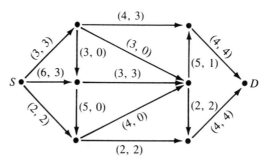

Figure 7-22.

(b) The value of this flow is 8.

6. First, remove all edges of C to obtain a graph H. Then define a set X of vertices v of G such that there is a directed path from S to v in H. The complement $\overline{X}$ of X is all vertices of G that cannot be reached via a directed path from S in H. Thus, $D \in \overline{X}$ since C is an S-D disconnecting set. The set of edges $(X,\overline{X})$ is therefore a cut. Now observe that $(X,\overline{X}) \subseteq C$. For if $(x,y) \in (X,\overline{X})$ and $(x,y) \notin C$, then in H, there is a directed path from S to y. This directed path is formed by a directed path S to x, followed by the edge (x,y). However, by definition of X, this implies $y \in X$, a contradiction. Hence, $(x,y) \in C$ and $(X,\overline{X}) \subseteq C$.

7. (a) By Exercise 6, any S-D disconnecting set C contains a cut $(X,\overline{X})$. If C is, in fact, an edge cut-set then no proper subset of C is a disconnecting set. But the cut $(X,\overline{X})$ is a disconnecting set. Thus, $C = (X,\overline{X})$.

(b) In Figure 7-16 let $X = \{S,e\}$. Then $(X,\overline{X}) = \{(S,a),(S,b),(e,g)\}$ is a cut but not an edge cut-set since the proper subset, $\{(S,a),(S,b)\}$ is a disconnecting set.

(c) Suppose $(X,\overline{X})$ is a minimal cut. If some proper subset of $(X,\overline{X})$ is a disconnecting set C, then C contains another cut $(Y,\overline{Y})$. The capacity of $(Y,\overline{Y})$ must be less than the capacity of $(X,\overline{X})$ since $(Y,\overline{Y}) \subset (X,\overline{X})$ and the capacity is the sum of a subset of the same numbers. This contradicts the fact that $(X,\overline{X})$ was a minimal cut. Thus, $(X,\overline{X})$ has no proper disconnecting subsets, and therefore must be an edge cut-set.

8. In general, the minimum of a set of numbers is less than or equal to the minimum of a subset of those numbers. Let $(X_0,\overline{X}_0)$ be a minimal S-D cut. By Exercise 7(c) $(X_0,\overline{X}_0)$ is also an S-D edge cut-set. Hence, any minimal cut is contained in E. Thus, $m_2 \geq m_1$.

7.4   THE MAX FLOW-MIN CUT THEOREM

Theorem 7.3.1 asserts that the value of any flow, in particular the value of a maximal flow, is at most the capacity of a minimal cut. Corollary 7.3.1 reveals that if ever there is a flow F with value equal to the capacity of some cut $(X,\overline{X})$, then F is a maximal flow and $(X,\overline{X})$ is a minimal cut. Corollary 7.3.2 shows that if $|F| = k(X,\overline{X})$, then edges in $(X,\overline{X})$ must be saturated and edges in $(\overline{X},X)$ must have zero flow. All of these results reveal properties about maximal flows but do not show how to find a maximal flow, and, for that matter, they do not guarantee that a maximal flow even exists. All we know at this point is that if certain conditions prevail then we have a maximal flow, but we do not know whether these conditions must prevail always.

The well-known Max Flow-Min Cut theorem of Ford and Fulkerson comes to our rescue.

Theorem 7.4.1. The Max Flow-Min Cut Theorem. In any transport network, the value of any maximal flow is equal to the capacity of a minimal cut.

This theorem asserts the following:

(1) the existence of a minimal cut $(X,\overline{X})$,
(2) the existence of a maximal flow F, and
(3) the equality of $|F|$ and $k(X,\overline{X})$ for any maximal flow F and any minimal cut $(X,\overline{X})$.

In other words, the inequality stated in Theorem 7.3.1 becomes an equality for a maximal flow and a minimal cut. Moreover, the conditions stated in Corollary 7.3.1 and 7.3.2 do, in fact, always hold for a maximal flow and a minimal cut.

First, let us clear the deck of the first two assertions.

The existence of a minimal cut is an easy thing to observe; the proof is essentially the idea expressed in Example 7.3.3. For a transport network with n intermediate vertices, there are 2^n possible subsets of these n intermediate vertices. By combining each of these sets with the source S we have 2^n possible sets X not containing the sink D. There are, therefore, 2^n S-D cuts $(X, \overline{X})$. Each of these cuts has an associated capacity, and the set of capacities of all S-D cuts obviously contains a smallest number. Any cut with this minimum capacity is a minimal cut. Of course, it is possible that there are several different minimal cuts.

The existence of a flow with maximal value is only slightly more difficult, but the proof requires a few more facts such as the following:

(i) A set A of numbers with an upper bound has a least upper bound L and this number L is the limiting value of a sequence $\{a_n\}$ of numbers in A, that is, $\lim_{n \to \infty} a_n = L$, and

(ii) If $\{a_n\}$ and $\{b_n\}$ are two sequences and $a = \lim_{n \to \infty} a_n$ and $b = \lim_{n \to \infty} b_n$, then $a + b = \lim_{n \to \infty} (a_n + b_n)$.

If these facts are not familiar, the reader must take for granted the existence of a maximal flow and proceed to the proof of (3).

Here we will sketch the proof of (2). Since the value of all flows is less than or equal to the capacity of any cut in G, the set $\mathcal{F}$ of all values of flows in (G, k) is bounded above. Let V be the least upper bound of $\mathcal{F}$. Let $F_1, F_2, \ldots, F_n, \ldots$ be a sequence of flows so that the limit of the values $|F_n|$ is V.

We indicate how to construct a flow with value V. For each edge (x, y) consider the flows $F_n(x, y)$ for all possible n. The sequence of numbers $\{F_n(x, y)\}$ may not approach a limit, but a subsequence of them does. Since there are only finitely many edges in (G, k), we can choose one subsequence such that the set of numbers $\{F_n(x, y)\}$ has a limiting value for all edges (x, y) in G. Then for an edge (x, y) let $F(x, y)$ equal to the limit of $F_n(x, y)$ as n approaches infinity, where F_n ranges over the flows in the chosen subsequence. It is fairly easy to see that F is a flow whose value is V. Thus, F is a maximal flow.

Two Basic Ways to Increase the Value of Flows

Before proceeding to the proof of the final assertion of the max flow-min cut theorem, let us discuss some ideas that will make the proof clearer.

There are two basic ways to increase the value of flows, that is, two basic ways to move more of a commodity toward the sink.

(i) If an edge is not being used to capacity, we could try to send more of the commodity through it.
(ii) If an edge is working against us by sending some of the commodity back toward the source, we could try to reduce the flow along this edge and redirect in a more practical direction.

Let us illustrate by considering an example.

Example 7.4.1. Consider the network with indicated capacities and flow F shown in Figure 7-23.

Suppose that we decide to increase the flow along the edges (S,a), (a,d) and (d,D). In other words, we are increasing the flow in the path S-a-d-D by a certain amount t.

We must decide how large this number t can be. Since the flow is at most the capacity, the flow of (S,a) cannot be increased by more than

$$k(S,a) - F(S,a) = 5 - 3 = 2.$$

Similarly, the flow from a to d and from d to D cannot be increased by more than

$$k(a,d) - F(a,d) = 6 - 0 = 6$$

and

$$k(d,D) - F(d,D) = 6 - 3 = 3,$$

Figure 7-23.

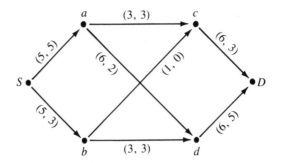

Figure 7-24.

respectively. Since we wish to increase the flows along all three edges by the same amount t, the largest possible value for t is the smallest of these three numbers, that is, $t = \min\{2,6,3\} = 2$. Increasing the flow of each edge in the path S-a-d-D by 2 yields the flow F_1 shown in Figure 7-24.

Notice that the value of F_1 is 8 while the value of F was 6. We know then that F was not a maximal flow. Is F_1 a maximal flow? We answer this by attempting to increase the flow along unsaturated edges.

Consider the path S-b-c-D. Each edge along this path is unsaturated, or in other words, each edge along this path has positive slack.

The slack of (S,b), (b,c) and (c,d) is respectively:

$$k(S,b) - F_1(S,b) = 5 - 3 = 2$$
$$k(b,c) - F_1(b,c) = 1 - 0 = 1$$
$$k(c,D) - F_1(c,D) = 6 - 3 = 3$$

Increasing the flow of each edge on the path S-b-c-D by $\min\{2,1,3\} = 1$ then yields the flow F_2 indicated in Figure 7-25.

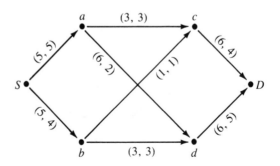

Figure 7-25.

Now there is no path from S to D that does not traverse a saturated edge. In fact, this last flow is a maximal flow since $|F_2| = 9$ and the capacity of the cut $(X,\overline{X})$, where $X = \{S,b\}$, is also 9. Thus, Corollary 7.3.1 implies that F_2 is a maximal flow.

This example suggests the following "greedy" method for increasing the value of the flow F:

(1) Choose any simple directed path $P = \{e_1,e_2,\ldots,e_k\}$ from source S to the sink D where each edge e_i is unsaturated.
(2) Calculate the slack of each edge e_i in P, and let t be the minimum of these numbers.
(3) Increase the value of the flow of each edge e_i of the path P by t.

Repeated application of this "greedy" method produced a maximal flow in Example 7.4.1. But unfortunately this technique does not always produce a maximal flow, as indicated in the following example. In fact, in the next example, there are no S-D paths made up of unsaturated edges, so in this case, we take an alternative approach at increasing the value of the flow by actually *decreasing* the flows along certain edges that may be working against the total flow.

Example 7.4.2. Let F denote the flow indicated in the network shown in Figure 7-26.

We should not, of course, decrease $F(S,a)$, $F(S,b)$, $F(a,D)$, $F(b,D)$ because doing so would decrease the flow out of S or the flow into D (that is, it would decrease the value of F). Our only choice, therefore, is to

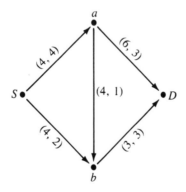

Figure 7-26.

attempt to decrease $F(a,b)$ by some amount t. Since

$$F(S,b) + F(a,b) = F(b,D) \text{ and}$$
$$F(S,a) = F(a,b) + F(a,D),$$

if we decrease $F(a,b)$ by t, we must also increase $F(a,D)$ and $F(S,b)$ by t to avoid decreasing $F(b,D)$ and $F(S,a)$. Hence, the value of the flow can be increased by increasing $F(S,b)$ by t, decreasing $F(a,b)$ by t, and increasing $F(a,D)$ by t. Now

$$k(S,b) - F(S,b) = 4 - 2 = 2 \text{ and } k(a,D) - F(a,D) = 6 - 3 = 3,$$

so (S,b) and (a,D) can be increased by the minimum of $\{2,3\}$. Hence, $t \le 2$. Moreover, since $F(a,b) = 1$ and a flow cannot take on negative values, $F(a,b)$ cannot be decreased by more than 1. Thus, we see that we can choose $t = 1$. Then we obtain the new flow F_1 indicated in Figure 7-27. It follows that F_1 is a maximal flow because $|F_1| = 7 = k(X,\overline{X})$ where $X = \{S,b\}$.

In order to generalize the process illustrated in the last example, let us reconsider the edges whose flows were changed. Notice that the edges (S,b), (a,b), and (a,D) do not form a directed path from S to D. But, of course, these edges do form a nondirected S-D path. In particular, notice that if we *reverse* the direction of the edge whose flow we decreased, we would obtain a new edge that together with (S,b) and (a,D) form a directed S-D path.

Perhaps another example will be instructive.

Figure 7-27.

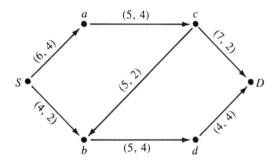

Figure 7-28.

Example 7.4.3. Consider the indicated flow F shown in Figure 7-28.

Consider the path S-a-c-D; the slacks of (S,a), (a,c), and (c,D) are, respectively 2, 1, and 5. Increase the flow by 1 along these edges to obtain the flow F_1 (Figure 7-29).

Now every directed S-D path contains a saturated edge. But the nondirected path S-b-c-D is such that the direction of the edge (c,b) could be reversed to form a directed S-D path with the other 2 edges (S,b) and (c,D). The 2 units of flow along (c,b) we view as working against the total flow. Note that 2 is the minimum of the slack of (S,b), the flow of (c,b), and the slack of (c,D), so we increase the flow along (S,b) and (c,D) by 2 units and decrease the flow along (c,b) by 2 units to obtain the flow F_2 (Figure 7-30). Now $|F_2| = 9 = k(X,\overline{X})$, where $X = \{S,a,b\}$ so that F_2 is a maximal flow.

Figure 7-29.

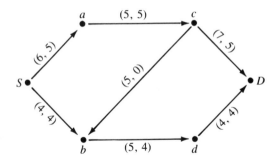

Figure 7-30.

These examples suggest the following definition.

Definition 7.4.1. Suppose that (G,k) is a transport network with source S and sink D. Suppose, moreover, that F is a flow in the network. Recall that a nondirected path P from S to D is a sequence of edges $e_1,e_2,\ldots,e_n$ and a sequence of vertices $S = v_0,v_1,\ldots,v_n = D$ where the edge e_i in G is directed from v_{i-1} to v_i or from v_i to v_{i-1}. The edge e_i is called a **forward edge** of P if it is directed from v_{i-1} to v_i; otherwise, e_i is a **reverse** or **backward edge** of P.

A path P is defined to be an F-**augmenting path** (or a flow-augmenting path for F) if

(i) for each forward edge e of P, $F(e) < k(e)$, and
(ii) for each backward edge e of P, $F(e) > 0$.

In other words, a path P is an F-augmenting path iff all forward edges are unsaturated and all backward edges have positive flow.

In order to state the next lemma let us formulate the definition of an augmenting path in yet another form.

For any path P, let

$$\epsilon_i(P) = \begin{cases} k(e_i) - F(e_i) = \text{the slack } s(e_i) \\ \qquad\quad \text{if } e_i \text{ is a forward edge of } P. \\ F(e_i) \qquad \text{if } e_i \text{ is a backward edge of } P. \end{cases}$$

With P, associate the number $\epsilon(P) = \min \epsilon_i(P)$. Then note that $\epsilon(P)$ is always a nonnegative number. In particular, the path P is an F-augmenting path iff $\epsilon(P) > 0$.

For example, if F is the first flow of Example 7.4.1, then the path P_1: S-a-c-D is not F-augmenting because $\epsilon(P_1) = 0$ since the edge (a,c) is saturated. But, on the other hand, the path P_2: S-b-c-D is F-augmenting because $\epsilon(P_2) = 1$ (here each edge was an unsaturated forward edge of P_2).

If F is the flow in Example 7.4.2, then the path P: S-b-a-D is F-augmenting because the two forward edges (S,b) and (a,D) are unsaturated and the backward edge (a,b) has positive flow. The number $\epsilon(P)$

$$= \min \{s(S,b), F(a,b), s(a,D)\}$$
$$= \min \{2,1,3\} = 1.$$

The path P: S-a-c-D, in Example 7.4.3 was F-augmenting because $\epsilon(P) = 1$. The path P_1: S-b-c-D is F_1-augmenting because $\epsilon(P_1) = 2$.

Now we have the machinery and terminology to state and prove the following lemma.

Lemma 7.4.1. Suppose that (G,k) is a transport network with flow F. If P is an F-augmenting path in G from S to D, then the function $\hat{F}$ defined by

$$\hat{F}(e) = \begin{cases} F(e) + \epsilon(P) & \text{if } e \text{ is a forward edge of } P \\ F(e) - \epsilon(P) & \text{if } e \text{ is a backward edge of } P \\ F(e) & \text{if } e \text{ is an edge of } G \text{ not in } P \end{cases}$$

is a flow with value $|F| + \epsilon(P)$.

Note that $\hat{F}$ is obtained by increasing the flow by $\epsilon(P)$ along forward edges of P, decreasing the flow along backward edges of P, and leaving the flow unchanged on all other edges of G.

Proof. Let P be the sequence of edges $e_1, e_2, \ldots, e_n$ through the vertices $S = v_0, v_1, \ldots, v_n = D$.

To prove $\hat{F}$ is a flow with value $|F| + \epsilon(P)$, we must show that

(a) $0 \le \hat{F}(e) \le k(e)$ for all edges e in G and G_1.

(b) $\hat{F}(x, V(G)) - \hat{F}(V(G), x) = \begin{cases} |F| + \epsilon(P) & \text{if } x = S \\ -|F| - \epsilon(P) & \text{if } x = D \\ 0 & \text{if } x \ne S \text{ or } D. \end{cases}$

To see that (a) holds, note that $F(e_i) + \epsilon(P) \le k(e_i)$ for each forward edge e_i of P since by definition $\epsilon(P) \le s(e_i) = k(e_i) - F(e_i)$. Likewise, $F(e_i) - \epsilon(P) \ge 0$ for each backward edge e_i since $\epsilon(P) \le F(e_i)$.

To prove (b), observe that we need only check for vertices x on P because for all other vertices $\hat{F}$ and F are equal. Furthermore, if $x = v_i$ is on P, recall that only on the two edges e_i and e_{i+1} touching v_i do we change the flow. If the edge $e_{i+1} = (v_i, v_{i+1})$ is a forward edge of P, then $\hat{F}(e_{i+1}) = F(e_{i+1}) + \epsilon(P)$. On the other hand, if $e_{i+1} = (v_{i+1}, v_i)$ is a backward edge of P, then $\hat{F}(e_{i+1}) = F(e_{i+1}) - \epsilon(P)$. In either case, the result of replacing F by $\hat{F}$ on this edge e_{i+1} is to increase the **net flow** out of v_i by $\epsilon(P)$.

Similarly, the changes with regard to the edge e_i joining v_{i-1} to v_i end up decreasing the net flow out of v_i by $\epsilon(P)$. Thus, the net flow out of v_i is $0 + \epsilon(P) - \epsilon(P)$, which is zero, if $i \neq 0$ or n; the net flow out of $v_0 = S$ is $|F| + \epsilon(P)$ and out of $v_n = D$ is $-|F| - \epsilon(P)$. Thus, $\hat{F}$ is a flow and its value is $|F| + \epsilon(P)$. $\square$

It follows from Lemma 7.4.1 that if a flow F admits an augmenting path, then F is not a maximal flow. The next lemma shows that the converse holds. The proof of this lemma contains the main idea of the proof of the max-flow min-cut theorem and contains a proof of assertion (3) that we mentioned before.

Lemma 7.4.2.   If a transport network (G,k) has a flow F with value $|F|$, then the network either contains an F-augmenting path or a cut $(X, \overline{X})$ such that the capacity $k(X, \overline{X}) = |F|$.

Proof.   First, we construct a collection of subsets of $V(G)$. Define $X_0 = \{S\}$. Then define X_1 as the set of all vertices $y \in V(G)$ such that there is a directed edge (S,y) in $E(G)$ such that $F(S,y) < k(S,y)$. Recall that in the definition of flow there is no edge (y,S) with positive flow. In other words, X_1 is the set of vertices y such that the sequence of vertices S,y could be the start of an F-augmenting path.

Suppose next that $X_1, \ldots, X_{k-1}$ have been defined. Then define X_k as the set of all vertices y such that either there is an edge (x,y) such that $F(x,y) < k(x,y)$ or else there is an edge (y,x) such that $F(y,x) > 0$ for some $x \in X_{k-1}$ and such that y has not already been chosen in $X_0, X_1, \ldots, X_{k-1}$.

This process of construction must terminate because the number of vertices is finite. Either we will find D belongs to X_n for some positive integer n or we will not.

Case 1.   If $D \in X_n$, then D arose from some $x_{n-1} \in X_{n-1}$, because $f(x_{n-1}, D) < k(x_{n-1}, D)$. (Recall that in the definition of flow there can be no outgoing edge from D with positive flow.) Similarly x_{n-1} arose from some $x_{n-2} \in X_{n-2}$ where either $F(x_{n-2}, x_{n-1}) < k(x_{n-2}, x_{n-1})$ or $F(x_{n-1}, x_{n-2}) > 0$. Continuing in this manner, we eventually reach $S = x_0$. Then the sequence of vertices $S = x_0, x_1, \ldots, x_n = D$ is the set of vertices of an F-augmenting path.

Case 2. If D does not belong to X_n for any positive integer n, let $X = \cup_{n=1}^{\infty} X_n$. Since $S \in X$ and $D \notin X$, the set of edges $(X,\overline{X})$ form an S-D cut. Moreover, if $(x,y) \in (X,\overline{X})$, then $F(x,y) = k(x,y)$ (since otherwise y would be in X). If (y,x) is any edge of $(\overline{X},X)$, then $F(y,x) = 0$. Consequently, $F(X,\overline{X}) - F(\overline{X},X) = k(X,\overline{X})$. But Theorem 7.3.1 implies $|F| = F(X,\overline{X}) - F(\overline{X},X)$, and Corollary 7.3.1 gives the conclusion that F is a maximal flow and $(X,\overline{X})$ is a minimal cut. $\square$

Now we have all the machinery to make the proof of assertion (3) of the max flow-min cut theorem of Ford and Fulkerson easy.

Suppose F is a maximal flow. By Lemma 7.4.1, F admits no augmenting path, so by Lemma 7.4.2, the network contains a cut $(X,\overline{X})$ such that $|F| = k(X,\overline{X})$, which means, of course, that $(X,\overline{X})$ is a minimal cut. This concludes the proof of assertion (3), and the proof of the theorem is complete. $\square$

In the course of proving Theorem 7.4.1 we proved two lemmas that together give another characterization of maximal flows. We state the combination of Lemmas 7.4.1 and 7.4.2 as a corollary to Theorem 7.4.1.

Corollary 7.4.1. A flow F is maximal iff there is no F-augmenting path from S to D.

If F is a maximal flow, then there are no flow-augmenting paths from S to D. Nevertheless, there may be intermediate vertices v and paths P from S to v such that $\epsilon(P) > 0$. In other words, while it is not possible to augment the flow from S to D, it is possible to augment the flow from S to v. In fact, the set X defined in Case 2 in the proof of Lemma 7.4.2 is the set of all such vertices $v \neq D$ for which the flow can be augmented from S to v. The proof concludes that for this definition of X, the set $(X,\overline{X})$ is a minimal cut.

Consider, for example, the maximal flow in Example 7.4.3. The set X defined above is, in this case, $\{S,a\}$, for the edge (S,a) is unsaturated and constitutes a flow-augmenting path from S to a. For the maximal flow of Example 7.3.4, the set X is $\{S,a,b\}$.

Corollary 7.4.1 is of fundamental importance in the study of network flow because it says, in essence, that in order to increase the value of a flow we need only look for improvements of the two restricted kinds mentioned earlier in the section or in the definition of an augmenting path. The combination of Theorem 7.4.1 and Corollary 7.3.2 gives a useful characterization of minimal cuts.

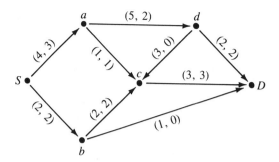

Figure 7-31.

Corollary 7.4.2.   A cut $(X,\overline{X})$ is a minimal cut iff every maximal flow F is such that

(i) $F(e) = k(e)$ for each edge $e \in (X,\overline{X})$, and
(ii) $F(e) = 0$ for each edge $e \in (\overline{X},X)$.

Suppose that $(X,\overline{X})$ is a minimal cut and F is a maximal flow. Then if an edge (x,y) has positive slack and positive flow, it must be that x and y are in the same "half" of any minimal cut $(X,\overline{X})$, that is, x and y are both in X or both in $\overline{X}$.

Example 7.4.4.   In the network with indicated flow shown in Figure 7-31, find a maximal flow and all minimal cuts.

Consider the augmenting path S-a-d-c-b-D. We increase the flow by 1 along all forward edges and decrease the flow by 1 along the backward edge (c,b) to obtain the flow shown in Figure 7-32.

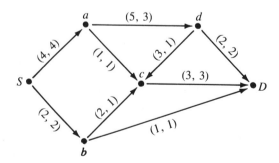

Figure 7-32.

Clearly this flow F is maximal because the value of this flow is 6 and the capacity of the cut $(X,\overline{X})$ where $X = \{S\}$ is also 6.

To find all minimum cuts we use Corollary 7.4.2. If $(Y,\overline{Y})$ is any minimal cut, then all edges of $(Y,\overline{Y})$ must be saturated by the maximal flow F and all edges of $(\overline{Y},Y)$ must have 0 flow. Therefore, none of the edges (a,d), (d,c), (b,c) can be either in $(Y,\overline{Y})$ or in $(\overline{Y},Y)$. Thus, if $a \in Y$, then b, c, and d must also be in Y. Similarly, if $b \in Y$, then so are a, d, c in Y; likewise for c and d. Therefore, $\{a,b,c,d\}$ is in either Y or $\overline{Y}$ for any minimal cut $(Y,\overline{Y})$. Thus, there are only two possibilities for minimal cuts:

(i) $(\{S,a,b,c,d\},\{D\})$ and
(ii) $(\{S\},\{a,b,c,d,D\})$.

Since each cut has capacity 6, each of these two cuts is minimal, therefore these are the only minimal cuts.

Construction of a Maximal Flow

The proof of the max flow-min cut theorem, and especially Lemma 7.4.1 as well as Examples 7.4.1, 7.4.2, and 7.4.3, suggests the following algorithm for increasing the value of the flow in a network.

Algorithm 7.4.1.

Input: a transport network with a given flow F.
Output: a maximal flow.

Repeat steps 1 and 2 below until step 1 finds no augmenting path.

1. Choose an F-augmenting path if possible.
2. Form a new flow of higher value using F and the augmenting path.

If there is ever a stage where, in fact, there are no more flow-augmenting paths, then the flow constructed by the above algorithm will be a maximal flow by Lemma 7.4.2. In other words, if the algorithm does terminate then the output will be a maximal flow. The question as to whether or not the algorithm terminates then is a crucial one.

In their book, *Flows in Networks* [48], Ford and Fulkerson produced an example of a network with irrational capacities for which the above algorithm does not terminate. Nevertheless, we shall show that the algorithm does terminate if the capacity of each edge is an integer. There is a modification of Algorithm 7.4.1, due to Edmonds and Karp, that terminates even for irrational capacities and thus constructs a maximal flow even in the more general case.

Theorem 7.4.2.   If the capacity of each edge in a transport network (G,k) is an integer, then Algorithm 7.4.1 terminates after at most $\Sigma_{e \in E(G)} k(e)$ steps.

Proof.   Start by constructing a sequence of flows $F_0, F_1, \ldots$, such that $F_i(e)$ is an integer for each flow, $|F_j|$ is an integer, and the sequence of values is strictly increasing. First, we let F_0 be the zero flow, that is, $F_0(e) = 0$ for each edge e. Suppose next that we have constructed F_k. If there is an F_k-augmenting path P, then $\epsilon(P)$, being the minimum of a set of positive integers, is a positive integer. The new flow F_{k+1} guaranteed by Lemma 7.4.1 has value $|F_k| + \epsilon(P)$, and $F_{k+1}(e)$ is an integer for each edge e since $F_{k+1}(e)$ is defined to be either $F_k(e) + \epsilon(P)$, $F_k(e) - \epsilon(P)$, or $F_k(e)$ according to whether e is a forward or backward edge of P or an edge not in P. Since

$$|F_k| < |F_k| + 1 \le |F_{k+1}| = |F_k| + \epsilon(P) \le k(X, \overline{X})$$

for any cut $(X, \overline{X})$ and $k(X, \overline{X}) \le \Sigma_{e \in E(G)} k(e)$, we can find at most $\Sigma_{e \in E(G)} k(e)$ new flows. $\square$

A corollary of this proof follows.

Theorem 7.4.3. (Integrality of Flows Theorem).   If each edge of a transport network (G,k) has an integer capacity, then there is a maximal flow F such that $F(e)$ is an integer for each edge e of G.

Let us apply Algorithm 7.4.1 to the following example.

Example 7.4.5.   Find a maximal flow for the network shown in Figure 7-33.

First we look for an augmenting path where, if possible, all edges are forward edges (and they have to be unsaturated). The path P_1: S-c-d-D is

Figure 7-33.

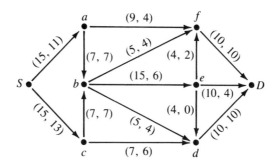

Figure 7-34.

such a path; the minimum slack of the edges along P_1 is 3. By increasing the flow by 3 units along the edges of P_1, we obtain the flow F_1 shown in Figure 7-34.

Now there are no other directed S-D paths with unsaturated edges. Thus, we look for augmenting paths with some backward edges. (Recall that backward edges must have positive flow, and, of course, the forward edges must be unsaturated.)

Sequentially we augment the flow by 1 along the path S-c-d-b-e-D, by 2 along the path S-a-f-e-D, and by 2 along the path S-a-f-b-e-D to obtain the flows shown in Figures 7-35, 7-36, and 7-37 respectively.

Note the decrease in the flow by 1 unit along edge (b,d) in Figure 7-35.

Note the decrease in the flow by 2 units along the edge (e,f) in Figure 7-36.

Note the decrease in the flow by 2 units along the edge (b,f) in Figure 7-37.

There are no more augmenting paths from S to D because (S,a) is

Figure 7-35.

Figure 7-36.

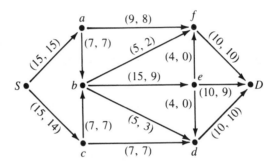

Figure 7-37.

saturated and both edges out of c are saturated. Thus, the last indicated flow is maximal. Of course, we could also argue that this flow is maximal by observing that the value of this final flow is 29 and that the capacity of the cut $(X,\overline{X})$, where $X = \{S,c\}$ is also 29.

In fact, let us determine all minimal cuts as we did in Example 7.4.6. If $(X,\overline{X})$ is a minimal cut, then none of the edges (a,f), (b,f), (b,e), (e,D), (b,d), (S,c) are in $(X,\overline{X}) \cup (\overline{X},X)$. In fact, since $D \in \overline{X}$, it follows that $e \in \overline{X}$, $b \in \overline{X}$, $f \in \overline{X}$, and $a \in \overline{X}$. Moreover, since $S \in X$ and $(S,c) \notin (X,\overline{X}) \cup (\overline{X},X)$, it follows that $c \in X$. Thus, if $X = \{S,c\}$, then $(X,\overline{X})$ is the only minimal cut.

A Labeling Algorithm*

To establish that a given flow F is maximal, we must verify that there are no flow-augmenting paths in the transport network (G,k). For graphs with few vertices and edges we usually can determine by inspection that

---

*The discussion from here to the end of the section may be omitted.

no flow-augmenting paths exist. But for larger graphs, inspection alone is not sufficient. For this reason, we introduce a labeling algorithm.

Basically, we label vertices in agreement with the definition of the sets $X_0, X_1, \ldots, X_k$ used in the proof of Lemma 7.4.2. In other words, first we label the source S and then we label other vertices recursively according to the following scheme.

Suppose a vertex x has been labeled and a vertex y is unlabeled where there is an edge e in the transport network connecting x and y. Then there are two rules we follow:

Forward Labeling: If $e = (x,y)$ (so that e is directed from x to y), then a labeling for y is possible if $F(e) < k(e)$. In other words, y receives a label if the edge e is a forward edge in a flow augmenting path.

Backward Labeling: If, on the other hand, $e = (y,x)$ and $F(e) > 0$ (so that (y,x) is a backward edge), then y receives a label.

In either case, assign y the label x, the name of the previously labeled vertex to which y is adjacent. (Other information could be included in the label for y such as an indicator signifying whether e is a forward or a backward edge, but we shall be content with the simpler label.)

The information this labeling succeeds in conveying is this: since x has been labeled there is a F-augmenting path from S to x, and the label x on the vertex y indicates that a flow-augmenting path from S to y can be constructed by first using the flow-augmenting path from S to x and then using the edge e between x and y as a forward or a backward edge whatever the case may be. In other words, x is a predecessor of y on a flow-augmenting path from S to y. We use the notation $\text{Pred}(y) = x$ to indicate that y has been labeled with the label x and a flow-augmenting path exists from S to y through x. By referring back to the proof of Lemma 7.4.2, we see that since x is labeled, x belongs to some set X_{k-1} for some k and then the possibility of labeling y means that $y \in X_k$.

As in the proof of Lemma 7.4.2 two cases present themselves: either the sink D receives a label or not. If D receives a label, then a flow-augmenting path from S to D is possible and the label on D indicates the predecessor of D on such a path. A flow-augmenting path can be found by using the vertices $W_0 = D$, $W_1 = \text{Pred}(D)$, $W_2 = \text{Pred}(W_1) = \text{Pred}(\text{Pred}(D))$, and so on until eventually we find $W_k = S = \text{Pred}(W_{k-1})$ as the predecessor of some vertex W_{k-1} in the sequence. Of course, should we discover a flow-augmenting path P we can easily determine the amount $\epsilon(P)$ by which we augment the flow F to produce a flow of larger value.

If, on the other hand, it is impossible to give D a label, then the present flow F is maximal and the set X of all labeled vertices is such that $(X, \overline{X})$ is a minimal cut.

The algorithm as presently described will successfully discover a flow-augmenting path or determine that F is a maximal flow, but the algorithm is nondeterministic as it stands. Let us incorporate an indexing of the vertices so as to make the labeling algorithm deterministic. Say $S = v_0, v_1, \ldots, v_n = D$ is an indexing of the vertices where S is first and D is last in the ordering. This will establish a priority by which we examine the vertices in order to label them.

Now let us give a formal description of the labeling algorithm.

Algorithm 7.4.2.   Determining Whether a Flow-Augmenting Path Exists.

Input: a transport network (G,k) with source S and sink D and an indexing of the vertices $S = v_0, v_1, \ldots, v_n = D$, and a given flow F in (G,k).

Output: a flow-augmenting path or the conclusion that none exists.

1. Label the source S with ϕ (to indicate that S has no predecessor).
2. If the sink D is labeled, go to Step 5.
3. (Choose a vertex to be examined.) If no vertices have been examined, choose the source v_0; otherwise, choose the vertex v_i of smallest index i which has not been examined. Set $v_i = x$ and go to Step 4. If no unexamined vertices are available, stop.
4. (Examination of a vertex.) For $j = 1$ to n, do the following: If v_j is labeled, continue. But if v_j is not labeled and can be, label v_j with the label x. Set $\text{Pred}(v_j) = x$. Return to Step 2.
5. If $\text{Pred}(D) = v$, let $W_0 = D$, $W_1 = v$, $W_2 = \text{Pred}(v)$. Continue until $W_k = S$ for some k. Then a flow-augmenting path is $S = W_k, W_{k-1}, \ldots, W_1, W_0 = D$.

Example 7.4.6.   Let us apply the labeling algorithm to determine whether or not a flow-augmenting path exists in the network shown in Figure 7-38.

Let us order the vertices $v_0 = S_1$, $v_1 = a$, $v_2 = b$, $v_3 = c$, $v_4 = d$, $v_5 = D$. First we label S with ϕ. Then at Step 3 of the labeling algorithm, we see that S is the lowest labeled vertex not yet examined. At Step 4, we consider vertices a and c. We cannot label a, but we can label c with the label S. Thus, $\text{Pred}(c) = S$.

Returning to Step 2, since the sink D is not labeled we proceed to Step 3. The labeled vertex with smallest index not yet examined is c. At Step 4, we consider the vertices a, b, and d. The vertices a and d cannot receive labels, but b can; label b with the label c so that $\text{Pred}(b) = c$.

Returning to Step 2, since the sink is not labeled, again we proceed to Step 3. The labeled vertex of lowest index not yet examined is b. At Step

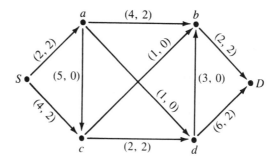

Figure 7-38.

4, we observe that a, d, and D are unlabeled and adjacent to b. Now a can be labeled by backward labeling with the label b, but d and D cannot receive labels. We return to Step 2.

Again, since the sink is not labeled, we go to Step 3. In this case, a is chosen as the vertex to examine. Only d is unlabeled and adjacent to a, and d can receive the forward label a. Return to Step 2 and then to Step 3. Choose d as the labeled vertex of smallest index with unlabeled adjacent vertices. At Step 4, label D with the forward label d. Return to Step 2.

Now at Step 2, D has received a label d and by Step 5, $W_0 = D$, $W_1 = \text{Pred}(D) = d$, $W_2 = \text{Pred}(d) = a$, $W_3 = \text{Pred}(a) = b$, $W_4 = \text{Pred}(b) = c$, $W_5 = \text{Pred}(c) = S$ determine a flow-augmenting path $P: S - c - b - a - d - D$.

Of course, we can now determine by inspection that $\epsilon(P) = 1$, and we can augment the present flow and erase all labels and start the labeling process all over again.

Exercises for Section 7.4

    1. Find a maximal flow and a minimal cut in the following networks:
       (a)

(b)

(c)

(d)

(e)

(f)   (g)

(h)

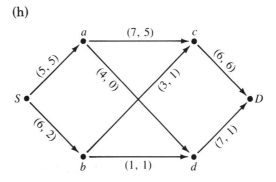

2. Find two maximal flows for the network shown in Figure 7-39 with indicated capacities.

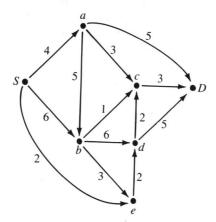

Figure 7-39.

3. Vertices a, b, and c in Figure 7-40 have supplies of 30, 30, and 20 units respectively, and vertices i and j have demands 30 and 50 units. Find a flow satisfying the demands (if possible).

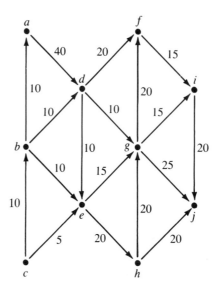

Figure 7-40.

4. Cities a, b, and c can supply 15, 25, and 45 units of merchandise, while cities f, g, and h require 20, 25, and 15 units, respectively. Transportation links are illustrated in Figure 7-41.
   (a) Draw a transport network that reflects total supply and demand.
   (b) Find a maximal flow.
   (c) How many units of merchandise should cities a, b, and c send and cities f, g, and h receive so that as much merchandise as possible is transported?

5. Use Corollary 7.4.2 to show that
   (a) $(X, \overline{X})$, where $X = \{S, c\}$, is the only minimal cut in the network of Example 7.4.5.

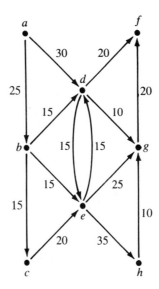

Figure 7-41.

(b) $(X,\overline{X})$, where $X = \{S,b,d\}$, is the only minimal cut of Exercise 1(a) in Section 7.4.

(c) $(X,\overline{X})$ and $(Y,\overline{Y})$, where $X = \{S,a,b,c,d,e,f\}$, and $Y = \{S,a,b,c,d\}$, are the only minimal cuts of the network of Exercise 1(b) of Section 7.4.

(d) $(X,\overline{X})$, $(Y,\overline{Y})$, and $(Z,\overline{Z})$ are the only minimal cuts of the network of Exercise 1(e) of Section 7.4, where $X = \{S,a,b,c\}$, $Y = \{S,a,c\}$, $Z = \{S,a\}$.

6. Find all minimal cuts in the following networks:
   (a)

(b)

(c)

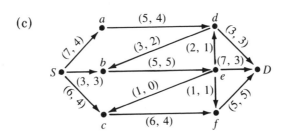

7. Suppose that 7 kinds of equipment are to be flown to a destination by 5 cargo planes. There are 4 units of each kind of equipment and the 5 planes can carry 7, 7, 6, 4, and 4 units, respectively. Determine a loading plan so that no 2 units of the same kind are on the same plane. Hint: see Exercise 5 of Section 7.2; use the algorithm to determine a maximal flow.

8. Use Corollary 7.4.2 to prove that if $(X,\overline{X})$ and $(Y,\overline{Y})$ are minimal cuts, then $(X \cup Y, \overline{X \cup Y})$ and $(X \cap Y, \overline{X \cap Y})$ are also minimal cuts.

9. Suppose that F is a maximal flow in a transport network (G,k). Let $(Y,\overline{Y})$ be any minimal cut and let X be the set defined in Case 2 of the proof of Lemma 7.4.2. Prove that $X \subseteq Y$. Hint: use Exercise 8 and Corollary 7.4.2.

10. A set X of residents belong to various clubs C_1, C_2, C_3, C_4, and to four disjoint political parties P_1, P_2, P_3, and P_4 as depicted by Figure 7-42. Suppose that each club must choose one of its members to represent it, and no person can represent more than one club, no matter how many clubs he belongs to. How should one choose a system of distinct representatives A so that $|A \cap P_j| = 1$ for each $j = 1, 2, 3$? Hint: set up a flow problem.

11. Let (G,k) be a transport network and suppose that $(X,\overline{X})$ is a minimal cut in G. Prove or disprove:
    (a) If F_1 and F_2 are flows in (G,k) such that $F_1(e) = F_2(e)$ for each edge e in $(X,\overline{X}) \cup (\overline{X},X)$, then F_1 and F_2 are maximal flows.

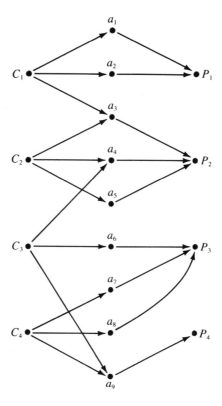

Figure 7-42.

(b) If F_1 and F_2 are maximal flows, then $F_1(e) = F_2(e)$ for each edge $e \in (X, \overline{X}) \cup (\overline{X}, X)$.

12. Suppose that (G, k) is a transport network with integer edge capacities. Suppose further that $e_1 = (v_0, v_1)$, $e_2 = (v_1, v_2)$, ...,$e_k = (v_{k-1}, v_k)$ is a directed k-cycle where $v_k = v_0$. Prove or disprove that there is a maximal flow F in G where $F(e_i) = 0$ for some i.

13. Prove or disprove. If F is a flow in a transport network (G, k) and $(X, \overline{X})$ is a cut such that $k(X, \overline{X}) > |F|$, then $(X, \overline{X})$ is not a minimal cut and F is not a maximal flow.

14. Four instructors A, B, C, and D are available for teaching four courses: algebra, biology, chemistry, and physics. Instructor A is trained in biology and chemistry; B is qualified to teach algebra and physics; C can teach algebra, biology, and chemistry; and D can teach chemistry.

(a) Model the conditions as a transport network.

(b) Use the concept of maximal flow to determine whether or not the four instructors can be assigned to the four courses so that no instructor will be obliged to teach a subject for which he is not trained.

15. True or false? Suppose F is a flow and $(X,\overline{X})$ is an S-D cut.

(a) If F is maximal, then every edge with nonzero capacity will have nonzero flow along it.

(b) If $F(X,\overline{X})$ is equal to $k(X,\overline{X})$, then $(X,\overline{X})$ is a minimal cut.

(c) If $|F| = k(X,\overline{X})$ for a cut $(X,\overline{X})$, then $(X,\overline{X})$ is a minimal cut.

(d) The flow $F(e)$ along an edge can be negative if the edge goes backward.

(e) If $F(X,\overline{X}) = k(X,\overline{X})$, then $F(\overline{X},X) = 0$.

(f) If $|F| = k(X,\overline{X})$ for a cut $(X,\overline{X})$, then $F(\overline{X},X) = 0$.

(g) If there is more than one maximal flow in a network, then there is more than one minimal cut.

(h) If $F(X,\overline{X}) = F(\overline{X},X)$, then F is a maximal flow.

(i) If $|F| = 0$, then $F(e) = 0$ for all edges e.

(j) If e runs directly from S to D, then $F(e) = k(e)$ whenever F is maximal.

(k) A cut $(X,\overline{X})$ may be minimal for some flows but not for others.

(l) The flow across a cut can be larger than the capacity of the cut if there is some flow back from $\overline{X}$ to X.

(m) If F is a maximal flow and $(X,\overline{X})$ is any cut, then $F(e) = 0$ for any edge $e \in (\overline{X},X)$.

(n) If all the edges of a transport network have the same capacity, then all S-D cuts have the same capacity.

(o) If $e = (a,b)$ is an unsaturated edge and $F(e) > 0$ for some maximal flow F, then a,b are both in X for any minimal S-D cut $(X,\overline{X})$.

(p) If all edges of a network are saturated, then the flow is maximal.

(q) If $(X,\overline{X})$ is a minimal cut, then all edges in $(X,\overline{X})$ are saturated.

(r) If $(X,\overline{X})$ is a minimal cut, then all edges in $(\overline{X},X)$ have zero flow.

Selected Answers for Section 7.4

1. (a) Use the augmenting paths S-a-c-D, S-b-d-D, and S-b-c-D shown in Figure 7-43. The value of the flow is 9 and a minimal cut is $(X,\overline{X})$ where $X = \{S,b,d\}$.

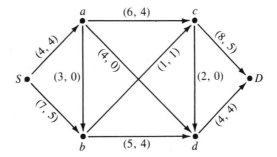

Figure 7-43.

(b) Using the augmenting paths S-a-c-e-D, S-a-b-d-f-D, S-b-d-f-D, and S-a-c-e-f-D, we obtain the flow shown in Figure 7-44. The value of this flow is 8, and the capacity of $(X,\overline{X})$ is also 8, where $X = \{S,a,b,c,d,e,f\}$.

(c) Consider the augmenting path S-b-d-a-c-D. Increase the flow by 1 along the forward edges and decrease the flow by 1 along the edge (a,d) to obtain a flow of value 8. A minimal cut is determined by $X = \{S,a,b,c,d\}$.

(d) Augment the flow by 2 units along S-c-e-D and then by 1 unit along S-a-d-b-e to obtain a flow of value 16. Let $X = \{S,a,b,c,d,e\}$, then $(X,\overline{X})$ is a minimal cut.

(e) Augment by 1 unit along the path S-a-c-D and then by 1 unit along S-b-d-D to obtain a flow of value 2. A minimal cut is determined by $X = \{S,a,b,c\}$.

3. Not possible, a maximal flow has value 50. For example, use paths S-a-d-f-i-D, S-a-d-g-i-D, S-a-d-e-g-i-D, S-b-e-h-j-D, S-c-e-h-j-D, S-b-a-d-e-h-j-D.

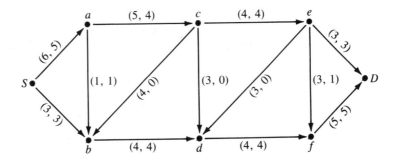

Figure 7-44.

5. (b) If $(Y,\overline{Y})$ is a minimal cut, then using the maximal flow indicated in answer 1(a), note that none of the edges (a,c), (b,d), (c,D), (S,b) can be in $(Y,\overline{Y})$ or in $(\overline{Y},Y)$. Since $S \in Y$, it follows that $b \in Y$ and $d \in Y$. Since $D \in \overline{Y}$, $c \in \overline{Y}$ and $a \in \overline{Y}$. Thus, $Y = \{S,b,d\}$ and $\overline{Y} = \{a,c,D\}$.

(d) Using the maximal flow of answer 1(d), we see that for any minimal cut $(Y,\overline{Y})$ (since $S \in Y$ and (S,a) is not saturated), it must be that $a \in Y$. Likewise, $D \in \overline{Y}$ and (d,D) unsaturated implies that $d \in \overline{Y}$. The cut $(Z,\overline{Z})$ where $Z = \{S,a\}$ satisfies the conclusion of Corollary 7.4.2, so it must be a minimal cut. We determine the other minimal cuts by considering the 2 cases whether b or c is in Y. If $b \in Y$, then c must also be in Y since the edge (b,c) is unsaturated. Then $c \in Y$ and $b \in Y$ is the only other case.

9. See Ford and Fulkerson's book, *Flows in Networks* [48] page 13.

10. Introduce a source and a sink; make the capacity of each edge 1. Find a maximal flow. $A = \{a_1,a_3,a_6,a_9\}$ is such a set.

7.5   APPLICATIONS: MATCHING AND HALL'S MARRIAGE THEOREM

In the exercises and examples of Chapter 7 we have encountered several problems like the ones that follow.

The Assignment Problem. A company has n positions to fill and m applicants. Each applicant has a list of qualifications which make him suitable for certain positions. Is it possible to assign each applicant to a position to which he or she is suited? If not, what is the largest number of people that can be assigned to the positions? And finally: How should these assignments be made?

The Committee Problem (or System of Distinct Representatives Problem). In a certain organization there are n people $\{p_1,p_2,\ldots,p_n\}$ belonging to m committees $\{C_1,C_2,\ldots,C_m\}$. (The committees need not be disjoint.) When is it possible to select a chairperson for each committee so that no member chairs more than one committee? In other words, when is it possible to select a system of m distinct representatives for the different committees? (A system of distinct representatives is a set of m distinct people $\{p_{j_1},p_{j_2},p_{j_3},\ldots,p_{j_m}\}$ where $p_{j_i} \in C_i$.)

These problems can be given another twist. This version is known as

The Marriage Problem. Given m girls and n boys, under what conditions can all the girls be married provided no girl is married to a boy she does not know?

All of these problems can be dressed up in many different costumes, but there are certain common elements in all of them. Let us attempt to get at the essence of the problems with the following definitions.

Definition 7.5.1. A directed bipartite graph is a directed graph G whose set of vertices is the disjoint union of two subsets A and B such that each edge of G is from vertices of A to vertices of B.

A **matching** for G is a set of edges M with no vertices in common. A **maximal matching** for G is a matching M with a maximal number of edges. A **complete matching** for G is a matching M having the property that for each $a \in A, (a,b) \in M$ for some $b \in B$. Or to put it another way: A matching M is complete iff the number of edges in M equals the number of vertices in A.

In other words, the set of edges M is a matching for G if there is a subset A_0 of A and a subset B_0 of B such that

(i) for each vertex $a \in A_0$ there is exactly one vertex $b \in B_0$ such that the edge $(a,b) \in M$. Moreover,

(ii) for each vertex $b \in B_0$, there is exactly one vertex $a \in A_0$ such that $(a,b) \in M$.

Then, of course, M is a complete matching iff $A_0 = A$.

The assignment problem can be modeled with a directed bipartite graph G with vertices in A representing the applicants and the vertices in B representing the positions. Furthermore, an edge (a,b) signifies that applicant a is qualified for position b. The assignment problem calls for determining whether or not there is a complete matching. But failing that, we are asked to determine the number of edges in a maximal matching. Finally, we are to exhibit a maximal matching.

The other problems can also be modeled by a bipartite graph. For the committee problem we let the vertices of A represent the committees and the vertices of B the people. Then an edge (a,b) signifies that person b belongs to committee a. In this context, a system of distinct representatives is nothing more than a complete matching.

Of course, in the marriage problem the vertices of A represent the girls, the vertices of B represent the boys, and an edge (a,b) conveys the information that girl a knows boy b.

Basic Strategy for Solutions of the Problems

While we have been able to model each of these problems with a directed bipartite graph, we have not yet determined a solution for any of them. Of course, having solved simpler versions of these problems in the exercises, we have already gained insight into how we will proceed. Basically, our strategy will be to extend our bipartite graph to a transport network and then apply the maximal flow-minimal cut theorem.

We design the transport network as follows: We adjoin a source S and a sink D and edges (S,a) and (b,D) for each $a \in A$ and each $b \in B$. Moreover, we assign the capacity 1 to all edges. This resulting network we call a **matching network.** Let us illustrate one such matching network where $|A| = 4$ and $|B| = 5$ (Figure 7-45).

The next theorem relates matching networks and flows.

Theorem 7.5.1. Let G be a directed bipartite graph whose vertex set is the disjoint union of subsets A and B such that each edge of G is from vertices in A to vertices in B. Let (G^*,k) denote the associated matching network. Then

(a) A flow in the network (G^*,k) gives a matching for G. A vertex $a \in A$ is matched to a vertex $b \in B$ iff the flow across edge (a,b) is 1.

(b) A maximal flow gives a maximal matching for G.

(c) A flow whose value is $|A|$ is a complete matching for G.

Proof. Let F be a flow in (G^*,k). Suppose that $F(a,b) = 1$, where $a \in A$ and $b \in B$. Then the only edge into a is (S,a) and it must be that

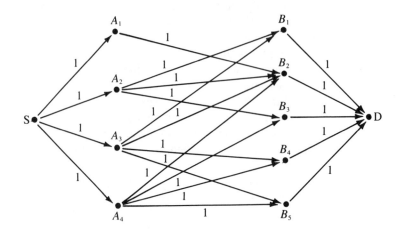

Figure 7-45.

$F(S,a) = 1$, or in other words, the flow into a is 1. Moreover, since the flow out of a is 1 and $F(a,b) = 1$, (a,b) is the only edge of the form (a,x) for which $F(a,x) = 1$. Similarly the only edge of the form (x,b) such that $F(x,b) = 1$ is the edge (a,b). Thus, if $M = \{$edges (a,b) where $a \in A$ and $b \in B$ and $F(a,b) = 1\}$, then the edges of M have no vertices in common and M is a matching for G.

Parts (b) and (c) follow immediately from the fact that the number of vertices of A matched with vertices of B is the value of the flow F.

Of course, since a maximal flow produces a maximal matching, we can apply the flow-augmenting path algorithm to the matching network and determine a maximal matching.　　□

Complete Matchings

A complete matching in a directed bipartite graph is obtained precisely when the value of the flow in the associated matching network is $|A|$. Let us show how to use this fact to produce a proof of a famous theorem known as *Hall's Marriage Theorem*. First we need some terminology.

Suppose G is a directed bipartite graph with vertex set equal to the disjoint union of A and B where each edge of G is from vertices of A to vertices of B. Then if $C \subseteq A$, let $R(C)$ be the subset of B of all vertices incident from vertices in C, that is, $R(C) = \{b \in B \,|\, (a,b)$ is an edge in G for some $a \in C\}$. Of course, if G has a complete matching, then we must have $|C| \leq |R(C)|$ for all subsets $C \subseteq A$. In other words, each subset of k elements of A is incident to at least k elements of B if G has a complete matching. We refer to the condition $|C| \leq |R(C)|$, for all subsets C of A, as the **matching condition**. For the assignment problem, the matching condition says that for all m applicants to be assigned, then for each subset of k applicants there must be at least k job positions for which they collectively qualify.

P. Hall proved the marriage theorem in 1935. Later it was discovered that Konig and Egervary had proved an equivalent version in 1931, but both theorems were immediate from an earlier theorem that Menger proved in 1927. Though others had priority, we continue to refer to the theorem by its familiar name.

Theorem 7.5.2. Hall's Marriage Theorem. Let G be a directed bipartite graph whose vertex set is the disjoint union of subsets A and B where each edge of G is from vertices in A to vertices in B. There exists a complete matching for G iff $|C| \leq |R(C)|$ for each subset C of A.

Proof. We have already observed that if there is a complete matching then the matching condition holds.

Suppose that $|C| \le |R(C)|$ for each subset C of A. Let $A = \{a_1, \ldots, a_m\}$, $B = \{b_1, \ldots, b_n\}$, and let (G^*, k) be the associated matching network. Finally, assume $(X, \overline{X})$ is a minimal cut in (G^*, k). Obviously, the value of any flow cannot exceed m. Thus, if we can show that $k(X, \overline{X}) \ge m$, then a maximal flow will have value m and the matching corresponding to this flow will be a complete matching.

After a change of notation we may write $X = \{S, a_1, \ldots, a_s, b_1, \ldots, b_t\}$. Then (consult Figure 7-46), $(X, \overline{X})$ has three types of edges.

I $m-s$ edges from S to $a_{s+1}, \ldots, a_m$,
II t edges from $b_1, \ldots, b_t$ to D,

and

III   edges from $a_1, \ldots, a_s$ to $b_{t+1}, \ldots, b_n$.

We need only estimate the number of edges of type III. Let $C = \{a_1, \ldots, a_s\}$. Then by the matching condition $s = |C| \le |R(C)|$. Now $R(C)$ is the disjoint union of W_1 and W_2, where $W_1 = R(C) \cap \{b_1, \ldots, b_t\}$ and $W_2 = $

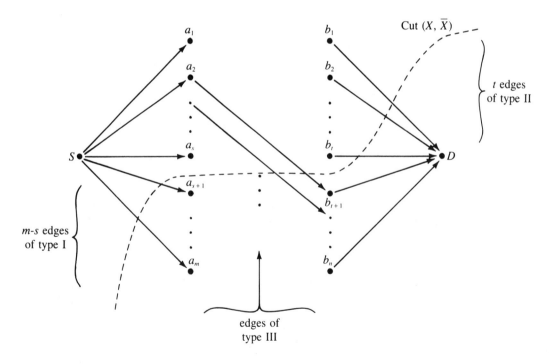

Figure 7-46.

$R(C) \cap \{b_{t+1}, \ldots, b_n\}$. The number of edges of type III is just $|W_2|$. Note $|W_1| \leq t$ and $|W_1| + |W_2| = |R(C)| \geq s$ implies $|W_2| \geq s - t$. Therefore, $k(X, \overline{X}) \geq m$ since the sum of the number of edges of types I, II, and III is greater than or equal to $(m - s) + t + (s - t) = m$. $\square$

The following corollary gives a sufficient condition for the existence of a complete matching.

Corollary 7.5.1. In a bipartite graph G whose vertex set is the disjoint union of A and B where each edge of G is from vertices of A to vertices of B, there exists a complete matching for G if there is an integer k such that

(i) every vertex in A is adjacent to k or more vertices in B and
(ii) every vertex in B is adjacent to k or less vertices in A.

Proof. Given a subset C of A, there are at least $k|C|$ edges from C to B. Since at most k edges go to any element of B, the number of elements of B can be reached from C is at least $k|C|(1/k) = |C|$. Thus, $|R(C)| \geq |C|$. $\square$

Thus, in the committee problem, if each committee has at least k members and no individual is permitted to belong to more than k committees, then it is always possible to find a separate chairperson for each committee. Likewise, in the marriage problem, if every girl knows at least k boys and every boy is known by at most k girls, then each girl can be married to a boy she knows.

Exercises for Section 7.5

1. A law firm has 6 partners and 6 important clients. Each partner has a good relationship with two of the clients, and each client has a good relationship with 2 of the partners. Is it possible to assign each partner to a different client with whom he or she has a good relationship? Explain.

2. Andy, Bob, Cindy, David, and Myrtie wish to form a car pool to drive their children to school. Andy can drive Mondays or Tuesdays, Bob can drive Mondays or Wednesdays, Cindy can drive Tuesdays or Thursdays, David can drive Wednesdays or Fridays, and Myrtie can drive Thursdays or Fridays.
   (a) Draw a graph to model this situation.
   (b) Use Corollary 7.5.1 to show that a complete matching exists.
   (c) Find one such complete matching.

3. Applicant A is qualified for jobs J_1, J_2, and J_6; B is qualified for jobs J_3, J_4, J_5, and J_6; C is qualified for jobs J_1 and J_5; D is qualified for J_1, J_3, and J_6; E is qualified for J_1, J_2, J_4, J_6, and J_7; F is qualified for J_4 and J_6; and G is qualified for J_3, J_5, and J_7.
   (a) Model this problem as a matching network.
   (b) Find a maximal matching.
   (c) Is there a complete matching?

4. Five students have agreed to fix an old house as a church charity project. The house must be painted, wallpapered, and cleaned. In addition, furniture must be moved and new curtains made. Suppose that Anna can paint and move furniture; Jody can make new curtains; Brian can paint, wallpaper, and move furniture; Myrtie can clean and wallpaper; and Joe can paint and wallpaper. Assuming that each does a different job, give an assignment of tasks that will match as many people as possible with jobs that they can do.

5. A set of code words $\{bcd, aefg, abef, abdf, abc, cdeg\}$ is to be transmitted. Is it possible to represent each word by one of the letters in the word so that the words will be uniquely represented? If so, how?

6. Five students, S_1, S_2, S_3, S_4, and S_5, are members of four committees, C_1, C_2, C_3, and C_4. The members of C_1 are S_1, S_3, and S_4; C_2 has members S_3 and S_5; S_1, S_4, and S_5 belong to C_3; and S_1, S_2, S_3, and S_5 belong to C_4. Each committee is to send a representative to a banquet. No student can represent two committees.
   (a) Model this problem as a matching network.
   (b) What is the interpretation of a maximal matching?
   (c) What is the interpretation of a complete matching?
   (d) Find a maximal matching.
   (e) Is there a complete matching?

7. Six senators, S_1, S_2, S_3, S_4, S_5, and S_6, are members of 5 committees. The committees are $C_1 = \{S_2, S_3, S_4\}$, $C_2 = \{S_1, S_5, S_6\}$, $C_3 = \{S_1, S_2, S_5, S_6\}$, $C_4 = \{S_1, S_2, S_4, S_6\}$, and $C_5 = \{S_1, S_2, S_3\}$. The activities of each committee are to be reviewed by a senator who is not on the committee. Can 5 distinct senators be selected for the reviewing tasks? If so, how?

8. There are mn couples at a dance. The men are divided into m groups with n men in each group according to their ages. The women are also divided into n groups with m women in each group according to their heights. Show that m couples can be chosen so that every age and every height will be represented.

9. Six persons $P_1, P_2, \ldots, P_6$ are held in a foreign prison. The prison warden wishes to keep the prisoners separated into cells so that inmates cannot understand each other. Suppose that P_1 speaks

Chinese, French, and Hebrew; P_2 speaks German, Hebrew, and Italian; P_3 speaks English and French; P_4 speaks Chinese and Spanish; P_5 speaks English and German; and P_6 speaks Italian and Spanish. Could these 6 prisoners be locked in 2 cells such that no inmates in the same cell would be able to understand a language the others speak?

10. Suppose m applicants apply for n jobs where $m > n$. Without knowing anything about the qualifications, can you determine whether or not there is a complete matching? Explain.

11. If every girl in school has k boyfriends and every boy in school has k girlfriends, is it possible for each girl to go to the school dance with one of her boyfriends and for each boy to go to the dance with one of his girlfriends? Explain.

12. Give an example of a directed bipartite graph that has a complete matching but does not satisfy the conditions (i) and (ii) of Corollary 7.5.1.

13. Prove or disprove. Any matching is contained in a maximal matching.

14. There are n computers and n disk drives. Each computer is compatible with m disk drives and each disk drive is compatible with m of the computers. Is it possible to match each computer with a compatible disk drive?

15. An $r \times n$ Latin rectangle, where $r \leq n$, is an $r \times n$ matrix that has the numbers $1, 2, \ldots, n$ as entries such that no number appears more than once in the same row or the same column. A Latin square is an $n \times n$ Latin rectangle. If $r < n$ show that it is always possible to append $n - r$ rows to an $r \times n$ Latin rectangle L to form an $n \times n$ Latin square. Hint: do the following:
    (a) Let $A = \{a_1, \ldots, a_n\}$ denote the set of columns of L and let $B = \{1, \ldots, n\}$. Design a directed bipartite graph G with edges (a_i, k) where k does not appear in column a_i of L.
    (b) Show that there is a complete matching for the matching network determined by G.
    (c) Interpret what a complete matching means in this context.

16. Let G be a directed bipartite graph with vertex set V equal to the disjoint union of subset A and B where each edge of G is from vertices of A to vertices of B. Define the *deficiency of G* as $d(G) = \max \{\|C\| - |R(C)| \text{ such that } C \subseteq A\}$.
    (a) Show that G has a complete matching iff $d(G) = 0$.
    (b) Obtain the matching network (G^*, k) for G. Show that the capacity of any cut in (G^*, k) is greater than or equal to $|A| - d(G)$.

(c) Conclude that the maximum number of vertices of A that can be matched with vertices of B is $|A| - d(G)$.

17. Let G be a directed bipartite graph whose vertex set is the disjoint union of subsets A and B where each edge of G is from vertices of A to vertices of B. Suppose that there are 4 or more edges incident from each vertex of A and, moreover, suppose that there are 7 or fewer edges incident to each vertex in B. Prove that the deficiency $d(G) \leq 6$ if $|A| \leq 14$.

18. Let G be a directed bipartite graph with vertex sets A and B as in Exercise 17. Suppose that there are 4 or more edges incident from each vertex of A and 5 or fewer edges incident to each vertex in B. Prove that $d(G) \leq 2$ if $|A| \leq 10$.

19. A telephone switching network was built to route phone calls through incoming lines to outgoing trunks. The 60 incoming lines are partitioned into 3 groups I, II, and III of 20 lines each. A group I line is connected so that it can be switched to one of eight outgoing trunks. A group II line can be switched to one of four outgoing trunks, and a group III line can be switched to one of two outgoing trunks. There are 48 outgoing trunks in all. Moreover, the number of incoming lines that can be switched to an outgoing trunk is at most 6.
   (a) Describe a directed bipartite model for G for this problem.
   (b) Determine an upper bound for $d(G)$. See problem 15 for the definition of $d(G)$.
   (c) How many of the 60 incoming lines will be connected to outgoing trunks?
   (d) At least how many calls will be routed to outgoing trunks when there are calls at all 60 incoming lines?

20. Thirty manufacturers use a certain type of miniature circuit in their microcomputers. There are 24 electronics companies that make this type of circuit. Because transportation costs are high, the computer manufacturers would like to buy their circuits from nearby electronics companies.

   The computer manufacturers are partitioned into 3 transportation categories with 10 members each. Manufacturers in category I are within 50 miles of 4 electronics companies, category II manufacturers are within 50 miles of 2 electronics companies, and category III manufacturers are within 50 miles of only 1 electronics company. No electronics company is within 50 miles of more than 3 manufacturers and no electronics company is able to supply more than 1 manufacturer.
   (a) Describe a directed bipartite graph model for this problem.
   (b) Show that $d(G) \leq 10$.

(c) Conclude that at least 20 supply contracts can be signed between the computer manufacturers and electronics companies that are within 50 miles of each other.

21. Let (G,k) be a transport network with $k(e) = 1$ for each edge e of G.

    (a) Show that the value of a maximal flow is equal to the maximum number of edge-disjoint directed S-D paths in G.

    (b) Show that the capacity of a minimal cut in (G,k) is equal to the minimum number of edges whose deletion destroys all directed S-D paths.

    (c) (*Menger's Theorem.*) Let x and y be two vertices in a directed graph G. Prove that the maximum number of edge-disjoint directed x-y paths in G is equal to the minimum number of edges whose deletion destroys all directed x-y paths in G.

    (d) (*Konig's Theorem.*) Let G be a directed bipartite graph whose vertex set is the disjoint union of subsets A and B where each edge of G is from vertices of A to vertices of B. Then the number of edges in a maximum matching is equal to the minimum number of vertices that cover all edges. (A set of vertices *cover* all the edges of G if each edge is incident with at least one of them.) Hint: Obtain Konig's Theorem as a restatement of part (c).

22. Given an $m \times n$ matrix, M, construct a bipartite graph G with $a_1, a_2, \ldots, a_m$ in A corresponding to the rows of M and with $b_1, b_2, \ldots, b_n$ of B corresponding to the columns of M. In the graph G, let there be an edge from a_i to b_j iff the entry in the ith row and j^{th} column of M is nonzero.

    (a) Interpret what a matching means in this context.

    (b) Expand to a matching network (G^*,k) and interpret the meaning of a minimal cut in (G^*,k).

    (c) Prove the Konig-Egervary Theorem: In any matrix, the maximum number of nonzero entries such that no two are in the same row or column is equal to the minimum number of rows and columns that between them contain all the nonzero entries.

Selected Answers for Section 7.5

5. Design a matching network with $A = \{$code words$\}$, $B = \{a,b,c,d,e,f,g\}$, and edges from a code word to a letter if the letter is in the code word. Obtain a maximal flow.

11. Apply Corollary 7.5.1.

15. See Liu [26] p. 286.

18. For any $C \subseteq A$ consider the subgraph of G consisting of the vertices in C, those in $R(C)$ and edges between the two. Let $e =$ the number of edges in this subgraph. Then $4|C| \le e \le 5|R(C)|$. Thus, $|C| - |R(C)| \le 1/5|C| \le 2$ since $|C| \le 10$. Since $d(G) = \max\{|C| - |R(C)|\}$, $d(G) \le 2$.

19. See Liu [26] p. 289.

21. See Bondy and Murty [4] pp. 203–204.

REVIEW FOR CHAPTER 7

1. In the network with indicated flow shown in Figure 7-47, determine:
   (a) $|F|$,
   (b) $F(X,\overline{X})$ and $k(X,\overline{X})$ for $X = \{S,a\}$, and
   (c) the value of a maximal flow.

2. There are two routes from city A to city D, one passes through city B and the other passes through city C. During the lunch hour traffic, the average trip times are:

   A to B: 15 minutes
   A to C: 30 minutes
   B to D: 15 minutes
   C to D: 15 minutes

   The maximum capacities of the routes are:

   A to B: 1000 vehicles
   A to C: 2000 vehicles
   B to D: 3000 vehicles
   C to D: 2000 vehicles

Figure 7-47.

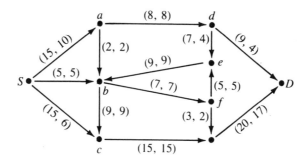

Figure 7-48.

Represent the flow of traffic from A to D during the lunch hour as a transport network.

3. If a transport network has 10 vertices including the source S and sink D, how many S-D cuts are there?

4. In the transport network shown in Figure 7-48:
   (a) Find a maximal flow.
   (b) List all flow-augmenting paths P that you find to increase the value of the flow, showing the value $\epsilon(P)$ for each path.
   (c) Find a minimal cut $(X,\overline{X})$.
   (d) Determine all minimal cuts.

5. Suppose that $(X,\overline{X})$ is an S-D cut in a transport network (G,k). Prove that $(X,\overline{X})$ and $(\overline{X},X)$ contain an equal number of edges in common with any directed circuit in G.

6. There are 50 girls and 50 boys in the senior class at Leon High School. If each girl has dated exactly 5 of the boys and each boy has dated exactly 4 girls, determine whether or not it is possible for each girl to go to the school banquet with a boy she has dated and each boy to go with a girl he has dated.

7. For each of the following collection of sets, determine whether or not there is a system of distinct representatives. If no such system exists, explain why not.
   (a) $C_1 = \{a,b,c\}$, $C_2 = \{a,d\}$, $C_3 = \{a\}$, $C_4 = \{a,d\}$.
   (b) $C_1 = C_2 = C_3 = \{b,d,e\}$, $C_4 = C_5 = \{a,b,c,d,e\}$.

8. The six state universities have, respectively, 13, 7, 7, 7, 7, and 7 graduates with training in computer science and economics. Ten state corporations wish to hire a total of 48 graduates with these qualifications, but company policy at each of the corporations dictates that no 2 newly hired professionals may be hired from the same university. Explain in terms of Hall's Marriage Theorem

whether or not all 48 graduates will be hired by the 10 corporations.

9. Let G be a directed bipartite graph whose vertex set is the disjoint union of subsets A and B where edges in G are from vertices in A to vertices in B. Suppose that there are 3 or more edges incident from each vertex in A, and moreover, suppose that there are 7 or fewer edges incident to each vertex in B. If $|A| \leq 50$, find an upper bound for the deficiency of G.

8

Representation and Manipulation of Imprecision

8.1 FUZZY SETS

8.1.1 Fuzzy Sets—Notation and Terminology

Fuzzy set theory, introduced by Zadeh in 1965, is a generalization of abstract set theory.

A **fuzzy set** consists of objects and their respective grades of membership in the set. The **grade of membership** of an object in the fuzzy set is given by a subjectively defined **membership function.** The value of the grade of membership of an object can range from 0 to 1 where the value of 1 denotes full membership, and the closer the value is to 0, the weaker is the object's membership in the fuzzy set.

In our discussion of representation and manipulation of fuzzy sets we follow closely the excellent presentations in Zemankova-Leech and Kandel [51] and Zadeh [50], as well as the interesting example of an Approximate Reasoning Inference Engine (ARIES) by Appelbaum and Ruspini [46].

Definition 8.1.1. Let U be the universe of discourse, with the generic element of U denoted by u. A *fuzzy subset F of U* is characterized by a *membership function* $m_F: U \rightarrow [0,1]$, which associates with each element u of U a number $m_F(u)$ representing the *grade of membership* of u in F. F is denoted as $\{(u, m_F(u)) \mid u \in U\}$.

Other widely used notations are:

$$F = \int_U m_F(u)/u \qquad \text{when } U \text{ is a continuum} \qquad (8.1.1)$$

$$\begin{aligned} F &= m_F(u_1)/u_1 + \cdots + m_F(u_n)/u_n \\ &= \sum_{i=1}^{n} m_F(u_i)/u_i \end{aligned} \qquad (8.1.2)$$

when U is a finite or countable set of n elements.

Definition 8.1.2. The **support** of F is the set of points in U at which $m_F(u)$ is positive.

Definition 8.1.3. The **height** of F is the least upper bound of $m_F(u)$ over U,

$$hgt(F) = \operatorname*{lub}_{u \in U} m_F(u). \qquad (8.1.3)$$

Definition 8.1.4. A fuzzy set F is said to be **normal** if its height is unity, that is, if

$$\operatorname*{lub}_{u \in U} m_F(u) = 1.$$

Otherwise F is **subnormal**. (It may be noted that a subnormal fuzzy set F may be normalized by dividing m_F by $hgt(F)$.)

Definition 8.1.5. Let A be a fuzzy subset of U, and let B be another fuzzy or nonfuzzy (ordinary) subset of U. A is a *subset* of B or is *contained* in B iff $m_A(u) \leq m_B(u)$ for all elements u of U, that is,

$$A \subseteq B \leftrightarrow m_A(u) \leq m_B(u), \quad u \in U. \qquad (8.1.4)$$

Example 8.1.1. Let the universe of discourse U be the interval $[-20,110]$ with u interpreted as temperature in degrees Fahrenheit. To define a fuzzy subset A of U labeled WARM, it is convenient to be able to express the membership function m_A as a standard function whose parameters may be adjusted to reflect the subjectivity of the characteristic function. A standard function of this type that maps a fuzzy subset of a real line onto the interval $[0,1]$ is the S-function, which is a piecewise quadratic function defined as follows:

$$
\begin{aligned}
S(u; a,b,c) &= 0 & &\text{for } u \le a \\[4pt]
&= 2\left(\frac{u-a}{c-a}\right)^2 & &\text{for } a \le u \le b \\[4pt]
&= 1 - 2\left(\frac{u-c}{c-a}\right)^2 & &\text{for } b \le u \le c \\[4pt]
&= 1 & &\text{for } u \ge c
\end{aligned}
\tag{8.1.5}
$$

where the parameters a and c are the lower and upper fuzzy subset interval bounds, respectively, and the parameter $b = (a + c)/2$ is the **crossover** point, that is, the value of u at which $S(u; a,b,c) = 0.5$.

Using the S-function, the fuzzy subset WARM can be subjectively characterized by the membership function $m_{A_1}(u) = S(u; 60,70,80)$ or $m_{A_2}(u) = S(u; 45,60,75)$ where $m_{A_1}(u)$ was given by a person from Florida whereas a person from Maine expressed the concept of "warm" by $m_{A_2}(u)$. Figure 8-1 shows the graphs of membership functions $m_{A_1}(u)$ and $m_{A_2}(u)$.

Notice that $m_{A_1}(70) = 0.5$ whereas $m_{A_2}(70) = 0.9$ with the interpretation that the degree of agreement with the statement "70°F is warm" would be 0.5 for the person from Florida as compared to 0.9 for the person from Maine.

The crossover points for $m_{A_1}(u)$ and $m_{A_2}(u)$ are 70 and 60, respectively.

The support of A_1 is $(60,110]$ whereas A_2 has $m_{A_2}(u) > 0$ on $(45,110]$.

Both A_1 and A_2 are normal since $hgt(A_1) = hgt(A_2) = 1$.

In general, $m_{A_1}(u) \le m_{A_2}(u)$ for all u in U, hence A_1 is a subset of A_2.

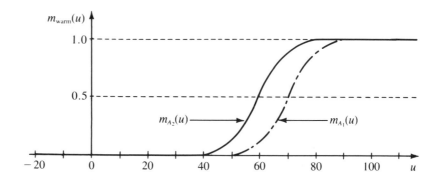

Figure 8-1. Membership function of a fuzzy subset WARM.

Example 8.1.2.   A fuzzy subset B of the universe U (from Example 8.1.1) labeled COLD can be defined in terms of the S-function given below and the corresponding graph in Figure 8-2.

$$m_B(u) = 1 - S(u; 30,45,60)$$

$$
\begin{aligned}
m_B(u) &= 1 && \text{for } u \le 30 \\
&= 1 - 2\left(\frac{u-30}{30}\right)^2 && \text{for } 30 \le u \le 45 \\
&= 2\left(\frac{u-60}{30}\right)^2 && \text{for } 45 \le u \le 60 \\
&= 0 && \text{for } u \ge 60.
\end{aligned}
$$

Example 8.1.3.   A fuzzy subset C labeled PLEASANT (temperature) can be characterized by a membership function of the normal class, that is, $C = \{$set of temperatures $\approx t\}$ (where $\approx$ means approximately equal to) with

$$m_C(u) = \exp\left(-((u-t)/b)^2\right) \quad \text{for all real } u. \tag{8.1.6}$$

The parameter b provides the degree of fuzziness of $\approx$.

Let us consider membership functions given by a Floridian and a person from Maine, respectively which are graphed in Figure 8-3:

$$m_{C_1}(u) = \exp\left(-((u-74)/6)^2\right) \quad \text{and}$$
$$m_{C_2}(u) = \exp\left(-((u-68)/12)^2\right).$$

It can be observed from Figure 8-3 that the Floridian not only likes

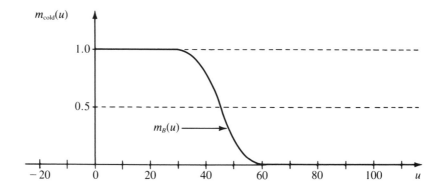

Figure 8-2. Membership function of a fuzzy subset COLD.

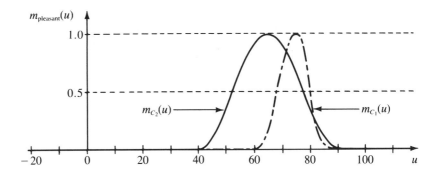

Figure 8-3. Membership function of a fuzzy subset PLEASANT.

warmer temperatures but also has a low tolerance for the deviation from the preferred "pleasant" temperature of 74°F, whereas the person from Maine considers 68°F as ideal and accepts a broader range of temperatures as "pleasant."

Example 8.1.4. Let $U = \{a,b,c,d,e\}$, or equivalently $U = a + b + c + d + e$. In this case a fuzzy subset A of U may be represented as $A = .3/a + .6/b + .9/c + .5/d$. Here the support of $A = \{a,b,c,d\}$, $hgt(A) = .9$, hence A is subnormal.

It is convenient to represent a fuzzy set definition in a tabular form:

u	$m_A(u)$
a	.3
b	.6
c	.9
d	.5

A:

Example 8.1.5. Let $U = \{u \mid u \in I, \ u \geq 0\}$ where I is the set of integers; a fuzzy set A labeled SMALL may be expressed using a function

$$m_A(u) = \sum_a^c \left[1 + \left(\frac{u}{a+b} \right)^2 \right]^{-1} \Big/ u. \qquad (8.1.7)$$

Here the support is the interval $[a,c]$. $m_A(u)$ is a monotonically decreasing function with $m_A(a) = 1$, $\lim_{c \to \infty} m_A(c) = 0$, and the crossover point being b.

Letting $a = 0$, $c = \infty$ and $b = 10$ yields the following definition of the fuzzy set SMALL:

$$SMALL = \sum_{0}^{\infty} \left[1 + \left(\frac{u}{10}\right)^2\right]^{-1} / u,$$

with sample values given below:

u	0	1	3	5	7	10	12	15
SMALL	1	.99	.91	.80	.67	.50	.41	.31

Example 8.1.6. Let $U = \{u \,|\, 0 \leq u \leq 100\}$ with u representing age in years. In this case U is a continuum, and a fuzzy set A labeled OLD may be expressed in terms of a function

$$m_A(u) = \int_a^c \left[1 + \left(\frac{u - c}{b - a}\right)^{-2}\right]^{-1} / u. \tag{8.1.8}$$

$m_A(u)$ is a monotonically increasing function over the support interval $[a,c]$ with $m_A(a) = 0$, $\lim_{c \to \infty} m_A(c) = 1$ and the crossover point is at $u = b$.

Letting the support be the interval $[50,100]$ and the crossover point be at $u = 55$, the definition of OLD becomes

$$OLD = \int_{50}^{100} \left[1 + \left(\frac{u - 50}{5}\right)^{-2}\right]^{-1} / u.$$

We can find various types of membership functions which are advantageous in other applications of fuzzy sets.

8.1.2   Operations on Fuzzy Sets

Throughout the forthcoming discussion, the symbols # and & stand for max and min, respectively; thus, for any real a,b

$$\begin{aligned} a \# b = \max(a,b) &= a \quad \text{if } a \geq b \\ &= b \quad \text{if } a < b \end{aligned} \tag{8.1.9}$$

and

$$\begin{aligned} a \,\&\, b = \min(a,b) &= a \quad \text{if } a \leq b \\ &= b \quad \text{if } a > b \end{aligned} \tag{8.1.10}$$

Consistent with this notation, the symbol $\#_z$ means "least upper bound over the values of z" and $\&_z$ should be read as "greatest lower bound over the values of z" where $z \in Z$.

Among the basic operations that can be performed on fuzzy sets are the following:

Definition 8.1.6. The **complement** of a fuzzy set A is denoted by $\tilde{}A$ (or by $\overline{A}$) and is defined by

$$\tilde{}A = \int_U [1 - m_A(u)]/u. \tag{8.1.11}$$

The operation of complementation is equivalent to negation. Hence, if A is a label for a fuzzy set, then NOT A would be interpreted as $\tilde{}A$.

Definition 8.1.7. The **union** of fuzzy sets A and B is denoted by $A + B$ (or by $A \cup B$) and is defined by

$$A + B = \int_U [m_A(u) \,\#\, m_B(u)]/u. \tag{8.1.12}$$

The union corresponds to the connective OR. Thus, if A and B are labels of fuzzy sets, then A OR B is expressed as $A + B$.

Definition 8.1.8. The **intersection** of fuzzy sets A and B is denoted by $A \cap B$ and is defined by

$$A \cap B = \int_U [m_A(u) \,\&\, m_B(u)]/u. \tag{8.1.13}$$

The intersection corresponds to the connective AND. Hence A AND B is interpreted as $A \cap B$.

Comment. It should be pointed out that $\#$ and $\&$ are not the only operators used as interpretations of the union and intersection, respectively. In particular, when AND is identified with $\&$ (i.e., min), it represents a "hard" AND in the sense that no trade-off is allowed between its operands. By contrast, an AND that is interpreted in terms of the arithmetic product of the operands, acts as a "soft" AND. Which of these or other possible interpretations is more appropriate depends on the applications in which OR and AND are used.

Definition 8.1.9. The **product** of fuzzy sets A and B is denoted by AB and is defined by

$$AB = \int_U m_A(u) \cdot m_B(u)/u. \tag{8.1.14}$$

Thus, A^p, where p is any positive number, is defined by

$$A^p = \int_U [m_A(u)]^p/u. \qquad (8.1.15)$$

Similarly, if w is any nonnegative real number such that $w \cdot hgt(A) \leq 1$, then

$$wA = \int_U w \cdot m_A(u)/u. \qquad (8.1.16)$$

Two operations that are defined as special cases of (8.1.15) are useful in the representation of linguistic hedges.

The operation of **concentration** is denoted by $CON(A)$ and is defined by

$$CON(A) = A^2. \qquad (8.1.17)$$

The concentration is an interpretation of VERY. Thus, if A is a label of a fuzzy set, then VERY A corresponds to $CON(A)$.

The operation of **dilation** is denoted by $DIL(A)$ and is expressed by

$$DIL(A) = A^{0.5}. \qquad (8.1.18)$$

If A is a label of a fuzzy set, then APPROXIMATELY A is interpreted as $DIL(A)$.

Example 8.1.7.   Given the universe of discourse

$$U = 1 + 2 + \cdots + 8,$$

and fuzzy subsets

$$A = .8/3 + 1/5 + .6/6 \quad \text{and} \quad B = .7/3 + 1/4 + .5/6, \qquad (8.1.19)$$

then

$$\begin{aligned}
\tilde{A} &= 1/1 + 1/2 + .2/3 + 1/4 + .4/6 + 1/7 + 1/8 \\
A + B &= .8/3 + 1/4 + 1/5 + .6/6 \\
A \cap B &= .7/3 + .5/6 \\
AB &= .56/3 + .3/6 \\
A^3 &= .512/3 + 1/5 + .216/6 \\
.6B &= .42/3 + .6/4 + .3/6 \\
CON(A) &= .64/3 + 1/5 + .36/6 \\
DIL(B) &= .84/3 + 1/4 + .71/7
\end{aligned} \qquad (8.1.20)$$

Definition 8.1.10. If $A_1, \ldots, A_n$ are fuzzy subsets of $U_1, \ldots, U_n$, respectively, the **Cartesian product** of $A_1, \ldots, A_n$ is denoted by $A_1 \times \cdots \times A_n$ and is defined as a fuzzy subset of $U_1, \ldots, U_n$ whose membership function is expressed by

$$m_{A_1} \times \cdots \times m_{A_n}(u_1, \ldots, u_n) = m_{A_1}(u_1) \ \& \ \cdots \ \& \ m_{A_n}(u_n). \quad (8.1.21)$$

Thus $A_1 \times \cdots \times A_n$ can be written as

$$\int_{U_1 \times \cdots \times U_n} [m_{A_1}(u_1) \ \& \ \cdots \ \& \ m_{A_n}(u_n)]/(u_1, \ldots, u_n) \quad (8.1.22)$$

The concept of the Cartesian product will be further referenced in the discussion of fuzzy relations.

Example 8.1.8. Given $U_1 = U_2 = 1 + 2 + 3$,

$$A_1 = .5/1 + 1./2 + .6/3 \quad \text{and} \quad A_2 = 1./1 + .6/2,$$

then

$$A_1 \times A_2 = .5/(1,1) + 1./(2,1) + .6/(3,1)$$
$$+ .5/(1,2) + .6/(2,2) + .6/(3.2) \quad (8.1.23)$$

Definition 8.1.11. If $A_1, \ldots, A_n$ are fuzzy subsets of $U_1, \ldots, U_n$ (not necessarily distinct), and $w_1, \ldots, w_n$ are nonnegative weights such that $\Sigma_{i=1}^n w_i = 1$, then the **convex combination** of $A_1, \ldots, A_n$ is a fuzzy set A whose membership function is defined by

$$m_A = w_1 m_{A_1} + \cdots + w_n m_{A_n} \quad (8.1.24)$$

where $+$ denotes the arithmetic sum. The concept of a convex combination is useful in the representation of linguistic hedges such as *essentially, typically*, etc., which modify the weights associated with components of a fuzzy set. The weights can also be interpreted as coefficients of importance of the components of a fuzzy set A "built" from fuzzy sets $A_1, \ldots, A_n$.

Example 8.1.9. Let $U_1 = \{u_1 : 10 \le u_1 \le 250\}$ with u_1 representing weight in kilograms. A fuzzy set A_1 labeled as HEAVY may be expressed as

$$\text{HEAVY} = \int_{40}^{100} \left[1 + \left(\frac{u_1 - 40}{30} \right)^{-2} \right]^{-1} / u_1 \quad (8.1.25)$$

with $m_{\text{HEAVY}}(70) = 0.5$. Let $U_2 = \{u_2: 50 \leq u_2 \leq 220\}$ with u_2 as height in centimeters. A fuzzy set A_2 labeled TALL may be defined as

$$\text{TALL} = \int_{140}^{190} \left[1 + \left(\frac{u_2 - 140}{30} \right)^{-2} \right]^{-1} \Big/ u_2 \qquad (8.1.26)$$

with $m_{\text{TALL}}(170) = 0.5$. Then a fuzzy set A labeled BIG may be defined as a convex combination of fuzzy sets HEAVY and TALL, that is, A_1 and A_2:

$$\text{BIG} = 0.6 \text{ HEAVY} + 0.4 \text{ TALL} \qquad (8.1.27)$$

$$m_A(u_1,u_2) = \int_{40}^{100} \int_{140}^{200} [.6 \, m_{A_1}(u_1) + .4 \, m_{A_2}(u_2)]/(u_1,u_2).$$

For example, if we denote the product by *,

$$m_{\text{BIG}}(70,170) = 0.6 \, m_{\text{HEAVY}}(70) + 0.4 \, m_{\text{TALL}}(170)$$
$$= 0.6 * 0.5 + 0.4 * 0.5 = \underline{0.5}$$
$$m_{\text{BIG}}(80,170) = 0.6 \, m_{\text{HEAVY}}(80) + 0.4 \, m_{\text{TALL}}(170)$$
$$= 0.6 * 0.64 + 0.4 * 0.5 = \underline{0.584}$$
$$m_{\text{BIG}}(70,180) = 0.6 \, m_{\text{HEAVY}}(70) + 0.4 \, m_{\text{TALL}}(180)$$
$$= 0.6 * 0.5 + 0.4 * 0.64 = \underline{0.556}$$

Notice that variations in weight have stronger influence on the values of the membership function of the fuzzy set BIG than variations in height. This is due to greater "importance" of the component HEAVY in determining the fuzzy set BIG using the convex combination of HEAVY and TALL.

Definition 8.1.12. If A is a fuzzy subset of U, then a **t-level set** of A is a nonfuzzy set denoted by A_t which comprises all elements of U whose grade of membership in A is greater or equal to t. In symbols,

$$A_t = \{u: m_A(u) \geq t\}. \qquad (8.1.28)$$

A fuzzy set A may be decomposed into its level-sets through the **resolution identity**

$$A = \int_0^1 t A_t \qquad (8.1.29)$$

or

$$A = \sum_t tA_t \qquad (8.1.30)$$

where tA_t is the product of a scalar t with the set A_t [in the sense of (8.1.16)], and $\int_0^1$ (or Σ_t) is the union of the A_t of the A_t sets, with t ranging from 0 to 1.

Since (8.1.1) or (8.1.2) may be interpreted as a representation of a fuzzy set as a union of its constituent fuzzy singletons (m_i/u_i), it follows from the definition of the union (8.1.12) that if in the representation of A we have $u_i = u_j$, then we can make the substitution expressed by

$$m_i/u_i + m_j/u_j = (m_i \# m_j)/u_i. \qquad (8.1.31)$$

For example, given

$$A = .4/a + .7/a + .6/b + .3/b,$$

A may be rewritten as

$$A = (.4 \# .7)/a + (.6 \# .3)/b = .7/a + .6/b.$$

Or conversely,

$$m_i/u_i = (\underset{m_j \in [t, m_i]}{\#} m_j)/u_i, \quad 0 \le t \le m_i. \qquad (8.1.32)$$

For example,

$$.4/a = (.1 \# .2 \# .3 \# .4)/a, \quad t = .1.$$

Thus the resolution identity may be viewed as the result of combining together those terms in (8.1.1) or (8.1.2) which fall into the same level-set.

More specifically, suppose that A is represented in the form

$$A = .1/a + .3/b + .5/c + .9/d + 1/e.$$

Then by using (8.1.32), A can be rewritten as

$$\begin{aligned}
A = .1/a + &.1/b + .1/c + .1/d + .1/e \\
+ &.3/b + .3/c + .3/d + .3/e \\
&+ .5/c + .5/d + .5/e \\
&+ .9/d + .9/e \\
&+ 1/e
\end{aligned}$$

or

$$A = .1\,(1/a + 1/b + 1/c + 1/d + 1/e) + $$
$$.3\,(1/b + 1/c + 1/d + 1/e) + $$
$$.5\,(1/c + 1/d + 1/e) + $$
$$.9\,(1/d + 1/e) + $$
$$1\,(1/e).$$

which is in the form (8.1.30). Using (8.1.28), the level-sets are given by

$$A_{.1} = a + b + c + d + e$$
$$A_{.3} = b + c + d + e$$
$$A_{.5} = c + d + e$$
$$A_{.9} = d + e$$
$$A_1 = e.$$

As will be seen later, the resolution identity provides a convenient way of generalizing various concepts associated with nonfuzzy sets to fuzzy sets. The level-sets of a fuzzy set will be used as the basis in establishing the response set to a given query given a threshold of acceptance, or t-level.

Definition 8.1.13.   The operation of **fuzzification** can be used in transforming a nonfuzzy set into a fuzzy set by assigning proper grades of membership to its elements. Thus, a **fuzzifier** F applied to a fuzzy subset A of U produces a fuzzy subset $F(A;K)$ which is expressed by

$$F(A;K) = \int_U m_A(u)K(u), \qquad (8.1.33)$$

where the fuzzy set $K(u)$ is the **kernel of F,** i.e., the result of applying F to a singleton $1/u$:

$$K(u) = F(1/u;K). \qquad (8.1.34)$$

Here, $m_A(u)K(u)$ represents the product of a scalar $m_A(u)$ and the fuzzy set $K(u)$ [see (8.1.16)], and $\int$ is the union of the family of fuzzy sets $m_A(u)K(u)$, $u \in U$. In effect, (8.1.33) can be viewed as an integral representation of a linear operator, with $K(u)$ being the counterpart of the impulse response.

Example 8.1.10.   Let U, A and $K(u)$ be defined by

$$U = a + b + c + d$$
$$A = .8/a + .6/b$$
$$K(1) = 1/a + .4/b$$
$$K(2) = 1/b + .4/a + .4/c.$$

Then

$$F(A;K) = .8 (1/a + .4/b) + .6 (1/b + .4/a + .4/c)$$
$$= .8/a + .32/b + .6/b + .24/a + .24/c$$
$$= (.8 \ \# \ .24)/a + (.32 \ \# \ .6)/b + .24/c$$
$$= .8/a + .6/b + .24/c.$$

The operation of fuzzification finds an important application in the definition of linguistic hedges such as MORE OR LESS, SOMEWHAT, SLIGHTLY, MUCH, etc. For example, if a set A of positive numbers is labeled as POSITIVE, then SLIGHTLY POSITIVE is a label for a fuzzy subset of the real line. In this case, SLIGHTLY is a fuzzifier of the term POSITIVE.

8.1.3   Fuzzy Relations

Definition 8.1.14.   If U is a Cartesian product of n universes of discourse $U_1,\ldots,U_n$, then an *n-ary fuzzy relation, R, in U* is a fuzzy subset of U. R can be expressed as the union of its constituent fuzzy singletons $m_R(u_1,\ldots,u_n)/(u_1,\ldots,u_n)$, that is,

$$R = \int_{U_1 \times \cdots \times U_n} m_R(u_1,\ldots,u_n)/(u_1,\ldots,u_n) \qquad (8.1.35)$$

where m_R is the membership function of R.

For the sake of simplicity, further discussion will be restricted to binary relations. In this case, $m_R(u,v)$ is a measure of the strength of the link between u and v, which are elements of universes of discourse U and V, respectively. Common examples of binary fuzzy relations are MUCH GREATER THAN, IS SIMILAR TO, IS RELEVANT TO, IS CLOSE TO, IS INFLUENCED BY, etc.

Example 8.1.11.   Let $U = V = 1 + 2 + 3 + 4$. Then the relation MUCH GREATER THAN may be defined by a relation matrix (Table 8-1).

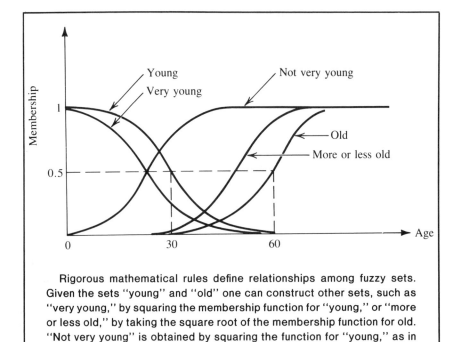

Rigorous mathematical rules define relationships among fuzzy sets. Given the sets "young" and "old" one can construct other sets, such as "very young," by squaring the membership function for "young," or "more or less old," by taking the square root of the membership function for old. "Not very young" is obtained by squaring the function for "young," as in "very young," and then subtracting the result from one, to represent "not."

Figure 8-4. Construction of fuzzy sets from given fuzzy sets.

Table 8-1.

R	1	2	3	4
1	0	0.3	0.8	1
2	0	0	0	0.8
3	0	0	0	0.3
4	0	0	0	0

Example 8.1.12. If $U = V = (-\infty, \infty)$, the fuzzy relation IS CLOSE TO may be defined as

$$\text{IS CLOSE TO} = \int_{U \times V} e^{\dfrac{-|u - v|}{a}} \Big/ (u,v),$$

where $a > 0$ is a scale factor.

Fuzzy relations generalize ordinary relations; therefore, they can be composed as shown in the following definition.

Definition 8.1.15. If R and S are fuzzy relations on $U \times V$ and $V \times W$ respectively, then the composition of R and S is a fuzzy relation on $V \times W$ denoted by $R \circ S$ and defined by

$$R \circ S = \int_{U \times W} \#_V [m_R(u,v) \; \& \; m_S(v,w)]/(u,w) \qquad (8.1.36)$$

If U, V and W are finite sets, then the relational matrix for $R \circ S$ is the min-max product (i.e., operations of addition and multiplication are replaced by min and max, respectively) of the relation matrices for R and S.

Comment. It should be mentioned that $\&$ in (8.1.36) can be min or a product or the algebraic operation used in the definition of the intersection of fuzzy sets.

Definition 8.1.16. A fuzzy binary relation R on $U \times V$ is said to be:

reflexive	iff $m_R(u,u) = 1$, $\forall u \in U$	(8.1.37)
symmetrical	iff $m_R(u,u') = m_R(u',u)$, $\forall u,u' \in U$	(8.1.38)
anti-symmetrical	iff $m_R(u,u') > 0$ and $m_R(u',u) = 0$, $\forall u,u' \in U$, $u \neq u'$	(8.1.39)
@-transitive	iff $m_R(u,u'') \geq m_R(u,u') \; @ \; m_R(u',u'')$ $\forall u,u'$, $u'' \in U$	(8.1.40)

where @ can be replaced by min, a product or other algebraic operation.

When R is transitive, its **transitive closure,** R^+, is defined by

$$R^+ = R + R^2 + \cdots + R^m + \cdots, \text{ where } R^m = R \circ R^{m-1} \quad (8.1.41)$$

in the sense of max-min composition. Due to the reflexivity of R, R^+ exists and is min-transitive.

Reflexive and symmetrical fuzzy relations are called **proximity** relations.

Reflexive, symmetrical and min-transitive fuzzy relations are called **similarity relations.** They are obtained as transitive closures of proximities.

Reflexive, symmetrical and product-transitive fuzzy relations (Example 8.1.12) are weaker than similarity relations.

8.1.4   Cardinality of a Fuzzy Set

Definition 8.1.17.   Let U be a finite universe and A be a fuzzy set on U. Its fuzzy **cardinality,** denoted by $c(A)$, is defined by

$$c(A) = \int_{t \in (0,1]} t / |A_t| \qquad (8.1.42)$$

where $|A_t|$ is the number of elements in the t-level set of A, A_t, as defined in (8.1.28). In other words, $c(A)$ can be defined as a fuzzy set of natural numbers, N, such that

$$\forall\, n \in N, \; m_{c(A)}(n) = \text{lub}\{t \in (0,1] : |A_t| = n\}. \qquad (8.1.43)$$

The concept of a fuzzy set cardinality can be applied when answering a question of the type "How many long streets are there?" or "Are there some states with pleasant climate?"

8.1.5   The Extension Principle

The extension principle is one of the most important concepts in the fuzzy set theory. Application of this principle transforms any mathematical relation between nonfuzzy elements to deal with fuzzy entities.

Definition 8.1.18.   The Extension Principle. Let $A_1, \ldots, A_n$ be fuzzy sets over $U_1, \ldots, U_n$ respectively, with their Cartesian product defined by (8.1.22). Let f be a function from $U_1 \times \cdots \times U_n$ to Y. The fuzzy image B of $A_1, \ldots, A_n$ through f has a membership function:

$$m_B(y) = \text{lub}_{(u_1,\ldots,u_n) \in U_1 \times \ldots \times U_n} \; \min_{i=1,\ldots,n} m_{A_i}(u_i), \quad \forall\, y \in Y \qquad (8.1.44)$$

under the constraint $y = f(u_1, \ldots, u_n)$ and the additional condition

$$m_B(y) = 0 \text{ when } f^{-1}(y) = \{u_1, \ldots, u_n : y = f(u_1, \ldots, u_n)\} = 0 \qquad (8.1.45)$$

The extension principle can be applied in the composition of fuzzy sets of functions, in the algebra of fuzzy numbers, in defining the fuzzy maximum value of a function over a fuzzy domain, in multivalued logic, and in other important areas.

8.1.6   Applications of Fuzzy Set Theory

Fuzzy set theory has gained its importance as a tool for vagueness modeling. It is being applied in the fields of fuzzy systems, fuzzy grammars and languages, and fuzzy algorithms; in models of natural language and approximate reasoning; and in economics, operations research, decision analysis, pattern recognition, system control, medical diagnosis, theory of learning, information retrieval, game theory and other areas of artificial intelligence and computer science. Specific applications in the field of artificial intelligence (AI) and expert systems (ES) will be discussed in Section 8.3.

8.2   POSSIBILITY THEORY

The following example is used to explain the relation between fuzziness and possibility.

Example 8.2.1.   Consider a nonfuzzy statement (or proposition p):

$$p = X \text{ is an integer in the interval } [0,5].$$

The interpretation of the proposition p asserts that:

(i) it is possible for any integer in the interval [0,5] to be a value of X, and
(ii) it is not possible for any integer outside of the interval [0,5] to be a value of X.

In other words, p induces a possibility distribution Π_X which associates with each integer $u \in [0,5]$ the possibility that u could be a value of X. Hence,

$$\Pi_X = \begin{cases} \text{Poss } \{X = u\} = 1 & \text{for } 0 \leq u \leq 5 \\ \text{Poss } \{X = u\} = 0 & \text{for } u < 0 \text{ or } u > 5 \end{cases}$$

where Poss $\{X = u\}$ stands for "The possibility that X may assume the value of u."

Now, "fuzzify" the proposition p:

$$p_F = X \text{ is a small integer}$$

where "small integer" is a fuzzy set defined in the universe of positive integers as

$$\text{SMALL INTEGER} = 1/0 + 1/1 + .9/2 + .7/3 + .5/4 + .2/5$$

where .7/3 signifies that the grade of membership of the integer 3 in the fuzzy set SMALL INTEGER—or, equivalently, the compatibility of the statement that 3 is a SMALL INTEGER—is 0.7.

Consequently, the interpretation of the proposition q can be fuzzified:

> It is possible for any integer to be a SMALL INTEGER with the possibility of X taking a value of u being equal to the grade of membership of u in the fuzzy set SMALL INTEGER.

In other words, p_F induces a possibility distribution Π_X which associates with each integer the possibility that u could be a value of X equal to the grade of membership of u in the fuzzy set SMALL INTEGER. Thus,

$$\text{Poss}\,\{X = 0\} = \text{Poss}\,\{X = 1\} = 1$$
$$\text{Poss}\,\{X = 2\} = .9$$
$$\text{Poss}\,\{X = 3\} = .7$$
$$\text{Poss}\,\{X = 4\} = .5$$
$$\text{Poss}\,\{X = 5\} = .2$$
$$\text{Poss}\,\{X = u\} = 0 \qquad \text{for } u < 0 \text{ or } u > 5.$$

More generally, the above interpretation of a fuzzy proposition can be stated in the following postulate.

Definition 8.2.1. Possibility Postulate. If X is a variable that takes values in U and F is a fuzzy subset of U characterized by a membership function m_F, then the proposition

$$q = X \text{ is } F \tag{8.2.1}$$

induces a **possibility distribution** Π_X which is equal to F, that is,

$$\Pi_X = F \tag{8.2.2}$$

implying that

$$\text{Poss}\,\{X = u\} = m_F(u) \qquad \text{for all } u \in U. \tag{8.2.3}$$

In essence, the possibility distribution of X engenders a fuzzy set which serves to define the possibility that X could take any value u in U. It is important to note that since $\Pi_X = F$, the possibility distribution depends on the definition of F and hence is purely subjective in nature.

Correspondingly, the **possibility distribution function** associated with X (or the possibility distribution function of Π_X) is denoted by p_X and is defined to be numerically equal to the membership function of F, that is,

$$p_X = m_F. \tag{8.2.4}$$

Thus, $p_X(u)$, the **possibility** that $X = u$, is postulated to be equal to $m_F(u)$.

Equation (8.2.2) is referred to as a **possibility assignment equation** because it signifies that the proposition "X is F" translates into the assignment of a fuzzy set F to the possibility distribution of X, or

$$q = X \text{ is } F \rightarrow \Pi_X = F \tag{8.2.5}$$

where $\rightarrow$ stands for "translates into."

More generally, the possibility assignment equation corresponding to a proposition of the form

$$\text{``}N \text{ is } F\text{''} \tag{8.2.6}$$

where F is a fuzzy subset of a universe of discourse U, and N is the name of (i) a variable, (ii) a fuzzy set, (iii) a proposition, or (iv) an object, may be expressed as

$$\Pi_{X(N)} = F \tag{8.2.7}$$

or, more simply,

$$\Pi_X = F.$$

X is either N itself (when N is a variable) or a variable that is explicit or implicit in N, with X taking values in U. Note that also "nested" propositions can be translated into (8.2.7).

Example 8.2.2. Consider the proposition

$$q = \text{Tom is old.}$$

Here $N = \text{Tom}$, $X = \text{Age(Tom)}$, and "old" is a fuzzy set defined on $U = \{u: 0 \le u \le 100\}$ with u signifying Age and characterized by a membership function m_{OLD}. Hence,

$$q = \text{Tom is old} \rightarrow \Pi_{\text{Age(Tom)}} = \text{OLD.}$$

It is clear that the concept of a possibility distribution is closely related to fuzzy sets; therefore, possibility distributions can be manipulated by the rules applicable to fuzzy sets.

The notion of a possibility distribution bears also a close relation to the concept of a **fuzzy restriction**. Hence the mathematical apparatus for fuzzy restrictions, the calculus of fuzzy restrictions, can be used as a basis for the manipulations of possibility distributions.

First, let us define the concept of a fuzzy restriction.

Definition 8.2.2. Let X be a variable which takes values in a universe of discourse U and let $X = u$ signify that X is assigned the value of u where u is an element of U. Let F be a fuzzy subset of U which is characterized by a membership function m_F.

Then F is a **fuzzy restriction on X** (or **associated with X**) if F acts as an elastic constraint on the value that may be assigned to X. In other words, the assignment of a value of u to X has the form

$$X = u : m_F(u) \tag{8.2.8}$$

where $m_F(u)$ is interpreted as the degree to which the constraint represented by F is satisfied when u is assigned to X, or $m_F(u)$ can be thought of as a degree of compatibility of u with F. It follows that $[1 - m_F(u)]$ is the degree to which the constraint represented by F is not satisfied, or it is the degree to which the constraint must be stretched so that the assignment of u to X is possible.

It is important to note that a fuzzy set per se is not a fuzzy restriction. To be a fuzzy restriction, it must be acting as a constraint on the values of a variable. In other words, a variable X can assume values in U depending on the definition on the fuzzy set F.

8.2.1   Possibility-Probability Relationship

Both the concepts of possibility and probability are means of representing and manipulating uncertainty or imprecision. A simple example is used to illustrate the difference between probability and possibility.

Example 8.2.3. Suppose that a 7-member family owns a 4-seat car. Now let us consider how many passengers can ride in the car. This corresponds to the statement "X passengers ride in a car" where X takes on values in $U = \{1, 2, \ldots, 7\}$. We can associate a possibility distribution with X by interpreting $\text{Poss}\{X = u\}$ as the degree of ease with which u passengers can ride in a car. Let us also associate a probability distribution with X where $\text{Prob}\{X = u\}$ stands for the probability that u people will ride in the car. The values of $\text{Poss}\{X = u\}$ and $\text{Prob}\{X = u\}$ are assessed subjectively and are shown in Table 8-2.

Table 8-2. Possibility and Probability Distributions Associated with X.

u	1	2	3	4	5	6	7
Poss$\{X = u\}$	1	1	1	.8	.4	.1	0
Prob$\{X = u\}$	.3	.4	.2	.1	0	0	0

Some intrinsic difference between possibility and probability can be observed from this table. While the probabilities have to sum up to 1 over U, the possibility values are not so restricted. Also notice that the possibility that 3 passengers will ride in the car is 1, while the probability is quite small, that is, 0.2. Thus, a high possibility does not imply a high probability, nor does a low degree of probability imply a low degree of possibility. However, lessening the possibility of an event tends to lessen its probability, but the converse is not true. Furthermore, if an event is impossible, it is bound to be improbable. This heuristic relationship between possibilities and probabilities is called the **possibility/probability consistency principle**.

Example 8.2.4. As another example of the difference between probability and possibility, consider again the statement "Tom is old" with the translation having the form

$$\text{Tom is old} \rightarrow R(\text{Age}(\text{Tom})) = \text{OLD}.$$

Let us assume that the fuzzy set OLD is subjectively defined on $U = \{u: 0 \leq u < 100\}$. Hence $\Pi_{A(X)}(u)$ may have sample values given in Table 8-3.

Table 8-3. Possibility Distribution Associated with OLD.

u	10	20	25	30	35	40	50	60	. . .
$\Pi_{A(X)}(u)$	0	.2	.3	.5	.8	.9	1	1	. . .

To associate a probability distribution with $A(X)$, it is necessary to translate the fuzzy concept "old" into a "hard" (nonfuzzy or nonelastic) concept. Let us adopt the definition that any age over or equal to 60 is "old." Then, the probabilities assume the values as shown in Table 8-4.

Table 8-4. Probability Distribution Associated with OLD.

u	40	50	60	70	80	90	100
Prob$_{A(X)}(u)$	0	0	.2	.2	.2	.2	.2

These tables demonstrate some differences between the concepts and application of possibility and probability, where it appears that the imprecision that is intrinsic in natural languages is possibilistic in nature.

8.3   APPLICATION OF FUZZY SET THEORY TO EXPERT SYSTEMS

8.3.1   Soft Expert Systems—Introduction

The media blitz is on, and hype is coming from all sides. Artificial intelligence (AI)—especially in the form of Knowledge-Based Expert Systems—is now a practical reality, and expert systems (ES) are the hottest glamour item in today's high-tech boom. A new programming technology has been created, based upon "production rules" and "knowledge representation." The result is a new class of enormously powerful high-order languages (HOLs) implemented through "inference engines" which form the heart of "knowledge engineering tools."

These new HOLs make it easier to create, debug, and maintain complex programs through the suppression of explicit control statements, use of English-like syntax, and other notational devices.

Expert Systems (sometimes also called Knowledge-Based Consultant Systems) are structured representations of data, experience, inferences, and rules that are implicit in the human expert. Expert systems draw conclusions from a store of task-specific knowledge, principally through logical or plausible inference, not by calculation.

The objective of an expert system is to help the user choose among a limited set of options, actions, conclusions, or decisions, within a specific context, on the basis of information that is likely to be qualitative rather than quantitative.

The creation of an expert system primarily revolves around the task of putting specific domain knowledge into the system. Expert systems support this task by explicitly separating the domain knowledge from the rest of the system into what is commonly called the *knowledge base*.

The everyday usage of an expert system requires access by the end user to the knowledge base. This is accomplished by a software system called the *inference engine*. It interacts with the user through a user-interface subsystem as shown in Figure 8-5.

Each production rule in an expert system implements an autonomous chunk of expertise that can be developed and modified independently of other rules. When thrown together and fed to the inference engine, the set of rules behaves synergistically, yielding effects that are "greater than the sum of its parts."

Reflecting human expertise, much of the information in the knowledge base of a typical expert system is imprecise, incomplete, or not totally reliable. For this reason, the answer to a question or the advice rendered by an expert system is usually qualified with a "certainty factor" (CF), which gives the user an indication of the degree of confidence that the system has in its conclusion. To arrive at the certainty factor, existing expert systems such as MYCIN—medical diagnosis of infectious blood diseases (Stanford University), PROSPECTOR—location of mineral

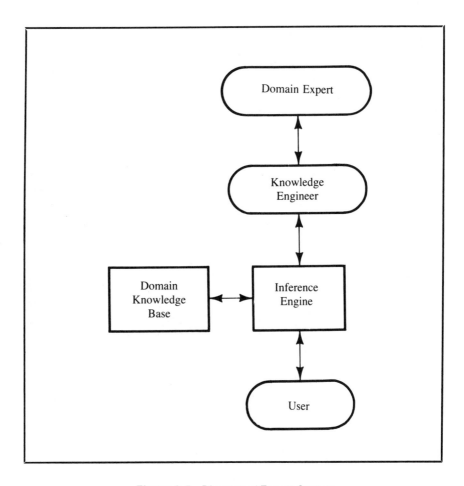

Figure 8-5. Diagram of Expert System.

deposits (Stanford Research Institute), and others employ what are essentially probability-based methods. However, since much of the uncertainty in the knowledge base of a typical expert derives from the fuzziness and incompleteness of data rather than from its randomness, the computed values of the certainty factor are frequently lacking in reliability.

By providing a single inferential system for dealing with the fuzziness, incompleteness, and randomness of information in the knowledge base, fuzzy logic furnishes a systematic basis for the computation of certainty factors in the form of fuzzy numbers. The numbers may be expressed as linguistic probabilities or fuzzy quantifiers, for example, "likely," "very unlikely," "almost certain," "most," "almost all," "frequently." In this perspective, fuzzy set theory is an effective tool in the development of a "soft" expert system (SES).

A number of SESs have been implemented, especially in the context of medical decision support systems. Indeed, the design of expert systems may well prove to be one of the most important applications of fuzzy logic in knowledge engineering and information technology. Although fuzziness is usually viewed as undesirable, the elasticity of fuzzy sets gives them a number of advantages over conventional sets. First, they avoid the rigidity of conventional mathematical reasoning and computer programming. Second, fuzzy sets simplify the task of translating between human reasoning, which is inherently elastic, and the rigid operation of digital computers. In particular, in common-sense reasoning, humans tend to use words rather than numbers to describe how systems behave. Finally, fuzzy sets allow computers to use the type of human knowledge called common sense. Common-sense knowledge exists mainly in the form of statements that are usually, but not always, true. Such a statement can be termed a *disposition*. A disposition contains an implicit fuzzy quantifier, such as "most," "almost always," "usually," and so on.

In fuzzy logic, fuzzy quantifiers are treated as fuzzy numbers that represent, imprecisely, the absolute or relative count of elements in a fuzzy set. Thus the proposition "Most As are Bs" means that the proportion of these elements is represented by the fuzzy quantifier "most" (see Figure 8-6).

Once the fuzzy quantifiers are made explicit, various syllogisms in fuzzy logic can be employed in order to arrive at a conclusion. For example, from the statements, "Most students are undergraduates" and "Most undergraduates are young," one could deduce that "most^2 students are young." In this case, "most^2" represents the product of the fuzzy quantifier "most" with itself. As should be expected, the quantifier "most^2" in the conclusion is less specific than the quantifier "most" in

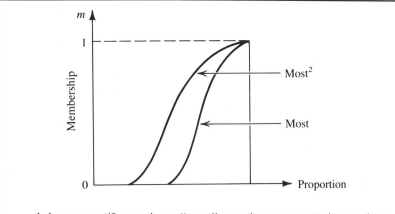

A fuzzy quantifier such as "most" may be represented as a fuzzy number—a fuzzy set that defines the degree to which any given proportion matches the definition of "most." The fuzzy quantifier "most" acts by finding the degree to which a proportion fits into "most" and then finding the degree to which that degree fits into "most." Thus, the value of "most" for a given proportion will be equal to the value of "most" for the square root of that proportion.

Figure 8-6. The fuzzy quantifier "most."

the premises. Most kinds of common-sense knowledge can be processed in a similar fashion through fuzzy reasoning.

A key problem in the application of fuzzy set theory to SESs is devising effective methods of reasoning with chains of fuzzy syllogisms. For example, to derive conclusions from common-sense knowledge, a computer must be able to deduce from the information that most students are undergraduates and most undergraduates are young, the conclusion that "most^2" students are young.

If we could take advantage of the tolerance for imprecision and devise programs that could quickly, if approximately, solve such problems, we could perhaps achieve algorithms that approach the ease and speed of humans in their everyday communication.

In the existing expert systems, uncertainty is dealt with through a combination of predicate logic and probability-based methods. A serious shortcoming of these methods is that they are not capable of coming to grips with the pervasive fuzziness of information in the knowledge base, and, as a result, are mostly ad hoc in nature. An alternative approach to the management of uncertainty is based on the use of fuzzy logic, which is the logic underlying approximate or, equivalently, fuzzy reasoning. A feature of fuzzy logic which is of particular importance to the management of uncertainty in expert systems is that it provides a systematic framework for dealing with fuzzy quantifiers. In this way fuzzy logic subsumes both predicate logic and probability theory and makes it possible to deal with different types of uncertainty within a single conceptual framework.

As a simple illustration of this point, consider the fact "Jim has duodenal ulcer (CF = 0.8)," where CF of 0.8 is a certainty factor of 0.8. Since "has duodenal ulcer" is a fuzzy predicate, so that Jim may have it to a degree, the meaning of the certainty factor becomes ambiguous. More specifically, does CF = 0.8 mean (a) that Jim has duodenal ulcer to the degree 0.8; or (b) that the probability of the fuzzy event "Jim has duodenal ulcer" is 0.8? Note that in order to make the latter interpretation meaningful, it is necessary to be able to define the probability of a fuzzy event. This can be done in fuzzy logic, but not in classical probability theory.

The computation of certainty factors in nonfuzzy ES is carried out when two or more rules are combined through conjunction, disjunction, or chaining. In the case of chaining, in particular, the standard inference rule—*modus ponens*—loses much of its validity and must be replaced by the more general compositional rule of inference. Furthermore, the transitivity of implication, which forms the basis for both forward and backward chaining in most expert systems, is a brittle property which must be applied with great caution.

However, fuzzy logic provides a natural conceptual framework for knowledge representation and inference from knowledge bases which are imprecise, incomplete, or not totally reliable. Generally, the use of fuzzy logic reduces the problem of inference to that of solving a nonlinear program and leads to conclusions whose uncertainty is a cumulation of the uncertainties in the premises from which the conclusions are derived. As a consequence, the conclusions, as well as the certainty factor, are fuzzy sets which are characterized by their possibility distributions.

8.3.2   Fuzzy Relational Knowledge Base

Much of human reasoning deals with imprecise, incomplete, or vague information. Therefore, there is a need for information systems that allow representation and manipulation of imprecise information in order to model human reasoning.

The Fuzzy Relational Knowledge Base (FRKB) model, based on the research in the fields of relational data bases and theories of fuzzy sets and possibility, is designed to satisfy the need for individualization and imprecise information processing.

The FRKB model design addresses the following:

1. Representation of imprecise information.
2. Derivation of possibility/certainty measures of acceptance.
3. Linguistic approximations of fuzzy terms in the query language.
4. Development of fuzzy relational operators (IS,AS . . . AS, GREATER,. . .).
5. Processing of queries with fuzzy connectors and truth quantifiers.
6. Null value handling using the concept of the possibilistic expected value.
7. Modification of the fuzzy term definitions to suit the individual user.

Such a knowledge base, in the form of an FRKB or a modified version thereof, is the basic unit of the "soft" expert system (SES).

A fuzzy relational knowledge base is a collection of fuzzy, time-varying relations which may be characterized by tables or functions and manipulated by recognition (retrieval) algorithms or translation rules.

The organization of the FRKB can be divided into three parts:

1. Value knowledge base (VKB),
2. Explanatory knowledge base (EKB), and
3. Translation rules.

The VKB is used to store actual data values, whereas the EKB consists of a collection of relations or functions (similarity, proximity, general fuzzy relations, and fuzzy set definitions) that "explain" how to compute the degree of compliance of a given data value with the user's query. This part of the knowledge base definition can be used to reflect the subjective knowledge profile of a user.

Data Types   The *domains* in the FRKB can be of the following types:

1. Discrete scalar set (e.g., COLOR = green, yellow, blue).
2. Discrete number sets, finite or infinite (limited by the maximum computer word size and precision).
3. The unit interval [0,1].

The *attribute values* are:

1. Single scalars or numbers.

2. A sequence (list) of scalars or numbers.
3. A possibilistic distribution of scalar or numeric domain values.
4. A real number from the unit interval [0,1] (membership or possibility distribution function value).
5. Null value.

In general, if A_i is an imprecise attribute with a domain D_i, then an attribute value can be a possibilistic distribution specified on D_i, denoted by Π_A.

Relations defined in the EKB are used in translation of fuzzy propositions. In essence, they relax the dependence of relational algebra operators on the regular relational operators ($=, \neq, <, >, \leq, \geq$). These include:

(i) *Similarity relation:* Let D_i be a scalar domain, $x, y \in D_i$.
    Then $s(x,y) \in [0,1]$ is a similarity relation with the following properties:

Reflexivity: $s(x,x) = 1$

Symmetry: $s(x,y) = s(y,x)$

θ-transitivity: where θ is most commonly specified as max-min transitivity. If $x, y, z \in U$, then

$$s(x,z) \geq \max_{y \in D} \{\min(s(x,y), s(y,z))\}.$$

(ii) *Proximity Relation:* Let D_i be a numerical domain and $x, y, z \in D_i$; $p(x,y) \in [0,1]$ is a proximity relation that is reflexive, symmetric with transitivity of the form

$$p(x,y) \geq \max_{y \in D_i} \{p(x,y) * p(y,z)\}.$$

The generally used form of the proximity relation is

$$p(x,y) = e^{\beta|x-y|}, \text{ where } \beta > 0.$$

This form assigns equal degrees of proximity to equally distant points.

A general fuzzy relation (link) can be defined in either VKB or EKB. A link may have any of the following properties:

nonreflexive: $g(x,x) = 0$

ϵ-reflexive: $g(x,x) = \epsilon, \epsilon > 0$

reflexive: $g(x,x) = 1$

nonsymmetric: $\exists\, x,y$ such that $g(x,y) \neq g(y,x)$

$\hat{g}$-transitive: generalized max-min transitivity.

A link can be used to express relationships that are not necessarily reflexive and symmetrical, but may obey a specific transitivity, or *transitivity improvement*. Typical relations that can be represented by a link are friendship or influence among the members of a group.

More complex queries in the FRKB system are evaluated by applying rules of:

1. Fuzzy modifiers.
2. Fuzzy relational operators.
3. Composition.
4. Qualified propositions.

The Relational Knowledge Base structure combined with the theory of fuzzy sets and possibility provides the solid theoretical foundation. The query language permits "natural-language-like" expressions that are easily understood by users and can be further developed to incorporate fuzzy inferences or production rules. The Value Knowledge Base appears to be an adequate schema for imprecise data representation. The Explanation Knowledge Base provides the means of individualization, and may be used to extend the query vocabulary by defining new fuzzy sets in terms of previously defined sets. This feature can become very useful in compounding knowledge, and it can be projected that the underlying structure can be utilized in knowledge extrapolation, or learning. Hence, it can be concluded that the FRKB system can serve as the data base in the implementation of such "soft" expert systems.

8.3.3   An Example of an Approximate Reasoning Inference Engine for an SES

In this section we present an example of a general purpose approximate reasoning inference engine (ARIES) used in an SES and developed by Appelbaum and Ruspini [46]. ARIES provides mechanisms for the representation and manipulation of multiple degrees of truth of propositions, as defined by several criteria of truth, belief, or likelihood (or desirability, preference, etc.). Further, ARIES has facilities for the definition of either default or proposition/operation-specific truth function formulas. ARIES has been implemented primarily as a general purpose approximate inference subsystem capable of being easily incorporated into a wide variety of expert systems or other information systems that rely on automated deductive reasoning to a substantial degree.

Degrees of truth, which may be interpreted in several ways, are typically represented in ARIES as either classical or fuzzy intervals of the [0,1] interval of the real line. These intervals represent constraints on the possible truth values of propositions (facts) or conditional propositions (approximate inference rules). Truth values are combined using user-defined truth-functional formulas (i.e., quasi-truth functional formulas that propagate interval constraints on truth values). Such formulas can be defined independently (and for each fact or rule) for the logical operations of negation, conjunction, disjunction, and implication.

In the context of ARIES deductive processes, inference consists of the determination of optimal or near-optimal hyperedge that link vertices representing available evidence with those representing hypothetical propositions in the AND-OR hypergraph that describes a propositional rule system (rule domain).

The operation of ARIES as an approximate reasoning inference engine is better understood when described in terms of hypergraph representations of rule domains.

Hypergraphs, also called AND/OR graphs are generalizations of standard graphs that are used to represent the relationships between propositions in a rule domain. In AND/OR graphs, such as the one illustrated in Figure 8-7, each vertex represents a different proposition in the rule domain being described. Directed edges in the graph (usually drawn to indicate their direction using arrows or, as in Figure 8-7, by assuming a bottom to top direction of edges) are used to connect propositions that appear in rule antecedents with the corresponding consequent proposition. The nature of the logical connectives joining propositions in the antecedent is also indicated by means of symbols drawn at the end of the directed edges incident on the consequent vertex. An arc drawn across several edges indicates that the corresponding propositions are joined by the logical connective AND in the antecedent of the rule, while absence of such an arc indicates that the corresponding propositions are joined by the logical connective OR.

In hypergraphs used for approximate reasoning, values (not shown in Figure 8-7) are also associated with vertices and edges to indicate, respectively, the a priori truth values assigned by users (on the basis of observations of the real world) to the propositions and inferential relations in the rule domain. These values can be combined by ARIES to arrive at estimates of the a posteriori truth value of a goal proposition.

In the hypergraph, each path leading to the goal represents a possible combination of knowledge about the truth of propositions and inference rules that provides a certain amount of support to the truth of the goal proposition or, as discussed below, to its logical negation. Among all such possible paths, ARIES usually seeks two paths having certain optimal properties which are further discussed below.

The paths examined by ARIES in this optimization process are not classical paths as conventionally defined in graph theory. Rather, they are hyperpaths, as several edges (rather than one) incident at the same

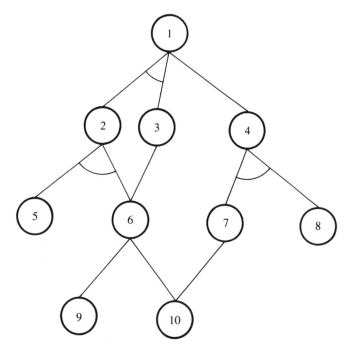

Figure 8-7. Example of AND/OR graph.

vertex may be part of the subgraph (hyperedge) that must be considered to compute the desired truth value. As discussed below, at any AND-VERTEX (i.e., a vertex where some of the incident arcs correspond to propositions that are joined by the connective AND in the antecedent of the rule), all of those edges (and their predecessors) are part of the hyperedge that must be evaluated. Depending on the numerical formalism (calculus of evidence) used to propagate truth values along the hypergraph, at an OR-VERTEX some or all of the edges corresponding to propositions that are connected by the logical connective OR (and their predecessors) might also require inclusion in the hyperedge, although in many formalisms only one of them requires such inclusion.

ARIES is a backward-chaining inference engine, that is, hyperedges leading to the goal proposition are searched and evaluated backward starting from the known final vertex (i.e., the goal) and establishing predecessor subgoals at each step of the search.

The numerical values that ARIES combines during its search and optimization process represent the degree of support provided by certain bodies of evidence for the truth of propositions in the rule domain as correct statements about the state of a real world system. A priori support values provided by users for facts and propositions are an indication of the extent of the support provided by external observations of the real world. Rules in the knowledge base, on the other hand, constrain the truth values that may be attained by factual statements on the basis of the truth values of related statements. Information about system behavior provided by rules can therefore be used by systems such as ARIES to further determine the support provided by the a priori truth values of the propositional system, as a whole, to the truth of the goal proposition.

The actual interpretation of the word support and the rationale for the use of certain formulas to combine evidence depends on the particular characteristics and needs of the approximate reasoning application being considered.

Fuzzy logic is an example of an excellent formalism for the manipulation of this type of evidential truth quantification. The input to such a "soft" expert system as ARIES falls into three categories: a list of facts, a list of rules, and a set of combination functions. For example, the format for a fact is usually given as

(((PROPOSITION) (TRUTH-VAL)) SUPPORT-PLAUSIBILITY).

The PROPOSITION is any statement in the base rule domain (i.e., a propositional statement or a clause in the antecedent or consequent of a rule). The TRUTH-VAL term is either an interval $[b,B]$ which represents the a priori evidential lower and upper bounds for the truth values of the proposition or a call to a function, with optional arguments, that generates those bounds. For example:

((Jim is much older than 30) (0.6,0.8))

assigns the interval [0.6,0.8] to the degree of membership of Jim to the set of people much older than 30. The proposition

((Jim is much older than 30) (age-factor Jim-age))

on the other hand, computes the interval as a function of Jim's age. The SUPPORT-PLAUSIBILITY term, which has the same format as TRUTH-VAL, is bound to the results produced by ARIES (if available). These results are the lower and upper bounds of an interval indicating the degree of support and plausibility provided by consideration of the body of evidence. This interval, if any has been computed, is therefore the a posteriori estimate of possible truth values for PROPOSITION. Its lower bound (support) measures the extent by which the propositions in the domain confirm the truth of PROPOSITION, while its upper bound measures the extent by which the same evidence refutes it. For example:

(((Jim is much older than 30) (age-factor Jim-age)) (0.5,0.9)) asserts that the evidence in the base domain (facts and rules) indicates that possible truth values for the proposition "Jim is much older than 30" may range from a lowest value of 0.5 to a highest (near certainty) value of 0.9.

The INFERENCE-FUNCTION requires as input the AND-VERTEX representing the clauses in the antecedent of a rule and the interval truth value (produced by the conjunctive formula of CONJUNCTION-FUNCTION) of their conjunction returning the truth value of the conclusion. The final set of input functions supplied by the user of ARIES is used to control the optimal path search process. When more than one rule has the same PROPOSITION as conclusion, a strategy must be chosen to determine which antecedents will be expanded. The input to these functions is always an OR-VERTEX specification, while their outputs are two sets of predecessor AND-VERTICES.

ARIES output is an interval (b, B), where b equals the greatest (among hypergraph paths) lower bound for the possible truth value (support) of the goal hypothesis, and where B equals the least upper bound (plausibility). The program also produces the solution (i.e., optimizing) graphs for b and B. These graphs can be retrieved using the explanation (WHY) capability of ARIES. Systems such as ARIES also provide graphics capabilities that allow display of portions of the hypergraph representing the rule domain, or of specific nodes and their predecessors/successors. On input of a PROPOSITION identifier along one of the optimal paths, they are also capable of displaying rules having PROPOSITION as a conclusion, their antecedents, and their truth values.

While many applications of fuzzy set theory are still in an early stage of development, it seems probable that in the next decade fuzzy logic will become routinely applied in many areas of artificial intelligence where communication with people or imitation of their thought processes is involved. This may help to bridge the gap between the analogic and flexible thinking of humans and the rigid framework of present computers.

Exercises for Chapter 8

1. "The union of fuzzy sets A and B is the smallest fuzzy set containing both A and B," which is also equivalent to "If D is any fuzzy set which contains both A and B, then it also contains the union of A and B." Prove the above statement.

2. Prove that the definition of intersection of two fuzzy sets is equivalent to the following definition: "The intersection of fuzzy sets A and B is the largest fuzzy set which is contained in both A and

B," which is also equivalent to "If D is any fuzzy set which is contained in both A and B, then it is also contained in the intersection of A and B."

3. For any fuzzy set A prove that
   (a) $A \cup \phi = A$
   (b) $A \cap \phi = \phi$
   (c) $A \cup U = U$
   (d) $A \cap U = A$
   (e) $A \cup \overline{A} = U$
   (f) $A \cap \overline{A} = \phi$.

4. Show that the following relations hold for ordinary sets.
   (a) $\phi - A = \phi$
   (b) $A - B = A \cap \overline{B}$
   (c) $A - (A - B) = A \cap B$
   (d) $C \cap D = A - [(A - C) \cup (A - D)]$
   (e) $C \cup D = A - [(A - C] \cap (A - D)]$.

5. The symmetrical difference of two fuzzy sets A and B with membership functions u_A and u_B, denoted by $A \Delta B$, is a fuzzy set whose membership function $u_{A\Delta B}$ is related to those of A and B by $u_{A\Delta B} = |u_A - u_B|$. If all the relative complement operations are changed to the symmetrical difference operation Δ defined above, do the above relations (problem 4) hold for fuzzy sets?

6. Let A and B be fuzzy sets. Prove that $A \Delta B = (A \cup B) \Delta (B \cap A)$.

7. *Definition:* Let $\mathbb{R}^R$ be the set of n-tuples $(y_1, y_2, \ldots, y_n)$ where y_i is a real number. A fuzzy set X is convex if $x_1, x_2 \in X, \forall \lambda \in [0, 1]$,

$$u_A[\lambda x_1 + (1 - \lambda)x_2] \geq \min [u_A (x_1), u_A(x_2)].$$

The following are some of the well-known continuous probability density functions.

The Uniform Distribution

$$f_U(x) = 1/a [u(x - \alpha) - u(x - \alpha - a)]$$

The Normal Distribution

$$f_N(x) = \frac{1}{(2\pi\sigma^2)^{1/2}} \exp \left[\frac{-(x - u)^2}{2\sigma^2} \right], \quad -\infty < x < \infty$$

The Gamma Distribution

$$f_\gamma(x) = \begin{cases} \dfrac{1}{\alpha!\beta^{\alpha+1}} \, x^\alpha \exp\left(-\dfrac{x}{\beta}\right), & 0 < x < \infty \\[2mm] 0 \text{ elsewhere} \end{cases}$$

where $\alpha > -1$ and $\beta > 0$.

The Beta Distribution

$$f_\beta(x) = \begin{cases} \dfrac{(\alpha + \beta + 1)!}{\alpha!\,\beta!} \, x^\alpha(1 - x)^\beta, & 0 < x < 1 \\[2mm] 0 \text{ elsewhere} \end{cases}$$

where α and β must both be greater than -1.

The Cauchy Density Function

$$f_c(x) = \frac{1}{\pi} \frac{\alpha}{1 + \alpha^2(x - u)^2}, \quad -\infty < x < \infty$$

Show that any fuzzy set described by one of these functions is convex.

Selected Answers for Chapter 8

6. $u_{A \triangle B} = |u_A - u_B| = |\max(u_A, u_B) - \min(u_A, u_B)|$.

Bibliography

1. Aho, A. V., J. E. Hopcroft, and J. D. Ullman, *Design and Analysis of Computer Algorithms,* Reading, Mass., Addison-Wesley, 1974.

2. Beckenbach, E., *Applied Combinatorial Mathematics,* New York, John Wiley and Sons, 1964.

3. Behzad, M., G. Chartrand, and L. Lesniak-Foster, *Graphs and Digraphs,* Boston: Prindle, Weber and Schmidt, 1979.

4. Bondy, J. A. and U. S. R. Murty, *Graph Theory with Applications,* New York: North-Holland, 1976.

5. Berman, G. and K. D. Fryer, *Introduction to Combinatorics,* New York: Academic Press, 1972.

6. Brualdi, R. A., *Introductory Combinatorics,* New York: North-Holland, 1977.

7. Buck, R. C., *Mathematical Induction and Recursive Definitions,* American Mat. Monthly, 70, 128–135, 1963.

8. Capobianco, M., and J. C. Mollezzo, *Examples and Counterexample in Graph Theory,* New York, North-Holland, 1978.

9. Chartrand, G., *Graphs as Mathematical Models,* Boston, Prindle, Weber and Schmidt, 1977.

10. Cohen, D. I. A., *Basic Techniques of Combinatorial Theory,* New York: John Wiley and Sons, 1978.

11. Dinkines, F., *Introduction to Mathematical Logic,* New York, Appleton-Century-Crofts, 1964.

12. Eisen, M., *Elementary Combinatorial Analysis,* New York, Gordon and Breach, 1969.

13. Even, S., *Algorithmic Combinatorics,* New York, Macmillan, 1973.

14. Gill, A., *Applied Algebra for the Computer Sciences.* Englewood Cliffs, New Jersey, Prentice-Hall, 1976.

15. Hayden, S. and J. Kennison, *Zermelo-Fraenkel Set Theory,* Columbus, Ohio, Charles Merrill, 1968.

16. Henkin, L., *On Mathematical Induction,* American Math. Monthly, 67, 323–337, 1960.

17. Hopcroft, J. E. and J. D. Ullman, *Introduction to Automata Theory, Languages, and Computation,* Reading, Mass., Addison-Wesley, 1979.

18. Horowitz, E. and S. Sahni, *Fundamentals of Data Structures,* Woodland Hills, Calif., Computer Science Press, 1976.

19. Horowitz, E. and S. Sahni, *Fundamentals of Computer Algorithms,* Woodland Hills, Calif., Computer Science Press, 1978.

20. Honsberger, R., *Mathematical Gems,* The American Mathematical Association of America, 1973.

21. Kenelly, John W., *Informal Logic,* Boston, Allyn and Bacon, 1967.

22. Korfhage, R. R., *Discrete Computational Structures,* New York, Academic Press, 1974.

23. Knuth, D. E., *The Art of Computer Programming, Vol. I: Fundamental Algorithms,* Reading, Mass., Addison-Wesley, 1968 (2nd ed. 1973).

24. Levy, H. and F. Lessman, *Finite Difference Equations,* New York, Macmillan, 1961.

25. Liu, C. L., *Elements of Discrete Mathematics,* New York: McGraw-Hill, 1977.

26. Liu, C. L., *Introduction to Combinatorial Mathematics,* New York, McGraw-Hill, 1968.

27. Mano, M. M., *Computer System Architecture,* Englewood Cliffs, New Jersey, Prentice-Hall, 1976.

28. Ore, O., *Graphs and Their Uses,* New York, Random House, 1963.

29. Page, E. S. and L. B. Wilson, *An Introduction to Computational Combinatorics,* New York, Cambridge University Press, 1979.

30. Prather, R., *Discrete Mathematical Structures for Computer Science,* Boston, Houghton Mifflin, 1976.

31. Preparata, F. P. and P. T. Yeh, *Introduction to Discrete Structures,* Reading, Mass., Addison-Wesley, 1973.

32. Polya, G., *How to Solve It,* Garden City, N.Y., Doubleday and Company, 1957.

33. Polya, G., *Mathematics and Plausible Reasoning, Vol. I. Induction and Analogy in Mathematics, Vol. II., Patterns of Plausible Inference,* Princeton, New Jersey, Princeton University Press, 1954.

34. Polya, G., *Mathematical Discovery,* New York, John Wiley and Sons, 1962.

35. Reingold, E., J. Nievergelt, and N. Deo, *Combinatorial Algorithms,* Englewood Cliffs, New Jersey, Prentice-Hall, 1977.

36. Riordan, J., *Combinatorial Identities,* New York, John Wiley and Sons, 1968.

37. Stanat, D. F. and D. F. McAllister, *Discrete Mathematics in Computer Science,* Englewood Cliffs, New Jersey, Prentice-Hall, 1977.

38. Standish, T. A., *Data Structure Techniques,* Reading, Mass., Addison-Wesly, 1980.

39. Tremblay, J. T. and R. P. Manohar, *Discrete Mathematical Structures with Applications to Computer Science,* New York, McGraw-Hill, 1975.

40. Tucker, A., *Applied Combinatorics,* New York, John Wiley and Sons, 1980.

41. Tutte, W. T., *Non-Hamiltonian Planar Maps,* in *Graph Theory and Computing,* edited by R. Read, Academic Press, 1972.

42. Welsh, D. J. A. and M. B. Powell, *An upper bound to the chromatic number of a graph and its applications to time tabling problems,* Comput. Journal 10, 85–86, 1967.

43. Whitworth, W. A., *Choice and Chance,* 5th ed. (1901), New York, Hafner Press, 1965.

44. Woodall, D. R. and R. J. Wilson, *The Appel-Haken Proof of the Four-Color Problem,* in *Selected Topics in Graph Theory,* edited by L. W. Beinke and R. J. Wilson, New York, Academic Press, 1978.

45. Youse, B. K., *Mathematical Induction,* Englewood Cliffs, New Jersey, Prentice-Hall, 1964.

46. Appelbaum, L. and E. H. Ruspini, *ARIES—An Example of an Approximate Reasoning Inference Engine,* in *Approximate Reasoning in Expert Systems,* edited by M. M. Gupta, A. Kandel, W. Bandler, and J. B. Kiszka, North Holland, 1985.

47. Ford, L. R. Jr. and D. R. Fulkerson, *Flows in Networks,* Princeton, N.J., Princeton University Press, 1962.

48. Goldberg, S., *Introduction to Difference Equations,* New York, John Wiley and Sons, 1958.

49. Zadeh, L. A., *Fuzzy Sets,* Information and Control, 8, 338–353, 1965.

50. Zadeh, L. A., *The Role of Fuzzy Logic in the Management of Uncertainty in Expert Systems,* Fuzzy Sets and Systems, 11, 199–227, 1983.

51. Zemankova-Leech, M. and A. Kandel, *Fuzzy Relational Data Bases—A Key to Expert Systems,* Köln, Verlag Tüv Rheinland, 1984.

Index